MW00784219

Introduction to Quantitative Finance

Introduction to Quantitative Finance

A Math Tool Kit

Robert R. Reitano

The MIT Press
Cambridge, Massachusetts
London, England

MIT Press books may be purchased at special quantity discounts for business or sales promotional use. For information, please email special_sales@mitpress.mit.edu or write to Special Sales Department, The MIT Press, 55 Hayward Street, Cambridge, MA 02142.

This book was set in Times New Roman on 3B2 by Asco Typesetters, Hong Kong and was printed and bound in the United States of America.

Library of Congress Cataloging-in-Publication Data

Reitano, Robert R., 1950–
Introduction to quantitative finance : a math tool kit / Robert R. Reitano.
    p.   cm.
Includes index.
ISBN 978-0-262-01369-7 (hardcover : alk. paper) 1. Finance—Mathematical models. I. Title.
HG106.R45   2010
332.01′5195—dc22   2009022214

10 9 8 7 6 5 4 3 2 1

to Lisa

Contents

List of Figures and Tables

Figures

Introduction

This book provides an accessible yet rigorous introduction to the fields of mathematics that are needed for success in investment and quantitative finance. The book's goal is to develop mathematics topics used in portfolio management and investment banking, including basic derivatives pricing and risk management applications, that are essential to quantitative investment finance, or more simply, investment finance. A future book, *Advanced Quantitative Finance: A Math Tool Kit*, will cover more advanced mathematical topics in these areas as used for investment modeling, derivatives pricing, and risk management. Collectively, these latter areas are called quantitative finance or mathematical finance.

The mathematics presented in this book would typically be learned by an undergraduate mathematics major. Each chapter of the book corresponds roughly to the mathematical materials that are acquired in a one semester course. Naturally each chapter presents only a subset of the materials from these traditional math courses, since the goal is to emphasize the most important and relevant materials for the finance applications presented. However, more advanced topics are introduced earlier than is customary so that the reader can become familiar with these materials in an accessible setting.

My motivation for writing this text was to fill two current gaps in the financial and mathematical literature as they apply to students, and practitioners, interested in sharpening their mathematical skills and deepening their understanding of investment and quantitative finance applications. The gap in the mathematics literature is that most texts are focused on a single field of mathematics such as calculus. Anyone interested in meeting the field requirements in finance is left with the choice to either pursue one or more degrees in mathematics or expend a significant self-study effort on associated mathematics textbooks. Neither approach is efficient for business school and finance graduate students nor for professionals working in investment and quantitative finance and aiming to advance their mathematical skills. As the diligent reader quickly discovers, each such book presents more math than is needed for finance, and it is nearly impossible to identify what math is essential for finance applications. An additional complication is that math books rarely if ever provide applications in finance, which further complicates the identification of the relevant theory.

The second gap is in the finance literature. Finance texts have effectively become bifurcated in terms of mathematical sophistication. One group of texts takes the recipe-book approach to math finance often presenting mathematical formulas with only simplified or heuristic derivations. These books typically neglect discussion of the mathematical framework that derivations require, as well as effects of assumptions by which the conclusions are drawn. While such treatment may allow more discussion of the financial applications, it does not adequately prepare the student who will inevitably be investigating quantitative problems for which the answers are unknown.

The other group of finance textbooks are mathematically rigorous but inaccessible to students who are not in a mathematics degree program. Also, while rigorous, such books depend on sophisticated results developed elsewhere, and hence the discussions are incomplete and inadequate even for a motivated student without additional class-room instruction. Here, again, the unprepared student must take on faith referenced results without adequate understanding, which is essentially another form of recipe book.

With this book I attempt to fill some of these gaps by way of a reasonably eco-nomic, yet rigorous and accessible, review of many of the areas of mathematics needed in quantitative investment finance. My objective is to help the reader acquire a deep understanding of relevant mathematical theory and the tools that can be ef-fectively put in practice. In each chapter I provide a concluding section on finance applications of the presented materials to help the reader connect the chapter's math-ematical theory to finance applications and work in the finance industry.

What Does It Take to Be a "Quant"?

In some sense, the emphasis of this book is on the development of the math tools one needs to succeed in mathematical modeling applications in finance. The imagery implied by "math tool kit" is deliberate, and it reflects my belief that the study of mathematics is an intellectually rewarding endeavor, and it provides an enormously flexible collection of tools that allow users to answer a wide variety of important and practical questions.

By tools, however, I do not mean a collection of formulas that should be memo-rized for later application. Of course, some memorization is mandatory in mathe-matics, as in any language, to understand what the words mean and to facilitate accurate communication. But most formulas are outside this mandatorily memorized collection. Indeed, although mathematics texts are full of formulas, the memoriza-tion of formulas should be relatively low on the list of priorities of any student or user of these books. The student should instead endeavor to learn the mathematical frameworks and the application of these frameworks to real world problems.

In other words, the student should focus on the thought process and mathematics used to develop each result. These are the "tools," that is, the mathematical methods of each discipline of explicitly identifying assumptions, formally developing the needed insights and formulas, and understanding the relationships between formulas and the underlying assumptions. The tools so defined and studied in this book will equip the student with fairly robust frameworks for their applications in investment and quantitative finance.

Despite its large size, this book has the relatively modest ambition of teaching a very specific application of mathematics, that being to finance, and so the selection of materials in every subdiscipline has been made parsimoniously. This selection of materials was the most difficult aspect of developing this book. In general, the selection criterion I used was that a topic had to be either directly applicable to finance, or needed for the understanding of a later topic that was directly applicable to finance. Because my objective was to make this book more than a collection of mathematical formulas, or just another finance recipe book, I devote considerable space to discussion on how the results are derived, and how they relate to their mathematical assumptions. Ideally the students of this book should never again accept a formulaic result as an immutable truth separate from any assumptions made by its originator.

The motivation for this approach is that in investment and quantitative finance, there are few good careers that depend on the application of standard formulas in standard situations. All such applications tend to be automated and run in companies' computer systems with little or no human intervention. Think "program trading" as an example of this statement. While there is an interesting and deep theory related to identifying so-called arbitrage opportunities, these can be formulaically listed and programmed, and their implementation automated with little further analyst intervention.

Equally, if not more important, with new financial products developed regularly, there are increased demands on quants and all finance practitioners to apply the previous methodologies and adapt them appropriately to financial analyses, pricing, risk modeling, and risk management. Today, in practice, standard results may or may not apply, and the most critical job of the finance quant is to determine if the traditional approach applies, and if not, to develop an appropriate modification or even an entirely new approach. In other words, for today's finance quants, it has become critical to be able to think in mathematics, and not simply to do mathematics by rote.

The many finance applications developed in the chapters present enough detail to be understood by someone new to the given application but in less detail than would be appropriate for mastering the application. Ideally the reader will be familiar with some applications and will be introduced to other applications that can, as needed, be enhanced by further study. On my selection of mathematical topics and finance applications, I hope to benefit from the valuable comments of finance readers, whether student or practitioner. All such feedback will be welcomed and acknowledged in future editions.

Plan of the Book

The ten chapters of this book are arranged so that each topic is developed based on materials previously discussed. In a few places, however, a formula or result is introduced that could not be fully developed until a much later chapter. In fewer places, I decided to not prove a deep result that would have brought the book too far afield from its intended purpose. Overall, the book is intended to be self-contained, complete with respect to the materials discussed, and mathematically rigorous. The only mathematical background required of the reader is competent skill in algebraic manipulations and some knowledge of pre-calculus topics of graphing, exponentials and logarithms. Thus the topics developed in this book are interrelated and applied with the understanding that the student will be motivated to work through, with pen or pencil and paper or by computer simulation, any derivation or example that may be unclear and that the student has the algebraic skills and self-discipline to do so.

Of course, even when a proof or example appears clear, the student will benefit in using pencil and paper and computer simulation to clarify any missing details in derivations. Such informal exercises provide essential practice in the application of the tools discussed, and analytical skills can be progressively sharpened by way of the book's formal exercises and ultimately in real world situations. While not every derivation in the book offers the same amount of enlightenment on the mathematical tools studied, or should be studied in detail before proceeding, developing the habit of filling in details can deepen mathematical knowledge and the understanding of how this knowledge can be applied.

I have identified the more advanced sections by an asterisk (*). The beginning student may find it useful to scan these sections on first reading. These sections can then be returned to if needed for a later application of interest. The more advanced student may find these sections to provide some insights on the materials they are already familiar with. For beginning practitioners and professors of students new to the materials, it may be useful to only scan the reasoning in the longer proofs on a first review before turning to the applications.

There are a number of productive approaches to the chapter sequencing of this book for both self-study and formal classroom presentation. Professors and practitioners with good prior exposure might pick and choose chapters out of order to efficiently address pressing educational needs. For finance applications, again the best approach is the one that suits the needs of the student or practitioner. Those familiar with finance applications and aware of the math skills that need to be developed will focus on the appropriate math sections, then proceed to the finance applications to better understand the connections between the math and the finance. Those less familiar with finance may be motivated to first review the applications section of each chapter for motivation before turning to the math.

Some Course Design Options

This book is well suited for a first-semester introductory graduate course in quantitative finance, perhaps taken at the same time as other typical first-year graduate courses for finance students, such as investment markets and products, portfolio theory, financial reporting, corporate finance, and business strategy. For such students the instructor can balance the class time between sharpening mathematical knowledge and deepening a level of understanding of finance applications taken in the first term. Students will then be well prepared for more quantitatively focused investment finance courses on fixed income and equity markets, portfolio management, and options and derivatives, for example, in the second term.

For business school finance students new to the subject of finance, it might be better to defer this book to a second semester course, following an introductory course in financial markets and instruments so as to provide a context for the finance applications discussed in the chapters of this book.

This book is also appropriate for graduate students interested in firming up their technical knowledge and skills in investment and quantitative finance, so it can be used for self-study by students soon to be working in investment or quantitative finance, and by practitioners needing to improve their math skill set in order to advance their finance careers in the "quant" direction. Mathematics and engineering departments, which will have many very knowledgeable graduate and undergraduate students in the areas of math covered in this book, may also be interested in offering an introductory course in finance with a strong mathematical framework. The rigorous math approach to real world applications will be familiar to such students, so a balance of math and finance could be offered early in the students' academic program.

For students for whom the early chapters would provide a relatively easy review, it is feasible to take a sequential approach to all the materials, moving faster through the familiar math topics and dwelling more on the finance applications. For non-mathematical students who risk getting bogged down by the first four chapters in their struggle with abstract notions, and are motivated to learn the math only after recognizing the need in a later practical setting, it may be preferable to teach only a subset of the math from chapters 1 through 4 and focus on the intuition behind these chapters' applications. For example, an instructor might provide a quick overview of logic and proof from chapter 1, choose selectively from chapter 2 on number systems, then skip ahead to chapter 4 for set operations. After this topical tour the instructor could finally settle in with all the math and applications in chapter 5 on sequences and then move forward sequentially through chapters 6 to 10. The other mathematics topics of chapters 1 through 4 could then be assigned or taught as required to supplement the materials of these later chapters. This approach and pace could keep the students motivated by getting to the more meaningful applications sooner, and thus help prevent math burnout before reaching these important applications.

Chapter Exercises

Chapter exercises are split into practice exercises and assignment exercises. Both types of exercises provide practice in mathematics and finance applications. The more challenging exercises are accompanied by a "hint," but students should not be constrained by the hints. The best learning in mathematics and in applications often occurs in pursuit of alternative approaches, even those that ultimately fail. Valuable lessons can come from such failures that help the student identify a misunderstanding of concepts or a misapplication of logic or mathematical techniques. Therefore, if other approaches to a problem appear feasible, the student is encouraged to follow at least some to a conclusion. This additional effort can provide reinforcement of a result that follows from different approaches but also help identify errors and misunderstandings when two approaches lead to different conclusions.

Solutions and Instructor's Manuals

For the book's practice exercises, a *Solutions Manual* with detailed explanations of solutions is available for purchase by students. For the assignment exercises, solutions are available to instructors as part of an *Instructor's Manual*. This Manual also contains chapter-by-chapter suggestions on teaching the materials. All instructor materials are also available online.

Organization of Chapters

Few mathematics books today have an introductory chapter on mathematical logic, and certainly none that address applications. The field of logic is a subject available to mathematics or philosophy students as a separate course. To skip the material on logic is to miss an opportunity to acquire useful tools of thinking, in drawing appropriate conclusions, and developing clear and correct quantitative reasoning.

Simple conclusions and quantitative derivations require no formality of logic, but the tools of truth tables and statement analysis, as well as the logical construction of a valid proof, are indispensable in evaluating the integrity of more complicated results. In addition to the tools of logic, chapter 1 presents various approaches to proofs that follow from these tools, and that will be encountered in subsequent chapters. The chapter also provides a collection of paradoxes that are often amusing and demonstrate that even with careful reasoning, an argument can go awry or a conclusion reached can make no sense. Yet paradoxes are important; they motivate clearer thinking and more explicit identification of underlying assumptions.

Finally, for completeness, this chapter includes a discussion of the axiomatic formality of mathematical theory and explains why this formality can help one avoid paradoxes. It notes that there can be some latitude in the selection of the axioms, and that axioms can have a strong effect on the mathematical theory. While the reader should not get bogged down in these formalities, since they are not critical to the understanding of the materials that follow, the reader should find comfort that they exist beneath the more familiar frameworks to be studied later.

The primary application of mathematical logic to finance and to any field is as a guide to cautionary practice in identifying assumptions and in applying or deriving a needed result to avoid the risk of a potentially disastrous consequence. Intuition is useful as a guide to a result, but never as a substitute for careful analysis.

Chapter 2, on number systems and functions, may appear to be on relatively trivial topics. Haven't we all learned numbers in grade school? The main objective in reviewing the different number systems is that they *are* familiar and provide the foundational examples for more advanced mathematical models. Because the aim of this book is to introduce important concepts early, the natural numbers provide a relatively simple example of an axiomatic structure from chapter 1 used to develop a mathematical theory.

From the natural numbers other numbers are added sequentially to allow more arithmetic operations, leading in turn to integers, rational, irrational, real, and complex numbers. Along the way these collections are seen to share certain arithmetic structures, and the notions of group and field are introduced. These collections also provide an elementary context for introducing the notions of countable and uncountable infinite sets, as well as the notion of a "dense" subset of a given set. Once defined, these number systems and their various subsets are the natural domains on which functions are defined.

While it might be expected that only the rational numbers are needed in finance, and indeed the rational numbers with perhaps only 6 to 10 decimal point representations, it is easy to exemplify finance problems with irrational and even complex number solutions. In the former cases, rational approximations are used, and sometimes with reconciliation difficulties to real world transactions, while complex numbers are avoided by properly framing the interest rate basis. Functions appear everywhere in finance—from interest rate nominal basis conversions, to the pricing functions for bonds, mortgages and other loans, preferred and common stock, and forward contracts, and to the modeling of portfolio returns as a function of the asset allocation.

The development of number system structures is continued in chapter 3 on Euclidean and other spaces. Two-dimensional Euclidean space, as was introduced in chapter 2, provided a visual framework for the complex numbers. Once defined, the vector space structure of Euclidean space is discussed, as well as the notions of the standard norm and inner product on these spaces. This discussion leads naturally to the important Cauchy–Schwarz inequality relating these concepts, an inequality that arises time and again in various contexts in this book. Euclidean space is also the simplest context in which to introduce the notion of alternative norms, and the l_p-norms, in particular, are defined and relationships developed. The central result is the generalization of Cauchy–Schwarz to the Hölder inequality, and of the triangle inequality to the Minkowski inequality.

Metrics are then discussed, as is the relationship between a metric and a norm, and cases where one can be induced from the other on a given space using examples from the l_p-norm collection. A common theme in mathematics and one seen here is that a general metric is defined to have exactly the essential properties of the standard and familiar metric defined on $\mathbb{R}^2$ or generalized to $\mathbb{R}^n$. Two notions of equivalence of two metrics is introduced, and it is shown that all the metrics induced by the l_p-norms are equivalent in Euclidean space. Strong evidence is uncovered that this result is fundamentally related to the finite dimensionality of these spaces, suggesting that equivalence will not be sustained in more general forthcoming contexts. It is also illustrated that despite this general l_p-equivalence result, not all metrics are equivalent.

For finance applications, Euclidean space is seen to be the natural habitat for expressing vectors of asset allocations within a portfolio, various bond yield term structures, and projected cash flows. In addition, all the l_p-norms appear in the calculation of various moments of sample statistical data, while some of the l_p-norms, specifically $p = 1, 2$, and ∞, appear in various guises in constrained optimization problems common in finance. Sometimes these special norms appear as constraints and sometimes as the objective function one needs to optimize.

Chapter 4 on set theory and topology introduces another example of an axiomatic framework, and this example is motivated by one of the paradoxes discussed in chapter 1. But the focus here is on set operations and their relationships. These are important tools that are as essential to mathematical derivations as are algebraic manipulations. In addition, basic concepts of open and closed are first introduced in the familiar setting of intervals on the real line, but then generalized and illustrated making good use of the set manipulation results. After showing that open sets in $\mathbb{R}$ are relatively simple, the construction of the Cantor set is presented as an exotic example of a closed set. It is unusual because it is uncountable and yet, at the same time, shown to have "measure 0." This result is demonstrated by showing that the Cantor set is what is left from the interval $[0, 1]$ after a collection of intervals are removed that have total length equal to 1!

The notions of open and closed are then extended in a natural way to Euclidean space and metric spaces, and the idea of a topological space is introduced for completeness. The basic aim is once again to illustrate that a general idea, here topology, is defined to satisfy exactly the same properties as do the open sets in more familiar contexts. The chapter ends with a few other important notions such as accumulation point and compactness, which lead to discussions in the next chapter.

For finance applications, constrained optimization problems are seen to be naturally interpreted in terms of sets in Euclidean space defined by functions and/or norms. The solution of such problems generally requires that these sets have certain topological properties like compactness and that the defining functions have certain regularity properties. Function regularity here means that the solution of an equation can be approximated with an iterative process that converges as the number of steps increases, a notion that naturally leads to chapter 5. Interval bisection is introduced as an example of an iterative process, with an application to finding the yield of a security, and convergence questions are made explicit and seen to motivate the notion of continuity.

Sequences and their convergence are addressed in chapter 5, making good use of the concepts, tools, and examples of earlier chapters. The central idea, of course, is that of convergence to a limit, which is informally illustrated before it is formally defined. Because of the importance of this idea, the formal definition is discussed at some length, providing both more detail on what the words mean and justification as to why this definition requires the formality presented. Convergence is demonstrated to be preserved under various arithmetic operations. Also an important result related to compactness is demonstrated: that is, while a bounded sequence need not converge, it must have an accumulation point and contain a subsequence that convergences to that accumulation point. Because such sequences may have many—indeed infinitely many—such accumulation points, the notions of limit superior and limit inferior are introduced and shown to provide the largest and smallest such accumulation points, respectively.

Convergence of sequences is then discussed in the more general context of Euclidean space, for which all the earlier results generalize without modification, and metric spaces, in which some care is needed. The notion of a Cauchy sequence is next introduced and seen to naturally lead to the question of whether such sequences converge to a point of the space, as examples of both convergence and nonconvergence are presented. This discussion leads to the introduction of the idea of completeness of a metric space, and of its completion, and an important result on completion is presented without proof but seen to be consistent with examples studied.

Interval bisection provides an important example of a Cauchy sequence in finance. Here the sequence is of solution iterates, but again the question of convergence of the associated price values remains open to a future chapter. With more details on this process, the important notion of continuous function is given more formality.

Although the convergence of an infinite sequence is broadly applicable in its own right, this theory provides the perfect segue to the convergence of infinite sums addressed in chapter 6 on series and their convergence. Notions of absolute and conditional convergence are developed, along with the implications of these properties for arithmetic manipulations of series, and for re-orderings or rearrangements of the series terms. Rearrangements are discussed for both single-sum and multiple-sum applications.

A few of the most useful tests for convergence are developed in this chapter. The chapter 3 introduction to the l_p-norms is expanded to include l_p-spaces of sequences and associated norms, demonstrating that these spaces are complete normed spaces, or Banach spaces, and are overlapping yet distinct spaces for each p. The case of $p = 2$ gets special notice as a complete inner product space, or Hilbert space, and implications of this are explored. Power series are introduced, and the notions of radius of convergence and interval of convergence are developed from one of the previous tests for convergence. Finally, results for products and quotients of power series are developed.

Applications to finance include convergence of price formulas for various perpetual preferred and common stock models with cash flows modeled in different functional ways, and various investor yield demands. Linearly increasing cash flows provide an example of double summation methods, and the result is generalized to polynomial payments. Approximating complicated pricing functions with power series is considered next, and the application of the l_p-spaces is characterized as providing an accessible introduction to the generalized function space counterparts to be studied in more advanced texts.

An important application of the tools of chapter 6 is to discrete probability theory, which is the topic developed in chapter 7 starting with sample spaces and probability measures. By discrete, it is meant that the theory applies to sample spaces with a finite or countably infinite number of sample points. Also studied are notions of conditional probability, stochastic independence, and an n-trial sample space construction that provides a formal basis for the concept of an independent sample from a sample space. Combinatorics are then presented as an important tool for organizing and counting collections of events from discrete sample spaces.

Random variables are shown to provide key insights to a sample space and its probability measure through the associated probability density and distribution functions, making good use of the combinatorial tools. Moments of probability density functions and their properties are developed, as well as moments of sample data drawn from an *n*-trial sample space. Several of the most common discrete probability density functions are introduced, as well as a methodology for generating random samples from any such density function.

Applications of these materials in finance are many, and begin with loss models related to bond or loan portfolios, as well as those associated with various forms of insurance. In this latter context, various net premium calculations are derived. Asset allocation provides a natural application of probability methods, as does the modeling of equity prices in discrete time considered within either a binomial lattice or binomial scenario model. The binomial lattice model is then used for option pricing in discrete time based on the notion of option replication. Last, scenario-based option pricing is introduced through the notion of a sample-based option price defined in terms of a sampling of equity price scenarios.

With chapter 7 providing the groundwork, chapter 8 develops a collection of the fundamental probability theorems, beginning with a modest proof of the uniqueness of the moment-generating and characteristic functions in the case of finite discrete probability density functions. Chebyshev's inequality, or rather, Chebyshev's inequalities, are developed, as is the *weak law of large numbers* as the first of several results related to the distribution of the sample mean of a random variable in the limit as the sample size grows. Although the weak law requires only that the random variable have a finite mean, in the more common case where the variance is also finite, this law is derived with a sleek one-step proof based on Chebyshev.

The *strong law of large numbers* requires both a finite mean and variance but provides a much more powerful statement about the distribution of sample means in the limit. The strong law is based on a generalization of the Chebyshev inequality known as Kolmogorov's inequality. The De Moivre–Laplace theorem is investigated next, followed by discussions on the normal distribution and the *central limit theorem* (CLT). The CLT is proved in the special case of probability densities with moment-generating functions, and some generalizations are discussed.

For finance applications, Chebyshev is applied to the problem of modeling and evaluating asset adequacy, or capital adequacy, in a risky balance sheet. Then the binomial lattice model for stock prices under the real world probabilities introduced in

chapter 7 is studied in the limit as the time interval converges to zero, and the probability density function of future stock prices is determined. This analysis uses the
methods underlying the De Moivre–Laplace theorem and provides the basis of the
next investigation into the derivation of the Black–Scholes–Merton formulas for the
price of a European put or call option. Several of the details of this derivation that
require the tools of chapters 9 and 10 are deferred to those chapters. The final application is to the probabilistic properties of the scenario-based option price introduced
in chapter 7.

The calculus of functions of a single variable is the topic developed in the last two
chapters. Calculus is generally understood as the study of functions that display various types of "smoothness." In line with tradition, this subject is split into a differentiation theory and an integration theory. The former provides a rigorous framework
for approximating smooth functions, and the latter introduces in an accessible framework an important tool needed for a continuous probability theory.

Chapter 9 on the calculus of differentiation begins with the formal introduction
of the notion of continuity and its variations, as well the development of important
properties of continuous functions. These basic notions of smoothness provide the
beginnings of an approximation approach that is generalized and formalized with
the development of the derivative of a function. Various results on differentiation follow, as does the formal application of derivatives to the question of function approximation via Taylor series. With these tools important results are developed related to
the derivative, such as classifying the critical points of a given function, characteriz-
ing the notions of convexity and concavity, and the derivation of Jensen's inequality.
Not only can derivatives be used to approximate function values, but the values of
derivatives can be approximated using nearby function values and the associated
errors quantified. Results on the preservation of continuity and differentiability under
convergence of a sequence of functions are addressed, as is the relationship between
analytic functions and power series.

Applications found in finance include the continuity of price functions and their
application to the method of interval bisection. Also discussed is the continuity of
objective functions and constraint functions and implications for solvability of constrained optimization problems. Deriving the minimal risk portfolio allocation is
one application of a critical point analysis. Duration and convexity of fixed income
investments is studied next and used in an application of Taylor series to price function approximations and asset-liability management problems in various settings.

Outside of fixed income, the more common sensitivity measures are known as the
"Greeks," and these are introduced and shown to easily lend themselves to Taylor
series methods. Utility theory and its implications for risk preferences are studied
as an application of convex and concave functions and Jensen's inequality, and then

applied in the context of optimal portfolio allocation. Finally, details are provided for the limiting distributions of stock prices under the risk-neutral probabilities and special risk-averter probabilities needed for the derivation of the Black–Scholes–Merton option pricing formulas, extending and formalizing the derivation begun in chapter 8. The risk-averter model is introduced in chapter 8 as a mathematical artifact to facilitate the final derivation, but it is clear the final result only depends on the risk-neutral model.

The notion of Riemann integral is studied in chapter 10 on the calculus of integration, beginning with its definition for a continuous function on a closed and bounded interval where it is seen to represent a "signed" area between the graph of the function and the x-axis. A series of generalizations are pursued, from the weakening of the continuity assumption to that of bounded and continuous "except on a set of points of measure 0," to the generalization of the interval to be unbounded, and finally to certain generalizations when the function is unbounded. Properties of such integrals are developed, and the connection between integration and differentiation is studied with two forms of the fundamental theorem of calculus.

The evaluation of a given integral is pursued with standard methods for exact valuation as well as with numerical methods. The notion of integral is seen to provide a useful alternative representation of the remainder in a Taylor series, and to provide a powerful tool for evaluating convergence of, and estimating the sum of or rate of divergence of, an infinite series. Convergence of a sequence of integrals is included. The Riemann notion of an integral is powerful but has limitations, some of which are explored.

Continuous probability theory is developed with the tools of this chapter, encompassing more general probability spaces and sigma algebras of events. Continuously distributed random variables are introduced, as well as their moments, and an accessible result is presented on discretizing such a random variable that links the discrete and continuous moment results. Several continuous distributions are presented and their properties studied.

Applications to finance in chapter 10 include the present and accumulated value of continuous cash flow streams with continuous interest rates, continuous interest rate term structures for bond yields, spot and forward rates, and continuous equity dividends and their reinvestment into equities. An alternative approach to applying the duration and convexity values of fixed income investments to approximating price functions is introduced. Numerical integration methods are exemplified by application to the normal distribution.

Finally, a generalized Black–Scholes–Merton pricing formula for a European option is developed from the general binomial pricing result of chapter 8, using a "continuitization" of the binomial distribution and a derivation that this continuitization converges to the appropriate normal distribution encountered in chapter 9. As another application, the Riemann–Stieltjes integral is introduced in the chapter exercises. It is seen to provide a mathematical link between the calculations within the discrete and continuous probability theories, and to generalize these to so-called mixed probability densities.

Acknowledgments

I have had the pleasure and privilege to train under and work with many experts in both mathematics and finance. My thesis advisor and mentor, Alberto P. Calderón (1920–1998), was the most influential in my mathematical development, and to this day I gauge the elegance and lucidity of any mathematical argument by the standard he set in his work and communications. In addition I owe a debt of gratitude to all the mathematicians whose books and papers I have studied, and whose best proofs have greatly influenced many of the proofs presented throughout this book.

I also acknowledge the advice and support of many friends and professional associates on the development of this book. Notably this includes (alphabetically) fellow academics Zvi Bodie, Laurence D. Booth, F. Trenery Dolbear, Jr., Frank J. Fabozzi, George J. Hall, John C. Hull, Blake LeBaron, Andrew Lyasoff, Bruce R. Magid, Catherine L. Mann, and Rachel McCulloch, as well as fellow finance practitioners Foster L. Aborn, Charles L. Gilbert, C. Dec Mullarkey, K. Ravi Ravindran and Andrew D. Smith, publishing professionals Jane MacDonald and Tina Samaha, and my editor at the MIT Press, Dana Andrus.

I thank the students at the Brandeis University International Business School for their feedback on an earlier draft of this book and careful proofreading, notably Amidou Guindo, Zhenbin Luo, Manjola Tase, Ly Tran, and Erick Barongo Vedasto. Despite their best efforts I remain responsible for any remaining errors.

Last, I am indebted to my parents, Dorothy and Domenic, for a lifetime of advice and support. I happily acknowledge the support and encouragement of my wife Lisa, who also provided editorial support, and sons Michael, David, and Jeffrey, during the somewhat long and continuing process of preparing my work for publication.

I welcome comments on this book from readers. My email address is *rreitano@ brandeis.edu.*

Robert R. Reitano
International Business School
Brandeis University

Introduction to Quantitative Finance

1 Mathematical Logic

1.1 Introduction

Nearly everyone thinks they know what logic is but will admit the difficulty in formally defining it, or will protest that such a formal definition is not necessary because its meaning is obvious. For example, we all like to stop an adversary in an argument with the statement "that conclusion is illogical," or attempt to secure our own victory by proclaiming "logic demands that my conclusion is correct." But if compelled in either instance, it may be difficult to formalize in what way logic provides the desired conclusion.

A legal trial can be all about attempts at drawing logical conclusions. The prosecution is trying to prove that the accused is guilty based on the so-called facts. The defense team is trying to prove the improbability of guilt, or indeed even innocence, based on the same or another set of facts. In this example, however, there is an asymmetry in the burden of proof. The defense team does not have to prove innocence. Of course, if such a proof can be presented, one expects a not guilty verdict for the accused. The burden of proof instead rests on the prosecution, in that they must prove guilt, at least to some legal standard; if they cannot do so, the accused is deemed not guilty.

Consequently a defense tactic is often focused not on attempting to prove innocence but rather on demonstrating that the prosecution's attempt to prove guilt is faulty. This might be accomplished by demonstrating that some of the claimed facts are in doubt, perhaps due to the existence of additional facts, or by arguing that even given these facts, the conclusion of guilt does not necessarily follow "logically." That is, the conclusion may be consistent with but not compelled by the facts. In such a case the facts, or evidence, is called "circumstantial."

What is clear is that the subject of logic applies to the drawing of conclusions, or to the formulation of inferences. It is, in a sense, the science of good reasoning. At its simplest, logic addresses circumstances under which one can correctly conclude that "B follows from A," or that "A implies B," or again, "If A, then B." Most would informally say that an inference or conclusion is logical if it makes sense relative to experience. More specifically, one might say that a conclusion follows logically from a statement or series of statements if the truth of the conclusion is guaranteed by, or at least compelled by, the truth of the preceding statement or statements.

For example, imagine an accused who is charged with robbing a store in the dark of night. The prosecution presents their facts: prior criminal record; eyewitness account that the perpetrator had the same height, weight, and hair color; roommate testimony that the accused was not home the night of the robbery; and the accused's inability to prove his whereabouts on the evening in question. To be sure, all these facts are consistent with a conclusion of guilt, but they also clearly do not compel such a conclusion. Even a more detailed eyewitness account might be challenged, since this crime occurred at night and visibility was presumably impaired. A fact that would be harder to challenge might be the accused's possession of many expensive items from the store, without possession of sales receipts, although even this would not be an irrefutable fact. "Who keeps receipts?" the defense team asserts!

The world of mathematical theories and proofs shares features with this trial example. For one, a mathematician claiming the validity of a result has the burden of proof to demonstrate this result is true. For example, if I assert the claim,

For any two integers N and M, it is true that $M + N = N + M$,

I have the burden of demonstrating that such a conclusion is compelled by a set of facts. A jury of my mathematical peers will then evaluate the validity of the assumed facts, as well as the quality of the logic or reasoning applied to these facts to reach the claimed conclusion. If this jury determines that my assumed facts or logic is inadequate, they will deem the conclusion "not proved." In the same way that a failed attempt to prove guilt is not a proof of innocence, a failed proof of truth is not a proof of falsehood. Typically there is no single judge who oversees such a mathematical process, but in this case every jury member is a judge.

Imagine if in mathematics the burden of proof was not as described above but instead reversed. Imagine if an acceptable proof of the claim above regarding N and M was: "It must be true because you cannot prove it is false." The consequence of this would be parallel to that of reversing the burden of proof in a trial where the prosecution proclaims: "The accused must be guilty because he cannot prove he is innocent." Namely, in the case of trials, many innocent people would be punished, and perhaps at a later date their innocence demonstrated. In the case of mathematics, many false results would be believed to be true, and almost certainly their falsity would ultimately be demonstrated at a later date. Our jails would be full of the innocent people; our math books, full of questionable and indeed false theory.

In contrast to an assertion of the validity of a result, if I claim that a given statement is false, I simply need to supply a single example, which would be called a "counterexample" to the statement. For example, the claim,

For any integer A, there is an integer B so that $A = 2B$,

can be proved to be false, or disproved, by the simple counterexample: $A = 3$.

What distinguishes these two approaches to proof is not related to the asserted statement being true or false, but to an asymmetry that exists in the approach to the presentation of mathematical theory. Mathematicians are typically interested in whether a general result is always true or not always true. In the first case, a general proof is required, whereas in the second, a single counterexample suffices. On the other hand, if one attempted to prove that a result is always false, or not always false, again in the first case, a general proof would be required, whereas in the second, a single counterexample would suffice. The asymmetry that exists is that one rarely sees propositions in mathematics stated in terms of a result that is always false, or not always false. Mathematicians tend to focus on "positive" results, as well as counterexamples to a positive result, and rarely pursue the opposite perspective. Of course, this is more a matter of semantic preference than theoretical preference. A mathematician has no need to state a proposition in terms of "a given statement is always false" when an equivalent and more positive perspective would be that "the negative of the given statement is always true." Why prove that "$2x = x$ is always false if $x \neq 0$" when you can prove that "for all $x \neq 0$, it is true that $2x \neq x$."

What distinguishes logic in the real world from the logic needed in mathematics is that in the real world the determination that A follows from B often reflects the human experience of the observers, for example, the judge and jury, as well as rules specified in the law. This is reinforced in the case of a criminal trial where the jury is given an explicit qualitative standard such as "beyond a reasonable doubt." In this case the jury does not have to receive evidence of the guilt of the accused that convinces with 100 percent conviction, only that the evidence does so beyond a reasonable doubt based on their human experiences and instincts, as further defined and exemplified by the judge.

In mathematics one wants logical conclusions of truth to be far more secure than simply dependent on the reasonable doubts of the jury of mathematicians. As mathematics is a cumulative science, each work is built on the foundation of prior results. Consequently the discovery of any error, however improbable, would have far-reaching implications that would also be enormously difficult to track down and rectify. So not surprisingly, the goal for mathematical logic is that every conclusion will be immutable, inviolate, and once drawn, never to be overturned or contradicted in the future with the emergence of new information. Mathematics cannot be built as a house of cards that at a later date is discovered to be unstable and prone to collapse.

In contrast, in the natural sciences, the burden of proof allowed is often closer to that discussed above in a legal trial. In natural sciences, the first requirement of a theory is that it be consistent with observations. In mathematics, the first requirement of a theory is that it be consistent, rigorously developed, and permanent. While it is always the case that mathematical theories are expanded upon, and sometimes become more or less in vogue depending on the level of excitement surrounding the development of new insights, it should never be the case that a theory is discarded because it is discovered to be faulty. The natural sciences, which have the added burden of consistency with observations, can be expected to significantly change over time and previously successful theories even abandoned as new observations are made that current theories are unable to adequately explain.

1.2   Axiomatic Theory

From the discussion above it should be no surprise that structure is desired of every mathematical theory:

1. Facts used in a proof are to be explicitly identified, and each is either assumed true or proved true given other assumed or proved facts.

2. The rules of inference, namely the logic applied to these facts in proofs, are to be "correct," and the definition of correct must be objective and immutable.

3. The collection of conclusions provable from the facts in item 1 using the logic in item 2 and known as **theorems**, are to be **consistent**. That is, for no statement P will the collection of theorems include both "statement P is true" and "the negation of statement P is true."

4. The collection of all **theorems** is to be **complete**. That is, for every statement P, either "statement P is a theorem" or "the negation of statement P is a theorem." A related but stronger condition is that the resulting theory is **decidable**, which means that one can develop a procedure so that for any statement P, one can determine if P is true or not true in a finite number of steps.

It may seem surprising that in item 1 the "truth" of the assumed facts was not the first requirement, but that these facts be explicitly identified. It is natural that identification of the assumed facts is important to allow a mathematical jury to do its review, but why not an absolute requirement of "truth"? The short answer is, there are no facts in mathematics that are "true" and yet at the same time dependent on no other statements of fact. One cannot start with an empty set of facts and somehow derive, with logic alone, a collection of conclusions that can be demonstrated to be true.

Consequently some basic collection of facts must be **assumed** to be true, and these will be the **axioms** of the theory. In other words, all mathematical theories are **axiomatic theories**, in that some basic set of facts must be assumed to be true, and based on these, other facts proved. Of course, the axioms of a theory are not arbitrary. Mathematicians will choose the axioms so that in the given context their truth appears undeniable, or at least highly reasonable. This is what ensures that the theorems of the mathematical theory in item 3, that is, the facts and conclusions that follow from these axioms, will be useful in that given context.

Different mathematical theories will require different sets of axioms. What one might assume as axioms to develop a theory of the integers will be different from the axioms needed to develop a theory of plane geometry. Both sets will appear undeniably true in their given context, or at least quite reasonable and consistent with experience. Moreover, even within a given subject matter, such as geometry, there may be more than one context of interest, and hence more than one reasonable choice for the axioms.

For example, the basic axioms assumed for **plane geometry**, or the geometry that applies on a "flat" two-dimensional sheet, will logically be different from the axioms one will need to develop **spherical geometry**, which is the geometry that applies on the surface of a sphere, such as the earth. Which axioms are "true"? The answer is both, since both theories one can develop with these sets of axioms are useful in the given contexts. That is, these sets of axioms can legitimately be claimed to be "true" because they imply theories that include many important and deep insights in the given contexts.

That said, in mathematics one can and does also develop theories from sets of axioms that may seem abstract and not have a readily observable context in the real world. Yet these axioms can produce interesting and beautiful mathematical theories that find real world relevance long after their initial development.

The general requirements on a set of axioms is that they are:

1. Adequate to develop an interesting and/or useful theory.

2. Consistent in that they cannot be used to prove both "statement P is true" and "the negation of statement P is true."

3. Minimal in that for aesthetic reasons, and because these are after all "assumed truths," it is desirable to have the simplest axioms, and the fewest number that accomplish the goal of producing an interesting and/or useful theory.

It is important to understand that the desirability, and indeed necessity, of framing a mathematical theory in the context of an axiomatic theory is by no means a modern invention. The earliest known exposition is in the *Elements* by **Euclid of Alexandria** (ca. 325–265 BC), so Euclid is generally attributed with founding the axiomatic method. The *Elements* introduced an axiomatic approach to two- and three-dimensional geometry (called **Euclidean geometry**) as well as number theory. Like the modern theories this treatise explicitly identifies axioms, which it classifies as "common notions" and "postulates," and then proceeds to carefully deduce its theorems, called "propositions." Even by modern standards the *Elements* is a masterful exposition of the axiomatic method.

If there is one significant difference from modern treatments of geometry and other theories, it is that the *Elements* defines all the basic terms, such as point and line, before stating the axioms and deducing the theorems. Mathematicians today recognize and accept the futility of attempting to define all terms. Every such definition uses words and references that require further expansion, and on and on. Modern developments simply identify and accept certain notions as undefined—the so-called primitive concepts—as the needed assumptions about the properties of these terms are listed within the axioms.

1.3   Inferences

Euclid's logical development in the *Elements* depends on "rules of inference" but does not formally include logic as a theory in and of itself. A formal development of the theory of logic was not pursued for almost two millennia, as mathematicians, following Euclid, felt confident that "logic" as they applied it was irrefutable. For instance, if we are trying to prove that a certain solution to an equation satisfies $x < 100$, and instead our calculation reveals that $x < 50$, without further thought we would proclaim to be done. Logically we have:

"$x < 50$ implies that $x < 100$" is a true statement.

"$x < 50$" is a true statement by the given calculation.

"$x < 100$" is a true statement, by "deduction."

Abstractly: if $P \Rightarrow Q$ and P, then Q. Here we use the well-known symbol $\Rightarrow$ for "implies," and agree that in this notation, all statements displayed are "true." That is, if $P \Rightarrow Q$ and P are true statements, then Q is a true statement. This is an example of the **direct method of proof** applied to the **conditional statement**, $P \Rightarrow Q$, which is also called an **implication**.

In the example above note that even as we were attempting to implement an objective logical argument on the validity of the conclusion that $x < 100$, we would likely have been simultaneously considering, and perhaps even biased by, the intuition we had about the given context of the problem. In logic, one attempts to strip away all context, and thereby strip away all intuition and bias. The logical conclusion we drew about x is true if and only if we are comfortable with the following logical statement in every context, for any meanings we might ever ascribe to the statements P and Q:

If $P \Rightarrow Q$ and P, then Q.

In logic, it must be all or nothing. The rule of inference summarized above is known as **modus ponens**, and it will be discussed in more detail below.

Another logical deduction we might make, and one a bit more subtle, is as follows:

"$x < 50$ implies that $x < 100$" is a true statement.

"$x < 100$" is not a true statement by demonstration.

"$x < 50$" is not a true statement, by deduction.

Again, abstractly: if $P \Rightarrow Q$ and $\sim Q$, then $\sim P$. Here we use the symbol $\sim Q$ to mean "the negation of Q is true," which is "logic-speak" for "Q is false." This is similar to the "direct method of proof," but applied to what will be called the **contrapositive** of the conditional $P \Rightarrow Q$, and consequently it can be considered an **indirect method of proof**. Again, we can apply this logical deduction in the given context if and only if we are comfortable with the following logical statement in every context:

If $P \Rightarrow Q$ and $\sim Q$, then $\sim P$.

The rule of inference summarized above is known as **modus tollens**, and will also be discussed below.

Clearly, the logical structure of an argument can become much more complicated and subtle than is implied by these very simple examples. The theory of mathematical logic creates a formal structure for addressing the validity of such arguments within which general questions about axiomatic theories can be addressed. As it turns out, there are a great many rules of inference that can be developed in mathematical logic, but *modus ponens* plays the central role because other rules can be deduced from it.

1.4   Paradoxes

One may wonder when and why mathematicians decided to become so formal with the development of a mathematical theory of logic, collectively referred to as **mathematical logic**, requiring an axiomatic structure and a formalization of rules of inference. An important motivation for increased formality has been the recognition that even with early efforts to formalize, such as in Euclid's *Elements*, mathematics has not always been formal enough, and the result was the discovery of a host of **paradoxes** throughout its history. A paradox is defined as a statement or collection of statements which appear true but at the same time produce a contradiction or a conflict with one's intuition. Some mathematical paradoxes in history where solved by later developments of additional theory. That is, they were indicative of an incomplete or erroneous understanding of the theory, often as a consequence of erroneous assumptions. Others were more fatal, in that they implied that the theory developed was effectively built as a house of cards and so required a firmer and more formal theoretical foundation.

Of course, paradoxes also exist outside of mathematics. The simplest example is the **liar's paradox**:

This statement is false.

The statement is paradoxical because if it is true, then it must be false, and conversely, if false, it must be true. So the statement is both true and false, or neither true nor false, and hence a paradox.

Returning to mathematics, sometimes an apparent paradox represents nothing more than sleight of hand. Take, for instance, the "proof" that $1 = 0$, developed from the following series of steps:

$a = 1,$

$a^2 = 1,$

$a^2 - a = 0,$

$a(a - 1) = 0,$

$a = 0,$

$1 = 0.$

The sleight of hand here is obvious to many. We divided by $a - 1$ before the fifth step, but by the first, $a - 1 = 0$. So the paradoxical conclusion is created by the illegitimate division by 0. Put another way, this derivation can be used to confirm the illegitimacy of division by zero, since to allow this is to allow the conclusion that $1 = 0$.

Sometimes the sleight of hand is more subtle, and strikes at the heart of our lack of understanding and need for more formality. Take, again, the following deduction that $1 = 0$:

$A = 1 - 1 + 1 - 1 + 1 - 1 + 1 - \cdots$

$\quad = (1 - 1) + (1 - 1) + (1 - 1) + \cdots$

$\quad = 0.$

$$A = 1 - (1 - 1) - (1 - 1) - (1 - 1) - \cdots$$
$$= 1,$$

so once more, $A = 1 = 0$. The problem with this derivation relates to the legitimacy of the grouping operations demonstrated; once grouped, there can be little doubt that the sum of an infinite string of zeros must be zero. Because we know that such groupings are fine if the summation has only finitely many terms, the problem here must be related to this example being an infinite sum. Chapter 6 on numerical series will develop this topic in detail, but it will be seen that this infinite alternating sum cannot be assigned a well-defined value, and that such grouping operations are mathematically legitimate only when such a sum is well-defined.

An example of an early and yet more complex paradox in mathematics is **Zeno's paradox**, arising from a mythical race between Achilles and a tortoise. **Zeno of Elea** (ca. 490–430 BC) noted that if both are moving in the same direction, with Achilles initially behind, Achilles can never pass the tortoise. He reasoned that at any moment that Achilles reaches a point on the road, the tortoise will have already arrived at that point, and hence the tortoise will always remain ahead, no matter how fast Achilles runs. This is a paradox for the obvious reason that we observe faster runners passing slower runners all the time. But how can this argument be resolved?

Although this will be addressed formally in chapter 6, the resolution comes from the demonstration that the infinite collection of observations that Zeno described between Achilles and the tortoise occur in a finite amount of time. Zeno's conclusion of paradox implicitly reflected the assumption that if in each of an infinite number of observations the tortoise is ahead of Achilles, it must be the case that the tortoise is ahead for all time. A formal resolution again requires the development of a theory in which the sum of an infinite collection of numbers can be addressed, where in this case each number represents the length of the time interval between observations.

Another paradox is referred to as the **wheel of Aristotle**. **Aristotle of Stagira** (384–322 BC) imagined a wheel that has inner and outer concentric circles, as in the inner and outer edges of a car tire. He then imagined a fixed line from the wheel's hub extending through these circles as the wheel rotates. Aristotle argued that at every moment, there is a one-to-one correspondence between the points of intersection of the line and the inner wheel, and the line and the outer wheel. Consequently the inner and outer circles must have the same number of points and the same circumference, a paradox. The resolution of this paradox lies in the fact that having a $1:1$ correspondence between the points on these two circles does not ensure that they have equal lengths, but to formalize this required the development of the theory of infinite sets many hundreds of years later. At the time of Aristotle it was not understood how two sets could be put in $1:1$ correspondence and not be "equivalent" in their size or measure, as is apparently the case for two finite sets. Chapter 2 on number systems will develop the topic of infinite sets further.

The final paradox is unlike the others in that it effectively dealt a fatal blow to an existing mathematical theory, and made it clear that the theory needed to be redeveloped more formally from the beginning. It is fair to say that the paradoxes above didn't identify any house of cards but only a situation that could not be appropriately explained within the mathematical theory or understanding of that theory developed to that date. The next paradox has many forms, but a favorite is called the **Barber's paradox**. As the story goes, in a town there is a barber that shaves all the men that do not shave themselves, and only those men. The question is: Does the barber shave himself? Similar to the liar's paradox, we conclude that the barber shaves himself if and only if he does not shave himself. The problem here strikes at the heart of set theory, where it had previously been assumed that a set could be defined as any collection satisfying a given criterion, and once defined, one could determine unambiguously whether or not a given element is a member of the set. Here the set is defined as the collection of individuals satisfying the criterion that they don't shave themselves, and we can get no logical conclusion as to whether or not the barber is a member of this set.

An equivalent form of this paradox, and the form in which it was discovered by **Bertrand Russell** (1872–1970) in 1901 and known as **Russell's paradox**, makes this set theory connection explicit. Let X denote the set of all sets that are not elements of themselves. The paradox is that one concludes X to be an element of itself if and only if it is not an element of itself. This discovery was instrumental in identifying the need for, and motivating the development of, a more careful axiomatic approach to set theory. Of course, the need for the development of a more formal axiomatic theory for all mathematics was equally compelled, since if mathematics went astray by defining an object as simple and intuitive as a set, who could be confident that other potential crises didn't loom elsewhere?

1.5   Propositional Logic

1.5.1   Truth Tables

Much of mathematical logic can be better understood once the concept of **truth table** is introduced and basic relationships developed. The starting point is to define a **statement** in a mathematical theory as any declarative sentence that is either true or false, but not both. For example, "today the sky is blue" and "$5 < 7$" are statements. An expression such as "$x < 7$" is not a statement because we cannot assign T or F to it without knowing what value the variable x assumes. Such an expression will be called a **formula** below. While a formula is not a statement because the variable x is a **free variable**, it can be made into a statement by making x a **bound variable**. The most common ways of accomplishing this is with the **universal quantifier**, $\forall$, and **existential quantifier**, $\exists$, defined as follows:

- $\forall x$ denotes: "for all x."
- $\exists x$ denotes: "there exists an x such that."

For example, $\forall x\ (x < 7)$ and $\exists x\ (x < 7)$ are now statements. The first, "for all x, x is less than 7" is assigned an F; the second, "there exists an x such that x is less than 7" is a T.

A truth table is a mechanical device for deciphering the truth or falsity of a complicated statement based on the truth or falsity of its various substatements. Complicated statements are constructed using **statement connectives** in various combinations. Of course, from the discussion above it should be no surprise that the initial collection of true statements for a given mathematical theory would be the "assumed facts" or axioms of the theory. Truth tables then provide a mechanism for determining the truth or falsity of more complicated statements that can be formulated from these axioms and, as we will see, also provide a framework within which one can evaluate the logical integrity of a given inference one makes in a proof.

If P and Q are statements, we define the following statement connectives and present the associated truth tables. Negation is a **unary** or **singulary connective**, whereas the others are **binary connectives**. In each case the truth table identifies all possible combinations of T or F for the given statements, denoted P or Q, and then assigns a T or F to the defined statements.

1. **Negation:** $\sim P$ denotes the statement "not P."

P	$\sim P$
T	F
F	T

2. **Conjunction:** $P \wedge Q$ denotes the statement "P and Q."

P	Q	$P \wedge Q$
T	T	T
T	F	F
F	T	F
F	F	F

3. Disjunction: $P \vee Q$ denotes the statement "P or Q" but understood as "P and/or Q."

P	Q	$P \vee Q$
T	T	T
T	F	T
F	T	T
F	F	F

4. Conditional: $P \Rightarrow Q$ denotes the statement "P implies Q."

P	Q	$P \Rightarrow Q$
T	T	T
T	F	F
F	T	T
F	F	T

5. Biconditional: $P \Leftrightarrow Q$ denotes the statement "P if and only if Q."

P	Q	$P \Leftrightarrow Q$
T	T	T
T	F	F
F	T	F
F	F	T

In other words, we have the following truth assignments, which are generally consistent with common usage:

- $\sim P$ has the **opposite** truth value as P.
- $P \wedge Q$ is true only when **both** P and Q are true.
- $P \vee Q$ is true when **at least one of** P and Q are true.
- $P \Rightarrow Q$ is true **unless** P is T, and Q is F.
- $P \Leftrightarrow Q$ is true when P and Q have the **same truth values**.

There may be two surprises here. First off, in mathematical logic the disjunctive "or" means "and/or." In common language, "P or Q" usually means "P or Q but not both." If you are told, "your money or your life," you do not expect an unfavorable outcome after handing over your wallet. Obviously, if the thief is a mathematician, there could be an unpleasant surprise.

An important consequence of this interpretation, which would not be true for the common language notion, is that there is a logical symmetry between conjunction and disjunction when negation is applied:

$$\sim(P \wedge Q) \Leftrightarrow (\sim P) \vee (\sim Q),$$

$$\sim(P \vee Q) \Leftrightarrow (\sim P) \wedge (\sim Q).$$

That is, the statement "$P \wedge Q$" is false if and only if "either P is false or Q is false," and the statement "$P \vee Q$" is false if and only if "both P is false and Q is false."

The equivalence of these statements follows from a truth table analysis that utilizes the basic properties above. For example, the truth table for the first statement is:

P	Q	$\sim(P \wedge Q)$	$(\sim P) \vee (\sim Q)$	$\sim(P \wedge Q) \Leftrightarrow (\sim P) \vee (\sim Q)$
T	T	F	F	T
T	F	T	T	T
F	T	T	T	T
F	F	T	T	T

This demonstrates that the two statements always have the same truth values.

The second surprise relates to the conditional truth values in the last two rows of the table, when P is false. Then, whether Q is true or false, the conditional $P \Rightarrow Q$ is declared true. For example, let

P : There is a mispricing in the market,

Q : I will attempt to arbitrage.

So $P \Rightarrow Q$ is a statement I might make:

"If there is a mispricing in the market, then I will attempt to arbitrage."

The question becomes, How would you evaluate whether or not my statement is true? The truth table declares this statement true when P and Q are both true, and so would you. In other words, if there was a mispricing and I attempted to arbitrage, you would judge my statement true. Similarly, if P was true and I did not make this attempt, you would judge my statement false, consistent with the second line in the truth table.

Now assume that there was not a mispricing in the market today, and yet I was observed to be attempting an arbitrage. Would my statement above be judged false? What if in the same market, I did not attempt to arbitrage, would my statement be deemed false? The truth table for the conditional states that in both cases my original statement would be deemed true, although in the real world the likely conclusion would be "not apparently false." In other words, in these last two cases my actions do not present evidence of the falsity of my statement, and hence the truth table deems my statement "true." Simply said, the truth table holds me truthful unless proved untruthful, or innocent unless proved guilty.

A consequence of this truth table assignment for the conditional is that

$$(P \Rightarrow Q) \Leftrightarrow \sim(P \wedge \sim Q).$$

In other words, $P \Rightarrow Q$ has exactly the same truth values as does $\sim(P \wedge \sim Q)$. The associated truth table is as follows:

P	Q	$P \Rightarrow Q$	$\sim(P \wedge \sim Q)$	$(P \Rightarrow Q) \Leftrightarrow \sim(P \wedge \sim Q)$
T	T	T	T	T
T	F	F	F	T
F	T	T	T	T
F	F	T	T	T

This truth table analysis and the one above were somewhat tedious, especially when all the missing columns are added in detail, but note that they were entirely mechanical. No intuition was needed; we just apply in a methodical way the logic rules as defined by the truth tables above.

These truth tables have another interpretation, and that is, for any statements P and Q, and any truth values assigned, the statement

$$\sim(P \wedge Q) \Leftrightarrow (\sim P) \vee (\sim Q),$$

is a **tautology**, which is to say that it is always true. The same can be said for the biconditional statements illustrated above. Tautologies will be seen to form the foundation for developing and evaluating rules of inference, and more specifically, the logical integrity of a given proof.

There are many other tautologies possible, in fact infinitely many. One reason for this is that there is redundancy in the list of connectives above:

$$\sim, \wedge, \vee, \Rightarrow, \Leftrightarrow.$$

In a formal treatment of mathematical logic, only $\sim$ and $\Rightarrow$ need be introduced, and the others are then defined by the following statements, all of which are tautologies in the framework above:

$$P \vee Q \Leftrightarrow \sim P \Rightarrow Q,$$

$P \wedge Q \Leftrightarrow \sim(P \Rightarrow \sim Q),$

$(P \Leftrightarrow Q) \Leftrightarrow (P \Rightarrow Q) \wedge (Q \Rightarrow P).$

Note that the last statement can in turn be expressed in terms of only $\sim$ and $\Rightarrow$ using the second tautology.

There is also redundancy between the universal and existential quantifiers. In formal treatments one introduces the universal quantifier $\forall$ and defines the existential quantifier $\exists$ by

$\exists x P(x) \Leftrightarrow \sim \forall x(\sim P(x)).$

In other words, "there exists an x so that statement $P(x)$ is true" is the same as "it is false that for all x the statement $P(x)$ is false."

Admittedly, such definitional connections require one to pause for understanding, and one might wonder why all the terms are simply not defined straightaway instead of in the complicated ways above. The reason was noted earlier in the discussion on axioms. One goal of an axiomatic structure is to be minimal, or at least parsimonious. The cost of this goal is often apparent complexity, as one might spend considerable effort proving a statement that virtually everyone would be more than happy just accepting as another axiom. But the goal of mathematical logic is not the avoidance of complexity by adding more axioms; it is the illumination of the theory and the avoidance of potential paradoxes by minimizing the number of axioms needed. The fewer the axioms, the more transparent the theory becomes, and the less likely the axioms will be in violation of another important goal of an axiomatic structure. And that is consistency.

1.5.2 Framework of a Proof

In later chapters various statements will be made under the heading **proposition**, which is the term used in this book for the more formal sounding **theorem**. These terms are equivalent in mathematics, and the choice reflects style rather than substance. In virtually all cases, a "proof" of the statement will be provided. A **lemma** is yet another name for the same thing, although it is generally accepted that a **lemma** is considered a relatively minor result, whereas a proposition or theorem is a major result. Some authors distinguish between proposition and theorem on the same basis, with theorem used for the most important results.

This terminology is by no means universally accepted. For example, students of finance will undoubtedly encounter *Ito's lemma*, and soon discover that in the theory underlying the pricing of financial derivatives like options, this lemma is perhaps the most important theoretical result in quantitative finance.

Now the typical structure for the statement of a proposition is

If P, then Q.

The statement P is the **hypothesis** of the proposition, and in some cases it will be a complex statement with many substatements and connectives, while the statement Q is the **conclusion**. The goal of this and the next section is to identify logical frameworks for such proofs.

First off, a proof of the statement "If P, then Q" is not equivalent to a proof of the statement '$P \Rightarrow Q$' despite their apparent equivalence in informal language. Specifically,

"If P, then Q" means "if statement P is true, then statement Q is true,"

whereas

'$P \Rightarrow Q$' means "the statement P implies Q is true."

Of course, one is hardly interested in proving statements such as '$P \Rightarrow Q$' unless Q can be asserted to be a true statement. That is the true goal of a proposition, to achieve the conclusion that Q is true. However, the statement $P \Rightarrow Q$ was seen to be true in three of the four cases displayed in the truth table above, and in only one of these three cases is Q seen to be true. Namely the truth of '$P \Rightarrow Q$' assures the truth of Q only when P is true. Consequently, if we want to prove the typical propositional structure above, which is to say that we can infer the truth of statement Q from the truth of statement P, we can prove the following:

If P and $P \Rightarrow Q$, then Q.

If this statement is written in the notation of logic, it is in fact a tautology, and always true. That is, in the truth table of

$$P \wedge (P \Rightarrow Q) \Rightarrow Q, \tag{1.1}$$

we have that for any assignment of the truth values to P and Q, this statement has constant truth value of "true."

This statement is the central **rule of inference** in logic, and it is known as **modus ponens**. It says that:

If statement P is true, and the statement $P \Rightarrow Q$ is demonstrated as true, then Q must be true.

This is the formal basis of many mathematical proofs of "If P, then Q." Of course, the language of the proof usually focuses on the development of the truth of the implication: $P \Rightarrow Q$, while the truth of the statement P, which is the hypothesis of the theorem, is simply implied. Moreover, if P were false, the demonstration of the truth of $P \Rightarrow Q$ would be for naught, since in this case Q could be true or false, as the truth table above attests.

In the next section we investigate proof structures in more detail. The central idea is every logical structure for a valid proof must be representable as a tautology, such as the *modus ponens* structure in (1.1). As we have seen, it is straightforward and mechanical, though perhaps tedious, to verify that a given proof structure, however complicated, is indeed a tautology. Here are a few other possible proof structures that are tautologies intuitively, as well as relatively easy to demonstrate in a truth table. Each is simply related to a single line on one of the basic truth tables given for the connectives:

$$P \wedge (P \wedge Q) \Rightarrow Q,$$

$$(P \vee Q) \wedge \sim Q \Rightarrow P,$$

$$(P \Leftrightarrow Q) \wedge \sim Q \Rightarrow \sim P.$$

For example, on the truth table for $P \wedge Q$, the only row where both P and $P \wedge Q$ are true is the row where Q is also true. In any other row, one or both of P and $P \wedge Q$ are false, and hence the conjunction $P \wedge (P \wedge Q)$ is false, assuring that the conditional $P \wedge (P \wedge Q) \Rightarrow Q$ is true. That is exactly how this statement becomes a tautology, and this logic will be seen to hold in all such cases. Specifically, when the hypothesis of the proposition is a conjunction, as is typically the case, we only really have to evaluate the case where all substatements are true, and assure that the conclusion is then true in this case. In all other cases the conjunction will be false and the conditional automatically true.

1.5.3 Methods of Proof

With *modus ponens* in the background, the essence of virtually any mathematical proof is a demonstration of the truth of the implication $P \Rightarrow Q$. To this end, the first choice one has is to prove the **direct conditional** statement $P \Rightarrow Q$, or its **contrapositive** $\sim Q \Rightarrow \sim P$. These statements are logically equivalent, which is to say that they have the same truth values in all cases. In other words, the statement

$$(P \Rightarrow Q) \Leftrightarrow (\sim Q \Rightarrow \sim P) \tag{1.2}$$

is a tautology, in that for any assignment of the truth values to P and Q, this statement has constant truth value of "true."

If *modus ponens* is applied to this contrapositive, we arrive at

$$\sim Q \wedge (\sim Q \Rightarrow \sim P) \Rightarrow \sim P. \tag{1.3}$$

However, because of (1.2), this can also be written as

$$\sim Q \wedge (P \Rightarrow Q) \Rightarrow \sim P, \tag{1.4}$$

which is a rule of inference known as **modus tollens** and exemplified in section 1.2 on axiomatic theory. It is not an independent rule of inference, of course, as it follows from *modus ponens*. In words, (1.4) states that if $P \Rightarrow Q$ is true, and $\sim Q$ is true, meaning Q is false, then $\sim P$ is also true, or P false.

In some proofs, the direct statement lends itself more easily to a proof, in others, the contrapositive works more easily, while in others still, both are easy, and in others still yet, both seem to fail miserably. The only general rule is, if the method you are attempting is failing, try the other. Experience with success and failure improves the odds of identifying the more expedient approach on the first attempt.

For example, assume that we wish to prove $P \Rightarrow Q$, where

$$P : a = b,$$

$$Q : a^2 = b^2.$$

The direct proof might proceed as

$$a = b \Rightarrow [a^2 = ab \text{ and } ab = b^2] \Rightarrow a^2 = b^2.$$

The contrapositive proof proceeds by first identifying the statement negations

$$\sim P : a \neq b,$$

$$\sim Q : a^2 \neq b^2,$$

and constructing the proof as

$$\sim Q \Rightarrow a^2 - b^2 \neq 0$$

$$\Rightarrow (a+b)(a-b) \neq 0$$

$$\Rightarrow [(a+b) \neq 0 \text{ and } (a-b) \neq 0]$$

$$\Rightarrow a \neq b.$$

In the last statement we also can conclude that $a \neq -b$, but this is extra information not needed for the given demonstration.

Once a choice is made between the direct statement and its contrapositive, there are two common methods for proving the truth of the resulting implication. To simplify notation, we denote the implication to be proved as $A \Rightarrow C$, where A denotes either P or $\sim Q$, and C denoted either Q or $\sim P$, respectively.

The Direct Proof

The first approach is what we often think of as the use of "deductive" reasoning, whereby if we cannot prove $A \Rightarrow C$ in one step, we may take two or more steps. For example, proving that for some statement B that $A \Rightarrow B$ and $B \Rightarrow C$, it would seem transparent that $A \Rightarrow C$. One expects that such a partitioning of the demonstration ought to be valid, independent of how many intermediate implications are developed, and indeed this is the case. It is based on a result in logic that is called a **syllogism** and forms the basis of what is known as a **direct proof**. Specifically, we have that

$$(A \Rightarrow B) \wedge (B \Rightarrow C) \Rightarrow (A \Rightarrow C) \tag{1.5}$$

is a tautology. That is, for any assignment of the truth values to A, B, and C, this statement has constant truth value of "true."

This direct method is very powerful in that it allows the most complicated implications to be justified through an arbitrary number of smaller, and more easily proved, implications. In the proof above that $P \Rightarrow Q$, this method was in fact used without mention as follows:

$A : a = b,$

$B : a^2 = ab \wedge ab = b^2,$

$C : a^2 = b^2.$

Proof by Contradiction

The second approach to proving an implication is considered an **indirect proof**, and is also known as **reductio ad absurdum**, as well as **proof by contradiction**. In its simplest terms, proof by contradiction proceeds as follows:

To prove P, assume $\sim P$. If $R \wedge \sim R$ is derived for any R, deduce P.

In other words,

If $\sim P \Rightarrow (R \wedge \sim R)$, then P.

If $\sim P \Rightarrow (R \wedge \sim R)$ is true, then since $R \wedge \sim R$ is always false, it must be the case that $\sim P$ is also false, and hence P is true. The logical structure of this is the tautology

$$[\sim P \Rightarrow (R \wedge \sim R)] \Rightarrow P. \tag{1.6}$$

Remark 1.1   *It is often the case that in a given application, what is called a proof by contradiction appears as*

If $\sim P \Rightarrow R$, and R is known to be false, then P. $\tag{1.7}$

For example, one might derive that $\sim P \Rightarrow R$, where R is the statement $1 \neq 1$. Implicitly, the truth of the statement $\sim R$, that $1 = 1$, does not need to be explicitly identified, but is understood. Also note that the truth of a statement like $1 = 1$ does not need to "follow" in some sense from the statement $\sim P$. That (1.7) is a valid conclusion can also be formalized by explicitly identifying the truth of $\sim R$ in the tautology

$$[(\sim P \Rightarrow R) \wedge \sim R] \Rightarrow P,$$

which except for notation is equivalent to modus tollens in (1.4). This approach also justifies the terminology of a reductio ad absurdum, namely from the assumed truth of $\sim P$ one deduces an absurd conclusion, R, such as $1 \neq 1$.

The indirect method of proof may appear complex, but with some practice, it is quite simple. The central point is that for any statement R, it is the case that $R \wedge \sim R$ is always false. This is because its negation, $\sim R \vee R$, is always true and

$$\sim(R \wedge \sim R) \Leftrightarrow \sim R \vee R \tag{1.8}$$

is a tautology. That is, for any statement R, either R is true or $\sim R$ is true. This is known as the **law of the excluded middle**.

Before formalizing this further, let's apply this approach to the earlier simple example, taking careful steps:

Step 1   State what we seek to prove: $a = b \Rightarrow a^2 = b^2$.

Step 2   Develop the negation of this implication. Looking at the truth table for the conditional, an implication $A \Rightarrow C$ is false only when A is true, and C is false. So the negation of what we seek to prove is

$$a = b \quad \text{and} \quad a^2 \neq b^2.$$

Step 3   What can we conclude from this assumed statement? This amounts to "playing" with some mathematics and seeing what we get:

$a^2 \neq b^2 \Leftrightarrow a^2 - b^2 \neq 0$

$\qquad \Leftrightarrow (a+b)(a-b) \neq 0$

$\qquad \Leftrightarrow a+b \neq 0 \quad \text{and} \quad a-b \neq 0,$

whereas

$a = b \Leftrightarrow a - b = 0.$

Step 4 Identify the contradiction: we have concluded that both $a - b = 0$ and $a - b \neq 0$.

Step 5 Claim victory: $a = b \Rightarrow a^2 = b^2$ is true.

Admittedly, this may look like an ominous process, but with a little practice the logical sequence will become second nature. The payoff to practicing this method is that this provides a powerful and frequently used alternative approach to proving statements in mathematics as will be often seen in later chapters.

Summarizing, we can rewrite (1.6) in the way it is most commonly used in mathematics, and that is when the statement P is in fact an implication $A \Rightarrow C$. To do this, we use the result from step 2 as to the logical negation of an implication. That is,

$\sim(A \Rightarrow C) \Leftrightarrow A \wedge \sim C.$

It is also the case that the most common contradiction one arrives at in (1.6) is not a general statement R, but as in the example above, it is a contradiction about A. We express this result first in the common form:

If $(A \wedge \sim C) \Rightarrow \sim A$, then $A \Rightarrow C$. $\hfill (1.9)$

Tautology: $[(A \wedge \sim C) \Rightarrow \sim A] \Rightarrow (A \Rightarrow C)$.

In the more general case,

If $(A \wedge \sim C) \Rightarrow R \wedge \sim R$, then $A \Rightarrow C$. $\hfill (1.10)$

Tautology: $[(A \wedge \sim C) \Rightarrow (R \wedge \sim R)] \Rightarrow (A \Rightarrow C)$.

Remark 1.2 *As in remark 1.1 above, (1.10) can also be applied in the context of $(A \wedge \sim C) \Rightarrow R$, where R is known to be false. The conclusion of the truth of $A \Rightarrow C$ again follows.*

Proof by Induction

A **proof by induction** is an approach frequently used when the statement to be proved encompasses a (countably) infinite number of statements (more on countably infinite sets in chapter 2 on number systems). A somewhat complicated example is the statement in the introduction: For any two integers M and N, we have that $M + N = N + M$. This is complicated because this statement involves two general quantities, and each can assume an infinite number of values. In other words, this statement is an economical way of expressing an infinite number of equalities ($1 + 9 = 9 + 1$, $-4 + 37 = 37 + (-4)$, etc.).

A simpler example involving only one such quantity is as follows:

$$\text{If } N \text{ is a positive integer, then } 1 + 2 + \cdots + N = \frac{N(N+1)}{2}. \tag{1.11}$$

This has the form of an equality, $P = Q$, but neither P nor Q is a simple declarative statement. Instead, both are indexed by the positive integers. That is, we seek to prove

$$\forall N, P(N) = Q(N), \tag{1.12}$$

where we define

$$P(N) = 1 + 2 + \cdots + N,$$

$$Q(N) = \frac{N(N+1)}{2}.$$

Obviously, for any fixed value of N, the proof requires no general theory, and the result can be demonstrated or contradicted by a hand or computer calculation. A proof by induction provides an economical way to demonstrate the validity of (1.12) for all N. The idea can be summarized as follows:

$$\text{If} \quad P(1) = Q(1),$$

$$\text{and} \quad [P(N) = Q(N)] \Rightarrow [P(N+1) = Q(N+1)], \tag{1.13}$$

$$\text{then} \quad \forall N, P(N) = Q(N).$$

In other words, proof by induction has two steps:

Step 1 (Initialization Step)   Show the statement to be true for the smallest value of N needed, say $N = 1$ (sometimes $N = 0$).

Step 2 (Induction Step)   Show that if the result is true for a given N, it must also be true for $N + 1$.

The logic is self-evident. From the initialization step, the induction step assures the truth for $N = 2$, which when applied again assures the truth of $N = 3$, and so forth.

Example 1.3 *To show (1.11), we see that the result is apparently true for* $N = 1$. *Next, assuming the result is true for* N, *we get*

$$1 + 2 + \cdots + N + (N + 1) = \frac{N(N + 1)}{2} + N + 1$$

$$= \frac{N(N + 1)}{2} + \frac{2(N + 1)}{2}$$

$$= \frac{(N + 1)(N + 2)}{2},$$

which is the desired result.

*1.6 Mathematical Logic

Mathematical logic is one of the most abstract and symbolic disciplines in mathematics. This is quite deliberate. As exemplified above, the goal of mathematical logic is to define and develop the properties of deductive systems that are context free. We cannot be certain that a given logical development is correct if our assessment of it is encumbered by our intuition in a given application to a field of mathematics. So the goal of mathematical logic is to strip away any hint of a context, eliminate all that is familiar in a given theory, and study the logical structure of a general, and unspecified, mathematical theory.

To do this, mathematical logic must first erase all familiar notations that imply a given context. Also its symbolic structure needs to be very general so that it allows application to a wide variety of mathematical disciplines or contexts. As a result mathematical logic is highly symbolic, highly stylized, leaving the logician with nothing to guide her except the rules allowed by the structure. This way every deduction can be verified mechanically, effectively as an appropriately structured computer program. This program then declares a symbolic statement to be "true" if and only if it is able to construct a symbol sequence, using only the axioms or assumed facts and rules of inference that results in the deductive construction of the statement. No context is assumed, and no intuition is needed or desired.

The preceding section's informal introduction to the mathematical logic of statements, which is referred to as **statement calculus** or **propositional logic**, is a small subset of the discipline of mathematical logic. The axiomatic structure of statement calculus includes:

1. Certain **formal symbols** made up of **logical operators** ($\sim$ and $\Rightarrow$, but excluding $\forall$ and $\exists$), **punctuation marks** (e.g., parentheses), and other **symbols** that are undefined, but in terms of which other needed concepts such as **variable, predicate, formula, operation, statement**, and **theorem** are defined.

2. Axioms that identify the basic formula structures that will be assumed true.

3. A **rule of inference**: *modus ponens*.

The resulting theory can then be shown to be **complete** because it is **decidable**. The algorithm for determining if a given statement is true or not is the construction of the associated truth table, any one of which requires only a finite number of steps to develop. The key to this result is that a statement is a theorem in statement calculus, meaning it can be deduced from the axioms with *modus ponens* if and only if the statement is a tautology in the sense of the associated truth table.

For many areas of mathematics, however, statement calculus is insufficient in that it excludes statements of the form

$$\forall x P(x) \quad \text{or} \quad \exists x P(x)$$

that are central to the statements in most areas of mathematics. The mathematical theory developed to accommodate these notions is called **first-order predicate calculus**, or simply **first-order logic**.

Landmark results in first-order logic are **Gödel's incompleteness theorems**, published in 1931 by **Kurt Gödel** (1906–1978). Although far beyond the boundaries of this book, the informal essence of Gödel's first theorem is this: In any consistent first-order theory powerful enough to develop the basic theory of numbers, one can construct a true statement that is not provable in this system. In other words, in any such theory one cannot hope to confirm or deny every statement that can be made within the theory, and hence every such theory is "incomplete."

The informal essence of Gödel's second theorem is this: In any consistent first-order theory powerful enough to develop the basic theory of numbers, it is impossible to prove consistency from within the theory. In other words, for any such theory the proof of consistency will of necessity have to be framed outside the theory.

1.7   Applications to Finance

The applications of mathematical logic discussed in this chapter to finance are both specific and general. First off, there are many specific instances in finance when one has to develop a proof of a given result. Typically the framework for this proof is not a formally stated theorem as one sees in a research paper. The proof is more or less an application of, and sometimes the adaptation of, a given theory to a situation not explicitly anticipated by the theory, or entirely outside the framework anticipated.

Alternatively, one might be developing and testing the validity of a variety of hypothetical implications that appear reasonable in the given context. In such specific applications the investigation pursued often requires a very formal process of derivation, logical deduction, and proof, and the tools described in the sections above can be helpful in that they provide a rigorous, or at least semi-rigorous, framework for such investigations.

More specifically, a truth table can often be put to good use to investigate the validity of a subtle logical derivation involving a series of implications and, based on the various identities demonstrated, to provide alternative approaches to the desired result. For example, a proof by contradiction applied to the contrapositive of the desired implication can be subtle in the language provided by the context of the problem. Just as in mathematics, isolating the logical argument from the context provides a better framework for assessing the former without the necessary bias that the latter might convey. In addition, when the investigation ultimately reduces to the proof of a given implication, as often arises in an attempt to evaluate the truth of a reasonable and perhaps even desired implication, the various methods of proof provide a framework for the attack.

There is also a general application of the topics in this chapter to finance, and more broadly, any applied mathematical discipline, and that is as a cautionary tale. All too often the power and rigor of mathematics is interpreted to imply a certain robustness. That is, one assumes that the true results in mathematics are "so true" that they are robust enough to remain true even when one alters the hypotheses a bit, or is careless in their application to a given situation. Actually nothing could be further from the truth.

The most profound thought on this point I recall was made long ago by my thesis advisor and mentor, **Alberto P. Calderón** (1920–1998), during a working visit made to his office. What he said on this point, as perhaps altered by less than perfect recall, was: "*The most interesting and powerful theorems in mathematics are just barely true.*" In other words, the conclusions of the "best theorems" in mathematics are both solid in their foundation and yet fragile; they represent a delicate relationship between the assumed hypothesis and the proved conclusion. In the "best" theorems the hypothesis is in a sense very close to the minimal assumption needed for the conclusion, or said another way, the conclusion is very close to the maximal result possible that follows from the given hypothesis. The "more true" a theorem is, in the sense of excessive hypotheses or suboptimal conclusions, the less interesting and important it is. Such theorems are often revisited in the literature in search of a more refined and economical statement.

The implication of this cautionary tale is that it is insufficient to simply memorize a general version of the many results in mathematics without also paying close attention to the assumptions made to prove these results. A slight alteration of the assumptions, or an attempt to broaden the conclusions, can and will lead to periodic disasters. But more than just the need to carefully utilize known results, it is important to understand the proof of how the given hypotheses provide the given conclusions since, in practice, the researcher is often attempting to alter one or the other, and evaluate what part of the original conclusion may still be valid.

The snippets of mathematical history alluded to in this chapter, and the paradoxes, support this perspective of the fragility of the best results, and the care needed to get them right and in balance. As careful as mathematicians were in the development of their subjects, pitfalls were periodically identified and ultimately had to be overcome. And perhaps it is obvious, but a great many of these mathematicians were intellectual giants, and leaders in their mathematical disciplines. The pitfalls were far less a reflection of their abilities than a testament to the subtlety of their discipline.

As a simple example of this cautionary tale, it is important that in any mathematical pursuit, any quantitative calculation, and any logical deduction, one must keep in mind that the truth of statement Q as promised by *modus ponens*, depends on **both** the truth of the hypothesis P and the truth of the implication $P \Rightarrow Q$. The truth of the latter relies on the careful application of many of the principles discussed above, and it is often the focus of the investigation. But *modus ponens* cautions that equally important is to do what is often the more tedious part of the derivation, and that is to check and recheck the validity of the assumptions, the validity of P.

A simple example is the principle of **arbitrage**, which tends to fascinate new finance students. In an arbitrage, one is able to implement a market trade at no cost, that is risk free over some period of time, and with positive likelihood of producing a profit at the end of the period and no chance of loss. Invariably, students will perform long, detailed, and very creative calculations that identify arbitrages in the financial markets. In other words, they are very detailed and creative in their derivations of the truths of the statements $P \Rightarrow Q$, where in their particular applications, P is the statement "I go long and short various instruments at the market prices I see in the press or online," and Q is the statement "I get embarrassingly rich as the profits come rolling in."

Of course, the poorly trained students make mistakes in this proof of $P \Rightarrow Q$, using the wrong collection of instruments, or not identifying the risks that exist post trade. But the better students produce perfect and sometimes subtle trade analyses. Invariably the finance professor is left the job of bursting bubbles with the question: "How sure are you that the securities are tradable at the prices assumed?" In other words, how sure are you that P is true?

The answer to this question comes from a logical analysis of the following argument using syllogism and *modus tollens*:

If finance students' arbitrages worked,
there would be numerous, embarrassingly rich finance students.
If finance students could trade at the assumed prices,
their arbitrages would work.
There are not numerous embarrassingly rich finance students.

Exercises

Practice Exercises

1. Create truth tables to evaluate if the following statements, $A \Leftrightarrow B$, are tautologies:

(a) $P \vee Q \Leftrightarrow \sim P \Rightarrow Q$

(b) $(P \vee Q) \vee (P \Rightarrow Q) \Leftrightarrow P \wedge Q$

(c) $(P \Leftrightarrow Q) \Leftrightarrow (P \Rightarrow Q) \wedge (Q \Rightarrow P)$

(d) $[P \Rightarrow (Q \vee R)] \wedge [Q \Rightarrow (P \vee R)] \Leftrightarrow R$

2. It was noted that the truth of $P \Rightarrow Q$ does not necessarily imply the truth of Q. Confirm this with a truth table by showing that $(P \Rightarrow Q) \Rightarrow Q$ is not a tautology. Create real world applications by defining statements P and Q illustrating a case where $(P \Rightarrow Q) \Rightarrow Q$ is true, and one where it is false.

3. The contrapositive provides an alternative way to demonstrate the truth of the implication $P \Rightarrow Q$. Confirm that $(P \Rightarrow Q) \Leftrightarrow (\sim Q \Rightarrow \sim P)$ is a tautology. Give a real world example.

4. Confirm that the structure of the proof by contradiction,

$$[(A \wedge \sim C) \Rightarrow \sim A] \Rightarrow (A \Rightarrow C),$$

is a tautology.

5. Comedically, the logical deduction

$$[(P \Rightarrow Q) \wedge Q] \Rightarrow P \qquad (1.14)$$

is known as **modus moronus**. Show that this statement is not a tautology, and provide a real world example of statements P and Q for which the hypothesis is true and conclusion false.

6. Show by mathematical induction that for any integer $n \geq 0$:

$$\sum_{i=0}^{n} 2^i = 2^{n+1} - 1.$$

7. Develop a direct proof of the formula in exercise 6. (*Hint*: Define $S = \sum_{i=0}^{n} 2^i$, consider the formula for $2S$, and then subtract.)

8. Develop a proof by contradiction in the form of (1.6) of the formula in exercise 6. (*Hint*: The formula is apparently true for $n = 0, 1, 2$, and other values of n. Let N be the first integer for which it is false. From the truth for $n = N - 1$, and falsity for $n = N$, conclude that $2^N \neq 2^N$ and recall the remark after (1.6).)

9. It is often assumed that the initialization step in mathematical induction is unnecessary, and that only the induction step need be confirmed. Show that the formula

$$\sum_{i=0}^{n} 2^i = 2^{n+1} + c$$

satisfies the induction step for any c, but that only for $c = -1$ does it satisfy the initialization step.

10. Show by mathematical induction that

$$\sum_{j=1}^{n} j^2 = \frac{n(n+1)(2n+1)}{6}.$$

11. A bank has made the promise that for some fixed $i > 0$, an investment with it will grow over every one-year period as $F_{j+1} = F_j(1 + i)$, where F_j denotes the fund at time j in years. Prove by mathematical induction that if an investment of F_0 is made today, then for any $n \geq 1$,

$$F_n = F_0(1 + i)^n.$$

12. Develop a proof using *modus tollens* in the structure of (1.4) that if at some time n years in the future, the bank communicates $F_n \neq F_0(1+i)^n$, then the bank at some point must have broken its promise of one-year fund growth noted in exercise 11. (*Hint*: Define $P : F_{j+1} = F_j(1+i)$ for all j; $Q : F_n = F_0(1+i)^n$ for all $n \geq 1$. What can you conclude from $(P \Rightarrow Q) \wedge \sim Q$?)

Assignment Exercises

13. Create truth tables to evaluate if the following statements, $A \Leftrightarrow B$ or $A \Rightarrow B$, are tautologies:

(a) $P \wedge Q \Leftrightarrow \sim(P \Rightarrow \sim Q)$

(b) $(P \vee Q) \wedge \sim Q \Rightarrow P$

(c) $(P \Rightarrow Q) \wedge (P \wedge R) \Rightarrow Q \wedge R$

(d) $\sim P \vee (Q \wedge R) \Leftrightarrow (\sim R \vee \sim Q) \wedge P$

14. *Modus ponens* identifies the necessary additional fact to convert a proof of the truth of the implication, $P \Rightarrow Q$, into a proof of the conclusion, Q. Confirm that $P \wedge (P \Rightarrow Q) \Rightarrow Q$ is a tautology. Demonstrate by real world examples as in exercise 2 that while $(P \Rightarrow Q) \Rightarrow Q$ can be true or false, $P \wedge (P \Rightarrow Q) \Rightarrow Q$ is always true.

15. Show that *modus ponens* combined with the contrapositive yields $\sim Q \wedge (P \Rightarrow Q)$ $\Rightarrow \sim P$, and show directly that this statement is a tautology. Give a real world example.

16. Identify and label (A, B, etc.) the statements in the argument at the end of this chapter, convert the argument to a logical structure, and demonstrate what conclusion can be derived using syllogism and *modus tollens*.

17. Show by mathematical induction that for $i > 0$ and integer $n \geq 1$,

$$\sum_{j=1}^{n}(1+i)^{-j} = \frac{1-(1+i)^{-n}}{i}.$$

18. Develop a direct proof of the formula in exercise 17. (*Hint*: See exercise 7.)

19. Show by mathematical induction that

$$\sum_{j=1}^{n} j^3 = \left[\sum_{j=1}^{n} j\right]^2.$$

20. A bank has made the promise that for some fixed $i > 0$, an investment with it will grow over every one-year period as $F_{j+1} = F_j(1+i)$, where F_j denotes the fund at time j in years. Develop a proof by contradiction in the form of (1.9) that for any $n \geq 1$,

$$F_n = F_0(1 + i)^n.$$

(*Hint*: Define $A : F_{j+1} = F_j(1 + i)$ for all $j \geq 0$; $C : F_n = F_0(1 + i)^n$ for all $n \geq 1$. If $A \wedge {\sim}C$ and N is the smallest n that fails in C, what can you conclude about F_N, which provides a contradiction, and about the conclusion $A \Rightarrow C$?)

2 Number Systems and Functions

2.1 Numbers: Properties and Structures

2.1.1 Introduction

In this chapter some of the detailed proofs on number systems are omitted. The reason is that to provide a rigorous framework for the fundamental properties of number systems summarized below would require the development of both subtle and detailed mathematical tools for which we will have no explicit use in subsequent chapters. The mathematics involved, however, gives beautiful examples of the extraordinary power and elegance of mathematics, and provides an intuitive context for many of the generalizations in later chapters.

This statement of the "power and elegance" of this theory might surprise a reader who is tempted to think that the power of mathematics is only revealed in the development of new and complex theory. However, the development of a rigorous framework to prove statements about properties of numbers that we have been taught as "true" since pre-school can be even more complex. For example, how would one set out to prove that for any integers n and m,

$$n + m = m + n?$$

Who but a mathematician would think that such an "obvious" statement would require proof, and who but a mathematician would commit to the effort of developing the necessary tools and mathematical framework to allow this and other such statements an objective and critical analysis?

As discussed in chapter 1, such a framework must introduce certain undefined terms, the **formal symbols**. It must also explicitly address what will be assumed within the **axioms** about these terms and symbols and the system of numbers under study. It will need to ensure that despite the strong belief system people have about properties of numbers learned since childhood, all demonstrations of statements within theory rely explicitly and exclusively on axioms, or on other results that follow from these axioms. Such provable statements are then called the **theorems** or **propositions** of the theory (terms used interchangeably), and the rigorous demonstrations of these statements' validity are called the **proofs** of the theory.

The modern axiomatic approach to natural numbers was introduced by **Giuseppe Peano** (1858–1932) in 1889, when he developed what has come to be known as **Peano's axioms**, which simplified a 1888 axiomatic treatment by **Richard Dedekind** (1831–1916).

2.1.2   Natural Numbers

Perhaps the simplest collection of numbers is that of **natural numbers** or **counting numbers**, denoted $\mathbb{N}$, and defined as

$$\mathbb{N} = \{1, 2, 3, \ldots\} \quad \text{or} \quad \{0, 1, 2, 3, \ldots\}.$$

To give a flavor for the axiomatic structure for $\mathbb{N}$, we introduce Peano's axioms in the framework that provides the basic arithmetic structure. The formal symbols are self-evident except for the symbol $'$. Intuitively, for any natural number n, the symbol n' denotes its successor, which in concrete terms can be thought of as $n + 1$.

1. Formal Symbols: $=, ', +, \cdot, 0$

2. Axioms:

- **A1:** $\forall m \forall n (m' = n' \Rightarrow m = n)$
- **A2:** $\forall m (m' \neq 0)$
- **A3:** $\forall m (m + 0 = m)$
- **A4:** $\forall m \forall n (m + n' = (m + n)')$
- **A5:** $\forall m (m \cdot 0 = 0)$
- **A6:** $\forall m \forall n (m \cdot n' = m \cdot n + m)$
- **A7:** For any formula $P(m)$: $[P(0) \wedge \forall m (P(m) \Rightarrow P(m'))] \Rightarrow \forall m P(m)$

We note that the formal symbols include the familiar addition $(+)$, multiplication $(\cdot)$, and equality $(=)$ symbols, as well as one numerical constant 0. There is also the prime symbol $(')$, which, as can be inferred from the axioms, is meant to denote "successor." In layman's terms, m' stands for $m + 1$, but in the more abstract axiomatic setting, m' simply denotes the successor of m.

Axiom 1 says that the "successor" is unique; two different elements of $\mathbb{N}$ cannot have the same successor, while axiom 2 formally puts 0 at the front of the successor chain. Axioms 3 and 4 form the foundation for how addition works while axioms 5 and 6 do the same for multiplication. Also axiom 6 reveals our layman understanding that $m' = m + 1$. To deduce this formally, we need to define $1 = 0'$, then prove that $m = 1 \cdot m$, as well as prove that we can factor $m \cdot n + m = m \cdot (n + 1)$. Finally, axiom 7 is the "induction" axiom, which provides a framework to prove general formulas about $\mathbb{N}$. Namely, if one proves that a formula is true for 0, and that its truth for m implies truth for m', then the formula is true for all m. This idea was introduced in chapter 1 as "proof by induction." We will not pursue this formal axiomatic development further.

Returning to the informal setting, we note that the natural numbers are useful primarily for counting and ordering objects. There are an infinite number of elements of the set $\mathbb{N}$, of course, and to distinguish this notion of infinity, we say that the set $\mathbb{N}$ is **countable** or **denumerable**. More generally, a collection X is said to denumerable if there is a **1:1 correspondence** between X and $\mathbb{N}$, denoted

$$X \leftrightarrow \mathbb{N},$$

meaning that there exists an **enumeration** of the elements of X,

$$X = \{x_1, x_2, x_3, \ldots\},$$

that includes all of the elements of X exactly once. Alternatively, each element of X can be paired with a unique element of N.

Note, however, that to prove that a set is countable, it is sometimes easier to explicitly demonstrate a correspondence that contains multiple counts where all elements of X are counted at least once. Such a demonstration implies the desired result, of course, and oftentimes there will be no reason to refine the argument to get an explicit correspondence "which includes all of the elements of X exactly once."

Proposition 2.1   *If the collections X_i are countable for $i = 1, 2, \ldots, n$, then $X = \{x \mid x \in X_i \text{ for some } i\}$ is also countable.*

Proof   The necessary correspondence $X \leftrightarrow \mathbb{N}$ is defined by associating the elements of each $X_i \equiv \{x_{i1}, x_{i2}, x_{i3}, \ldots, x_{ij}, \ldots\}$ with $\{i + (j-1)n \mid j = 1, 2, \ldots\}$. In other words, the first elements of the $\{X_i\}$ are counted sequentially, then the second elements, etc.   ∎

Remark 2.2   *In the next chapter we introduce sets and operations on sets such as unions and intersections, but for those already familiar with these concepts, it is apparent that X above is defined as the union of the X_i. It is the case that the proposition above holds even if there are a countable number of X_i. A proof of this statement will be seen below when it is demonstrated that the rational numbers are countable.*

As a collection the natural numbers are **closed under addition and multiplication**, meaning that these operations produce results that are again natural numbers,

$$n_1, n_2 \in \mathbb{N} \Rightarrow n_1 + n_2 \in \mathbb{N} \quad \text{and} \quad n_1 \cdot n_2 \in \mathbb{N},$$

but are not closed under subtraction or division. An important property of $\mathbb{N}$ under multiplication ($\cdot$), and one known to the ancient Greeks, is that of **unique factorization**. We first set the stage.

Definition 2.3   *A number $n \in \mathbb{N}$ is **prime** if $n > 1$ and*

$n = n_1 \cdot n_2$ *implies* $n_1 = 1$ *and* $n_2 = n$,   *or conversely.*

*A number $n > 1$ is **composite** if it is not prime. That is, $n = n_1 \cdot n_2$ and neither factor n_j, equals* 1.

Note that $n = 1$ is neither prime nor composite by this definition. That is a matter of personal taste, and one can define it to be prime without much consequence, other than needing to be a bit more careful in the definition of "unique factorization," which will be discussed below.

Proposition 2.4   *The collection of primes is infinite.*

Proof   Following **Euclid of Alexandria** (ca. 325–265 BC), who presented the proof in **Euclid's** *Elements*, we use the method of proof by contradiction. If the conclusion were false and $n_1, n_2, n_3, \ldots, n_N$ were the only primes, then define $n = n_1 \cdot n_2 \cdot n_3 \cdot \ldots \cdot n_N + 1$. So either n is prime, which would be a contradiction as it is clearly bigger than any of the original primes, or it is composite, meaning that it is evenly divisible by one of the original set of primes. But this too is impossible given the formula for n, since 1 is not evenly divisible by any prime.   ∎

We now return to the notion of unique factorization. By this we simply mean that every natural number can be expressed as a product of prime numbers in only one way.

Definition 2.5   *The set $\mathbb{N}$ satisfies **unique factorization** if for every n, there exists a collection of primes $\{p_j\}_{j=1}^{N}$ so that $n = \Pi p_j$, and if there exist collections of primes $\{p_j\}_{j=1}^{N}$ and $\{q_k\}_{k=1}^{M}$ so that*

$n = \Pi p_j = \Pi q_k;$

then $N = M$, and when these primes are arranged in nondecreasing order, $p_j = q_j$ for all j.

Remark 2.6

1. *In the definition above, Πp_j is shorthand for the product*

$\Pi p_j = p_1 p_2 p_3 \ldots p_N,$

and analogously for Πq_k. When necessary for clarity, this product will be expressed as $\prod_{j=1}^{N} p_j$.

2. *The notion here of a nondecreasing arrangement seems awkward at first. We tend to think of increasing and decreasing as opposites, so we expect a nondecreasing arrangement to be an increasing one. But this definition must allow for cases where the primes are not all distinct, and hence the arrangement can not be truly "increasing." In other contexts, the notion of "nonincreasing" will have the same intent.*

3. *If the natural number 1 is defined above to be a prime number, the definition of unique factorization would have to be a bit more complicated to allow for any number of factors equaling 1.*

Proposition 2.7 (*Fundamental Theorem of Arithmetic*) $\mathbb{N}$ *satisfies unique factorization.*

Proof The complexity of this proof lies in the proof of a much simpler idea: if a prime divides a composite number, then given any factorization of that number, this prime must divide at least one of its factors. This is known as **Euclid's lemma** (after **Euclid of Alexandria**), which we discuss below. Once this lemma is demonstrated, the proof then proceeds by induction. The proposition is clearly true for $n = 2$, which is prime. Assume next that it is true for all $n < N$, and that N has been factored: $N = \Pi p_j = \Pi q_j$, where, for definitiveness, the primes have been arranged in nondecreasing order. Of course, we can assume that N is composite, since all primes satisfy unique factorization by definition. Now by Euclid's lemma, if p_1 divides $N = \Pi q_j$, it must divide one of the factors. Because the q_j are prime, it must be the case that $p_1 = q_i$ for some i. Similarly, because q_1 must divide Πp_j and the p_j are prime, it must be the case that $q_1 = p_k$ for some k. Consequently, by the assumed arrangements of primes, we must have $q_1 = p_1$, and this common factor can be eliminated from the expressions by division. We now have two prime factorizations for $N/p_1 = N/q_1$, a number which is less that N. Hence by the induction step, unique factorization applies, and the result follows. ∎

Remark 2.8

1. *Euclid's Lemma* *The modern idea behind Euclid's lemma, in contrast to the original proof, is that if p and a are natural numbers that have no common factors, one can find natural numbers x and y so that*

$$1 = \pm(px - ay).$$

*In other words, if p and a have no common factors, one can find multiples of these numbers that differ by 1. This result is a special case of **Bézout's identity**, named for **Étienne Bézout** (1730–1783), and discussed below. Assuming this lemma, if p is a* *prime that divides $n = ab$ but does not divide a, we know that p and a have no common factors, so the identity above holds. Multiplying through by b, we conclude that*

$$b = \pm(bpx - aby),$$

and hence p divides b, since it clearly divides bpx, and also divides $aby = ny$, since p divides n by assumption.

2. *Bézout's Identity* *Bézout's identity states that given any natural numbers a and b, if d denotes the **greatest common divisor**, $d = \gcd(a, b)$, then there are natural numbers x and y so that*

$$d = \pm(ax - by).$$

In other words, one can find multiples of these numbers that differ by the greatest common division of these numbers. If a and b have no common factors, then $d = 1$, and this becomes Euclid's lemma utilized above. The proof of this result comes from another very neat construction of Euclid.

3. *Euclid's Algorithm* *Euclid's algorithm provides an efficient process for finding d, the greatest common divisor of a and b. To understand the basic idea, let's assume $b > a$, and write*

$$b = q_1 a + r_1,$$

where q_1 is a natural number including 0, and r_1 is a natural number satisfying $0 \leq r_1 < a$. Euclid's critical observation is that any number that divides a and b must also divide r_1, since $r_1 = b - q_1 a$. Consequently the number $\gcd(a, b)$ must also divide r_1, and hence

$$\gcd(a, b) = \gcd(r_1, a).$$

We now repeat the process with a and r_1:

$$a = q_2 r_1 + r_2,$$

$$r_1 = q_3 r_2 + r_3,$$

$$r_2 = q_4 r_3 + r_4, \ldots,$$

where in each step, $0 \leq r_{j+1} < r_j$. We continue in this way until a remainder of 0 is obtained, which must happen because the remainders must decrease. The second to last remainder must then be d because of the critical observation above. In other words, we eventually get to the last two steps:

$$r_{n-1} = q_{n+1}r_n + r_{n+1},$$

$$r_n = q_{n+2}r_{n+1} + 0.$$

Since $\gcd(a,b) = \gcd(r_{n+1},0) = r_{n+1}$, *it must be the case that* $r_{n+1} = d$. *We then obtain* x *and* y *by reversing the steps above. For example, assume that the process stops with a remainder of* 0 *on the third step so that* $r_3 = 0$ *and* $r_2 = d$. *Then*

$$d = a - q_2 r_1$$

$$= a - q_2(b - q_1 a)$$

$$= (1 + q_2 q_1)a - q_2 b.$$

Example 2.9   *To show that* $\gcd(68013, 6172) = 1$:

$$68013 = 11 \cdot 6172 + 121,$$

$$6172 = 51 \cdot 121 + 1,$$

$$121 = 121 \cdot 1 + 0.$$

Reversing the steps obtains

$$1 = 6172 - 51 \cdot 121$$

$$= 6172 - 51 \cdot (68013 - 11 \cdot 6172)$$

$$= -51 \cdot 68013 + 562 \cdot 6172.$$

2.1.3   Integers

The set of **integers**, denoted $\mathbb{Z}$, and defined as

$$\mathbb{Z} = \{\ldots, -3, -2, -1, 0, 1, 2, 3, \ldots\},$$

is closed under both addition and subtraction, as well as multiplication. In fact, under the operation of $+$, the integers have the structure of a **commutative group**, $(\mathbb{Z}, +)$, which we state without proof.

Definition 2.10   *A set* X *is a **group under the operation** $\star$, denoted* $(X, \star)$ *if:*

1. X *is closed under* $\star$: *that is,* $x, y \in X \Rightarrow x \star y \in X$.

2. X *has a **unit**: there is an element* $e \in X$ *so that* $e \star x = x \star e = x$.

3. X *contains **inverses**: for any* $x \neq e$, *there is* $x^{-1} \in X$ *so that* $x^{-1} \star x = x \star x^{-1} = e$.

4. $\star$ *is **associative**: for any* $x, y, z \in X$: $((x \star y) \star z) = (x \star (y \star z))$.

Definition 2.11 $(X, \star)$ *is an* **abelian** *or* **commutative group** *if X is a group and for all* $x, y \in X$,

$$x \star y = y \star x.$$

Of course, in $(\mathbb{Z}, +)$, the unit $e = 0$, and the inverses $x^{-1} = -x$.

Also the set $\mathbb{Z}$ **is denumerable**, since it is the union of three denumerable sets, the natural numbers and their negatives, and $\{0\}$. It is also the case that unique factorization holds in $\mathbb{Z}$ once one accounts for the possibility of products of ± 1, since we clearly must allow for examples such as $2 \cdot 3 = (-2) \cdot (-3)$. In other words, the Fundamental Theorem of Arithmetic holds for both positive and negative natural numbers, but for prime factorization the conclusion must allow for the possibility that

$$p_j = \pm q_j \qquad \text{for all } j.$$

Finally, one sometimes sees the notation $\mathbb{Z}^+$ and $\mathbb{Z}^-$ to denote the positive and negative integers, respectively, although there is not a reliable convention as to whether $\mathbb{Z}^+$ contains 0, which is similar to the case for $\mathbb{N}$.

2.1.4   Rational Numbers

The group $\mathbb{Z}$ is not closed under division, but it can be enlarged to the collection of **rational numbers**, denoted $\mathbb{Q}$, and defined as

$$\mathbb{Q} = \left\{ \frac{n}{m} \,\middle|\, n, m \in \mathbb{Z}, m \neq 0 \right\}.$$

The collection $\mathbb{Q}$ is a group under both addition $(+)$ and multiplication $(\cdot)$. In $(\mathbb{Q}, +)$, as in $(\mathbb{Z}, +)$, the unit is $e = 0$ and inverses are $x^{-1} = -x$, whereas in $(\mathbb{Q}, \cdot)$, $e = 1$ and $x^{-1} = 1/x$. In fact $(\mathbb{Q}, +, \cdot)$ has the structure of a field.

Definition 2.12   *A set X under the operations $+$ and $\cdot$ is a* **field**, *denoted $(X, +, \cdot)$, if:*

1. $(X, +)$ *is a commutative group.*
2. $(X, \cdot)$ *is a commutative group.*
3. $(\cdot)$ *is* **distributive** *over $(+)$: for any $x, y, z \in X$: $x \cdot (y + z) = x \cdot y + x \cdot z$.*

The set $\mathbb{Q}$ is **denumerable** as can be demonstrated by a famous construction of **Georg Cantor** (1845–1918). Express all positive rational numbers in a grid such as

$\frac{1}{1} \quad \frac{1}{2} \quad \frac{1}{3} \quad \frac{1}{4} \quad \cdots$

$\frac{2}{1} \quad \frac{2}{2} \quad \frac{2}{3} \quad \frac{2}{4} \quad \cdots$

$$\frac{3}{1} \quad \frac{3}{2} \quad \frac{3}{3} \quad \frac{3}{4} \quad \cdots$$

$$\frac{4}{1} \quad \frac{4}{2} \quad \frac{4}{3} \quad \frac{4}{4} \quad \cdots$$

$$\vdots \quad \vdots \quad \vdots \quad \vdots \quad \vdots$$

It is clear that this is a listing of all positive rational numbers, with all rationals counted infinitely many times. However, even with this redundancy, these numbers can be enumerated by starting in the upper left-hand cell, and weaving through the table in diagonals:

$$\frac{1}{1} \mapsto \frac{1}{2} \mapsto \frac{2}{1} \mapsto \frac{3}{1} \mapsto \frac{2}{2} \mapsto \frac{1}{3} \mapsto \frac{1}{4} \mapsto \cdots.$$

All rationals are then countable as the union of countable sets: positive and negative rationals and $\{0\}$.

Remark 2.13   *As noted above, this demonstration applies to the more general state-ment that the **union of a countable number of countable collections is countable**, since these collections can be displayed as rows in the table above and the enumeration defined analogously.*

While closed under the arithmetic operations of $+, -, \cdot, \div$, the set of rationals $\mathbb{Q}$ is not closed under exponentiation of positive numbers. In other words,

$$x > 0 \quad \text{and} \quad y \in \mathbb{Q} \not\Rightarrow x^y \in \mathbb{Q},$$

where "$\not\Rightarrow$" is shorthand here for "does not necessarily imply." The simplest demon-stration that there exist numbers that are not rational comes from Greece around 500 BC, some 200 years before Euclid's time. The original result was that $\sqrt{2}$ was not rational. The general result is that only perfect squares of natural numbers have ra-tional square roots, only perfect cubes have rational cube roots, and so forth. We demonstrate the square root result on natural numbers next.

Proposition 2.14   *If $n \in \mathbb{N}$ and $n \neq m^2$ for any $m \in \mathbb{N}$, then $\sqrt{n} \notin \mathbb{Q}$.*

Proof   Again, using proof by contradiction, assume that $\sqrt{n}$ is rational, with $\sqrt{n} = \frac{a}{b} \in \mathbb{Q}$. Then $nb^2 = a^2$. Now if $a = \Pi p_j$ and $b = \Pi q_k$ are the respective unique fac-torizations, we get

$$n\Pi q_k^2 = \Pi p_j^2.$$

However, since nb^2 also has unique factorization, it must be the case that the collection of primes on the left and right side of this equality are identical, which means that after cancellation, there is a remaining set of primes so that $n = \Pi r_j^2$. That is, $n = m^2$ for $m = \Pi r_j$, contradicting the assumption that $n \neq m^2$ for any m.   ∎

This proposition can be generalized substantially, with exactly the same proof. Specifically, if $r \in \mathbb{N}$ and $r > 1$, the only time the rth root of a rational number is rational is in the most obvious case, when both the numerator and denominator are rth powers of natural numbers.

Proposition 2.15   *Let $\frac{n'}{m'} \in \mathbb{Q}$, expressed with no common divisors and $\frac{n'}{m'} \neq 0$. If $\frac{n'}{m'} \neq \frac{n^r}{m^r}$ for some $n, m \in \mathbb{N}$, and $r \in \mathbb{N}$, $r > 1$, then $\sqrt[r]{\frac{n'}{m'}} \notin \mathbb{Q}$.*

Proof   Follow the steps of the special case above.   ∎

The set $\mathbb{Q}$ has four interesting, and perhaps not surprising, properties that provide insight to the ultimate expansion below to the real numbers. As will be explained in chapter 4, these properties can be summarized to say that within the collection of real numbers, the rational numbers are a **dense subset**, as is the collection of numbers that are not rational, called the **irrational numbers**. However, these number sets will later be seen to differ in a dramatic and surprising way.

Proposition 2.16

1. *For any $q_1, q_2 \in \mathbb{Q}$ with $q_1 < q_2$, there is a $q \in \mathbb{Q}$ with $q_1 < q < q_2$.*
2. *For any $q_1, q_2 \in \mathbb{Q}$ with $q_1 < q_2$, there is an $r \notin \mathbb{Q}$ with $q_1 < r < q_2$.*
3. *For any $r_1, r_2 \notin \mathbb{Q}$ with $r_1 < r_2$, there is a $q \in \mathbb{Q}$ with $r_1 < q < r_2$.*
4. *For any $r_1, r_2 \notin \mathbb{Q}$ with $r_1 < r_2$, there is an $r \notin \mathbb{Q}$ with $r_1 < r < r_2$.*

Proof   The first statement is easy to justify by construction, by letting $q = 0.5(q_1 + q_2)$, or more generally, $q = p(q_1 + q_2)$ for any rational number p, $0 < p < 1$. For the second statement we demonstrate with a proof by contradiction. Assume that all such r are in fact rational numbers. Then it is also the case that for any $p \in \mathbb{Q}$, we have that all r with $q_1 + p < r < q_2 + p$ are also rational, since rationals are closed under addition. Choosing $p = -q_1$, we arrive at a contradiction as follows: The proposition above shows that if $n \neq m^2$ for any m, then $\sqrt{n} \notin \mathbb{Q}$, and hence $\frac{1}{\sqrt{n}} \notin \mathbb{Q}$. However, we clearly have values of $\frac{1}{\sqrt{n}}$ satisfying $0 < \frac{1}{\sqrt{n}} < q_2 - q_1$. The third statement has the same demonstration as the second. Specifically, if we assume that all such q are irrational, then we can translate this collection by a rational number p, to conclude that all numbers q with $r_1 + p < q < r_2 + p$ are not rational (it is an easy exercise that the sum of a rational number and an irrational number is again irratio-nal). But then we can easily move this range to capture an integer, or any rational number of our choosing. Finally, the fourth statement follows from the observation that the construction for the third statement can produce two rationals between r_1 and r_2, to which we can apply the second statement.   ∎

Consequently the collection of rational numbers can be informally thought of as being "infinitely close," with no "big holes," but at the same time, containing infi-nitely many "small holes" that are also infinitely close. The same is true for the col-lection of irrational numbers. One might guess that this demonstrates that there are an equal number of rational and irrational numbers. In other words, we might guess that the above proposition implies that both sets are denumerable. We will see shortly that this guess would be wrong.

2.1.5 Real Numbers

The rational numbers can be expanded to the **real numbers**, denoted $\mathbb{R}$, which includes the rationals and irrationals, although the actual construction is subtle. This construction of $\mathbb{R}$ was introduced by **Richard Dedekind** (1831–1916) in a 1872 paper, using a method that has come to be known as **Dedekind cuts**. Although we will discuss "sets" in chapter 4, we note that $\emptyset$ is the universal symbol for the "empty set," or the set with no elements.

The idea in this construction is to capitalize on the one common property that rationals and irrationals share, which follows from the proposition above as gener-alized in exercises 2 and 17. That is, for any $r \in \mathbb{Q}$ or $r \notin \mathbb{Q}$ there is a sequence of ra-tional numbers, $q_1, q_2, q_3, \ldots$ so that q_n gets "arbitrarily close" to r as n increases without bound, denoted $n \to \infty$.

Definition 2.17   *A Dedekind cut is a subset* $\alpha \subset \mathbb{Q}$ *with the following properties:*

1. $\alpha \neq \emptyset$, *and* $\alpha \neq \mathbb{Q}$.
2. *If* $q \in \alpha$ *and* $p \in \mathbb{Q}$ *with* $p < q$, *then* $p \in \alpha$.
3. *There is no* $p \in \alpha$ *so that* $\alpha = \{q \in \mathbb{Q} \mid q \leq p\}$.

That is, a cut can neither be the empty set nor the set of all rationals, it must con-tain all the rationals smaller than any member rational, and it contains no largest ra-tional. Dedekind's idea was to demonstrate that the collection of cuts form a field, denoted $\mathbb{R}$, that contains the field $\mathbb{Q}$. Of course, he also needed to create an identifi-cation between cuts and real numbers. That identification was

$$r \in \mathbb{R} \leftrightarrow \alpha_r = \{q \in \mathbb{Q} \mid q < r\}.$$

Put another way, each real number r is identified with the **least upper bound** (or **l.u.b.**) of the cut α_r, defined as the minimum of all upper bounds:

$$r = \text{l.u.b.}\{p \mid p \in \alpha_r\}$$

$$= \min\{q \in \mathbb{Q} \mid q > p \text{ for all } p \in \alpha_r\}.$$

Intuitively, this minimum is an element of $\mathbb{Q}$ if and only if $r \in \mathbb{Q}$. For example,

$$\frac{1}{2} = \text{l.u.b.}\{p \mid p \in \alpha_{1/2}\}$$

$$= \min\{q \in \mathbb{Q} \mid q > p \text{ for all } p \in \alpha_{1/2}\},$$

$$\sqrt{2} = \text{l.u.b.}\{p \mid p \in \alpha_{\sqrt{2}}\}$$

$$= \min\{q \in \mathbb{Q} \mid q > p \text{ for all } p \in \alpha_{\sqrt{2}}\}.$$

In 1872 **Augustin Louis Cauchy** (1789–1857) introduced an alternative construction of $\mathbb{R}$, using the notion of Cauchy sequences studied in chapter 5, and showed that the field of real numbers could be identified with a field of Cauchy sequences of rational numbers. In effect, each real number is identified with the limit of such a sequence. To make this work, Cantor needed to "identify as one sequence" all sequences with the same limit, but then for the purpose of the identification with elements of $\mathbb{R}$, any sequence from each association class worked equally well.

Like $\mathbb{Q}$, the set $\mathbb{R}$ is also a field that is closed under $+$, $-$, $\cdot$, $\div$, and while closed under exponentiation if applied to positive reals, it is not closed under exponentiation if applied to negative reals. Also unlike $\mathbb{Q}$, the set $\mathbb{R}$ is **not countable**.

Proposition 2.18   *There exists no enumeration* $\mathbb{R} = \{r_n\}_{n=1}^{\infty}$.

Proof   The original proof was discovered by **Georg Cantor** (1845–1918), published in 1874, and proceeds by contradiction as follows. It has come to be known as **Cantor's diagonalization argument**. Assume that such an enumeration was possible, and that the reals between 0 and 1 could be put into a table:

$0.a_{11}a_{12}a_{13}a_{14}a_{15}a_{16}\cdots$

$0.a_{21}a_{22}a_{23}a_{24}a_{25}a_{26}\cdots$

$0.a_{31}a_{32}a_{33}a_{34}a_{35}a_{36}\cdots$

$0.a_{41}a_{42}a_{43}a_{44}a_{45}a_{46}\cdots$

$0.a_{51}a_{52}a_{53}a_{54}a_{55}a_{56}\cdots$

$0.a_{61}a_{62}a_{63}a_{64}a_{65}a_{66}\cdots$

$\vdots$

Here each digit, a_{ij}, is an integer between 0 and 9. Cantor's idea was that the enumeration above could not be complete. His proof was that one could easily construct many real numbers that could not be on this list. Simply define a real number a by

$$a = \tilde{a}_{11}\tilde{a}_{22}\tilde{a}_{33}\tilde{a}_{44}\tilde{a}_{55}\ldots,$$

where each digit of the constructed number $\tilde{a}_{jj}$, denotes any number other than the a_{jj} found on the listing above. For each j, we then have nine choices for $\tilde{a}_{jj}$, and infinitely many combinations of choices, and none of these constructed real numbers will be on the list. Consequently the listing above cannot be complete and hence $\mathbb{R}$ is not countable.   ∎

On first introduction to this notion of a **nondenumerably infinite**, or an **uncountably infinite** collection, it is natural to be at least a bit skeptical. Perhaps it would be easier to use a number base other than decimal, with fewer digits, so that we could be more explicit about this listing. Naturally, since any number can be written in any base, the question of countability or uncountability is also independent of this base.

Standard decimal expansions, also called **base-10 expansions**, represent a real number $x \in [0, 1]$ as

$$x = 0.x_1x_2x_3x_4x_5x_6\ldots$$

$$= \frac{x_1}{10} + \frac{x_2}{10^2} + \frac{x_3}{10^3} + \frac{x_4}{10^4} + \cdots,$$

where each $x_j \in \{0, 1, 2, \ldots, 9\}$. Similarly a **base-$b$ expansion** of x is defined, for b a positive integer, $b \geq 2$:

$$x_{(b)} = 0.a_1a_2a_3a_4a_5a_6\ldots$$

$$\equiv \frac{a_1}{b} + \frac{a_2}{b^2} + \frac{a_3}{b^3} + \frac{a_4}{b^4} + \frac{a_5}{b^5} + \cdots, \tag{2.1}$$

where each $a_j \in \{0, 1, 2, \ldots, b-1\}$. Each a_j is defined iteratively by the so-called **greedy algorithm** as the largest multiple of $\frac{1}{b^j}$ that is less than or equal to what is left after the prior steps. That is, the largest multiple less than or equal to $x - \sum_{k=1}^{j-1} \frac{a_k}{b^k}$.

Other real numbers $x \in \mathbb{R}$ are accommodated by applying this algorithm using both positive and negative powers of b in the expression, as is done for base-10.

In particular, with $b = 2$, the **base-2** or **binary system** is produced, and all $a_j \in \{0, 1\}$, so one easily imagines a well-defined and countable listing of the real numbers $x \in [0, 1]$ by an explicit ordering as follows:

0.000000000000 . . .

0.100000000000 . . .

0.010000000000 . . .

0.110000000000 . . .

0.001000000000 . . .

0.011000000000 . . .

0.101000000000 . . .

0.111000000000 . . . ,

and so forth. It seems apparent that such a careful listing represents all possible reals, and hence the reals are countable.

Unfortunately, the logic here comes up short. Since every number on this list has all 0s from a fixed binary position forward, every such number is a finite summation of the form $\sum_{k=1}^{n} \frac{a_k}{2^k}$, with $a_k \in \{0, 1\}$, and hence is rational. So we have simply developed a demonstration that this proper subset of the rationals is countable. It is a proper subset, since it does not contain $\frac{1}{3}$, for instance, which has no such finite expansion in base-2. Once infinite binary expansions are added to the listing, we can again apply the Cantor diagonalization argument as before and find infinitely many missing real numbers.

An interesting observation is that despite the analysis in the section on rational numbers that seemed to imply that rational and irrational numbers are effectively interspersed, the rational numbers are countable, and yet the irrational numbers are uncountable; otherwise, the real numbers would be countable as well. This observation will have interesting and significant implications in later chapters.

*2.1.6   Complex Numbers

The real numbers form a field, $(\mathbb{R}, +, \cdot)$, that is closed under the algebraic operations of $+, -, \cdot, \div$, as well as exponentiation, x^y, if $x > 0$, but it is not closed under exponentiation of negative reals. The simplest case is $\sqrt{-1}$, since the square of every real number is nonnegative. More generally, not all polynomials with real coefficients have solutions in $\mathbb{R}$, again the simplest example being

$$x^2 + 1 = 0.$$

Remarkably, one only needs to augment $\mathbb{R}$ by the addition of the so-called **imaginary unit**, denoted $\iota = \sqrt{-1}$, in an appropriate way, and all polynomials are then solvable.

Definition 2.19 *The collection of **complex numbers**, denoted $\mathbb{C}$, is defined by*

$$\mathbb{C} = \{z \mid z = a + b\iota; a, b \in \mathbb{R}, \iota = \sqrt{-1}\}.$$

*The term a is called the **real part** of z, denoted $\mathrm{Re}(z)$, and the term b is called the **imaginary part** of z, and denoted $\mathrm{Im}(z)$. Also the **complex conjugate** of z, denoted $\bar{z}$, is defined as*

$$\bar{z} = a - b\iota, \quad if \ z = a + b\iota.$$

*The **absolute value** of z, denoted $|z|$, is defined as*

$$|z| = \sqrt{a^2 + b^2} = \sqrt{z\bar{z}}, \tag{2.2}$$

where the positive square root is taken by convention.

It is common to identify the complex "number line" with the two-dimensional real space, also known as the **Cartesian plane**, denoted $\mathbb{R}^2$ (see chapter 3):

$$z \leftrightarrow (a, b).$$

This way $\mathrm{Re}(z)$ is plotted along the traditional x-axis, and $\mathrm{Im}(z)$ is plotted along the y-axis. The **absolute value** of z can then be seen to be a natural generalization of the **absolute value** of x, $|x|$, for real x:

$$|x| = \sqrt{x^2} = \begin{cases} x, & x \geq 0, \\ -x, & x < 0, \end{cases} \tag{2.3}$$

again with the positive square root taken by convention.

This absolute value can be interpreted as the distance from x to the origin, 0. Likewise $|z|$ is the distance from the point $z = (a, b)$ to the origin, $(0, 0)$, by the **Pythagorean theorem** applied to a right triangle with side lengths $|a|$ and $|b|$. For example, in **figure 2.1** is displayed the case where $a > 0$ and $b > 0$.

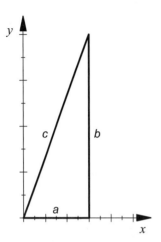

Figure 2.1
Pythagorean theorem: $c = \sqrt{a^2 + b^2}$

Another interesting connection between $\mathbb{C}$ and the Cartesian plane comes by way
of the so-called **polar coordinate representation** of a point $(a, b) \in \mathbb{R}^2$. The identification is $(a, b) \leftrightarrow (r, t)$, where r denotes the distance to the origin, and t is the "radian"
measure of the angle α that the "ray" from $(0, 0)$ to (a, b) makes with the positive x-
axis, measured counterclockwise. By convention, the measurement of α is limited to
one revolution so that $0° \leq \alpha < 360°$, or in the usual **radian measure**, $0 \leq t < 2\pi$. The
connection between an angle of $\alpha°$ and the associated "radian measure of t" is that
the radian measure of an angle equals the arc length of the sector on a circle of radius
1, with internal angle $\alpha°$. Such a circle is commonly called a **unit circle**. Numerically,
canceling the degrees units obtains $t = \frac{2\pi\alpha}{360}$.

The polar coordinate representation is then defined as

$$(a, b) = (r \cos t, r \sin t), \tag{2.4a}$$

$$r = \sqrt{a^2 + b^2}, \tag{2.4b}$$

$$t = \begin{cases} \arctan \frac{b}{a}, & 0 \leq \theta < 2\pi, \, a \neq 0, \\ \frac{\pi}{2}, & a = 0, \, b > 0, \\ \frac{3\pi}{2}, & a = 0, \, b < 0. \end{cases} \tag{2.4c}$$

In **figure 2.2** is shown a graphical depiction of these relationships when $a > 0$ and
$b > 0$. For $a = b = 0$, t can be arbitrarily defined. In other words, $(0, 0) \leftrightarrow (0, t)$ for
all t.

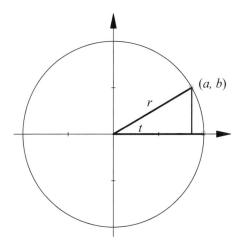

Figure 2.2
$a = r \cos t$, $b = r \sin t$

By this idea it is natural to also associate the complex number $z = a + bi = |z|(\cos t + i \sin t)$. However, an even more remarkable result is known as **Euler's formula**, after **Leonhard Euler** (1707–1783). He derived this formula based on methods of calculus presented in chapter 9. Specifically, for $z = a + bi$,

$$e^z = e^a(\cos b + i \sin b), \tag{2.5}$$

which for $z = bi$ implies that $|e^{bi}| = 1$ for all b. This is because by (2.2), $|e^{bi}|^2 = \cos^2 b + \sin^2 b = 1$.

In addition, when applied to $z = \pi i$, this formula provides the most remarkable, and perhaps most famous, identity in all of mathematics. It is called **Euler's identity**, and follows from (2.5), since $\cos \pi = -1$, and $\sin \pi = 0$:

$$e^{\pi i} = -1. \tag{2.6}$$

More generally, Euler's formula has other interesting trigonometric applications (see exercise 5), and it is a "lifesaver" for those of us who struggled with the memorization of the many complicated formulas known as "identities" in trigonometry.

We next show that for either (2.2) or (2.3) the so-called **triangle inequality** is satisfied.

Proposition 2.20   *Under either (2.2) or (2.3), we have that*

$$|x + y| \le |x| + |y|. \tag{2.7}$$

Proof We will demonstrate (2.7) by using the definition of absolute value in (2.2), which is equivalent to (2.3) for real numbers x and y. We then have

$$|x + y|^2 = (x + y)(\bar{x} + \bar{y})$$

$$= x\bar{x} + x\bar{y} + y\bar{x} + y\bar{y}$$

$$= |x|^2 + 2\,\mathrm{Re}(x\bar{y}) + |y|^2$$

$$\leq |x|^2 + 2|x|\,|y| + |y|^2$$

$$= (|x| + |y|)^2.$$

Note that in the third step it was used that $y\bar{x} = \overline{x\bar{y}}$, and that $z + \bar{z} = 2\,\mathrm{Re}(z)$, whereas for the fourth, $\mathrm{Re}(x\bar{y}) \leq |x\bar{y}| = \sqrt{x\bar{y}\overline{x\bar{y}}} = \sqrt{x\bar{x}y\bar{y}} = |x|\,|y|$. ∎

As it turns out, $(\mathbb{C}, +, \cdot)$ is a field under the usual laws of arithmetic because $\imath^2 = -1$. For example, multiplication proceeds as

$$(a + b\imath) \cdot (c + d\imath) = (ac - bd) + (ad + bc)\imath. \tag{2.8}$$

The one item perhaps not immediately obvious is the multiplicative inverse for $z \in \mathbb{C}$, where $z \neq 0$. It is easy to check that with

$$z^{-1} = \frac{\bar{z}}{|z|^2} = \frac{a - b\imath}{a^2 + b^2},$$

we have $zz^{-1} = 1$.

With these definitions, we can identify the real number field $\mathbb{R}$ as a "subfield" of the field $\mathbb{C}$:

$$\mathbb{R} \leftrightarrow \{(a, b) \mid b = 0\},$$

completing the list of inclusions

$$\mathbb{N} \subset \mathbb{Z} \subset \mathbb{Q} \subset \mathbb{R} \subset \mathbb{C}.$$

Remarkably, as alluded to above, $\mathbb{C}$ is the end of the number field "chain" for the vast majority of mathematics, at least in part due to a result first proved (in his doctoral thesis!) by **Johann Carl Friedrich Gauss** (1777–1855) in 1799 after more than 200 years of study by other great mathematicians. We state this result without proof, and mention that there are numerous demonstrations of this result using many different mathematical disciplines.

Proposition 2.21 (*Fundamental Theorem of Algebra*)   *Let $P(z)$ be an* n-*degree poly-nomial with complex coefficients*

$$P(z) = \sum_{j=0}^{n} c_j z^j.$$

Then the equation $P(z) = 0$ has exactly n complex roots, $\{w_j\} \subset \mathbb{C}$, counting multiplicities, and $P(z)$ can be factored:

$$P(z) = c_n \prod_{j=0}^{n} (z - w_j).$$

Remark 2.22

1. *The expression, "counting multiplicities," means that the collection of roots is not necessarily distinct, and that some may appear more than once. An example is $P(z) \equiv z^2 - 2z + 1 = (z - 1)^2$, which has two roots, 1 and 1, counting multiplicities.*

2. *This important theorem is often expressed with the assumption that $P(z)$ has a leading coefficient, $c_n = 1$, which then eliminates the coefficient in the factorization above.*

3. *If $P(z)$ has real coefficients, then the complex roots, namely those with $w = a + bi$ and $b \neq 0$, come in* **conjugate pairs**. *That is,*

$$P(w) = 0 \quad \text{iff} \quad P(\overline{w}) = 0,$$

where the abbreviation **iff** *is mathematical shorthand for "if and only if." It denotes the fact that the two statements are both true, or both false, and in this respect is the common language version of the logical symbol $\Leftrightarrow$ of chapter 1. The complete logical statement is that*

$$P(w) = 0 \quad \text{if} \quad P(\overline{w}) = 0 \quad \text{and only if} \quad P(\overline{w}) = 0.$$

This result on conjugate pairs is easily demonstrated by showing that for real coefficients, $\overline{P(w)} = P(\overline{w})$ because conjugation satisfies the following properties:

- *If $w = w_1 + w_2$, then $\overline{w} = \overline{w}_1 + \overline{w}_2$,*
- *If $w = w_1 \cdot w_2$, then $\overline{w} = \overline{w}_1 \cdot \overline{w}_2$.*

2.2   Functions

Definition 2.23   *A* **function** *is a rule by which elements of two sets of values are associated. There is only one restriction on this association and that is that each element of* *the first set of values, called the **domain**, must be identified with a **unique** element of a second set of values, called the **range**.*

For many applications of interest in this book, both the domain and range of a function are subsets of the real numbers or integers, but these may also be defined on more general sets as will be seen below. The rule is then typically expressed by a formula such as

$$f(x) = x^2 + 3.$$

Here x is an element of the domain of the function f, while $f(x)$ is an element of the range of f. Functions are also thought of and "visualized" as mappings between their domain and range, whereby x is mapped to $f(x)$, and this imagery is intuitively helpful at times. In this context one might use the notation

$$f : X \to Y,$$

where X denotes the domain of f, and Y the range. It is also common to write $f(x)$ for both the function, which ought to be denoted only by f, and the value of the function at x. This bit of carelessness will rarely cause confusion. Finally, $\mathrm{Dmn}(f)$ and $\mathrm{Rng}(f)$ are commonly used as abbreviations for the domain and range of the function.

In many applications, f will be a **multivariate function**, also called a **function of several variables**, meaning that the domain of f is made up of **n-tuples** of variables: $(x_1, x_2, \ldots, x_n)$, where each of the variables x_j, is defined on the reals, or complexes, and so forth. For example, $f(x, y, z) = 1 - xy + yz$ is a function of three variables, and illustrates the notational convention that when n is small, the n-tuple is denoted as (x, y), or (x, y, z), avoiding subscripts. To distinguish the special case of 1-variable functions, such functions are sometimes called **univariate**.

In general mathematical language, the word "function" typically implies that the range of f, or Y, is a subset of one of the number systems defined above. When $Y \subset \mathbb{R}$, the function f is called a **real-valued** function, and one similarly defines the notions of **complex-valued** function, **integer-valued** function, and so forth. This terminology applies to both multivariate and univariate functions. Similarly, if $X \subset \mathbb{R}$, the function f is referred to as a function of a **real variable**, and one similarly defines the notion of a function of a **complex variable**, and so forth. When necessary, this terminology might be modified to **univariate function of a real variable**, or **multivariate function of a real variable**, for example, but the context of the discussion is usually adequate to avoid such cumbersome terminology. In the more general case, where X and Y are collections of n-tuples, perhaps with different values of n, f is typically referred to as a **transformation** from X to Y.

It is important to note that while the definition of a function requires that $f(x)$ be unique for any x, it is not required that x be unique for any $f(x)$. For instance, the function, $f(x) = x^2 + 3$, above has $f(x) = f(-x)$ for any $x > 0$. Another way of expressing this is that a function can be a **many-to-one rule**, or a **one-to-one rule**, but it cannot be a **one-to-many rule**. A function that is in fact one to one has the special property that it has an "inverse" that is also a function.

Definition 2.24 *If f is a one-to-one function, $f : X \to Y$, the **inverse function**, denoted f^{-1}, is defined by*

$$f^{-1} : Y \to X, \tag{2.9a}$$

$$f^{-1}(y) = x \quad \text{iff} \quad f(x) = y. \tag{2.9b}$$

The example, $f(x) = x^2 + 3$, above has no inverse if defined as a function with domain equal to all real numbers where it is many to one, but the function does have an inverse if the domain is restricted to any subset of the nonnegative or nonpositive real numbers, since this then makes it one to one.

Naturally, a function can also relate nonnumerical sets of values. For example, the domain could be the set of all strings of heads (H) and tails (T) that arise from 10 flips of a fair coin. A function f could then be defined as the rule that counts the number of heads for a given string. So this is a function

$$f : \{\text{strings of 10 } Ts \text{ and/or } Hs\} \to \{0, 1, 2, \ldots, 9, 10\},$$

where $f(\text{string}) = $ number of Hs in the string.

2.3 Applications to Finance

2.3.1 Number Systems

This may seem too obvious, but ultimately finance is all about money, in one or several currencies, and money is all about numbers. One hardly needs to say more on this point. Admittedly, finance would seem to be only about rational numbers, since who ever earned a profit on an investment of $\$\sqrt{200}$? On the other hand, when one is dealing with rates of return or solving financial problems and their equations, the rational numbers are inadequate, and this is true even if all the inputs to the problem, or terms in the resulting equations, are in fact rational numbers.

For example, if one had an investment that doubled in n years, the implied annual return is irrational for any natural number $n > 1$. For $n = 5$ and an initial investment of $\$1000$, say, one solves the equation

$$1000(1 + r)^5 = 2000,$$

$$r = \sqrt[5]{2} - 1.$$

Well that's the theory, but no one in the market would quote a return of $100(\sqrt[5]{2} - 1)\%$. It would be rounded to a rational return of 14.87%, or if one wanted to impress, 14.869836%. Most people would be satisfied with the former answer, and yet if we use a rational approximation, and the dollar investment is large enough, we begin to see differences between the actual return and the approximated return using the approximate rational yield.

For example, using the return $r = 0.1476$, we would have a positive error of \$14.30 or so with a \$1 million investment. Such discrepancies are commonly observed in the financial markets. Not a big deal, perhaps, for so-called retail investors with modest sums to invest, but for institutional investors with millions or billions, this rounding error creates ambiguities and the need for conventions. It is also important to note that as one uses rational approximations in the real world, it comes at a cost: rounding errors begin to appear in our calculations. In other words, if we solve an equation and use a rational approximation to the solution, this solution will not exactly reproduce the desired result unless amounts are so small that the rounding error is less than the minimum currency unit. Our theoretical calculations don't balance with the real world in other cases. When complex calculations are performed, the error can be big enough to complicate our debugging of the computer program, since we need to determine if the discrepancy is a rounding problem or an as yet undiscovered error.

But are even the real numbers all that is needed? We are all likely to say so because of an inherent suspicion of the complex numbers that is certainly reinforced by lack of familiarity and compounded by the unfortunate terminology of "imaginary" numbers versus "real" numbers. But consider that some investment strategies can produce a negative final fund balance, even though this may be disguised if the investor has posted margin.

For example, if a hedge fund manager with \$100 million of capital is leveraged 10:1 by borrowing \$1 billion, and investing the \$1.1 billion in various strategies, one of which loses \$20 million in an investment of \$10 million, what is the fund return to the capital investors on this strategy? Naturally the broker would require margin for such a strategy, so the negative final fund balance would be reflected in the reduction of the margin account and overall fund capital. One can similarly develop investment strategies in the derivatives market directly, by going long and/or short futures contracts on commodities or other "underlying" investments, or implementing long/short strategies in the options markets. One invests \$100, say, and has a final balance on delivery or exercise date of $-\$100$, again in reality observed by a reduction in margin balances of $100.

For the period, we could argue that the return was -200%, or a period return of $r = -2.00$. On the other hand, if one desires to put this return on an **annual rate** basis, difficulties occur. For example, if this investment occurred over a month, the annual return satisfies

$$100(1 + r)^{1/12} = -100,$$

$$(1 + r)^{1/12} = -1,$$

which has no solutions in $\mathbb{R}$ but, as it turns out, 12 distinct solutions in $\mathbb{C}$. Note that exponentiation provides an illusory escape from $\mathbb{C}$:

$$(1 + r) = (-1)^{12},$$

$$r = 0.$$

However, while $r = 0$ solves the algebraically transformed equation, it does not solve the original equation. Such a solution is sometimes called a **spurious solution**.

Alternatively, if this return occurred over a year and we sought to determine the return for this investment on a **monthly nominal rate basis** discussed below, we obtain

$$100\left(1 + \frac{r}{12}\right)^{12} = -100,$$

$$1 + \frac{r}{12} = \sqrt[12]{-1},$$

$$r = 12[\sqrt[12]{-1} - 1],$$

a decidedly complex return, and as above, it has 12 distinct solutions in $\mathbb{C}$. On the other hand, by squaring the original equation, we can again produce the spurious solution of $r = 0$. But this, of course, will not work if substituted into the equation above.

So what is the correct answer? Despite possible discomfort, any one of the 12 possible values of $r = 12[\sqrt[12]{-1} - 1]$ is the actual complex return on a monthly nominal basis, since each solves the required equation, and there are correspondingly 12 possible complex returns that can be articulated on an annual basis.

To be sure, the market can always avoid this problem by simply using the language that the return was $r = -200\%$ "over the period."

2.3.2   Functions

The other major area of application for this chapter is related to functions. Functions are everywhere! Not just in finance but in every branch of the natural sciences, as well as in virtually every branch of the social sciences, and indeed in every human endeavor. This is because virtually every branch of human inquiry contains recipes, or formulas, that describe relationships between quantities that are either provable in theory or based on observations and considered approximate models of a true underlying theory. It is these formulas that help us understand the theories by revealing relationships in the theories. We note a truism:

Every formula is a function in disguise.

The difference between a formula and a function is simply based on the objective of the user. For example, if we seek the area of a circle of radius, $r = 2$, we recall or look up the formula, which is

area equals π times radius squared,

and with the approximation $\pi \approx 3.1416$, we estimate that $A \approx 12.5664$. On the other hand, if we seek to understand the relationship between area and radius, the preferred perspective is one of a function:

$$A(r) = \pi r^2.$$

We can now see clearly that if the radius doubles, the area quadruples. We can also easily determine that a large 17-inch pizza has just about the same area as two small 12-inch pizzas, an important observation when thinking about feeding the family. This is especially useful given that a large pizza is often much less expensive than two small pizzas, which is an application to finance, of course.

Returning to other areas of finance, we consider several examples. In every case it is purely a matter of taste and purpose which of the parameters in the given formula are distinguished as variables of the associated function. The general rule of thumb is that one wants to frame each function in as few variables as possible, but sufficiently many to allow the intended analysis.

Present Value Functions
If a payment of $100 is due in five years, the value today, or **present value**, can be represented as a function of the assumed annual interest rate, r:

$$V(r) = 100(1 + r)^{-5},$$

which easily generalizes to a payment of F due in n years as

$$V(r) = F(1+r)^{-n} = Fv^n. \qquad v = \text{discount factor}, \; v = (1+r)^{-1} \qquad (2.10)$$

The present value function in (2.10) is often written in the shorthand of $V(r) = Fv^n$, where v is universally understood as the discount factor for one period, so here $v = (1+r)^{-1}$.

More generally, if a series of payments of amount F are due at the end of each of the next n years, the present value can be represented as a function of an assumed annual rate:

$$V(r) = F \sum_{j=1}^{n} (1+r)^{-j}, \quad = F \sum_{j=1}^{n} \left(\tfrac{1}{1+r}\right)^{j} = F \cdot \frac{\left(\tfrac{1}{1+r}\right)\left[1 - \left(\tfrac{1}{1+r}\right)^n\right]}{1 - \left(\tfrac{1}{1+r}\right)} \qquad \sum_{j=1}^{n}(a)^j = \frac{a - a^{n+1}}{1-a}$$

$$= F \cdot \frac{\left(\tfrac{1}{1+r}\right)\left[1 - \left(\tfrac{1}{1+r}\right)^n\right]}{\tfrac{r}{1+r}} \qquad = a(1-a^n)$$

$$= F \frac{1 - (1+r)^{-n}}{r}. \qquad\qquad\qquad\qquad = \frac{a(1-a^n)}{1-a}$$

This last formula is derived in exercises 17 and 18 of chapter 1.

Because this present value factor is so common in finance, representing the present value of an **annuity** of n fixed payments, it warrants a special notation:

$$a_{n;r} \equiv \frac{1 - (1+r)^{-n}}{r} = \frac{1 - v^n}{r}. \qquad \text{Since } a_{n;r} := \sum_{j=1}^{n}(1+r)^{-j} \qquad (2.11)$$

Note that $a_{n;r}$ is a function of n and r, and could equally well have been denoted $a(n,r)$.

Accumulated Value Functions

If an investment of F at time 0 is accumulated for n years at an assumed annual interest rate r, the **accumulated value** at time n is given as

$$A(r) = F(1+r)^n. \qquad (2.12)$$

The accumulated value at time n of a series of investments of amount F at the end of each of the next n years can be represented as

$$A(r) = F \sum_{j=0}^{n-1} (1+r)^j, \quad = F \cdot \sum_{j=1}^{n}(1+r)^{j-1} = \frac{F}{(1+r)} \sum_{j=1}^{n}(1+r)^{j}$$

$$= \frac{F}{(1+r)} \cdot \frac{(1+r)\left[1 - (1+r)^n\right]}{(1 - (1+r))}$$

$$= F \frac{(1+r)^n - 1}{r}, \qquad\qquad = F \cdot \frac{1 - (1+r)^n}{-r} = F \cdot \frac{(1+r)^n - 1}{r}$$

where this last formula is derived with the same trick as was used for (2.11). Again, as this accumulated factor is so commonly used in finance, it is often accorded the special notation:

$$s_{n;r} \equiv \frac{(1+r)^n - 1}{r},$$
(2.13)

and as a function of n and r it could equally well have been denoted $s(n,r)$.

Although one could formally identify $V(r)$ with the multivariate function $V(r,F)$, and similarly for $A(r)$, there is little point to this formalization since the dependence of the valuation on F is fairly trivial. However, there are applications whereby the functional dependence on n is of interest, and one sees this notation explicitly in the $a_{n;r}$ and $s_{n;r}$ functions.

Nominal Interest Rate Conversion Functions

The financial markets require the use of interest rate bases for which the compounding frequency is other than annual. The conventional system is that of **nominal interest rates**, whereby rates are quoted on an **annualized basis**, but calculations are performed in the following way, generalizing the monthly nominal rate example above.

In the same way that an annual rate of r means that interest is accrued at $100r\%$ per year, if r is a semiannual rate, interest is accrued at the rate of $100\left(\frac{r}{2}\right)\%$ per half year, while a monthly rate is accrued at $100\left(\frac{r}{12}\right)\%$ per month, and so forth. In each case the numerical value quoted pertains to an annual period, as it is virtually never the case in finance that an interest rate is quoted on the basis of a period shorter or longer than a year. An interest rate of 6% on a monthly basis, or simply 6% monthly, **does not mean** that 6% is paid or earned over one month; rather, it is the market convention for expressing that 0.5% will be paid or earned over one month. Similarly 8% semiannual means 4% per half year, and so forth. Consequently one can introduce the notion of a rate r, on an *mthly* nominal basis, meaning that $100\left(\frac{r}{m}\right)\%$ is paid or accrued every $\frac{1}{m}$th of a year.

Nominal interest rates simplify the expression and calculation of present and accumulated values where payments are made other than annually. For example, a bond's payments are typically made semiannually in the United States. If payments of F are made semiannually for n years, the present value is expressible in terms of an annual rate by

$$V(r) = F \sum_{j=1}^{2n} (1+r)^{-j/2},$$

or more simply in terms of a semiannual rate

$$V(r) = F \sum_{j=1}^{2n} \left(1 + \frac{r}{2}\right)^{-j}$$

$$= F a_{2n;\, r/2},$$

$$\star\ a_{n;r} = \sum_{j=1}^{n} (1+r)^{-j}$$

$$\star\ a_{2n;\frac{r}{2}} = \sum_{j=1}^{2n} \left(1 + \frac{r}{2}\right)^{-j} = \frac{1 - v^{2n}}{r/2} = 2 \cdot \frac{1 - v^{2n}}{r}$$

making the application of the present value and accumulated value functions in (2.11) and (2.13) more flexible.

Finally, one can introduce the notion of **equivalence of nominal rates**, meaning that accumulating or present-valuing payments using equivalent rates produces the same answer. If r_m is on an *mthly* nominal basis, and r_n is on an *nthly* nominal basis, in order for the present value of F payable at time N years to be the same with either rate requires

$$F\left(1 + \frac{r_m}{m}\right)^{-Nm} = F\left(1 + \frac{r_n}{n}\right)^{-Nn},$$

and we immediately conclude that the notion of equivalence is independent of the cash flow F and time period N. The resulting identity between r_n and r_m equals that produced by contemplating accumulated values rather than present values. Of course, this identity between r_n and r_m can be converted to a function such as $r_m(r_n)$. This tells us that for any r_n on an *nthly* nominal basis, the equivalent r_m on an *mthly* nominal basis is given as

$$r_m(r_n) = m\left[\left(1 + \frac{r_n}{n}\right)^{n/m} - 1\right].$$

$$\star\ v = (1 + r)^{-1}$$

$$\star\ v_{i/2} = \left(1 + \frac{i}{2}\right)^{-1} \tag{2.14}$$

Bond-Pricing Functions

The application of the formulas and functions above to fixed income instruments such as bonds and mortgages is relatively straightforward. For example, under the US convention of semiannual coupons quoted at a semiannual rate r, the coupon paid is $F\frac{r}{2}$ per half year, where F denotes the bond's **par value**. If the bond has a maturity of n years, the price of the bond at semiannual yield i is given by

$$P(i) = \left(F\frac{r}{2} a_{2n;\, i/2}\right) + F v_{i/2}^{2n} = F \cdot \left(\frac{r}{2}\right) \cdot \left(\frac{1 - v^{2n}}{i/2}\right) + F. \tag{2.15}$$

Here $v_{i/2}$ again denotes the the discount factor for one period, $v = \left(1 + \frac{i}{2}\right)^{-1}$, but with a subscript for notational consistency. Sometimes this yield is expressed as i_n to emphasize that this is the yield on an n-year bond.

This formula allows a simple analysis of the relationship between $P(i)$ and F, or price and par. From (2.11) applied to $a_{2n;\,i/2}$ we derive that $v_{i/2}^{2n} = 1 - \frac{i}{2}a_{2n;\,i/2}$. When substituted into (2.15), this price function becomes

$$P(i) = F\left[1 + \frac{1}{2}(r - i)a_{2n;\,i/2}\right]. \tag{2.16}$$

From this expression we conclude the following:

- $P(i) > F$, and the bond **sells at a premium**, iff $r > i$.
- $P(i) = F$, and the bond **sells at par**, iff $r = i$.
- $P(i) < F$, and the bond **sells at a discount**, iff $r < i$.

Notice that the bond price function as expressed in either (2.15) or (2.16) can be thought of as a function of time. Identifying the given formulas as the price today when the bond has n years to maturity, and denoted $P_0(i)$, the price at time $\frac{j}{2}$, immediately after the jth coupon, denoted $P_{j/2}(i)$, is given by

$$P_{j/2}(i) = F\left[1 + \frac{1}{2}(r - i)a_{2n-j;\,i/2}\right], \tag{2.17}$$

using the format of (2.16), with a similar adjustment to express this in the format of (2.15). This formula is correct at time 0, of course, as well as at time n, where it reduces to F. In other words, immediately after the last coupon, the bond has value equal to the outstanding par value then payable.

The price of this bond between coupons, for instance, at time t, $0 < t < \frac{1}{2}$, can be derived prospectively, as the present value of remaining payments at that time, or retrospectively, in terms of the value required by the investor to ensure that a return of i is achieved. In either case one derives $P_t(i) = \left(1 + \frac{i}{2}\right)^{2t} P_0(i)$, which generalizes to

$$P_{(j/2)+t}(i) = \left(1 + \frac{i}{2}\right)^{2t} P_{j/2}(i), \qquad 0 \le t < \frac{1}{2}, \tag{2.18}$$

which demonstrates that for fixed yield rate i, the price of a bond varies "smoothly" between coupon dates and abruptly at the time of a coupon payment. In the language of chapter 9, this price function is **continuous** between coupon payments and **discontinuous** at coupon dates.

More generally, we may wish to express P as a function of $2n$ yield variables, allowing each cash flow to be discounted by the appropriate semiannual **spot rate**, in which case we obtain

$$P(i_{0.5}, i_1, \ldots, i_n) = F\frac{r}{2} \sum_{j=1}^{2n} \left(1 + \frac{i_{j/2}}{2}\right)^{-j} + F\left(1 + \frac{i_n}{2}\right)^{-2n}. \tag{2.19}$$

The domain of all these bond-pricing functions would logically be understood to be real numbers with $0 \le i < 1$ or $0 \le i_j < 1$ for most applications, although the functions are mathematically well defined for $1 + \frac{i}{m} > 0$, where i is an *mthly* nominal yield.

Mortgage- and Loan-Pricing Functions

The same way that bonds often have a semiannual cash flow stream, mortgages and other consumer loans are often repaid with monthly payments, and consequently rate quotes are typically made on a monthly nominal basis. If a loan of L is made, to be repaid with monthly payments of P over n years, the relationship between L and P depends on the value of the loan rate, r. Specifically, the loan value must equal the present value of the payments at the required rate. Using the tools above, this becomes

$$L = Pa_{12n; r/12},$$

producing a monthly repayment of

$$P(r, n) = \frac{Lr}{12(1 - v_{r/12}^{12n})}. \tag{2.20}$$

Here the monthly repayment is expressed as a function of both r and n. In some applications, where n is fixed, the notation is simplified to $P(r)$.

Note that the identity between the value of the loan and the remaining payments can also be used to track the progress of the loan's **outstanding balance** over time either immediately after a payment is made, as in (2.17), or in between payment dates, as in (2.18) (see exercise 13).

Preferred Stock-Pricing Functions

A so-called **perpetual preferred stock** is effectively a bond with maturity $n = \infty$. That is, there is a par value, F, a coupon rate, r, that is typically quoted on a semiannual basis and referred to as the preferred's **dividend rate**, but the financial instrument has no maturity and hence no repayment of par. At a given semiannual yield of i, the price of this instrument can be easily inferred from (2.15) by considering what happens to each of the present value functions as the term of the bond, n, grows without bound. This subject of "limits" will be addressed more formally in chapters 5 and 6, but here we present an informal but compelling argument.

Since it is natural to assume that the market yield rate $i > 0$, it is apparent that $1 + \frac{i}{2} > 1$, and hence $v_{i/2}^{2n}$ decreases to 0 as n increases to ∞. Using (2.11) modified to a semiannual yield, it is equally apparent that as $v_{i/2}^{2n}$ decreases to 0, the annuity factor $a_{2n;i/2}$ increases to $\frac{1}{i/2}$, which can be denoted $a_{\infty;i/2}$. Combining, and canceling the $\frac{1}{2}$ terms, we have that the pricing function for a perpetual preferred stock, is given by

$$P(i) = \frac{Fr}{i}. \tag{2.21}$$

From (2.21) we see that when the dividend rate and yield rate are both on a semiannual basis, the price does not explicitly reflect this basis. Generalizing, the same price would be obtained if r and i were quoted on any common nominal basis.

It is also clear that a perpetual preferred will be priced at a premium, par or at a discount in exactly the same conditions as was observed above for a given bond, and that was if $r > i$, $r = i$, or $r < i$, respectively.

Common Stock-Pricing Functions

The so-called **discounted dividend model** for evaluating the price of a common stock, often shortened to DDM, is another function of several variables. The basic idea of this model is that the price of the stock equals the present value of the projected dividends. Since a common stock has no "par" value, the dividends are quoted and modeled in dollars or the local currency, although it is common to unitize the calculation to a "per share" basis.

If D denotes the annual dividend just paid (per share), and it is assumed that annual dividends will grow in the future at annual rate of g, and investors demand an annual return of r, then in its most general notational form, the price of the stock can be modeled as a function of all these variables:

$$V(D, g, r) = D\frac{1 + g}{r - g}, \qquad r > g. \tag{2.22}$$

The derivation of (2.22) is similar to that for the preferred stock above, but with a small trick. That is, the present value of the dividends can be written

$$D\sum_{j=1}^{\infty}(1 + r)^{-j}(1 + g)^{j},$$

and since $(1 + r)^{-j}(1 + g)^{j} = \left(1 + \frac{r-g}{1+g}\right)^{-j}$, this present value becomes a preferred stock with dividend D, valued with a yield of $\frac{r-g}{1+g}$. Consequently (2.22) follows from

(2.21), and where the requirement that $r > g$ is simply to ensure that in (2.11), $\left(1 + \frac{r-g}{1+g}\right)^{-n}$ decreases to 0 as n increases to ∞.

In many applications one thinks of this price function as a function of a single variable. For example, if we think of D and r as fixed, we can express the stock value as a function of the assumed growth rate, $V(g)$, and so forth. This illustrates the important point noted above. The functional representation of a quantity is usually not uniquely defined; it is typically best defined based on the objectives of the user. As was the case for the price of a bond, one could also allow either g and/or r to vary by year, further expanding the multivariate nature of this price function, or modify this derivation to allow for dividends payable other than annually.

Portfolio Return Functions

If the return on asset A_1 is projected to be r_1, and that of A_2 projected to be r_2, we can define a function $f(w)$ to represent the projected return on a portfolio of both assets, with $100w\%$ allocated to A_1, and $100(1-w)\%$ to A_2. We then have

$$f(w) = wr_1 + (1-w)r_2$$

$$= r_2 + w(r_1 - r_2).$$

While this may be initially modeled with the understanding that $0 \le w \le 1$, it is a perfectly sensible function outside this domain by understanding a "negative investment" to represent a **short sale**.

A short sale is one whereby the investor borrows and sells an asset for the cash proceeds, with the future obligation to repurchase the asset in the open market to **cover the short**, which is to say, return the asset to the original owner. Such short sales require the posting of collateral in a margin account, typically in addition to the cash proceeds or the securities purchased with these proceeds.

This model is easily generalized to the case of n assets, whereby our asset choices are $\{A_j\}_{j=1}^n$ with projected returns of $\{r_j\}_{j=1}^n$ and asset allocations of $\{w_j\}_{j=1}^n$ with $0 \le w_j \le 1$ and $\sum_{j=1}^n w_j = 1$. One then sees that the projected portfolio return is a function of these asset allocation weights:

$$f(w_1, w_2, \ldots, w_n) = \sum_{j=1}^n w_j r_j. \tag{2.23}$$

Once again, with short sales allowed, the domain of this function can be expanded beyond the original restricted domains of $0 \le w_j \le 1$ for all j.

As a final comment, it may seem odd that with 2 assets, f was a function of 1 variable, yet with n assets, f is a function of n variables. This provides another example of the flexibility one has in such representations. As currently expressed, it must be remembered in the analysis that logically $\sum_{j=1}^{n} w_j = 1$, and hence these n variables are **constrained**, meaning that the domain of this function is not the "n-dimensional cube," $\{(w_1, w_2, \ldots, w_n) \mid 0 \leq w_j \leq 1 \text{ for all } j\}$, but a subset of this cube, $\{(w_1, w_2, \ldots, w_n) \mid 0 \leq w_j \leq 1 \text{ for all } j \text{ and } \sum_{j=1}^{n} w_j = 1\}$. To eliminate the need to remember this constraint, it can be built into the definition of the function, as was done in the 2-asset model. For example, writing $w_n = 1 - \sum_{j=1}^{n-1} w_j$, we can rewrite the projected return function as a function of $n - 1$ variables:

$$f(w_1, w_2, \ldots, w_{n-1}) = r_n + \sum_{j=1}^{n-1} w_j(r_j - r_n).$$

The domain of this function is now defined to either preclude or allow short sales.

Naturally this functional representation also makes sense when the r_j values are not initially defined as constants but instead represent values that will only be revealed at the end of the period. This perspective then lends itself to thinking about these returns as **random variables**, as will be discussed in chapter 7 on probability theory. Within that framework, good analysis can be done with this function, and the asset allocation will be seen to influence properties of the randomness of the portfolio return.

Forward-Pricing Functions

As a final example, consider a **forward contract** on an equity, with current price S_0. A forward contract is a contract that obligates the **long** position to purchase the equity, and the **short** position to sell the equity, at forward time $T > 0$, measured in years say, and at a price agreed to today, denoted F_0. No cash changes hands at time 0, whereas at time T one share of the stock is exchanged for F_0. The natural question is, What should be the value of F_0 and on what variables should it depend?

As it turns out, the long position can **replicate** this contract in theory, which means that the long can implement a trade at time 0 that provides the obligation to "buy" the stock at time T, and this can be done without finding another investor that is willing to take on the short position. Similarly a short position can be replicated, so an investor can implement this contract without finding another investor that is willing to take on the long position.

The replication of the long position is accomplished by purchasing the equity today for a price of S_0, and acquiring the cash to do so by **short-selling** a T-period Treasury bill. Imagine for clarity that the equity is placed in the margin account required for the short position, along with other investor funds, so the investor doesn't actually have possession of it at the time of this trade. At time T, the short sale will be **covered** at a cost of $S_0(1 + r_T)^T$, the value of the T-bill to the original owner at that time, where r_T denotes the annual return on the T-period T-bill, and T is in units of years. Because the short position has been covered, the margin account is released and the investor takes possession of the stock, implicitly for the price of covering the short.

Similarly a short forward can be replicated with a short position in the stock and an investment in T-bills, and the same cost of $S_0(1 + r_T)^T$ is derived. In both cases the position is replicated with no out-of-pocket cost at time 0 for the investor.

So in either case we conclude that the forward price, F_0, that makes sense today with no money now changing hands, if it is to be agreed to by independent parties each of whom could in theory replicate their positions, is a function of 3 variables:

$$F_0(S_0, r_T, T) = S_0(1 + r_T)^T. \tag{2.24}$$

In some applications one might think of one or two of these variables as fixed, and the forward price function expressed with fewer variables. The reason this is the "correct price" is that if forwards were offered at a different price, it would be possible for investors to make riskless profits by committing to forwards and then replicating the opposite position (see exercise 15).

Once the forward contract is negotiated and committed to, there arises the question of the value of the contract to the long and to the short at time t where $0 < t \leq T$. For definitiveness, let F_0 denote the price agreed to at time $t = 0$. At time t, we know from the formula above that the forward price will be

$$F_t(S_t, r_{T-t}, T - t) = S_t(1 + r_{T-t})^{T-t}. \tag{2.25}$$

So the long position is committed to buy at time T at price F_0, but today's market indicates that the right price is F_t. That's good news for the long if $F_0 \leq F_t$, and bad news otherwise. The sentiments of the short position are opposite. So the value at time t is "plus or minus" the present value of the difference between the two prices F_0 and F_t, that is, $\pm[F_t - F_0](1 + r_{T-t})^{-(T-t)}$, which for the long position can be expressed as

$$V_t(S_t, r_{T-t}, T - t) = S_t - F_0(1 + r_{T-t})^{-(T-t)}. \tag{2.26}$$

The function representing the value of this contract to the short position is simply the negative of the function in (2.26).

Exercises

Practice Exercises

1. Apply Euclid's algorithm to the following pairs of integers to find the greatest common divisor (g.c.d.), and express the g.c.d. in terms of Bezout's identity:

(a) 115 and 35

(b) 4531 and 828

(c) 1915 and 472

(d) 46053 and 3042

2. In a remark after the proof of the existence of nonrational numbers, or irrational numbers, it was demonstrated that between any two rational numbers is a rational number and an irrational number. Prove by construction, or by contradiction, that in both cases there are infinitely many rationals and irrationals between the two given rationals. (*Hint*: For intermediate irrationals, note that for $n \neq m^2$, we know that $\sqrt{n} \notin \mathbb{Q}$, and hence $\frac{1}{\sqrt{n}} \notin \mathbb{Q}$. Note also that $\frac{1}{\sqrt{n}} \to 0$ as $n \to \infty$.)

3. Prove that the irrationals are uncountable. (*Hint*: Consider a proof by contradiction based on the countability of the rationals and uncountability of the reals.)

4. Express the following real numbers in the indicated base using the greedy algorithm either exactly or to four digits to the right of the "decimal point":

(a) 100.4 in base-6

(b) 0.1212121212... in base-2

(c) 125,160.256256256... in base-12

(d) −127.33333333... in base-7

5. Demonstrate that if a number's decimal expansion either terminates, or ends with an infinite repeating cluster of digits such as $12.12536363636 \equiv 12.125\overline{36}$, then this number is rational. (*Hint*: If the number in this example is called x, compare $1000x$ and $100,000x$. Generalize.)

6. Euler's formula gives a practical and easy way to derive many of the trigonometric identities involving the sine and cosine trigonometric functions. Verify the following (*Hint*: $e^{2ai} = (e^{ai})^2$):

(a) $\cos 2a = \cos^2 a - \sin^2 a$

(b) $\sin 2a = 2 \sin a \cos a$

7. If an annual payment annuity of 100 is to be received from time 8 to time 20, show that the value of this **7-year deferred, 13-year annuity** can be represented in either of the following ways:

(a) $100(a_{20;r} - a_{7;r})$

(b) $100(1+r)^{-7}a_{13;r}$

8. What is the domain and range of the following functions? Note that the domain may include real numbers that would not make sense in a finance application.

(a) Annuity present value: $V(r) = F\sum_{j=1}^{n}(1+r)^{-j}$ (If this is written in the equivalent form $V(r) = F\frac{1-(1+r)^{-n}}{r}$, the domain initially looks different. Convince yourself by numerical calculation, or analysis, that $r = 0$ is not really a problem for this function even in the second form, since the r in the denominator "cancels" an r in the numerator, much like $3r/r$.)

(b) Bond price: $P(i) = F\frac{r}{2}a_{2n;i/2} + Fv_{i/2}^{2n}$

(c) Loan repayment: $P(r,n) = \frac{L(r/12)}{1-v_{r/12}^{12n}}$

9. Use the nominal equivalent yield formula and demonstrate numerically for annual "rates" $r_1 = 0.01, 0.10, 0.25, 1.00$, that as $m \to \infty$, the equivalent yield $r_m(r_1)$ gets closer and closer to $\ln(1+r_1)$. Consider m up to 1000, say. Show algebraically that if this limiting result is true for all r_1, and n and r_n are fixed, then as $m \to \infty$, the equivalent yield, $r_m(r_n)$, again gets closer and closer to $\ln(1+r_1)$ where r_1 is the annual rate equivalent to r_n. (*Note*: These results can be proved with the tools of chapter 5, once the notion of the limit of a sequence is formally introduced, and chapter 9, which provides Taylor series approximations to the function $\ln x$.)

10. Complete the rows of the following table with equivalent nominal rates:

r_1	r_2	r_4	r_{12}	r_{365}
0.05				
	0.10			
		0.0825		
			0.0450	
				0.0775

11. You are given a 5-year and a 30-year bond, each with a par of 1000 and a semiannual coupon rate of 8%. Calculate the price of each at an 8% semiannual yield, and graph each price function over the range of semiannual yields $0\% \le i \le 16\%$ on the same set of axes. What pattern do you notice between the graphs?

12. For the 5-year bond in exercise 11, start with prices calculated at 6% and 10%:

(a) Develop graphs of these bond prices over time using (2.18)

(b) Show that in the case of the 6% valuation, that the successive ratios of the bond's **write downs**, defined as the quantities $P_{j/2}(0.06) - P_{(j+1)/2}(0.06)$, have a constant ratio of 1.03.

(c) Show similarly that for the 10% valuation, the successive ratios of the bond's **write ups**, defined as the quantities $P_{(j+1)/2}(0.10) - P_{j/2}(0.10)$, have a constant ratio of 1.05.

(d) Derive algebraically using (2.16), the general formula for a write up or write down and show that the common ratio is $1 + \frac{i}{2}$, where i denotes the investor's yield.

13. You are considering a 10-year loan for $100,000 at a monthly nominal rate of 7.5%.

(a) Calculate the monthly payment for this loan.

(b) Calculate the outstanding balance of this loan over the first year immediately following each of the required 12 payments as well as the changes in these balances, called **loan amortizations**. (*Hint*: recall that the loan balance equals the present value of remaining payments)

(c) Confirm that the ratio of successive amortizations are in constant ratio of $1 + \frac{0.075}{12}$.

(d) Derive algebraically the general formula for the loan amortizations and confirm that the ratio of successive values is a constant $1 + \frac{i}{12}$.

(e) Demonstrate that given the formula derived for the values of the amortizations, they indeed add up to the original loan value, L.

14. What is the DDM price for a common stock with quarterly dividends, where the last dividend of 2.50 was paid yesterday:

(a) If dividends are assumed to grow at a quarterly nominal rate of 9% and the investor requires a return of 15% quarterly?

(b) If dividends are assumed to grow at a quarterly nominal rate of 9% only for 5 years, and then to a grow at a rate of 4%, again on a quarterly basis? (*Hint*: Show that the dividends can be modeled as a 5-year annuity at one rate, followed by a 5-year deferred **perpetuity** [i.e., an infinite annuity] at another rate, where by "deferred" means the first payment is one-quarter year after $t = 5$. See also exercise 7.).

15. A common stock trades today at $S_0 = 15$, and the risk free rate is 6% on a semi-annual basis.

(a) What is the forward price of this stock for delivery in one year?

(b) Replicate a long position in this forward contract with a portfolio of stock and T-bills, giving details on the initial position as well as trade resolution in 1 year.

(c) If the market traded long and short 1-year forwards on this stock with a price of 15.10, develop an arbitrage to take advantage of this mispricing, giving details on the initial position as well as trade resolution in 1 year. (*Hint*: Go long the forward if this price is low, and short if this price is high. Offset the risk with replication.)

(d) If an investor goes short the forward in part (a), what is the investor's gain or loss at 3 months' time when the contract is "offset" in the market (i.e., liquidated for the then market value) if the stock price has fallen to 13.50, and the 9-month risk-free rate is 7.50% (semiannual)?

Assignment Exercises

16. Apply Euclid's algorithm to the following pairs of integers to find the greatest common divisor (g.c.d.), and express the g.c.d. in terms of Bezout's identity:

(a) 697 and 221

(b) 7500 and 2412

(c) 21423 and 3441

(d) 79107 and 32567

17. (See exercise 2.) In a remark after the proof of the existence of nonrational numbers, or irrational numbers, it was demonstrated that between any two irrational numbers is a rational and an irrational. Prove by construction, or by contradiction, that in both cases there are infinitely many rationals and irrationals between the two irrational numbers.

18. Express the following real numbers in the indicated base using the greedy algorithm either exactly or to four digits to the right of the "decimal" point:

(a) 25.5 in base-2

(b) 150.151515... in base-5

(c) 237,996.1256 in base-12

(d) $-2,399.27$ in base-9

19. (See exercise 5.) Explain why it is the case that if a number is rational, its decimal expansion either terminates or, after a certain number of digits, ends with an infinite repeating cluster of digits such as $12.125\overline{36}$. Specifically, explain that if this rational number is given by $\frac{n}{m}$ where n and m have no common divisors, then the decimal expansion will terminate by the mth decimal digit, or there will be repeating cluster that will begin on or before the mth decimal digit, and in this case, the repeating cluster can contain at most $m - 1$ digits. (*Hint*: Think about the remainders you get at each division step.)

20. Euler's formula gives a practical and easy way to derive many of the trigonometric identities involving the sine and cosine trigonometric functions. Verify the following (*Hint*: $e^{(a+b)i} = e^{ai}e^{bi}$):

(a) $\cos(a + b) = \cos a \cos b - \sin a \sin b$

(b) $\sin(a + b) = \cos a \sin b + \cos b \sin a$

21. (See exercise 7.) If an annual payment annuity of 100 is to be received from time $n + 1$ to time $n + m$, show that the value of this n-**year deferred**, m-**year annuity** can be represented as either of the following:

(a) $100(a_{n+m;r} - a_{n;r})$

(b) $100(1 + r)^{-n} a_{m;r}$

22. What is the domain and range of the following functions? Note that the domain may include real numbers that would not make sense in a finance application:

(a) Nominal equivalent rate: $r_m(r_n) = m\left[\left(1 + \frac{r_n}{n}\right)^{n/m} - 1\right]$

(b) Common stock price: $V(D, g, r) = D\frac{1+g}{r-g}$

(c) Forward price: $F_t(S_t, r_{T-t}, T - t) = S_t(1 + r_{T-t})^{T-t}$

23. Complete the rows of the following table with equivalent nominal rates:

r_1	r_2	r_4	r_{12}	r_{365}
0.16				
	0.045			
		0.0955		
			0.0150	
				0.025

24. A \$25 million, 10-year commercial mortgage is issued with a rate of 8% on a monthly nominal basis.

(a) What is the monthly repayment, P, over the term of the mortgage?

(b) If B_j denotes the outstanding balance on this loan immediately after the jth payment, with $B_0 = 25$ million, show that

$$B_j = Pa_{(120-j);0.08/12}$$

$$= [B_0 - Pa_{j;0.08/12}]\left(1 + \frac{0.08}{12}\right)^j.$$

(c) If P_j denotes the principal portion of the jth payment, show that

$$P_j = P - \frac{0.08}{12} B_{j-1}.$$

(d) Show that $P_{j+1} = \left(1 + \frac{0.08}{12}\right) P_j$ for $j \geq 1$.

(e) From part (d), confirm that $\sum P_j = 25$ million.

25. A common stock trades today at $S_0 = 50$, and the risk-free rate is 5% on a semi-annual basis.

(a) What is the forward price of this stock for delivery in 6 months?

(b) Replicate a long position in this forward contract with a portfolio of stock and T-bills, giving details on the initial position as well as the trade resolution in 6 months.

(c) If the market traded long and short 6-month forwards on this stock with a price of 53, develop an arbitrage to take advantage of this mispricing, giving details on the initial position as well as the trade resolution in 6 months.

(d) If an investor goes long the forward in part (a), how much does the investor make or lose at 3 months' time when the contract is offset in the market if the stock price has risen to 52, and the 3-month risk-free rate is at 4.50% (semiannual)?

3 Euclidean and Other Spaces

3.1 Euclidean Space

3.1.1 Structure and Arithmetic

The notion of a **Euclidean space of dimension** n is a generalization of the two-dimensional plane and three-dimensional space studied by Euclid in the *Elements*.

Definition 3.1 *Denoted* $\mathbb{R}^n$ *or sometimes* E^n, n-**dimensional Euclidean space**, *or* **Euclidean** n-**space**, *is defined as the collection of* n-**tuples** *of real numbers, referred to as* **points**:

$$\mathbb{R}^n \equiv \{(x_1, x_2, \ldots, x_n) \mid x_j \in \mathbb{R} \ for \ all \ j\}. \tag{3.1}$$

Arithmetic operations of **pointwise addition** *and* **scalar multiplication** *in* $\mathbb{R}^n$ *are defined by*

1. $\mathbf{x} + \mathbf{y} = (x_1 + y_1, x_2 + y_2, \ldots, x_n + y_n)$.
2. $a\mathbf{x} = (ax_1, ax_2, \ldots, ax_n)$, *where* $a \in \mathbb{R}$.

In other words, addition and multiplication by so-called **scalars** $a \in \mathbb{R}$, are defined componentwise. Because points in $\mathbb{R}^n$ have n components and are thought of as generalizing the corresponding notion in familiar two- and three-dimensional space, they are typically referred to as **points** and sometimes **vectors**, and are either notated in **boldface**, $\mathbf{x}$, as will be used in this book, or with an overstrike arrow, $\vec{x}$. The components of these points, the $\{x_j\}$, are called **coordinates**, and a given x_j is referred to as the jth coordinate.

The terminology of n-tuple may seem a bit strange at first. It is but a generalization of the typical language for such groupings whereby, following "twin" and "triplet," one says quadruple, quintuple, sextuple, and so forth. For specific values of n, the language would be 2-tuple, 3-tuple, and on and on.

Note that the notation for Euclidean space, $\mathbb{R}^n$, is more than just a fanciful play on the notation for the real numbers, $\mathbb{R}$. This notation rather stems from that for a **product space** defined in terms of a **direct** or **Cartesian product**:

Definition 3.2 *If* X *and* Y *are two collections, the* **direct** *or* **Cartesian product** *of* X *and* Y, *denoted:* $X \times Y$ *is defined as*

$$X \times Y = \{(x, y) \mid x \in X, y \in Y\}. \tag{3.2}$$

That is, $X \times Y$ *is the collection of* **ordered pairs**, *which is to say that* $X \times Y \neq Y \times X$ *in general, and the order of the terms in the product matter. One similarly defines* $X \times Y \times Z$, *etc., and refers to all such constructions as* **product spaces**.

When $X = Y$, it is customary to denote $X \times X$ by X^2, $X \times X \times X$ by X^3, etc. Consequently the notation for Euclidean space, which is the original example of a product space, is consistent with this notational convention:

$$\mathbb{R}^n \equiv \mathbb{R} \times \mathbb{R} \times \cdots \times \mathbb{R}, \quad \text{with } n \text{ factors.}$$

One similarly defines $\mathbb{C}^n$, n-**dimensional complex space**; $\mathbb{Z}^n$, n-**dimensional integer space** or the n-**dimensional integer lattice**; and so forth.

In general, Euclidean space does not have the structure of a field as was the case for $\mathbb{Q}$, $\mathbb{R}$, and $\mathbb{C}$ in chapter 2. This reason is not related to the "addition" in $\mathbb{R}^n$ but to the problem of defining a multiplication of vectors with the required properties. However, Euclidean space has the structure of a **vector space**, and it is easily demonstrated that $\mathbb{R}^n$ is a **vector space over the real field** $\mathbb{R}$. In this book we will almost exclusively be interested in **real vector spaces** that are defined by $\mathcal{F} = \mathbb{R}$:

Definition 3.3   *A collection of **points** or **vectors**, X, is a **vector space over a field** $\mathcal{F}$, if:*

1. *X is **closed** under pointwise addition and scalar multiplication:*

If $\mathbf{x}, \mathbf{y} \in X$ and $a \in \mathcal{F}$, then $\mathbf{x} + \mathbf{y} \in X$ and $a\mathbf{x} \in \mathcal{F}$.

2. *There is a **zero vector**: $\mathbf{0} = (0, 0, \ldots, 0) \in X$ such that*

$$\mathbf{x} + \mathbf{0} = \mathbf{0} + \mathbf{x} = \mathbf{x} \quad \text{for all } \mathbf{x} \in X.$$

3. *Point addition is **commutative** and **associative**: Given $\mathbf{x}, \mathbf{y}, \mathbf{z} \in X$,*

$$\mathbf{x} + \mathbf{y} = \mathbf{y} + \mathbf{x},$$

$$\mathbf{x} + (\mathbf{y} + \mathbf{z}) = (\mathbf{x} + \mathbf{y}) + \mathbf{z}.$$

4. *Scalar multiplication satisfies the **distributive law** over addition: For $\mathbf{x}, \mathbf{y} \in X$ and $a \in \mathcal{F}$,*

$$a(\mathbf{x} + \mathbf{y}) = (\mathbf{x} + \mathbf{y})a = a\mathbf{x} + a\mathbf{y}.$$

As was noted in chapter 2, one can define a multiplication and a field structure on $\mathbb{R}^2$ by the identification with the complex numbers:

$$\mathbb{R}^2 \leftrightarrow \mathbb{C} : (a, b) \leftrightarrow a + b\imath.$$

Then multiplication is defined using (2.8):

$$(a, b) \cdot (c, d) = (ac - bd, ad + bc),$$

and multiplicative inverses follow from the formula for z^{-1}:

$$(a,b)^{-1} = \left(\frac{a}{a^2 + b^2}, \frac{-b}{a^2 + b^2} \right).$$

It is natural to wonder if such an identification can be made for $\mathbb{R}^n$, with $n > 2$, and other fields produced. The answer is that yes, identifications do exist for some $n > 2$, but these do not produce the structure of fields.

For example, the first of these identifications was discovered by **Sir William Rowan Hamilton** (1805–1865) in 1843, and called the **quaternions**. The quaternions can be identified with $\mathbb{R}^4$, and have the appearance of "generalized" complex numbers. That is, having a "real" component and three "imaginary" components i, j, k, and the identification is

$$(a,b,c,d) \leftrightarrow a + bi + cj + dk,$$

$$i^2 = j^2 = k^2 = ijk = -1.$$

The resulting structure falls short of a field structure because multiplication is not commutative. This follows from $ijk = -1$, which implies that $ij = -ji$. The resulting structure is called an **associative normed division algebra**.

The quaternions can in turn be generalized and an identification made with $\mathbb{R}^8$, known as the **octonions**, which were independently discovered by **John T. Graves** (1806–1870) in 1843 and **Arthur Cayley** (1821–1895) in 1845. Although octonions form a normed division algebra, in contrast to the quaternions, multiplication in the octonions is neither commutative nor associative. Further generalizations to $\mathbb{R}^{2^n}$ are possible for all n, each successive term in the sequence derived from the former term through what is known as the **Cayley–Dickson construction**, also after **Leonard Eugene Dickson** (1874–1954).

3.1.2   Standard Norm and Inner Product for $\mathbb{R}^n$

Besides an arithmetic on $\mathbb{R}^n$, there is the need for a notion of **length**, or magnitude, of a point. In mathematics this notion is called a "norm."

Definition 3.4   *The **standard norm** on $\mathbb{R}^n$, denoted $|\mathbf{x}|$ or $\|\mathbf{x}\|$, is defined by*

$$|\mathbf{x}| \equiv \sqrt{\sum_{j=1}^{n} x_i^2}, \tag{3.3}$$

where the positive square root is implied.

This norm generalizes the **Pythagorean theorem** and the notion of the length of a vector in the plane or in 3-space, which in turn generalizes the notion of length on the real line or 1-space achieved by the absolute value of x: $|x|$, defined in (2.3).

Another useful notion on $\mathbb{R}^n$ that generalizes to other vector spaces is that of an **inner product**, whose formula generalizes the notion of a **dot product** of vectors in the plane and 3-space:

Definition 3.5   *The **standard inner product** on $\mathbb{R}^n$, denoted $\mathbf{x} \cdot \mathbf{y}$ or $(\mathbf{x}, \mathbf{y})$, is defined for $\mathbf{x}, \mathbf{y} \in \mathbb{R}^n$ as*

$$\mathbf{x} \cdot \mathbf{y} = \sum_{j=1}^{n} x_i y_i. \tag{3.4}$$

Inner products are intimately connected with norms. As may be apparent from the definitions above, the standard norm for $\mathbb{R}^n$ satisfies

$$|\mathbf{x}| = (\mathbf{x} \cdot \mathbf{x})^{1/2}, \quad \text{or} \quad |\mathbf{x}|^2 = |\mathbf{x} \cdot \mathbf{x}|. \tag{3.5}$$

Remark 3.6   *The notion of an inner product is one that will reappear in later chapters and studies in a variety of contexts. As it turns out, there are many possible inner products on $\mathbb{R}^n$ that satisfy the same critical properties as the standard inner product above. Here we identify these defining properties and leave their verification for the standard inner product as an exercise. Note that item 4 below follows from properties 2 and 3, but is listed for completeness.*

Definition 3.7   *An **inner product** on a real vector space X, is a real-valued function defined on $X \times X$ with the following properties:*

1. $(\mathbf{x}, \mathbf{x}) \geq 0$ *and* $(\mathbf{x}, \mathbf{x}) = 0$ *if and only if* $\mathbf{x} = \mathbf{0}$.
2. $(\mathbf{x}, \mathbf{y}) = (\mathbf{y}, \mathbf{x})$.
3. $(a\mathbf{x}_1 + b\mathbf{x}_2, \mathbf{y}) = a(\mathbf{x}_1, \mathbf{y}) + b(\mathbf{x}_2, \mathbf{y})$ *for* $a, b \in \mathbb{R}$.
4. $(\mathbf{x}, a\mathbf{y}_1 + b\mathbf{y}_2) = a(\mathbf{x}, \mathbf{y}_1) + b(\mathbf{x}, \mathbf{y}_2)$ *for* $a, b \in \mathbb{R}$.

Definition 3.8   *If $(\mathbf{x}, \mathbf{y})$ is an inner product on a real vector space X, the **norm associated with this inner product** is defined by (3.5).*

*3.1.3   Standard Norm and Inner Product for $\mathbb{C}^n$

We note for completeness that in order to appropriately generalize (2.2) to an n-dimensional complex vector space, the inner product and norm definitions are modified when the space involved, such as $\mathbb{C}^n$, and its underlying field, have complex values. We provide the definition here:

Definition 3.9   *The **standard inner product** on $\mathbb{C}^n$, denoted $\mathbf{x} \cdot \mathbf{y}$ or $(\mathbf{x}, \mathbf{y})$ is defined for $\mathbf{x}, \mathbf{y} \in \mathbb{C}^n$,*

$$\mathbf{x} \cdot \mathbf{y} = \sum_{j=1}^{n} x_i \bar{y}_i, \tag{3.6}$$

*where $\bar{y}_i$ denotes the complex conjugate of y_i. The **standard norm** for $\mathbb{C}^n$ is defined as*

$$|\mathbf{x}| = (\mathbf{x} \cdot \mathbf{x})^{1/2} \quad \text{or} \quad |\mathbf{x}|^2 = |\mathbf{x} \cdot \mathbf{x}|. \tag{3.7}$$

Remark 3.10   *In the context of a complex space, there are again many possible inner products satisfying the critical properties of the standard inner product above. These properties are identical to those listed for $\mathbb{R}^n$, with the necessary adjustments for the complex conjugate on the second term. As before, 5 follows from 3 and 4, and also here 1 follows from 3, but these properties are listed for completeness.*

Definition 3.11   *An inner product on a complex vector space X, is a complex-valued function defined on $X \times X$ with the following properties:*

1. $(\mathbf{x}, \mathbf{x}) \in \mathbb{R}$ *for all $\mathbf{x}$.*
2. $(\mathbf{x}, \mathbf{x}) \geq 0$ *and $(\mathbf{x}, \mathbf{x}) = \mathbf{0}$ if and only if $\mathbf{x} = \mathbf{0}$.*
3. $(\mathbf{x}, \mathbf{y}) = \overline{(\mathbf{y}, \mathbf{x})}$.
4. $(a\mathbf{x}_1 + b\mathbf{x}_2, \mathbf{y}) = a(\mathbf{x}_1, \mathbf{y}) + b(\mathbf{x}_2, \mathbf{y})$ *for $a, b \in \mathbb{C}$.*
5. $(\mathbf{x}, a\mathbf{y}_1 + b\mathbf{y}_2) = \bar{a}(\mathbf{x}, \mathbf{y}_1) + \bar{b}(\mathbf{x}, \mathbf{y}_2)$ *for $a, b \in \mathbb{C}$.*

3.1.4   Norm and Inner Product Inequalities for $\mathbb{R}^n$

An important property of inner products is the **Cauchy–Schwarz inequality**, which was originally proved in 1821 in the current finite-dimensional context by **Augustin Louis Cauchy** (1759–1857), and generalized 25 years later to all "inner product spaces" by **Hermann Schwarz** (1843–1921).

Throughout this section, results on inner products are derived in the context of the "standard" inner products in (3.4) or (3.6) for specificity. However, it should be noted that the proofs of these results rely only on the properties identified above for general inner products, and consequently these results will remain true for all inner products once defined.

Proposition 3.12 (*Cauchy–Schwarz Inequality*)   *With* $\mathbf{x} \cdot \mathbf{y}$ *defined as in (3.4) or (3.6)*,

$$|\mathbf{x} \cdot \mathbf{y}| \le |\mathbf{x}|\,|\mathbf{y}|. \tag{3.8}$$

In other words, the absolute value of an inner product is bounded above by the product of the vector norms.

Proof   Consider $\mathbf{x} - a\mathbf{y}$. By definition of a norm, we have for any real number a:

$$|\mathbf{x} - a\mathbf{y}| \ge 0.$$

However, a calculation produces

$$|\mathbf{x} - a\mathbf{y}|^2 = (\mathbf{x} - a\mathbf{y}, \mathbf{x} - a\mathbf{y})$$

$$= \mathbf{x} \cdot \mathbf{x} - 2a\mathbf{x} \cdot \mathbf{y} + a^2 \mathbf{y} \cdot \mathbf{y}$$

$$= |\mathbf{x}|^2 + a^2 |\mathbf{y}|^2 - 2a\mathbf{x} \cdot \mathbf{y}.$$

Choosing $a = \frac{\mathbf{x} \cdot \mathbf{y}}{|\mathbf{y}|^2}$, and combining, we get

$$|\mathbf{x}|^2 - \frac{(\mathbf{x} \cdot \mathbf{y})^2}{|\mathbf{y}|^2} \ge 0,$$

and the result follows.   ■

Remark 3.13   *We can remove the absolute values from* $\mathbf{x} \cdot \mathbf{y}$, *and the result remains true since, by definition,* $\mathbf{x} \cdot \mathbf{y} = \pm|\mathbf{x} \cdot \mathbf{y}| \le |\mathbf{x} \cdot \mathbf{y}|$. *We use this below.*

The general notion of a norm is a fundamental tool in mathematics and is formalized as follows:

Definition 3.14   *A **norm on a real vector space** X, is a real-valued function on X with values, denoted $|\mathbf{x}|$ or $\|\mathbf{x}\|$, satisfying:*

1. $|\mathbf{x}| \in \mathbb{R}$.
2. $|\mathbf{0}| = 0$, *and* $|\mathbf{x}| > 0$ *for* $\mathbf{x} \ne \mathbf{0}$.
3. $|a\mathbf{x}| = |a|\,|\mathbf{x}|$ *for* $a \in \mathbb{R}$.
4. **(Triangle inequality)** $|\mathbf{x} + \mathbf{y}| \le |\mathbf{x}| + |\mathbf{y}|$.

Definition 3.15   *A **normed vector space** is any real vector space, X, on which there is defined a norm, $|\mathbf{x}|$. For specificity, a normed space is sometimes denoted $(X, |\mathbf{x}|)$ or $(X, \|\mathbf{x}\|)$.*

Remark 3.16 *Item 4 is known as the **triangle inequality** because it generalizes the result in (2.7) that the length of any side of a triangle cannot exceed the sum of the lengths of the other two sides. Also note that item 4 is easily generalized by an iterative application to*

$$\left| \sum_{j=1}^{n} \mathbf{x}_i \right| \le \sum_{j=1}^{n} |\mathbf{x}_i|. \tag{3.9}$$

Remark 3.17 *A norm can be equally well defined on a vector space over a general field $\mathcal{F}$, such as the complex field $\mathbb{C}$, where $|a|$ denotes the norm of $a \in \mathcal{F}$. But we will have no need for this generalization.*

The general definition of a norm was intended to capture the essential properties known to be true of the standard norm $|\mathbf{x}|$ defined on $\mathbb{R}^n$. Not surprisingly, we therefore have:

Proposition 3.18 $|\mathbf{x}|$ *defined in (3.3) is a **norm on** $\mathbb{R}^n$.*

Proof Only the triangle inequality needs to be addressed as the others follow immediately from definition. From (3.5) we have that

$$|\mathbf{x} + \mathbf{y}|^2 = (\mathbf{x} + \mathbf{y}, \mathbf{x} + \mathbf{y})$$

$$= \mathbf{x} \cdot \mathbf{x} + 2\mathbf{x} \cdot \mathbf{y} + \mathbf{y} \cdot \mathbf{y}$$

$$\le |\mathbf{x}|^2 + 2|\mathbf{x}|\,|\mathbf{y}| + |\mathbf{y}|^2$$

$$= (|\mathbf{x}| + |\mathbf{y}|)^2,$$

and the result follows. Note that in the third step, the Cauchy–Schwarz inequality was used because it implies that $\mathbf{x} \cdot \mathbf{y} \le |\mathbf{x}|\,|\mathbf{y}|$. ∎

*3.1.5 Other Norms and Norm Inequalities for $\mathbb{R}^n$

It turns out that there are many norms that can be defined on $\mathbb{R}^n$ in addition to the standard norm in (3.3).

Example 3.19

1. *For any p with $1 \le p < \infty$, the so-called l_p-norm, pronounced "lp-norm," is defined by*

$$\|\mathbf{x}\|_p \equiv \left(\sum_{j=1}^{n}|x_i|^p\right)^{1/p}.$$

(3.10)

2. *Extending to* $p = \infty$, *the so-called* l_∞-***norm**, *pronounced "l infinity norm," is defined by*

$$\|\mathbf{x}\|_\infty = \max_i |x_i|.$$

(3.11)

Remark 3.20   *We still have to prove that these l_p-norms are true norms by the defini-tion above, but note that for $p = 2$, the l_2-norm is identical to the standard norm defined in (3.3). So the l_p-norms can be seen to generalize the standard norm by gen-eralizing the power and root used in the definition. Also, as will be seen below, while appearing quite differently defined, the l_∞-norm will be seen to be the "limit" of the l_p-norms as p increases to ∞.*

The challenge of demonstrating that these examples provide true norms is to show the triangle inequality to be satisfied, since the other needed properties are easy to verify. For the l_∞-norm in (3.11) the triangle inequality follows from (2.7), since the l_∞-norm is a maximum of absolute values. That is, $|x_i + y_i| \le |x_i| + |y_i|$ for any i by (2.7), and we have that

$$\max_i |x_i + y_i| \le \max_i(|x_i| + |y_i|) \le \max_i |x_i| + \max_i |y_i|.$$

Similarly the l_1-norm again satisfies the triangle inequality due to (2.7), since the l_1-norm is a sum of absolute values, and

$$\sum_{j=1}^{n}|x_i + y_i| \le \sum_{j=1}^{n}|x_i| + \sum_{j=1}^{n}|y_i|.$$

For the l_p-norm with $1 < p < \infty$, the proof will proceed in a somewhat long series of steps that should be simply scanned on first reading, focusing instead on the flow of the logic. The proof proceeds in steps:

1. First off, the triangle inequality in this norm is called the **Minkowski inequality or Minkowski's inequality**, and was derived by **Hermann Minkowski** (1864–1909) in 1896. The proof of this inequality requires a generalization of the Cauchy–Schwarz inequality, which is called the **Hölder inequality** or **Hölder's inequality**, derived by **Otto Hölder** (1859–1937) in 1884 in a more general context than presented here.

2. To derive Hölder's inequality, we require **Young's inequality**, which was derived by **W. H. Young** (1863–1942) in 1912.

Reversing the steps to a proof, we begin with Young's inequality. It introduces a new notion that arises often in the study of l_p-norms, and that is the notion of an index q being the **conjugate index** to p. Specifically, given $1 < p < \infty$, the index q is said to be conjugate to p if $\frac{1}{p} + \frac{1}{q} = 1$. It is then easy to see that $q = \frac{p}{p-1}$ also satisfies $1 < q < \infty$, and that p is also conjugate to q. In some cases this notion of conjugacy is extended to $1 \le p \le \infty$, where one defines $\frac{1}{\infty} \equiv 0$, and hence $p = 1$ and $q = \infty$ are conjugate. This notion highlights the uniqueness of the index $p = 2$, namely that this is the only index conjugate to itself, a fact that will later be seen to be quite significant.

Before turning to the statement and proof of Young's inequality, note that the natural logarithm is a **concave function**, which means that for any $x, y > 0$,

$$t \ln x + (1 - t) \ln y \le \ln(tx + (1 - t)y) \qquad \text{for } 0 \le t \le 1. \tag{3.12}$$

Graphically, for given points $x, y > 0$, say $y > x > 0$ for definiteness, the straight line connecting the points $(x, \ln x)$ and $(y, \ln y)$ never exceeds the graph of the function $f(z) = \ln z$ for $x \le z \le y$. This line in fact is always below the graph of this function except at the endpoints, where the curve and line intersect. This is a property called "strictly concave."

This property is difficult to prove with the tools thus far at our disposal, but as will be seen in chapter 9, the tools there will make this an easy derivation. At this point we simply note that the inequality in (3.12) is equivalent to the **arithmetic mean–geometric mean inequality** whenever t is a rational number. This familiar inequality, which is also developed in chapter 9, states that for any collection of positive numbers, $\{x_i\}_{i=1}^n$, that AM $\ge$ GM, or notationally,

$$\frac{1}{n} \sum_{i=1}^n x_i \ge \left(\prod_{i=1}^n x_i \right)^{1/n}. \tag{3.13}$$

If $t = \frac{a}{b}$, a rational number in $[0, 1]$, apply (3.13) with a of the x_i equal to x, and $b - a$ of the x_i equal to y, producing

$$\frac{a}{b} x + \left(1 - \frac{a}{b} \right) y \ge x^{a/b} y^{1 - (a/b)}.$$

Taking logarithms of this inequality is equivalent to (3.12) for rational $t \in [0, 1]$. While it is compelling that (3.12) is proved true for all rational t, the tools of chapter 9 are still needed to extend this to all $t \in [0, 1]$. For now, we assume (3.12) and defer a proof to chapter 9.

Proposition 3.21 (*Young's Inequality*)   *Given p, q so that $1 < p, q < \infty$, and $\frac{1}{p} + \frac{1}{q} = 1$, then for all $a, b > 0$,*

$$ab \leq \frac{a^p}{p} + \frac{b^q}{q}. \tag{3.14}$$

Proof   Assuming the concavity of $\ln x$, and with $t = \frac{1}{p}$ in (3.12), we derive

$$\ln(ab) = \frac{\ln a^p}{p} + \frac{\ln b^q}{q}$$

$$\leq \ln\left(\frac{a^p}{p} + \frac{b^q}{q}\right).$$

The result in (3.14) follows by exponentiation.   ■

Remark 3.22   *The notion of **concave function** in (3.12) makes sense for any function $f : X \to \mathbb{R}$, and not just where X is the one-dimensional real line. All that is required is that X is a vector space over $\mathbb{R}$ so that the addition of vectors in the inequality makes sense. In other words, a function f is concave if for $\mathbf{x}, \mathbf{y} \in X$,*

$$tf(\mathbf{x}) + (1 - t)f(\mathbf{y}) \leq f(t\mathbf{x} + (1 - t)\mathbf{y}) \qquad \text{for } 0 \leq t \leq 1. \tag{3.15}$$

As noted above, the next result generalizes the Cauchy–Schwarz inequality, which is now seen as the special case: $p = q = 2$.

Proposition 3.23 (*Hölder's Inequality*)   *Given p, q so that $1 \leq p, q \leq \infty$, and $\frac{1}{p} + \frac{1}{q} = 1$, where notationally, $\frac{1}{\infty} \equiv 0$, we have that*

$$|\mathbf{x} \cdot \mathbf{y}| \leq \|\mathbf{x}\|_p \|\mathbf{y}\|_q. \tag{3.16}$$

In other words, the absolute value of the standard inner product is bounded above by the product of the l_p- and l_q-norms of the vectors, if (p, q) are a conjugate pair of indexes.

Proof   First, if $p = 1$ and $q = \infty$ or conversely, then by the triangle inequality for absolute value in (2.7) applied to (3.4),

$$|\mathbf{x} \cdot \mathbf{y}| \leq \sum_{i=1}^{n} |x_i y_i| \leq \max_i |x_i| \sum_{i=1}^{n} |y_i| = \|\mathbf{x}\|_\infty \|\mathbf{y}\|_1.$$

Otherwise, we apply Young's inequality n-times to each term of the summation with $a_i \equiv \frac{|x_i|}{\|\mathbf{x}\|_p}$, and $b_i \equiv \frac{|y_i|}{\|\mathbf{y}\|_q}$, which produces

$$\sum_{i=1}^{n} \frac{|x_i|}{\|\mathbf{x}\|_p} \cdot \frac{|y_i|}{\|\mathbf{y}\|_q} \leq \frac{1}{p} \sum_{i=1}^{n} \frac{|x_i|^p}{\|\mathbf{x}\|_p^p} + \frac{1}{q} \sum_{i=1}^{n} \frac{|y_i|^q}{\|\mathbf{y}\|_q^q} = \frac{1}{p} + \frac{1}{q} = 1,$$

and consequently, $\sum_{i=1}^{n} |x_i| \, |y_i| \leq \|\mathbf{x}\|_p \|\mathbf{y}\|_q$. Now since $|\mathbf{x} \cdot \mathbf{y}| \leq \sum_{i=1}^{n} |x_i| \, |y_i|$ by the triangle inequality, the result follows.   ∎

Finally, the goal of this series of results, that the l_p-norms satisfy the triangle inequality, can now be addressed:

Proposition 3.24 (*Minkowski's Inequality*)   *Given p with $1 \leq p \leq \infty$,*

$$\|\mathbf{x} + \mathbf{y}\|_p \leq \|\mathbf{x}\|_p + \|\mathbf{y}\|_p. \tag{3.17}$$

Proof   The cases of $p = 1, \infty$, were handled above, so we assume that $1 < p < \infty$. We then have by (2.7),

$$\|\mathbf{x} + \mathbf{y}\|_p^p = \sum_{i=1}^{n} |x_i + y_i|^{p-1} |x_i + y_i|$$

$$\leq \sum_{i=1}^{n} |x_i + y_i|^{p-1} |x_i| + \sum_{i=1}^{n} |x_i + y_i|^{p-1} |y_i|.$$

We can now apply Hölder's inequality to the last two summations:

$$\sum_{i=1}^{n} |x_i + y_i|^{p-1} |x_i| \leq \|\mathbf{x}\|_p \left(\sum_{i=1}^{n} |x_i + y_i|^{(p-1)q} \right)^{1/q} = \|\mathbf{x}\|_p \|\mathbf{x} + \mathbf{y}\|_p^{p/q},$$

$$\sum_{i=1}^{n} |x_i + y_i|^{p-1} |y_i| \leq \|\mathbf{y}\|_p \left(\sum_{i=1}^{n} |x_i + y_i|^{(p-1)q} \right)^{1/q} = \|\mathbf{y}\|_p \|\mathbf{x} + \mathbf{y}\|_p^{p/q},$$

since $(p-1)q = p$. Combining, we get

$$\|\mathbf{x} + \mathbf{y}\|_p^p \leq (\|\mathbf{x} + \mathbf{y}\|_p^{p/q})(\|\mathbf{y}\|_p + \|\mathbf{x}\|_p),$$

and the result follows by division by $\|\mathbf{x} + \mathbf{y}\|_p^{p/q}$ since $p - \frac{p}{q} = 1$.   ∎

Admittedly, quite a lot of machinery was needed to demonstrate that the definition above for $\|\mathbf{x}\|_p$ produced a true norm. However, there will be a significant payoff in later chapters as these norms are the basis of important spaces of series, and in later studies, important spaces of functions.

Remark 3.25   *Note that despite its appearance the l_∞-norm, $\|\mathbf{x}\|_\infty$, is the limit of the l_p-norms $\|\mathbf{x}\|_p$ as $p \to \infty$. That is,*

$$\|\mathbf{x}\|_p \to \|\mathbf{x}\|_\infty \qquad as\ p \to \infty.$$

To see this, assume that the l_∞-norm of $\mathbf{x}$ satisfies $\|\mathbf{x}\|_\infty = |x_j|$. That is, no component is larger in absolute value than the jth element. Then

$$\frac{\|\mathbf{x}\|_p}{\|\mathbf{x}\|_\infty} = \left(\sum_{i=1}^{n} \frac{|x_i|^p}{\|\mathbf{x}\|_\infty^p} \right)^{1/p} = \left(\sum_{i=1}^{n} \lambda_i^p \right)^{1/p}.$$

Now, since $\lambda_j = 1$ and all other $\lambda_i \leq 1$, we have $1 \leq \sum_{j=1}^{n} \lambda_i^p \leq n$, and hence the pth root of this sum approaches 1 as $p \to \infty$.

3.2   Metric Spaces

3.2.1   Basic Notions

An important application of the notion of a norm is that it provides the basis for defining a **distance function** or a **metric**, which will be seen to have many applications. On $\mathbb{R}^n$, the **standard metric** is defined in terms of the standard norm by

$$d(\mathbf{x}, \mathbf{y}) \equiv |\mathbf{x} - \mathbf{y}|. \tag{3.18}$$

Just as the general definition of norm was intended to capture the essential properties of the standard norm $|\mathbf{x}|$ defined on $\mathbb{R}^n$, so too is the general definition of distance or metric intended to capture the essential properties of $|\mathbf{x} - \mathbf{y}|$ defined on $\mathbb{R}^n$. The connection between norms and metrics is discussed below, but note that in order for a set X to have a norm defined on it, this set must have an arithmetic structure so that quantities like $\mathbf{x} + \mathbf{y}$, and $a\mathbf{x}$ make sense. Consequently norms are defined on vector spaces that allow such an arithmetic structure. On the other hand, a metric can be defined on far more general sets than vector spaces.

Definition 3.26   *A **distance function** or **metric** on an arbitrary set X is defined as a real-valued function on $X^2 \equiv X \times X$, and denoted $d(\mathbf{x}, \mathbf{y})$ or $d(x, y)$, with the following properties:*

1. *$d(\mathbf{x}, \mathbf{x}) = 0$.*

2. *$d(\mathbf{x}, \mathbf{y}) > 0$ if $\mathbf{x} \neq \mathbf{y}$.*

3. $d(\mathbf{x}, \mathbf{y}) = d(\mathbf{y}, \mathbf{x})$.

4. *(Triangle inequality)* $d(\mathbf{x}, \mathbf{y}) \leq d(\mathbf{x}, \mathbf{z}) + d(\mathbf{z}, \mathbf{y})$ *for any* $\mathbf{z} \in X$.

*If X is a vector space over $\mathcal{F}$, a distance function is called **translation invariant** if for any $\mathbf{z} \in X$:*

5. $d(\mathbf{x}, \mathbf{y}) = d(\mathbf{x} + \mathbf{z}, \mathbf{y} + \mathbf{z})$.

*A distance function is called **homogeneous** if for any $a \in \mathcal{F}$:*

6. $d(a\mathbf{x}, a\mathbf{y}) = |a| d(\mathbf{x}, \mathbf{y})$.

Definition 3.27   *A **metric space** is any collection of points X on which there is defined a distance function or metric $d(\cdot, \cdot)$. For clarity, a metric space may be denoted (X, d).*

Remark 3.28   *The name "triangle inequality" will be momentarily shown to be consistent with the same notion defined in the context of norms.*

Proposition 3.29   *If $d(x, y)$ is a given metric, then:*

1. $d'(x, y) \equiv \lambda d(x, y)$ *is a metric for any real $\lambda > 0$.*

2. $d'(x, y) \equiv \frac{d(x, y)}{1 + d(x, y)}$ *is a metric.*

Proof   The first statement follows easily from the definition, and in this case, the new metric d' can be thought of as measuring distances in a different set of units. For example, if d measures distances in units of meters, then with $\lambda = 100$, d' provides distances in centimeters. For the second statement, only the triangle inequality requires examination. To show that

$$\frac{d(x, y)}{1 + d(x, y)} \leq \frac{d(x, z)}{1 + d(x, z)} + \frac{d(z, y)}{1 + d(z, y)},$$

we simply cross-multiply, since all denominators are positive, and cancel common terms.   ■

This second metric is interesting because under this definition, the distance between any two points of X is less than 1. More specifically, for any λ, $0 \leq \lambda < 1$,

$$d'(x, y) = \lambda \quad \text{if and only if} \quad d(x, y) = \frac{\lambda}{1 - \lambda}, \tag{3.19a}$$

$$d(x, y) = \lambda \quad \text{if and only if} \quad d'(x, y) = \frac{\lambda}{1 + \lambda}. \tag{3.19b}$$

3.2.2   Metrics and Norms Compared

Because the definitions of norm and metric appear so related, it is natural to wonder about the connection between the two concepts. Can we make norms out of metrics and metrics out of norms? First, we have to be careful because, as noted above, norms are always defined on vector spaces while a metric can be defined on an arbitrary set. Norms require an arithmetic structure on the set X, since one item in the definition required that $|\mathbf{0}| = 0$, and hence we needed to have $\mathbf{0} \in X$ well defined. Given $\mathbf{x}, \mathbf{y} \in X$ and $a \in \mathbb{R}$, we also require in the definition of norm that $\mathbf{x} + \mathbf{y} \in X$ and $a\mathbf{x} \in X$ be well defined. So, by definition, a normed space must have this minimal arithmetic structure, and the vector space structure is a natural requirement as noted in the norm definition.

On the other hand, a metric can be defined on any set, as long as the distance function $d(\mathbf{x}, \mathbf{y})$ satisfies the required properties. There are no arithmetic operations on the elements of X as part of the definition of metric. So the better question is, Given a vector space X, can we make norms out of metrics and metrics out of norms?

The following shows that if the metric satisfies the additional properties 5 and 6 above, that a norm can be constructed.

Proposition 3.30   *If $d(\mathbf{x}, \mathbf{y})$ is a metric on a vector space X that is homogeneous and translation invariant, then $\|\mathbf{x}\| \equiv d(\mathbf{x}, \mathbf{0})$ is a norm and is said to be **induced by the metric** d.*

Proof   Property 1 in the norm definition, that $|\mathbf{x}| \in \mathbb{R}$, follows from a metric being a real-valued function, while norm property 2, that $|\mathbf{0}| = 0$, and $|\mathbf{x}| > 0$ for $\mathbf{x} \neq \mathbf{0}$, follows from 1 and 2 in the metric definition. Finally, norm property 3, that $|a\mathbf{x}| = |a|\,|\mathbf{x}|$ for $a \in \mathbb{R}$, follows from the assumed homogeneity of d, while norm property 4, that $|\mathbf{x} + \mathbf{y}| \leq |\mathbf{x}| + |\mathbf{y}|$ is a consequence of translation invariance and homogeneity. Specifically,

$$|\mathbf{x} + \mathbf{y}| = d(\mathbf{x} + \mathbf{y}, \mathbf{0}) = d(\mathbf{x}, -\mathbf{y}) \leq d(\mathbf{x}, \mathbf{0}) + d(\mathbf{0}, -\mathbf{y}) = |\mathbf{x}| + |\mathbf{y}|. \qquad \blacksquare$$

The reverse implication is easier: on a vector space, a norm always gives rise to a distance function.

Proposition 3.31   *If $\|\mathbf{x}\|$ is a norm on a vector space X, then*

$$d(\mathbf{x}, \mathbf{y}) \equiv \|\mathbf{x} - \mathbf{y}\|, \tag{3.20}$$

*is a metric on X, and in particular, (X, d) is a metric space. The metric d is said to be **induced by the norm** $\|\ \|$.*

Proof Only distance property **4**, which is again called the **triangle inequality**, requires comment. Rewriting, we seek to prove that

$$d(\mathbf{x}, \mathbf{y}) \le d(\mathbf{x}, \mathbf{z}) + d(\mathbf{z}, \mathbf{y}),$$

$$\|\mathbf{x} - \mathbf{y}\| \le \|\mathbf{x} - \mathbf{z}\| + \|\mathbf{z} - \mathbf{y}\|.$$

Letting $\mathbf{x}' = \mathbf{x} - \mathbf{z}$, and $\mathbf{y}' = \mathbf{z} - \mathbf{y}$, we have that $\mathbf{x}' + \mathbf{y}' = \mathbf{x} - \mathbf{y}$, and this inequality for d is equivalent to the triangle inequality for the associated norm applied to $\mathbf{x}'$, $\mathbf{y}'$ and $\mathbf{x}' + \mathbf{y}'$. ∎

Corollary 3.32 $d(\mathbf{x}, \mathbf{y}) \equiv |\mathbf{x} - \mathbf{y}|$ *defined in (3.3) is a metric on* $\mathbb{R}^n$, *and consequently* $(\mathbb{R}^n, d)$ *is a metric space. In addition* $d(x, y) \equiv |x - y|$ *defined in (2.2) is a metric on* $\mathbb{C}$, *and consequently* $(\mathbb{C}, d)$ *is a metric space.*

Proof The proof follows immediately from the proposition above. ∎

The corollary above provides the "natural" metric on $\mathbb{R}^n$, but there are many more that are definable in terms of the various l_p-norms:

Corollary 3.33 *Given any* l_p*-norm* $\|\mathbf{x}\|_p$ *for* $1 \le p \le \infty$ *on* $\mathbb{R}^n$, *then*

$$d_p(\mathbf{x}, \mathbf{y}) \equiv \|\mathbf{x} - \mathbf{y}\|_p, \qquad 1 \le p \le \infty, \tag{3.21}$$

is a metric on $\mathbb{R}^n$, *and consequently* $(\mathbb{R}^n, d_p)$ *is a metric space.*

Proof The proof follows immediately from the proposition above, since $\mathbb{R}^n$ is a vector space. ∎

Remark 3.34 *Of course,* $d_2(\mathbf{x}, \mathbf{y})$ *in this corollary is just the standard metric* $d(\mathbf{x}, \mathbf{y})$ *on* $\mathbb{R}^n$ *defined in (3.3). The metrics defined in (3.21) are referred to as* l_p***-metrics****, or* ***metrics induced by the*** l_p***-norms****.*

To understand the structure of these l_p-metrics, $d_p(\mathbf{x}, \mathbf{y})$, we investigate $\mathbb{R}^2$ where visualization is simple but instructive. Specifically, it is instructive to graph the **closed** l_p**-ball of radius** 1 **about 0**,

$$\bar{B}_1^p(\mathbf{0}) = \{\mathbf{x} \in \mathbb{R}^2 \mid d_p(\mathbf{x}, \mathbf{0}) \equiv \|\mathbf{x}\|_p \le 1\}, \tag{3.22}$$

for various values of p, $1 \le p \le \infty$. Analogously, one can define the **closed** l_p**-ball of radius** r **about y** by

$$\bar{B}_r^p(\mathbf{y}) = \{\mathbf{x} \in \mathbb{R}^2 \mid d_p(\mathbf{x}, \mathbf{y}) \equiv \|\mathbf{x} - \mathbf{y}\|_p \le r\}. \tag{3.23}$$

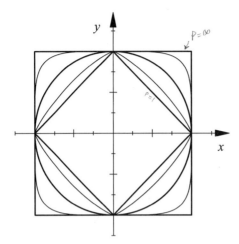

Figure 3.1
l_p-Balls: $p = 1, 1.25, 2, 5, \infty$

The corresponding **open l_p-ball of radius 1 about 0** is defined as

$$B_1^p(\mathbf{0}) = \{\mathbf{x} \in \mathbb{R}^2 \mid d_p(\mathbf{x}, \mathbf{0}) \equiv \|\mathbf{x}\|_p < 1\}, \tag{3.24}$$

and the **open l_p-ball of radius r about $\mathbf{y}$** by

$$B_r^p(\mathbf{y}) = \{\mathbf{x} \in \mathbb{R}^2 \mid d_p(\mathbf{x}, \mathbf{y}) \equiv \|\mathbf{x} - \mathbf{y}\|_p < r\}. \tag{3.25}$$

Note that all these l_p-ball definitions makes sense in any $\mathbb{R}^n$. Of course, for $p = 2$, the closed l_2-ball of diameter 1 is truly a "2-dimensional ball," and it represents the familiar circle of radius 1, including its interior. In $\mathbb{R}^3$, it is indeed a ball, or sphere of radius 1, again including its interior. The corresponding open balls are just the interiors of these closed balls.

For other values of p, these figures do not resemble any ball we would ever consider playing with, but mathematicians retain the familiar name anyway. For example, l_p-balls about $\mathbf{0}$ for $p = 1, 1.25, 2, 5$, and ∞ in $\mathbb{R}^2$ are seen in **figure 3.1**. These can be understood to be open or closed balls depending on whether or not the "boundary" of the ball is included.

For $p = 1$, this innermost "ball" has corners at its intersection points with the coordinate axes, while for $p > 1$, these corners round out, approaching a circle as $p \to 2$. For $p > 2$, these balls again begin to square off in the direction of the diagonal lines in the plane, $y = \pm x$. It is clear from this figure that these balls very quickly converge to the l_∞-ball, which is the square with sides parallel to the axes, and four corners at $(\pm 1, \pm 1)$.

Even more generally, given any metric space (X, d) or normed space, $(X, \|\mathbf{x}\|)$, one can define the **closed ball of radius r about $\mathbf{y}$** by

$$\bar{B}_r(\mathbf{y}) = \{\mathbf{x} \in \mathbf{X} \mid d(\mathbf{x}, \mathbf{y}) \le r\}, \tag{3.26}$$

or

$$\bar{B}_r(\mathbf{y}) = \{\mathbf{x} \in \mathbf{X} \mid \|\mathbf{x} - \mathbf{y}\| \le r\}, \tag{3.27}$$

as well as the associated **open ball of radius r about $\mathbf{y}$**, denoted $B_r(\mathbf{y})$, using strict inequality $<$, rather than the inequality $\le$.

One thing that each of these balls has in common with a true ball, if $1 \le p \le \infty$, is that they are all **convex sets**. This means that if $\mathbf{x}_1, \mathbf{x}_2 \in \bar{B}_r^p(\mathbf{y})$, then the straight line segment joining these points also lies in $\bar{B}_r^p(\mathbf{y})$. That is,

If $\mathbf{x}_1, \mathbf{x}_2 \in \bar{B}_r^p(\mathbf{y})$, then $t\mathbf{x}_1 + (1 - t)\mathbf{x}_2 \in \bar{B}_r^p(\mathbf{y})$ for $0 \le t \le 1$. $\tag{3.28}$

The same is true for a closed ball in a general normed space, as well as in a metric space X that is also a vector space, so in (3.28), $t\mathbf{x}_1 + (1 - t)\mathbf{x}_2$ makes sense. And similarly open balls are convex:

If $\mathbf{x}_1, \mathbf{x}_2 \in B_r^p(\mathbf{y})$, then $t\mathbf{x}_1 + (1 - t)\mathbf{x}_2 \in B_r^p(\mathbf{y})$ for $0 \le t \le 1$. $\tag{3.29}$

Use of this terminology and of the word "convex" is related to the notion of a concave function defined in (3.12). Analogously, the l_p-ball above and the general normed ball are convex because a norm, interpreted as a function $f(\mathbf{x}) = \|\mathbf{x}\|$, is a **convex function**. That is, given $\mathbf{x}_1$, $\mathbf{x}_2$,

$$\|t\mathbf{x}_1 + (1 - t)\mathbf{x}_2\| \le t\|\mathbf{x}_1\| + (1 - t)\|\mathbf{x}_2\| \quad \text{for } 0 \le t \le 1. \tag{3.30}$$

This inequality follows directly from the triangle inequality. Stated more generally, a function $f(\mathbf{x})$ is a **convex function** if for $\mathbf{x}_1, \mathbf{x}_2 \in X$,

$$f(t\mathbf{x}_1 + (1 - t)\mathbf{x}_2) \le tf(\mathbf{x}_1) + (1 - t)f(\mathbf{x}_2) \quad \text{for } 0 \le t \le 1. \tag{3.31}$$

Note that here the inequality is reversed compared to the definition of concave function in (3.15) above.

Graphically, when X is the real line and $x < y$, the inequality in (3.31) states that on the interval $[x, y]$, the value of the function never rises above the line segment connecting $(x, f(x))$ and $(y, f(y))$. This insight on convexity provides a geometric

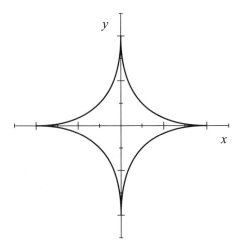

Figure 3.2
l_p-Ball: $p = 0.5$

interpretation of the implication of the triangle inequality as required in the defini-tion of norm. That is, the triangle inequality assures that all balls defined by norms are convex sets. Also the reason why no attempt was made to define an l_p-norm for $0 < p < 1$ is that in these cases the triangle inequality is not satisfied and geometri-cally, as is easily demonstrated, the associated l_p-balls are not convex.

For example, with $p = 0.5$, we have $\bar{B}_1^{0.5}(\mathbf{0})$ in **figure 3.2**. If we choose $\mathbf{x}_1 = (1, 0)$ and $\mathbf{x}_2 = (0, 1)$, it is clear that $\|t\mathbf{x}_1 + (1 - t)\mathbf{x}_2\|_{0.5} = \|(t, 1 - t)\|_{0.5} > 1$ for $0 < t < 1$, and this point is outside the ball. However, $t\|\mathbf{x}_1\|_{0.5} + (1 - t)\|\mathbf{x}_2\|_{0.5} = 1$. Conse-quently this ball is not convex by definition, as is also visually apparent.

*3.2.3   Equivalence of Metrics

Two metrics on a metric space X, say d_1 and d_2, may produce different numerical values of distance between arbitrary points $x, y \in X$, but they may be fundamentally "equivalent" in terms of conclusions that might be drawn from certain observations on the space. A trivial example on $\mathbb{R}$ would be where $d_1(x, y) = |x - y|$, the standard metric, and $d_2(x, y) = \lambda d_1(x, y)$, where λ is a positive real number. As noted above, d_2 is a metric for any positive number λ. Also, while all such metrics produce differ-ent numerical values of distance, such as miles and kilometers, they are fundamen-tally the same in many ways.

For this example, if $\{x_n, y\} \subset X$ is a collection of points so that $d_1(x_n, y) \to 0$ as $n \to \infty$, we would observe the same property under d_2 for any positive λ. Correspondingly $d_2(x_n, y) \to 0$ as $n \to \infty$ would imply the same thing about d_1. Note that a formal definition of what $d_2(x_n, y) \to 0$ means will be presented in the chapter 5, but the intuition for this idea is adequate for our purposes here.

In general, two metrics are defined as equivalent when this simultaneous convergence property is satisfied. The following definition provides a neat way of ensuring this conclusion:

Definition 3.35   *Two metrics, d_1 and d_2, on a metric space X are **Lipschitz equivalent** if there exists positive real constants λ_1 and λ_2 so that for all $x, y \in X$,*

$$\lambda_1 d_1(x, y) \le d_2(x, y) \le \lambda_2 d_1(x, y). \tag{3.32}$$

Lipschitz equivalence is named for **Rudolf Lipschitz** (1832–1903), who introduced a related notion of Lipschitz continuity that will be studied in chapter 9.

It is clear from this definition that the original objective is satisfied. That is, it would seem clear that

$$d_1(x_n, y) \to 0 \quad \text{iff} \quad d_2(x_n, y) \to 0,$$

based on our current informal understanding of the definition of convergence. But logically, and this will be made rigorous in chapter 5, the result is forced by the incqualities in (3.32).

Note that every metric is Lipschitz equivalent to itself, and also it is easy to see that this notion of Lipschitz metric equivalence is **symmetric**. That is, if (3.32), then

$$\frac{1}{\lambda_2} d_2(x_n, y) \le d_1(x_n, y) \le \frac{1}{\lambda_1} d_2(x_n, y). \tag{3.33}$$

This notion is also **transitive**: if d_1 and d_2 are Lipschitz equivalent, and d_2 and d_3 are Lipschitz equivalent, then d_1 and d_3 are Lipschitz equivalent.

An important concept in mathematics is one of an **equivalence relation**, defined on an arbitrary set. The simplest equivalence relation is equality, where xRy denotes $x = y$.

Definition 3.36   *An equivalence relation on a set X, denoted xRy or $x \sim y$ as shorthand for "x is related to y," is a binary relation on X; that is:*

1. ***Reflexive***:   *xRx for all $x \in X$.*

2. ***Symmetric***:   *xRy if and only if yRx.*

3. ***Transitive***:   *if xRy and yRz, then xRz.*

The importance of equivalence relations is that one can form **equivalence classes** of elements of X. An equivalence class is a collection of elements related to each other under R. It is defined so that any two elements from a given class are equivalent, while any two elements from different classes are not equivalent.

For example, the collections of Lipschitz equivalent metrics on a given space X are equivalence classes. For many applications it matters not which element of the class is used. For example, continuing with some informality, if we define $x_n \to y$ by $d(x_n, y) \to 0$ for a given metric d, we could equally well define $x_n \to y$ relative to any metric in the equivalence class of d. That is, the notion $x_n \to y$ depends not so much on d as on the equivalence class of d. If this property is true for a given d, it is also true for an other d' that is Lipschitz equivalent, $d \sim_L d'$, while if this property is false for a given d, it is also false for an other d' with $d \sim_L d'$. However, in neither case can one draw a conclusion about the truth or falsity of this property for metrics outside the given equivalence class.

Proposition 3.37   *If $d(x, y)$ is a metric on X, then:*

1. $\lambda d(x, y) \sim_L d(x, y)$ *for any real $\lambda > 0$.*
2. $d'(x, y) \equiv \frac{d(x,y)}{1+d(x,y)} \sim_L d(x, y)$ *if and only if $d(x, y) \leq M$ for all $x, y \in X$.*

Proof   In defining $d_2(x, y) = \lambda d(x, y)$ and $d_1(x, y) = d(x, y)$, it is apparent that (3.32) is satisfied with $\lambda_1 = \lambda_2 = \lambda$, proving part 1. The second statement is initially less obvious, but it follows directly from the one-to-one correspondence between d and d' distances in (3.19). With $d_2(x, y) = d'(x, y)$ and $d_1(x, y) = d(x, y)$, we derive from (3.19b) that $d'(x, y) \leq d(x, y)$, which is consistent with $\lambda_2 = 1$ in (3.32). For the other inequality we have from (3.19b) that if $d(x, y) \leq M$, then $d'(x, y) \leq \frac{M}{M+1}$, which is algebraically equivalent to $\frac{1}{1-d'(x,y)} \leq 1 + M$. Then from (3.19a),

$$d(x, y) = \frac{d'(x, y)}{1 - d'(x, y)} \leq (1 + M)d'(x, y),$$

and so the second inequality in (3.32) is satisfied with $\lambda_1 = \frac{1}{M+1}$. If $d(x, y)$ is unbounded, there can be no λ_1 for which $\lambda_1 d(x, y) \leq d'(x, y)$, since $d'(x, y) \leq 1$.   ∎

In addition to these examples of equivalent metrics, it may be surprising but it turns out that the various l_p-norms, for $1 \leq p \leq \infty$, are equivalent in $\mathbb{R}^n$.

Proposition 3.38   *On $\mathbb{R}^n$, all distances given by the l_p-norms in (3.21) for $1 \leq p \leq \infty$ are Lipschitz equivalent.*

Proof   We first show that if $1 \leq p < \infty$, that the l_p-distance is Lipschitz equivalent to the l_∞-distance. For given $\mathbf{x} = (x_1, x_2, \ldots, x_n)$ and $\mathbf{y} = (y_1, y_2, \ldots, y_n)$, we have that

$$\max_i |x_i - y_i|^p \le \sum_{i=1}^{n} |x_i - y_i|^p \le n \max_i |x_i - y_i|^p.$$

That is, taking pth roots:

$$d_\infty(\mathbf{x}, \mathbf{y}) \le d_p(\mathbf{x}, \mathbf{y}) \le \mathbf{n}^{1/p} d_\infty(\mathbf{x}, \mathbf{y}),$$

and so every l_p-distance is Lipschitz equivalent to the l_∞-distance if $1 \le p < \infty$. Since Lipschitz equivalence is transitive, we conclude that $d_p(\mathbf{x}, \mathbf{y})$ is equivalent to $d_{p'}(\mathbf{x}, \mathbf{y})$ for any $1 \le p, p' \le \infty$. In fact, using (3.32) and (3.33), we can infer bounds between $d_p(\mathbf{x}, \mathbf{y})$ and $d_{p'}(\mathbf{x}, \mathbf{y})$:

$$n^{-1/p'} d_{p'}(\mathbf{x}, \mathbf{y}) \le d_p(\mathbf{x}, \mathbf{y}) \le \mathbf{n}^{1/p} d_{p'}(\mathbf{x}, \mathbf{y}). \tag{3.34}$$

■

Remark 3.39

1. *Note that the λ_1 and λ_2 bounds between $d_p(\mathbf{x}, \mathbf{y})$ and $d_\infty(\mathbf{x}, \mathbf{y})$ are **sharp** in that these bounds can be achieved by examples and hence cannot be improved upon. The left-hand bound is attained, for example, with $\mathbf{x} = (x, 0, \ldots, 0)$ and $\mathbf{y} = (y, 0, \ldots, 0)$, or with $\mathbf{x}$ and $\mathbf{y}$ being similarly defined to be on the same "axis." We can in fact observe this equality in figure 3.1, where the five l_p-balls about $\mathbf{0}$ for $p = 1, 1.25, 2, 5, \infty$, are seen to intersect at the axes. On the other hand, the right-hand bound is attained for $\mathbf{x} = (x, x, \ldots, x)$ and $\mathbf{y} = (y, y, \ldots, y)$, as well as other point combinations with $|x_i - y_i| = c > 0$—that is, on the "diagonals" of $\mathbb{R}^n$, which is again seen on figure 3.1. However, the inequalities between $d_p(\mathbf{x}, \mathbf{y})$ and $d_{p'}(\mathbf{x}, \mathbf{y})$ in (3.34) are not sharp, as is easily verified by considering the case $p = p'$. With a more detailed analysis using the tools of multivariate calculus, we would obtain the sharp bounds with $1 \le p \le p' \le \infty$,*

$$d_{p'}(\mathbf{x}, \mathbf{y}) \le d_p(\mathbf{x}, \mathbf{y}) \le \mathbf{n}^{(p'-p)/pp'} d_{p'}(\mathbf{x}, \mathbf{y}),$$

and these bounds would again be seen to be achieved on the axes and diagonals of $\mathbb{R}^n$, respectively.

2. *Note also that the Lipschitz equivalence of $d_p(\mathbf{x}, \mathbf{y})$ and $d_\infty(\mathbf{x}, \mathbf{y})$, and more generally, of $d_p(\mathbf{x}, \mathbf{y})$ and $d_{p'}(\mathbf{x}, \mathbf{y})$, depends on the dimension of the space n in a way that precludes any hope that this equivalence will be preserved as $n \to \infty$ (as will be formalized in chapter 6 on series). In other words, an informal consideration of the notion of an $\mathbb{R}^\infty$ suggests that the various l_p-distances will not be Lipschitz equivalent.*

3. *Not all metrics are Lipschitz equivalent to those in this proposition. For example, define*

$$d(\mathbf{x}, \mathbf{y}) = \begin{cases} 0, & \mathbf{x} = \mathbf{y} \\ 1, & \mathbf{x} \neq \mathbf{y} \end{cases}.$$

It is easy to show that this is indeed a metric on $\mathbb{R}^n$ that is not Lipschitz equivalent to the l_p-distances.

4. *It was noted above that every norm on a vector space induces a metric on that space. Consequently it is common to say that two such **norms are Lipschitz equivalent** if the respective induced metrics are equivalent in the above-described sense.*

As a final comment regarding Lipschitz equivalence of metrics, we note that there is a simple and natural geometric interpretation of this concept. First, we introduce a more general notion of metric equivalence, sometimes called topologically equivalent. The term "topology" will be addressed in chapter 4, and is related to the notion of open sets in a space.

Definition 3.40   *Two metrics on a metric space X, say d_1 and d_2, are **equivalent**, and sometimes **topologically equivalent** for specificity, if for any $\mathbf{x} \in X$ and $r > 0$, $B_r^{(2)}(\mathbf{x})$ defined relative to d_2 both contains an open d_1-ball and is contained in an open d_1-ball. That is, there are real numbers r_1, r_2, both formally functions of r and $\mathbf{x}$, so that*

$$B_{r_1}^{(1)}(\mathbf{x}) \subset B_r^{(2)}(\mathbf{x}) \subset B_{r_2}^{(1)}(\mathbf{x}), \tag{3.35}$$

where $B_r^{(j)}(\mathbf{x})$ denotes an open ball defined relative to d_j, and $A \subset B$ denotes "set inclusion" and means that every point in A is also contained in B.

Proposition 3.41   *In a metric space X, if d_1 and d_2 are Lipschitz equivalent as in (3.32), then they are topologically equivalent as in (3.35).*

Proof   If we are given $\mathbf{x} \in X$ and $r > 0$, and $B_r^{(2)}(\mathbf{x}) = \{\mathbf{y} \,|\, d_2(\mathbf{x}, \mathbf{y}) < r\}$, by (3.32) we conclude that for any $\mathbf{y} \in B_r^{(2)}(\mathbf{x})$,

$$\lambda_1 d_1(\mathbf{x}, \mathbf{y}) \leq d_2(\mathbf{x}, \mathbf{y}) \leq \lambda_2 d_1(\mathbf{x}, \mathbf{y}),$$

so (3.35) is satisfied with $r_2 = r/\lambda_1$ and $r_1 = r/\lambda_2$.   ∎

This geometric statement is simple to see in **figure 3.1**. Notice that any l_p-ball can be envisioned as containing, and being contained in, two $l_{p'}$-balls for any p'. A more specific example is seen in **figure 3.3** where the l_2-ball of radius 1 contains the l_1-ball

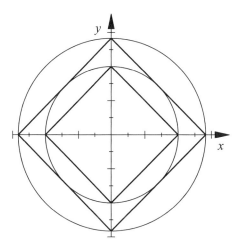

Figure 3.3
Equivalence of l_1- and l_2-metrics

of radius 1, and is contained in the l_1-ball of radius $\sqrt{2}$, and this l_1-ball in turn is contained in the l_2-ball of radius $\sqrt{2}$.

Remark 3.42 *The notion of metric equivalence, or "topological equivalence," is more general than Lipschitz equivalence, since it allows the relationship between these metrics to vary with $\mathbf{x} \in \mathbf{X}$ since the numbers r_1, r_2 depend on $\mathbf{x}$. For Lipschitz equivalence this relationship is fixed for all $\mathbf{x}$, as noted in the proof above.*

3.3 Applications to Finance

3.3.1 Euclidean Space

Euclidean space provides a natural framework in any discipline in which one is trying to solve problems that involve several parameters, and such problems exist in many areas of finance. For example, in asset allocation problems one is attempting to divide a given total investment fund between certain available asset classes, however defined, and the solution to such a problem can naturally be identified with a point, or allocation vector, in a Euclidean space. The dimension of this space is logically equal to the number of available asset classes. In the fixed income markets the very notion of a yield curve, which is defined in terms of the yields on a collection of reference bonds of increasing maturities, compels the interpretation of a yield curve vector in an appropriately dimensioned Euclidean space. Such yield vectors can then be translated to spot rate or forward rate vectors as needed by the given application, or used in a price risk analysis. Finally, a given security or portfolio of securities can be modeled in terms of projected cash flows, and these cash flow vectors, whether fixed or variable, can then be used in a variety of portfolio modeling applications.

Asset Allocation Vectors

An **asset allocation problem** involves determining a vector of dollar amounts: $(x_1, x_2, \ldots, x_n)$, where n denotes the number of available assets, x_i denotes the dollar investment in the ith asset, and $\sum x_i = A$, the total amount to be invested. In certain applications, all x_i satisfy $x_i \geq 0$ and represent long positions, but we can allow $x_i < 0$ in cases where short-selling is possible. Equivalently, we can parametrize the solution to the problem in percentage units so that x_i denotes the proportion of the portfolio to be invested in the ith asset, again long or short, and then $\sum x_i = 1$.

Alternatively, the n-tuple $(x_1, x_2, \ldots, x_n)$ might represent a **portfolio trade**, whereby $x_i > 0$ implies a purchase and $x_i < 0$ a sale of $|x_i|$ units of the ith asset, and now $\sum x_i = 0$ unless the trade is intended to also increase or decrease the portfolio balance due to net deposits or redemptions. In all such cases it is only natural to think of the feasible n-tuples as residing in some collective structure such as $\mathbb{R}^n$. This is especially true in the trading model, since the vector space arithmetic properties of $\mathbb{R}^n$ exactly reflect arithmetic operations for such trades. Scalar multiplication by 2, say, which doubles the trading done, doubles each individual trade, which is to say, is reflected componentwise in the trade vector. If one trade is implemented after another, the net trade is equivalent to the componentwise sum of the trade vectors.

However, this may appear to be a case of overkill. Admittedly, in all such cases the real world **feasible solution space** is a finite collection of points, which clearly $\mathbb{R}^n$ is not. The real world provides a finite solution set because first, no portfolio can be arbitrarily large, nor can a trade be implemented in arbitrarily large volumes. Second, even the maximally detailed solution cannot be implemented in units of less than \$.01 in the United States, or 1¥ in Japan, or .01€ in the European Union, and so there are only finitely many portfolio allocations, or trades, to consider. More realistically, assets cannot be acquired in such units. For instance, we cannot acquire an extra \$.01 of a given US asset, and so the feasible solution set is far cruder than this maximally detailed solution set implies.

Ironically, most problems in finance are harder to solve if one explicitly recognizes the finiteness of this solution set. That is, if the objective of the asset allocation or portfolio trade is to optimize a given function, referred to as the **objective function**, it can be very difficult to solve this problem over the finite "grid" of feasible solutions, other than by a brute-force search. The difficulty arises because despite its finiteness, the feasible solution set can be quite large. In most cases it is far easier to make believe that one can trade any amount of any asset and solve the problem at hand using the methods of later chapters that take advantage of the structure of $\mathbb{R}^n$. It is then reasonable to assume that the approximate implementation in the real world of this too-detailed solution will be quite close to that which would have been obtained had the finite feasibility set been explicitly recognized at the outset.

That is, by interpreting our problem in an artificially refined setting of $\mathbb{R}^n$, we simplify the solution, but we are then required to assume that the approximate implementation of the exact solution is close to the exact solution obtained had we begun with the finite feasible solution set. In many cases this assumption can be checked. That is, once we solve the more detailed problem, we can investigate to what extent its approximate implementation is an optimal or near-optimal solution among feasible alternatives. Even this analysis can be simpler than searching for a best solution on the grid at the outset.

Interest Rate Term Structures

There are three common bases for describing the **term structures of interest rates**, where by "structure" is meant the functional dependence of rates on the term of the implied loan. In practice, the most readily available data for loans exist in the bond markets. The three term structure bases are:

1. Bond Yields: The interest rates that equate each coupon bond's price to the present value of the bond's scheduled cash flows.

2. Spot Rates: The bond yields on real or hypothetical zero coupon bonds.

3. Forward Rates: The bond yields on "forward" zero coupon bonds, which is to say, the yield today for future investments in zero coupon bonds.

The bond market provides insights to these structures, but for the term structure to be meaningful, it is important that as many of the bond characteristics as possible are controlled for, so that only the dependency on the bond's terms remain.

For example, it is common to group bonds by currency and credit quality, avoiding when possible unusual cash flow structures that get special pricing, or bonds with embedded options. One special class in every major currency is the class of all risk-free Treasury bonds issued by the country's central government. Bonds at the next highest credit rating, often denoted AAA or Aaa, are then grouped, as are the next level of AA or Aa, and so forth. With enough bonds in a given group, a term structure can be inferred in any of the three bases. When bond data are sparse, interpolation techniques are often used to estimate missing data.

For a bond yield or spot rate, there is one implied time parameter determined by the maturity of the bond. For forward rates, there are two time parameters: one establishes the time of the investment in the forward zero coupon bond, and the second determines the time of maturity of this bond.

To illustrate the calculation of these term structures, we assume that bonds have semiannual coupons and that there are bonds available at all maturities from 0.5 to n-years. As noted above, interpolation is often necessary to infer information at maturities that have no market representatives. We also implement all calculations with semiannual nominal rates, but note that these calculations can be implemented in any nominal basis.

Bond Yields   Using (2.15), bond yields at each maturity are derived by solving the following equations for $\{i_j\}$, the semiannual bond yields:

$$P_j = F_j \frac{r_j}{2} a_{2j;\, i_j/2} + F_j v_{i_j/2}^{2j}, \qquad j = 0.5, 1.0, 1.5, \ldots, n. \tag{3.36}$$

Here j denotes the term of the bond in years; $\{P_j\}$ are the bonds' prices, $\{r_j\}$ the semiannual coupon rates, and $\{F_j\}$ the bonds' par values. It is typical to fix $F_j = 100$, and so P_j denotes the price per 100 par. The result is the bond yield term structure: $(i_{0.5}, i_1, \ldots, i_n)$, which can be envisioned as a vector in $\mathbb{R}^{2n}$.

One numerical approach to solving these equations, called **interval bisection**, is discussed in chapters 4 and 5.

Spot Rates   From the same data used to determine the bond yield term structure, one can in theory calculate the spot rate structure, since a coupon bond is nothing but a portfolio of zero coupon bonds. Using (2.19), the price P_j must reflect spot rates: $(s_{0.5}, s_1, \ldots, s_j)$, each appropriate to discount a single cash flow of the bond:

$$P_j = F_j \frac{r_j}{2} \sum_{k=1}^{2j} \left(1 + \frac{s_{k/2}}{2}\right)^{-k} + F_j \left(1 + \frac{s_j}{2}\right)^{-2j}. \tag{3.37}$$

Notation 3.43   *In this summation the present value of the cash flow at time k-years is calculated with the factor*

$$\left(1 + \frac{s_k}{2}\right)^{-2k},$$

but then the summation above would be expressed in the nonstandard notation as

$$\sum_{k=0.5}^{j} \left(1 + \frac{s_k}{2}\right)^{-2k},$$

where it would be hoped that the reader understood that the index values must be incremented by 0.5. To avoid this notational ambiguity, we use standard natural number indexing, and consequently we need to halve the index values to obtain the correct result.

Forward Rates   As noted above, forward rates are functions of two time parameters, defining the investment date in the zero coupon bond and the maturity date. In other words, a forward can be denoted, $f_{j,k}$, where $j, k \in \{0, 0.5, 1.0, 1.5, \ldots, n\}$, with $k > j$. In this notation, $f_{j,k}$ denotes the yield today for a $(k - j)$-year zero coupon bond, which is to be acquired at time j-years. Consequently $f_{0,k} = s_k$. The forward rate $f_{j,k}$ would be described as the $(k - j)$-year forward rate at time j-years.

In the same way that s_k is appropriate for discounting a cash flow from time k-years to time 0, the forward rate $f_{j,k}$ is appropriate for discounting a cash flow from time k-years to time j-years. With this interpretation, it must be the case that one can discount from time k-years to time 0 either with the spot rate s_k, or a sequence of forward rates:

$$f_{0,0.5}, f_{0.5,1.0}, f_{1.0,1.5}, \ldots, f_{k-0.5,k}.$$

Of course, if k is an integer, one could also use the forward rates:

$$f_{0,1.0}, f_{1.0,2.0}, \ldots, f_{k-1,k}.$$

Specifically, using the first sequence, and recalling the notational comment above, obtains

$$\left(1 + \frac{s_{k/2}}{2}\right)^{-k} = \prod_{i=1}^{k} \left(1 + \frac{f_{(i-1)/2, i/2}}{2}\right)^{-1}. \tag{3.38}$$

So the price of a bond can be written in the messy but unambiguous notation

$$P_j = F_j \frac{r_j}{2} \sum_{k=1}^{2j} \prod_{i=1}^{k} \left(1 + \frac{f_{(i-1)/2, i/2}}{2}\right)^{-1} + F \prod_{i=1}^{2j} \left(1 + \frac{f_{(i-1)/2, i/2}}{2}\right)^{-1}. \tag{3.39}$$

In general, forward rates are calculated in series for applications, since from these any forward $f_{j,k}$ can be calculated in the same way one calculates spot rates. Reverting to the original notation with $j, k \in \{0.5, 1.0, 1.5, \ldots, n\}$ obtains

$$\left(1+\frac{f_{j,k}}{2}\right)^{-2(k-j)} = \prod_{i=2j+1}^{2k}\left(1+\frac{f_{(i-1)/2,\,i/2}}{2}\right)^{-1}. \tag{3.40}$$

Equivalence of Term Structures   What is apparent from the three bond pricing formulas (3.36), (3.37), and (3.39) is that if a term structure is given in any of the three bases, all coupon bonds can be priced. What is also apparent is that these term structures must be consistent and produce the same prices, or else risk-free arbitrage is possible.

For example, the price of zeros must be consistent with the pricing of coupon bonds of the same issuer, since a coupon bond is a portfolio of zeros, and hence, in theory, one can buy coupon bonds and sell zeros, or sell coupon bonds and buy zeros. The first transaction is called **coupon stripping**, and the second, **bond reconstitution**.

Similarly forward bond prices must be consistent with zero coupon pricing, since by (3.38), a zero coupon bond is equivalent to a series of forward bonds. For example, one could invest 100 in a 3-year zero, or invest this money in a 0.5-year zero, and at the same time commit to a forward contract from time 0.5 to time 1.0 years, and another from time 1.0 to 1.5 years, and so forth. The investment amount for each forward contract would be calculated as the original 100 compounded with the interest earned to that time, which is known. For example, if the 0.5-year spot rate is 2%, and the 0.5-year forward rate at time 0.5 is 2.2%, the investment amount for the time 0.5-year forward contract would be 101, and the investment amount for the time 1-year forward contract would be 102.11 to 2 decimals.

There is also a direct way to "replicate" a forward on a zero with a long/short market trade in zero coupon bonds.

Example 3.44   *Assume that a 5-year zero has semiannual yield 4%, and a 2-year zero has yield 2%. To create a "long" forward contract from time 2 to time 5 years, meaning an investment opportunity, we proceed as follows: In order to be able to invest 100 at time 2 years, we "short" $100(1.01)^{-4}$ of the 2-year zero, and go long an equivalent amount of the 5-year zero. So at time 0, no out-of-pocket money is required other than perhaps a margin account deposit, which is not a cost. At time 2 years, we "cover the short" position with an "investment" of 100. At time 5 years, we mature the original 5-year zero for $100(1.01)^{-4}(1.02)^{10}$, or 117.14 to 2 decimals. It is easy to show that if all decimals are carried, then the rate obtained on this 100 investment at time 2 is exactly equal to the 3-year forward rate at time 2 years, or 5.344%, implied by (3.40) and (3.38):*

$$\left(1+\frac{f_{j,k}}{2}\right)^{-2(k-j)} = \frac{\left(1+\frac{s_k}{2}\right)^{-2k}}{\left(1+\frac{s_j}{2}\right)^{-2j}}. \tag{3.41}$$

So spot rates and forward rates must be equivalent because one can transact to create zeros from forwards and forwards from zeros. Mathematically the associated rates must satisfy (3.38), to create spot rates from forward rates, and (3.41), to create forward rates from spot rates.

To convert between bond yields and spot rates is done as follows:

1. Spot Rates to Bond Yields: This is the easier direction, since spot rates provide bond prices by (3.37), and one then calculates the associated bond yields by solving (3.36) for i_j (see **interval bisection** in chapters 4 and 5).

2. Bond Yields to Spot Rates: This methodology is known as **bootstrapping** or the **bootstrap method**. First, all bond prices can be calculated from the bond yields using (3.36). To derive the spot rates, the bootstrap method is an iterative procedure whereby one spot rate is calculated at a time using (3.37). Specifically, one starts with $j = 0.5$, and this produces

$$P_{0.5} = F_{0.5}\left(1+\frac{r_{0.5}}{2}\right)\left(1+\frac{s_{0.5}}{2}\right)^{-1},$$

from which $s_{0.5}$ is easily calculated. One next calculates s_1 from P_1 using

$$P_1 = F_1\frac{r_1}{2}\sum_{k=1}^{2}\left(1+\frac{s_{k/2}}{2}\right)^{-k} + F_1\left(1+\frac{s_1}{2}\right)^{-2},$$

which can be solved since $s_{0.5}$ is known from the first step. This process continues in that once $(s_{0.5}, s_1, \ldots, s_j)$ is calculated, (3.37) is used to calculate $s_{j+0.5}$ from $P_{j+0.5}$, which is straightforward as this is then the only unknown in this equation.

Bond Yield Vector Risk Analysis

Besides portfolio allocation vectors, or trade vectors, another natural application of n-tuples in finance is where $(x_1, x_2, \ldots, x_n)$ represents one of the term structures of interest rates discussed above. For example, these might be the yields of a collection **of benchmark bonds** at certain maturities in increasing order, with interpolation used for the other yields, or a complete collection of bond yields or spot rates, or a sequence of forwards.

The prices of other bonds might then be modeled as a function:

$$P(x_1, x_2, \ldots, x_n).$$

Within this model, one then envisions moment-to-moment changes in the term structure as vector increments to this initial yield curve:

$$\Delta \mathbf{x} = (\Delta x_1, \Delta x_2, \ldots, \Delta x_n).$$

In turn, as this yield curve evolves over time, so too does the price of the portfolio, and the change in this price can be modeled:

$$\Delta P(x_1, x_2, \ldots, x_n) \equiv P(x_1 + \Delta x_1, x_2 + \Delta x_2, \ldots, x_n + \Delta x_n) - P(x_1, x_2, \ldots, x_n).$$

In practice, a spot rate structure is sometimes the most transparent approach. This is because the connection between $\Delta \mathbf{x}$ and ΔP is then clearly visible for option-free bonds. But there is far less transparency for bonds with embedded options. Also, although spot rates can be readily calculated, they are not typically visible in market trades, so a model that better connects ΔP with market observations might be a bond yield model, whereby the mathematics needed to transform $\Delta \mathbf{x}$ on a bond yield basis to $\Delta \mathbf{y}$ say, on a spot rate basis needed for pricing, is just part of the computer model calculations, and then ΔP is modeled in terms of $\Delta \mathbf{x}$.

Within this model, price sensitivities and hedging strategies can be evaluated. Formal methods for this risk analysis will be introduced in chapters 9 and 10.

Again, using an $\mathbb{R}^n$-based model for such yield curve analyses is overkill formally, as yields are rarely if ever quoted with even six decimal precision, which is equivalent to "hundreths of a basis point" (1 **basis point** $= 0.01\% = 0.0001$). However, just as in the case of portfolio allocation and trading, most problems are easier to solve within the framework of $\mathbb{R}^n$ than the discrete framework of feasible yield curves and yield curve changes.

Cash Flow Vectors and ALM

As another example, the vector $(x_1, x_2, \ldots, x_n)$ might represent the **period-by-period cash flows** in a fixed income security such as a bond or a mortgage-backed security (MBS). Because of the prepayment options afforded borrowers in MBS and callable bonds, there can be significant variability in future cash flow which reflects the evolution of future interest rates, among other factors. Similarly, even a simple bullet bond with no call option, where cash flows are, in theory, known with certainty at issue, may experience variability due to the presence of **credit risk** and the potential for default and loss.

At a portfolio level, one might model the cash flow vectors representing the assets and liabilities of a firm such as a life insurance or property and casualty insurance company, commercial or investment bank, or pension plan. The liabilities could reflect explicit contractual obligations of the firm, or implicit liabilities associated with short positions in investment securities or financial derivatives. In any such case, these cash flows may contain embedded options or credit risks, as well as changes due to the issuance of new liabilities and portfolio management of assets.

Once so modeled, the firm is in a better position to evaluate its **asset–liability management risk**, or **ALM risk**, which is the residual risk to firm capital caused by any risks in assets and liabilities that are not naturally offsetting or otherwise hedged. Interest rate risk noted in the last section is often a major component of ALM risk.

In each case, one can embed the possible cash flow structures in $\mathbb{R}^n$ and begin the risk analysis and evaluation of hedging strategies with the advantage of the structure this space affords.

3.3.2 Metrics and Norms

Truthfully, the most prevalent norms and metrics in finance are of l_p-type for $p = 1, 2,$ and ∞. However, it is no easier to develop the necessary theory for these three needed cases than it is to develop the general l_p theory. So rather than expend the effort to develop three special cases and leave the reader thinking that these are isolated and special metrics, this book takes the position that for the given effort, it is better to understand that $p = 1, 2,$ and ∞ are simply three special points in a continuum of metrics spanning: $1 \leq p \leq \infty$.

And who knows, you may discover a natural application in finance of a different l_p-metric, and you will be ready with all the necessary tools.

One exception to the $p = 1, 2,$ and ∞ rule is for the analysis of sample data.

Sample Statistics

Of the given three common l_p-norms, l_2 is the most frequently used. As is well known and will be further developed in the chapter 7 on statistics, the most common measure of risk in finance is defined in terms of the measure known as **variance**, and its square root, **standard deviation**, and both reflect an l_2-type measurement. These are special cases of what are known as the **moments** of the sample, and in general, sample statistics utilize the full range of l_p-norms for integer p.

For example, assume that $\mathbf{x} = (x_1, x_2, \ldots, x_n)$ represents a "sample" of observations of a random variable of interest. In finance, a common example would be observations of sequential period returns of an asset or portfolio of interest. For example, the monthly returns of a given common stock, or a benchmark portfolio such as the S&P 500 Index, would be natural candidates for analysis. Alternatively, these observations might reflect equally spaced observations of a currency exchange rate, or interest rate, or price of a given commodity. In any such case, the variable of interest might be the actual observation, or the change in the observed value measured in absolute or relative percentage units. The so-called moments of the sample are all defined in a way which can be seen to be equivalent to an l_p-norm:

1. Moments about the Origin

Mean: The mean of the sample is defined as

$$\hat{\mu} = \frac{1}{n}\sum_{j=1}^{n} x_j. \tag{3.42}$$

If all observations $x_j \geq 0$, the sample mean is equivalent to an l_1-norm, $\hat{\mu} = \frac{1}{n}\|\mathbf{x}\|_1$. In general, however, this is not true as the "sign" of x_j is preserved in the definition of a mean, but not preserved in the definition of an l_1-norm.

Higher Moments: For r a positive integer, the so-called rth moment of the sample is defined as

$$\hat{\mu}_r' = \frac{1}{n}\sum_{j=1}^{n} x_j^r, \; = \frac{1}{n}\left(\|\vec{x}\|_r\right)^r \tag{3.43}$$

so we see that $\hat{\mu} = \hat{\mu}_1'$. Also, when the observations are nonnegative, or in the general case where r is an even integer, this moment is related to the l_r-norm, and we have that $\hat{\mu}_r' = \frac{1}{n}\|\mathbf{x}\|_r^r$.

Notation: To distinguish between the moments of the sample and those of the unknown theoretical distribution of all such data, of which the sample is just a subset, one sometimes sees the notation of m or $\bar{x}$ for the sample mean, and m_r' for the rth sample moment about the origin, with μ and μ_r' preserved as notation for the moments of the theoretical distributions. A caret over a variable, such as $\hat{\mu}$, is also standard notation to signify that its value is based on a sample estimate and not the theoretical distribution.

2. Moments about the Mean

Variance and Standard Deviation: The "unbiased" variance of the sample is denoted $\hat{\sigma}^2$, and the standard deviation is the positive square root, denoted $\hat{\sigma}$, where

$$\hat{\sigma}^2 = \frac{1}{n-1}\sum_{j=1}^{n}(x_j - \hat{\mu})^2. \tag{3.44}$$

In some applications (see chapter 7), $\hat{\sigma}^2$ is defined with a divisor of n rather than $n-1$. If we denote by $\hat{\boldsymbol{\mu}}$ the vector with constant components equal to $\hat{\mu}$,

$$\hat{\boldsymbol{\mu}} = (\hat{\mu}, \hat{\mu}, \ldots, \hat{\mu}),$$

the variance is related to the l_2-norm, and we have that $\boxed{\hat{\sigma}^2 = \frac{1}{n-1} \|\mathbf{x} - \hat{\boldsymbol{\mu}}\|_2^2.}$

General Moments: The rth moment about the mean is denoted $\hat{\mu}_r$ and defined by

$$\hat{\mu}_r = \frac{1}{n} \sum_{j=1}^{n} (x_j - \hat{\mu})^r \tag{3.45}$$

so that $\hat{\sigma}^2 = \frac{n-1}{n} \hat{\mu}_2$. When r is an even integer, we have that $\boxed{\hat{\mu}_r = \frac{1}{n} \|\mathbf{x} - \hat{\boldsymbol{\mu}}\|_r^r.}$

Notation: As noted above, to distinguish between the moments of the sample and those of the unknown theoretical distribution of all such data, of which the sample is a subset, one sometimes sees the notation of s^2 for the variance, and s for the standard deviation. There is no standard notation for the rth moment about the mean, although analogous to the notational comment above, m_r would be a logical choice.

Constrained Optimization

It turns out that many mathematical problems in finance, especially those related to optimizing an objective function given certain constraints, are more easily solvable within an l_2-type measurement framework for reasons related to the tools of multivariate calculus, although these constraints may in fact be more accurately represented in terms of other norms.

Optimization with an l_1-Norm   An example of an l_1-norm occurs within a trading model. Assume that we have a portfolio within which we are trying to change some portfolio attribute through a trade. Typically there are infinitely many trades that can provide the desired objective. What is clear is that trading can be expensive due to the presence of **bid–ask spreads** as well as other direct costs. If one evaluates the portfolio value after a trade represented by the n-tuple $\mathbf{x} = (x_1, x_2, \ldots, x_n)$, whereby $x_i > 0$ implies a purchase, and $x_i < 0$ a sale of a dollar amount of $|x_i|$ of the ith asset and $\sum x_i = 0$, the portfolio value after the trade can be represented as

$$P(x_1, x_2, \ldots, x_n) = P - e \sum |x_i|.$$

Here e denotes the average cost per currency unit of trading, and P the current portfolio market value.

Consequently one problem to be solved can be stated:

Of all $(x_1, x_2, \ldots, x_n)$ that achieve portfolio objectives,

Minimize: $\displaystyle\sum |x_i| = \|\mathbf{x}\|_1.$

Typically the condition of achieving portfolio objectives can also be expressed in terms of an equation involving the terms $(x_1, x_2, \ldots, x_n)$. For example, if β denotes the current **portfolio beta** value, and β' the desired value, the constraint on traded assets to achieve the target could be expressed as

$$\beta + \frac{\sum x_i \beta_i}{P} = \beta', \tag{3.46}$$

where β_i denotes the beta of the ith asset traded.

Summarizing, we see that this trading problem becomes one of finding a solution of this equation with minimal l_1-norm. That is, rewriting objectives results in

Minimize: $\|\mathbf{x}\|_1$ given

$$(\mathbf{x}, \boldsymbol{\beta}) = P(\beta' - \beta), \tag{3.47}$$

$$\sum x_i = 0.$$

Here $\boldsymbol{\beta}$ denotes the vector of tradable asset betas, and we used inner product notation $(\mathbf{x}, \boldsymbol{\beta}) = \sum x_i \beta_i$. This is an example of a **constrained optimization** in that we are optimizing, and in this case minimizing, the l_1-norm with the constraint that the solution satisfies two given equations.

We can envision the problem in (3.47) geometrically. Of the set of all $\mathbf{x}$ that satisfy the given constraints, find the value that is closest to the origin in terms of the l_1-norm.

Optimization with an l_∞-Norm   An example of the same type but with an l_∞-norm occurs when one is trying to control the total amount of any asset traded. Such a constraint may occur because of illiquid markets and the desire to avoid a trade that moves prices, or because one has an investment policy constraint on the concentration in any given asset. In the simplest form, where all traded assets have the same limitation, the objective would be one of finding a solution to equation (3.46) with l_∞-norm bounded by this common limit: $\|(x_1, x_2, \ldots, x_n)\|_\infty \leq L$.

More realistically, one is generally not so much interested in limiting the maximum trade as the maximum portfolio exposure post-trade. Consequently one would instead look for solutions to equation (3.46) with the limit: $\|(p_1, p_2, \ldots, p_n) + (x_1, x_2, \ldots, x_n)\|_\infty \leq L$, where x_i is the amount of each trade, and p_i the initial portfolio exposure. Since there are potentially many such solutions, an optimization is still possible, and the problem becomes

Minimize: $\|\mathbf{x}\|_1$ given

$$(\mathbf{x}, \boldsymbol{\beta}) = (\beta' - \beta)P,$$

$$\sum x_i = 0,$$ (3.48)

$$\|\mathbf{p} + \mathbf{x}\|_\infty \leq L,$$

where $\mathbf{p} = (p_1, p_2, \ldots, p_n)$.

Optimization with an l_2-Norm   Although in both types of problems above the use of l_1-norms and l_∞-norms is more natural, one might actually solve the problem using an l_2-norm instead. The reason relates to the tools of multivariate calculus and is one of mathematical tractability. That is, explicit solutions to such problems with an l_2-norm can often be derived analytically, whereas with other norms one must typically utilize numerical procedures. Obviously, given the prevalence and power of computers today, one could hardly imagine that obtaining an explicit mathematical expression, rather than a numerical solution, would be worth much. However, the popularity of l_2-norm methods was certified long before the "computer age," and still has merit.

The advantage of representing the solution as an explicit mathematical expression is that the functional relationship between the problem's inputs and the output solution is explicitly represented in a form that allows further analysis. For example, one can easily perform a sensitivity analysis that quantifies the dependence of the solution on changes to various constraints, and the addition of constraints. Such analyses are also possible with numerical solutions, but they require the development of solutions over a "grid" of input assumptions from which sensitivities can be estimated.

Tractability of the l_p-Norms: An Optimization Example
A simple example of the mathematical tractability of l_2-norms is as follows: Assume we are given a collection of data points $\{x_i\}_{i=1}^n$, which we may envision either as distributed on the real line $\mathbb{R}$ or as a point $\mathbf{x} = (x_1, x_2, \ldots, x_n)$ in $\mathbb{R}^n$. The goal is to find a single number a_p that best approximates these points in the l_p-norm, where $p \geq 1$. That is,

Find a_p so that $\|(x_1 - a_p, x_2 - a_p, \ldots, x_n - a_p)\|_p$ is minimized.

Assume that for notational simplicity that we have arranged the data points in increasing order: $x_1 \leq x_2 \leq \cdots \leq x_n$. This problem can be envisioned as a problem in $\mathbb{R}$, such as for $p < \infty$,

Minimize: $f(a) = \left(\sum_{i=1}^{n} |x_i - a|^p \right)^{1/p}$,   (3.49)

or as a problem in $\mathbb{R}^n$, for any p,

Minimize: $f(\mathbf{a}) = \|\mathbf{x} - \mathbf{a}\|_p$,   (3.50)

where $\mathbf{x} \equiv (x_1, x_2, \ldots, x_n)$ and $\mathbf{a} \equiv (a, a, \ldots, a)$ is a point on the "diagonal" in $\mathbb{R}^n$.

Geometrically, for the problem statement in $\mathbb{R}^n$, we seek the smallest l_p-ball centered on $\mathbf{x}$, $\bar{B}_r^p(\mathbf{x})$, that intersects this diagonal. The point or points of intersection are then the values of $\mathbf{a}_p$ that minimize $f(\mathbf{a})$, and the "radius" of this minimal ball is the value of $f(\mathbf{a}_p)$.

In either setting, minimizing the stated functions, or their pth powers to eliminate the pth-root, are equivalent, since $p \geq 1$ and hence $g(y) = y^p$ is an increasing function on $[0, \infty)$. Consequently, if $y \equiv f(a)$ and $y' \equiv f(a')$, then $y \leq y'$ if and only if $g(y) \leq g(y')$.

What is easily demonstrated is that any solution must satisfy $x_1 \leq a \leq x_n$. For example, if $a > x_n$,

$$f(a)^p = \sum |x_i - a|^p = \sum (a - x_i)^p,$$

which is an increasing function on $[x_n, \infty)$, so we must have $a \leq x_n$. Similarly, for $a < x_1$, we have that $f(a)^p = \sum (x_i - a)^p$ which is a decreasing function on $(-\infty, x_1]$, and so $a \geq x_1$.

The analytical solution of this general problem is somewhat difficult and with three exceptions requires the tools of calculus from chapter 9. In fact, at this point, it is not even obvious that in the general case a solution exists, or if it does, that it is unique. However, in the special cases of $p = 1, 2$, and ∞, this problem can be solved with elementary methods, and this is easiest when $p = 2$, which we address first. In chapter 9, the other cases will be addressed.

l_2-**Solution**   Given the points $\{x_i\}_{i=1}^n$, define the simple average consistently with the sample mean in (3.42):

$$\bar{x} = \frac{1}{n} \sum x_i.$$

By writing,

$$x_i - a = (x_i - \bar{x}) + (\bar{x} - a),$$

a simple algebraic calculation leads to

$$f(a) \equiv \sum (x_i - a)^2 = \sum (x_i - \bar{x})^2 + n(\bar{x} - a)^2,$$

where $f(a)$ denotes now the l_2-norm squared. So it is clear that $a_2 = \bar{x}$ gives the l_2-norm minimizing point, since then $n(\bar{x} - a)^2 = 0$.

In other words, the sample mean of a collection of points minimizes the l_2-norm in the sense of (3.49). Since this l_2-norm is related to the sample variance in (3.44), this result can be restated. Considering $\hat{\mu}$ in the definition of sample variance as undefined for a moment, the analysis above implies that the value of the sample variance is minimized when $\hat{\mu}$ equals the sample mean, which it does.

l_1-Solution The case of $p = 1$ is more difficult but still tractable. Because $x_1 \leq x_2 \leq \cdots \leq x_n$, we can relabel these to be distinct points $y_1 < y_2 < \cdots < y_m$. Now, letting n_i denote the number of occurrences of y_i, so that $\sum_{i=1}^{m} n_i = n$, we write

$$f(a) \equiv \sum |x_i - a| = \sum n_i |y_i - a|.$$

We know that if $y_j \leq a \leq y_{j+1}$, then $|y_i - a| = y_i - a$ for $i \geq j+1$, and $|y_i - a| = a - y_i$ for $i \leq j$. Consequently

$$f(a) = c_j - \left(n - 2 \sum_{i=1}^{j} n_i \right) a,$$

where c_j is a constant in each interval, and specifically, $c_j = \sum_{i=j+1}^{n} n_i y_i - \sum_{i=1}^{j} n_i y_i$. So the graph of $f(a)$ is linear between any consecutive distinct points, is decreasing if $n - 2 \sum_{i=1}^{j} n_i > 0$, is increasing if $n - 2 \sum_{i=1}^{j} n_i < 0$, and is constant if $n - 2 \sum_{i=1}^{j} n_i = 0$. We can therefore conclude:

1. If $n = 2m + 1$ is odd, then there is no value of j for which $n - 2 \sum_{i=1}^{j} n_i = 0$, and hence there is a unique value of j with $n - 2 \sum_{i=1}^{j} n_i > 0$ and $n - 2 \sum_{i=1}^{j+1} n_i < 0$. Consequently $a_1 = x_{j+1}$ is the l_1-norm minimizing point, since $f(a)$ is decreasing when $a < x_{j+1}$ and increasing when $a > x_{j+1}$. When all $n_i = 1$, then $a_1 = x_{m+1}$.

2. If $n = 2m$, then the solution will be unique if there is no value of j for which $n - 2 \sum_{i=1}^{j} n_i = 0$, and in this case the value of a_1 is calculated as above. However, if there is a value of j for which $n - 2 \sum_{i=1}^{j} n_i = 0$, then any a_1 with $y_j \leq a_1 \leq y_{j+1}$ will gives the same value for $f(a_1)$, namely c_j, so the solution will not be unique. When all $n_i = 1$, then the solution is never unique, and any a_1 with $x_m \leq a_1 \leq x_{m+1}$ is a solution.

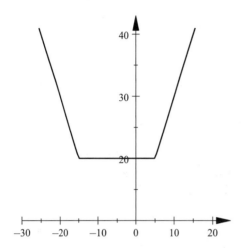

Figure 3.4
$f(a) = |5 - a| + |-15 - a|$

As a simple graphical illustration of non-uniqueness in the even case when all $n_i = 1$, let $x_1 = 5$, and $x_2 = -15$. The graph of $f(a)$ as a function on $\mathbb{R}$ is seen in **figure 3.4**.

Considered as a problem of in $\mathbb{R}^2$, the minimal l_1-Ball centered on $(5, -15)$ that intersects the diagonal in $\mathbb{R}^2$ is presented in **figure 3.5**. As can be seen, this minimal l_1-ball intersects the diagonal line over the same range of x-values that minimize the function in **figure 3.4**.

Remark 3.45   *The earlier l_1-norm trading problem is similar to this problem. However, the "admissible" set of solutions there is not defined as the $\mathbb{R}^n$ diagonal, unless we wish to trade the same amount in all assets, an unlikely scenario. The admissible set is instead the collection of points that satisfy the beta-constraint in (3.46). In addition, rather than look for the point on the admissible set that is "closest" to some initial point $\mathbf{x} = (x_1, x_2, \ldots, x_n)$, we seek the trading point on the admissible set that is closest to $\mathbf{0} = (0, 0, \ldots, 0)$ in the l_1-norm.*

l_∞**-Solution**   The case $p = \infty$ is considered next, and in this special example the solution is immediate, though often this is not the case. Here the goal is to determine a that minimizes

$$f(a) = \max\{|x_i - a|\},$$

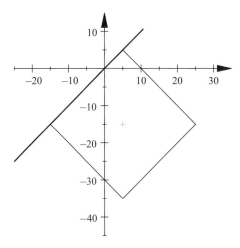

Figure 3.5
$|x - 5| + |y + 15| = 20$

and this is easily seen to be $a = (x_n - x_1)/2$, the midpoint of the interval $x_1 \le x \le x_n$. This is because the l_∞-norm must be attained at one of the end points, so to minimize this distance, the interval midpoint is optimal.

General l_p-Solution   In general, framing this l_p-norm problem as a problem in $\mathbb{R}$ or $\mathbb{R}^n$ are identical problems, but the intuitive framework differs between the two Euclidean settings. The geometry and intuition in $\mathbb{R}$ can be exemplified by a simple graph. Here we illustrate the problem with x_i values of 5, and -15, and $p = 3$ in **figure 3.6**.

The function we aim to minimize is graphed in a bold line, and equals the cube of the l_3-norm. This function is seen to equal the sum of the two component functions defined as $f_i(a) = |x_i - a|^3$, graphed in light lines. Clearly, the minimum appears to be at $a = -5$, and this is easily confirmed. Letting $a = -5 + b$, and assuming $b < 10$ say to make the absolute value unambiguous, we get that $f(a) = 2000 + 60b^2$, which is minimized when $b = 0$.

In $\mathbb{R}^2$ this problem can be written as one of minimizing the cube of the l_3-norm between $(5, -15)$ and (a, a):

Minimize: $\|(5, -15) - (a, a)\|_3^3.$

Geometrically, we are looking for a point on the diagonal of $\mathbb{R}^2$: $\{(x, y) \mid x = y\}$ that is closest to the point $(5, -15)$ in the l_3-norm. Intuitively, we imagine l_3-balls centered

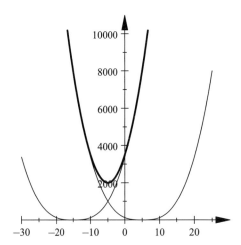

Figure 3.6
$f(a) = |5 - a|^3 + |-15 - a|^3$

on $(5, -15)$ of various radius values, and seek the smallest one that intersects this diagonal. Graphically, the solution is seen in **figure 3.7**. If the radius of this ball is less than $\sqrt[3]{2000} \approx 12.6$, there is no intersection, while if is greater, there are two points of intersection.

Without more powerful tools, however, we are not able to confirm that such problems have solutions for general p and n, nor if they do, if and when such solutions are unique. Even if a solution is known to be unique, there may be no "closed-form solution" to the problem whereby the value of a_p can be expressed as an explicit function of p and the initial collection $\{x_i\}$.

In the cases $p = 1, 2$, and ∞, it was shown that the solution of the problems in (3.49) and (3.50) were always uniquely and explicitly solvable, except in the case where $p = 1$ and n is even, where although explicitly solvable, there could be infinitely many solutions.

General Optimization Framework
Optimization problems are everywhere in finance, and they usually take the following form:

Problem 3.46   *Of all values of* $\mathbf{x} = (x_1, x_2, \ldots, x_n)$ *that satisfy*

$$f(\mathbf{x}) = c,$$

find the value that optimizes (i.e., minimizes or maximizes)

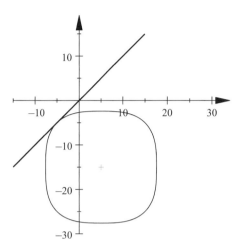

Figure 3.7
$|x - 5|^3 + |y + 15|^3 = 2000$

$$\|\mathbf{x} - \mathbf{a}\|_p,$$

where c is a constant, and $\mathbf{a}$ *is a point, perhaps* $\mathbf{0}$, *and p is typically* 1, 2, *or* ∞.

In the more general case, the norm minimization is replaced:

Problem 3.47   *Of all values of* $\mathbf{x} = (x_1, x_2, \ldots, x_n)$ *that satisfy*

$$f(\mathbf{x}) = c,$$

find the value that optimizes (i.e., minimizes or maximizes)

$$g(\mathbf{x}),$$

where $g(\mathbf{x})$ *is a given function.*

Note that in both cases the problem is known as a **constrained optimization** and is defined by:

• One or more **constraint functions**: The function that provides constraints on the solution.

• An **objective function**: The function that is to be optimized.

Depending on the application, one or both of these functions may reflect one or more l_p-norms, as well as a variety of other financial functions of interest.

Exercises

Practice Exercises

1. Calculate the l_p-norms of the following vectors in $\mathbb{R}^n$, for $p = 1, 2, 5$, and ∞ and a a positive real number:

(a) $\mathbf{a} = (\pm a, \pm a, \dots, \pm a)$

(b) $\mathbf{a} = (\pm a, 0, 0, \dots, 0)$

(c) $\mathbf{a} = (a_1, a_2, \dots, a_n)$ where one $a_j = \pm a$, all others are 0.

2. Calculate the inner product of the following pairs of vectors and confirm Hölder's inequality in (3.16) (which is the Cauchy–Schwarz inequality for $p = 2$) for $p = 1, 2, 5, 10$, and ∞:

(a) $\mathbf{x} = (-5, 3)$ and $\mathbf{y} = (-2, -8)$

(b) $\mathbf{x} = (-1, 2, 3)$ and $\mathbf{y} = (-1, -1, 20)$

(c) $\mathbf{x} = (2, 12, -3, -3)$ and $\mathbf{y} = (-10, 3, 2, 0)$

(d) $\mathbf{x} = (-3, -3, -5, -10, -1)$ and $\mathbf{y} = (2, 5, 10, 20, 1)$

3. For the vector pairs in exercise 2, verify the Minkowski inequality in (3.17) for $p = 1, 2, 5, 10$, and ∞.

4. For the vector pairs in exercise 2:

(a) Calculate the l_p-distances for $p = 1, 2, 5, 10$, and ∞.

(b) Demonstrate explicitly that for each pair of vectors, the l_p-distance gets closer to the l_∞-distance as p increases without bound. (*Hint*: Recall remark 3.25 following the proof of the Minkowski inequality.)

5. Develop graphs of the l_p-balls in $\mathbb{R}^2$, $\bar{B}_r^p(0)$, for $p = 1, 2$, and ∞, and r-values $r = 0.10, 0.5$ and 1. Evaluate the relationship between the different balls for various r by comparing l_1- and l_2-balls, then l_2- and l_∞-balls.

(a) Demonstrate the equivalence of the l_1- and l_2-norms by showing how one can choose the associated r-values to verify (3.35).

(b) Demonstrate the equivalence of the l_2- and l_∞-norms by showing how one can choose the associated r-values to verify (3.35).

6. Show that if $(\mathbf{x}, \mathbf{y})$ is an inner product on a real vector space X, all the properties of a norm are satisfied by $|\mathbf{x}|$ as defined by (3.5), and hence the terminology "norm associated with this inner product" is justified.

7. If $\mathbf{x}$ and $\mathbf{y}$ are two vectors in $\mathbb{R}^n$, $n = 2, 3$ and $\mathbf{z} \equiv \mathbf{y} - \mathbf{x}$:

(a) Demonstrate $\|\mathbf{z}\|_2^2 = \|\mathbf{x}\|_2^2 + \|\mathbf{y}\|_2^2 - 2\mathbf{x} \cdot \mathbf{y}$. (*Hint*: Use (3.5) and properties of inner products.)

(b) Show that if $\theta < \pi$ denotes the angle between $\mathbf{x}$ and $\mathbf{y}$, then

$$\cos \theta = \frac{\mathbf{x} \cdot \mathbf{y}}{\|\mathbf{x}\|_2 \|\mathbf{y}\|_2}. \tag{3.51}$$

(*Hint*: the **law of cosines** from trigonometry states that

$$c^2 = a^2 + b^2 - 2ab \cos \theta, \tag{3.52}$$

where a, b, c are the sides of a triangle, and θ is the radian measure of the angle between sides a and b. Now create a triangle with sides $\mathbf{x}$, $\mathbf{y}$, and $\mathbf{z}$.)

(c) Show that if $\theta < \pi$ denotes the angle between $\mathbf{x}$ and $\mathbf{y}$, then $\mathbf{x} \cdot \mathbf{y} = \mathbf{0}$ iff $\theta = 90°$, so $\mathbf{x}$ and $\mathbf{y}$ are "perpendicular." (*Note*: The usual terminology is that $\mathbf{x}$ and $\mathbf{y}$ are **orthogonal**, and this is often denoted $\mathbf{x} \perp \mathbf{y}$.)

Remark 3.48 *Note that for $n > 3$, the formula in (3.51) is taken as the definition of the cosine of the angle between $\mathbf{x}$ and $\mathbf{y}$, and logically represents the true angle between these vectors in the two-dimensional plane in $\mathbb{R}^n$ that contain them. As was noted in the section on inner products, the derivations in (a) and (b) remain true for a general inner product and associated norm, and hence the notion of "orthogonality" can be defined in this general context.*

8. Show that if $\{\mathbf{x}_j\}_{j=1}^n$ is a collection of mutually orthogonal, unit vectors in $\mathbb{R}^n$, namely $\mathbf{x}_j \cdot \mathbf{x}_k = 0$ for $j \neq k$, and $|\mathbf{x}_j|^2 = \mathbf{x}_j \cdot \mathbf{x}_j = 1$ for all j, then for a vector $\mathbf{y} \in \mathbb{R}^n$ that can be expressed as a linear combination of these vectors

$$\mathbf{y} = \sum_{j=1}^n a_j \mathbf{x}_j, \tag{3.53}$$

the constants a_j must satisfy $a_j = \mathbf{x}_j \cdot \mathbf{y}$. (*Hint*: Consider an inner product of each side with $\mathbf{x}_j$.)

Remark 3.49 *The usual terminology is that the collection of vectors, $\{\mathbf{x}_j\}_{j=1}^n$, are **orthonormal**. With the tools of linear algebra, it can be shown that all vectors $\mathbf{y} \in \mathbb{R}^n$ can be represented as in (3.53).*

9. Given a vector of sample data: $\mathbf{x} = (x_1, x_2, \ldots, x_n)$, demonstrate that $\hat{\sigma}^2 = \hat{\mu}_2' - \hat{\mu}^2$, where here $\hat{\sigma}^2$ is defined with n rather than $n - 1$.

10. Given semiannual coupon bond data with prices expressed per 100 par:

Term	0.5 years	1.0 years	1.5 years	2.0 years
Coupon	2.0%	2.2%	2.6%	3.0%
Price	99.5	100.0	100.5	101.0

(a) Bootstrap this data to determine semiannual market spot rates for 0.5, 1.0, 1.5, and 2.0 years.

(b) What is the semiannual forward rate between 0.5 and 1.5 years?

11. Demonstrate that the forward rate in exercise 10(b) can be realized by an investor desiring to invest $1 million between time 0.5 and 1.5 years, by constructing an appropriate portfolio of long and short zero coupon bonds. Assume that these zeros are trading with the spot rates from 10(a).

12. Given a portfolio of three stocks with market values in $millions of 350, 150, and 500, and respective betas of 1.0, 0.9, and 1.1:

(a) Calculate the **beta** of the portfolio, where $\beta = \sum x_i \beta_i / \sum x_i$ and x_i denotes the amount invested in stock i.

(b) Find the trade in $\mathbb{R}^3$ that changes the portfolio beta to 1.08 that has the lowest transaction fee, assuming that this fee is proportional to the market value bought and sold, and that all final positions must be long. (*Hint:* See (3.47), but note that while the constraint $\sum x_i = 0$ allows you to analytically consider this a problem in $\mathbb{R}^2$, because $x_3 = -x_1 - x_2$, the norm minimization in $\mathbb{R}^2$ will not work in general.)

(c) Repeat part (b) but now with a beta target of 0.935, and where final positions can be long or short.

(d) Achieve the same objective in part (c), but adding the constraint that the investment policy maximum for any stock is 600 on a long or short basis.

Assignment Exercises

13. Calculate the inner product of the following pairs of vectors, and confirm Hölder's inequality in (3.16) (which is the Cauchy–Schwarz inequality for $p = 2$) for $p = 1, 2, 5, 10$, and ∞:

(a) $\mathbf{x} = (11, -133)$ and $\mathbf{y} = (12, 28)$

(b) $\mathbf{x} = (10, -2, 13)$ and $\mathbf{y} = (-10, 101, 30)$

(c) $\mathbf{x} = (1, -24, 3, 13)$ and $\mathbf{y} = (-1, -23, 21, 10)$

(d) $\mathbf{x} = (10, 53, -53, -10, 21)$ and $\mathbf{y} = (1, -15, -10, 25, 11)$

14. For the vector pairs in exercise 13, verify the Minkowski inequality in (3.17) for $p = 1, 2, 5, 10$, and ∞.

15. For the vector pairs in exercise 13:

(a) Calculate the l_p-distances for $p = 1, 2, 5, 10$, and ∞.

(b) Demonstrate explicitly that for each pair of vectors, the l_p-distance gets closer to the l_∞-distance as p increases without bound (*Hint*: Recall remark 3.25 following the proof of the Minkowski inequality.)

16. Develop a graph of the l_p-balls in $\mathbb{R}^2$, $\bar{B}_r^p(\mathbf{0})$, for $p = 1, 5$, and ∞, and r-values $r = 0.10, 0.5$, and 1. Evaluate the relationship between the different balls for various r by comparing l_1- and l_5-balls, then l_5- and l_∞-balls.

(a) Demonstrate the equivalence of the l_1- and l_5-norms by showing how one can choose the associated r-values to verify (3.35).

(b) Demonstrate the equivalence of the l_5- and l_∞-norms by showing how one can choose the associated r-values to verify (3.35).

17. For fixed $a, b > 0$, say $a = 3$, $b = 5$, develop a graph of the function for $1 < p < \infty$:

$$f(p) = \frac{a^p}{p} + \frac{b^q}{q},$$

where $q = \frac{p}{p-1}$ is conjugate to p. Confirm Young's inequality that $ab \le f(p)$ for all p. What happens if $a = b$?

18. Not all metrics are equivalent to the l_p-metrics. Show that

$$d(\mathbf{x}, \mathbf{y}) = \begin{cases} 0, & \mathbf{x} = \mathbf{y} \\ 1, & \mathbf{x} \ne \mathbf{y} \end{cases},$$

is a metric on $\mathbb{R}^n$ that is not equivalent to the l_p-metrics.

19. Given a portfolio of $100{,}000$ par of 6% semiannual (s.a.) coupon, 10-year bonds, and $250{,}000$ par of 4.5% s.a. coupon, 3-year bonds, let $(i, j) \in \mathbb{R}^2$ denote the market yield vector, where i is the s.a. yield for the 3-year bond, and j the s.a. yield for the 10-year bond.

(a) Develop the formula for the portfolio price function, $P(i, j)$, using (2.15) or an equivalent formulation, and calculate the initial portfolio market value assuming that $(i_0, j_0) = (0.04, 0.055)$.

(b) Assume that the initial yield vector shifts, $(i_0, j_0) \to (i, j)$, where $(i, j) = (i_0 + \Delta i, j_0 + \Delta j)$. Consider only shifts, $(\Delta i, \Delta j)$, that have the same l_p-norm as the shift vector $(0.01, 0.01)$ for $p = 1, 2, \infty$. Show by examples that the portfolio gain/loss $P(i_0 + \Delta i, j_0 + \Delta j) - P(i_0, j_0)$ is not constant in any of these norms. (*Hint*: Consider shifts on the l_p-balls in $\mathbb{R}^2$, $\bar{B}_r^p(0.04, 0.055)$ where $r = \|(0.01, 0.01)\|_p$. Try shift vectors, $(\Delta i, \Delta j)$ where $\Delta i = \pm \Delta j$, or where one or the other is 0, to get started.)

(c) For each p-value, estimate numerically the yield shift vectors that provide the largest portfolio gain and loss.

20. For the portfolio in exercise 19, implement a **market-value neutral trade** at the initial yields, selling 75,000 par of the 10-year and purchasing an equivalent dollar amount of the 3-year bonds.

(a) Express this trade as a vector-shift in $\mathbb{R}^{20}$, where the initial vector $\mathbf{C}_0$ is the original cash flow vector, and $\mathbf{C}$ the vector after the trade.

(b) Repeat exercise 19(b) and 19(c) for the traded portfolio, comparing results.

21. Given semiannual coupon bond data with prices expressed per 100 par:

Term	0.5 years	1.0 years	1.5 years	2.0 years
Coupon	3.0%	2.8%	2.4%	2.0%
Price	100.0	100.5	101.0	101.5

(a) Bootstrap this data to determine semiannual market spot rates for 0.5, 1.0, 1.5, and 2.0 years.

(b) What is the semiannual forward rate between 1.0 and 2.0 years?

22. Demonstrate that the forward rate in exercise 21(b) can be realized by an investor desiring to borrow $100 million between time 1.0 and 2.0 years, by constructing an appropriate portfolio of long and short zero coupon bonds. Assume that these zeros are trading with the spot rates from 21(a).

23. Given a portfolio of 3 bonds with market values in $millions of: 200, 450, and 350, and respective durations of 3.5, 5.0, and 8.5.

(a) Calculate the **duration** of the portfolio, where $D = \sum x_i D_i / \sum x_i$ and x_i denotes the amount invested in bond i.

(b) Find the trade in $\mathbb{R}^3$ that changes the portfolio duration to 4.0 that has the lowest transaction fee, assuming that this fee is proportional to the market value bought and sold, and that all final positions must be long. (*Hint*: See (3.47), but note that while the constraint $\sum x_i = 0$ allows you to analytically consider this a problem in $\mathbb{R}^2$, because $x_3 = -x_1 - x_2$, the norm minimization in $\mathbb{R}^2$ will not work in general.)

(c) Repeat part (b) but now with a duration target of 6.5, and where final positions can be long or short.

(d) Achieve the same objective in part (c), but adding the constraint that the investment policy maximum for any bond is 462 on a long or short basis.

4 Set Theory and Topology

4.1 Set Theory

4.1.1 Historical Background

In this section we formalize the notion of sets and their most common operations. Ironically, the definition of a "set" is more complex than it first appears. Before the early 1900s, a set was generally accepted as being definable as any collection of objects that satisfy a given property,

$$X = \{a \mid a \text{ satisfies property } P\},$$

and an axiomatic structure was developed around this basic concept. This approach has come to be known perhaps unfairly as **naive set theory**, despite the fact that it was developed within a formal axiomatic framework.

In 1903 **Bertrand Russell** (1872–1970) published a paradox he discovered in 1901, which has come to be known as **Russell's paradox**, by proposing as a "set" the following:

$$X = \{R \mid R \text{ is a set, and } R \notin R\}.$$

In other words, X is the "set" of all sets which are not a member of themselves.

The paradox occurs in attempting to answer the question

Is $X \in X$?

If $X \in X$, then by the defining property above it is a set that is **not** an element of itself. However, if we posit that $X \notin X$, then again by definition, X should be one of the sets R that are included in X. In summary,

$$X \in X \quad \text{iff} \quad X \notin X.$$

This is a situation that gives mathematicians great anxiety and rightfully so! What is causing this unexpected result? Are there others? Could such paradoxes be avoided? How? Defining a set as a "collection satisfying a property" certainly works fine most of the time, but apparently not this time.

What was needed was an even more careful and formal articulation of the axioms of set theory and the fundamental properties that would be assumed. With this, mathematicians would be able to develop a theory that was, on the one hand, "familiar," but on the other, paradox free. This approach has come to be known as **axiomatic set theory**.

A number of axiomatic approaches have been developed. The first approach was introduced by **Ernst Zermelo** (1871–1953) in 1908, called the **Zermelo axioms**, and

produced **Zermelo set theory**. This axiomatic structure was later improved upon by
Adolf Fraenkel (1891–1965) in 1922, and produced the **Zermelo–Fraenkel axioms**,
and the **Zermelo–Fraenkel set theory**, or **ZF set theory**. This is the approach largely
used to this day.

In essence, sets are defined as those collections that can be constructed based on
the 10 or so ZF axioms, and the paradox above is resolved because it is not possible
to construct the Russell collection X as a set within this axiomatic structure. It is also
not possible to construct the set of all sets, which underlies another paradox. However, these axioms have been shown to be adequate to construct virtually all of the
types of sets one needs in mathematics, and that for these sets, set manipulations
can proceed just as if these sets were defined via naive set theory, as collections of
objects which satisfy given criteria.

*4.1.2   Overview of Axiomatic Set Theory

To give a flavor for the axiomatic structure of set theory, we introduce the Zermelo–
Fraenkel axioms, including the so-called **axiom of choice**, which collectively produce
what is referred to as **ZFC set theory**. This structure is presented below in a simplified
framework that omits many of the quantifiers necessary to make statements formal,
and is presented in both plain and informal English and approximately formal symbolic language.

In this structure it will be noted that the intuitive notions of "set" and "element"
are formalized as relative terms, not absolute terms. A set may be an element of another set, and an element of a set may itself be a set that contains elements. In addition the expression $P(x)$ will denote a statement that may be true or false for any
given set x, and $P(X)$ will denote that the statement is true for a given set X. For
example, if

$P(x) : x$ contains an integer as an element,

then $P(\mathbb{N})$. Also $P(x, y)$ will denote a conditional statement in that given a set x,
there is a unique set y so that $P(x, y)$ is true, and then $P(X, Y)$ denotes that the
statement is true for X, Y. For example,

$P(x, y) : y$ contains the elements of x plus the integers as elements.

Finally, we recall the logical symbols: $\forall$ (for all), $\exists$ (there exists), $\sim$ (not), $\ni$ (such
that), $\vee$ (or), $\wedge$ (and), $\Rightarrow$ (implies), and $\Leftrightarrow$ (if and only if).

1. Formal Symbols: $\emptyset, \in, \{,\}, X, Y, Z, \ldots.$

2. Axioms

- **ZF1 (Extensionality):**   Two sets are equal means they contain the same elements,

$$X = Y \Leftrightarrow (Z \in X \Leftrightarrow Z \in Y).$$

- **ZF2 (Empty Set):**   There exists a set with no elements,

$$\exists \emptyset = \{ \ \}.$$

- **ZF3 (Pairing):**   Given any two sets, there exists a set that contains these as elements,

$$X, Y \Rightarrow \exists Z = \{X, Y\}.$$

- **ZF4 (Union):**   Given two sets, there exists a set that contains as elements exactly the elements of the original sets,

$$X, Y \Rightarrow \exists Z \ni: W \in Z \Leftrightarrow W \in X \vee W \in Y.$$

- **ZF5 (Infinity):**   There exists a set with an infinite number of elements, in that it contains the empty set as an element, and for any element Y that it contains, it also contains the element $\{Y, \{Y\}\}$,

$$\exists X \ni: \emptyset \in X \wedge (Y \in X \Rightarrow \{Y, \{Y\}\} \in X).$$

- **ZF6 (Subset):**   Given any set and any statement, there is a set that contains all the elements of the original set for which the statement is true,

$$X, P(x) \Rightarrow \exists Y \ni: Z \in Y \Leftrightarrow Z \in X \wedge P(Z).$$

- **ZF7 (Replacement):**   Given any set and conditional statement, there is a set that contains as elements the unique sets associated with the elements of the original set as defined by the conditional statement,

$$X, P(x, y) \Rightarrow \exists Y \ni: Z \in Y \Leftrightarrow \exists W \in X \wedge P(W, Z).$$

- **ZF8 (Power Set):**   For any set, there is a set that contains as elements any set that contains elements of the original set. In other words, this new set, called the **power set**, contains all the "subsets" of the original set,

$$X \Rightarrow \exists Y \ni: Z \in Y \Leftrightarrow (W \in Z \Rightarrow W \in X).$$

- **ZF9 (Regularity):**   Any set that is not empty contains an element that has no elements in common with the original set,

$$X \neq \emptyset \Rightarrow \exists Y \ni: Y \in X \wedge \sim\exists W(W \in X \wedge W \in Y),$$

where $\sim\exists$ is shorthand for "there does not exist."

· **ZF10 (Axiom of Choice):**   For any set, there is a set that contains as elements an element from each nonempty element of the original set,

$$X \Rightarrow \exists Y \ni: \forall Z \in X(Z \neq \emptyset)\exists W \in Y \wedge W \in Z.$$

These axioms fall into four categories.

1. Axiom 1 introduces the notion of equality of "sets," and indirectly provides a context for the undefined term $\in$. Although the notion of subset is not explicitly defined, we see that this is implicitly referenced in axiom 8, which suggests that the condition on Z is one of "subset":

$$Z \subset X \Leftrightarrow (W \in Z \Rightarrow W \in X).$$

2. Axioms 2 and 5 are existence axioms, on the one hand, declaring the existence of an empty set and, on the other, the existence of a set with an infinite number of elements.

3. All the other axioms except axiom 9 identify how one can make new sets from old sets, or from sets and statements. For example, axiom 3 states that a set can be formed to include as members two other sets, while axiom 4 states that the union of sets is a set. Axioms 6 and 7 state that sets can be formed from sets and statements. A simple application of axiom 6 is that the intersection of X and Y must be a set since we can use the statement: $P(Z) : (Z \in Y)$. Axiom 8 introduces the power set, or the set of all subsets of a given set, and axiom 10 states that there is a set that contains one element from every nonempty element of a given set. In other words, from the elements of X, we can form a set which "chooses" one element from each such element, and hence the name, "axiom of choice."

4. Finally, axiom 9 puts a limit on what a set can be, and can be shown to preclude the "set of all sets" from being a set in this theory. It states that any nonempty set contains an element that is disjoint from the original set.

In what follows, we will treat sets as if definable as collections of objects that satisfy certain statements or formulaic properties, and this can generally be justified by axiom 6. More specifically, the ZFC set theory states that defining a set as a collection of objects that satisfy certain properties will avoid paradoxes if the original collection of objects is itself a set or a subcollection of a set. That is, if A is a set,

$$\{x \,|\, x \in A \text{ and } P(x)\},$$

is a set for any "statement" P, by axiom 6. However, although beyond the scope of this introduction to set theory, one needs to be careful as to exactly what kinds of "statements" are appropriate in this axiom, as it can be shown that for a general property P, paradoxes are still possible.

4.1.3   Basic Set Operations

As a collection of objects, and with the axiomatic structure in the background, we distinguish between the notions: "element of," "subset of," and "equal to":

1. Membership: "x is an element of A," denoted $x \in A$, is only defined indirectly in the axioms, but understanding this notion in terms of the heuristic

$$A \equiv \{x \mid x \in A\}$$

is consistent with the axioms and operationally efficient.

2. Subset: "B is a subset of A," denoted $B \subset A$, and defined by $x \in B \Rightarrow x \in A$.

3. Equality: "B equals A," denoted $A = B$, and defined by $B \subset A$ and $A \subset B$.

Given sets A and B, the basic set operations are:

1. Union: $A \cup B = \{x \mid x \in A \text{ and/or } x \in B\}$.

2. Intersection: $A \cap B = \{x \mid x \in A \text{ and } x \in B\}$.

3. Complement: $\tilde{A} = \{x \mid x \notin A\}$. A^c is an alternative notation, especially if A is a complicated expression. Note that $\tilde{\tilde{A}} = A$.

4. Difference: $A \sim B = \{x \mid x \in A \text{ and } x \notin B\}$. Note that $A \sim B = A \cap \tilde{B}$.

Union and intersection are similarly defined for any indexed collection of sets: $\{A_\alpha \mid \alpha \in I\}$, where I denotes any indexing set which may be finite, or denumerably or uncountably infinite (recall chapter 2):

$$\bigcup_\alpha A_\alpha = \{x \mid x \in A_\alpha \text{ for some } \alpha \in I\},$$

$$\bigcap_\alpha A_\alpha = \{x \mid x \in A_\alpha \text{ for all } \alpha \in I\}.$$

It is straightforward to justify the so-called **De Morgan's laws**, named after **Augustus De Morgan** (1806–1871), who formalized a system of "relational algebra" in 1860. Examples are:

1. $\widetilde{\bigcup_\alpha A_\alpha} = \bigcap_\alpha \tilde{A}_\alpha$.

2. $\widetilde{\bigcap_\alpha A_\alpha} = \bigcup_\alpha \tilde{A}_\alpha$.

3. $B \cap [\bigcup_\alpha A_\alpha] = \bigcup_\alpha [A_\alpha \cap B]$.

4. $B \cup [\bigcap_\alpha A_\alpha] = \bigcap_\alpha [A_\alpha \cup B]$.

To demonstrate the first example in detail, we use the definitions above:

$$x \in \widetilde{\bigcup_\alpha A_\alpha}$$

$$\Leftrightarrow x \notin \bigcup_\alpha A_\alpha$$

$$\Leftrightarrow x \notin A_\alpha \qquad \text{for all } \alpha$$

$$\Leftrightarrow x \in \widetilde{A_\alpha} \qquad \text{for all } \alpha$$

$$\Leftrightarrow x \in \bigcap_\alpha \widetilde{A_\alpha}.$$

4.2   Open, Closed, and Other Sets

4.2.1   Open and Closed Subsets of $\mathbb{R}$

The reader is undoubtedly familiar with the notion of an **interval** in $\mathbb{R}$, as well as the various types of intervals. First off, an interval is a subset of $\mathbb{R}$ that has "no holes."

Definition 4.1   *An **interval** I is a subset of $\mathbb{R}$ that has the property:*

If x, $y \in I$, then for all $z : x \leq z \leq y$ we have that $z \in I$.

There are four types of intervals, as we list next. Interval notation is universal.

1. Open: $(a,b) = \{x \mid a < x < b\}$.

2. Closed: $[a,b] = \{x \mid a \leq x \leq b\}$.

3. Semi-open or **Semi-closed:** $(a,b]$ and $[a,b)$.

In some applications, where it is unimportant if the interval contains its endpoints, the "generic interval" will be denoted: $\langle a, b \rangle$, meaning that it can be any of the four examples above without consequence in the given statement.

Any of these interval types may be **bounded**—meaning that $-\infty < a, b < \infty$—and that all but the closed interval may be **unbounded**. For example,

$$(a, \infty), \quad (-\infty, b), \quad (-\infty, \infty), \quad (-\infty, b], \quad [a, \infty).$$

Each of these characteristics of an interval: open, closed, bounded, and unbounded, can be generalized, and each is important in mathematics for reasons that will emerge over coming chapters. The notions of open and closed subsets of $\mathbb{R}$ are generalized next.

Definition 4.2   *Given $x \in \mathbb{R}$, a **neighborhood of x of radius** r, or **open ball about x of radius** r, denoted $B_r(x)$, is defined as*

$$B_r(x) = \{ y \in \mathbb{R} \mid |x - y| < r \}. \tag{4.1}$$

*A subset $G \subset \mathbb{R}$ is **open** if given $x \in G$, there is an $r > 0$ so that $B_r(x) \subset G$. A subset $F \subset \mathbb{R}$ is **closed** if the complement of F, $\tilde{F}$, is open.*

Intuitively, an open set only contains "interior" points, in that every point can be surrounded by an open ball that fits entirely inside the set. In contrast, a closed set will contain at least one point that is not interior to the set. In other words, no matter how small an open ball one constructs that contains this point, this ball will always contain points outside the set. But while, by definition, the existence of such a point is ensured for a closed set, the existence of such a point does not ensure that the set is closed, and hence the need to define closed in terms of the complement of the set being open. The problem is that a set can be neither open nor closed.

A useful exercise is to think through how an interval like $(-1, 1)$ is open by this definition, whereas the interval $[-1, 1]$ is closed. On the other hand, the interval $[-1, 1)$ has one exceptional point that prevents it from being open, yet this set is also not closed since $(-\infty, -1) \cup [1, \infty)$ is not open.

That open and closed sets are fundamentally different can be first appreciated by observing how differently they behave under set operations.

Proposition 4.3   *If $\{G_\alpha\}$ is any collection of open sets, $G_\alpha \subset \mathbb{R}$, with $\alpha \in I$ an arbitrary indexing set, then*

$\bigcup G_\alpha$ *is an open set.*

If this collection is finite, then $\bigcap G_\alpha$ is an open set. If $\{F_\alpha\}$ is any collection of closed sets, $F_\alpha \subset \mathbb{R}$, then

$\bigcap F_\alpha$ *is a closed set.*

If this collection is finite, then $\bigcup F_\alpha$ is a closed set.

Proof   If $x \in \bigcup G_\alpha$, then $x \in G_\alpha$ for some α. Since each G_α is open, there is an $r > 0$ so that $B_r(x) \subset G_\alpha \subset \bigcup G_\alpha$, proving the first statement. If the collection is finite, and

$x \in \bigcap G_n$, then for every n there is an r_n so that $B_{r_n}(x) \subset G_n$, and therefore $B_r(x) \subset \bigcap G_n$, where $r = \min r_n$. The second statement on closed sets follows from De Morgan's laws and the first result. That is, the complement of this general intersection is open, since

$$\widetilde{\bigcap F_\alpha} = \bigcup \tilde{F_\alpha},$$

which is a union of open sets by assumption. Similarly, if the collection is finite, the complement of the union is an intersection of a finite collection of open sets, which is open.   ∎

This proposition cannot be extended to a statement about the general intersection of open sets, or the general union of closed sets. For example,

$$G_n = \left(-\frac{1}{n}, 1 + \frac{1}{n} \right)$$

has intersection equal to $[0, 1]$, whereas

$$F_n = \left[\frac{1}{n}, 1 - \frac{1}{n} \right]$$

has union $(0, 1)$ (see exercise 3).

Other examples are easy to generate where openness and closeness are preserved, or where semi-openness/closeness is produced (see exercise 15). In other words, anything is possible when an infinite collection of open sets are intersected or closed sets are unioned.

It turns out that open sets in $\mathbb{R}$ can be characterized in a simple and direct way, but not so closed sets.

Proposition 4.4 $G \subset \mathbb{R}$ *is an open set iff there is a countable collection of disjoint open intervals, $\{I_n\}$, so that $G = \bigcup I_n$.*

Proof   Clearly, if G is a countable union of open intervals, it is open by the proposition above. On the other hand, for any $x \in G$, let $\{I_{(a,b)}(x)\}$ be the collection of open intervals that contain x and that are contained in G. This family is not empty, since by definition of open, $I_{(x-r, x+r)}(x) \equiv B_r(x)$ is in this collection for some $r > 0$. Define $I(x) = \bigcup I_{(a,b)}(x)$. By the proposition above, $I(x)$ is an open set. But also we have that $I(x)$ must be an open interval: $I(x) = I_{(a',b')}(x)$. To show this, let $y, z \in I(x)$, with $y < z$ for definitiveness. We must show that $[y, z] \subset I(x)$. Now since $y \in I_{(a,b)}(x)$ for some (a,b), all points between x and y are also in $I_{(a,b)}(x)$. Similarly all points between x and z are in some other interval, $I_{(c,d)}(x)$, say. So we conclude that

$$[y, z] \subset I_{(a,b)}(x) \cup I_{(c,d)}(x) \subset I(x).$$

Finally, to show that $\{I(x)\}$ can be collected into disjoint intervals, assume that for some $x \neq y$, $I(x) \cap I(y) \neq \emptyset$. That is, assume that two such open intervals have non-empty intersection. Then it must be the case that $I(x) = I(y)$, since otherwise, $I(x) \cup I(y)$ would be a larger interval for each of x and y, contradicting the maximality of the individual intervals. That this collection is countable follows from the observation that each of the disjoint open intervals constructed must contain a rational number. ∎

From this result we can redefine closed sets by reverse reasoning:

$F \subset \mathbb{R}$ is closed iff $\tilde{F}$ is a countable collection of disjoint open intervals.

Unlike an open set, which is always a union of a finite or countably infinite number of disjoint open intervals, closed sets can differ greatly. Any singleton set, $\{x\}$, is closed, as is any finite set, $\{x_j\}_{j=1}^n$. Countably infinite closed sets can be sparsely spaced in $\mathbb{R}$, like the integers, or with accumulation points, such as $\{m + \frac{1}{n} \mid m, n \in \mathbb{Z}, n > 0\} \cup \mathbb{Z}$. A closed set can even contain uncountably many points, and yet contain no interval. A famous example is the **Cantor ternary set**, named for its discoverer **Georg Cantor** (1845–1918).

The Cantor set, K, is a subset of the interval $[0, 1]$ and is defined as the intersection of a countable number of closed sets, $\{F_n\}$, so $K = \bigcap F_n$ and K is closed. Each successive closed set is defined as the prior set, with the open "middle third" intervals removed. For example,

$$F_0 = [0, 1],$$

$$F_1 = [0, 1] \sim \left(\frac{1}{3}, \frac{2}{3}\right),$$

$$F_2 = F_1 \sim \left\{ \left(\frac{1}{9}, \frac{2}{9}\right) \cup \left(\frac{7}{9}, \frac{9}{9}\right) \right\},$$

$$\vdots$$

Interestingly, the total length of the open intervals removed is 1, the length of the original interval $[0, 1]$. This can be derived by noting that in the first step, one interval of length one-third is removed, then two intervals of length one-ninth, then four of length one-twenty-seventh, and so forth. The total length of these intervals can be expressed as

$$\sum_{n=0}^{\infty} \frac{2^n}{3^{n+1}} = \frac{1}{3} \sum_{n=0}^{\infty} \left(\frac{2}{3}\right)^n = 1.$$

This last summation is accomplished using the informal methodology introduced in chapter 2 in the applications section for pricing a preferred stock. Recall, if $S \equiv \sum_{n=0}^{\infty} \left(\frac{2}{3}\right)^n$, then $\frac{2}{3} S = \sum_{n=1}^{\infty} \left(\frac{2}{3}\right)^n$. Subtracting, we conclude that $\frac{1}{3} S = 1$, and the result follows. (See also the chapter 6 discussion of geometric series for a formal justification.)

Because the complement of the Cantor ternary set in $[0, 1]$ has length 1, the Cantor ternary set is said to be a **set of measure 0**. The intuition, which will be formalized in chapter 10 is that a set of measure zero can be contained in, or "covered by" a collection of intervals, the total length of which is as small as desired. In this case the closed sets F_n provide just such a sequence of sets, as each is a collection of intervals, each covers K, and by the analysis above, the total length of the intervals in F_n is $1 - \sum_{j=0}^{n-1} \frac{2^j}{3^{j+1}}$, which is as small as we want by taking n large enough.

That the Cantor ternary set is in fact uncountable is not at all obvious, since it is easy to believe that all that will be left in this set are the endpoints of the intervals removed, and these form a countable collection. The demonstration of uncountability relies on the base-3 expansion of numbers in the interval $[0, 1]$, introduced in chapter 2. Paralleling the decimal expansion, the base-3 expansion uses the digits 0, 1, and 2:

$$x_{(3)} = 0.a_1 a_2 a_3 a_4 \ldots$$

$$\equiv \sum_{j=1}^{\infty} \frac{a_j}{3^j}, \qquad \text{where } a_j = 0, 1, 2.$$

It turns out that the removal of the "middle thirds" is equivalent to eliminating the possibility of $a_j = 1$, so the Cantor ternary set is made up of all numbers in $[0, 1]$ with base-3 expansions using only 0s and 2s. This at first seems counterintuitive because $\frac{1}{3} \in K$, and yet the base-3 expansion of $\frac{1}{3}$ is 0.1. The same is true for the left endpoints of the leftmost intervals removed at each step, which are all of the form $\frac{1}{3^j}$. But these can all be rewritten as

$$\frac{1}{3^j} = \sum_{n=j+1}^{\infty} \frac{2}{3^n},$$

as can be verified using the derivation above.

By dividing each a_j term by 2, all such expansions can then be identified in a 1:1 way with the base-2 expansions of all numbers in $[0, 1]$, which are uncountable as was seen in section 2.1.5. Specifically, the identification is

$$\text{If } \sum_{n=1}^{\infty} \frac{a_j}{3^j} \in K, \quad \text{then } \sum_{n=1}^{\infty} \frac{a_j}{3^j} \leftrightarrow \sum_{n=1}^{\infty} \frac{a_j/2}{2^j}.$$

4.2.2   Open and Closed Subsets of $\mathbb{R}^n$

Generalizing the ideas from $\mathbb{R}$ in the natural way to $\mathbb{R}^n$, we have the following:

Definition 4.5   *Given* $\mathbf{x} \in \mathbb{R}^n$, *a **neighborhood of** $\mathbf{x}$ **of radius** r, or **open ball about** $\mathbf{x}$ **of radius** r, denoted $B_r(\mathbf{x})$, is defined as*

$$B_r(\mathbf{x}) = \{\mathbf{y} \in \mathbb{R}^n \mid |\mathbf{x} - \mathbf{y}| < r\}, \tag{4.2}$$

where $|\mathbf{x}|$ *denotes the standard norm on* $\mathbb{R}^n$. *A subset* $G \subset \mathbb{R}^n$ *is **open** if, given* $\mathbf{x} \in G$, *there is an* $r > 0$ *so that* $B_r(\mathbf{x}) \subset G$. *A subset* $F \subset \mathbb{R}^n$ *is **closed** if the complement of* F, $\tilde{F}$, *is open.*

The proposition above on unions and intersections of open and closed sets in $\mathbb{R}$ carries over to $\mathbb{R}^n$ without modification. We state this result without proof.

Proposition 4.6   *If* $\{G_\alpha\}$ *is any collection of open sets,* $G_\alpha \subset \mathbb{R}^n$, *then*

$\bigcup G_\alpha$ *is an open set.*

If this collection is finite, then $\bigcap G_\alpha$ *is an open set. If* $\{F_\alpha\}$ *is any collection of closed sets,* $F_\alpha \subset \mathbb{R}^n$, *then*

$\bigcap F_\alpha$ *is a closed set.*

If this collection is finite, then $\bigcup F_\alpha$ *is a closed set.*

It is also the case that one cannot generalize this result to arbitrary intersections of open sets, nor arbitrary unions of open sets, and the examples above easily generalize to this setting (see exercise 16).

Remark 4.7   *Note that "open" was defined in terms of open balls, and in turn by the standard metric in* $\mathbb{R}^n$, *also called the* l_2*-metric in chapter 3. However, as might be guessed from that chapter, we could have used any metric equivalent to the standard metric and obtained the same open and closed sets due to (3.35). We formalize this observation in the following:*

Proposition 4.8   *Let $d'(x, y)$ be any metric on $\mathbb{R}^n$ equivalent to the standard metric $d(x, y) = |\mathbf{x} - \mathbf{y}|$ given in (3.18), and let open sets be defined relative to open d'-balls. Then $G \subset \mathbb{R}^n$ is open relative to d' iff it is open relative to d.*

Proof   We demonstrate one implication only, as the other is analogous. Assume that G is open relative to d', and let $\mathbf{x} \in G$. Then, by definition, there is an $r' > 0$ so that $B'_{r'}(\mathbf{x}) \subset G$. By (3.35), there is an $r > 0$ so that

$$B_r(\mathbf{x}) \subset B'_{r'}(\mathbf{x}).$$

and hence $B_r(\mathbf{x}) \subset G$ and so G is open relative to $d(x, y)$.   ∎

It is important to note that this proposition cannot be expanded arbitrarily. If d and d' are metrics that are not equivalent, it will generally be the case that the associated notions of open and closed will also not be equivalent.

Remark 4.9   *Because as proved in proposition 3.41, Lipschitz equivalence of metrics implies equivalence, any result stated concerning equivalent metrics is automatically true for Lipschitz equivalent metrics.*

*4.2.3   Open and Closed Subsets in Metric Spaces

The definition of a neighborhood, or open ball about $\mathbf{x} \in \mathbb{R}^n$, is fundamentally a metric notion. Namely an open ball of radius r about $\mathbf{x}$ is defined to be equal to all points within a distance of r from $\mathbf{x}$. Consequently, for any metric space, whether familiar like $\mathbb{C}$ or an exotic construction, we can likewise define open ball, and open and closed sets, in terms of the distance function—or metric—that defines the space.

Definition 4.10   *Given $\mathbf{x} \in X$, where (X, d) is a metric space, a **neighborhood of** $\mathbf{x}$ or **open ball about** $\mathbf{x}$ **of radius** r, denoted $B_r(\mathbf{x})$ is defined as*

$$B_r(\mathbf{x}) = \{\mathbf{y} \,|\, d(\mathbf{x}, \mathbf{y}) < \mathbf{r}\}. \tag{4.3}$$

*A subset $G \subset X$ is **open**, and sometimes d-open, if given $\mathbf{x} \in G$, there is an $r > 0$ so that $B_r(\mathbf{x}) \subset G$. A subset $F \subset \mathbb{R}$ is **closed** if the complement of F, $\tilde{F}$, is open.*

For example, let $X = \mathbb{C}$, the complex numbers under the metric defined by the norm in (2.2), and let $B_r(\mathbf{x})$ be defined as in (4.3). Then, if $x = a + bi$ and $y = c + di$, we have $y \in B_r(\mathbf{x})$ iff $|x - y| < r$. That is, by (2.2),

$$[(a - c)^2 + (b - d)^2]^{1/2} < r.$$

Note that under the identification $\mathbb{C} \leftrightarrow \mathbb{R}^2$, $a + b\iota \leftrightarrow (a, b)$, we can define $\mathbf{y} \in B_r(\mathbf{x})$ on $\mathbb{C}$ iff $\mathbf{y} \in B_r(\mathbf{x})$ defined on $\mathbb{R}^2$ under this identification. That is, the identification $\mathbb{C} \leftrightarrow \mathbb{R}^2$ preserves the metrics defined on these respective spaces, as well as the notions of open and closed.

We note that in the general context of a metric space, as was demonstrated for $\mathbb{R}$, $\mathbb{C}$, and $\mathbb{R}^n$, the concept of an open set is not as metric-dependent as it first appears.

Proposition 4.11   *Let X be a metric space under two equivalent metrics, d_1 and d_2. Then a set $G \subset X$ is open in (X, d_1) iff G is open in (X, d_2).*

Proof   The proof, based on (3.35), is identical to that above in $\mathbb{R}^n$.   ∎

*4.2.4   Open and Closed Subsets in General Spaces

In a more general space without a metric, one can specify the open sets of X by defining a so-called **topology** on X as follows:

Definition 4.12   *Given a space X, a **topology** is a collection of subsets of X, $\Im$, which are the **open sets**, with the following properties:*

1. $\emptyset, X \in \Im$,

2. *If $\{G_\alpha\} \subset \Im$, then $\bigcup G_\alpha \in \Im$,*

3. *If $\{G_n\} \subset \Im$, a finite collection, then $\bigcap G_n \in \Im$.*

Hence a topology identifies the collection of open sets and demands that this collection behaves the same way under union and intersection as we have shown open sets to behave in the familiar settings of $\mathbb{R}$, $\mathbb{C}$, $\mathbb{R}^n$ or a general metric space X. In particular, in any of these special spaces, if we define $\Im$ as the collection of open sets under the definition of open as a metric space, then $\Im$ is a topology by the above definition. Such a topology is said to be **induced by the metric d**.

Closed sets are then defined by

$$F \subset X \quad \text{is closed iff} \quad \tilde{F} \in \Im,$$

and we see that this collection of closed sets again behaves in a familiar way under unions and intersections, based on De Morgan's laws.

Equivalent topologies can then be defined as follows:

Definition 4.13   *Two topologies $\Im_1$ and $\Im_2$ on a space X are equivalent if for any $G_1 \subset \Im_1$, there is a $G_2 \subset \Im_2$ with $G_2 \subset G_1$, and conversely, for any $G_2 \subset \Im_2$, there is a $G_1 \subset \Im_1$ with $G_1 \subset G_2$.*

Not surprisingly, especially given the terminology, we have immediately from the above proposition in a general metric space:

Corollary 4.14   *Let X be a metric space under two equivalent metrics, d_1 and d_2. Then the topologies induced by d_1 and d_2 are equivalent.*

Remark 4.15   *This corollary provides the motivation for the use of the language as noted in chapter 3, that d_1 and d_2 are "topologically equivalent," as an alternative to the terminology, d_1 and d_2 are "equivalent." The point is, such metrics provide the equivalent topologies on the space.*

Finally, we note that if a space X has a topology, $\Im$, and $Y \subset X$ is a subset, then there is a natural topology on Y called the **relative topology** or **induced topology**, denoted $\Im_Y$, which is defined as

$$\Im_Y = \{Y \cap G \,|\, G \in \Im\}.$$

For example, if we consider $\mathbb{R}$ as a topological space with open sets defined by the standard metric, and $Y = [0, 1]$, then the induced topology on Y contains sets of the form $[0, b)$, (a, b), $(b, 1]$, where $0 < a < b < 1$, as well as $[0, 1]$.

4.2.5   Other Properties of Subsets of a Metric Space

In the preceding sections it was clear that the notions of open and closed could be defined in any metric space using nearly identical definitions, the only difference related to the particular space's notion of distance as given by that space's metric. In this section, rather than repeat the same development for other important properties of sets from an initial definition in $\mathbb{R}$, to one in $\mathbb{R}^n$, to a general metric space X, we introduce the definitions directly in a general metric space, and leave it to the reader to reformulate these definitions in the other special cases.

Many of these notions also have meaning in a general topological space, but we will not have need for this development.

Definition 4.16   *In a metric space X with metric d:*

1. *If $\mathbf{x} \in X$, the **closed ball about** $\mathbf{x}$ **of radius** $r > 0$ is defined by*

$$\bar{B}_r(\mathbf{x}) = \{\mathbf{y} \,|\, d(\mathbf{x}, \mathbf{y}) \leq \mathbf{r}\}. \tag{4.4}$$

2. *If $E \subset X$, then $\mathbf{x} \in X$ is a **limit point of** E, a **cluster point of** E, or an **accumulation point of** E, if for any $r > 0$, $B_r(\mathbf{x}) \cap E \neq \emptyset$. So every $\mathbf{x} \in E$ is a limit point, but if there is an $r > 0$ with $B_r(\mathbf{x}) \cap E \equiv \mathbf{x}$, the point $\mathbf{x}$ is also said to be an **isolated point** of E. We denote by $\bar{E}$ the **set of limit points** of E, or the **closure of** E, and note that $E \subset \bar{E}$.*

3. $E \subset X$ is **dense in** X if every $\mathbf{x} \in X$ is a limit point of E.

4. $E \subset X$ is **bounded** if for any $\mathbf{x} \in X$, there is a number $r = r(\mathbf{x})$ so that $E \subset B_r(\mathbf{x})$, and is **unbounded** otherwise. In the special case of $X = \mathbb{R}$, one also has the notion of **bounded from above** and **bounded from below**. In the former case, there exists x^{max} so that $x < x^{max}$ for all $x \in E$, whereas in the latter case, there exists x^{min} so that $x > x^{min}$ for all $x \in E$.

5. Given $E \subset X$, a collection of open sets, $\{G_\alpha\}$, is an **open cover of** E if $E \subset \bigcup_\alpha G_\alpha$.

6. $E \subset X$ is **compact** if given any open cover of E, $\{G_\alpha\}$, which may be uncountably infinite, there is a finite subcollection, $\{G_j\}_{j=1}^m$ so that

$$E \subset \bigcup_{j \leq m} G_j.$$

7. $E \subset X$ is **connected** if given any two open sets, G_1 and G_2, with $E \subset G_1 \cup G_2$, we have $G_1 \cap G_2 \neq \emptyset$. $E \subset X$ is **disconnected** if there exists open sets, G_1 and G_2, with $E \subset G_1 \cup G_2$, and $G_1 \cap G_2 = \emptyset$.

Several of the important properties related to these notions are summarized in the following proposition, stated in the general metric space context. However, on first reading the intuition may be more easily developed if one envisions $\mathbb{R}$ as the given metric space, rather than X.

Proposition 4.17   *Let X be a metric space, then:*

1. *If $E \subset X$ is closed and $\mathbf{x}$ is a limit point of E, then $\mathbf{x} \in E$, and hence $\bar{E} = E$. Conversely, if $\bar{E} = E$, then E is closed.*

2. *If $\mathbf{x} \in X$ is a limit point of $E \subset X$ that is not an isolated point, then for any $r > 0$ there is a countable collection $\{\mathbf{x}_n\} \subset B_r(\mathbf{x}) \cap E$, with $\mathbf{x}_n \neq \mathbf{x}$.*

3. *If $E \subset X$ is compact, then E is closed and bounded.*

4. *(**Heine–Borel theorem**) $E \subset \mathbb{R}^n$ is compact iff E is closed and bounded.*

5. *If $\{\mathbf{x}_\alpha\} \subset E$ is a countable or uncountable infinite set, and E is compact, then $\{\mathbf{x}_\alpha\}$ has a limit point $\mathbf{x} \in E$.*

Proof   We prove each statement in turn:

1. If $E \subset X$ is closed, and $\mathbf{x} \notin E$, then $\mathbf{x} \in \tilde{E}$, which is open, and hence by definition, there is an $r > 0$ so that $B_r(\mathbf{x}) \subset \tilde{E}$. So it must be the case that $B_r(\mathbf{x}) \cap E = \emptyset$, and therefore $\mathbf{x}$ cannot be a limit point of E. Hence, if $\mathbf{x}$ is a limit point of E, we must have $\mathbf{x} \in E$ and so $\bar{E} = E$. Conversely, if $\bar{E} = E$, and $\mathbf{x} \in \tilde{E} = \tilde{\bar{E}}$, then since $\mathbf{x}$ is not a limit point, there is an $r > 0$ so that $B_r(\mathbf{x}) \cap E = \emptyset$. That is, $\tilde{E}$ is open and hence E is closed.

2. Choose a sequence $r_n \to 0$. Then by assumption that $\mathbf{x} \in X$ is a limit point of E that is not isolated, $B_{r_n}(\mathbf{x}) \cap E \neq \emptyset$ for all n, and each such intersection contains at least one point other than $\mathbf{x}$. Choose $\mathbf{x}_n \in B_{r_n}(\mathbf{x}) \cap E$ with $\mathbf{x}_n \neq \mathbf{x}$. Then $\{\mathbf{x}_n\}$ must be countably infinite, since for any n, there is $r_N < \min_{j \le n} d(\mathbf{x}, \mathbf{x}_j)$, and hence $\mathbf{x}_N$ must be distinct from $\{\mathbf{x}_j\}_{j=1}^n$.

3. If $E \subset X$ is compact, it is bounded, since we can define an open cover of E by $\{B_1(\mathbf{x}) \mid \mathbf{x} \in E\}$. Then by compactness, there is a finite collection $\{B_1(\mathbf{x}_j) \mid j = 1, \dots, n\}$. Let $D = \max d(\mathbf{x}_j, \mathbf{x}_k)$. Next, given any $\mathbf{x} \in \mathbf{X}$, if $\mathbf{y} \in E$, then $\mathbf{y} \in B_1(\mathbf{x}_k)$ for some k, and we can derive from the triangle inequality that

$$d(\mathbf{x}, \mathbf{y}) \le d(\mathbf{x}, \mathbf{x}_1) + d(\mathbf{x}_1, \mathbf{x}_k) + d(\mathbf{x}_k, \mathbf{y})$$

$$\le d(\mathbf{x}, \mathbf{x}_1) + D + 1,$$

and hence $E \subset B_R(\mathbf{x})$ for $R = d(\mathbf{x}, \mathbf{x}_1) + D + 1$ and E is bounded. To show that E is closed, we demonstrate that $\tilde{E}$ is open. To this end, let $\mathbf{x} \in \tilde{E}$. Then for any $\mathbf{y} \in E$, let $e(\mathbf{y}) = d(\mathbf{x}, \mathbf{y})/2$ and construct $B_{e(y)}(\mathbf{y})$. Clearly, by construction, $\{B_{e(y)}(\mathbf{y})\}$ is an open cover of E. Since E is compact, let $\{B_{e(y_n)}(\mathbf{y}_n)\}$ be the finite subcollection, which is again a cover of E, and define $e = \frac{1}{2} \min e(\mathbf{y}_n)$. By construction, $B_e(\mathbf{x}) \cap (\bigcup B_{e(y)}(\mathbf{y}_n)) = \emptyset$. So since $E \subset \bigcup B_{e(y)}(\mathbf{y}_n)$, we get that $B_e(\mathbf{x}) \subset \tilde{E}$, and hence $\tilde{E}$ is open and E closed.

4. From step 3 we only have to prove the "only if" part, that in $\mathbb{R}^n$, closed and bounded implies compact. To that end, assume that $E \subset \mathbb{R}^n$ is closed and bounded. Since it is bounded, we have that for some $R > 0$, $E \subset B_R(\mathbf{0})$. Also $B_R(\mathbf{0}) \subset \bar{\mathbf{C}}_R(\mathbf{0})$, the **closed cube about 0 of diameter 2R** defined by

$$\bar{\mathbf{C}}_R(\mathbf{0}) = \{\mathbf{x} \mid -R \le x_j \le R, \text{ all } j\}. \tag{4.5}$$

We will prove below that the closed cube, $\bar{C}_R(\mathbf{0})$, is compact for any R, and this will prove that E is compact as follows. Given any open cover of E, it can be augmented to become an open cover of $\bar{C}_R(\mathbf{0})$ by addition of the open set $C_{R+1}(\mathbf{0}) \sim E$. Here $C_{R+1}(\mathbf{0})$ is the open cube defined as in (4.5) but with strict inequalities, and since E is closed, $C_{R+1}(\mathbf{0}) \sim E = C_{R+1}(\mathbf{0}) \cap \tilde{E}$ is open. Now once $\bar{C}_R(\mathbf{0})$ is proved to be compact, this cover will have a finite subcover that then covers E without the added set $C_{R+1}(\mathbf{0}) \sim E$, and hence E is compact. We now prove that $\bar{C}_R(\mathbf{0})$ is compact by contradiction—assuming that $\bar{C}_R(\mathbf{0})$ is not compact. Then there is an open cover $\{G_j\}$ for which no finite subcover exists. Subdivide $\bar{C}_R(\mathbf{0})$ into 2^n closed cubes by halving each axis,

$$\bar{C}_R(\mathbf{0}) = \bigcup_{j=1}^{2^n} \bar{C}_j,$$

where each $\bar{C}_j$ is defined by one of the 2^n combinations of positive and negative coordinates:

$$\bar{C}_j = \{\mathbf{x} \,|\, \text{for each } i, 0 \le x_i \le R \text{ or } -R \le x_i \le 0\}.$$

Then at least one of these $\bar{C}_j$ has no finite subcover from $\{G_j\}$, for if all did, then $\bar{C}_R(\mathbf{0})$ would have a finite cover and hence be compact. Choose this $\bar{C}_j$ and subdivide it into 2^n closed cubes,

$$\bar{C}_j = \bigcup_{k=1}^{2^n} \bar{C}_{jk},$$

by again halving each axis, and choose any one of these cubes that has no finite subcover. Continuing in this way, we have an infinite collection of closed cubes: $\bar{C}_R(\mathbf{0}) \supset \bar{C}_j \supset \bar{C}_{jk} \supset \bar{C}_{jkl} \supset \cdots$, none of which have a finite subcover from $\{G_j\}$. By construction, the intersection of all such cubes is a single point $\mathbf{x}$, but since $\mathbf{x} \in G_j$ for some j, and G_j is open, there is a $B_r(\mathbf{x}) \subset G_j$. Beyond a given point this ball must then contain all the subcubes in the sequence above, since at each step the sides of the cube are halved and decrease to 0. This contradicts that no subcube has a finite subcover, and hence all such cubes have a finite subcover and $\bar{C}_R(\mathbf{0})$ is compact.

5. Assume that $\{\mathbf{x}_\alpha\} \subset E$, and E is compact, but that $\{\mathbf{x}_\alpha\}$ has no limit point in E. Then for any α there is an open ball $B_{r_\alpha}(\mathbf{x}_\alpha)$ that contains no other point in the sequence than $\mathbf{x}_\alpha$. Indeed, if there was such an $\mathbf{x}_\alpha$ so that $B_r(\mathbf{x}_\alpha)$ always contained at least one other point for any $r \to 0$, then this $\mathbf{x}_\alpha$ would be a limit point of the sequence by definition. Now $\{B_{r_\alpha}(\mathbf{x}_\alpha)\}$ is an infinite collection of open sets, to which we can add the open set $A \equiv X \sim [\bigcup B_{r_\alpha/2}(\mathbf{x}_\alpha)]$, which is open since the complement of A in X is the closed set $[\bigcup B_{r_\alpha/2}(\mathbf{x}_n)]$. We hence have an open cover of E with no finite subcover by construction, contradicting the compactness of E.   ∎

Note that in the proof of the Heine–Borel theorem there is a construction that can easily be generalized to demonstrate:

Corollary 4.18   *If X is a metric space, $E \subset X$ is compact and $F \subset E$ is closed, then F is also compact.*

Proof   If $\{G_j\}$ is an open cover of F, then $\{G_j\} \cup \tilde{F}$ is an open cover of E that has a finite subcover by compactness. This finite subcover, excluding the set $\tilde{F}$, is then a finite subcover of F.   ∎

Corollary 4.19 (*Heine–Borel Theorem*) $E \subset \mathbb{C}$ *is compact iff* E *is closed and bounded.*

Proof We have seen that the identification $\mathbb{C} \Leftrightarrow \mathbb{R}^2$ preserves the respective metrics in these spaces, and hence the closed and open balls defined in (4.4) and (4.3) are identical in both spaces. In $\mathbb{R}^2$ we have shown that the closed cube is compact, and by the corollary above, any closed ball within this cube is also compact. Consequently every closed ball in $\mathbb{C}$ is also compact and the above proof can be streamlined. If a closed and bounded $E \subset \mathbb{C}$ had an open cover with no finite subcover, then this cover could be augmented with the open set $B_{R+1}(\mathbf{0}) \sim E = B_{R+1}(\mathbf{0}) \cap \tilde{E}$; here, as above, we assume that $E \subset B_R(\mathbf{0})$. We have now constructed an open cover of $\bar{B}_R(\mathbf{0})$ with no finite subcover, contradicting the compactness of this closed ball.

∎

The Heine–Borel theorem is named after **Eduard Heine** (1821–1881) and **Émile Borel** (1871–1956). Borel formalized the earlier work of Heine in an 1895 publication that applied to the notion of compactness, which was then defined in terms of countably infinite open covers. Specifically, compact meant that every countable open cover had a finite subcover. This in turn was generalized by **Henri Lebesgue** (1875–1941) in 1898 to the notion of compactness defined in terms of an arbitrary infinite open cover, and this is the definition now used.

Remark 4.20 *The reader reviewing the propositions above may notice a glaring omission. On the one hand, in every metric space a compact set is closed and bounded. On the other hand, the subject of the Heine–Borel theorem, that closed and bounded implies compact, is only stated as true in $\mathbb{R}^n$ and $\mathbb{C}$. While Heine–Borel is also true in $\mathbb{C}^n$, we do not prove this as we have no need for this result. But it is only natural to wonder if Heine–Borel can be extended to all metric spaces. The answer is no, although the development of such an example will take us too far afield to be justified given that we will not make use of this in what follows.*

4.3   Applications to Finance

4.3.1   Set Theory

In general, knowledge of the axiomatic structure of set theory, or even the need for an axiomatic structure, is not directly applicable to finance except as a cautionary tale, as was discussed in chapter 1. While one's intuition can be a valuable facilitator in the development of an idea, or the pursuit of a solution to a problem, it is rarely adequate in and of itself even when the topic at hand appears elementary, and caution seems unwarranted. The ideal approach to problems in finance is where the development is mathematically formal but enlightened with intuition.

In finance as in all mathematical applications, one sometimes has a compelling intuitive argument as to how a problem ought to be solved, and then perhaps struggles to make this intuition precise. On the other hand, one sometimes discovers (or stumbles upon) a formal mathematical relationship and then struggles with an intuitive understanding. Both approaches are common, and both are valuable. The key is that until one has both, mathematical rigor and intuition, one hasn't really solved the problem. That is, a true "solution" requires a quantitative derivation of the solution to the problem as well as an intuitive understanding of why this solution works.

Of course, the tools of set theory are necessary and important simply because many problems in finance can be articulated in terms of sets, and so call for formal understanding and working knowledge of the set operations as well as their properties.

4.3.2   Constrained Optimization and Compactness

The constrained optimization problems discussed in chapter 3 on Euclidean spaces can be posed in terms of sets. For example, consider the constrained maximization problem in $\mathbb{R}^n$:

$$\max g(\mathbf{x}), \quad \text{given that} \quad f(\mathbf{x}) = c.$$

Now define the sets

$$A = \{\mathbf{x} \in \mathbb{R}^n \mid f(\mathbf{x}) = c\},$$

$$B = \{g(\mathbf{x}) \mid \mathbf{x} \in A\}.$$

Then $A \subset \mathbb{R}^n$ is clearly the constraint set, and $B \subset \mathbb{R}$ is the set of values the objective function takes on this constraint set. For example, A might denote the portfolio allocations that provide a given level of "risk" appropriately defined, and B then evaluates the average or return "expected" from these allocations.

Now, if B is unbounded from above, then the constrained optimization obviously has no solution. Hence, within this framework, solvability is seen to depend at the minimum on conditions on A and $g(\mathbf{x})$, which assure that B is bounded from above. Of course, if we seek a minimum, we need B to be bounded from below.

However, while boundedness is necessary, it is not sufficient. If B is an open subset of $\mathbb{R}$, it will not contain its minimum or maximum points. This comes from the definition of open, which is to say, if $x \in E$ an open set, then there is an $r > 0$ so that

$B_r(x) \subset E$, and no x can be a maximum or a minimum. Hence within this framework, solvability is also seen to depend at the minimum on conditions on A and $g(\mathbf{x})$, which assure that B is bounded and closed—which is to say, by the Heine–Borel theorem, that B is compact. In that case, if $x^{opt} \in B$ is the optimized value, either the maximum or the minimum, then by definition there is an $\mathbf{x}^{opt} \in A$ so that $g(\mathbf{x}^{opt}) = x^{opt}$. Hence, if B is compact, there is in theory a solution to the constrained optimization problem. Uniqueness will then depend on conditions on $g(\mathbf{x})$.

Logically, the condition(s) on A and $g(\mathbf{x})$ that assure compactness of B are in fact conditions on the constraint function $f(\mathbf{x})$ and the objective function $g(\mathbf{x})$. More generally, the constraint set A may be defined as

$$A = \{\mathbf{x} \,|\, f(\mathbf{x}) \in C\},$$

where C is a given constraint set, $C \subset \mathbb{R}$. Alternatively, A may be defined in terms of multiple constraints, as the intersection of sets of the form $\{\mathbf{x} \,|\, f_i(\mathbf{x}) \in C_i\}$:

$$A = \{\mathbf{x} \,|\, f_i(\mathbf{x}) \in C_i \text{ for all } 1 \leq i \leq m\}.$$

So we see that in this general case the compactness of the objective function's range B reflects conditions on the functions f and g, as well as the constraint set C. Notationally, if f is one-to-one, we can express $A = f^{-1}(C)$ and $B = g(A)$, and hence

$$B = g(f^{-1}(C)).$$

So we seek conditions on C, f, and g that ensure that B is compact.

When f is not one-to-one, we seek conditions on g and A to ensure that $g(A)$ is compact, and in turn conditions on f and C to ensure that the needed conditions on A are satisfied.

To explore this, we need to study additional properties of functions that will provide answers to these and related questions. The first steps will be taken in chapter 9 on calculus I, which addresses differential calculus on $\mathbb{R}$, but this will not be enough for the question above despite the fact that both B and C are subsets of $\mathbb{R}$. The problem, of course, is that in going from C to B we need to "travel" through $A \subset \mathbb{R}^n$, so for a complete answer, multivariate calculus is required.

That said, there is still the issue of determining a solution. The analysis above would provide what in mathematics is known as a **qualitative theory and solution** to the constrained optimization problem. What is meant by "qualitative" is that the theory demonstrates that a solution exists and whether or not it is unique. There is then the question of developing a **quantitative theory and solution**. That is, either an explicit formula or procedure that provides the answer, or a numerical algorithm that will "converge" to the given solution after infinitely many iterations. In the latter case, since we only have finitely much time, our goal would be to perform enough iterations to assure accuracy to some level of tolerance.

This raises the question of "convergence" and rate of convergence, issues introduced in the next two chapters on sequences and series. This discussion will then be expanded in chapter 9, where the relationship between properties of sets and properties of functions on $\mathbb{R}$ and related questions will be addressed.

4.3.3 Yield of a Security

In chapter 2 a number of formulas were derived for the prices of various securities, expressed as functions of variables that define the security's cash flow characteristics as well as of the investors' required yields. Put another way, given the cash flow structure, price can be thought of as a function of yield. One application of these formulas is to determine the price investors are willing to pay, given their yield requirements. Oftentimes in the financial markets, however, an investor faces a different question, and that is, given the market price of a security, what is the implied investor yield.

Such questions can arise in terms of a **bid price**, the price that a dealer is willing to pay on a purchase from an investor, or an **offer** (or **ask**) **price**, the price that a dealer requires on a sale to an investor. In both cases the investor is interested in the yield implications of the trade.

The offer price is always more, of course, and hence the offer yield is less than the associated bid yield. It is common to be interested in the so-called **bid–ask spread**, or **bid–offer spread**, defined as the difference between the higher bid yield and the lower ask–offer yield. This yield differential provides information to the investor on the **liquidity** of the security. A small bid–ask spread is usually associated with high liquidity, and increasingly larger spreads are associated with increasingly less liquidity.

In this context the notion of liquidity implies the commonly understood meaning as a measure of "ease of sale," since the dealer can encourage or discourage investor sales by favorable or unfavorable pricing. Narrow spreads are associated with deep markets of actively traded securities, and wide spreads with thin or narrow markets. In effect, a wide spread is compensation to the dealer for the expected delayed offsetting transaction, and the risks or hedge costs incurred during the intervening period.

But more important, liquidity is a measure of the fairness of the transaction's price. A small spread implies pricing is fair, since dealers are willing to transact either way at similar prices, whereas a wide spread implies that an investor sale may be well below a fair price, and purchase well above a fair price. Of course, fairness is like beauty; it is in the eye of the investor. Nonetheless, all market participants agree that the size of the spread tells a lot about both the ease of transacting and the fairness of pricing.

If $P(i)$ denotes the pricing function for a given security, and P_0 the price quoted, the security's **implied yield**, or in the case of a fixed income security, **implied yield to maturity**, is the solution i_0 to the equation

$$P(i) = P_0. \tag{4.6}$$

In this section we informally introduce the method of **interval bisection** in solving (4.6) for i_0, and return to this methodology with greater formality in later chapters.

First off, one can do a qualitative analysis of this equation to determine if a solution is feasible. In virtually all markets one expects that all yields on securities are greater than 0%, and less than 100%, so a very simple qualitative assessment for the existence of a solution is that

$$P(1.0) \leq P_0 \leq P(0),$$

where $i = 0$ and 1.0 mean that the respective discount factors in the pricing formula, $v = (1 + i)^{-1}$, are 1 and $\frac{1}{2}$. From this assessment we can posit that $i_0 \in [0, 1] \equiv F_0$. In practice, this first step could well produce a much smaller initial solution interval, such as $[0.05, 0.1]$, but for notational simplicity, we ignore this refinement.

Next we could evaluate $P(0.5)$, or in general, the price function at the midpoint of the initial interval. We then have either

$$P(1.0) \leq P_0 \leq P(0.5) \quad \text{or} \quad P(0.5) \leq P_0 \leq P(0).$$

From this we conclude that either $i_0 \in [0.5, 1]$ or $i_0 \in [0, 0.5]$, respectively, and choose the appropriate interval and label it F_1. Of course, if $P_0 = P(0.5)$, we are done.

Continuing in this way, one of two things happens:

1. We develop a sequence of closed intervals F_n, with $i_0 \in F_n$ for all n, and lengths $|F_n| = \frac{1}{2^n}$.

2. Or the process serendipitiously stops in a finite number of steps, since i_0 is an endpoint of one of the F_n.

Assuming the process does not stop, we can identify "approximate" solutions to the equation in (4.6) by simply choosing the midpoints of the respective intervals. Specifically, defining i_n as the midpoint of F_n, it must be the case, since $i_0 \in F_n$ and $|F_n| = \frac{1}{2^n}$, that

$$|i_n - i_0| < \frac{1}{2^{n+1}}.$$

Also, since $\{F_n\}$ are closed sets, and the length $|F_n|$ decreases to 0, then $\bigcap F_n$ is closed and hence must be a single point. That is, it must be the case that

$$i_0 = \bigcap F_n.$$

Or does it?

If $F_n = [a_n, b_n]$, all we know is that $P(b_n) \leq P_0 \leq P(a_n)$ for all n, and that $b_n - a_n = \frac{1}{2^n}$. But how do we know that there is a unique value in this interval, which we denote i_0, on which P_0 is achieved by $P(i)$? Let's summarize the assumptions made to draw this desired conclusion:

1. We implicitly assumed that the price function was **decreasing**, or more generally, that it is **monotonic** (increasing or decreasing), since at each step we assumed, in the notation above, that the value at the midpoint was "between" the endpoint values:

$$P(b_n) \leq P(i_n) \leq P(a_n).$$

Then we could choose one or the other of the subintervals $[a_n, i_n]$ or $[i_n, b_n]$ in the next step. We know, or at least intuit, that this is true for most pricing functions in finance, and this can be demonstrated with the tools of chapter 9. But in a more general application, $P(i_n)$ could exceed either endpoint value, or be less than either. In such a case there could be more than one solution, and we would have to choose which interval(s) we search to find them.

2. We implicitly assumed that the price function varies in a smooth and predictable way, a property that will be called **continuity** in chapter 9. Specifically, we know from the values of $|F_n|$ that the intersection of these closed sets will produce a unique point, call it i_0. We also know that by construction, $P(b_n) \leq P_0 \leq P(a_n)$, and given the assumption of monotonicity, that $P(b_n) \leq P(i_n) \leq P(a_n)$, where i_n is the midpoint of F_n. But to conclude from $i_n \to i_0$ that $P(i_n) \to P(i_0)$ requires an assumption on the **continuity** of the price function $P(i)$, which thankfully, is true for pricing functions as will also be seen.

We will return to the analysis of interval bisection in chapter 5.

Exercises

Practice Exercises

1. Russell's paradox is equivalently formulated as the Barber Paradox: *In a town there is a barber who shaves all the men that do not shave themselves, and only those men.* Define "set" A as the set of all men in the town that the barber shaves.

(a) Is the barber a member of this set? Show that the barber's set membership cannot be determined as we would conclude the paradox that the barber shaves himself if and only if he does not shave himself.

(b) Note that the paradox works because at the time, the barber was assumed to be male. Show that if the barber were female, we could conclude that the barber is not a member of this set, whether she shaves or not.

2. Prove De Morgan's laws 2 to 4, using the operational definitions.

3. Demonstrate the following, using operational definitions:

(a) If $G_n \equiv \left(-\frac{1}{n}, 1+\frac{1}{n}\right)$, then $\bigcap G_n = [0,1]$, so the intersection of open sets can be closed.

(b) If $F_n \equiv \left[\frac{1}{n}, 1-\frac{1}{n}\right]$, then $\bigcup F_n = (0,1)$, so the union of closed sets can be open.

4. Show that if a set A contains n-elements, that the **power set** of A, defined as the set of all subsets of A, contains 2^n elements. (*Hint:* Label the elements of A as $x_1, x_2, \ldots, x_n$, and define a **chooser function** on the power set that produces decimal expansions as follows:

$$f(B) = 0.a_1 a_2 a_3 a_4 \ldots a_n,$$

where

$$a_j = \begin{cases} 0, & x_j \notin B, \\ 1, & x_j \in B. \end{cases}$$

Show that $f(B) = f(B')$ iff $B = B'$, and that the range of f has 2^n elements.)

5. Generalize exercise 4 to the case where A contains a countably infinite number of elements, and show that with an abuse of notation, there are 2^∞ elements in the power set, where this symbol denotes the uncountable infinity of real numbers in the interval $[0,1]$. (*Hint:* Use the construction in exercise 4, and recall the binary expansions of reals $x \in [0,1]$.)

6. Demonstrate that the points in the Cantor set can be identified with base-3 expansions, $0.a_1 a_2 a_3 a_4 \ldots$, where $a_j = 0,2$. (*Hint:* Show that points in F_1 all begin as $0.0a_2 a_3 a_4 \ldots$ or $0.2a_2 a_3 a_4 \ldots$, then show that points in F_2 are of the form: $0.a_1 a_2 a_3 a_4 \ldots$, where $a_1 a_2 = 00, 02, 20, 22$, etc.)

7. Develop another proof that the Cantor set has measure 0, using the fact that if we denote by $|F_n|$ the total length of the intervals in F_n, then by the construction, $|F_{n+1}| = \frac{2}{3}|F_n|$. (*Hint:* $K = \bigcap_{j=0}^{\infty} F_j$, but $\bigcap_{j=0}^{n} F_j = F_n$.)

8. Show that the following sets have measure 0 by constructing a covering with intervals that have arbitrarily small total length. (*Hint:* Recall that $\sum_{n=1}^{\infty} \frac{1}{2^n} = 1$.)

(a) The integers $\mathbb{Z}$

(b) $\{\frac{1}{n} \mid n = 1, 2, 3, \ldots\}$

(c) The rationals $\mathbb{Q} \cap (0, 1)$

9. For $1 \le p \le \infty$, define a set, $G \subset \mathbb{R}^2$ to be "l_p-open" if for any $x \in G$ there is an $r > 0$ so that $B_r^{(p)}(\mathbf{x}) \subset \mathbf{G}$ where $B_r^{(p)}(\mathbf{x}) = \{\mathbf{y} \mid \|\mathbf{x} - \mathbf{y}\|_p < \mathbf{r}\}$, and where $\|\mathbf{x} - \mathbf{y}\|_p$ denotes the l_p-norm. The usual definition of open is then l_2-open.

(a) Show that G is open if and only if it is l_1-open. (*Hint*: Recall the graphs of equivalent metrics in chapter 3.)

(b) Generalize part (a) to show that G is open if and only if it is l_p-open for all p.

10. Define a set $G \subset \mathbb{R}^n$ to be open if for any $x \in G$, there is an $r > 0$ so that $B_r^{(d)}(\mathbf{x}) \subset \mathbf{G}$, where $B_r^{(d)}(\mathbf{x}) = \{\mathbf{y} \mid d(\mathbf{x}, \mathbf{y}) < \mathbf{r}\}$.

(a) Exercise 18 of chapter 3 introduced a metric on $\mathbb{R}^n$ that was not equivalent to the l_p-metrics. Specifically,

$$d(\mathbf{x}, \mathbf{y}) = \begin{cases} 0, & \mathbf{x} = \mathbf{y}, \\ 1, & \mathbf{x} \ne \mathbf{y}. \end{cases}$$

Determine all the open sets in $\mathbb{R}^n$.

(b) Define:

$$d(\mathbf{x}, \mathbf{y}) = \{0, \quad \text{all } \mathbf{x}, \mathbf{y}.$$

Prove that there is only one open set, and determine what it is.

11. The Heine–Borel theorem assures that a set is compact in $\mathbb{R}^n$ if and only if it is closed and bounded. Explain how to choose the finite subcovers of the following open covers of the given sets:

(a) $F = [0, 1] \subset \bigcup B_r(x_j)$, where $\{x_j\}$ is an arbitrary enumeration of the rational numbers in the interval and $r > 0$ is an arbitrary constant.

(b) $F = [0, 1] \subset \bigcup B_{r_j}(x_j)$, where $\{x_j\}$ is an arbitrary enumeration of the rational numbers in the interval, and $r_j > 0$ are arbitrary values. If $r_j > r > 0$, this can be solved as in part (a), so assume that 0 is an accumulation point of $\{r_j\}$.

(c) $F = \bar{C}_R(\mathbf{0}) \subset \bigcup C_r(\mathbf{x}_j)$ in $\mathbb{R}^2$, where $\bar{C}_R$ denotes the closed 2-cube, or square, about $\mathbf{0}$ of diameter $2R$, and $C_r(\mathbf{x}_j)$ denotes the open cubes about points $\mathbf{x}_j$, with rational coordinates, of fixed diameter $2r > 0$.

12. Show that the interval $(0, 1)$ is not compact by constructing an infinite open cover for which there is no finite subcover. (*Hint*: Construct an open cover sequentially, with $I_j \subset I_{j+1}$.)

13. Use the method of interval bisection to determine the yields of the following securities to four decimals (i.e., to basis points). Solve each in the appropriate nominal rate basis:

(a) A 10 year bond with 5% semiannual coupons, with a price of 98.75 per 100 par.

(b) An annual dividend common stock, last dividend of $10 paid yesterday and assumed to grow at 8% annually, selling for $115.00.

(c) A 5-year monthly pay commercial mortgage, with loan amount $5 million and amortization schedule developed with a monthly rate of 6%, selling in the secondary market for $5.2 million.

Assignment Exercises

14. Simplify the following expressions by applying De Morgan's laws, and then demonstrate that the expression derived is correct using the operational definitions.

(a) $(A \cap \tilde{B})^c \cup C$

(b) $(B \cap [\bigcup_\alpha A_\alpha])^c$

(c) $(\bigcup_\alpha A_\alpha)^c \cup (\bigcap_\beta \tilde{B}_\beta)^c$

Recall that $(C)^c$ denotes $\tilde{C}$.

15. Generalize exercise 3:

(a) Provide an example of a countably infinite collection of open $G_n \subset \mathbb{R}$ so that $\bigcap G_n$ is open.

(b) Repeat part (a) so that $\bigcap G_n$ is neither open nor closed.

(c) Provide an example of a countably infinite collection of closed $F_n \subset \mathbb{R}$ so that $\bigcup F_n$ is closed.

(d) Repeat part (c) so that $\bigcup F_n$ is neither open nor closed.

16. Develop examples in $\mathbb{R}^2$ of the results illustrated in:

(a) Exercise 3

(b) Exercise 15

Can your constructions in parts (a) and (b) be applied in $\mathbb{R}^n$?

17. Generalize exercise 5 and show that if A is a set of any "cardinality," the power set of A has greater cardinality; that is to say, its elements cannot be put into one-to-one correspondence with the elements of A. (*Hint*: Assume there is such a correspondence, and define $f(a)$ as the $1:1$ function that connects A and its power set. In other words, $f(a) = A_a$, the unique subset of A associated with a. Consider the set $A' = \{a \mid a \notin A_a\}$. Then there is an $a' \in A$ so that this collection is produced by $f(a')$; that is, $A' = A_{a'}$. Show that $a' \in A_{a'}$ iff $a' \notin A_{a'}$.)

Remark 4.21 *In Cantor's theory of infinite cardinal numbers, where "cardinal" is intended as a generalization of the idea of the "number" of elements in a set, the symbol $\aleph_0$ and read "Aleph-naught," denotes the cardinality of the integers, or "countably infinite." Then $\aleph_1$ denotes the next greater cardinality, $\aleph_2$ the next, and so forth, and Cantor proved with the construction of this exercise that there is an infinite sequence of cardinal numbers so that no one-to-one correspondence could be produced between any two sets with different cardinalities. For example, we have already seen that a set of cardinality $\aleph_0$ cannot be put into one-to-one correspondence with the set of real numbers, so the cardinality of the reals must exceed $\aleph_0$. Now the cardinality of the power set of a set of cardinality $\aleph_0$ is the same as the cardinality of the collection of all functions from a set of cardinality $\aleph_0$, to the 2-element set, $\{0,1\}$. This follows from the construction in exercise 5, since every set in the power set implies a function that has value 1 on every element in this set, and value 0 on every element not in this set. The notation used for the cardinality of this class of functions is $2^{\aleph_0}$ and exercise 5 assures that $\aleph_0 < 2^{\aleph_0}$ and that $2^{\aleph_0} = c$, the uncountable infinity of the real numbers, also called the* **continuum**. *The power set of a set of cardinality $\aleph_1$ again has greater cardinality by exercise 17, equal to the $2^{\aleph_1}$, and so $\aleph_1 < 2^{\aleph_1}$. This process continues, in turn producing an infinite sequence of increasingly large infinite cardinals, since for all j, $\aleph_j < 2^{\aleph_j}$. The* **continuum hypothesis**, *which is a statement that has been proved to be independent of ZFC set theory (the 10 Zermelo–Fraenkel axioms with the axiom of choice), is that there is no cardinal strictly between $\aleph_0$ and $c = 2^{\aleph_0}$, and hence the next greater cardinal $\aleph_1$ is $2^{\aleph_0}$. In other words, $\aleph_1 = 2^{\aleph_0}$. The* **generalized continuum hypothesis** *states that there is no cardinal strictly between $\aleph_j$ and $2^{\aleph_j}$ for any j and so $\aleph_{j+1} = 2^{\aleph_j}$. It has been proved that this hypothesis is also independent of the ZFC set theory, and hence can neither be proved nor disproved in that theory. In other words, mathematicians have the option to add these hypotheses or their negative to the theory, and in each case derive a consistent theory of cardinals.*

18. Denote the Cantor set developed in this chapter by $K_{2/3}$ to signify that in each step, each closed interval from the prior step is divided equally into three-subintervals, and the second open subinterval is removed. Define a **generalized Cantor set**, denoted $K_{m/n}$, for n, m integers, $n \geq 3$, $m = 1, 2, \ldots, n$, analogously. That is, at each step, each closed interval of the form $\left[\frac{k}{n^j}, \frac{k+1}{n^j}\right]$ from the prior step is divided equally into n-subintervals, and the mth open subinterval removed.

(a) Defining $K_{m/n}$ as the intersection of all the sets produced in these steps, confirm that $K_{m/n}$ is closed.

(b) Show that $K_{m/n}$ has measure 0 using the approach of exercise 7. Note the complexity of proving this result by considering the sum of the lengths of the intervals removed.

(c) Show that $K_{m/n}$ is uncountable by identifying points in this set with base-n expansions, but without the digit $m - 1$. (*Hint*: Identify these expansions with base-$(n - 1)$ expansions of all real numbers in $[0, 1]$.)

19. Demonstrate that the exercise 18(c) construction does not work if $n = 2$.

(a) Show that $K_{m/2}$ is a closed set of measure 0.

(b) Prove that $K_{m/2}$ is countable and identify explicitly the elements of these two sets, where $m = 1$ or $m = 2$.

20. Generalizing exercise 8, show that the following sets in $\mathbb{R}^2$ have measure 0, which means that the set can be covered by a collection of balls with total area as small as we choose.

(a) The "integer lattice": $\{(n, m) \mid n, m \in \mathbb{Z}\}$

(b) $\left\{\left(\frac{1}{n}, \frac{1}{m}\right) \mid n, m \in \mathbb{Z}, n, m \neq 0\right\}$

(c) $\{(q, r) \mid q, r \in \mathbb{Q}\}$

21. Generalize exercise 9 to $\mathbb{R}^n$. (*Hint*: Recall (3.34).)

22. Show that the following sets are not compact by constructing an infinite open cover for which there is no finite subcover.

(a) $\{(x, y) \subset \mathbb{R}^2 \mid |x| + |y| < 1\}$

(b) $\{(x, y) \subset \mathbb{R}^2 \mid x^2 + y^2 < R\}$ for $R > 0$.

(c) $\{\mathbf{x} \subset \mathbb{R}^n \mid x_1 \neq 0\}$ where $\mathbf{x} = (x_1, x_2, \ldots, x_n)$ (*Hint*: Try $n = 2$ first.)

23. Prove that:

(a) $Q_1 \subset \mathbb{R}^n$ defined as $Q_1 = \{\mathbf{x} \subset \mathbb{R}^n \mid x_j \in \mathbb{Q} \text{ for all } j\}$ is dense for any n.

(b) For any $k \in \mathbb{N}$, the set $Q_k \subset \mathbb{R}^n$ defined as $Q_k = \{\mathbf{x} \subset \mathbb{R}^n \mid x_j^k \in \mathbb{Q} \text{ for all } j\}$ is dense for any n. (*Hint*: Show $Q_1 \subset Q_k$.)

24. Use the method of interval bisection to determine the yields of the following securities to four decimal places (i.e., to basis points). Solve each in the appropriate nominal rate basis:

(a) A 15-year bond with 3% semiannual coupons, with a price of 92.50 per 100 par.

(b) A semiannual dividend common stock, last dividend of $6 paid yesterday and assumed to grow at a 5% semiannual rate, selling for $66.00.

(c) A perpetual preferred stock with quarterly dividends at a quarterly dividend rate of 7%, priced at 105.25 per 100 par.

5 Sequences and Their Convergence

5.1 Numerical Sequences

5.1.1 Definition and Examples

The mathematical concept of a numerical sequence is deceptively simple, and yet its study provides a solid foundation for a great many deep and useful results as we will see in coming chapters.

Definition 5.1 *A **numerical sequence**, denoted $\{x_n\}$, $\{z_j\}$, and so forth, is a countably infinite collection of real or complex numbers for which a numerical ordering is specified:*

$$\{x_n\} \equiv x_1, x_2, x_3, \ldots.$$

*For specificity, the sequence may be called a **real sequence** or a **complex sequence**. A numerical sequence is said to be **bounded** if there is a number B so that $|x_n| \le B$ for all n. A **subsequence of a numerical sequence** is a countably infinite subcollection that preserves order. That is, $\{y_m\}$ **is a subsequence of** $\{x_n\}$ if*

$$y_m = x_{n_m} \quad and \quad n_{m+1} > n_m \quad for\ all\ m.$$

Remark 5.2 *In some applications a numerical sequence is indexed as $\{x_n\}_{n=0}^{\infty}$ rather than $\{x_n\}_{n=1}^{\infty}$.*

Note that the notion of a numerical sequence requires both a countable infinite collection of numbers as well as an ordering on this collection. For example, while the collection of rational numbers is, as we have seen, a countably infinite collection of real numbers, it is not a numerical sequence until an ordering has been imposed. One such ordering was introduced in section 2.1.4 on rational numbers to prove countability, although this ordering counted each rational infinitely many times. However, there are infinitely many other orderings, in fact uncountably many.

Order is particularly important because one is generally interested in whether or not the numerical sequence "converges" as $n \to \infty$. For example, even without a formal definition of convergence, it is intuitively clear that the following sequences behave as indicated:

Example 5.3

1. $y_m \equiv \frac{1}{m}$ *converges to 0 as $m \to \infty$.*
2. $x_n \equiv \frac{(-1)^n}{n}$ *converges to 0 as $n \to \infty$.*
3. $a_j \equiv \frac{j-1}{j}$ *converges to 1 as $j \to \infty$.*

4. $c_j \equiv (-1)^j \frac{j-1}{j}$ *does not converge as* $j \to \infty$.

5. $z_n \equiv \frac{2n-5}{4n+1000} + \frac{3n^3}{5n^3+6} \imath$ *converges to* $0.5 + 0.6\imath$ *as* $n \to \infty$.

6. $b_n \equiv \begin{cases} m, & n = 2m \\ -m, & n = 2m+1 \end{cases}$ *does not converge as* $n \to \infty$.

7. $w_k \equiv k$ *diverges to* ∞ *as* $k \to \infty$.

8. $u_j = -j^2$ *diverges to* $-\infty$ *as* $j \to \infty$.

On an intuitive level, cases 1 and 3 of example 5.3 not only converge, but **converge monotonically**, which is to say that both sequences get closer to their respective limits at each increment of the index. Case 2 also converges but not monotonically because of the alternating signs. Case 4 "almost" converges, in that "half" of the sequence is converging to a limit of $+1$, while the other half is converging to a limit of -1. Specifically, case 4 has two convergent subsequences:

$$\{y_n\} \equiv \{c_{2n}\} \to 1,$$

$$\{y_n'\} \equiv \{c_{2n-1}\} \to -1.$$

That case 5 converges is made more transparent by rewriting the rational functions, for example,

$$\frac{2n-5}{4n+1000} = \frac{2 - \frac{5}{n}}{4 + \frac{1000}{n}},$$

which converges to $\frac{1}{2}$. Cases 6, 7, and 8 all "explode" in a sense, but cases 7 and 8 seem to be reasonable candidates for a definition of converge to ∞, or converge to $-\infty$, for which we will use the language **diverge to** $\pm\infty$.

These examples provide a range of sample behaviors for numerical sequences. After formalizing the definition of convergence that will capture the intuition of all convergent examples, we will develop several properties of numerical sequences and see that the comment above on case 4 generalizes. That is, any bounded numerical sequence has at least one convergent subsequence.

5.1.2   Convergence of Sequences

The following definition of convergence of a numerical sequence is formal, and will be discussed below to provide additional intuition. But at this point, we note the key intuitive idea that this formality is attempting to capture. The notion of convergence $x_n \to x$ means more than just "as n increases, there are terms x_n that get arbitrarily close to x." This is a notion that is weaker than convergence and will be addressed below. The stronger property defined here is that "as n increases, **all** terms x_n get arbitrarily close to x." More precisely:

Definition 5.4   *A numerical sequence $\{x_n\}$ **converges to the limit** x as $n \to \infty$ if for any $\epsilon > 0$ there is an $N \equiv N(\epsilon)$ so that*

$$|x_n - x| < \epsilon \quad whenever \quad n \geq N. \tag{5.1}$$

In this case we write

$$\lim_{n \to \infty} x_n = x \quad or \quad x_n \to x.$$

*In (5.1) the notation $|x_n - x|$ is to be interpreted in terms of the standard norm in $\mathbb{R}$ and $\mathbb{C}$ given in (2.3) and (2.2), respectively. A real sequence $\{x_n\}$ **diverges to** ∞ as $n \to \infty$ if for any $M > 0$ there is $N \equiv N(M)$ so that*

$$x_n \geq M \quad whenever \quad n \geq N,$$

*and **diverges to** $-\infty$ as $n \to \infty$ if for any $M > 0$ there is $N \equiv N(M)$ so that*

$$x_n < -M \quad whenever \quad n \geq N.$$

In these cases we write, as appropriate,

$$\lim_{n \to \infty} x_n = \pm\infty \quad or \quad x_n \to \pm\infty,$$

*In all other cases we say that $\{x_n\}$ **diverges as** $n \to \infty$, or simply, **does not converge**.*

Definition 5.5   *A real sequence $\{x_n\}$ is **monotonic** if any of the following conditions are satisfied:*

*$x_n < x_{n+1}$ for all n: **strictly increasing***

*$x_n \leq x_{n+1}$ for all n: **increasing, or nondecreasing***

*$x_n > x_{n+1}$ for all n: **strictly decreasing***

*$x_n \geq x_{n+1}$ for all n: **decreasing, or nonincreasing***

*A real sequence $\{x_n\}$ **converges monotonically to the limit** x as $n \to \infty$ if $\{x_n\}$ is **monotonic and converges to the limit** x as $n \to \infty$.*

Note that while convergence of a complex sequence is easily defined with the same notation as that for a real sequence, as was noted in section 2.1.6 on complex numbers, there is no ordering of $\mathbb{C}$ as there is in $\mathbb{R}$, and hence one does not have the notion of a monotonic complex sequence or that of monotonic convergence. Note also that again with the exception of monotonicity, these definitions generalize without change to vector sequences $\mathbf{x}_n \in \mathbb{R}^n$, only where (2.3) is replaced by the standard norm in (3.3). Moreover this notion of convergence only depends on the norm up to equivalence. So, if $\mathbf{x}_n \to \mathbf{x}$ under the standard norm, it will also converge relative to the l_p-norms for $1 \le p \le \infty$, or any other equivalent norm. This more general notion will be discussed below.

Remark 5.6 *The concept in the definition above, that "for any $\epsilon > 0$ there is an $N \equiv N(\epsilon)$," can be a difficult one to grasp initially. But this theme is repeated time and again in the following chapters, so we pause a moment here to develop it a bit further. The difficulty some have is that the intuitive notion of a limit, that*

"x_n gets closer to x as n gets large"

seems simple enough. But the detail that needs to be addressed is:

- *Does convergence mean that we can find values of x_n that get arbitrarily close to x?*
- *Or does convergence mean that all values of x_n eventually get arbitrarily close to x?*

  *For some purposes, the former weaker definition may suffice, and this idea is essentially captured in the notion of **accumulation point** or **limit point** introduced in section 4.2.5. But for many applications we want the stronger definition of convergence in that not just some x_n get arbitrarily close to x as $n \to \infty$, but all x_n get arbitrarily close to x as $n \to \infty$. This is the reason to insist that $|x_n - x| < \epsilon$ for all $n \ge N$.*

The formal definition of convergence may seem to suggest that we can randomly generate any ϵ, and as long as there is an associated N with the needed property, we are done and have proved convergence. Actually the terminology "for any $\epsilon > 0$ there is an $N \equiv N(\epsilon)$" is not to be interpreted as if ϵ is arbitrarily selected by the mathematician. The idea is instead that the mathematician wants to be sure that there is a sequence of epsilons $\epsilon_j \to 0$, for example, $\epsilon_j = \frac{1}{j}$, so that for every term in that sequence, an associated $N_j \equiv N(\epsilon_j)$ can be found, resulting in $|x_n - x| < \epsilon_j$ whenever $n \ge N_j$. In other words, for any such ϵ_j there is an N_j so that all terms of the sequence from term x_{N_j} onward are closer to x than ϵ_j. Logically, as $\epsilon_j \to 0$, we expect to have that $N_j \to \infty$. That is, as one insists that sequence values be increasingly close to their limit, it may be necessary to exclude more and more of the sequence's initial terms. So a good intuitive model for the expression "for any $\epsilon > 0$ there is an $N \equiv N(\epsilon)$ so that..." is that "there is a sequence of epsilons, $\epsilon_j \to 0$, and associated $N_j \equiv N(\epsilon_j)$, so that...."

The payoff from this definition is that one immediately has error bounds

$$-\epsilon_j < x - x_n < \epsilon_j$$

as long as $n \geq N_j$, so any such x_n could be used as an approximation to x with the error bounded as noted.

Example 5.7   *Let's prove the convergence of cases 3 and 5 in example 5.3 above to the intuited limits of* 1 *and* $0.5 + 0.6i$. *First off, for case 3,*

$$|a_j - 1| = \frac{1}{j}.$$

Given $\epsilon > 0$, *to have* $|a_j - 1| < \epsilon$ *then requires that* $j > \frac{1}{\epsilon}$. *So N is chosen as any integer that exceeds this value. For case 5 of example 5.3, we use the triangle inequality, and recalling that* $|i| = 1$, *we write*

$$|z_n - (0.5 + 0.6i)| = \left| \frac{-555}{4n + 1000} - \frac{0.36}{5n^3 + 6}i \right|$$

$$\leq \frac{555}{4n + 1000} + \frac{0.36}{5n^3 + 6}$$

$$< \frac{556}{4n + 1000}.$$

This last inequality follows since $5n^3 + 6 > 4n + 1000$ *for* $n > 10$, *say, and this is good enough. Given* $\epsilon > 0$, *to have* $|z_n - (0.5 + 0.6i)| < \epsilon$ *requires that* $n > \frac{556 - 1000\epsilon}{4\epsilon}$. *So N is chosen to exceed this value.*

5.1.3   Properties of Limits

The first observation about the definition of convergence, which is not true for the weaker notion of accumulation point, is that if a numerical sequence converges, the limit must be unique.

Proposition 5.8   *If* $\lim_{n \to \infty} x_n = x$ *and* $\lim_{n \to \infty} x_n = x'$, *then* $x = x'$.

Proof   This result is obvious if $x = \pm\infty$: by definition, a sequence cannot have both a finite limit and diverge to $\pm\infty$, nor can it have both ∞ and $-\infty$ as limits. If x and x' are both finite, then for any $\epsilon > 0$, there is an $N \equiv N(\epsilon)$ so that $|x_n - x| < \epsilon$ and $|x_n - x'| < \epsilon$ for $n \geq N$. Actually the definition of limit assures the existence of N_1 and N_2, one for each limit, so we simply define $N = \max(N_1, N_2)$. By the triangle inequality,

$|x - x'| \leq |x - x_n| + |x_n - x'| < 2\epsilon.$

As this is true for any $\epsilon > 0$, we conclude that $x = x'$.   ∎

The next observation concerning convergence is that convergence implies boundedness.

Proposition 5.9   *Let $\{x_n\}$ be a convergent numerical sequence with $x_n \to x$; then $\{x_n\}$ is bounded.*

Proof   Fix any $\epsilon > 0$, for example, $\epsilon = 1$, and let N be the associated integer so that $|x_n - x| < 1$ whenever $n \geq N$. Then by the triangle inequality,

$|x_n| = |x_n - x + x| < 1 + |x|$ for $n \geq N$.

For $n < N$, $|x_n| \leq \max_{n \leq N}|x_n|$, which is also finite. So all $|x_n|$ are bounded by the larger of $1 + |x|$ and $\max_{n \leq N}|x_n|$.   ∎

Remark 5.10   *Note that case 4 of example 5.3 above shows that boundedness does not guarantee convergence.*

It is relatively easy to show that the notion of convergence is preserved under arithmetic operations:

Proposition 5.11   *Let $\{x_n\}$ and $\{y_n\}$ be convergent numerical sequences with $x_n \to x$, and $y_n \to y$, and let a be a real or complex number. Then:*

1. $ax_n \to ax.$

2. $x_n + y_n \to x + y.$

3. $x_n y_n \to xy.$

4. $\frac{1}{y_n} \to \frac{1}{y}$ *as long as $y \neq 0$, and $y_n \neq 0$ for all n.*

5. $\frac{x_n}{y_n} \to \frac{x}{y}$ *as long as $y \neq 0$, and $y_n \neq 0$ for all n.*

Proof   In each case we show that convergence is guaranteed by convergence of the original sequences:

1. $|ax_n - ax| = |a|\,|x_n - x|$ by either (2.3) or (2.2), so assuming $a \neq 0$, $|ax_n - ax| < \epsilon$ if $|x_n - x| < \frac{\epsilon}{|a|}$. If $a = 0$, there is nothing to prove.

2. $|(x_n + y_n) - (x + y)| \leq |x_n - x| + |y_n - y|$ by the triangle inequality in (2.7), so $|(x_n + y_n) - (x + y)| < \epsilon$ if each of the absolute values on the right-hand side are bounded by $\frac{\epsilon}{2}$.

3. Again, by the triangle inequality, $|x_n y_n - xy| \leq |x_n y_n - x_n y| + |x_n y - xy| = |x_n||y_n - y| + |y||x_n - x|$. So if $y \neq 0$, $|x_n y_n - xy| < \epsilon$ if $|y_n - y| < \frac{\epsilon}{2B}$, where B is an upper bound for $\{|x_n|\}$, and $|x_n - x| < \frac{\epsilon}{2|y|}$. If $y = 0$, the second term drops out.

4. $\left|\frac{1}{y_n} - \frac{1}{y}\right| = \left|\frac{y_n - y}{y y_n}\right|$. Now, since $y \neq 0$ and $y_n \neq 0$ for all n, we can take $\epsilon = 0.5|y|$. We know that by convergence $y_n \to y$, there is an N so that $|y_n - y| < 0.5|y|$ for $n > N_0$. Now for $n > N_0$, $|y_n| > 0.5|y|$, and so $|y_n y| > 0.5|y|^2$ and $\left|\frac{1}{y_n} - \frac{1}{y}\right| < \frac{2|y_n - y|}{|y|^2}$. Given arbitrary $\epsilon > 0$, we have that $\left|\frac{1}{y_n} - \frac{1}{y}\right| < \epsilon$ for $n \geq \max(N, N_0)$, if N is chosen to have $|y_n - y| < 0.5|y|^2 \epsilon$.

5. This follows from parts 3 and 4, since $\frac{x_n}{y_n} = x_n \left(\frac{1}{y_n}\right)$.   ∎

While we have seen by example that boundedness does not guarantee convergence, we have the following result that boundedness assures the existence of a convergent subsequence, generalizing case 4 of example 5.3 above.

Proposition 5.12   *Let $\{x_n\}$ be a bounded numerical sequence. Then there is a subsequence $\{y_m\} \subset \{x_n\}$ and y so that $y_m \to y$.*

Proof   Because both $\mathbb{R}$ and $\mathbb{C}$ are metric spaces under the standard norms defined in (2.3) and (2.2), we have by proposition 5.9 that there is a closed ball in $\mathbb{R}$ or $\mathbb{C}$ so that $\{x_n\} \subset \bar{B}_R(0)$ for some R. By the Heine–Borel theorem, closed balls are compact in both $\mathbb{R}$ and $\mathbb{C}$, so we can apply proposition 4.17 that any infinite collection of points in a compact set must have an accumulation point. That is, $\{x_n\}$ has an accumulation point $y \in \bar{B}_R(0)$. So for any $r > 0$, $B_r(y) \cap \{x_n\} \neq \emptyset$. Next we choose $r_m \to 0$, and for each m choose an arbitrary $y_m \in B_{r_m}(y) \cap \{x_n\}$. Then $y_m \to y$, since for any $\epsilon > 0$ we can choose any $r_N < \epsilon$, and by construction, $y_m \in B_{r_N}(y)$ for all $m \geq N$. That is, $|y_m - y| < \epsilon$ for all $m \geq N$.   ∎

The apparent arbitrariness in this proof implied by "choose an arbitrary $y_m \in B_{r_m}(y) \cap \{x_n\}$" may surprise the reader. However, not only will there be for a given y many sequences $\{y_m\}$ with $y_m \to y$, but there may also be many such accumulation points y. For example, every point of the sequence can be an accumulation point, and moreover the total number of such accumulation points may be uncountably infinite.

Example 5.13   *Let $\{x_n\}$ be an arbitrary enumeration of the rational numbers in $[0, 1]$. Then every $y \in [0, 1]$ is an accumulation point. This is easily seen by taking an arbitrary $y = 0.d_1 d_2 d_3 \ldots$ as a decimal expansion. If y is a rational number ending in all 0s, we first rewrite this as an equivalent decimal ending in all 9s. For example, $0.5 = 0.49999 \ldots$ The subsequence is then formed by looking at the rational truncations of r:*

$0.d_1, 0.d_1d_2, 0.d_1d_2d_3, 0.d_1d_2d_3d_4, \ldots.$

Define $y_1 = 0.d_1$. Clearly, $0.d_1 = x_{n_1}$ for some n_1. The next term of the subsequence, y_2, is the first decimal truncation, $0.d_1d_2d_3 \ldots d_m$, so that $0.d_1d_2d_3 \ldots d_m = x_{n_2}$, where $n_2 > n_1$. Continuing in this way, we obtain a subsequence $\{y_m\}$ with $y_m \to y$.

*5.2   Limits Superior and Inferior

The preceding example illustrates that a bounded numerical sequence not only has an accumulation point as well as a subsequence convergent to that accumulation point, but that it may have a great many such accumulation points. For this reason the notions of **limit superior** and **limit inferior** of a sequence have been introduced. These are defined to equal the **least upper bound** or **l.u.b.**, and **greatest lower bound**, or **g.l.b.**, respectively, of the collection of accumulation points, although unfortunately, not in an immediately transparent way. A small but important application of these notions will be seen in chapter 6 in the statement of the ratio test for series convergence.

In addition these notions of limits have great utility in the advanced topic of real analysis. But rather than deferring their introduction to that more abstract context, we introduce limits superior and inferior here where the essence of these ideas is more transparent.

Before defining formally and justifying the interpretations of limits superior and inferior, we first define the l.u.b. and g.l.b. and introduce alternative notation.

Definition 5.14   *Let $\{x_\alpha\}$ be a collection of real numbers. The **least upper bound** or **supremum** is defined by*

$$\text{l.u.b.}\{x_\alpha\} = \sup\{x_\alpha\} \equiv \min\{x \mid x \geq x_\alpha \text{ for all } \alpha\}. \tag{5.2}$$

*If $\{x_\alpha\}$ is unbounded from above, we define $\text{l.u.b.}\{x_\alpha\} = \sup\{x_\alpha\} \equiv \infty$. The **greatest lower bound** or **infimum** is defined by*

$$\text{g.l.b.}\{x_\alpha\} = \inf\{x_\alpha\} \equiv \max\{x \mid x \leq x_\alpha \text{ for all } \alpha\}. \tag{5.3}$$

If $\{x_\alpha\}$ is unbounded from below, we define $\text{g.l.b}\{x_\alpha\} = \inf\{x_\alpha\} \equiv -\infty$.

Notation 5.15   *It is common to write l.u.b. as lub and g.l.b. as glb.*

Next we state the formal definitions of the limits superior and inferior, and then work toward the demonstration that these achieve the stated objective concerning the g.l.b. and l.u.b. of accumulation points of the given sequence.

Unfortunately, this is another example of where a lot of carefully positioned words are needed to define an idea that has a relatively simple intuitive meaning.

Definition 5.16 *Let $\{x_n\}$ be a numerical sequence. If $\sup\{x_n\} = \infty$, meaning there exists no U so that $x_n \leq U$ for all n, then we define the **limit superior of** $\{x_n\}$ to be ∞, and denote this as*

$$\limsup_{n\to\infty} x_n = \infty.$$

If there exists a U so that $x_n \leq U$ for all n, let $U_n = \sup_{m\geq n}\{x_m\}$ and define

$$\limsup_{n\to\infty} x_n = \lim_{n\to\infty} U_n. \tag{5.4}$$

*Similarly, if $\inf\{x_n\} = -\infty$, meaning there exists no L so that $L \leq x_n$ for all n, then we define the **limit inferior of** $\{x_n\}$ to be $-\infty$, and denote this as*

$$\liminf_{n\to\infty} x_n = -\infty.$$

If there exists an L so that $L \leq x_n$ for all n, let $L_n = \inf_{m\geq n}\{x_m\}$ and define

$$\liminf_{n\to\infty} x_n = \lim_{n\to\infty} L_n. \tag{5.5}$$

Notation 5.17 *In some mathematical references, the **limit superior of** $\{x_n\}$ is denoted by $\overline{\lim}_{n\to\infty} x_n$, and the **limit inferior of** $\{x_n\}$ is denoted by $\underline{\lim}_{n\to\infty} x_n$, but throughout this book we will use the more explicit notation above.*

Before demonstrating that these rather abstract definitions provide the l.u.b. and the g.l.b. of the collection of accumulation points of the sequence, we address a technicality within the definition above. That is, both the definition of lim sup in (5.4) and that of lim inf in (5.5) involve limits of sequences as $n \to \infty$. It is natural to wonder why such limits exist when nothing but one-sided boundedness is assumed of the original sequence $\{x_n\}$.

The following proposition provides the missing detail because both sequences, U_n and L_n, are monotonic as can be demonstrated by

$$U_n = \sup_{m\geq n}\{x_m\} \geq \sup_{m\geq n+1}\{x_m\} = U_{n+1}, \tag{5.6a}$$

$$L_n = \inf_{m\geq n}\{x_m\} \leq \inf_{m\geq n+1}\{x_m\} = L_{n+1}. \tag{5.6b}$$

Consequently U_n is monotonically decreasing, and L_n monotonically increasing, although in neither case must this monotonicity be strict.

The next result is that a monotonic sequence either converges, or diverges to $\pm\infty$, depending on whether it is bounded or unbounded.

Proposition 5.18   *If $\{x_n\}$ is monotonically decreasing, then $\lim_{n\to\infty} x_n = -\infty$ if this sequence is unbounded from below; otherwise, there is an x such that $\lim_{n\to\infty} x_n = x$. Similarly, if $\{x_n\}$ is monotonically increasing, we have that $\lim_{n\to\infty} x_n = \infty$ or $\lim_{n\to\infty} x_n = x$, depending on whether this sequence is unbounded from above or bounded, respectively.*

Proof   The unbounded cases are straightforward. For example, if unbounded from below, we have for any positive integer M there is an N so that $x_N \leq -M$, but by the decreasing monotonicity assumption, we conclude that

$$x_n \leq -M \quad \text{whenever} \quad n \geq N,$$

and we have $\lim_{n\to\infty} x_n = -\infty$. If bounded, we know from proposition 5.12 that $\{x_n\}$ has an accumulation point x and a subsequence $\{y_m\}$ so that $y_m \to x$. By definition of this convergence, we have that for any $\epsilon > 0$ there is an $N \equiv N(\epsilon)$ so that $|y_m - x| < \epsilon$ when $m \geq N$. We now show that x is in fact the limit of the original sequence, and indeed $\lim_{n\to\infty} x_n = x$. First, choose N' defined by $x_{N'} = y_{N+1}$. Next, if $\{x_n\}$ is monotonically decreasing, for any $n \geq N'$ choose $y_{m(n)}$ and $y_{m(n)+1}$ so that and $y_{m(n)+1} \leq x_n \leq y_{m(n)}$. Then

$$|x_n - x| \leq |y_{m(n)} - x| < \epsilon,$$

since by assumption $m(n) \geq N$. The result is analogously proved in the opposite monotonicity case, except that we have $y_{m(n)} \leq x_n \leq y_{m(n)+1}$ and

$$|x_n - x| \leq |y_{m(n)+1} - x| < \epsilon. \qquad \blacksquare$$

We now return to the relationship between limits superior and inferior, and the accumulation points of the sequence $\{x_n\}$. Given the formality in the definitions, it may not be apparent how the definition of limit superior and limit inferior captures the intention set out earlier, that being, to define the g.l.b. and the l.u.b. of all the accumulation points of $\{x_n\}$. The next proposition establishes this connection.

Proposition 5.19   *Given a sequence $\{x_n\}$, let $\{z_k\}$ denote the set of accumulation points. Then*

$$\limsup_{n\to\infty} x_n = \text{l.u.b.}\{z_k\}, \tag{5.7a}$$

$$\liminf_{n\to\infty} x_n = \text{g.l.b.}\{z_k\}. \tag{5.7b}$$

Proof First off, if the sequence $\{x_n\}$ is unbounded from above, then by definition, there is a subsequence $\{y_n\}$ so that $y_n \to \infty$ and hence $\infty \in \{z_k\}$, but also $\limsup_{n\to\infty} x_n = \infty$. Similarly, if unbounded below, there is a subsequence $\{y_n'\}$ so that $y_n' \to -\infty$, and we conclude that $-\infty \in \{z_k\}$, but also $\liminf_{n\to\infty} x_n = -\infty$. So in these cases the intended goal regarding the collection of accumulation points is achieved. On the other hand, if bounded above, then since the sequence $\{U_n\}$ must be monotonically decreasing, it has a finite limit or diverges to $-\infty$ by the proposition above. If $U_n \to U'$, a finite limit, we claim that U' is the supremum or l.u.b. of all accumulation points. To see this, we have by definition of $U_n \to U'$, that for any $\epsilon > 0$ there is an N so that $|U_n - U'| < \epsilon$ for $n \geq N$. Now, since $U_n = \sup_{m \geq n}\{x_m\}$, we can find a value of $x_{m(n)}$ so that $|U_n - x_{m(n)}| < \frac{1}{n}$, say. Define $y_n \equiv x_{m(n)}$. Then we have that $y_n \to U'$, since by the triangle inequality,

$$|y_n - U'| \leq |y_n - U_n| + |U_n - U'| < \epsilon + \frac{1}{n},$$

and hence $U' \in \{z_k\}$. Also there can be no subsequence $\{y_n'\}$ so that $y_n' \to U''$ with $U'' > U'$, since by definition of U_n we have $U_n \geq \sup\{y_j' \mid y_j' = x_m \text{ and } m \geq n\}$. Hence, since $U_n \to U$ we cannot have $y_n' \to U''$ with $U'' > U'$.

The cases where $U_n \to -\infty$, $L_n \to L' < \infty$, and $L_n \to \infty$ are reasoned similarly. ∎

Example 5.20 *Define the sequence*

$$x_n = \begin{cases} 3 - (-1/n)^n, & n = 3m, \\ (-1)^n((n+1)/n), & n = 3m+1, \ m = 0, 1, 2, \ldots, \\ (-3/4)^n, & n = 3m+2. \end{cases}$$

This sequence has four accumulation points. The subsequence with $n = 3m$ converges to 3, the subsequence with $n = 3m + 1$ has two subsequences that converge to -1 and $+1$, and the subsequence with $n = 3m + 2$ converges to 0. So we conclude that by the proposition above, it must be the case that $\limsup_{n\to\infty} x_n = 3$ and $\liminf_{n\to\infty} x_n = -1$. Now

$$U_n = \sup_{m \geq n}\{x_m\} = 3 + \left(\frac{1}{n'}\right)^{n'},$$

$$L_n = \inf_{m \geq n}\{x_m\} = -\frac{n'' + 1}{n''},$$

where $n' = \min\{3m \mid 3m \geq n$ and $3m$ is even$\}$ and $n'' = \min\{3m + 1 \mid 3m + 1 \geq n$ and $3m + 1$ is odd$\}$. We see that each of $\{U_n\}$ and $\{L_n\}$ are convergent monotonic sequences, and that $U_n \to 3$ and $L_n \to -1$.

In summary, we conclude from this proposition that the limit superior equals the supremum of all accumulation points, and the limit inferior the infimum of all accumulation points of $\{x_n\}$. Based on this result, the following proposition's conclusion cannot be a surprise. In theoretical applications this result can provide a useful and powerful way of finding the limit of a convergent sequence, since it is sometimes the case that the limits superior and inferior are easier to estimate than the actual limit itself, as each allows one to focus on what is often a more manageable subsequence.

Proposition 5.21 *Let $\{x_n\}$ be a numerical sequence. Then, for $-\infty \leq x \leq \infty$, $\lim_{n \to \infty} x_n = x$ if and only if*

$$\liminf_{n \to \infty} x_n = \limsup_{n \to \infty} x_n = x.$$

Proof We consider three cases. The proof is a good example of "following the definition" to the logical conclusion:

1. For $x = \infty$, if $x_n \to \infty$, then for any M there is an N so that $x_n \geq M$ for $n \geq N$. Hence $\{x_n\}$ is unbounded from above and $\limsup_{n \to \infty} x_n = \infty$. Also $L_n = \inf_{m \geq n}\{x_m\} \geq M$, for $n > N$, so $L_n \to \infty$ as $n \to \infty$. That is, $\liminf_{n \to \infty} x_n = \infty$. Conversely, if $\liminf_{n \to \infty} x_n = \limsup_{n \to \infty} x_n = \infty$, then $L_n = \inf_{m \geq n}\{x_m\} \to \infty$ as $n \to \infty$. That is, for any M there is an N so that $L_n \geq M$ for $n \geq N$. Hence, by definition of $L_n, x_n \geq M$ for $n \geq N$ and $x_n \to \infty$.

2. For $x = -\infty$, the argument is identical.

3. For $-\infty < x < \infty$, if $x_n \to x$, then for any ϵ there is an N so that $|x_n - x| < \epsilon$ for $n \geq N$. That is, $x - \epsilon < x_n < x + \epsilon$ for $n \geq N$, and hence $x - \epsilon < L_n$, $U_n < x + \epsilon$, and we conclude that $\liminf_{n \to \infty} x_n = \limsup_{n \to \infty} x_n = x$. Conversely, $\liminf_{n \to \infty} x_n = \limsup_{n \to \infty} x_n = x$ implies that for any ϵ there is an N so that $|L_n - x| < \epsilon$ and $|U_n - x| < \epsilon$ for $n \geq N$, and hence by the definition of U_n and L_n, we conclude that $|x_n - x| < \epsilon$ for $n \geq N$ and $x_n \to x$.   ∎

The next result says that the interval with endpoints equal to the limits superior and inferior, if expanded arbitrarily little, will contain all but finitely many values of the original sequence $\{x_n\}$.

Proposition 5.22   *If $L^S = \limsup_{n \to \infty} x_n$ and $L^I = \liminf_{n \to \infty} x_n$, then for any $\epsilon > 0$ there is an N so that for all $n \geq N$,*

$$L^I - \epsilon \leq x_n \leq L^S + \epsilon. \tag{5.8}$$

Proof   We proceed with a proof by contradiction, illustrating the upper inequality. Assume that for some $\epsilon > 0$ there are infinitely many sequence terms satisfying $x_j > L^S + \epsilon$. Then, for any n, $U_n = \sup_{m \geq n}\{x_m\} > L^S + \epsilon$, and hence $\limsup_{n \to \infty} x_n = \lim_{n \to \infty} U_n \geq L^S + \epsilon$, contradicting the definition of L^S.   ∎

Example 5.13 discussed above, on an arbitrary enumeration of rationals in $[0, 1]$, also introduces an issue that will play a critically important role in subsequent chapters. That being, if a sequence $\{x_n\} \subset X$, where X is a subset of $\mathbb{R}$ or $\mathbb{C}$ and where $x_n \to x$, is x necessarily an element of this subset? The answer is "no," and we provide two examples of what can happen.

Example 5.23

1. *If $X = (0, 1)$, then both $\{\frac{1}{n}\}$ and $\{1 - \frac{1}{n}\}$ converge, but not to a point in X. On the other hand, any convergent sequence $\{x_n\} \subset [a, b] \subset (0, 1)$ must converge to a point in X.*

2. *If $X = \mathbb{Q}$, the rational numbers, then as example 5.13 demonstrates, some sequences converge to a point in X and some converge to a point outside X.*

In the next section we generalize the notion of sequence to an arbitrary metric space where $x \in X$ becomes an explicit component of the criterion for convergence.

*5.3   General Metric Space Sequences

The preceding section focused on properties of numerical sequences. However, if one reviews the various proofs, it becomes clear that with one exception, no special property of $\mathbb{R}$ or $\mathbb{C}$ is used other than the existence of a metric or distance function, $d(x, y) = |x - y|$, which was used as a measure of "closeness." The one special property of $\mathbb{R}$ or $\mathbb{C}$ we used was the Heine–Borel theorem, which assures us that a bounded sequence lies in a compact set and hence has a convergent subsequence.

Consequently it should be expected that we can define sequences $\{\mathbf{x}_n\} \subset \mathbb{R}^n$ and their convergence under the standard metric, defined by (3.18), or under any one of the l_p-norms defined in (3.10). This notion of convergence would satisfy all the properties in the preceding section, since in this context we once again have the benefit of the Heine–Borel theorem. Moreover the notion of convergence under equivalent metrics d and d' are identical. Namely $\mathbf{x}_n \to \mathbf{x}$ under d if and only if $\mathbf{x}_n \to \mathbf{x}$ under d'.

More generally, if $\{\mathbf{x}_n\} \subset X$, where (X, d) is a general metric space, convergence can again be defined, and virtually all properties are satisfied. In this general context, however, the definition of convergence must explicitly require that $x \in X$. That is because for a general metric space if $\{\mathbf{x}_n\} \subset X$ and $\mathbf{x} \notin X$, the notion of $d(\mathbf{x}_n, \mathbf{x}) < \epsilon$ is not well defined. Also we note that we have two issues in this general metric space setting that do not exist in $\mathbb{R}$, $\mathbb{R}^n$, or $\mathbb{C}$:

1. In a general metric space, numerical operations like addition may not be defined. If they are defined, the proposition above on arithmetic operations on sequences with limits remains valid.

2. In a general metric space, we do not necessarily have the Heine–Borel theorem. That is, a closed and bounded set need not be compact (the converse is true as proved in proposition 4.17). Consequently a bounded sequence need not be contained in a compact set, and hence it need not have a convergent subsequence.

In this section we document definitions and properties, the latter generally without proof, which the reader can supply as an exercise by redeveloping the arguments above.

Definition 5.24   *Let (X, d) be a metric space. A **sequence**, denoted $\{\mathbf{x}_n\}$, $\{\mathbf{z}_j\}$, and so forth, is a countably infinite collection of elements of X for which a numerical ordering is specified:*

$$\{\mathbf{x}_n\} \equiv \mathbf{x}_1, \mathbf{x}_2, \mathbf{x}_3, \dots.$$

*A sequence is **bounded** if there is a number D and an element $\mathbf{y} \in X$ so that $d(\mathbf{y}, \mathbf{x}_n) \leq D$ for all n. A **subsequence of a sequence** is a countably infinite subcollection that preserves order. That is, $\{\mathbf{y}_m\}$ **is a subsequence of** $\{\mathbf{x}_n\}$ if*

$$\mathbf{y}_m = \mathbf{x}_{n_m} \quad and \quad n_{m+1} > n_m \quad for\ all\ m.$$

We begin by noting that in the definition of bounded, there is nothing special about the identified $\mathbf{y}$.

Proposition 5.25   *If $\{\mathbf{x}_n\} \subset X$, a metric space, and $\{\mathbf{x}_n\}$ is bounded, then for any $\mathbf{y}' \in X$ there is a $D(\mathbf{y}')$ so that $d(\mathbf{y}', \mathbf{x}_n) \leq D(\mathbf{y}')$ for all n.*

Proof   Let $\mathbf{y}$ and D be given as in the definition of bounded, and let $\mathbf{y}' \in X$ be arbitrary. Then by the triangle inequality,

$d(\mathbf{y}', \mathbf{x}_n) \le d(\mathbf{y}', \mathbf{y}) + d(\mathbf{y}, \mathbf{x}_n) \le d(\mathbf{y}', \mathbf{y}) + D.$

Hence $D(\mathbf{y}') = d(\mathbf{y}', \mathbf{y}) + D.$ ∎

Next we define convergence.

Definition 5.26   *A sequence* $\{\mathbf{x}_n\} \subset (X, d)$, *a metric space*, ***converges to a limit*** $\mathbf{x} \in X$ *as* $n \to \infty$ *if for any* $\epsilon > 0$ *there is an* $N \equiv N(\epsilon)$ *so that*

$$d(\mathbf{x}_n, \mathbf{x}) < \epsilon \quad \text{whenever} \quad n \ge N, \tag{5.9}$$

and in this case we write

$$\lim_{n \to \infty} \mathbf{x}_n = \mathbf{x} \quad \text{or} \quad \mathbf{x}_n \to \mathbf{x}.$$

If $\{\mathbf{x}_n\}$ *does not converge, we say it* ***diverges*** *as* $n \to \infty$, *or simply* ***does not converge***.

We note in the general context of a metric space, which of course includes $\mathbb{R}$, $\mathbb{C}$, and $\mathbb{R}^n$, that the concept of convergence is not as metric dependent as it first appears. We state the result for equivalent metrics, also called topologically equivalent, but recall this will also be true for Lipschitz equivalent metrics, since this latter notion implies the former by proposition 3.41.

Proposition 5.27   *Let* X *be a metric space under two equivalent metrics,* d_1 *and* d_2. *Then a sequence* $\{\mathbf{x}_n\} \subset X$ *converges to* $\mathbf{x}$ *in* (X, d_1) *iff* $\{\mathbf{x}_n\}$ *converges to* $\mathbf{x}$ *in* (X, d_2).

Proof   Since $\mathbf{x}_n \to \mathbf{x}$ in (X, d_1), we have that for any $\epsilon' > 0$ there is an $N \equiv N(\epsilon')$ so that $d_1(\mathbf{x}_n, \mathbf{x}) < \epsilon'$ whenever $n \ge N(\epsilon')$. In other words, $\{\mathbf{x}_n\}_{n=N(\epsilon')}^{\infty} \subset B_{\epsilon'}^{(1)}(\mathbf{x})$, the open ball about $\mathbf{x}$ of d_1-radius ϵ'. To show convergence in (X, d_2), let $\epsilon > 0$ be given. By (3.35) there is an ϵ' so that $B_{\epsilon'}^{(1)}(\mathbf{x}) \subset B_{\epsilon}^{(2)}(\mathbf{x})$. But from above, we have for this ϵ',

$$\{\mathbf{x}_n\}_{n=N(\epsilon')}^{\infty} \subset B_{\epsilon'}^{(1)}(\mathbf{x}) \subset B_{\epsilon}^{(2)}(\mathbf{x}),$$

so $d_2(\mathbf{x}_n, \mathbf{x}) < \epsilon$ for $n \ge N(\epsilon')$. The reverse demonstration is identical.   ∎

We now record these convergence results in this general context, where (X, d) is a given metric space.

Proposition 5.28   *If* $\{\mathbf{x}_n\} \subset X$ *is a convergent sequence with* $\lim_{n \to \infty} \mathbf{x}_n = \mathbf{x}$ *and* $\lim_{n \to \infty} \mathbf{x}_n = \mathbf{x}'$, *then* $\mathbf{x} = \mathbf{x}'$.

Proposition 5.29   *If* $\{\mathbf{x}_n\} \subset X$ *is a convergent sequence with* $\{\mathbf{x}_n\} \to \mathbf{x}$, *then* $\{\mathbf{x}_n\}$ *is bounded.*

The next proposition requires a caveat, because a general metric space need not have arithmetic operations. Recall that by definition, X can be any collection of points on which a metric is defined. However, many metric spaces of interest are vector spaces that at least allow addition and scalar multiplication, so we record this result without proof as the proof is identical to that above. These vector spaces are called **(real or complex) linear metric spaces**, depending on whether the vector space structure is over the real or complex numbers. Of course, $\mathbb{R}^n$ is the classic example of a real linear metric space, and correspondingly $\mathbb{C}^n$ is the classic example of a complex linear metric space.

Proposition 5.30   *Let $\{\mathbf{x}_n\}$ and $\{\mathbf{y}_n\}$ be convergent sequences in a linear metric space X with $\{\mathbf{x}_n\} \to \mathbf{x}$, and $\{\mathbf{y}_n\} \to \mathbf{y}$, and let a be a scalar. Then we have:*

1. $a\mathbf{x}_n \to a\mathbf{x}$.

2. $\mathbf{x}_n + \mathbf{y}_n \to \mathbf{x} + \mathbf{y}$.

As noted above, a bounded sequence in a general metric space need not be contained in a compact subset of that metric space. It will be contained in a closed and bounded subset, but in general, this does not necessarily imply compact. Hence, if this sequence is not contained in a compact set, it need not have an accumulation point and hence need not have a convergent subsequence. One approach to ensuring that every bounded sequence is contained in a compact subset is to introduce the notion of a **compact metric space**.

Definition 5.31   *A metric space (X, d) is **compact** if every open cover of X contains a finite subcover.*

Proposition 5.32   *Let $\{\mathbf{x}_n\} \subset \mathbb{R}^n$ be a bounded sequence, or $\{\mathbf{x}_n\} \subset X$ a general sequence in a compact metric space. Then there is a subsequence $\{\mathbf{y}_m\} \subset \{\mathbf{x}_n\}$ so that $\mathbf{y}_m \to \mathbf{y}$ where $\mathbf{y} \in \mathbb{R}^n$ in the first case, and $\mathbf{y} \in X$ in the second.*

Proof   In the first case, boundedness implies that $\{\mathbf{x}_n\} \subset \bar{B}_R(\mathbf{x})$ for any $\mathbf{x} \in \mathbb{R}^n$, where R in general depends on $\mathbf{x}$. Now in $\mathbb{R}^n$, $\bar{B}_R(\mathbf{x})$ is closed and bounded and hence compact by the Heine–Borel theorem, so an accumulation point exists in $\bar{B}_R(\mathbf{x})$ by proposition 4.17. Consequently a convergent subsequence can be constructed as in proposition 5.12. If X is compact, we argue by contradiction and assume that there is no such accumulation point. Then about each point $\mathbf{x}_n$, an open ball can be constructed, $B_{r_n}(\mathbf{x}_n)$, that contains no other point of the sequence. We define the set A by $A \equiv X \sim [\bigcup B_{r_n/2}(\mathbf{x}_n)]$, which is open since the complement of A in

X is the closed set $\overline{[\bigcup B_{r_n/2}(\mathbf{x}_n)]}$. With A and $\{B_{r_n}(\mathbf{x}_n)\}$ we now have an open cover of X that admits no finite subcover, since each $B_{r_n}(\mathbf{x}_n)$ contains only one point of X. This contradicts that X is compact, and hence $\{\mathbf{x}_n\}$ must have an accumulation point in X.   ■

It may not be surprising, at least on an intuitive level, that in a compact metric space a sequence has a subsequence that clusters around some point and "wants" to converge to this point. What should be surprising in this general case is that this subsequence converges to a point $\mathbf{y} \in X$. The question is, why can X have no "holes" so that the bounded sequence converges to the hole and not to a point in X?

Example 5.33 *Using the standard metric, imagine the "apparently compact" metric space $X \equiv [0, 1] \cap \mathbb{Q}$ made up of all rational numbers q with $0 \leq q \leq 1$. It is easy to produce a sequence in X that converges to a hole, which would be an irrational $y \in [0, 1]$, simply by defining this sequence in terms of the rational decimal approximations to y. This appears to contradict proposition 5.32, so it is best to evaluate our assumptions more carefully. Since X is clearly a metric space under the standard metric, it must be compactness that is in question. Is X compact?*

To be compact, it must be the case that any open cover of X admits a finite open subcover. So there must be an infinite open cover that cannot be so reduced. Recall how such a cover was constructed in exercise 12 of chapter 4 to show that $(0,1)$ was not compact. The trick was that since 0 did not need to be covered, a collection of slightly overlapping open intervals could be constructed that collectively covered all real numbers between 0 and 1, but no finite subcover accomplished this. That same trick works here, since we can split X using any irrational y as

$$X = [[0, y) \cap \mathbb{Q}] \cup [(y, 1] \cap \mathbb{Q}].$$

Now the construction of that exercise can be applied to $[0, y)$ and $(y, 1]$ since neither is compact, producing an open cover of $[0, y) \cup (y, 1]$ that has no finite subcover. As this is also now an open cover for X that has no finite subcover, we have demonstrated that X is not compact.

An alternative and simpler argument to show that a compact metric space can have no holes is to apply what we know from proposition 4.17, that a compact set is closed and hence it must contain all its limit points. It is apparent that X in the example above does not contain all its limit points, so it is not closed and cannot be compact.

5.4   Cauchy Sequences

5.4.1   Definition and Properties

In practice, given a sequence $\{x_n\} \subset X$, where X is Euclidean or a metric space, the principal challenge in applying the definition for convergence is that this definition requires knowledge of the limiting value x. The notion of a **Cauchy sequence**, named for **Augustin Louis Cauchy** (1759–1857), allows one to determine in many cases if a sequence converges without first knowing its limiting value. The key defining idea is that all *pairs* of points in the sequence will be found to be arbitrarily close if the index values are required to exceed some value. Specifically:

Definition 5.34 *A sequence $\{x_n\} \subset X$, where (X,d) is a metric space, is a **Cauchy sequence**, or satisfies the **Cauchy criterion**, if for any $\epsilon > 0$, there is an $N = N(\epsilon)$ so that*

$$d(x_n, x_m) < \epsilon \quad \text{whenever} \quad n, m \geq N. \tag{5.10}$$

Example 5.35

1. *Consider the sequence in case 3 of example 5.3: $a_j \equiv \frac{j-1}{j}$. Then by the triangle inequality,*

$$|a_n - a_m| = \left|\frac{n-m}{mn}\right| \leq \frac{1}{n} + \frac{1}{m}.$$

Consequently, to have $|a_n - a_m| < \epsilon$, choose $n, m > \frac{2}{\epsilon}$. In other words, define N as any integer which exceeds $\frac{2}{\epsilon}$.

2. *Consider the sequence defined by the **harmonic series**: $x_n = \sum_{j=1}^n \frac{1}{j}$. Then given m, consider $n = 2m$:*

$$|x_{2m} - x_m| = \sum_{j=m+1}^{2m} \frac{1}{j} > m\left(\frac{1}{2m}\right) = \frac{1}{2}.$$

In other words, no matter how large m is, the sum of the terms from m to $2m$ exceeds $\frac{1}{2}$, so this sequence is not a Cauchy sequence and cannot converge. Since this sequence is apparently monotonically increasing, we conclude that $x_n \to \infty$.

We note that in the general context of a metric space, which of course includes $\mathbb{R}$, $\mathbb{C}$, $\mathbb{R}^n$, and $\mathbb{C}^n$, the concept of a Cauchy sequence is not as metric-dependent as it first appears.

Proposition 5.36 *Let X be a metric space under two equivalent metrics, d_1 and d_2. Then a sequence $\{x_n\} \subset X$ is a Cauchy sequence in (X, d_1) iff $\{x_n\}$ is a Cauchy sequence in (X, d_2).*

Proof The proof is identical to that in proposition 5.27 for convergence of a sequence and is given as exercise 13(a). ∎

The definition of a Cauchy sequence is somewhat more complex than that of convergence to x because the condition in (5.10) applies to all pairs (n, m) of indexes that exceed N rather than the simpler statement concerning all single indexes that exceed N. This definition can be reframed in a logically more simple statement, although this is rarely if ever so noted. The proof of the equivalence of these definitions is assigned in exercise 7.

Definition 5.37 *A sequence $\{x_n\} \subset X$, where (X, d) is a metric space, is a **Cauchy sequence**, or satisfies the **Cauchy criterion**, if for any $\epsilon > 0$, there is an $N = N(\epsilon)$ so that*

$$d(x_N, x_n) < \epsilon \quad whenever \quad n \geq N. \tag{5.11}$$

We next investigate the relationship between the property of a sequence converging and the property of a sequence being a Cauchy sequence. First off, we show that just like convergent sequences, every Cauchy sequence in a metric space is bounded.

Proposition 5.38 *If (X, d) is a metric space and $\{x_n\} \subset X$ a Cauchy sequence, then $\{x_n\}$ is bounded.*

Proof Let $\epsilon > 0$ be arbitrarily chosen. Since $\{x_n\}$ is a Cauchy sequence, there is an N so that $d(x_n, x_m) < \epsilon$ whenever $n, m \geq N$. In particular, $d(x_n, x_N) < \epsilon$ whenever $n \geq N$. Now, if $B = \max_{n < N} d(x_n, x_N)$, then with $x = x_N$ we have $d(x_n, x) < \max(\epsilon, B)$ for all n, and hence $\{x_n\}$ is bounded. ∎

It is easy to show that every convergent sequence is in fact a Cauchy sequence:

Proposition 5.39 *If $\{x_n\} \subset X$, where X is a metric space and $x_n \to x$, then $\{x_n\}$ is a Cauchy sequence.*

Proof By the triangle inequality,

$$d(x_n, x_m) \leq d(x_n, x) + d(x, x_m).$$

Now, if $\epsilon > 0$ is given, choose N so that $d(x_n, x) < \frac{\epsilon}{2}$ for $n \geq N$. By the inequality above we then have $d(x_n, x_m) < \epsilon$ for $n, m \geq N$. ∎

While this last result is of interest, the result of greater value in applications has to do with the reverse implication. Namely, when does a Cauchy sequence converge? The answer can be readily seen to be: "not necessarily."

Example 5.40

*1. Let $\{x_n\} = \{\frac{1}{n}\}$ in the metric space $X = (0,1) \subset \mathbb{R}$ under the standard metric in (3.18). This is a Cauchy sequence, and one readily verifies that $d(x_n, x_m) < \epsilon$ whenever $n, m \geq N$ for any $N > \frac{1}{\epsilon}$. However, it is clear that this sequence does not converge in X. It is also clear that in this case X can be enlarged somewhat or **completed**, to its closure $\bar{X} = [0,1]$ in $\mathbb{R}$, and in this metric space we obtain convergence.*

*2. In example 5.33 was introduced $X = \mathbb{Q} \cap [0,1]$, under the standard metric, where it was shown that for any real number $y \in [0,1]$ there was a sequence $\{y_n\} \subset X$ so that $y_n \to y$. By the proposition above, all such sequences are Cauchy sequences. However, these sequences only converge in X if y is chosen to be rational. Again, we see that this metric space can be **completed** by enlarging it to $\bar{X} = [0,1]$, and then all these Cauchy sequences converge to a point in X.*

To motivate the result below, note that we have shown that if $\{x_n\} \subset X$ is a Cauchy sequence in any metric space, then it is bounded. So the question of convergence is closely related to the existence of an accumulation point, and we have seen from the above that such an accumulation point can be assured if $X = \mathbb{R}, \mathbb{C}, \mathbb{R}^n$ (as well as $\mathbb{C}^n$, though not proved) or if X is a compact metric space. Although the results below that rely on the Heine–Borel theorem are also true in $\mathbb{C}^n$, we will drop this reference since this theorem was not proved in this case, and we do not need this result in this book.

Proposition 5.41 *If $\{x_n\} \subset X$ is a Cauchy sequence, where $X = \mathbb{R}, \mathbb{C}, \mathbb{R}^n$, or X is a compact metric space, then there is an $x \in X$ so that $x_n \to x$.*

Proof In all cases we know that $\{x_n\}$ is bounded. Also for any $\epsilon > 0$ there is an N so that $|x_n - x_m| < \epsilon$ for $n, m \geq N$. That is,

$$\{x_n\}_{n=N}^{\infty} \in \bar{B}_{\epsilon}(x_N).$$

Choose $\epsilon_j = \frac{1}{j}$, and let N_j be the associated integer. Then as $j \to \infty$,

$$\{x_n\}_{n=N_j}^{\infty} \in \bar{B}_{1/j}(x_{N_j}).$$

We now claim that there is a unique $x \in X$ so that $\bigcap_j \bar{B}_{1/j}(x_{N_j}) = x$, and that $x_n \to x$. Of course, the latter conclusion follows from the existence of x, since we can conclude that for any ϵ_j, $x \in \bar{B}_{\epsilon_j}(x_{N_j})$ and hence for $n > N_j$,

$$d(x, x_n) \leq d(x, x_{N_j}) + d(x_n, x_{N_j}) < \frac{2}{j}.$$

To demonstrate the intersection claim, first note that every finite collection of these closed balls has a nonempty intersection, since all contain $\{x_n\}_{n=N}^{\infty}$ where $N = \max\{N_j\}$, and this maximum is finite for any finite collection. Also the intersection of all such balls cannot contain more than one point since the radius of these balls, $\epsilon_j = \frac{1}{j}$ converges to 0. To complete the proof, we show by contradiction that this infinite intersection cannot be empty, and hence it contains the unique point x. Assume that $\bigcap_j \bar{B}_{1/j}(x_{N_j}) = \emptyset$ and, in particular, $\{\bigcap_{j \geq 2} \bar{B}_{1/j}(x_{N_j})\} \cap \bar{B}_1(x_{N_1}) = \emptyset$. Then with $A^c \equiv \tilde{A}$, denoting the complement of A,

$$\bar{B}_1(x_{N_1}) \subset \left\{ \bigcap_{j \geq 2} \bar{B}_{1/j}(x_{N_j}) \right\}^c = \bigcup_{j \geq 2} \tilde{\bar{B}}_{1/j}(x_{N_j}),$$

by De Morgan's laws. Now the set $\bar{B}_1(x_{N_1})$ is compact either by Heine–Borel if $X = \mathbb{R}, \mathbb{C}, \mathbb{R}^n$ or as a closed set in the compact metric space X, and it is covered by a union of open sets $\{\tilde{\bar{B}}_{1/j}(x_{N_j})\}_{j \geq 2}$. It therefore has a finite subcover, so $\bar{B}_1(x_{N_1}) \subset \bigcup_{j \leq M} \tilde{\bar{B}}_{1/j}(x_{N_j})$ for some M. Again, using De Morgan's laws, we conclude that $\{\bigcap_{2 \leq j \leq M} \bar{B}_{1/j}(x_{N_j})\} \cap \bar{B}_1(x_{N_1}) = \emptyset$, contradicting the observation above that every finite collection of these balls has nonempty intersection. ∎

Unfortunately, many of the general metric spaces of interest are not compact. Hence we cannot, in general, conclude that Cauchy sequences converge to a point in the space. Of course, $\mathbb{R}$, $\mathbb{C}$, and $\mathbb{R}^n$ are also metric spaces of great interest, and are not compact, yet we have seen that in these cases Cauchy sequences do converge. So compactness is not a necessary condition for the convergence of Cauchy sequences, but it is a sufficient condition.

*5.4.2 Complete Metric Spaces

Because the property that Cauchy sequences converge to a point of the space is so important in mathematics, special terminology has been introduced for metric spaces that have this property.

Definition 5.42 *Let (X, d) be a metric space. Then X is said to be **complete** under d if every Cauchy sequence in X converges to a point in X.*

It should be noted that this notion of being complete is not just a property of the space X, but it is explicitly specified as "complete under d." This is because by the very definition in (5.10) or (5.11) above, the metric d determines which sequences are Cauchy sequences and therefore determines which sequences must converge in order to satisfy the completeness criterion. However, as was seen above, the dependence on the metric d is only up to metric equivalence. That is, X is complete under d if and only if it is complete under d' for any metric equivalent to d.

Example 5.43

1. *We have seen from the analysis above that* $\mathbb{R}$, $\mathbb{C}$, *and* $\mathbb{R}^n$ *are all complete under the standard metrics defined in (2.3), (2.2), and (3.3), respectively.*

2. $\mathbb{R}^n$ *is also complete under all the* l_p-*norms in (3.10) and (3.11), since these norms are equivalent to the standard metric.*

3. *Every compact metric space is complete under its metric.*

4. *The metric space* $\mathbb{Q}$ *is not complete under the standard metric, nor is* $\mathbb{Q} \cap [0, 1]$, *nor is any bounded open interval,* (a, b).

5. *The metric space* $\mathbb{Q}^n \subset \mathbb{R}^n$ *of rational n-tuples is not complete under the standard metric, nor is* $\mathbb{Q}^n \cap B_R(x)$ *for any R and x, nor is* $B_R(x)$.

Because completeness of a metric space is so important in applications, yet so often it is the case that a metric space of interest is not complete, it is of no surprise that the question of **completing a metric space** has received considerable attention. In the various examples above, it was obvious why the given spaces failed to be complete, and equally obvious how one could solve this problem by adding to the space the "missing" points that prevented the space from being complete in the first place.

For the examples above we note that what is interesting about these informal completions of the given spaces was that within the resulting completed spaces, the original spaces were **dense**. In addition distances between points of the original spaces were preserved in the completed spaces.

Alternatively, by looking at the incomplete space as a subspace of a larger space, we could interpret the completion of the original space as the closure of that space in the larger space that contained it. The completions in effect just added the original space's accumulation points. For example, (a, b) is not complete, but the closure of this interval in the metric space $\mathbb{R}$, which is $\overline{(a, b)} = [a, b]$, is complete. Similarly, while $\mathbb{Q}$ and $\mathbb{Q} \cap [0, 1]$ are not complete metric spaces, we can create their closures in $\mathbb{R}$, where $\overline{\mathbb{Q}} = \mathbb{R}$, and $\overline{\mathbb{Q} \cap [0, 1]} = [0, 1]$, and these are complete. We can do the same for $\mathbb{Q}^n$, $\mathbb{Q}^n \cap B_R(x)$, and $B_R(x)$ in $\mathbb{R}^n$.

The next proposition, which we state without proof, indicates that these examples illustrate the general case. Namely every metric space can be embedded in a complete metric space in a way that preserves distances, and where the original space is dense in the larger space. In addition, if the original space is already contained within a complete metric space, then this completion is equivalent to the closure of the original space.

Proposition 5.44 *Let (X, d) be a metric space. Then there is a complete metric space (X', d') so that (X, d) is **isometric** to a dense subset of (X', d'). That is, there is a dense subset $X'' \subset X'$ and a one-to-one identification $X'' \Leftrightarrow X$ so that for any $x'', y'' \in X''$, and identifications: $x'' \Leftrightarrow x$ and $y'' \Leftrightarrow y$, with $x, y \in X$, we have that*

$$d'(x'', y'') = d(x, y).$$

Also, if under d there is a complete metric space, Y, with $X \subset Y$, then X'' is isometric to $\overline{X}$, the closure of X in Y.

This proposition guarantees that any metric space (X, d) of interest can be completed in a way that does not change the original space very much, which is the meaning of the isometric identification. Also, if we are working with a space (X, d) that we know to be a subspace of a larger complete space Y, we can accomplish this completion by forming the closure of X in Y, as was seen to be the case in the earlier simpler examples.

5.5 Applications to Finance

The results of this chapter are to a large extent needed as an introduction to concepts that underlie applicable mathematics in later chapters. For example, the notion of convergence will be seen to be fundamental to much of what is to come. More directly, the notion of convergence of a sequence provides a context for understanding what it means for an iterative numerical calculation to converge to the correct answer, where in each step the calculation provides an approximate solution to a finance problem.

We return to the example of interval bisection next, extending the analysis originally introduced in section 4.3.3 for the evaluation of the yield to maturity of a bond or other security offered at a given price. Here we illustrate the general procedure with a detailed bond yield example.

5.5.1 Bond Yield to Maturity

Assume that we are offered a 1000 par, 10-year, 8% semiannual coupon bond at a price of 1050. First off, we easily confirm that the yield to maturity (YTM) is less than 8% on a semiannual basis because this bond is selling at a premium. The cash flows on this bond are 40 per half year for 10 years, with an extra payment of 1000 at time 10. So if r is the yield on a semiannual basis, we have from (2.16) that

$$P(r) = 1000 + 1000[0.5(0.08 - r)]a_{20;0.5r}.$$

From this equation it is apparent that in order to have $P(r_0) = 1050$, we need $r_0 < 0.08$.

We now detail an **interval bisection** approximation procedure and construct a sequence $\{r_j\}$, which we will prove is a Cauchy sequence. Consequently, without knowing to what value this sequence converges, we will be able to assert that this sequence will indeed converge because $\mathbb{R}$ is complete. Moreover, because of the nature of the approximation procedure, we will be able to calculate the rate at which convergence is achieved, and hence how many steps are needed for any given degree of accuracy. All this is doable without our ever knowing the exact answer.

To this end, for the first step we require two trial values of r, denoted r^+ and r^- so that

$$P(r^+) < 1050 < P(r^-).$$

In other words, since r^+ provides too small a price, $r^+ > r_0$, where r_0 is the desired exact value, and similarly $r^- < r_0$. That is,

$$r^- < r_0 < r^+.$$

For this step we choose somewhat arbitrarily, since this process will always converge, but not naively, since to do so increases the number of steps needed to get a good approximation. An example of a naive initial set of values is $r^+ = 1.00$ (i.e., 100%) and $r^- = 0$. We can with a moment of thought do better with $r^+ = 0.08$ and $r^- = 0.07$, producing $P(r^+) = 1000$, and $P(r^-) = 1071.0620165$. The first estimate of r_0 is then

$$r_1 = 0.5(r^+ + r^-),$$

which produces $r_1 = 0.075$.

For the second step, the process is to now evaluate $P(r_1)$. If $P(r_1) < 1050$, r_1 becomes the new r^+ and we retain the former r^-. Otherwise, r_1 becomes the new r^- and we retain the former r^+. In either case we calculate the second estimate of r_0 as

Table 5.1
Interval bisection for bond yield

Step	r^-	$P(r^-)$	r^+	$P(r^+)$	r_j	$r^+ - r^-$
1	7.0000%	1071.06202	8.00000%	1000.00000	7.50000%	1.00000%
2	7.0000%	1071.06202	7.50000%	1034.74051	7.25000%	0.50000%
3	7.2500%	1052.69870	7.50000%	1034.74051	7.37500%	0.25000%
4	7.2500%	1052.69870	7.37500%	1043.66959	7.31250%	0.12500%
5	7.2500%	1052.69870	7.31250%	1048.17157	7.28125%	0.06250%
6	7.2813%	1050.43198	7.31250%	1048.17157	7.29688%	0.03125%
7	7.2813%	1050.43198	7.29688%	1049.30099	7.28906%	0.01562%
8	7.2813%	1050.43198	7.28906%	1049.86629	7.28516%	0.00781%
9	7.2852%	1050.14908	7.28906%	1049.86629	7.28711%	0.00391%
10	7.2871%	1050.00767	7.28906%	1049.86629	7.28809%	0.00195%

$$r_2 = 0.5(r^+ + r^-),$$

and the process continues into the third step and beyond. If at any step we find that the calculated r_n serendipitously equals the exact answer, r_0, the process stops. However, this virtually never happens to anyone, so we have no need to dwell on this outcome.

The implementation of this algorithm to the bond yield problem yields the table of results in **table 5.1**, where for visual appeal, yields are presented in percentage units, on a semiannual nominal basis:

Now at each step, we have $r_n \in (r^-, r^+)$ by definition, and for any $r' \in (r^-, r^+)$,

$$|r' - r_n| \leq \frac{r^+ - r^-}{2}.$$

Since the lengths of these intervals halve at each step by construction, and for $n = 1$ we have $r^+ - r^- = 0.01$, we conclude that for any $r' \in (r^-, r^+)$ at the nth step,

$$|r' - r_n| \leq \frac{0.01}{2^n}.$$

From this estimate we demonstrate that the sequence $\{r_j\}$ is a Cauchy sequence, and hence because $\mathbb{R}$ is complete by the analysis above, we conclude that there is an $r_0 \in (r^-, r^+)$ for all such intervals and that $r_j \to r_0$.

To this end, let m and $n > m$ be given; then for $r' \in I_n \equiv (r^-, r^+)$ defined as the interval produced as of the nth step, we also have $r' \in I_m \equiv (r^-, r^+)$ defined as of the mth step since $I_n \subset I_m$. By the triangle inequality, with $r' \in I_n \cap I_m$,

$$|r_n - r_m| \leq |r_n - r'| + |r' - r_m|$$

$$\leq \frac{0.01}{2^n} + \frac{0.01}{2^m}.$$

From this estimate we can, for any ϵ, choose N so that $\frac{0.01}{2^N} < \frac{\epsilon}{2}$, and conclude that

$$|r_n - r_m| < \epsilon \qquad \text{for } n, m > N.$$

In other words, $\{r_j\}$ is a Cauchy sequence, and hence there is an $r_0 \in (r^-, r^+)$ for all such intervals with $r_j \to r_0$.

From the error estimate above, true for all $r' \in I_n$, we derive the error estimate for r_0 by letting $m \to \infty$:

$$|r_0 - r_n| \leq \frac{0.01}{2^n}. \tag{5.12}$$

From (5.12) we can choose n to provide any given level of accuracy. For example, to have k-decimal point accuracy, we need the error to be less than $5(10^{-k-1}) = \frac{10^{-k}}{2}$, that is,

$$\frac{0.01}{2^n} < \frac{10^{-k}}{2}.$$

From this point we conclude that n must be chosen so that $2^{n-1} > 10^{k-2}$, which is easily solved with logarithms.

This simple, yet powerful algorithm is known as the **interval bisection algorithm**. It has the property that the error decreases geometrically with a factor of $\frac{1}{2}$. Note that although the error in each step halves as is illustrated in the last column in **table 5.1**, it is not the case that the sequence of estimators, $\{r_j\}$, monotonically converge to r, as is seen from the second last column of this table. This conclusion is logical, since in each step one of the values of r^- and r^+ is replaced, and one is used in the next step. Consequently, if r^- is replaced in a given step, that step's estimate will exceed the prior step's estimate, and conversely.

5.5.2   Interval Bisection Assumptions Analysis

As was observed in section 4.3.3, the usefulness of this algorithm relies on subtle assumptions about the objective function, here $P(r)$, but in general, $f(x)$, where we are attempting to solve

$$f(x) = c$$

for some value c. The interval bisection algorithm produces a Cauchy sequence, $\{x_j\}$, which then has the property that $x_j \rightarrow x$ for some $x \in \mathbb{R}$ typically, where by construction, for every sequence point either $f(x_j) > c$ or $f(x_j) < c$.

The first subtlety in the application of interval bisection is that we are assuming that because $\{x_j\}$ is a Cauchy sequence, this implies that $\{f(x_j)\}$ is a convergent sequence. This appears to be the case for the bond yield example in **table 5.1**, but should this always be the case? Consider the next example where it is not initially feasible to produce a complete picture of what the graph of a given function looks like.

Imagine that it is a complicated function that has been programmed in terms of an iterative process. All that is possible is that by crunching the program for a given value of y, the value of $f(y)$ can be calculated. You are attempting to find a value of x so that $f(x) = c$. You know from sample calculations that c is within the range of sample values of $f(y)$ so far calculated. You proceed to program the interval bisection algorithm, and let it run. At each step, either $f(x_j) > c$ or $f(x_j) < c$, and it is apparent that $x_j \rightarrow x$ for some $x \neq 0$. However, it is also apparent that $f(x_j)$ is not converging. To see what is going wrong, a graphical depiction of this function must be laboriously estimated, and it appears to be given by

$$f(y) = \begin{cases} 1 - 2y, & y < x, \\ 1 + 2y, & y \geq x. \end{cases}$$

In this case a subsequence of $\{f(x_j)\}$ is approaching $1 - 2x$, another subsequence is approaching $1 + 2x$, and of course, $1 - 2x < c < 1 + 2x$.

The second subtle assumption needed for the usefulness of the interval bisection method is that if $x_j \rightarrow x$, and we observe $f(x_j)$ to be converging in that there is some c with

$$|f(x_j) - c| \rightarrow 0,$$

then it must be the case that $f(x) = c$. But this conclusion is really just another assumption about the behavior of the function, $f(x)$. That is, the assumption that $x_j \rightarrow x$ and $f(x_j) \rightarrow c$ implies that $f(x) = c$.

As it turns out, both assumptions are valid for an important, and fortuitously abundant and commonly encountered collection of functions, known as the **continuous functions**. These functions satisfy both properties needed. Namely, if $f(x)$ is continuous on an interval, and $\{x_j, x\}$ are contained in this interval, then from $x_j \rightarrow x$ we can conclude that:

1. $\{f(x_j)\}$ converges.
2. $\{f(x_j)\}$ converges to $f(x)$.

Continuous functions will be investigated in more detail, along with other important properties of functions, in chapter 9 on calculus I.

Exercises

Practice Exercises

1. Evaluate the convergence or lack of convergence of the following. In the cases of convergence, attempt to determine the formula for $N(\epsilon)$ for arbitrary $\epsilon > 0$, while for divergence to $\pm\infty$, do the same for $N(M)$. (*Hint*: The formulas for $N(\epsilon)$ and $N(M)$ do not have to be the "best possible," so estimate the results.)

(a) $c_n = \sqrt{n+1} - \sqrt{n}$ (*Hint*: Multiply by $\frac{\sqrt{n+1}+\sqrt{n}}{\sqrt{n+1}+\sqrt{n}}$.)

(b) $b_m = \frac{\sqrt{m+1}-\sqrt{m}}{\sqrt{m+3}}$

(c) $d_i = \frac{a^i}{i!}$, where $a > 1$ (*Hint*: $d_{i+1} = \frac{a}{i+1}d_i$.)

(d) $x_k = \frac{k^k}{k!}$ (*Hint*: Consider $\ln x_k$.)

(e) $z_j = \frac{4j}{j^2+\sqrt{j}}$

(f) $y_m = \frac{3m^2-5m}{8m^2+5m}$

2. Let $\{x_n\}$ be a convergent sequence and $\{y_n\}$ an arbitrary bounded sequence:

(a) Prove that if $x_n \to 0$, then $y_n x_n \to 0$.

(b) Show by example that if $x_n \to x \neq 0$, then $y_n x_n$ need not be convergent. (*Hint*: Consider y_n with alternating signs.)

(c) Repeat part (b), showing that we need not have $y_n x_n$ convergent even if all $y_n \geq 0$.

3. How does taking absolute values influence convergence?

(a) If $x_n \to x$ is convergent, must $|x_n|$ be convergent? Does the answer depend on whether $x = 0$ or $x \neq 0$?

(b) If $|x_n| \to x$ is convergent, must x_n be convergent? Does the answer depend on whether $x = 0$ or $x \neq 0$?

4. For $n = 0, 1, 2, 3, \ldots$, consider the sequence defined by

$$
y_m = \begin{cases}
\frac{1}{(n+1)!}, & m = 3n, \\
(-1)^n 10 + \frac{(-1)^{n+1}n}{2(n+1)}, & m = 3n+1, \\
(-1)^{n+1} + \frac{(-1)^n}{10(n+1)}, & m = 3n+2.
\end{cases}
$$

(a) Determine all the limit points of this sequence and the associated convergent subsequences.

(b) Determine the formula for U_n and L_n, as given in the definition of limits superior and inferior, and evaluate the limits of these monotonic sequences to derive $\limsup y_m$ and $\liminf y_m$, respectively.

(c) Confirm that the limit superior and limit inferior, derived in part (b), correspond to the l.u.b. and g.l.b. of the limit points in part (a).

5. Let $\{q_n\}$ denote an ordering of all rational numbers in $[0, 1]$.

(a) For the ordering implied by Cantor's construction in section 2.1.4, including or excluding multiple counts, demonstrate that for every n, $U_n = 1$, $L_n = 0$, and hence $\limsup q_m = 1$ and $\liminf q_m = 0$.

(b) Generalize the result on part (a) by showing that the same conclusion follows for an arbitrary ordering.

6. Demonstrate that the sequence in exercise 4 is not a Cauchy sequence, and draw the otherwise obvious conclusion that this sequence does not converge.

7. Prove that the two definitions given for Cauchy sequence in (5.10) and (5.11) are equivalent. (*Hint:* That (5.10) $\Rightarrow$ (5.11) is true follows by definition. For the reverse implication, express $d(x_n, x_m)$ using the triangle inequality.)

8. Identify which of the following sequences are Cauchy sequences and hence must converge, even in cases where their limiting values may be unknown.

(a) $d_n = \frac{n}{n+1}$

(b) $x_n = \frac{2n^2-4}{4n^2+10}$

(c) $y_n = \sum_{j=1}^{n}(-1)^{j+1}$

(d) $x_n = \sum_{j=1}^{n}(-1)^{j+1}2^{-j}$

(e) $f_n = \sum_{j=1}^{n}(-1)^{j+1}a^{-j}$, $a > 1$

(f) $c_k = k + \frac{1}{k}$

9. For the following securities, implement the interval bisection method to produce a tabular analysis as in table 5.1, and determine how many steps are needed to assure six decimal place yield accuracy.

(a) A 7-year, 3.5% s.a. coupon bond with a price of 92.50 per 100 par.

(b) A 2% annual dividend perpetual preferred stock with a price of 87.25 per 100 par.

(c) A \$1 million mortgage repayment loan, issued at 8% monthly, at a price of \$997,500.

Assignment Exercises

10. Evaluate the convergence or lack of convergence of the following. In the cases of convergence, attempt to determine the formula for $N(\epsilon)$ for arbitrary $\epsilon > 0$, while for divergence to $\pm\infty$, do the same for $N(M)$.

(a) $c_n = \sqrt[m]{n+1} - \sqrt[m]{n}$ for $m \in \mathbb{N}$, $m > 1$ (*Hint*: Confirm that

$$a^m - b^m = (a - b)\left(\sum_{j=0}^{m-1} a^j b^{m-1-j}\right),\tag{5.13}$$

and compare to exercise 1(a).)

(b) $z_j = \frac{j!+j}{(j+1)!}$

(c) $w_m = (-1)^{m+1} \ln\left(1 + \frac{1}{m}\right)$

(d) $x_n = (n+1)! + (-1)^{n+1} n!$

(e) $a_k = (-1)^{k+1} \frac{2^k}{10^k+k}$

(f) $b_i = (-1)^{i+1} (i^5 - i^3 + 10^i)$

(g) $u_n = \frac{(-1)^{n+1} n^p}{a^n}$, $p \in \mathbb{R}$, $a > 1$ (*Hint*: Consider the value of $\left|\frac{u_{n+1}}{u_n}\right|$.)

11. Consider the rational numbers in $[0, 1]$. Under an arbitrary enumeration, $\{q_n\}$, this set is a bounded sequence. Show that:

(a) As proposition 5.12 states, this sequence has a convergent subsequence.

(b) This sequence has a countably infinite number of convergent sequences.

(c) This sequence has an uncountably infinite number of convergent sequences.

(d) These results remain true if we require all sequences to be monotonic.

12. For $n = 0, 1, 2, 3, \ldots$, consider the sequence defined by

$$x_m = \begin{cases} \frac{(-1)^n}{n+1}, & m = 5n, \\ 1 + \frac{(-1)^n n}{2(n+1)}, & m = 5n + 1, \\ -1 + \frac{(-1)^n}{n+1}, & m = 5n + 2, \\ -n^2 + n, & m = 5n + 3, \\ 10e^{-n}, & m = 5n + 4. \end{cases}$$

(a) Determine all the limit points of this sequence and the associated convergent subsequences.

(b) Determine the formula for U_n and L_n, as given in the definition of limits superior and inferior, and evaluate the limits of these monotonic sequences to derive $\limsup x_m$ and $\liminf x_m$, respectively.

(c) Confirm that the limit superior and limit inferior, derived in part (b), correspond to the l.u.b. and g.l.b. of the limit points in part (a).

13. Consider the notion of Cauchy sequence under different metrics.

(a) Prove proposition 5.27 in the form: In a metric space X under two equivalent metrics, d_1 and d_2, a sequence $\{x_n\} \subset X$ is a Cauchy sequence in (X, d_1) iff $\{x_n\}$ is a Cauchy sequence in (X, d_2).

(b) Give an example of a metric on $\mathbb{R}^n$, d, so that sequences that are Cauchy under d are different than sequences that are Cauchy under the standard metric. (*Hint*: Consider a nonequivalent metric, like d in exercise 18 in chapter 3.)

14. Identify which of the following sequences are Cauchy sequences and hence must converge, even in cases where their limiting values may be unknown.

(a) $a_j = \sum_{n=1}^{j} \frac{1}{n!}$ (*Hint*: Show that $n! > 2^n$ for $n \geq 4$.)

(b) $a_j = \sum_{n=1}^{j} \frac{(-1)^{n+1}}{n!}$

(c) $y_n = \frac{(-1)^{n+1}}{n}$

(d) $b_k = \sum_{n=1}^{k} \frac{1}{n^2}$ (*Hint*: $n^2 > n(n-1)$.)

(e) $b_k = \sum_{n=1}^{k} \frac{(-1)^{n+1}}{n^2}$

(f) $\{z_n\} \subset \mathbb{R}$, increasing and bounded.

15. For the following securities, implement the interval bisection method to produce a tabular analysis as in table 5.1. Determine how many steps need to be implemented to assure six decimal place yield accuracy.

(a) A 10-year zero-coupon bond with a price of 66.75 per 100 par, priced with a semiannual yield.

(b) A 10-year, 4% annual coupon bond, with a "sinking fund" payment of 50% of par at time 5 years, with a price of 101 per 100 par.

(c) A $25 million, 30-year mortgage repayment loan, issued at 6% monthly, at a price of $25.525 million.

6 Series and Their Convergence

6.1 Numerical Series

6.1.1 Definitions

While a series can be defined in any space X that allows addition, and convergence defined in any such space that also has a metric, we will focus on numerical series defined on $\mathbb{R}$ or $\mathbb{C}$. More general definitions can be inferred now, and will be made in later chapters as needed.

Definition 6.1 *Given a numerical sequence $\{x_j\}$, the **infinite series associated with** $\{x_j\}$ is notationally represented by*

$$\sum_{j=1}^{\infty} x_j.$$

*For $\{x_j\} \subset \mathbb{R}$, if all $x_j > 0$, the series is called a **positive series**, if all $x_j < 0$, the series is called a **negative series**, whereas if the signs of the consecutive terms alternate, most commonly with $x_1 > 0$, the series is called an **alternating series**. The **partial sums** of a numerical series, denoted s_n, are defined as*

$$s_n = \sum_{j=1}^{n} x_j.$$

*The infinite series is said to **converge to a numerical value** s if the sequence of partial sums converges to s. That is, we define*

$$\sum_{j=1}^{\infty} x_j = s \quad \text{if and only if} \quad \lim_{n \to \infty} s_n = s.$$

*An infinite series that does not converge is said to **diverge** or be **divergent**.*

*A series is said to **converge absolutely** or be **absolutely convergent** if the series $\sum_{j=1}^{\infty} |x_j|$ converges, and is said to **converge conditionally** or be **conditionally convergent** if $\sum_{j=1}^{\infty} x_j$ converges yet $\sum_{j=1}^{\infty} |x_j|$ diverges. If a series diverges in the sense that $\lim_{n \to \infty} s_n = \pm \infty$, we will often write $\sum_{j=1}^{\infty} x_j = \pm \infty$ and say that $\sum_{j=1}^{\infty} x_j$ **diverges to** $\pm \infty$.*

Remark 6.2

1. *For some examples, an infinite series will be indexed as $\sum_{j=0}^{\infty} x_j$ rather than $\sum_{j=1}^{\infty} x_j$.*

2. *By definition, every convergent positive or negative series is absolutely convergent, but in general convergence does not imply absolute convergence (see cases 3 and 6 in examples 6.9 and 6.10 below).*

This definition implies that to be convergent it must be the case that $x_j \to 0$ as $j \to \infty$ (see exercise 1). This property alone is not enough to assure convergence as will be seen. However, while $x_j \to 0$ as $j \to \infty$ does not assure the convergence of $\sum_{j=1}^{\infty} x_j$ in general, it does assure convergence when the series is alternating, as will be demonstrated in proposition 6.20.

Applying the definition of convergence of a sequence to this series context, we have that:

Definition 6.3 $\sum_{j=1}^{\infty} x_j = s$ *if for any $\epsilon > 0$ there is an N so that $|s_n - s| < \epsilon$ whenever $n \geq N$. That is,*

$$\left| \sum_{j=n+1}^{\infty} x_j \right| < \epsilon \quad whenever \quad n \geq N.$$

In other words, a numerical series converges when it can be shown that by discarding a finite number of terms, here the first N terms, the residual summation can be made as small as desired. Alternatively, because a numerical sequence converges if and only if it is a Cauchy sequence, we can state that:

Definition 6.4 $\sum_{j=1}^{\infty} x_j = s$ *if for any $\epsilon > 0$ there is an N so that $|s_n - s_m| < \epsilon$ whenever $n, m \geq N$. That is, assuming $n > m$,*

$$\left| \sum_{j=m+1}^{n} x_j \right| < \epsilon \quad whenever \quad n, m \geq N.$$

6.1.2   Properties of Convergent Series

In this section three simple, useful results are presented. More subtle properties will be investigated in section 6.1.4 on rearrangements. The first result reinforces the intuitive conclusion that absolute convergence is a stronger condition than convergence. In the examples below we will see that this implication cannot, in general, be reversed.

Proposition 6.5 *If $\sum_{j=1}^{\infty} x_j$ is absolutely convergent, then it is convergent.*

Proof We show that $s_n = \sum_{j=1}^{n} x_j$ is a Cauchy sequence. By the assumption of absolute convergence, $s_n' = \sum_{j=1}^{n} |x_j|$ is Cauchy, and hence for any $\epsilon > 0$ there is an N

so that $|s'_n - s'_m| < \epsilon$ whenever $n, m \geq N$. Now, by the triangle inequality, say $n > m$ for specificity,

$$|s_n - s_m| = \left| \sum_{j=m+1}^{n} x_j \right| \leq \sum_{j=m+1}^{n} |x_j| = |s'_n - s'_m|,$$

so $|s_n - s_m| < \epsilon$ whenever $n, m \geq N$.   ∎

Next we see that convergent sequences combine well in terms of sums and scalar multiples.

Proposition 6.6   *Let $\sum_{j=1}^{\infty} x_j$ and $\sum_{j=1}^{\infty} y_j$ be convergent series with respective summations of s and s', then for any constants $a, b \in \mathbb{R}$, the series $\{ax_j + by_j\}$ is convergent, and $\sum_{j=1}^{\infty} (ax_j + by_j) = as + bs'$.*

Proof   The proof follows directly from the earlier result on sequences. The assumed convergence of the series implies that as sequences, $s_n \equiv \sum_{j=1}^{n} x_j$ and $s'_n \equiv \sum_{j=1}^{n} y_j$, converge to s and s', respectively; hence $as_n + bs'_n \rightarrow as + bs'$ from proposition 5.11.
                                                                                      ∎

Finally, we consider the termwise product sequence $\{x_j y_j\}$.

Proposition 6.7   *Let $\sum_{j=1}^{\infty} x_j$ and $\sum_{j=1}^{\infty} y_j$ be absolutely convergent series. Then for any a, b (real or complex):*

1. *$\sum_{j=1}^{\infty} [ax_j + by_j]$ is absolutely convergent.*
2. *$\sum_{j=1}^{\infty} x_j y_j$ is absolutely convergent.*

Proof   The first statement follows from the triangle inequality, since

$$\sum_{j=1}^{\infty} |ax_j + by_j| \leq |a| \sum_{j=1}^{\infty} |x_j| + |b| \sum_{j=1}^{\infty} |y_j|.$$

For the second, we show that $s_n \equiv \sum_{j=1}^{n} |x_j y_j|$ is a Cauchy sequence. Given $\epsilon > 0$, there is an N so that $\sum_{j=n}^{m} |x_j| < \epsilon$ and $\sum_{j=n}^{m} |y_j| < \epsilon$ for $n, m > N$. Now $\sum_{j=n}^{m} |x_j y_j| < \sum_{j=n}^{m} |x_j| \sum_{j=n}^{m} |y_j| < \epsilon^2$ for $n > N$, and the result follows.   ∎

Remark 6.8   *If the assumption on $\sum_{j=1}^{\infty} x_j$ and $\sum_{j=1}^{\infty} y_j$ is reduced to convergent, rather than absolutely convergent, then $\sum_{j=1}^{\infty} [ax_j + by_j]$ is convergent as noted in proposition 6.6, but $\sum_{j=1}^{\infty} x_j y_j$ need not be convergent. This will be assigned as exercise 21.*

6.1.3   Examples of Series

Example 6.9

1. *If* $x_n = a^n$, *a **geometric sequence**, then the associated **geometric series** converges if and only if* $|a| < 1$, *as can be demonstrated since the partial sums can be explicitly calculated. Specifically, if* $a \neq 1$, *from* $s_n = \sum_{j=1}^{n} a^j$ *and* $as_n = \sum_{j=2}^{n+1} a^j$, *we can solve for* s_n *by subtraction and obtain*

$$s_n = \frac{a^{n+1} - a}{a - 1}.$$

It is apparent that if $a > 1$, *then* $s_n \to \infty$, *and* a^{n+1} *grows without bound; while if* $a < -1$, *then* s_n *alternates sign between* $\pm$, *and* $|s_n| \to \infty$. *Similarly, if* $a = 1$, *then by the definition we have that* $s_n = n$, *which diverges, and if* $a = -1$, s_n *alternates between* -1 *and* 0. *Hence this series does not converge in any case for which* $|a| \geq 1$. *If* $|a| < 1$, *we conclude* $a^{n+1} \to 0$, *and hence*

$$\ast \quad \sum_{j=N}^{\infty} a^j = \frac{a^N}{1-a} \qquad \ast \quad \sum_{j=0}^{\infty} a^j = \frac{1}{1-a}$$

$$\sum_{j=1}^{\infty} a^j = \frac{a}{1-a}; \tag{6.1}$$

equivalently, $\sum_{j=0}^{\infty} a^j = \frac{1}{1-a}$. *Of course, this is exactly the calculation introduced in the pricing of perpetual preferreds in section 2.3.2, with* $a = (1+i)^{-1}$.

2. *If* $x_j = \frac{1}{j(j+1)}$, *then again by explicit calculation we can conclude that the sum* $\sum_{j=1}^{\infty} \frac{1}{j(j+1)}$ *converges. Since* $\frac{1}{j(j+1)} = \frac{1}{j} - \frac{1}{j+1}$, *we derive that* $s_n = \sum_{j=1}^{n} \frac{1}{j} - \sum_{j=2}^{n+1} \frac{1}{j}$, *which reduces to*

$$s_n = 1 - \frac{1}{n+1},$$

and hence $\sum_{j=1}^{\infty} \frac{1}{j(j+1)} = 1$.

3. *If* $x_j = \frac{1}{j}$, *the **harmonic series**, then surprisingly,* $\sum_{j=1}^{\infty} \frac{1}{j} = \infty$. *This result is justifiably the most surprising example of divergence of a series. The surprise stems from thinking about an arbitrarily large integer* N, *say the number of subatomic particles in the known universe. Then it is apparent that* $\sum_{j=1}^{N} \frac{1}{j}$ *is finite, and the next omitted term* $\frac{1}{N+1}$ *is an unimaginably small number, and the rest smaller yet. However, the divergence of the harmonic series implies that despite this unimaginable smallness,* $\sum_{j=N+1}^{\infty} \frac{1}{j}$ *is not finite. There are many proofs of this well-known fact; one seen in example 5.35 in chapter 5, but perhaps the simplest two are as follows:*

• *For an arbitrary integer $m > 1$, write*

$$\sum_{j=1}^{\infty} \frac{1}{j} = \sum_{j=1}^{m} \frac{1}{j} + \sum_{j=m+1}^{2m} \frac{1}{j} + \sum_{j=2m+1}^{3m} \frac{1}{j} + \cdots.$$

Now every summation on the right has m terms, and because the harmonic series is decreasing, each of these finite sums is strictly greater than m times the last term. That is,

$$\sum_{j=1}^{\infty} \frac{1}{j} > m\left(\frac{1}{m}\right) + m\left(\frac{1}{2m}\right) + m\left(\frac{1}{3m}\right) + \cdots$$

$$= \sum_{j=1}^{\infty} \frac{1}{j}.$$

So if $\sum_{j=1}^{\infty} \frac{1}{j}$ is finite, we can divide this inequality by this value to derive the absurd result $1 > 1$, or subtract to derive $0 > 0$. So via proof by contradiction we conclude that the harmonic series diverges.

• *Alternatively, we can manipulate this summation another way using a similar trick:*

$$\sum_{j=1}^{\infty} \frac{1}{j} = \sum_{j=1}^{m} \frac{1}{j} + \sum_{j=m+1}^{m^2} \frac{1}{j} + \sum_{j=m^2+1}^{m^3} \frac{1}{j} + \cdots$$

$$> m\left(\frac{1}{m}\right) + (m^2 - m)\left(\frac{1}{m^2}\right) + (m^3 - m^2)\left(\frac{1}{m^3}\right) + \cdots$$

$$= 1 + \left(1 - \frac{1}{m}\right) + \left(1 - \frac{1}{m}\right) + \left(1 - \frac{1}{m}\right) + \cdots,$$

from which the divergence is apparent since each term after the first equals the constant $1 - \frac{1}{m}$.

4. *If $x_j = \frac{1}{j^a}$ for $a > 1$, then the **power harmonic series**, $\sum_{j=1}^{\infty} \frac{1}{j^a}$, converges. Using the second trick above for the harmonic series, we create an upper bound with the first term of each group:*

$$\sum_{j=1}^{\infty} \frac{1}{j^a} = \sum_{j=1}^{m} \frac{1}{j^a} + \sum_{j=m+1}^{m^2} \frac{1}{j^a} + \sum_{j=m^2+1}^{m^3} \frac{1}{j^a} + \cdots$$

$$< m(1) + (m^2 - m)\frac{1}{(m+1)^a} + (m^3 - m^2)\frac{1}{(m^2+1)^a} + \cdots$$

$$< m + \frac{m^2 - m}{m^a} + \frac{m^3 - m^2}{m^{2a}} + \cdots$$

$$= m + (m-1) \sum_{j=1}^{\infty} m^{j(1-a)}.$$

The last summation is a convergent geometric series if $m^{1-a} < 1$. That is, if $a > 1$. Of course, as $a \to 1$, this last summation becomes increasingly large, as the given series approaches a summation of 1s, and the original series approaches the harmonic series.

In all these cases, note that the analysis done for the harmonic series was to infer divergence by manipulating the terms to produce a smaller and yet obviously divergent series, while the approach taken in the first two examples was to explicitly derive the summation. In many ways the harmonic series analysis is a more realistic example of analytics done in practice. The reason is that although there are many examples of series that can be evaluated explicitly, most of these require advanced methods of later chapters. In addition it is common to be confronted with a series that cannot be so evaluated even with more advanced techniques. In many of these cases this inability to find an exact value is not a problem since the primary question is related to the convergence or divergence of the series, and not to the exact value that the series converges to. If one can prove convergence, it is usually possible to develop a numerical approximation to the summation, or reasonable upper and lower bounds adequate for the purposes at hand.

There are many ways to prove convergence of series without an explicit evaluation of its summation. The most direct is the strategy employed for the geometric harmonic series, namely, to demonstrate that the series is smaller than one that apparently converges.

Example 6.10

5. *If $x_j = \frac{\ln j}{j^3}$, then $\sum_{j=1}^{\infty} x_j$ converges. To demonstrate this convergence without explicitly evaluating the actual summation, we show that this series is smaller than a simpler series that is readily seen to converge. First off, $\ln j < j$, and so $x_j < \frac{1}{j^2}$. Hence*

$$\sum_{j=1}^{\infty} \frac{\ln j}{j^3} < \sum_{j=1}^{\infty} \frac{1}{j^2} < \infty.$$

This second summation converges as in case 4 of example 5.9 with $a = 2$. Alternatively, by noting that $\frac{1}{j^2} < \frac{1}{j(j-1)}$ for $j \geq 2$, and with case 2 we conclude that this series converges to a value less than 2.

6. *If* $x_j = \frac{(-1)^{j+1}}{j}$, *the **alternating harmonic series**, then* $\sum_{j=1}^{\infty} x_j$ *converges. Taking this series in pairs, we obtain for* $n = 1, 3, 5, \ldots$ *that* $x_n + x_{n+1} = \frac{1}{n(n+1)}$, *which equals the odd terms of the series in case 2. Consequently*

$$\sum_{j=1}^{n} \frac{(-1)^{j+1}}{j} = \begin{cases} \sum_{j=1}^{m} \frac{1}{2j(2j-1)}, & n = 2m, \\ \sum_{j=1}^{m} \frac{1}{2j(2j-1)} - \frac{1}{2m+1}, & n = 2m + 1. \end{cases}$$

Therefore the even partial sums of the alternating harmonic series equal the partial sums of a subseries of the convergent series of case 2, while the odd partial sums equal this same convergent series but minus a term that converges to 0. The even and odd partial sums of this series must therefore converge to the same value. Yet, this series is only conditionally convergent, since the absolute value of this series is the harmonic series that diverges. As we will see as an application of a result from calculus in chapter 10, it turns out that $\sum_{j=1}^{\infty} \frac{(-1)^{j+1}}{j} = \ln 2$, *the natural logarithm of 2, which is approximately* 0.69315.

It is important to note that a subseries of a convergent series need not converge. The conclusion in case 6 is justified because the original convergent series in case 2 had all positive terms. More generally, what is needed is that the original series is absolutely convergent. An example of what can go wrong in the conditionally convergent case follows:

Example 6.11

7. *If* $x_j = \frac{(-1)^{j+1}}{j}$, *the (convergent) **alternating harmonic series**, then* $\sum_{j=1}^{\infty} x_{2j}$ *and* $\sum_{j=1}^{\infty} x_{2j-1}$ *both diverge. First off,*

$$\sum_{j=1}^{\infty} x_{2j} = -\frac{1}{2} \sum_{j=1}^{\infty} \frac{1}{j},$$

which is a multiple of the harmonic series. Similarly

$$\sum_{j=1}^{\infty} x_{2j-1} = \sum_{j=1}^{\infty} \frac{1}{2j - 1} > \sum_{j=1}^{\infty} \frac{1}{2j} = \frac{1}{2} \sum_{j=1}^{\infty} \frac{1}{j},$$

another multiple of the harmonic series.

Cases 3, 4, 5, 6, and 7 of the examples above present an application of the **comparison test** for a series. This and other tests are presented below in section 6.1.5 on tests of convergence. However, the next section provides two important results on absolutely versus conditionally convergent series.

*6.1.4   Rearrangements of Series

In attempting to evaluate the sum of a series or even to prove convergence, it is often desirable to be able to rearrange the order of the series. This is especially true for double series as will be seen below. But while a valid manipulation for finite sums, it is not always the case that an infinite sum can be rearranged without changing its value, or indeed changing whether or not it even converges. This section analyzes the relationship between convergence of a series and convergence of its rearrangements, as well as the associated summations.

To introduce the notion of a rearrangement formally, we introduce the notion of a **rearrangement function**, $\pi(n)$, defined on the index collection $J \equiv \{j\}_{j=0}^{\infty}$ or $J \equiv \{j\}_{j=1}^{\infty}$, with the property that $\pi : J \to J$ as a **one-to-one and onto function**. These words reflect three notions that can be reduced to the intuitive idea that π creates a "shuffle" of the set J:

• A "function" $J \to J$ means that for any $j \in J$, $\pi(j)$ is a unique element of J.

• "One-to-one" means that there cannot be $j, k \in J$ with $\pi(j) = \pi(k)$. Each j is mapped to a different point.

• "Onto" means that for any element $k \in J$, there is a $j \in J$, with $\pi(j) = k$.

Given a series $\{x_j\}$ the focus of this section has to do with the value of $\sum_{j=1}^{\infty} x_j$ versus the value of $\sum_{j=1}^{\infty} x_{\pi(j)}$ for an arbitrary rearrangement function π. Before presenting the results, let us consider two examples that highlight what can happen.

Example 6.12

1. *Recall the **alternating harmonic series** in example 6.11, $x_j = \frac{(-1)^{j+1}}{j}$, which converges but is not absolutely convergent. As was demonstrated, both $\sum_{j=1}^{\infty} x_{2j}$ and $\sum_{j=1}^{\infty} x_{2j-1}$ diverge, so the conditional convergence of this series occurs because of the cancellation that occurs between one subseries that is accumulating to $+\infty$, and the other subseries that is accumulating to $-\infty$. Intuition warns that rearranging this series could cause trouble. Indeed, if we simply rearrange the series with all the positive terms first, and all the negatives last, we arrive at a meaningless conclusion that $\sum_{j=1}^{\infty} x_j = \infty - \infty$, and we are justifiable cautious about concluding that this sum is 0. However, with a bit of ingenuity it is possible to rearrange this series so that the rearranged series converges conditionally to any real number, or even to $\pm\infty$. This seems impossible, but it is not too difficult to demonstrate. Let $r \in \mathbb{R}$ be given, and assume that $r \geq 0$. Choose N_1 to be the first integer so that $\sum_{j=1}^{N_1} x_{2j} > r$. Next choose M_1 to be the first integer so that $\sum_{j=1}^{N_1} x_{2j} + \sum_{j=1}^{M_1} x_{2j-1} < r$. Both choices are possible since the positive and nega-*

tive series grow without bound. Now choose $N_2 > N_1$ to be the first integer so that $\sum_{j=1}^{N_1} x_{2j} + \sum_{j=1}^{M_1} x_{2j-1} + \sum_{j=N_1+1}^{N_2} x_{2j} > r$, *and* $M_2 > M_1$ *to be the first integer so that* $\sum_{j=1}^{N_1} x_{2j} + \sum_{j=1}^{M_1} x_{2j-1} + \sum_{j=N_1+1}^{N_2} x_{2j} + \sum_{j=M_1+1}^{M_2} x_{2j-1} < r$, *and so forth. We can therefore show that this implied rearrangement of the series,*

$$x_2, \ldots, x_{2N_1}, x_1, \ldots, x_{2M_1-1}, x_{2(N_1+1)}, \ldots,$$

converges conditionally to r. For example, at the last step above, since M_2 was the first integer to produce the desired property, it is the case that

$$\sum_{j=1}^{N_1} x_{2j} + \sum_{j=1}^{M_1} x_{2j-1} + \sum_{j=N_1+1}^{N_2} x_{2j} + \sum_{j=M_1+1}^{M_2-1} x_{2j-1} > r,$$

and hence

$$\left| r - \left(\sum_{j=1}^{N_1} x_{2j} + \sum_{j=1}^{M_1} x_{2j-1} + \sum_{j=N_1+1}^{N_2} x_{2j} + \sum_{j=M_1+1}^{M_2} x_{2j-1} \right) \right| < |x_{2M_2}|.$$

In other words, at each step the difference between the partial summation and r is bounded by the absolute value of the last term added. Consequently, as these last added terms converge to 0 absolutely, conditional convergence is proved. If $r < 0$, the process is simply reversed. If $r = \pm\infty$, think about how this construction can be modified (answer is below in the proof of the Riemann series theorem).

2. *Consider an **alternating geometric series**, $x_j = (-1)^j a^j$, $j \geq 0$, where $0 < a < 1$. This series is absolutely convergent by example 6.9 above, so it is also convergent. Let the summation be denoted: $s = \sum_{j=0}^{\infty} (-1)^j a^j$. Then with $s_1 = \sum_{j=0}^{\infty} a^{2j} = \frac{1}{1-a^2}$ and $s_2 = \sum_{j=0}^{\infty} a^{2j+1} = as_1 = \frac{a}{1-a^2}$, we have $s = s_1 - s_2 = \frac{1}{1+a}$. Let π be a given rearrangement, and consider $\sum_{j=0}^{\infty} (-1)^{\pi(j)} a^{\pi(j)}$. The goal is to show that $\sum_{j=0}^{\infty} (-1)^{\pi(j)} a^{\pi(j)} = s$ and has the same value as the original series. To do so, for a given $\epsilon > 0$ we need to show that there is an N so that $|s - \sum_{j=0}^{n} (-1)^{\pi(j)} a^{\pi(j)}| < \epsilon$ for $n \geq N$. To this end, we focus on the positive and negative series separately. Since $s_1 = \sum_{j=0}^{\infty} a^{2j}$, choose N_1 so that $|s_1 - \sum_{j=0}^{n} a^{2j}| < \frac{\epsilon}{3}$ for $n \geq N_1$, and choose N_2 so that $|s_2 - \sum_{j=0}^{n} a^{2j+1}| < \frac{\epsilon}{3}$ for $n \geq N_2$. Also, since this series is absolutely convergent, we can apply the Cauchy criteria and choose N_3 so that $|\sum_{j=n}^{m} a^j| < \frac{\epsilon}{3}$ for $n, m > N_3$. Now note that for any n, $\{\pi(j)\}_{j=0}^{n}$ can be split into even and odd integers, and we choose N large enough so that $\{\pi(j)\}_{j=0}^{N}$ contains $\{j\}_{j=0}^{\max(N_j)}$. Then for $n \geq N$ we have by the triangle inequality,*

$$\left| s - \sum_{j=0}^{n} (-1)^{\pi(j)} a^{\pi(j)} \right|$$

$$= \left| [s_1 - s_2] - \left[\sum_{j=0}^{\max(N_j)} a^{2j} - \sum_{j=0}^{\max(N_j)} a^{2j+1} \right] + \sum_{\pi(j) \ge \max(N_j)} (-1)^{\pi(j)} a^{\pi(j)} \right|$$

$$\le \left| s_1 - \sum_{j=0}^{\max(N_j)} a^{2j} \right| + \left| s_2 - \sum_{j=0}^{\max(N_j)} a^{2j+1} \right| + \left| \sum_{\pi(j) \ge \max(N_j)} a^{\pi(j)} \right|$$

$$< \frac{\epsilon}{3} + \frac{\epsilon}{3} + \frac{\epsilon}{3} = \epsilon.$$

The following propositions summarize the results illustrated in the examples above. The proofs will be brief since they follow closely the developments given in these special cases. The first result is named for **Bernhard Riemann** (1826–1866).

Proposition 6.13 (Riemann Series Theorem)   *Let $\{x_j\}_{j=1}^{\infty}$ be a conditionally convergent series, $\sum_{j=1}^{\infty} x_j = s$. Then for any $r \in \mathbb{R}$, as well as $r = \pm\infty$, there is a rearrangement function π so that $\sum_{j=1}^{\infty} x_{\pi(j)} = r$.*

Proof   Since $\{x_j\}_{j=1}^{\infty}$ is not absolutely convergent, it must be the case that there are infinitely many terms in the series that are both positive and negative. This is because if either set was finite, say $\{x_j\}_{j=1}^{n}$ were the positive terms, then since $\sum_{j=1}^{\infty} x_j = \sum_{j=1}^{n} x_j + \sum_{j=n+1}^{\infty} x_j$, we derive that $\sum_{j=n+1}^{\infty} x_j = s - \sum_{j=1}^{n} x_j$. Now since all $x_j < 0$ for $j > n$, we have that $\sum_{j=n+1}^{\infty} |x_j| = \sum_{j=1}^{n} x_j - s$. This implies that $\sum_{j=1}^{\infty} |x_j| = 2 \sum_{j=1}^{n} x_j - s$, contradicting that $\{x_j\}_{j=1}^{\infty}$ is not absolutely convergent. So both positive and negative subseries are infinite. Next, denoting by $\{x_j^+\}_{j=1}^{\infty}$ and $\{x_j^-\}_{j=1}^{\infty}$ these infinite collections of positive and negative terms represented in their respective orderings, it must be the case that both $\sum_{j=1}^{\infty} x_j^+ = \infty$ and $\sum_{j=1}^{\infty} x_j^- = -\infty$. Again, if either were finite, the conditional convergence of $\{x_j\}_{j=1}^{\infty}$ would imply its absolute convergence, a contradiction. Now with these divergent positive and negative subseries, the proof is identical to the derivation above for the alternating harmonic series if $r \in \mathbb{R}$. In the case $r = \infty$, choose N_1 so that $\sum_{j=1}^{N_1} x_j^+ \ge 10|x_1^-|$, then choose N_2 so that $\sum_{j=N_1+1}^{N_2} x_j^+ \ge 10|x_2^-|$, and so forth. The rearrangement is $x_1^+, \ldots, x_{N_1}^+, x_1^-,$ $x_{N_1+1}^+, \ldots, x_{N_2}^+, x_2^-, \ldots$. By construction, the summation of each block of positives and one negative term, x_j^-, exceeds $9|x_j^-|$, and hence $\sum_{j=1}^{n} x_{\pi(j)}$ grows like $9 \sum_{j=1}^{m} |x_j^-|$, where m is the subscript of the largest N_j with $N_j \le n$. A similar type of construction produces the result for $r = -\infty$.   ∎

It is interesting to note that the rearrangements implied by this proposition have a special and initially not obvious property. Namely the collection of "forward shifts," $\{\pi(j) - j\}$, must be unbounded in the construction above for the summation of the series to shift from the original value of s to any new value r. In other words, in order to get the desired results, the rearrangement implied by this construction needs to map the elements of the index set $\{j\}$ farther and farther from their initial positions to new forward positions.

To investigate this, note that the construction in the proof above creates a series

$$x_1^+, \ldots, x_{N_1}^+, x_1^-, \ldots, x_{M_1}^-, x_{N_1+1}^+, \ldots, x_{N_2}^+, x_{M_1+1}^-, \ldots, x_{M_2}^-, \ldots$$

within which the forward shifts for positive terms appear to be unbounded, since they grow in relation to $\sum M_j$ as caused by the insertion of groups of negative terms. Similarly the forward shifts of negative terms appear unbounded as caused by the insertion of groups of positive terms.

But we need to be skeptical of this argument. The positive and negative terms were interspersed somehow initially, and perhaps interspersed similarly to what the construction called for. So this construction likely only changed the order a small amount, and not in the claimed unbounded way.

The next result shows in fact that if the rearrangement function only moves indexes by a limited amount, then the rearranged series converges to the original summation value and cannot be changed.

Proposition 6.14   *Let $\{x_i\}_{i=1}^{\infty}$ be a conditionally convergent series, $\sum_{j=1}^{\infty} x_j = s$, and π a rearrangement function with the property that for some integer P and all j, $\pi(j) \le j + P$. Then $\sum_{j=1}^{\infty} x_{\pi(j)} = s$.*

Proof   Consider the partial sums, $\sum_{j=1}^{n} x_j$ and $\sum_{j=1}^{n} x_{\pi(j)}$. By the given assumption on π that $\pi(j) \le j + P$, it must be the case that

$$\{x_{\pi(j)}\}_{j=1}^{n-P} \subset \{x_j\}_{j=1}^{n}.$$

It is also possible that some or all of $\{x_{\pi(j)}\}_{j=n-P+1}^{n}$ are also included in $\{x\}_{j=1}^{n}$, but this will not matter for the proof. So we can conclude that

$$\sum_{j=1}^{n} x_j - \sum_{j=1}^{n} x_{\pi(j)} = \sum_{j=n-P+1}^{n} x_j - \sum_{j=n-P+1}^{n} x_{\pi(j)},$$

where by assumption, $n - P + 1 \le \pi(j) \le n + P$ for $n - P + 1 \le j \le n$. Denoting $\{\pi(j)\}_{j=n-P+1}^{n}$ by $\{n - P + n_j\}_{j=1}^{P}$ for integers $1 \le n_j \le 2P$, we derive by the triangle inequality,

$$\left| \sum_{j=1}^{n} x_j - \sum_{j=1}^{n} x_{\pi(j)} \right| \le \sum_{j=1}^{P} |x_{n-P+j}| + \sum_{j=1}^{P} |x_{n-P+n_j}|.$$

Now, since $\{x_j\}_{j=1}^{\infty}$ is a convergent series, we have that $x_j \to 0$ as $j \to \infty$, so the sum of the $2P$ terms in this upper bound also converges to 0. More formally, for any $\epsilon > 0$, choose N so that $|x_j| < \frac{\epsilon}{2P}$ for $j > N$. Then choose n above so that $n - P + 1 > N$.   ∎

The implication of this result is that rearrangements of conditionally convergent series are allowable as long as the rearrangement is limited to index movements that are bounded in the sense above, whereby for all j, $\pi(j) \le P + j$ for some fixed P.

As an application, if a series is presented for evaluation of convergence, any number of rearrangements are possible within the rule that $\pi(j) \le P + j$ for some fixed P. If such manipulations then provide a basis for concluding convergence, then one can be assured that the original series converges to the same value. In other words, this result can be applied backward in that if a bounded rearrangement produces a convergent series, then the original series must be convergent to the same value. As the proposition demonstrated, however, with unbounded rearrangements, anything can happen.

The conclusion for absolutely convergent series is completely general, in that such a series can be rearranged in any way without changing the value of the sum.

Proposition 6.15   *Let $\{x_j\}_{j=1}^{\infty}$ be an absolutely convergent series, $\sum_{j=1}^{\infty} x_j = s$, and π any rearrangement function. Then $\sum_{j=1}^{\infty} x_{\pi(j)} = s$.*

Proof   The goal is to reproduce the proof used for the alternating geometric series in case 2 of example 6.12, but we first need to show that this series can be split into a positive and negative subseries, and that each of these converges to values that in turn sum to s. To this end, define $\{x_j^+\}_{j=1}^{\infty}$ and $\{x_j^-\}_{j=1}^{\infty}$ by

$$x_j^+ = \max\{x_j, 0\}, \quad x_j^- = \max\{-x_j, 0\}.$$

For the alternating geometric series above, this definition produces $x_{2j}^+ = a^{2j}$, $x_{2j-1}^- = a^{2j-1}$, and both subseries are 0 for other indexes. Now note that $x_j = x_j^+ - x_j^-$, and $|x_j| = x_j^+ + x_j^-$. Since this series is absolutely convergent, both subseries $x_j^+ = \frac{1}{2}[x_j + |x_j|]$ and $x_j^- = \frac{1}{2}[|x_j| - x_j]$ are absolutely convergent to s_1 and s_2, respectively. Therefore

$$\left| \sum_{j=1}^{n} x_j - (s_1 - s_2) \right| \le \left| \sum_{j=1}^{n} x_j^+ - s_1 \right| + \left| \sum_{j=1}^{n} x_j^- - s_2 \right|,$$

which implies that $\sum_{j=1}^{\infty} x_j = s_1 - s_2 = s$. With this setup the proof of this result for the alternating geometric series now can be implemented identically, by substituting x_j^+ and x_j^- in the roles of the positive and negative terms in example 6.12. ∎

Example 6.16 *Two common and important applications of this last result are:*

1. *If a series is given with only positive or negative terms, or one with only a finite number of terms of one sign and the remainder of the other, then such a series is convergent if and only if it is absolutely convergent. Consequently one can apply completely arbitrary rearrangements to the series in search of evidence of convergence because, once such evidence is found, one concludes absolute convergence, justifies the rearrangement by the proposition above, and knows that the original series must have the same summation as that developed for the rearrangement.*

2. *Since the rearrangement functions contemplated by the proposition above are completely general, one could in theory split such a series into a series of even terms followed by a series of odd terms, or in three collections*

$$x_1, x_4, \ldots, x_2, x_5, \ldots, x_3, x_6, \ldots$$

or any number of countably infinite subseries. An important application of this observation is to a "multiple" series, such as the double series,

$$\sum_{j=1}^{\infty} \sum_{i=1}^{n(j)} x_{ij},$$

where $n(j)$ is some function of j, or simply $n(j) = \infty$, for all j. A common example is $n(j) = j$. Of course, triple, quadruple, and higher order series are similarly defined, though less common in applications. These summations are always intended to be performed from the outer summation inward so that in the example above,

$$\sum_{j=1}^{\infty} \sum_{i=1}^{n(j)} x_{ij} = \sum_{i=1}^{n(1)} x_{i1} + \sum_{i=1}^{n(2)} x_{i2} + \sum_{i=1}^{n(3)} x_{i3} + \sum_{i=1}^{n(4)} x_{i4} + \cdots.$$

*One can envision these index points on the **positive integer lattice** in $\mathbb{R}^2$, where x_{ij} is defined at each point (i, j), $i, j > 0$ as in **figure 6.1**. The double summation is then envisioned as summing along rows, starting with $j = 1$ and summing the first row from $i = 1$ to $n(1)$, then the second row, from $i = 1$ to $n(2)$, and so forth. It is often convenient to be able to reverse the order of the summation, to in effect sum by columns first. For example,*

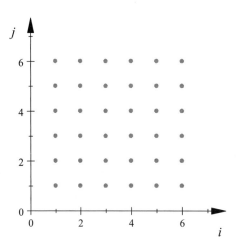

Figure 6.1
Positive integer lattice

$$\sum_{j=1}^{\infty}\sum_{i=1}^{\infty} x_{ij} \quad becomes \quad \sum_{i=1}^{\infty}\sum_{j=1}^{\infty} x_{ij},$$

$$\sum_{j=1}^{\infty}\sum_{i=1}^{j} x_{ij} \quad becomes \quad \sum_{i=1}^{\infty}\sum_{j=i}^{\infty} x_{ij}.$$

In the second summation, the integer lattice model simplifies the setting of the limits for the reversed summations by providing a visual representation. The question that arises is, can summations be switched in such a manner? Intuitively, if the series is only conditionally convergent, there is little hope of a positive conclusion, since it is apparent that such rearrangements move series terms by arbitrarily large distances. On the other hand, if the series has terms of one sign, or all but a finite number of one sign, then again it will be convergent if and only if absolutely convergent. In such cases, the result above on absolute convergence is again applied backward; that is, if one rearranges as necessary and convergence is justified, so too is absolute convergence. So we can conclude that the original multiple series has the same summation as the rearranged series.

6.1.5   Tests of Convergence

There are many tests of convergence for a series, and at first their large number may seem odd. Just how many tests does one need? The problem is that no test is stated in the unambiguous language:

The series $\sum_{j=1}^{\infty} x_j$ converges if and only if...;

that is, except the test in the definition itself, which then goes on to require for the Cauchy condition that

...for every $\epsilon > 0$, $\exists N$ with $|\sum_{j=1}^{m} x_j - \sum_{j=1}^{n} x_j| < \epsilon$ for $n, m > N$.

So the definition of convergence provides an "iff" test of convergence, but in many cases there is no easy way to demonstrate that there is a value of $N \equiv N(\epsilon)$ that will work.

The various tests of convergence provide the benefit of relative ease of implementation, but at the cost of so-called **indeterminate cases**. To be more precise, all tests provide the following schema, either explicitly or implicitly:

1. The series $\sum_{j=1}^{\infty} x_j$ converges if **condition** A is satisfied.

2. The series $\sum_{j=1}^{\infty} x_j$ diverges if **condition** B is satisfied.

3. No information on convergence is provided in other cases.

So every test divides the collection of all series $\{\sum_{j=1}^{\infty} x_j \mid x_j \in \mathbb{R} \text{ or } \mathbb{C}\}$, into these three groups according to that test's conditions. A given series may be in the indeterminate group for one test, and demonstrated to converge or diverge with another. Of course, it will never be the case that one test assures convergence, another divergence, or conversely.

The reason for the multitude of tests is that each varies in terms of ease of implementation for a given series, as well as in terms of the specific members of the group of series that remain indeterminate. Tests can be intuitively thought of as stronger if they provide a smaller indeterminate set, but there is no generally accepted ordering for the strength of such tests unless one test's indeterminate set is contained in another's.

So far, no test other than the definition itself has been discovered that has $\emptyset$, the empty set, as its indeterminate collection. In this section we identify a few of the best and easiest to implement tests. Also a very useful test will be added in chapter 10, using a method involving Riemann integrals. The first test is probably the most widely used because it affords the analyst a great deal of flexibility in its application.

Proposition 6.17 (*The Comparison Test*) *If $\sum_{j=1}^{\infty} x_j$ is absolutely convergent, and $\sum_{j=1}^{\infty} y_j$ is any series with $|y_j| \le |x_j|$, for $j \ge N$ for some N, then $\sum_{j=1}^{\infty} y_j$ converges absolutely. Conversely, if $\sum_{j=1}^{\infty} x_j$ and $\sum_{j=1}^{\infty} y_j$ are any series with $|y_j| \le |x_j|$, for some $j \ge N$, and $\sum_{j=1}^{\infty} |y_j| = \infty$, then $\sum_{j=1}^{\infty} |x_j| = \infty$.*

Proof For the convergence condition, if $s_n = \sum_{j=1}^{n} |y_j|$, then for $n \geq N$,

$$s_n \leq \sum_{j=1}^{N} |y_j| + \sum_{j=N+1}^{n} |x_j| \leq \sum_{j=1}^{N} |y_j| + \sum_{j=1}^{\infty} |x_j|.$$

In other words, the absolute partial sums of the series $\{y_j\}$ are both increasing with n, and bounded. Because these partial sums are bounded, they must have an accumulation point. So there is an s such that for any $\epsilon > 0$ there is an $M(\epsilon)$ with $|s_M - s| < \epsilon$. However, since the sequence $\{s_n\}$ is increasing, $|s_n - S| < \epsilon$ for $n \geq M$, and hence s is the limit of the partial sums. That is, $\sum_{j=1}^{\infty} |y_j| = s$. For the divergence condition, it is clear by assumption that the absolute partial sums, $s_n = \sum_{j=1}^{n} |y_j|$, are unbounded. Consequently, since all but a finite number of $|x_j|$ exceed $|y_j|$, the partial sums of this series must also be unbounded, and hence $\sum_{j=1}^{\infty} |x_j| = \infty$. ∎

Remark 6.18

1. Note that for the purpose of establishing convergence by the comparison test, or divergence, one can ignore any finite number of terms of the respective sequences. In other words, the relationship between $|y_j|$ and $|x_j|$, for $j \leq N$ and any fixed N, is irrelevant to the conclusions.

2. Note also that the assumption in the comparison test for convergence is that for some N and $j \geq N$,

$$-|x_j| \leq y_j \leq |x_j|.$$

That is, that all but finitely many terms of $\{y_j\}$ are bounded by two convergent series. This can be generalized. Namely, if there are two convergent series $\sum_{j=1}^{\infty} x_j$ and $\sum_{j=1}^{\infty} z_j$ so that

$$x_j \leq y_j \leq z_j \qquad \text{for } j \geq N \text{ for some } N,$$

then $\sum_{j=1}^{\infty} y_j$ is convergent. This is because $0 \leq z_j - y_j \leq z_j - x_j$, and since $\sum_{j=1}^{\infty}(z_j - x_j)$ converges by assumption, and hence converges absolutely because the terms are nonnegative, we conclude that $\sum_{j=1}^{\infty}(z_j - y_j)$ converges, and in fact converges absolutely. Subtracting the convergent $\sum_{j=1}^{\infty} z_j$ implies the result.

Example 6.19 *Consider $\sum_{n=1}^{\infty} \frac{1}{n!}$, where as usual, $n! \equiv n(n-1)(n-2)\cdots 2 \cdot 1$ is called n **factorial**. Note that for $n \geq 4$,*

$$\frac{n(n+1)}{n!} = \frac{n+1}{n-1} \frac{1}{(n-2)!} \leq \frac{5}{3} \cdot \frac{1}{2} < 1.$$

In other words, $\frac{1}{n!} < \frac{1}{n(n+1)}$ for $n \geq 4$, and consequently, $\sum_{n=1}^{\infty} \frac{1}{n!}$ converges by the comparison test because $\sum_{n=1}^{\infty} \frac{1}{n(n+1)}$ converges by case 2 in example 6.9.

The next test generalizes the result observed for the alternating harmonic series in example 6.10.

Proposition 6.20 (Alternating Series Convergence Test)   If $\sum_{j=1}^{\infty} x_j$ is an alternating series, and for some N we have $|x_{j+1}| \leq |x_j|$ for $j \geq N$ and $x_j \to 0$, then $\sum_{j=1}^{\infty} x_j$ converges. If s denotes the summation, we have the partial sum error estimate with $s_n = \sum_{j=1}^{n} x_j$:

$$|s_n - s| \leq |x_{n+1}| \qquad \text{for } n \geq N.$$

Proof   Since $\sum_{j=1}^{N-1} x_j = s'$ is finite, and for $n \geq N$,

$$s_n = \sum_{j=N}^{n} x_j + s',$$

we can ignore these exceptional terms and assume that $|x_j|$ monotonically decreases to 0 for all j. For specificity, assume that $x_1 > 0$. We first show that the odd partial sums form a decreasing sequence that is bounded below. This follows from

$$s_{2n+1} = s_{2n-1} + x_{2n} + x_{2n+1}$$

$$\leq s_{2n-1},$$

since $x_{2n} \leq 0 \leq x_{2n+1}$ and $|x_{2n+1}| \leq |x_{2n}|$ by the monotonicity assumption. In addition this sequence is bounded below by 0, since we have that every s_{2n+1} can be expressed as a summation of nonnegative terms by $s_{2n+1} = x_{2n+1} + \sum(x_{2j} + x_{2j-1})$, where the summation is from $j = 1$ to n.

Similarly the even partial sums form an increasing sequence that is bounded above. By proposition 5.18, both sequences are convergent, say to E and O for even and odd. But since $|s_{2n+1} - s_{2n}| = |x_{2n+1}| \to 0$, we have $E = O = s$ and $s_n \to s$. Now by this discussion,

$$s_{2n} \leq s \leq s_{2n+1} \qquad \text{for all } n,$$

so $0 \leq s - s_{2n} \leq s_{2n+1} - s_n = x_{2n+1}$. Similarly $0 \leq s_{2n+1} - s \leq s_{2n+1} - s_{2n+2} \leq x_{2n+2}$, and the error bounds follow.   ∎

Example 6.21   *As a simple application to the alternating harmonic series, if we desire an estimate of the summation that is within ϵ of the true sum, we simply choose N so* *that $\frac{1}{N+1} < \epsilon$. We then know from the proposition above that $s_N = \sum_{j=1}^{N} \frac{(-1)^{j+1}}{j}$ will be within ϵ of the correct answer. As noted above, using methods of calculus, we will derive that $s = \ln 2$, and we can conclude that*

$$\left| \ln 2 - \sum_{j=1}^{N} \frac{(-1)^{j+1}}{j} \right| \leq \frac{1}{N+1}.$$

To get the correct Mth decimal place of $\ln 2$, which is to say that we want an error of less that $\frac{0.5}{10^{M+1}}$, requires about $N \approx 2(10^{M+1})$ terms of this summation. In other words, although this series converges, it does so very slowly.

Next are two tests for convergence that depend on ratios. The first uses ratios of the given series' terms with those of an absolutely convergent series; the second uses ratios of consecutive terms from the given series.

Proposition 6.22 (*Comparative Ratio Test*)   *If $\sum_{j=1}^{\infty} x_j$ is an absolutely convergent series, and $\{y_j\}$ is a sequence so that $\lim_{j\to\infty} \frac{|y_j|}{|x_j|}$ exists, then $\sum_{j=1}^{\infty} y_j$ is absolutely convergent.*

Proof   The existence of this limit implies that $\left\{ \frac{|y_j|}{|x_j|} \right\}$ is a bounded sequence, and hence $|y_j| \leq B|x_j|$ for all j. Since $\sum_{j=1}^{\infty} Bx_j$ is absolutely convergent by assumption, the result follows by the comparison test in proposition 6.17.   ■

Remark 6.23   *This innocent looking result provides a powerful intuitive conclusion about convergence. First off, if $\sum_{j=1}^{\infty} x_j$ is an absolutely convergent series, it is apparent that $|x_j| \to 0$. Therefore for any $\epsilon > 0$ there is an N so that $|x_j| < \epsilon$ for $j \geq N$. The comparative ratio test says that if $\{y_j\}$ is any sequence that converges as fast or faster to 0, that is,*

$$\lim_{j\to\infty} \frac{|y_j|}{|x_j|} = C \geq 0,$$

then $\sum_{j=1}^{\infty} y_j$ is also absolutely convergent.

In other words, any absolutely convergent series provides a "speed benchmark" for the rate at which the absolute value of its terms converge to 0 in that every series that converges as fast or faster must also be absolutely convergent.

Although there are many other tests of convergence, we end with one of the most useful, as will be seen in the next section.

Proposition 6.24 (*Ratio Test*)   *If $\sum_{j=1}^{\infty} x_j$ is a series so that*

$$\limsup_{n \to \infty} \left\{ \frac{|x_{n+1}|}{|x_n|} \right\} = L < 1,$$

then $\sum_{j=1}^{\infty} x_j$ is absolutely convergent. On the other hand, if

$$\liminf_{n \to \infty} \left\{ \frac{|x_{n+1}|}{|x_n|} \right\} = L > 1,$$

then $\sum_{j=1}^{\infty} x_j$ diverges. If $L = 1$ in either case, no conclusion can be drawn.

Remark 6.25   *Recall the intuitive definition of limits superior and inferior. That is, consider all values of the sequential ratios $\left\{ \frac{|x_{n+1}|}{|x_n|} \right\}$, as well as all possible accumulation points. The ratio test says that if the largest such accumulation point is less than 1, the series must be absolutely convergent, and if the smallest is greater than 1, the series diverges. This test is powerful because it does not require the existence of the limit of these ratios, it only depends on values of the smallest and largest accumulation points.*

Of course, if the limit of these ratios exists, then the series converges absolutely or diverges according to whether the limit is less than or greater than 1. The indefinite case of $L = 1$ is easy to illustrate. From cases 3 and 4 of example 6.9, we know that $\sum \frac{1}{j}$ diverges and $\sum \frac{1}{j^2}$ converges, and yet for both, $L = 1$ as is easily verified.

Proof   In the first case where $\limsup_{n \to \infty} \left\{ \frac{|x_{n+1}|}{|x_n|} \right\} = L < 1$, by proposition 5.22, for any ϵ there is an N so that $\left\{ \frac{|x_{n+1}|}{|x_n|} \right\} < L + \epsilon$ for $n \geq N$. Choose $\epsilon < 1 - L$; then for any $m \geq 1$,

$$\frac{|x_{N+m}|}{|x_N|} = \frac{|x_{N+m}|}{|x_{N+m-1}|} \frac{|x_{N+m-1}|}{|x_{N+m-2}|} \cdots \frac{|x_{N+1}|}{|x_N|}$$

$$< (L + \epsilon)^m.$$

In other words, $|x_{N+m}| < (L + \epsilon)^m |x_N|$ for all $m \geq 1$, so $\{|x_{N+m}|\}$ is bounded above by a geometric series. Now, since $L + \epsilon < 1$ by construction, this geometric series must converge, and so too the original series by the comparison test. The limit inferior result is similar, only we conclude that $|x_{N+m}| > (L - \epsilon)^m |x_N|$, where ϵ is chosen as $\epsilon < L - 1$, so this sequence is bigger than a divergent geometric series as $L - \epsilon > 1$.   ∎

6.2   The l_p-Spaces

6.2.1   Definition and Basic Properties

The primary reason to introduce the notion of the l_p-**spaces** is that they represent an accessible introduction to an idea that will find more application with the notion of L_p function spaces studied in real analysis. In addition l_p-spaces provide an interesting and important counterpoint to the conclusion drawn in chapter 3, that all l_p-norms are equivalent in $\mathbb{R}^n$. We now study what happens to this conclusion when $n \to \infty$.

Notation 6.26   *While one can easily distinguish between l_p-space and L_p-space in writing, it is more difficult to do so in conversation, since both are pronounced "lp space." For this reason one sometimes hears "little lp space" and "big lp space" in a discussion.*

Definition 6.27   *For $1 \leq p \leq \infty$ the space l_p is defined by*

$$l_p = \{ \mathbf{x} = \{x_j\}_{j=1}^{\infty} \mid \|\mathbf{x}\|_p < \infty \},$$

where, consistent with the l_p-norms defined for Euclidean space,

$$\|\mathbf{x}\|_p \equiv \left(\sum |x_j|^p \right)^{1/p}, \qquad 1 \leq p < \infty, \tag{6.2a}$$

$$\|\mathbf{x}\|_{\infty} \equiv \sup_{j}\{|x_j|\}. \tag{6.2b}$$

Real l_p-space and complex l_p-space are defined according to whether $\{x_j\}_{j=1}^{\infty} \subset \mathbb{R}$ or $\{x_j\}_{j=1}^{\infty} \subset \mathbb{C}$. The absolute values $|x_j|$ in (6.2) are defined according to x_j being real or complex, as in (2.3) and (2.2), respectively.

Intuitively, one can imagine real l_p-space as an infinite Euclidean space, $\mathbb{R}^{\infty}$, under the previously defined l_p-norms. That is a good starting point for our intuition, in that we will see that the l_p-spaces are vector spaces just as was Euclidean space, and that the l_p-norms defined above are indeed norms in the sense of chapter 3.

There is a dramatic difference, however. Earlier we saw that all l_p-norms are equivalent in $\mathbb{R}^n$ for $1 \leq p \leq \infty$. Switching from one norm to another changed the numerical value of our norm measurements, but in every real sense the spaces were identical. By definition, the basic collection of points in $\mathbb{R}^n$ were the same, and the notions of open and closed, as well as convergence, were identical under any of these norms.

For example, $G \subset \mathbb{R}^n$ is open with respect to one l_p-norm if and only if it is open with respect to all l_p-norms. Similarly a sequence $\{\mathbf{x}_n\} \subset \mathbb{R}^n$ converges to $\mathbf{x} \in \mathbb{R}^n$ in one l_p-norm if and only if it converges with respect to all l_p-norms. Put another way, $\{\mathbf{x}_n\} \subset \mathbb{R}^n$ is a Cauchy sequence with respect to one l_p-norm if and only if it is a Cauchy sequence with respect to all l_p-norms.

On the other hand, the l_p-norms are not equivalent in $\mathbb{R}^\infty$. In fact, for any p with $1 \le p < \infty$, it is easy to produce a sequence $\{x_j\}_{j=1}^\infty$ so that $\{x_j\}_{j=1}^\infty \in l_{p'}$ for all p' with $p < p' \le \infty$ but $\{x_j\}_{j=1}^\infty \notin l_p$. The simplest example uses case 4 in example 6.9. For p given, define

$$\mathbf{x} = \{x_j\}_{j=1}^\infty \equiv \left\{ \left(\frac{1}{j} \right)^{1/p} \right\}_{j=1}^\infty.$$

Then in l_p, the norm of this point is the pth-root of the sum of the harmonic series, and hence it cannot have finite l_p-norm. However, by case 4, this point has finite $l_{p'}$-norm for any $p' > p$ with $p' < \infty$. In addition $\|\mathbf{x}\|_\infty = 1$. This generalizes to:

Proposition 6.28 *If $1 \le p < p' \le \infty$, then $l_p \subset l_{p'}$, and the inclusion is strict.*

Proof Let $\mathbf{x} = \{x_j\}_{j=1}^\infty \in l_p$ be given. Then the finiteness of $\|\mathbf{x}\|_p$ implies that all but a finite number of x_j satisfy $|x_j| < 1$. Now, if $p' > p$ and $p' < \infty$, then

$$\sum |x_j|^{p'} = \sum_{|x_j| < 1} |x_j|^{p'} + \sum_{|x_j| \ge 1} |x_j|^{p'} < \sum_{|x_j| < 1} |x_j|^p + C.$$

Consequently $\|\mathbf{x}\|_{p'}$ is finite and $\{x_j\}_{j=1}^\infty \in l_{p'}$. For $p' = \infty$, it is apparent that $\|\mathbf{x}\|_\infty = \sup_j\{|x_j|\}$ is finite since $\sum |x_j|^p$ is finite. Hence, in all cases, $l_p \subset l_{p'}$. That this inclusion is strict was exemplified in the case of the power harmonic series for $p' < \infty$. The case $p' = \infty$ is easily handled by the example $\mathbf{x} = \{x_j\}_{j=1}^\infty$, where all $x_j = 1$, say. Clearly, $\mathbf{x} \in l_\infty$ but in no other l_p-space for $p < \infty$.   ∎

More surprisingly, there exists an infinite collection of sequences that are in all the l_p-spaces and are in fact dense in all the l_p-spaces for $1 \le p < \infty$. So the differences between these spaces is caused entirely by the "completion" of the common collection of sequences in the various norms. To be more precise each l_p-space can be created by adding to this common collection of sequences the limiting values obtained by forming convergent sequences in the various norms.

To illustrate such a construction first in a more familiar setting, consider $\mathbb{Q}^n \subset \mathbb{R}^n$, defined as the n-tuples of rational numbers. That is,

$$\mathbb{Q}^n = \{\mathbf{x} \equiv (x_1, x_2, \ldots, x_n) \mid x_j \in \mathbb{Q} \text{ for all } j\},$$

which is a vector space over the rational numbers. Next, define $\mathbb{R}_p^n$ by

$$\mathbb{R}_p^n = \{\mathbf{x} \in \mathbb{R}^n \text{ so that } \exists \{\mathbf{x}_j\} \subset \mathbb{Q}^n \text{ with } \|\mathbf{x} - \mathbf{x}_j\|_p \to 0, 1 \le p \le \infty\}. \tag{6.3}$$

Proposition 6.29   *For any n, $\mathbb{R}_p^n = \mathbb{R}^n$ for all p, $1 \le p \le \infty$.*

Proof   By definition, $\mathbb{R}_p^n \subset \mathbb{R}^n$, so only the reverse inclusion need be proved. Let $\mathbf{x} \in \mathbb{R}^n$ be given. Define $\mathbf{x}_j \in \mathbb{Q}^n$ so that the integer parts, and the first j-decimals of the components of $\mathbf{x}_j$ agree with those of $\mathbf{x}$, and the decimal expansions of the components of $\mathbf{x}_j$ are all 0s past this jth position. Clearly, $\{\mathbf{x}_j\} \subset \mathbb{Q}^n$. It is also clear that for $p < \infty$, using the geometric series summation approach illustrated in example 6.9 above obtains

$$\|\mathbf{x} - \mathbf{x}_j\|_p \le 9\left(\sum_{k=j+1}^{\infty} 10^{-kp}\right)^{1/p} = 9\left(\frac{10^{-(j+1)}}{(1 - 10^{-p})^{1/p}}\right),$$

which converges to 0 as $j \to \infty$. For $p = \infty$, $\|\mathbf{x} - \mathbf{x}_j\|_\infty \le 9(10^{-(j+1)})$, which again converges to 0. Hence $\mathbb{R}^n \subset \mathbb{R}_p^n$ for all p, $1 \le p \le \infty$, and $\mathbb{R}_p^n = \mathbb{R}^n$.   ∎

In other words, starting with this common vector space, $\mathbb{Q}^n$, if we complete this space with respect to any of the l_p-norms, the same vector space arises, namely $\mathbb{R}^n$. Put yet another way, $\mathbb{Q}^n$ is dense in $\mathbb{R}^n$ with respect to every l_p-norm. We next show that there is also a common vector space that is dense in all the l_p-spaces, and that each l_p-space arises by completing this common space with respect to the associated norm. To this end, we introduce the following:

Definition 6.30 $\mathbb{R}^\infty$ *and* $\mathbb{C}^\infty$ *are formally defined as the collection of sequences:*

$$\mathbb{R}^\infty = \{\mathbf{x} \equiv (x_1, x_2, \ldots, x_n, \ldots) \mid x_j \in \mathbb{R} \text{ for all } j\}, \tag{6.4a}$$

$$\mathbb{C}^\infty = \{\mathbf{x} \equiv (x_1, x_2, \ldots, x_n, \ldots) \mid x_j \in \mathbb{C} \text{ for all } j\}. \tag{6.4b}$$

Similarly $\mathbb{R}_0^\infty$ *and* $\mathbb{C}_0^\infty$ *are formally defined as "truncated" sequences:*

$$\mathbb{R}_0^\infty = \{\mathbf{x} \in \mathbb{R}^\infty \mid x_j = 0 \text{ for all } j > N, \text{ some } N\}, \tag{6.5a}$$

$$\mathbb{C}_0^\infty = \{\mathbf{x} \in \mathbb{C}^\infty \mid x_j = 0 \text{ for all } j > N, \text{ some } N\}. \tag{6.5b}$$

Addition and scalar multiplication are defined pointwise:

$$\mathbf{x} + \mathbf{y} \equiv (x_1 + y_1, x_2 + y_2, \ldots, x_n + y_n, \ldots),$$

$$a\mathbf{x} \equiv (ax_1, ax_2, \ldots, ax_n, \ldots),$$

where $a \in \mathbb{R}$ for $\mathbb{R}^\infty$ and $\mathbb{R}_0^\infty$, and $a \in \mathbb{C}$ for $\mathbb{C}^\infty$ and $\mathbb{C}_0^\infty$.

Remark 6.31   *It is easy to see that $\mathbb{R}^\infty$ and $\mathbb{R}_0^\infty$ are vector spaces over $\mathbb{R}$, and that $\mathbb{C}^\infty$ and $\mathbb{C}_0^\infty$ are vector spaces over $\mathbb{C}$, based on definition 3.3 in chapter 3. Also, by the definition of the l_p-spaces, it is clear that for every p, $1 \le p \le \infty$, $\mathbb{R}_0^\infty \subset l_p \subset \mathbb{R}^\infty$ in the real case, and $\mathbb{C}_0^\infty \subset l_p \subset \mathbb{C}^\infty$ in the complex case. We study the l_p-spaces in the next section, but first demonstrate an interesting point. For conciseness, we limit the following statement to the real l_p-spaces, but it is equally valid in the complex case:*

Proposition 6.32   *The vector space $\mathbb{R}_0^\infty$ is dense in every l_p-space, $1 \le p < \infty$. That is, given any $\mathbf{x} \in l_p$, there is a sequence, $\{\mathbf{x}_n\} \subset \mathbb{R}_0^\infty$ so that $\|\mathbf{x} - \mathbf{x}_n\|_p \to 0$.*

Proof   Given $\mathbf{x} \equiv (x_1, x_2, \ldots, x_n, x_{n+1}, \ldots)$, define $\mathbf{x}_n = (x_1, x_2, \ldots, x_n, 0, 0, 0, \ldots)$. In other words, $\mathbf{x}_n$ is defined to have n nonzero components equal to the first n-components of $\mathbf{x}$. Now for $p < \infty$, $\mathbf{x} \in l_p$ implies that $\|\mathbf{x}\|_p^p = \sum_{j=1}^\infty |x_j|^p < \infty$. By definition, this implies that for any $\epsilon > 0$ there is an N so that $\sum_{j=n}^\infty |x_j|^p < \epsilon$ for $n > N$. However, $\|\mathbf{x} - \mathbf{x}_n\|_p^p = \sum_{j=n+1}^\infty |x_j|^p$, and hence $\|\mathbf{x} - \mathbf{x}_n\|_p^p \to 0$ as $n \to \infty$.   ■

It is important to note that this result does not extend to $p = \infty$, as a simple example demonstrates. If $\mathbf{x} = (1, 1, 1, 1, 1, \ldots)$, the constant vector, $\|\mathbf{x} - \mathbf{x}_n\|_\infty = \sup_{j>n}\{|x_j|\} = 1$, so no convergence occurs in the l_∞-norm.

*6.2.2   Banach Space

For l_p-spaces to be really useful, there are two as yet unanswered questions that need to be addressed:

1. While l_p-space is closed under addition and scalar multiplication as a vector space, is it closed as a normed space? In other words, if $\mathbf{x}, \mathbf{y} \in l_p$, must it be true that $\mathbf{x} + \mathbf{y} \in l_p$, and so $\mathbf{x} + \mathbf{y}$ has a finite l_p-norm?

2. Are the l_p-spaces complete? That is, if $\{\mathbf{x}_n\} \subset l_p$ is a Cauchy sequence, must there be an $\mathbf{x} \in l_p$ so that

$$\|\mathbf{x} - \mathbf{x}_n\|_p \equiv \left[\sum_{j=1}^\infty (x_j - x_{nj})^p\right]^{1/p} \to 0?$$

These questions are addressed in this section, and both are answered in the affirmative. First, the affirmative result on closure under addition.

Proposition 6.33   *Real l_p-space is a normed linear space over the real numbers, $\mathbb{R}$, and complex l_p-space is a normed linear space over the complex numbers, $\mathbb{C}$. In addition in both spaces we have the **Minkowski inequality**:*

$$\|\mathbf{x} + \mathbf{y}\|_p \leq \|\mathbf{x}\|_p + \|\mathbf{y}\|_p. \tag{6.6}$$

Proof   Because these collections are defined as subsets of the vector spaces $\mathbb{R}^\infty$ and $\mathbb{C}^\infty$, all that is left to prove is that these spaces are closed under the above-given definitions of addition and scalar multiplication, and that the l_p-norms defined in (6.2) are indeed norms in the sense of chapter 3. Of course, closure under scalar multiplication is immediate, since for any p, $\|a\mathbf{x}\|_p = |a|\,\|\mathbf{x}\|_p$. The more subtle question is addition, and for this, we demonstrate the Minkowski inequality. As in Euclidean space, the Minkowski inequality is the name given to the triangle inequality under the l_p-norm. This result is apparent for $p = \infty$ since

$$\sup_j \{|x_j + y_j|\} \leq \sup_j \{|x_j|\} + \sup_j \{|y_j|\},$$

and for $p = 1$ by the triangle inequality,

$$|x_j + y_j| \leq |x_j| + |y_j|,$$

which implies by summation that $\|\mathbf{x} + \mathbf{y}\|_1 \leq \|\mathbf{x}\|_1 + \|\mathbf{y}\|_1$. For $1 < p < \infty$ the subtle issue to address is the finiteness of $\|\mathbf{x} + \mathbf{y}\|_p$. If its finiteness is demonstrated, the proof of the inequality in (6.6) for $\mathbb{R}^n$ and $\mathbb{C}^n$ in proposition 3.24 in chapter 3, for which finiteness was guaranteed, goes through step by step.

To demonstrate the finiteness of $\|\mathbf{x} + \mathbf{y}\|_p$, we note that for $1 < p < \infty$, the function $f(x) = x^p$ is **convex**, which consistent with (3.31) means that

$$f(tz_1 + (1-t)z_2) \leq tf(z_1) + (1-t)f(z_2) \qquad \text{for } 0 \leq t \leq 1.$$

This function is also increasing for $t \in [0, \infty)$. This can be readily demonstrated with the tools in chapter 9 on calculus, although it is intuitively apparent from sample graphs. We will assume this result and let $z_1 = |x_j|$, $z_2 = |y_j|$, and $t = 0.5$. We get by the triangle inequality that $(0.5|x_j + y_j|)^p \leq (0.5|x_j| + 0.5|y_j|)^p$ since $p \geq 1$, and

$$(0.5|x_j| + 0.5|y_j|)^p \equiv f(0.5|x_j| + 0.5|y_j|).$$

By the convexity of $f(x)$ above,

$$f(0.5|x_j| + 0.5|y_j|) < 0.5(|x_j|^p + |y_j|^p).$$

That is, $(0.5|x_j + y_j|)^p \leq 0.5(|x_j|^p + |y_j|^p)$, and hence

$$\|\mathbf{x} + \mathbf{y}\|_p \leq (0.5)^{(1-p)/p}(\|\mathbf{x}\|_p^p + \|\mathbf{y}\|_p^p)^{1/p},$$

which is finite. Following the exact steps of the proof of proposition 3.24, we then derive the better estimate of the upper bound for $\|\mathbf{x} + \mathbf{y}\|_p$. Consequently l_p-space is closed under addition. Finally, the Minkowski inequality is also the critical step in proving that the l_p-norms are indeed norms in the sense of chapter 3, which is to say that the triangle inequality is satisfied, since the other norm requirements are immediate. ∎

Because l_p-space is a vector space, and $\|\ \|_p$ a norm, we can define a distance function or metric on l_p, the l_p-**metric**, consistent with this norm, just as it was defined in Euclidean space, $\mathbb{R}^n$ and complex space $\mathbb{C}^n$.

Definition 6.34 *The l_p-metric, $d_p(\mathbf{x}, \mathbf{y})$, is defined on l_p by*

$$d_p(\mathbf{x}, \mathbf{y}) \equiv \|\mathbf{x} - \mathbf{y}\|_p \qquad for\ 1 \leq p \leq \infty. \tag{6.7}$$

The final critical property of the l_p-spaces to verify is that they are complete in the sense of chapter 4. That is, every Cauchy sequence in l_p-space converges to a point in that l_p-space. In the proposition above it was proved that l_p-space is closed under addition, but this gives no insight to the completeness question.

A simple example is the space of rational numbers, $\mathbb{Q} \subset \mathbb{R}$. Clearly, $\mathbb{Q}$ is closed under addition, but equally clearly, as was seen in example 5.13, it is not complete. That is, while a Cauchy sequence in $\mathbb{Q}$ may well converge to a rational number, it is also possible that a sequence of rationals can converge to an irrational number. In fact, because $\mathbb{Q}$ is dense in $\mathbb{R}$, every number in $\mathbb{R}$ can be achieved by Cauchy sequences in $\mathbb{Q}$.

As it turns out, the l_p-spaces are complete for $1 \leq p \leq \infty$.

Proposition 6.35 *If $1 \leq p \leq \infty$, l_p is a complete normed linear space. That is, if $\{\mathbf{x}_n\} \subset l_p$ is a Cauchy sequence, then there exists $\mathbf{x} \in l_p$ so that $d_p(\mathbf{x}_n, \mathbf{x}) \equiv \|\mathbf{x}_n - \mathbf{x}\|_p \to 0$.*

Proof The assumption that $\{\mathbf{x}_n\}$ is a Cauchy sequence means that for any $\epsilon > 0$ there is an N so that $\|\mathbf{x}_n - \mathbf{x}_m\|_p < \epsilon$ for $n, m \geq N$. Now, if $p < \infty$, this means that $\sum_{j=1}^{\infty} |x_{n_j} - x_{m_j}|^p < \epsilon^p$, where x_{n_j} denotes the jth component of $\mathbf{x}_n$. This implies that $|x_{n_j} - x_{m_j}|^p < \epsilon^p$ for every j, and so the jth components of $\{\mathbf{x}_n\}$ form Cauchy sequences in $\mathbb{R}$ for every j. Since $\mathbb{R}$ is complete, there exists $x_j \in \mathbb{R}$ so that $x_{n_j} \to x_j$ for all j. A similar conclusion holds in the case of $p = \infty$ where the Cauchy property means that $\sup_j |x_{n_j} - x_{m_j}| < \epsilon$ for $n, m \geq N$. Defining $\mathbf{x} = (x_1, x_2, \ldots)$, the vector of componentwise limits, we must now show that $\mathbf{x} \in l_p$ and that $\|\mathbf{x}_n - \mathbf{x}\|_p \to 0$. The convergence of $\mathbf{x}_n$ to $\mathbf{x}$ is immediate from the Cauchy assumption, since for any $\epsilon > 0$ there is an N so that $\|\mathbf{x}_n - \mathbf{x}_m\|_p < \epsilon$ for $n, m \geq N$. Letting $m \to \infty$, we conclude that for any $\epsilon > 0$ there is an N so that $\|\mathbf{x}_n - \mathbf{x}\|_p < \epsilon$ for $n \geq N$. Finally, to show that $\mathbf{x} \in l_p$, note that $\|\mathbf{x}\|_p \leq \|\mathbf{x} - \mathbf{x}_N\|_p + \|\mathbf{x}_N\|_p$ by the Minkowski inequality, from which we derive that $\|\mathbf{x}\|_p \leq \epsilon + \|\mathbf{x}_N\|_p$ and so $\|\mathbf{x}\|_p$ is finite. That is, $\mathbf{x} \in l_p$.   ∎

The notion of complete normed linear space is so important in mathematics that it warrants a special name, after **Stefan Banach** (1892–1945), who first identified and studied properties of this special class of spaces:

Definition 6.36   *A normed linear space, $(X, \| \; \|)$, that is complete is called a **Banach** space.*

Remark 6.37   *To identify our list of Banach spaces so far, we include $\mathbb{R}^n$ and $\mathbb{C}^n$, under any of the l_p-norms, $1 \leq p \leq \infty$, as well as all the real and complex l_p-spaces, again for $1 \leq p \leq \infty$. In real analysis this list will be expanded to the function space counterparts to the l_p-spaces, denoted the L_p-spaces.*

*6.2.3   Hilbert Space

The preceding analysis shows that all the l_p-spaces are Banach spaces for $1 \leq p \leq \infty$, which is to say, complete normed linear spaces. As it turns out, there is one l_p-space that is more special than the rest. Specifically, l_2 has the additional property that its norm is given by an "inner product," and in that respect, l_2 is most like ordinary Euclidean space $\mathbb{R}^n$, or its complex counterpart $\mathbb{C}^n$, for which the same point was made concerning the "standard norm." Recall from chapter 3 that the inner product between two vectors can be defined as in (3.4) and (3.6), and that there is an intimate relationship between these inner products and the standard norms in these spaces as in (3.5) and (3.7), as summarized by

$$|\mathbf{x}| = (\mathbf{x} \cdot \mathbf{x})^{1/2}.$$

In the context of l_2-space we formally revise these inner product definitions by

$$\mathbf{x} \cdot \mathbf{y} = \sum_{j=1}^{\infty} x_i y_i, \qquad \mathbf{x}, \mathbf{y} \in l_2 \text{ (real)}, \tag{6.8}$$

$$\mathbf{x} \cdot \mathbf{y} = \sum_{j=1}^{\infty} x_i \bar{y}_i, \qquad \mathbf{x}, \mathbf{y} \in l_2 \text{ (complex)}. \tag{6.9}$$

To the extent these definitions can be shown to make sense, one has immediately as in (3.5) and (3.7) for the standard l_2-norms in $\mathbb{R}^n$ and $\mathbb{C}^n$, that in either real or complex l_2-space:

$$\|\mathbf{x}\|_2 = (\mathbf{x} \cdot \mathbf{x})^{1/2}. \tag{6.10}$$

This inner product construction can be implemented in l_2 and only in l_2. The subtlety, of course, is the demonstration that the inner products above actually converge, since in contrast to the case for $\mathbb{R}^n$ and $\mathbb{C}^n$, now $n = \infty$. If convergence is demonstrated, it will be straightforward to demonstrate that this inner product satisfies the same four properties as did the inner products in $\mathbb{R}^n$ and $\mathbb{C}^n$ highlighted in definitions 3.7 and 3.11 in chapter 3. That is, (6.8) satisfies the same properties as (3.4), while (6.9) satisfies the same properties as (3.6).

To this end, the critical insight to the convergence of the series in (6.8) and (6.9) is an inequality that was seen in chapter 3, and that was Hölder's inequality. In that chapter this inequality was demonstrated as one of the steps toward the proof of the Minkowski inequality. As noted above, the proof of the Minkowski inequality in l_p is identical to that in $\mathbb{R}^n$ and $\mathbb{C}^n$, subject only to the demonstration above that $\|\mathbf{x} + \mathbf{y}\|_p$ is in fact finite for $\mathbf{x}, \mathbf{y} \in l_p$, $1 \leq p \leq \infty$. Consequently, as a step in that proof, the Hölder inequality is also valid, and we state this without additional proof.

Proposition 6.38 (Hölder's Inequality)   *Given p, q so that $1 \leq p, q \leq \infty$, and $\frac{1}{p} + \frac{1}{q} = 1$, where notationally, $\frac{1}{\infty} \equiv 0$, then for $\mathbf{x} \in l_p$, $\mathbf{y} \in l_q$,*

$$|(\mathbf{x}, \mathbf{y})| \leq \|\mathbf{x}\|_p \|\mathbf{y}\|_q, \tag{6.11}$$

where $\mathbf{x} \cdot \mathbf{y}$ is defined in (6.8) or (6.9).

It is easy to see that this result highlights the special case of $p = 2$. That is, this is the only case where both $\mathbf{x}$ and $\mathbf{y}$ can be selected from the same l_p-space and an inner product defined. In this case the inner product is well defined, and has absolute value bounded by the product of the associated l_2-norms:

$$|(\mathbf{x}, \mathbf{y})| \leq \|\mathbf{x}\|_2 \|\mathbf{y}\|_2, \qquad \mathbf{x}, \mathbf{y} \in l_2. \tag{6.12}$$

Another important interpretation of (6.11) that is valuable in the future context of function spaces is that the componentwise product of two series from l_2 is a series in l_1. That is, if we momentarily define the componentwise product

$$\mathbf{x} * \mathbf{y} \equiv (x_1 y_1, x_2 y_2, x_3 y_3, \ldots), \tag{6.13}$$

then if $\mathbf{x}, \mathbf{y} \in l_2$ we have that $\mathbf{x} * \mathbf{y} \in l_1$ and by the Holder inequality

$$\|\mathbf{x} * \mathbf{y}\|_1 \leq \|\mathbf{x}\|_2 \|\mathbf{y}\|_2. \tag{6.14}$$

The power of having this inner product in l_2 is that is provides a basis for defining when two points are perpendicular, or in the language of such spaces, **orthogonal**. This is a natural generalization of this same notion in chapter 3 (see exercises 7 and 8 in that chapter):

Definition 6.39   *If* $\mathbf{x}, \mathbf{y} \in l_2$, *then we say* $\mathbf{x}$ *and* $\mathbf{y}$ *are* ***orthogonal****, denoted* $\mathbf{x} \perp \mathbf{y}$, *if* $(\mathbf{x}, \mathbf{y}) = 0$.

Of course, orthogonality is a generalization of the notion of perpendicularity in $\mathbb{R}^n$ and $\mathbb{C}^n$, in which $(\mathbf{x}, \mathbf{y}) = 0$ is also the defining relation using the standard inner product in those spaces. The classical collection of orthogonal vectors are those defined by the coordinate axes. For example, in $\mathbb{R}^n$ we have the set of n vectors

$$(1, 0, 0, \ldots, 0), (0, 1, 0, \ldots, 0), (0, 0, 1, 0, \ldots, 0) \ldots (0, 0, 0, \ldots, 0, 1),$$

denoted $\mathbf{e}_j$, for $j = 1, \ldots, n$, and it is apparent that these vectors are orthogonal and have unit norm or length

$$(\mathbf{e}_j, \mathbf{e}_k) = \begin{cases} 0, & j \neq k, \\ 1, & j = k, \end{cases}$$

where of course, $(\mathbf{e}_j, \mathbf{e}_j) = \|\mathbf{e}_j\|_2^2$, the square of the norm of $\mathbf{e}_j$.

Such a collection of vectors is said to be **orthonormal**. Here "ortho" is short for orthogonal, and "normal" means of unit length. In this case this collection is actually an **orthonormal basis** where by "basis" is meant that with these vectors, every other vector in $\mathbb{R}^n$ can be generated using linear combinations of these. In other words, we have for any vector $\mathbf{x} = (x_1, x_2, \ldots, x_n)$,

$$\mathbf{x} = \sum_{j=1}^{n} x_j \mathbf{e}_j,$$

where the coefficients, $\{x_j\}$ are used as scalars in what is called a **linear combination of vectors**.

This construction generalizes to l_2, for which an infinite sequence of vectors, $\{\mathbf{e}_j\}_{j=1}^{\infty}$ can be correspondingly defined. In l_2, however, the meaning given to the representation above for $\mathbf{x}$ is with $\mathbf{x}_n \equiv \sum_{j=1}^{n} x_j \mathbf{e}_j$:

$$\mathbf{x} = \sum_{j=1}^{\infty} x_j \mathbf{e}_j \qquad \text{iff } \|\mathbf{x} - \mathbf{x}_n\|_2 \to 0 \text{ as } n \to \infty. \tag{6.15}$$

In both cases, $\mathbb{R}^n$ and l_2, the norm of $\mathbf{x}$ can be derived from the scalar coefficients by

$$\|\mathbf{x}\|_2^2 = \sum_{j=1}^{\infty} x_j^2.$$

This perhaps feels a bit like a notational sleight of hand, as the orthonormal basis $\{\mathbf{e}_j\}_{j=1}^{\infty}$ is pretty trivial, and so is the expansion of $\mathbf{x}$ in terms of this basis and the corresponding identity for $\|\mathbf{x}\|_2^2$. But in reality, this is just the tip of the iceberg. It turns out that l_2-space has infinitely many orthonormal bases, although we do not prove this. The following is then a critical result on these bases.

Proposition 6.40   *If $\{\mathbf{e}_j\}_{j=1}^{\infty}$ is **any** orthonormal basis in $\mathbb{R}^n$, $\mathbb{C}^n$ or l_2-space, then for any $\mathbf{x}$ in the respective space defined by*

$$\mathbf{x} = \sum_{j=1}^{\infty} y_j \mathbf{e}_j, \tag{6.16}$$

the coefficients are given by

$$y_j = (\mathbf{x}, \mathbf{e}_j), \tag{6.17}$$

and

$$\|\mathbf{x}\|_2^2 = \sum_{j=1}^{\infty} |y_j|^2. \tag{6.18}$$

Proof   We focus on the l_2-space result, and leave $\mathbb{R}^n$ and $\mathbb{C}^n$ as an exercise. First off, the expression for y_j follows from (6.15), since by (6.12) we have as $n \to \infty$,

$$|(\mathbf{x} - \mathbf{x}_n, \mathbf{e}_j)| \le \|\mathbf{x} - \mathbf{x}_n\|_2 \|\mathbf{e}_j\|_2 \to 0,$$

and so $(\mathbf{x}_n, \mathbf{e}_j) \to (\mathbf{x}, \mathbf{e}_j)$. But then $(\mathbf{x}_n, \mathbf{e}_j) = y_j$ for $n \ge j$, using the orthonormal properties above, proving (6.17). Also, for (6.18), first note that (6.15) implies that $\|\mathbf{x}_n\|_2 \to \|\mathbf{x}\|_2$ as $n \to \infty$. That is, recalling that $\|\mathbf{x}_n\|_2^2 = (\mathbf{x}_n, \mathbf{x}_n)$,

$$\|\mathbf{x}_n\|_2^2 - \|\mathbf{x}\|_2^2 = \|\mathbf{x}_n - \mathbf{x}\|_2^2 + 2(\mathbf{x}, \mathbf{x}_n - \mathbf{x}).$$

So from (6.12) we have

$$\big|\|\mathbf{x}_n\|_2^2 - \|\mathbf{x}\|_2^2\big| \le \|\mathbf{x}_n - \mathbf{x}\|_2^2 + 2\|\mathbf{x}\|_2\|\mathbf{x}_n - \mathbf{x}\|_2,$$

and the result follows. Then, again using the orthonormal properties above, we derive (6.18), since

$$\|\mathbf{x}_n\|_2^2 = (\mathbf{x}_n, \mathbf{x}_n) = \sum_{j=1}^{n} |y_j|^2. \qquad \blacksquare$$

Remark 6.41

1. *The purpose of the absolute value in the identity in (6.18) is to indicate that in complex l_2-spaces, it is the square of the norms of these complex numbers that are summed.*

2. *The identity in (6.18) is known as **Parseval's identity**, after **Marc-Antoine Parseval** (1755–1836), who derived this identity in the more general content of L_2 function spaces. In that context, the collection of orthonormal functions used in (6.16) gave rise to what is known as the **Fourier series** representation of the "function" $\mathbf{x}$, named for **Jean Baptiste Joseph Fourier** (1768–1830), who studied such functional expansions.*

In real analysis this additional inner product structure in l_2 is repeated in the function space counterpart L_2, and this structure has important consequences there as well, similar to what was illustrated above.

The notion of complete normed linear space with a compatible inner product is so important in mathematics that it warrants a special name, after **David Hilbert** (1862–1943), who first identified and studied properties of this special class of infinite dimensional Euclidean spaces.

Definition 6.42   *A normed linear space, $(X, \| \ \|)$, that is complete and has a compatible inner product is called a **Hilbert space**.*

Remark 6.43   *To identify our list of Hilbert spaces so far, we include $\mathbb{R}^n$ and $\mathbb{C}^n$, under the standard or l_2-norm, as well as the real and complex l_2-spaces. There will be another identified later, but not until the study of real analysis, where we will be introduced to the function space counterpart to the l_2-spaces, denoted L_2-space.*

6.3   Power Series

In this section we introduce the notion of a power series that will justifiably get more attention in chapter 9 on calculus in the study of Taylor series. Here we focus on power series of a single variable, although one can imagine that multivariate versions are also possible, and as it turns out, important.

Definition 6.44 *Given a real numerical sequence, $\{c_n\}_{n=0}^{\infty}$, the **power series** associated with this sequence is notationally defined as a real function of x (or y, z, etc.), by*

$$f(x) = \sum_{n=0}^{\infty} c_n x^n. \tag{6.19}$$

In other words, a power series can be thought of as an infinite polynomial function of x, defined on $\mathbb{R}$. Not surprisingly, the central question to address here is the convergence of the expression given in (6.19), outside of the obvious point of convergence of $x = 0$ for which $f(0) = c_0$. In the later chapters on calculus, we will also address questions such as:

1. Given a function $f(x)$, when can this function be represented as in (6.19) for some sequence $\{c_n\}_{n=0}^{\infty}$?

2. Given a function $f(x)$, when can this function be approximated by a finite version of this series, and what is the nature of the error in this case?

Utilizing the results above on the convergence of numerical series, the following result is easily demonstrated.

Proposition 6.45 *Given the power series, $f(x) = \sum_{n=0}^{\infty} c_n x^n$, define*

$$L = \limsup_{n \to \infty} \left\{ \frac{|c_{n+1}|}{|c_n|} \right\}. \tag{6.20}$$

Then with $R = \frac{1}{L}$, this power series converges absolutely for $|x| < R$, diverges for $|x| > R$, and is indeterminate for $|x| = R$.

Proof By the ratio test, the requirement for absolute convergence is that

$$\limsup_{n \to \infty} \left\{ \frac{|c_{n+1} x^{n+1}|}{|c_n x^n|} \right\} < 1,$$

which occurs exactly when $|x| < R$ with R as defined. Similarly we conclude divergence when $|x| > R$ and that $|x| = R$ is an indeterminate case. ∎

Remark 6.46 *R is called the **radius of convergence** of the power series, and the interval, $|x| < R$ is called the **interval of convergence**.*

Example 6.47

1. *If* $f(x) = \sum_{n=0}^{\infty} \frac{x^n}{n!}$, *then* $L = \limsup_{n\to\infty} \left\{\frac{1}{n+1}\right\} = 0$. *Therefore* $R = \infty$, *and this power series converges for all* $x \in \mathbb{R}$. *In chapter 9 we will see that* $f(x) = e^x$.

2. *If* $f(x) = \sum_{n=0}^{\infty} (-1)^n \frac{x^n}{n+1}$, *then* $L = \limsup_{n\to\infty} \left\{\frac{n+1}{n+2}\right\} = 1$. *Therefore* $R = 1$, *and this power series converges for* $|x| < 1$. *This series diverges for* $x = -1$, *producing the harmonic series but converges for* $x = 1$ *by the alternating series test. In chapter 9 we will see that* $f(x) = \ln(1+x)$.

3. *If* $f(x) = \sum_{n=0}^{\infty} (-1)^n \frac{3^n x^n}{(n+1)^a}$, $a > 1$, *then* $L = \limsup_{n\to\infty} \left\{3\left(\frac{n+1}{n+2}\right)^a\right\} = 3$. *Therefore* $R = \frac{1}{3}$, *and this power series converges for* $|x| < \frac{1}{3}$. *It is also convergent for* $x = \frac{1}{3}$ *by the alternating series test, and for* $x = -\frac{1}{3}$, *producing a power harmonic series.*

4. *If* $f(x) = \sum_{n=0}^{\infty} x^n$, *then* $L = \limsup_{n\to\infty} \{1\} = 1$. *Therefore* $R = 1$, *and this power series converges for* $|x| < 1$. *This series is easily seen to diverge for* $x = 1$, *and not converge for* $x = -1$. *In chapter 9 we will see that* $f(x) = \frac{1}{1-x}$, *although this is easily derivable as follows. Since we have convergence for* $|x| < 1$, *we can infer that* $xf(x) = \sum_{n=1}^{\infty} x^n$ *and hence* $f(x) - xf(x) = 1$.

5. *If* $f(x) = \sum_{n=0}^{\infty} n! x^n$, *then* $L = \limsup_{n\to\infty} \{n+1\} \to \infty$. *Therefore* $R = 0$, *and this series converges only for* $x = 0$.

An alternative approach to power series convergence comes from the Comparison test.

Proposition 6.48   *Given the power series,* $f(x) = \sum_{n=0}^{\infty} c_n x^n$, *if* $f(x)$ *converges absolutely for* $x = a$, *then it converges absolutely for all* x *with* $|x| \leq |a|$.

Proof   If $|x| \leq |a|$, then it is obvious that $|c_n x^n| \leq |c_n a^n|$ for all n, and since $\sum_{n=0}^{\infty} |c_n a^n|$ converges, so does $\sum_{n=0}^{\infty} |c_n x^n|$ by the comparison test. That is, $f(x)$ is absolutely convergent.   ■

A simple application of this last result is that every absolutely convergent numerical series gives rise to a power series that is absolutely convergent for $|x| \leq 1$. To see this, assume that $\sum_{n=0}^{\infty} c_n$ is an absolutely convergent numerical series. Define the power series $f(x) = \sum_{n=0}^{\infty} c_n x^n$. By assumption, $f(1)$ is absolutely convergent, so the result follows.

Example 6.49   *It was demonstrated in case 4 of example 6.9 that if* $x_j = \frac{1}{j^a}$ *for* $a > 1$, *then the power harmonic series* $\sum_{j=1}^{\infty} \frac{1}{j^a}$ *converges, and since all terms are positive, it converges absolutely. Consequently it is immediate that the power series*

$$f(x) = \sum_{j=1}^{\infty} \frac{x^j}{j^a}$$

converges absolutely at least for $|x| \leq 1$. Calculating the radius of convergence from the previous proposition, we obtain $L = \limsup_{n \to \infty} \left\{ \left(\frac{n}{n+1} \right)^a \right\} = 1$, and $R = \frac{1}{L} = 1$. So in these cases the indeterminate case of $|x| = R$ converges, although this is not determinable by the ratio test.

As a final note, it will often be the case that the definition of power series requires a small adjustment for the applications coming in chapter 9 on calculus.

Definition 6.50 *Given a real numerical sequence $\{c_n\}_{n=0}^{\infty}$ and a constant a, the **power series centered on** a associated with this sequence is notationally defined as a real function of x, by*

$$f(x) = \sum_{n=0}^{\infty} c_n (x - a)^n. \tag{6.21}$$

The analysis above on power series convergence can be applied in this context, with one adjustment:

Proposition 6.51 *Given the power series $f(x) = \sum_{n=0}^{\infty} c_n (x - a)^n$, define*

$$L = \limsup_{n \to \infty} \left\{ \frac{|c_{n+1}|}{|c_n|} \right\}.$$

Then $f(x)$ converges absolutely for $|x - a| < R$, diverges for $|x - a| > R$, and is indeterminate for $|x - a| = R$, where $R = \frac{1}{L}$.

Proof The proof is an immediate application of proposition 6.45 above, or can be derived directly from the ratio test. ∎

In other words, for these power series the **radius of convergence** is independent of a, but the **interval of convergence** is shifted from being "centered on 0" with $|x| < R$, to being "centered on a" with $|x - a| < R$, justifying the name.

*6.3.1 Product of Power Series

The discussion in this section relates to the product of two functions given by power series. Obviously, if $f(x)$ and $g(x)$ are any two functions, the function $h(x) \equiv f(x)g(x)$ is well defined. The question here is, if $f(x)$ and $g(x)$ are given as convergent power series centered on a, with respective radii of convergence of R and R', what is the power series representation of $h(x)$ and what is its radius of convergence? The following proposition addresses this question:

Proposition 6.52   *Let $f(x)$ and $g(x)$ be given as convergent power series centered on a,*

$$f(x) = \sum_{n=0}^{\infty} b_n(x-a)^n, \quad g(x) = \sum_{n=0}^{\infty} c_n(x-a)^n,$$

with respective radii of convergence of R and R'. Then $h(x) \equiv f(x)g(x)$ is given by the power series

$$h(x) = \sum_{n=0}^{\infty} d_n(x-a)^n, \tag{6.22}$$

where

$$d_n = \sum_{j=0}^{n} b_j c_{n-j}. \tag{6.23}$$

Further the radius of convergence of $h(x)$ is $R'' = \min(R, R')$.

Proof   The formula for the coefficients in (6.23) follows immediately from the observation that when multiplying these series, the only way that the product of a $b_j(x-a)^j$ term from the expansion of $f(x)$ and a $c_k(x-a)^k$ term from the expansion of $g(x)$ can contribute to the coefficient of $(x-a)^n$ is to have $j+k=n$. So we see that this formula for d_n simply accounts for all such products. The question of convergence of (6.22) is the more difficult question which is addressed next. To simplify notation, let $f_m(x)$ denote the partial summation

$$f_m(x) = \sum_{n=0}^{m} b_n(x-a)^n,$$

and $\tilde{f}_m(x) = f(x) - f_m(x)$, which is given by the summation

$$\tilde{f}_m(x) = \sum_{n=m+1}^{\infty} b_n(x-a)^n.$$

Using similar notation for $g(x)$ and $h(x)$, and noting that the finite double summations such as $\sum_{n=0}^{m} \sum_{j=0}^{n}$ can be reversed to $\sum_{j=0}^{m} \sum_{n=j}^{m}$, we have for $|x - a| < R''$, due to the convergence of both $f(x)$ and $g(x)$,

$$h_m(x) = \sum_{n=0}^{m} \left[\sum_{j=0}^{n} (b_j(x-a)^j)(c_{n-j}(x-a)^{n-j}) \right]$$

$$= \sum_{j=0}^{m} b_j(x-a)^j \sum_{n=j}^{m} c_{n-j}(x-a)^{n-j}$$

$$= \sum_{j=0}^{m} b_j(x-a)^j g_{m-j}(x)$$

$$= g(x) \sum_{j=0}^{m} b_j(x-a)^j - \sum_{j=0}^{m} (b_j(x-a)^j)\tilde{g}_{m-j}(x).$$

Now $\sum_{j=0}^{m} b_j(x-a)^j \to f(x)$ as $m \to \infty$. If it can be shown that $\sum_{j=0}^{m} b_j(x-a)^j \tilde{g}_{m-j}(x) \to 0$ absolutely, the proof will be complete since then

$$\left| h_m(x) - g(x) \sum_{j=0}^{m} b_j(x-a)^j \right| \to 0.$$

Now since $\tilde{g}_n(x) \to 0$, for any $\epsilon > 0$ there is an N so that $|\tilde{g}_n(x)| < \epsilon$ for $n > N$. To have $|\tilde{g}_{m-j}(x)| < \epsilon$ requires $j < m - N$, and so for m large enough,

$$\left| \sum_{j=0}^{m} b_j(x-a)^j \tilde{g}_{m-j}(x) \right| \leq \sum_{j=0}^{m-N-1} |b_j(x-a)^j \tilde{g}_{m-j}(x)| + \sum_{j=m-N}^{m} |b_j(x-a)^j \tilde{g}_{m-j}(x)|$$

$$< \epsilon \sum_{j=0}^{\infty} |b_j(x-a)^j| + \sum_{j=m-N}^{m} |b_j(x-a)^j \tilde{g}_{m-j}(x)|$$

$$= K(x)\epsilon + \sum_{j=0}^{N} |b_{m-j}(x-a)^{m-j} \tilde{g}_j(x)|$$

$$\leq K(x)\epsilon + \max_{0 \leq j \leq N} |\tilde{g}_j(x)| \max_{0 \leq j \leq N} |b_{m-j}(x-a)^{m-j}|.$$

Note that the first summation converged to a finite value, $K(x)$ say, for any x, because the power series for $f(x)$ is absolutely convergent. Also the second term converges to 0 as $m \to \infty$ because the finite collection $\{\tilde{g}_j(x)\}_{j=0}^N$ is bounded for any x, and the maximum of the finite collection $\{|b_{m-j}(x-a)^{m-j}|\}_{j=0}^N$ converges to 0 as $m \to \infty$, again because the power series for $f(x)$ is absolutely convergent. ∎

*6.3.2   Quotient of Power Series

One important application of the proposition above is to generate the coefficients of the reciprocal of a power series, or the quotient of two power series. Specifically, the proposition above assures that if

$$f(x)g(x) = h(x),$$

$$\sum_{n=0}^{\infty} b_n(x-a)^n \sum_{n=0}^{\infty} c_n(x-a)^n = \sum_{n=0}^{\infty} d_n(x-a)^n,$$

then the coefficients $\{d_n\}$ satisfy (6.23). Consequently, if $f(x)$ and $h(x)$ are given, and if coefficients $\{c_n\}$ can be found that satisfy (6.23) and produce a convergent power series, then we can conclude that

$$\sum_{n=0}^{\infty} c_n(x-a)^n = \frac{h(x)}{f(x)}.$$

And in the special case where $h(x) \equiv 1$, the reciprocal of $f(x)$ is produced.

Of course, to have any hope that the resultant expansion is convergent in an interval of a, we require that $f(a) \neq 0$, which is equivalent to $b_0 \neq 0$. In such a case the equations in (6.23) can be solved for $\{c_n\}$ iteratively, producing after re-indexing for visual appeal,

$$c_0 = \frac{d_0}{b_0},$$

$$(6.24)$$

$$c_n = \frac{1}{b_0}\left[d_n - \sum_{j=0}^{n-1} b_{n-j}c_j\right], \qquad n \geq 1.$$

We now show that the condition $b_0 \neq 0$ is sufficient to ensure that $\sum_{n=0}^{\infty} c_n(x-a)^n$ is an absolutely convergent power series.

Proposition 6.53   *Let $f(x)$ and $h(x)$ be given as convergent power series centered on a:*

$$f(x) = \sum_{n=0}^{\infty} b_n (x-a)^n, \quad h(x) = \sum_{n=0}^{\infty} d_n (x-a)^n,$$

with common radius of convergence of R, and where $f(a) = b_0 \neq 0$. Then $g(x) \equiv \frac{h(x)}{f(x)}$ is given by the power series

$$g(x) = \sum_{n=0}^{\infty} c_n (x-a)^n,$$

where $\{c_n\}$ satisfy (6.24), and this series is absolutely convergent on $|x-a| < R'$ for some $R' > 0$.

Proof   We prove this proposition in two steps.

1. Assume that we can prove this result for $h(x) \equiv 1$, where $\{c_n'\}$ satisfy (6.24) with $d_0 = 1$ and $d_n = 0$ for all $n \geq 1$. In other words,

$$\frac{1}{f(x)} = \sum_{n=0}^{\infty} c_n' (x-a)^n$$

is absolutely convergent, where

$$c_n' = \begin{cases} \frac{1}{b_0}, & n = 0, \\ \frac{-1}{b_0} \sum_{j=0}^{n-1} b_{n-j} c_j', & n \geq 1. \end{cases} \tag{6.25}$$

Then by the proposition above, $g(x) = h(x) \frac{1}{f(x)}$ is well defined:

$$g(x) = \sum_{n=0}^{\infty} c_n (x-a)^n,$$

where by (6.23), stated in terms of $\{c_n'\}$ and $\{d_n\}$,

$$c_n = \sum_{j=0}^{n} d_j c_{n-j}' = \begin{cases} \frac{d_0}{b_0}, & n = 0, \\ \frac{1}{b_0} [d_n - \sum_{j=0}^{n-1} d_j \sum_{k=0}^{n-j-1} b_{n-j-k} c_k'], & n \geq 1. \end{cases}$$

We must now show that this definition of c_n is equivalent to (6.24). In this summation for $n \geq 1$, we define a new index variable $l = j + k$ and observe that given j, we have $j \leq l \leq n - 1$. Therefore

$$\sum_{j=0}^{n-1}\sum_{k=0}^{n-j-1} d_j b_{n-j-k} c_k' = \sum_{j=0}^{n-1}\sum_{l=j}^{n-1} b_{n-l} d_j c_{l-j}'$$

$$= \sum_{l=0}^{n-1} b_{n-l} \sum_{j=0}^{l} d_j c_{l-j}'$$

$$= \sum_{l=0}^{n-1} b_{n-l} c_l,$$

where we reversed the double summation $\sum_{j=0}^{n-1}\sum_{l=j}^{n-1} = \sum_{l=0}^{n-1}\sum_{j=0}^{l}$ in the second line. Substituting this final result into the definition above for c_n produces (6.24) as desired.

2. To prove (6.24) in the special case of $h(x) \equiv 1$, first note that we can assume that $b_0 = 1$, since this term can be factored out of the series without changing convergence properties, and factored back in as $\frac{1}{b_0}$ after the inversion. Now since the power series for $f(x)$ converges for $|x - a| = r < R$, its terms must converge to 0. Hence its terms are bounded, $|b_n| r^n \leq M$. Therefore

$$|b_n| \leq \frac{M}{r^n}.$$

For convenience below we take $M > 1$. With c_n' defined as the coefficients of $\frac{1}{f(x)}$ above with $b_0 = 1$, we now show by induction that

$$|c_n'| \leq 2^n \frac{M^n}{r^n}.$$

Since $c_0' = 1$, we assume this statement is true for n *and* evaluate c_{n+1}'. Then by (6.25),

$$|c_{n+1}'| = \left| \sum_{j=0}^{n} b_{n+1-j} c_j' \right|$$

$$\leq \sum_{j=0}^{n} \frac{M}{r^{n+1-j}} 2^j \frac{M^j}{r^j}$$

$$< \frac{M^{n+1}}{r^{n+1}} \sum_{j=0}^{n} 2^j$$

$$< 2^{n+1} \frac{M^{n+1}}{r^{n+1}}.$$

Since the power series coefficients for $\frac{1}{f(x)}$ are bounded in absolute value by a geometric series, we conclude that this series converges if

$$2^n \frac{M^n}{r^n} |x - a|^n < 1.$$

So the interval of absolute convergence contains

$$|x - a| < \frac{r}{2M}. \qquad \blacksquare$$

6.4   Applications to Finance

6.4.1   Perpetual Security Pricing: Preferred Stock

The most apparent application of numerical series to finance is the evaluation of the price of common stock or nonredeemable preferred stock, both "perpetual" securities. A preferred stock with par value of 1000 and dividend rate of 5% on an annual basis pays 50 per year to the investor in perpetuity. In general, with par value of F and dividend rate d on an annual basis, the investor receives Fd per year in perpetuity. For an investor desiring a fixed yield of r on an annual basis, and assuming the next dividend is one year into the future, the appropriate price function is given as

$$P(r) = Fd \sum_{j=1}^{\infty} (1 + r)^{-j}. \qquad (6.26)$$

From the methods above on numerical series we conclude that for any $r > 0$, this price function converges absolutely as noted in section 2.3.2, to

$$P(r) = \frac{Fd}{r}.$$

This model is easily generalized to different dividend payment frequencies and/or yield nominal bases.

A more general model of yearly varying yields is easily handled formally. Now the price is a function of a sequence of yields, $\{r_j\}$, and

$$P(\{r_j\}) = Fd \sum_{j=1}^{\infty} (1 + r_j)^{-j}. \tag{6.27}$$

But the question of convergence is more subtle. Clearly, if there is an $r > 0$ so that $r_j \geq r$ for all j, then by the comparison test, $P(\{r_j\})$ converges and $P(\{r_j\}) \leq P(r)$.

Consequently the only question is, if $r_j > 0$ for all j, but $r_j \to 0$, does this price converge? However, the question is not really about the stronger condition of convergence of $r_j \to 0$; it is only about the weaker condition of $\{r_j\}$ having 0 as a possible accumulation point. This can be problematic, since it is then possible that infinitely many terms in the summation are large enough to cause divergence. As was seen in section 5.2, this accumulation point condition can be expressed as $\liminf_{j \to \infty} r_j = 0$.

To investigate the question of convergence, we apply the ratio test to this series. The criterion for convergence is that

$$\limsup_{n \to \infty} \left\{ \frac{(1 + r_{j+1})^{-j-1}}{(1 + r_j)^{-j}} \right\} = L < 1.$$

By proposition 5.22 this condition is satisfied if and only if for any $\epsilon > 0$ there is an N so that $j \geq N$,

$$(1 + r_{j+1})^{j+1} \geq \frac{(1 + r_j)^j}{L + \epsilon}.$$

Choosing ϵ so that $L + \epsilon < 1$ and iterating, we derive that with $j = N + k$ and $k \geq 1$:

$$(1 + r_{N+k})^{N+k} \geq \frac{(1 + r_N)^N}{(L + \epsilon)^k}.$$

That is,

$$r_{N+k} \geq \frac{(1 + r_N)^{N/(N+k)}}{(L + \epsilon)^{k/(N+k)}} - 1,$$

which appears to be a bound on the rate at which $r_j \to 0$.

But closer inspection reveals more. As $k \to \infty$, it is clear that $\frac{k}{N+k} \to 1$ and hence $(L+\epsilon)^{k/(N+k)} \to L+\epsilon$. Also, $\frac{N}{N+k} \to 0$, and assuming that $r_N > 0$, we conclude that $(1+r_N)^{N/(N+k)} \to 1$. Consequently the lower bound for r_{N+k} converges to $\frac{1}{L+\epsilon} - 1 = \frac{1-L-\epsilon}{L+\epsilon}$, which exceeds 0 if $L+\epsilon < 1$.

Hence we obtain that as $k \to \infty$, the ratio test assures convergence of the preferred stock price only when for some $L' = L + \epsilon < 1$,

$$\liminf_{j \to \infty} r_j \geq \frac{1-L'}{L'},$$

and hence we return to the case where the sequence is bounded away from 0. That is, this condition implies that for any $\epsilon > 0$ there is an N so that $r_j > \frac{1-L'}{L'} - \epsilon$ for all $j \geq N$.

Of course, this does not prove that there is no sequence $\{r_j\}$ with $\liminf_{j \to \infty} r_j = 0$ for which the preferred stock price converges, it only proves that there is no such sequence for which convergence is verifiable by the ratio test.

With a similar analysis, one could anticipate the convergence of this pricing function for nonconstant dividends. Again, if these dividends are bounded from above, $d_j \leq d$ for all j, the price function $P(r) = F \sum_{j=1}^{\infty} d_j (1+r)^{-j}$ is easily seen to converge by the comparison test. For unbounded dividends, the answer is more subtle, but insights can often be developed with the aid of the ratio test.

6.4.2   Perpetual Security Pricing: Common Stock

A similar analysis can be implemented for the price of common stock under the discounted dividend model introduced in section 2.3.2. From (2.22) we have that the price—as a function of the last annual dividend assumed to have been just paid D, the annual dividend growth rate g, and the investor required yield r—is

$$V(D, g, r) = D \sum_{j=1}^{\infty} (1+r)^{-j} (1+g)^j$$

$$= D \sum_{j=1}^{\infty} \left(1 + \frac{r-g}{1+g} \right)^{-j}.$$

For fixed r and g, the analysis above for preferred stock indicates that this price converges as long as $r > g$, and in this case we have as in (2.22),

$$V(D,g,r) = D\frac{1+g}{r-g}, \qquad r > g.$$

To generalize this in contemplation of a growth rate sequence $\{g_j\}$ and yield sequence $\{r_j\}$, we apply the same considerations as for preferred stock. Convergence by the ratio test is assured if the effective discount rates are bounded away from 0, that is, $\frac{r_j - g_j}{1+g_j} \geq r > 0$. But this approach may be challenged if these rates converge to 0.

6.4.3   Price of an Increasing Perpetuity

In addition to pricing formulas for perpetuities with constant and geometrically increasing payments discussed above, we can apply double summations methods to value a linearly increasing payment stream with a fixed annual rate. Generalizations to this payment model are then discussed.

First, if the perpetuity payment at time j is $D_j = aj + b$ for constants a and b, then by linearity,

$$V(D_j, r) = a\sum_{j=1}^{\infty} j(1+r)^{-j} + b\sum_{j=1}^{\infty}(1+r)^{-j},$$

and only the first summation has not yet been evaluated. Writing $j = \sum_{i=1}^{j} 1$, we have

$$\sum_{j=1}^{\infty} j(1+r)^{-j} = \sum_{j=1}^{\infty}\sum_{i=1}^{j}(1+r)^{-j}$$

$$= \sum_{i=1}^{\infty}\sum_{j=i}^{\infty}(1+r)^{-j}.$$

Reversing the summations will be justified once we demonstrate convergence, which will imply absolute convergence.

Now

$$\sum_{j=i}^{\infty}(1+r)^{-j} = (1+r)^{-i+1}\sum_{j=1}^{\infty}(1+r)^{-j} = \frac{(1+r)^{-i+1}}{r}.$$

Substituting into the double sum, we obtain

$$\sum_{j=1}^{\infty} j(1+r)^{-j} = \frac{1+r}{r^2}. \tag{6.28}$$

The last answer makes sense because $\frac{1}{r}$ is the value of a perpetuity of 1s payable annually from $t = 1$ forward, so $\frac{1}{r^2}$ is the value of a perpetuity of perpetuities, whereby $\frac{1}{r}$ is paid annually from $t = 1$ forward. The first such perpetuity provides for a series of 1s annually from $t = 2$ forward, the second for a series of 1s annually from $t = 3$ forward, so it is clear that the total payment is growing linearly. However, $\frac{1}{r^2}$ starts payment one year later than desired, so the multiplicative factor of $1 + r$ adjusts for this. Combining results, we obtain

$$V(D_j, r) = \frac{a(1+r)}{r^2} + \frac{b}{r}, \qquad D_j = aj + b. \tag{6.29}$$

The double-summations approach can be generalized to present values of the form $P_n \equiv \sum_{j=1}^{\infty} j^n (1+r)^{-j}$. However, rather than obtaining an explicit formula as in the case of $n = 0, 1$, we derive an iterative formula whereby we give P_n in terms of $\{P_0, P_1, \ldots, P_{n-1}\}$. Of course, here $P_0 = \frac{1}{r}$, and $P_1 = \frac{1+r}{r^2}$.

There are two ways to develop this iterative formula. First, we can proceed as above and write multiple series:

$$\sum_{j=1}^{\infty} j^n (1+r)^{-j} = \sum_{j=1}^{\infty} \sum_{i=1}^{j^n} (1+r)^{-j}$$

$$= \sum_{i=1}^{\infty} \sum_{j=n(i)}^{\infty} (1+r)^{-j},$$

where $n(i) = k + 1$ for $k^n + 1 \le i \le (k+1)^n$ and for $k \ge 0$. In other words, we have

$$n(i) = \begin{cases} 1, & i = 1, \\ 2, & 2 \le i \le 2^n, \\ 3, & 2^n + 1 \le i \le 3^n, \\ \vdots & \vdots \\ k+1, & k^n + 1 \le i \le (k+1)^n. \end{cases}$$

Then the inner sums can be collected into groups for fixed $n(i)$, and the outer sum converted to index $k \ge 0$, producing

$$\sum_{j=1}^{\infty} j^n (1+r)^{-j} = \sum_{k=0}^{\infty} [(k+1)^n - k^n] \sum_{j=k+1}^{\infty} (1+r)^{-j}$$

$$= \sum_{k=1}^{\infty} \left[\sum_{i=0}^{n-1} \binom{n}{i} k^i \right] \frac{(1+r)^{-k}}{r} + \frac{1}{r}$$

$$= \frac{1}{r} \left[\sum_{i=0}^{n-1} \binom{n}{i} \sum_{k=1}^{\infty} k^i (1+r)^{-k} + 1 \right],$$

since $\sum_{j=k+1}^{\infty} (1+r)^{-j} = (1+r)^{-k} \sum_{j=1}^{\infty} (1+r)^{-j} = \frac{(1+r)^{-k}}{r}$.

Note that a bit of care is necessary for the k-summation, which is split as $\sum_{k=0}^{\infty} = \sum_{k=1}^{\infty} + \sum_{k=0}$, to avoid 0^0 in the second step where the **binomial theorem** (see chapter 8 for details) was used. This theorem states that with $n!$ ("n factorial") defined by $n! = n(n-1)(n-1)\dots(2)(1)$ and with $0! \equiv 1$,

$$(k+1)^n = \sum_{i=0}^{n} \binom{n}{i} k^i,$$

$$\binom{n}{i} \equiv \frac{n!}{i!(n-i)!}.$$

So rewriting, we obtain

$$P_n = \frac{1}{r} \left[\sum_{i=0}^{n-1} \binom{n}{i} P_i + 1 \right], \qquad n = 2, 3, \dots, \tag{6.30}$$

where

$$P_0 = \frac{1}{r}, \qquad P_1 = \frac{1+r}{r^2}.$$

This formula is also valid for $n = 1$, with only the initial value $P_0 = \frac{1}{r}$.

See exercise 15 for an alternative derivation.

6.4.4   Price of an Increasing Payment Security

The price of a security such as a bond or mortgage, or a fixed term annuity with linearly increasing payments, is now easily handled. Specifically, with $D_j = aj + b$ for $j = 1, \dots, n$,

$$V(D_j, r) = a \sum_{j=1}^{n} j(1+r)^{-j} + b \sum_{j=1}^{n} (1+r)^{-j}.$$

Now the second summation equals $a_{n:r}$ by (2.11), while

$$\sum_{j=1}^{n} j(1+r)^{-j} = \sum_{j=1}^{\infty} j(1+r)^{-j} - \sum_{j=n+1}^{\infty} j(1+r)^{-j}.$$

Here the first summation is the perpetuity above in (6.28), while the second splits as

$$\sum_{j=n+1}^{\infty} j(1+r)^{-j} = \sum_{j=1}^{\infty} (j+n)(1+r)^{-j-n}$$

$$= (1+r)^{-n} \left[n \sum_{j=1}^{\infty} (1+r)^{-j} + \sum_{j=1}^{\infty} j(1+r)^{-j} \right].$$

Combining and simplifying, and using notation from chapter 2, we derive for the first summation

$$\sum_{j=1}^{n} j(1+r)^{-j} = (1+r) \frac{a_{n:r}}{r} - \frac{n(1+r)^{-n}}{r}. \tag{6.31}$$

This formula can again be intuited from the component parts. The term $\frac{a_{n:r}}{r}$ provides a perpetuity that pays $a_{n:r}$ at each of times $1, 2, 3, \ldots$, each of which is in turn equivalent to a series of n payments of 1 starting one year later. So collectively this perpetuity provides for a payment stream that grows from 1 to n at times 2 to $n+1$, and is then frozen at level n from time $n+2$ forward. The $1+r$ factor puts these increasing payments at times 1 to n, and the frozen payments of n from time $n+1$ onward. The second term eliminates the payments of n from time $n+1$ onward, since $\frac{n}{r}$ is a perpetuity of n per year starting at time 1 and the $(1+r)^{-n}$ factor moves these payments to start at time $n+1$.

By defining $A_m = \sum_{j=1}^{n} j^m (1+r)^{-j}$, we can again split this increasing annuity as

$$A_m = \sum_{j=1}^{\infty} j^m (1+r)^{-j} - \sum_{j=n+1}^{\infty} j^m (1+r)^{-j}$$

$$= P_m - (1+r)^{-n} \sum_{j=1}^{\infty} (j+n)^m (1+r)^{-j}. \tag{6.32}$$

The binomial theorem can be applied to the second summation, producing a formula involving $\{P_j\}_{j=0}^{m}$ (see exercise 28).

6.4.5   Price Function Approximation: Asset Allocation

The primary application of power series in finance is to the problem of modeling and understanding the behavior of a complicated function $f(x)$ in a neighborhood of some fixed point $a \in \mathbb{R}$, or in the more general case, a multivariate function $f(\mathbf{x})$ in a neighborhood of $\mathbf{a} \in \mathbb{R}^n$. For example, $f(x)$ might denote the price of a bond when x is the bond's yield to maturity (YTM), and a denotes the yield today. Of course, as this price function is not very complicated, one could argue that to understand its behavior as the YTM changes from a to x, we simply can generate additional prices. However, this prospect becomes more daunting if one is managing a portfolio of such bonds, or if the price calculations are made more complex by the presence of embedded options like calls (i.e., early prepayment option for the issuer).

In more general multivariate cases, $f(\mathbf{x})$ might reflect a given bond's or bond portfolio's price as a function of a given yield curve, parametrized as a vector of values $\mathbf{x} \in \mathbb{R}^n$ as noted in section 3.3.1, with $f(\mathbf{a})$ the value on the current yield curve as parametrized by the vector $\mathbf{a}$. Prices of preferred stock, or common stock with the formulas above, can also be contemplated as a single variable or as multivariate functions. In each case the vector $\mathbf{a}$ denotes the collection of parameters that determine today's prices, and we are interested in approximating how prices change as these parameters change from $\mathbf{a}$ to $\mathbf{x}$.

A different kind of problem might be contemplated in the context of asset allocation. For example, for a given allocation vector $\mathbf{a}$ denoting the proportionate allocation to the various asset classes, one might develop a function $f(\mathbf{a})$ that quantifies return expectations, and another function $g(\mathbf{a})$ that quantifies risk expectations, given the current allocation vector. The analysis undertaken is one of understanding the behaviors of these functions in a neighborhood of $\mathbf{a}$ to investigate the possibilities of improving both return and risk through allocation changes, or at least to quantify the trade-off between risk and return.

In all such cases, as the complexity of the calculations increases, the utility and attractiveness of developing reasonable approximations also increases. To this end, methods discussed in chapter 9 on calculus will provide a basis for determining a sequence of coefficients, $\{c_j\}$, which may be finite or infinite. In the special infinite case, we will have

$$f(x) = \sum_{n=0}^{\infty} c_n (x - a)^n,$$

while in the finite case,

$$f(x) \approx \sum_{n=0}^{N} c_n (x-a)^n.$$

Both cases easily support approximations when x is "close to" a.

For example, assume that with either expansion above, with $N > 2$ say, that we attempt a **linear approximation**:

$$f(x) \approx c_0 + c_1(x-a).$$

Then in either case we conclude that the absolute error in this approximation, using the triangle inequality, is bounded by

$$|f(x) - [c_0 + c_1(x-a)]| \leq |c_2|(x-a)^2 \sum_{n=2}^{N} \left| \frac{c_n}{c_2}(x-a)^{n-2} \right|.$$

That is, as $x \to a$ the relative error satisfies

$$\frac{|f(x) - [c_0 + c_1(x-a)]|}{|c_2|(x-a)^2} \to 1. \tag{6.33}$$

This implies that for $x \sim a$, the absolute error is of order of magnitude $|c_2|(x-a)^2$.

Similarly one can show that the absolute error in the approximation $f(x) \approx c_0$ is of order of magnitude $|c_1||x-a|$, and similarly for approximations using higher order polynomials in $(x-a)$. Ultimately the ability to approximate $f(x)$ depends on how many terms in the series above the given function allows. When only finitely many terms are possible, approximation accuracy is limited but may still be adequate for applications. Otherwise, any given degree of accuracy is possible in theory as long as the analyst is willing to calculate additional terms in the approximating polynomial.

6.4.6 l_p-Spaces: Banach and Hilbert

The importance of these series spaces in finance is really that they provide an introduction to some subtle and important concepts in higher mathematics in an intuitive and accessible environment. The power of these concepts will only achieve full applicability in later studies on real analysis and stochastic processes, where these spaces are re-introduced as the L_p function spaces in general, and most important for stochastic processes, the special Hilbert space L_2. Consequently, while not of immediate application, these spaces and their properties, in addition to the examples of

Euclidean and complex spaces $\mathbb{R}^n$ and $\mathbb{C}^n$, will provide a solid foundation of examples that can be used to aid intuition in these admittedly more abstract settings.

In other words, the goal of this material in this chapter is to make the important but necessarily more remote setting of L_p-space better understood as a generalization of an accessible and familiar idea, than as an isolated and abstract construction.

Exercises

Practice Exercises

1. Show that if $\sum_{n=1}^{\infty} b_n$ is a convergent series, then as a sequence, $b_n \to 0$. (*Hint:* Consider the Cauchy criterion.)

2. Use the comparison test to demonstrate the absolute convergence of the following series, by comparing them to series shown to converge in this chapter:

(a) $\sum_{n=1}^{\infty} \frac{(\ln n)^2}{n^4}$

(b) $\sum_{n=1}^{\infty} \frac{\ln n}{n^p}$ for $p > 2$

(c) $\sum_{n=1}^{\infty} (-1)^{n+1} \frac{\sin(n)}{n^p}$, for $p > 1$, where for $\sin(n)$, n is understood in radians (i.e., π radians $= 180°$)

(d) $\sum_{n=1}^{\infty} (-1)^{n+1} a^n$ for $0 < a < 1$

3. Use the alternating series test or other means to demonstrate that the following converge and determine which converge absolutely:

(a) $\sum_{n=1}^{\infty} \frac{(-1)^{n+1}}{\ln(n+1)}$

(b) $\sum_{n=1}^{\infty} \frac{(-1)^{n+1} \ln(n!)}{n!}$

(c) $\sum_{n=1}^{\infty} \frac{(-1)^{n+1} \ln(n)}{n^p}$ for $p \geq 1$

(d) $\sum_{n=1}^{\infty} (-1)^{n+1} \ln\left(\frac{n+1}{n}\right)$

4. For each series in exercise 2, demonstrate absolute convergence using the comparative ratio test. In other words, in each case determine an absolutely convergent series $\sum_{i=1}^{\infty} c_n$ so that if a_n denotes the original series, $\frac{|a_n|}{|c_n|}$ converges as $n \to \infty$.

5. For the series in exercises 2 and 3, identify which would be declared as absolutely convergent using the ratio test, which would be not convergent, and which would be inconclusive.

6. Given a real number $x \in [0, 1]$, with decimal expansion $x = 0.a_1 a_2 a_3 \ldots$, where each $a_j \in \{0, 1, 2, \ldots, 9\}$, identify x with the sequence, $\mathbf{x} \in \mathbb{R}^\infty$ defined by $\mathbf{x} = (x_1, x_2, \ldots, x_j, \ldots)$, where $x_j = \frac{a_j}{10^j}$.

(a) Confirm that so defined, $\mathbf{x} \in l_p$ for all p, $1 \leq p \leq \infty$.

(b) Show that the truncated point sequence $\mathbf{x}_n \in \mathbb{R}_0^\infty$, defined by $\mathbf{x}_n = (x_1, x_2, \ldots, x_n, 0, 0, 0, \ldots)$, converges to $\mathbf{x}$ in the l_1-norm.

(c) Generalize part (b) to show that $\|\mathbf{x} - \mathbf{x}_n\|_p \to 0$ for all p, $1 \leq p \leq \infty$.

(d) Show that if the real number x is identified with the sequence $\mathbf{y} \in \mathbb{R}^\infty$, defined by $\mathbf{y} = (a_1, a_2, \ldots, a_j, \ldots)$, that $\mathbf{y} \in l_p$ only for $p = \infty$, yet even in this case, $\|\mathbf{y} - \mathbf{y}_n\|_\infty \nrightarrow 0$, where $\mathbf{y}_n = (a_1, a_2, \ldots, a_n, 0, 0, 0, \ldots)$ unless $\mathbf{y} \in \mathbb{R}_0^\infty$.

7. Using the Minkowski inequality, demonstrate that the following series are absolutely convergent:

(a) $\sum_{n=1}^\infty \left| \frac{n^{1.5}}{(n+2)^2} - \frac{(-1)^{n+1}}{\sqrt{n}} \right|^p$ for $p > 2$.

(b) $\sum_{k=1}^\infty \left| \frac{(-1)^{k+1}(k+1)}{k(k+10)} - \frac{\ln k}{k^2} - (0.5)^k \right|^p$ for $p > 1$.

8. Determine the radius of convergence and interval of convergence for the following power series:

(a) $u(z) = \sum_{n=1}^\infty \frac{z^n}{n}$

(b) $f(x) = \sum_{m=0}^\infty (-1)^m (x-1)^m$

(c) $g(y) = \sum_{n=1}^\infty (-1)^{n+1} n^p (y+2)^n$ for $p > 0$

(d) $h(z) = \sum_{k=1}^\infty \frac{(z-4)^k}{k}$

(e) $w(x) = \sum_{j=1}^\infty a^j (x+1)^j$, $a > 0$

(f) $v(y) = \sum_{m=0}^\infty \frac{(y-10)^m}{(m+1)(m+2)}$

(g) $k(y) = \sum_{n=1}^\infty n^n (y+4)^n$

(h) $m(u) = \sum_{n=1}^\infty \frac{2^n u^n}{n}$

9. With $f(x) = \sum_{n=0}^\infty \frac{x^n}{n!}$, develop the series expansion for $(f(x))^2$ using (6.23), and show that $(f(x))^2 = f(2x)$. (*Hint:* As will be demonstrated in (7.14) in chapter 7, $2^n = \sum_{j=0}^n \frac{n!}{j!(n-j)!}$ using the binomial theorem.)

10. Generalize exercise 9 to show that for all $n \in \mathbb{N}$, $(f(x))^n = f(nx)$. (*Hint:* Use induction.)

11. Confirm that for a preferred stock or common stock with nonconstant dividends $\{d_j\}$, where $d_j = a_1 j + a_0$, $a_1, a_0 \geq 0$, the price function $P(r) = \sum_{j=1}^\infty d_j (1+r)^{-j}$ is absolutely convergent for $r > 0$. (*Hint:* Consider the ratio test.)

12. Consider the preferred or common stock pricing function applied to the case of general nonconstant dividends $P(r) = \sum_{j=1}^\infty d_j (1+r)^{-j}$. Use the ratio test to develop bounds on the greatest rate of dividend growth allowable, which will ensure convergence for $r > 0$.

13. With a semi-annual yield rate of $r = 0.10$:

(a) Value a semiannual payment perpetuity that pays $10j + 15$ at time $j = 0.5, 1.0, 1.5 \ldots$.

(b) What is the semiannual payment increase for a 20 year $10 million semiannual payment mortgage where the borrower wants the payments to increase by equal amounts each payment and the first payment to be $0.25 million?

14. With an annual rate of 15%:

(a) Price a common stock with an annual dividend growth rate of 10% if the next dividend, due tomorrow, is expected to be $5.

(b) What is the price of the stock in part (a) if dividends are projected to grow for only 5 years at the 10% rate, then decrease to a growth rate of 5%?

15. With P_n defined as in the chapter by $P_n \equiv \sum_{j=1}^{\infty} j^n (1+r)^{-j}$:

(a) Derive (6.30),

$$P_n = \frac{1}{r}\left[\sum_{i=0}^{n-1}\binom{n}{i}P_i + 1\right].$$

(*Hint*: Note that

$$\sum_{j=1}^{\infty} j^n (1+r)^{-j} = (1+r)^{-1} + (1+r)^{-1}\sum_{j=1}^{\infty}(j+1)^n(1+r)^{-j}$$

and expand $(j+1)^n$ with the binimial theorem of (7.15) as seen in this chapter's derivation.)

(b) Develop explicit formulas for P_n, $n = 2, 3, 4, 5$ using $P_0 = \frac{1}{r}$ and $P_1 = \frac{1+r}{r^2}$.

16. Starting with the powers series, $f(x) = \sum_{n=0}^{\infty}\frac{x^n}{n!}$, consider the linear approximation $f_1(x) = 1 + x$. As indicated in (6.33),

$$\frac{f(x) - (1+x)}{\frac{x^2}{2}} \to 1,$$

as $x \to 0$. Demonstrate this result by calculating the power series for this ratio function, and confirming that it is absolutely convergent, which then justifies a substitution of $x = 0$.

Assignment Exercises

17. Use the comparison test to demonstrate the absolute convergence of the following series, by comparing them to series shown to converge in this chapter:

(a) $\sum_{n=1}^{\infty} c_n a^n$ for $0 < a < 1$ and any bounded sequence $\{c_n\}$

(b) $\sum_{j=1}^{\infty} (-1)^{j+1} \left(\frac{1}{j}\right)^q \ln j$ for $q > 2$

(c) $\sum_{k=1}^{\infty} \frac{(k+2)^2}{k(k+1)(k+10)^2}$

(d) $\sum_{k=1}^{\infty} \frac{(-1)^k (k+2)^3}{k!}$

18. Use the alternating series test to demonstrate that the following converge and determine which converge absolutely:

(a) $\sum_{n=1}^{\infty} \frac{(-1)^{n+1} \ln(n+1)}{n^4}$

(b) $\sum_{n=1}^{\infty} \frac{(-1)^{n+1} n^p}{a^n}$, $p \in \mathbb{R}$, $a > 1$

(c) $\sum_{n=1}^{\infty} \frac{(-1)^{n+1} n^2}{n^3+1}$

(d) $\sum_{n=1}^{\infty} \frac{(-1)^{n+1} n}{\ln[(n+1)^n]}$

19. For each series in exercise 17, demonstrate absolute convergence using the comparative ratio test. In other words, in each case determine an absolutely convergent series $\sum_{i=1}^{\infty} c_n$ so that if a_n denotes the original series, then $\frac{|a_n|}{|c_n|}$ converges as $n \to \infty$.

20. For the series in exercises 17 and 18, identify which would be declared as absolutely convergent using the ratio test, which would be not convergent, and which would be inconclusive.

21. Proposition 6.7 states that if $\sum_{j=1}^{\infty} x_j$ and $\sum_{j=1}^{\infty} y_j$ are absolutely convergent, then so too is $\sum_{j=1}^{\infty} x_j y_j$.

(a) Show that if $\sum_{j=1}^{\infty} x_j$ is absolutely convergent, and $\sum_{j=1}^{\infty} y_j$ conditionally convergent, then again $\sum_{j=1}^{\infty} x_j y_j$ is absolutely convergent.

(b) Give an example of conditionally convergent $\sum_{j=1}^{\infty} x_j$ and $\sum_{j=1}^{\infty} y_j$ for which $\sum_{j=1}^{\infty} x_j y_j$ is not convergent. (*Hint*: Can x_j and y_j be defined to satisfy the assumptions yet with $x_j y_j = \frac{1}{j}$?)

22. Prove that parts (c) and (d) of Exercise 6 have nothing to do with the base-10 assumption in the decimal expansion. In other words, if b is any positive integer, $b \geq 2$, and each such $x \in [0,1]$ is expanded in base-b so that $x = 0.a_1 a_2 a_3 \ldots$, where each $a_j \in \{0, 1, 2, \ldots, b-1\}$, then again:

(a) With $\mathbf{x} \in \mathbb{R}^\infty$ defined by $\mathbf{x} = (x_1, x_2, \ldots, x_j, \ldots)$, where $x_j = \frac{a_j}{b^j}$, and $\mathbf{x}_n \in \mathbb{R}_0^\infty$ is defined as before, we have that $\|\mathbf{x} - \mathbf{x}_n\|_p \to 0$ for all p, $1 \leq p \leq \infty$.

(b) With $\mathbf{y} \in \mathbb{R}^{\infty}$ defined by $\mathbf{y} = (a_1, a_2, \ldots, a_j, \ldots)$, where $x_j = a_j$, we have that $\mathbf{y} \in l_p$ only for $p = \infty$; yet even in this case $\|\mathbf{y} - \mathbf{y}_n\|_{\infty} \nrightarrow 0$, where $\mathbf{y}_n = (a_1, a_2, \ldots, a_n, 0, 0, 0, \ldots)$ unless $\mathbf{y} \in \mathbb{R}_0^{\infty}$.

23. Consider two sequences, $\mathbf{x} = (x_1, x_2, \ldots, x_j, \ldots)$, where $x_j = a^{-j}$, and $\mathbf{y}$ defined by $y_j = b^{-j}$, where $a, b > 1$:

(a) Confirm that $\mathbf{x}, \mathbf{y} \in l_p$ for all p, $1 \le p \le \infty$, and calculate the associated l_p-norms.

(b) Calculate the inner product $(\mathbf{x}, \mathbf{y})$, which is well defined.

(c) Develop the implication of Hölder's inequality, that for $1 \le p, q \le \infty$, with $\frac{1}{p} + \frac{1}{q} = 1$, where notationally, $\frac{1}{\infty} \equiv 0$, we have $|(\mathbf{x}, \mathbf{y})| \le \|\mathbf{x}\|_p \|\mathbf{y}\|_q$. Express the inequality in terms of one parameter, say with $q = \frac{p}{p-1}$.

(d) Express the inequality in part (c) in the special case of $p = q = 2$.

24. Determine the radius of convergence and interval of convergence for the following power series:

(a) $f(x) = \sum_{m=1}^{\infty} \frac{(-1)^m (x-5)^m}{m}$

(b) $g(y) = \sum_{n=1}^{\infty} n^p (y-6)^n$ for $p > 0$

(c) $h(z) = \sum_{k=1}^{\infty} \frac{(z-4)^k}{k!}$

(d) $t(z) = \sum_{k=1}^{\infty} (-1)^k \frac{(z+1)^k}{k!}$

(e) $w(x) = \sum_{j=1}^{\infty} a^{-j} (x-2)^j$, $a > 0$

(f) $v(y) = \sum_{m=0}^{\infty} \frac{(y+2)^m}{(m+1)^2}$

(g) $k(z) = \sum_{n=1}^{\infty} n!(z+1000)^n$

(h) $n(u) = \sum_{n=1}^{\infty} \frac{c^n u^n}{n}$, $c > 0$

25. Generalize exercise 11 to an arbitrary polynomial growth dividend model $d_j = \sum_{k=0}^{n} a_k j^k$, $a_k \ge 0$ for all k.

26. With an monthly rate of $r = 0.06$:

(a) Value a monthly payment perpetuity that pays $12j + 3$ at time j.

(b) What is the monthly payment increase for a 30-year, $5 million monthly payment mortgage, where the borrower wants the payments to increase by equal amounts each payment, and the first payment to be $10,000?

27. With an annual rate of 18%:

(a) Price a common stock with a semiannual nominal dividend growth rate of 8% if the next dividend, due tomorrow, is expected to be $15.

(b) What is the price of the stock in part (a) if dividends are projected to grow for only 3 years at the 8% rate and then increase to a growth rate of 12%?

28. Defining the increasing n-pay annuity, $A_m = \sum_{j=1}^{n} j^m (1+r)^{-j}$, use the formula in (6.32) and show that

$$A_m = P_m - (1+r)^{-n} \sum_{k=0}^{m} \binom{m}{k} n^{m-k} P_k,$$

where $\{P_k\}$ are given in exercise 15.

7 Discrete Probability Theory

7.1 The Notion of Randomness

In this chapter some basic ideas in probability theory are introduced and applied within a discrete distribution context. In chapter 10 these ideas will be generalized to continuous and so-called mixed distributions. The last step of the progression to "measurable" distributions will be deferred, since it requires the tools of real analysis.

Probability theory is the mathematical discipline that provides a framework for modeling and developing insights to the **random outcomes** of experiments developed in a laboratory or a staged setting or observed as natural or at least unplanned phenomenon. By **random** is meant that the outcome is not perfectly predictable, even when many of the features of the event are held constant or otherwise controlled and accounted for. By **discrete probability theory** is meant this theory as applied to situations for which there are only a finite or countably infinite number of outcomes possible. Later generalizations will extend these models and methods to situations for which an uncountable collection of outcomes are envisioned and accommodated.

It may seem surprising that the definition of "random" above states that the outcome is not perfectly predictable, rather than not predictable. This language is motivated by the fact that in many applications the outcome of an experiment or observation logically considered to be random may not be completely random in the stronger sense that we have no idea of what the outcome will be, but only random in the weaker sense that we have an imperfect idea of what the outcome will be.

For example, imagine that the observation to be made is the change in a major US stock market index, such as the S&P 500 Index, but simplified and reduced to a binary variable: -1 for a down market, and $+1$ for an up market. Most observers would agree that the result of this observation would appear to be a random outcome on a given day, at least as of the beginning of the day. However, just before the US market opens, stock markets in Japan and Asia have recently closed, Europe's trading day is half over, and based on their binary results it would appear that one could make a better guess of the subsequent US binary result than what would be possible without this information. Not a perfect prediction, of course, and the US result would still be considered random, but it would not be considered perfectly random.

Even more to the point, an hour before the US market closes, the binary result of this market remains random, but in a real sense, less random than at the opening bell because of the emergence of information throughout the trading hours. And this result one hour before market close is in turn apparently less random than the result as of the prior evening, before the Asian markets have traded.

So the definition of randomness given here allows all such observations to be modeled as random, until the moment in time when the outcome is perfectly predictable, which in this example, is moments after the "closing bell" when final trades are processed. Degrees of randomness is one of the ideas that can be quantified in probability theory. The notion of randomness here is admittedly informal, and it is to a large extent formalized only as a mathematical creation. But in the presence of the multitude of real world events that appear random, this informality is not fatal and the mathematical discipline of probability theory proves to be very useful.

For example, the flip of a "fair coin," by which is meant a coin for which it is equally likely to achieve a head, H, or a tail, T, is considered a standard model of randomness. On the assumption that the coin in question is perfectly fair, probability theory can address questions about a real or imagined experiment such as:

1. In 100 flips, how likely is it that exactly 80 Hs will occur?

2. In 10,000 flips, how likely is it that the number of Hs will exceed 5800?

3. In each case, what does "likely" mean?

In the absence of absolute knowledge of the fairness of the coin, probability theory can address questions on observations like:

1. In 10 flips, does 7 Hs provide "certain" evidence that the coin is "biased" and not fair?

2. In 10,000 flips, how large (or small) would the number of Hs have to be in order to be "certain" that the coin in question is not fair?

3. In each case, what does "certain" mean?

In real life one might think of the occurrences of car accidents, or untimely ends of life, as random outcomes within groups of individuals, though often not a perfectly random outcome in a given example. The modeling of these events is critical for property and casualty insurance and life insurance companies, respectively. In finance, virtually all observed market variables are also considered random, although generally not perfectly random. Prices of stock and bond market indexes, individual stocks and bonds, levels of interest rates, realized price or wage inflation indexes, currency exchange rates, commodity prices, and so forth, are all examples, as are events such as bond issuer defaults or bankruptcies or natural disasters.

Once mathematical models are produced for these variables, probability theory provides a framework for understanding the possible outcomes and answering questions such as those above, adapted to the given contexts.

7.2 Sample Spaces

7.2.1 Undefined Notions

As in every mathematical theory there must be some notions in probability theory that are considered "primitive" and hence will be left formally undefined. However, in the same way that most can work effectively in geometry without a formal definition of point, line, or plane, most can work effectively in probability theory without a formal definition of "sample space" or "sample points." In either case, the lack of formal definitions is made acceptable by the intuitive framework one can bring to bear on the subject.

For example, when one encounters point, line, or plane in geometry, a picture immediately comes to mind, and all statements about these terms understood, or at least interpreted, in the context of these pictures, however imperfectly. One's mental pictures of these terms in fact sharpen with time as their properties, developed in the context of the emerging theory, are revealed. So too for sample space and sample points, which are intended to provide a "set theory" structure to probability theory. In that context the sample space is understood as the "universe" of possible outcomes of a given experiment or natural phenomenon, and sample points understood to be the smallest possible units into which the sample space is decomposed, namely the individual outcomes or events. In this context the sample space can be viewed as a set of sample points, appropriately defined for the given application. By **discrete sample space** is meant, a sample space with a finite, or countably infinite, collection of sample points.

Example 7.1

1. *Returning to the coin flip examples above, if we are interested in understanding the possible outcomes of a* 10*-flip experiment, the sample space could be envisioned as the set of all* 10*-flip outcomes, and the sample points the individual sequences of* 10 *Hs and Ts. Similarly one could contemplate the sample space for the* 100*- and* 10,000*-flip questions.*

2. *In a different context with playing cards, one could envision a sample space of all 5-card hands that can be dealt from a single deck of cards, as would be relevant to a poker player. Similarly a sample space of all n-card hands that can be dealt from a multiple deck of cards, with point total less than 21, would be relevant in Black-jack. Especially relevant is the likelihood in any such case, that the $(n+1)$th card brings the point total above 21. The significance of the single deck versus multiple* *deck models is that the latter allows repeated cards in a single hand, whereas the former does not.*

3. *A related model for many probability problems is the "urn" problem, in which one envisions an urn that contains several colors of balls, with various numbers of each color. For example, the urn contains 25 balls: 2 red, 11 blue, and 12 green. One can then imagine an experiment where one selects 3 balls "at random" and forms the associated sample space of ball triplets. This sample space differs depending on how we assume that the 3 balls are selected:*

• *With replacement: Each of the 3 balls selected is returned to the urn after selection, so for each of the 3 draws, the urn contains the same 25 balls.*

• *Without replacement: Selected balls are not returned, so the balls in the urn for the second draw depend on the first ball drawn, and similarly for the third draw.*

For example, 3 red balls are a sample point of the sample space with replacement, but not in the space without replacement, since the urn contains only 2 red balls.

7.2.2   Events

We continue the set theory analogy. An **event** is defined to be a subset of the sample space. In the discrete models contemplated here, whereby one could feasibly list all possible sample points in the finite case, or produce a formula for the listing of all outcomes in the countably infinite case, the collection of events could be defined as the set of all subsets of the sample space. In other words, every subset of the sample space could be defined as an event. In later applications, beginning in chapter 10, where the idea of a sample space will be generalized, it will not be possible to allow all subsets of the sample space to qualify as events. Consequently we introduce ideas here, in a context where they are admittedly not strictly needed, in order to facilitate the generalization we will see later in chapter 10 and need in more advanced treatments. For subsets of the sample space to qualify as events, the specific question we need to address is: If the collection of events defined does not equal the collection of all subsets of the sample space, what minimal properties should this collection satisfy in order to be useful in applications?

The answer is as follows:

Definition 7.2   *Given a sample space, S, a collection of events, $\mathcal{E} = \{A \mid A \subset S\}$, is called a **complete collection** if it satisfies the following properties:*

1. $\emptyset, S \in \mathcal{E}$.

2. *If $A \in \mathcal{E}$ then $\tilde{A} \in \mathcal{E}$.*

3. *If $A_j \in \mathcal{E}$ for $j = 1, 2, 3, \ldots$, then $\bigcup_j A_j \in \mathcal{E}$.*

In other words, we require that a complete collection of events contain the "null event," $\emptyset$, and the "certain event," $\mathcal{S}$, the complement of any event, and that it be closed under countable unions. However, while item 3 is stated only for countable unions, it is also true for countable intersections because of item 2 and De Morgan's laws (see exercise 1). So it is also the case that $\bigcap_j A_j \in \mathcal{E}$. Similarly, if $A, B \in \mathcal{E}$, then $A \sim B \in \mathcal{E}$, where $A \sim B \equiv \{x \in \mathcal{S} \mid x \in A \text{ and } x \notin B\}$, since $A \sim B = A \cap \tilde{B}$.

Remark 7.3

1. *In a discrete sample space, $\mathcal{E}$ usually contains each of the sample points, and hence all subsets of $\mathcal{S}$, and is consequently always a complete collection. In other words, $\mathcal{E}$ is the **power set** of $\mathcal{S}$.*

2. *The use of the term "complete collection" is not standard but is introduced for simplicity. The three conditions in the definition above are general requirements for $\mathcal{E}$ to be a so-called **sigma algebra** as will be seen in chapter 10 and more advanced treatments.*

In discrete probability theory this extra formality may seem absurd, since we can so easily just list all possible events and work within this total collection in all applications. For example, in the sample space of 10 flips of a fair coin, the sample points are strings of 10 Hs and Ts, which we could list, even though there are 2^{10} such points. Also we could at least imagine the power set of this sample space, the collection of all subsets of sample points, of which there are $2^{2^{10}}$ (recall exercise 4 in chapter 4).

If the sample space is defined as the collection of Hs and Ts in n flips of a coin for all n, or defined as all sequences that emerge from flips that terminate on the occurrence of the first H, or the mth H, then these sample spaces have countably many sample points, and although significantly more complicated, one could envision the collection of all subsets as events.

However, if the sample space is defined as the collection of Hs and Ts in a countably infinite number of flips, this space has the same cardinality as the real numbers (recall exercise 5 in chapter 4), and the prospect of defining events as every subset of this space becomes hopeless, as can be proved using the tools of real analysis. Consequently the definition above is needed in such cases, and identifies the minimal properties for an event space for the next step, which is the introduction of event probabilities.

7.2.3   Probability Measures

The intuition behind the notion of the "probability" of an event is a simple one. One approach is sometimes deemed the "frequentist" interpretation. That is, the probability of an event is the long-term proportion of times the event would be observed in a repeated trials of an experiment that was designed to result in two outcomes:

Event *A* observed;

Event *A* not observed.

In this interpretation it is assumed that each trial is "independent" of the others, which is to say, that its outcome neither influences nor is influenced by the outcomes of the other trials.

Example 7.4   *In the 10-flip coin sample space S, define the event A as the subset of the sample space that has HH as the first two flips. Intuitively, a fair coin makes every sequence equally likely, and it is easy to see that 25% of the sequences in S begin with HH. So if we designed an experiment that flipped a coin 10 times, and recorded the results after many trials, the expectation would be that in 25% of the tests, A would be observed. The term "frequentist" probability comes from the idea that 25% is the relative frequency of event A in a long string of such trials. It is the relative frequency that would be observed in the long run.*

An alternative interpretation is related to games of chance, or gambling, which was a primary motivator for the original studies of probability by **Abraham de Moivre** (1667–1754), who published an early treatise on the subject in 1718 called ***The Doctrine of Chances***. The gambling perspective for this example can be phrased as: *For a $1 bet, what should the payoff be when event A occurs so that a gambler's wealth can be expected to not change in the long run?* Such a bet would be called a "fair bet." There is of course a frequentist flavor to this interpretation, since present are the notions of "repeated trials" and "in the long run."

So, if p denotes the probability of event A occurring and N is a large integer, then in N bets the gambler will bet $1 and lose about $(1 - p)N$ bets and $$(1 - p)N$, and the gambler will win about Np bets and $$Npw$ if w is the associated payoff or "winnings" for a $1 bet. This bet will be a fair bet if won and lost bets are equal, which happens when

$$w = \frac{1-p}{p}. \tag{7.1}$$

Example 7.5   *In the coin-flip example above, the gambler's winnings for a $1 bet, to ensure that it is a fair bet, must be $w = \$3$. That is, the gambler wins $3 if the coin-flip sequence is HH . . . , and he loses $1 otherwise.*

The formula for w in (7.1) really only makes sense for p values of $0 < p < 1$. Otherwise, the bet degenerates to a sure win or sure loss, and it cannot be made "fair" in the sense above. On this domain, $w = \frac{1}{p} - 1$ is seen to decrease as p increases, is unbounded as $p \to 0$, and decreases to 0 as $p \to 1$, consistent with intuition.

Note that (7.1) also encodes information about the "probability" we seek, and can be rewritten as

$$p = \frac{1}{w+1}. \tag{7.2}$$

Example 7.6 *Again in the coin-flip example, if participants agreed that the correct payoff was $w = 3$, then we would conclude that the probability of the sequence $HH\ldots$ is 0.25 or 25%.*

This intuitive framework provides a starting point for formalizing the notion of probability. Probabilities are logically associated with events and can therefore be identified with a function on the collection of events, denoted $\Pr(A)$ for $A \in \mathcal{E}$. Furthermore the value of this function must be between 0 and 1 for any event, and these extremes should be achieved on the null event, $\emptyset$, and the full sample space, $\mathcal{S}$, respectively. Finally, we expect this function to behave logically on the collection of events. For example, if $A \subset B$ are events, we want $\Pr(A) \leq \Pr(B)$, and if $A \cap B = \emptyset$, then $\Pr(A \cup B) = \Pr(A) + \Pr(B)$, and so forth.

We collect the necessary properties in the following, and note in advance that in a discrete sample space, $\Pr(s)$ is typically defined for all $s \in \mathcal{S}$ since $\mathcal{E}$ contains the individual sample points.

Definition 7.7 *Given a sample space, $\mathcal{S}$, and a complete collection of events, $\mathcal{E} = \{A \mid A \subset \mathcal{S}\}$, a **probability measure** is a function $\Pr : \mathcal{E} \to [0,1]$ that satisfies the following properties:*

1. $\Pr(\mathcal{S}) = 1$.

2. *If $A \in \mathcal{E}$, then $\Pr(A) \geq 0$ and $\Pr(\tilde{A}) = 1 - \Pr(A)$.*

3. *If $A_j \in \mathcal{E}$ for $j = 1, 2, 3, \ldots$ are **mutually exclusive events**, that is, with $A_j \cap A_k = \emptyset$ for all $j \neq k$, then $\Pr(\bigcup_j A_j) = \sum \Pr(A_j)$.*

*In this case the triplet $(\mathcal{S}, \mathcal{E}, \Pr)$ is called a **probability space**.*

Definition 7.8 *An event $A \in \mathcal{E}$ is a **null event under** $\Pr$ if $\Pr(A) = 0$. If A is a null event and every $A' \subset A$ satisfies $A' \in \mathcal{E}$, then the triplet $(\mathcal{S}, \mathcal{E}, \Pr)$ is called a **complete probability space**.*

Some properties of this probability measure are summarized next.

Proposition 7.9   *If* Pr *is a probability measure on a complete collection of events* $\mathcal{E}$, *then:*

1. $\Pr(\emptyset) = 0$.

2. *If* $A, B \in \mathcal{E}$, *with* $A \subset B$, *then* $\Pr(A) \leq \Pr(B)$.

3. *If* $A_j \in \mathcal{E}$ *for* $j = 1, 2, 3, \ldots$, *then*

$$\max_j\{\Pr(A_j)\} \leq \Pr\left(\bigcup_j A_j\right) \leq \sum_j \Pr(A_j).$$

4. *If* $A_j \in \mathcal{E}$ *for* $j = 1, 2, 3, \ldots$, *then*

$$\Pr\left(\bigcap_j A_j\right) \leq \min_j\{\Pr(A_j)\}.$$

Proof   See exercise 26.   ∎

Remark 7.10   *Note that in property* 2 *of the proposition above, it might be expected that if* $B \in \mathcal{E}$, *and* $A \subset B$, *then automatically it is true that* $A \in \mathcal{E}$. *In the special case of this chapter of discrete probability spaces, this is virtually always true in applications, since then* $\mathcal{E}$ *typically contains all the sample points and hence contains all possible subsets of* $\mathcal{S}$. *In the general case of what is called a "complete" collection of events, or generally a sigma algebra, subsets of events need not be events.*

7.2.4   Conditional Probabilities

Given a sample space $\mathcal{S}$, a complete collection of events $\mathcal{E} = \{A \mid A \subset \mathcal{S}\}$, and a probability measure $\Pr : \mathcal{E} \to [0, 1]$, there are many situations in which we are interested in probability values that reflect additional information. For example, if the sample space is the collection of all 10-flip sequences of a fair coin, we know that the probability of every one of the 2^{10} sample points is $\left(\frac{1}{2}\right)^{10}$. Similarly, if we define an event B as the collection of sample points with exactly 1-H and 9-Ts, then $\Pr(B) = 10\left(\frac{1}{2}\right)^{10}$ since we know there are exactly 10 such sequences.

Now imagine that we know that event B is true. How would that knowledge alter our calculation of the probabilities of all the events in $\mathcal{E}$? Perhaps simpler, how would that knowledge alter our calculation of the probabilities of all the sample points in $\mathcal{S}$? In other words, what is $\Pr(A$ conditional on the knowledge that B is true), where A denotes any sample point or event? In probability theory, this is called a **conditional probability**, and is written

$\Pr(A \mid B)$,

and read, "the probability of A given B," or "the probability of A conditional on B."

Example 7.11 *The sample points are somewhat easier to address first. Since we want $\Pr(\cdot \mid B)$ to be a genuine probability measure on $\mathcal{E}$, we need $\Pr(\mathcal{S} \mid B) = 1$, and since $\mathcal{S}$ is the disjoint union of its sample points, we must have that the sum of all the conditional probabilities of the sample points is also 1. Now, if A is any event with more or less than 1 H, it must be the case that $\Pr(A \mid B) = 0$. What about the 10 sample points, each with 1 H? Since each is equally likely in $\mathcal{E}$, it is logical to define $\Pr(A \mid B) = \frac{1}{10}$ for each such point. Similarly, if A is a general event that contains none of these 1-H points, we define $\Pr(A \mid B) = 0$, while if A contains j of these points, we define $\Pr(A \mid B) = \frac{j}{10}$.*

In this simple context the notion of conditional probability is somewhat transparent. The general definition is intended to formalize this idea to be more applicable in more complex situations, and provide a calculation that explicitly references the original probabilities of events under Pr.

Definition 7.12 *Given a discrete sample space $\mathcal{S}$, a complete collection of events $\mathcal{E} = \{A \mid A \subset \mathcal{S}\}$, a probability measure $\Pr : \mathcal{E} \to [0, 1]$, and an event $B \in \mathcal{E}$ with $\Pr(B) > 0$, then for any $A \in \mathcal{E}$, the **conditional probability of** A given B, denoted $\Pr(A \mid B)$, is defined by*

$$\Pr(A \mid B) = \frac{\Pr(A \cap B)}{\Pr(B)}, \qquad \Pr(B) \neq 0. \tag{7.3}$$

It is a straightforward exercise that for any such event B, that $\Pr(\cdot \mid B)$ defines a true probability measure on $\mathcal{S}$ as given in the definition above (see exercise 5). One can also review the example above in the formalized context of (7.3) and see that the respective intuitive results are reproduced.

Law of Total Probability
Another important application of these ideas is exemplified as follows:

Example 7.13 *Imagine an urn containing 10 balls, 5 each of red (R) and blue (B), from which 2 are to be selected. Let C_1 denote the color of the first ball drawn, and C_2 the color of the second. Then construct two sample spaces of the pair of balls drawn, (C_1, C_2): one space defined under the assumption that the draws are done with replacement, and the other reflecting no replacement. In the sample space with replacement, it is easy to see that $\Pr(C_2 \mid C_1) = \Pr(C_2)$. For example, $\Pr(R_2) \equiv \Pr(C_2 = R) = 0.5$, and $\Pr(R_2 \mid C_1) = 0.5$ whether $C_1 = R$ or $C_1 = B$.*

In the sample space without replacement, it is never the case that $\Pr(C_2 \mid C_1) = \Pr(C_2)$. *For example,* $\Pr(R_2 \mid R_1) = \frac{4}{9}$ *and* $\Pr(R_2 \mid B_1) = \frac{5}{9}$, *and we now show that* $\Pr(R_2) = 0.5$. *To this end, first note that* $\Pr(R_1 \mid R_2) \neq \frac{1}{2}$, *as might be expected given that* R_1 *happens "first" when there are five of each color. But that is not the meaning of* $\Pr(R_1 \mid R_2)$. *The question is, looking at the outcomes for which* $C_2 = R$, *what is the probability that* $C_1 = R$? *There are two such outcomes:*

$$\Pr(R_1 \cap R_2) = \frac{4}{18} \quad \text{and} \quad \Pr(B_1 \cap R_2) = \frac{5}{18},$$

from which we conclude that $\Pr(R_1 \mid R_2) = \frac{4}{9}$. *An application of (7.3) now shows that* $\Pr(R_2) = \frac{\Pr(R_1 \cap R_2)}{\Pr(R_1 \mid R_2)} = 0.5$. *This probability could have also been more easily calculated from the respective conditional probabilities using a method discussed next.*

Let $\{B_j\}$ be a collection of mutually exclusive events with $\bigcup B_j = S$. Then for any event A, $\{A \cap B_j\}$ are also mutually exclusive, and have union A. By the third property of the probability measure, we have that $\Pr(A) = \Pr(\bigcup[A \cap B_j]) = \sum \Pr(A \cap B_j)$. Also, by (7.3), $\Pr(A \cap B_j) = \Pr(A \mid B_j) \Pr(B_j)$. Combining, we get the **law of total probability**:

$$\Pr(A) = \sum_j \Pr(A \mid B_j) \Pr(B_j). \tag{7.4}$$

This law has widespread application because it is often easier to calculate conditional probabilities of an event than the direct probability because each "condition" provides a restriction on the sample points that need be considered.

Example 7.14   *In the urn problem of example 7.13 without replacement,* $\Pr(R_2) = 0.5$ *could have been more easily derived using this law of total probability. The mutually exclusive events* $\{B_j\}$ *are the events* $C_1 = R$ *and* $C_1 = B$, *and each of these events has probability equal to 0.5. Consequently, using the respective conditional probabilities, we can write*

$$\Pr(R_2) = \Pr(R_2 \mid R_1) \Pr(R_1) + \Pr(R_2 \mid B_1) \Pr(B_1),$$

again producing $\Pr(R_2) = 0.5$.

7.2.5   Independent Events

The notion of **stochastic independence** is a property of pairs of events under a given probability measure Pr. Intuitively we say that A and B are stochastically independent, or simply independent, if their probabilities are not changed by conditioning on each other. This idea is a simple one, except for the formality that in order for the various conditional probabilities to be defined, it is necessary that both events have nonzero probability.

To circumvent this technicality, observe that the desired condition: $\Pr(A \mid B) = \Pr(A)$, which requires that $\Pr(B) \neq 0$ to be well defined, is by (7.3) equivalent to $\Pr(A \cap B) = \Pr(A)\Pr(B)$, which does not require a condition on $\Pr(B)$ or $\Pr(A)$ to be well defined. This latter formulation of the idea of independence also has the immediate advantage of reflexivity; that is, A is independent of B iff B is independent of A. Formally, we state:

Definition 7.15 *Events $A_1, A_2 \in \mathcal{E}$ are **stochastically independent**, or simply **independent**, under the probability measure* $\Pr$, *if*

$$\Pr(A_1 \cap A_2) = \Pr(A_1)\Pr(A_2). \tag{7.5}$$

*More generally, a collection of events: $\{A_j\}_{j=1}^n$, where n may be ∞, are **mutually independent**, if for any integer subset $J \subset \{1, 2, \ldots, n\}$ we have that*

$$\Pr\left(\bigcap_J A_j\right) = \prod_J \Pr(A_j). \tag{7.6}$$

This definition makes sense even if A_k is a null event, $\Pr(A_k) = 0$ for some k. In either setting, we have from property 2 of the proposition above on probability measures that $\Pr(\bigcap_J A_j) = 0$ as well if $k \in J$. So formally, null sets are independent of all other sets.

In the case where one or both of A or B have nonzero probability, the notion of independence can be reformulated using conditional probabilities. For example, if A and B are independent, and $\Pr(B) \neq 0$, then

$$\Pr(A) = \Pr(A \mid B).$$

In other words, if A and B are independent, their probabilities are unaffected by knowledge of the occurrence of the other event.

In the urn examples above, with C_1 denoting the color of the first ball drawn and C_2 the color of the second, it was seen that in the sample space with replacement, these events were independent, whereas without replacement, these events are not independent.

7.2.6 Independent Trials: One Sample Space

One of the most important applications of the notion of independence is in the formalization of the idea of a **random sample** from a discrete sample space, or equivalently, a series of **independent trials** from a discrete sample space. Given a sample space S with associated probability measure Pr, a random sample of size n, or a sequence of n trials, is defined as a sample point in another sample space, S^n, which is formalized in:

Definition 7.16   *Given a discrete sample space S, a complete collection of events $\mathcal{E} = \{A \mid A \subset S\}$ containing the sample points, and a probability measure $\Pr : \mathcal{E} \to [0,1]$, the associated n-**trial sample space**, denoted S^n, is defined by*

$$S^n = \{(s_1, s_2, \ldots, s_n) \mid s_j \in S\}.$$

The collection of events, denoted $\mathcal{E}^n$, is defined by

$$\mathcal{E}^n = \{(A_1, A_2, \ldots, A_n) \mid A_j \in \mathcal{E} \text{ and by unions of such events}\}.$$

The associated probability measure, P_n, is defined on $\mathcal{E}^n$ by

$$P_n[(s_1, s_2, \ldots, s_n)] = \prod_{j=1}^{n} \Pr(s_j), \tag{7.7}$$

as extended additively to events, for $A \in \mathcal{E}^n$,

$$P_n(A) = \sum_{(s_1, s_2, \ldots, s_n) \in A} P_n[(s_1, s_2, \ldots, s_n)]. \tag{7.8}$$

The goal of the next proposition is to confirm that the collection of events in n-trial sample space is a complete collection, and that P_n is indeed a probability measure on S^n. Most important, we confirm that any event in $\mathcal{E}$ can be identified in a natural but not unique way with an event in $\mathcal{E}^n$, and that under this identification, n events in $\mathcal{E}$ are mutually independent as events in $\mathcal{E}^n$. This identification and associated independence result provides a formal meaning to the notion of independent trials, or independent draws, from a given sample space.

Before stating this proposition, we note that the multiplicative rule in (7.7) extends to events in $\mathcal{E}^n$. That is, with $A \equiv (A_1, A_2, \ldots, A_n)$, $A_j \in \mathcal{E}$,

$$P_n[A] = \sum_{(s_1, s_2, \ldots, s_n) \in A} \prod_{j=1}^{n} \Pr(s_j)$$

$$= \prod_{j=1}^{n} \left[\sum_{s_j \in A_j} \Pr(s_j) \right]$$

$$= \prod_{j=1}^{n} \Pr(A_j).$$

That is, for $\{A_j\}_{j=1}^{n} \subset \mathcal{E}$,

$$P_n[(A_1, A_2, \ldots, A_n)] = \prod_{j=1}^{n} \Pr(A_j). \tag{7.9}$$

Remark 7.17 *In the definition of n-trial sample space it is assumed that the event space $\mathcal{E}$ contained all the sample points. In fact, while this assumption is almost always true in discrete probability theory, it is more of a convenience here than a necessity. With this assumption, $\mathcal{E}^n$ then contains all the n-tuples of sample points, $(s_1, s_2, \ldots, s_n)$, whose probabilities are defined by (7.7), and the probability measure P_n is then easily generalized to all events in $\mathcal{E}^n$ by (7.8). In the more general case where $\mathcal{E}$ does not contain all the sample points, but is a complete collection of events as defined above, a similar construction is possible but more difficult. In this case $\mathcal{E}^n$ is defined as above to include all n-tuples of events, $(A_1, A_2, \ldots, A_n)$, and then expanded to include all unions of these n-tuples and their complements so that $\mathcal{E}^n$ becomes complete. The probability measure P_n is defined on n-tuples of events, $(A_1, A_2, \ldots, A_n)$, using (7.9) and then extended to all of $\mathcal{E}^n$. It is not possible to define this extension directly using a generalization of (7.8) because of a technicality that is avoided with our convenient assumption. And that technicality is, if an event $A \subset \mathcal{E}^n$ is a union of n-tuples of events, $\{(A_{k1}, A_{k2}, \ldots, A_{kn})\}_{k=1}^{N}$, where N may be ∞, these events need not be disjoint, and so a direct application of a formula such as (7.8) may involve multiple counts. This problem is avoided when $\mathcal{E}^n$ contains all n-tuples of sample points, $(s_1, s_2, \ldots, s_n)$. This general construction is subtle and developed in advanced studies using the tools of real analysis.*

Proposition 7.18 *Given a discrete sample space $\mathcal{S}$, a complete collection of events $\mathcal{E} = \{A \mid A \subset \mathcal{S}\}$ containing the sample points, and $\Pr$ a probability measure on $\mathcal{E}$, then:*

1. *Every event $A \subset \mathcal{E}$ can be identified with n-events in $\mathcal{E}^n$, any one if denoted $\bar{A}$, satisfies $P_n[\bar{A}] = \Pr(A)$.*

2. *Under the identification in 1, every collection of up to n-events in $\mathcal{E}$ can be identified with mutually independent events in $\mathcal{S}^n$. That is, for any collection of events in $\mathcal{E}$, $\{A_k\}_{k=1}^{n}$, there are associated $\{\bar{A}_k\}_{k=1}^{n} \subset \mathcal{E}^n$, so that for any $K \subset \{1, 2, \ldots, n\}$:*

$$P_n \left[\bigcap_{k \in K} \bar{A}_k \right] = \prod_{k \in K} P_n[\bar{A}_k] = \prod_{k \in K} P[A_k].$$

3. $\mathcal{E}^n$ *is a complete collection of events.*

4. P_n *defined in (7.7) and (7.8) is a probability measure on* $\mathcal{E}^n$.

Proof

1. The n identifications as noted above are simply $A \leftrightarrow (A, \mathcal{S}, \dots, \mathcal{S}), (\mathcal{S}, A, \mathcal{S}, \dots, \mathcal{S}) \dots (\mathcal{S}, \dots, \mathcal{S}, A)$, and for each identification by (7.9) we have $P_n[\bar{A}] = \text{Pr}(A)$, since $\text{Pr}(\mathcal{S}) = 1$.

2. Given $\{A_k\}_{k=1}^n$ we associate each with $\bar{A}_k$ where the event A_k is assigned to the kth component of $\bar{A}_k$, and $\mathcal{S}$ assigned to the other components as in 1 above. Now, if $K \subset \{1, 2, \dots, n\}$, $\bigcap_{k \in K} \bar{A}_k$ equals the event in $\mathcal{E}^n : (A'_1, A'_2, \dots, A'_n)$, where each A'_j equals A_j or $\mathcal{S}$, and the result follows from (7.9).

3. Both $\mathcal{S}^n \equiv (\mathcal{S}, \mathcal{S}, \dots, \mathcal{S})$ and $\emptyset \equiv (\emptyset, \emptyset, \dots, \emptyset)$ are elements of $\mathcal{E}^n$, by definition. Also, since $\mathcal{E}^n$ contains all n-tuples of sample points, $(s_1, s_2, \dots, s_n)$, if $A \in \mathcal{E}^n$, then also $\tilde{A} \in \mathcal{E}^n$. Similarly, if $A_j \in \mathcal{E}^n$, then $\bigcup A_k \in \mathcal{E}^n$.

4. By definition of P_n, we have $P_n[\emptyset] = 0$, and

$$P_n[\mathcal{S}^n] = \sum_{(s_1, s_2, \dots, s_n) \in \mathcal{S}^n} \left[\prod_{j=1}^n \text{Pr}(s_j) \right] = \left[\sum_{s_j \in \mathcal{S}} \text{Pr}(s_j) \right]^n = 1.$$

Now, if $A = \bigcup_{k=1}^M (s_{k1}, s_{k2}, \dots, s_{kn})$, then $A \cup \tilde{A} = \mathcal{S}^n$, and we can rewrite the identity above for $P_n[\mathcal{S}^n]$ as

$$1 = \sum_{(s_1, s_2, \dots, s_n) \in \mathcal{S}^n} \left[\prod_{j=1}^n \text{Pr}(s_j) \right]$$

$$= \sum_{(s_1, s_2, \dots, s_n) \in A} \left[\prod_{j=1}^n \text{Pr}(s_j) \right] + \sum_{(s_1, s_2, \dots, s_n) \in \tilde{A}} \left[\prod_{j=1}^n \text{Pr}(s_j) \right].$$

$$= P_n(A) + P_n(\tilde{A}).$$

Hence $P_n(\tilde{A}) = 1 - P_n(A)$. Finally, if $\{B_k\}_{k=1}^m$ are mutually exclusive events, meaning that for any $K \subset \{1, 2, \dots, m\}$,

$$\bigcap_{k \in K} B_k - \emptyset,$$

then by (7.8),

$$P_n(\bigcup B_k) = \sum_{(s_1, s_2, \ldots, s_n) \in \bigcup B_k} \prod_{j=1}^{n} \Pr(s_j)$$

$$= \sum_{k} \sum_{(s_1, s_2, \ldots, s_n) \in B_k} \prod_{j=1}^{n} \Pr(s_j)$$

$$= \sum_{k} P_n(B_k),$$

where the second equality is due to mutual exclusivity: $\sum_{(s_1, s_2, \ldots, s_n) \in \bigcup B_k} = \sum_{k} \sum_{(s_1, s_2, \ldots, s_n) \in B_k}$. ∎

*7.2.7   Independent Trials: Multiple Sample Spaces

The construction of an n-trial sample space $\mathcal{S}^n$, reflecting independent samples from a given sample space $\mathcal{S}$, is readily generalized to the notion of an n-trial sample space reflecting independent samples from a collection of different sample spaces. To this end, we start with a definition.

Definition 7.19 *Given a collection of discrete sample spaces $\{\mathcal{S}_j\}_{j=1}^{n}$, complete collections of events $\{\mathcal{E}_j\}_{j=1}^{n}$ where each $\mathcal{E}_j = \{A \mid A \subset \mathcal{S}_j\}$ contains all the sample points of $\mathcal{S}_j$, and associated probability measures $\Pr_j : \mathcal{E}_j \to [0, 1]$, the associated **generalized n-trial sample space**, denoted $\mathcal{S}^{(n)}$, is defined by*

$$\mathcal{S}^{(n)} = \{(s_1, s_2, \ldots, s_n) \mid s_j \in \mathcal{S}_j\}.$$

The collection of events, denoted $\mathcal{E}^{(n)}$, is defined by

$$\mathcal{E}^{(n)} = \{(A_1, A_2, \ldots, A_n) \mid A_j \in \mathcal{E}_j \text{ and unions of such events}\}.$$

The associated probability measure, $P_{(n)}$, is defined on $\mathcal{E}^{(n)}$ by

$$P_{(n)}[(s_1, s_2, \ldots, s_n)] = \prod_{j=1}^{n} \Pr_j(s_j), \tag{7.10}$$

as extended additively to events, for $A \in \mathcal{E}^{(n)}$:

$$P_{(n)}(A) = \sum_{(s_1, s_2, \ldots, s_n) \in A} P_{(n)}[(s_1, s_2, \ldots, s_n)]. \tag{7.11}$$

The proofs of the results in proposition 7.18 in the special case where $\mathcal{S}_j = \mathcal{S}$ and $\mathcal{E}_j = \mathcal{E}$ for all j carry over to this more general case without material change other than notational. This is because, with one exception, nowhere in the derivations above was it necessary to use the fact that the sample spaces, collections of events, and probability measures underlying the various components of an n-trial sample point were identical. The single exception is related to the identifications of events in $\mathcal{S}$ with events in $\mathcal{S}^n$. In the simpler case above, each event in $A \subset \mathcal{S}$ could be identified with n events in $\mathcal{S}^n$, all of which had the same probability under P_n, and this common probability equaled $\Pr(A)$, the probability in $\mathcal{S}$. In the general case it is natural to assume that the given sample spaces are ordered. Hence each event $A \subset \mathcal{S}_j$ is identified with a unique element $\bar{A} \subset \mathcal{S}^{(n)}$, and that is defined with A in the jth component, and the various $\mathcal{S}_k$ spaces used as events in the other components, in order. Of course the ordering is a convenience more than a necessity, and different orderings do not produce fundamentally different spaces.

As an example of how a result above generalizes to this setting, we note that (7.10) generalizes in the same way that (7.7) generalizes to (7.9). Specifically, with the same derivation, and for $A_j \in \mathcal{E}_j$,

$$P_{(n)}[(A_1, A_2, \ldots, A_n)] = \prod_{j=1}^{n} \Pr_j(A_j). \tag{7.12}$$

Finally, we state without proof the fundamental result that generalizes the proposition above to this setting, and note that remark 7.17 in that section, regarding the assumption that each $\mathcal{E}_j$ contains the sample points, applies here as well.

Proposition 7.20   *Given a collection of discrete sample spaces $\{\mathcal{S}_j\}_{j=1}^{n}$, complete collections of events $\{\mathcal{E}_j\}_{j=1}^{n}$ that contain the sample points, and associated probability measures $\Pr_j : \mathcal{E}_j \rightarrow [0, 1]$, then:*

1. *Every event $A \subset \mathcal{E}_j$ can be identified with a unique event in $\bar{A} \subset \mathcal{E}^{(n)}$ that satisfies $P_{(n)}[\bar{A}] = \Pr_j(A)$.*

2. *Under the identification in 1, every collection of events $A_k \subset \mathcal{E}_k$, $1 \leq k \leq n$, can be identified with mutually independent events in $\mathcal{S}^{(n)}$. That is, for any such collection of events $\{A_k\}_{k=1}^{n}$, there are associated $\{\bar{A}_k\}_{k=1}^{n} \subset \mathcal{E}^{(n)}$ so that for any $K \subset \{1, 2, \ldots, n\}$,*

$$P_{(n)}\left[\bigcap_{k \in K} \bar{A}_k\right] = \prod_{k \in K} P_{(n)}[\bar{A}_k] = \prod_{k \in K} P_k[A_k].$$

3. $\mathcal{E}^{(n)}$ *is a complete collections of events.*

4. $P_{(n)}$ *defined in (7.10) and (7.11) is a probability measure on* $\mathcal{S}^{(n)}$.

7.3   Combinatorics

To determine the values of $\Pr(A)$ in various sample space applications, it is often necessary to be able to efficiently count the sample points in the event A as well as those in the sample space $\mathcal{S}$, and such calculations can be both subtle and difficult. The mathematical discipline of **combinatorics**, or **combinatorial analysis**, provides a structured framework for addressing these types of problems, and we only scratch the surface of this discipline here with the most common applications.

7.3.1   Simple Ordered Samples

In many applications we require the number of ways that m items can be selected from a collection of $n \geq m$ distinguishable items. For example, an urn may contain n balls, all distinguishable by color or other markings, and we seek to determine how many distinct m-ball collections can be drawn from this urn. As we have seen from the examples above, we need to distinguish between whether this is an urn problem with replacement or without replacement.

With Replacement
On the first draw there are n possible outcomes, and due to replacement, each successive draw has the same number of possible outcomes. So we conclude that there are n^m total possibilities. This can be formalized by observing that for $m = 2$ we can explicitly enumerate the outcomes, and then proceed by induction. That is, we assume the truth of the formula for m, and verify the truth for $m + 1$ based on the explicit pairings of each m-tuple with each last draw.

Without Replacement
On the first draw there are again n possible outcomes, but since the first draw is not returned to the urn, the second draw has fewer possible outcomes, namely $n - 1$. This process continues to the mth draw for which there are $n - (m - 1) = n - m + 1$ possible outcomes. Using the same logic and proof as above, we see that there are $n(n-1)\ldots(n-m+1)$ possible outcomes. This sequential product is common in combinatorics, and it is worthwhile to note that it can easily be expressed in terms

of the **factorial function**. Recall that n **factorial** is defined $n! = n(n-1)(n-2)\cdots 2\cdot 1$, and so

$$n(n-1)\ldots(n-m+1) = \frac{n!}{(n-m)!}.$$

In some texts this partial factorial, which contains m terms, is denoted $(n)_m \equiv n(n-1)\ldots(n-m+1)$. Of course, in this notation, $(n)_n = n!$.

7.3.2   General Orderings

Here we seek an approach to determining how many distinguishable ways a given collection of n objects can be ordered. The answer depends on how many subset types are represented by the n objects, where all objects in each subset are identical. For example, if there is one subset type, and all n objects are identical, there is only one distinguishable ordering. If each of the objects are themselves distinguishable, which is n subset types, this is equivalent to the without replacement model and $m = n$, and we have from the section above that there are $n!$ distinguishable orderings.

Two Subset Types

Next assume that there are two subsets of indistinguishable objects, say n_1 of one type and $n_2 = n - n_1$ of the other. Envision a collection of n_1 1s and n_2 0s to be ordered, or n_1 red balls and n_2 blue balls. What distinguishes this example from that where all the objects differ is that here, the collection of all orderings will contain multiple counts. For example, if we start with the collection $\{1,2,3,4\}$, there are $4! = 24$ possible orderings, but if we begin with $\{1,1,1,4\}$, there are only 4 orderings. This is because we only have to choose the position for the one 4-digit, for the other digits will all be 1s. This can also be deduced by observing that in the 4! orderings of the 4 digits in this second set, each distinct outcome will be seen 3! times, reflecting the indistinguishable orderings of the three 1s.

Analogously in this general case, the number of orderings is

$$\frac{(n_1 + n_2)!}{n_1! n_2!} = \frac{n!}{n_1! n_2!}.$$

The logic of this formula, as will be analyzed in more detail next, is that the numerator reflects the number of orderings of the n objects, temporarily treating them as if all are distinguishable. The denominator then adjusts for multiple counts, since there will be $n_1!$ orderings with the n_1 objects of the first type in the same locations but with different orderings of these actual objects. Likewise for each of these orderings there will be n_2 objects of the second type in the same locations but with different orderings of these actual objects.

Binomial Coefficients

The formula above has many applications in mathematics, especially with respect to coin-flip and associated binomial models, where "binomial" means with two outcomes. The two outcomes represent the two subset types discussed above. Because of its prevalence, this formula has been given a special notation.

As a traditional binomial example, imagine that a coin is flipped n times. What is the total number of sample points in the associated sample space that have exactly m heads, for $m = 0, 1, 2, \ldots, n$? This question is identical to that of a general ordering of n objects, where there are m of one type, the Hs, and $n - m$ of the other type, the Ts. The analysis above shows that there will be $\frac{n!}{(n-m)!m!}$ such sample points, and the general notation is

$$\binom{n}{m} = \frac{n!}{(n - m)!m!}.$$ (7.13)

This factor is sometimes denoted $_nC_m$, and read, "n choose m," and we recall that by convention, $0! = 1$.

For any n, these constants, $\left\{\binom{n}{m}\right\}_{m=0}^{n}$, are known as **binomial coefficients**, for a reason that will be apparent below. The terminology "n choose m" is shorthand for "the number of ways of choosing m positions from n positions." In the example above, the m positions chosen are of course equal to the locations of the m-Hs, with the remaining positions filled with Ts.

Example 7.21 *As another example of an application of "n choose m," consider explicitly choosing all possible subsets of a set of n distinguishable objects. For any $m = 0, 1, 2, \ldots, n$, there are $\binom{n}{m}$ possible subsets that can be selected. This is just a reformulation of the earlier model in that we can envision these n objects as n positions, and the selection of a subset of m objects as equivalent to the selection of m of these positions. When $m = 0$, we are selecting the empty subset $\emptyset$, and there is only one way to do this. If we seek the total number of subsets of all sizes, which is the number of sets in the power set, the answer must therefore be equal to $\sum_{m=0}^{n} \binom{n}{m}$. But we also know from exercise 4 in chapter 4, that the number of sets in the power set of a set of n elements is 2^n. So we must have*

$$\sum_{m=0}^{n} \binom{n}{m} = 2^n.$$ (7.14)

The Binomial Theorem

Formula (7.14) is a special case of the so-called **binomial theorem**, which is yet another application of "n choose m." This theorem addresses the expansion of an integer power of a binomial, such as $(a+b)^n$. The problem posed is a "chooser" problem because in this multiplication we have to choose an a or a b from each of the n factors of $(a+b)$ and multiply the selected n terms. Consequently the general term in the product is of the form $a^m b^{n-m}$ for $m = 0, 1, 2, \ldots, n$. The question is, how many times will each such factor arise? Of course, the answer is $\binom{n}{m}$ times, since for each m there are $\binom{n}{m}$ ways of selecting the m a-factors from these n binomial factors. Consequently the binomial theorem states that

$$(a+b)^n = \sum_{m=0}^{n} \binom{n}{m} a^m b^{n-m}. \tag{7.15}$$

From (7.15), the special case of (7.14) is easily derived by setting $a = b = 1$.

Also of interest, for $a = -1$, $b = 1$, the sum of the alternating binomial coefficients is seen to equal 0:

$$\sum_{m=0}^{n} \binom{n}{m} (-1)^m = 0.$$

Finally, if $a + b = 1$, this theorem assures us that

$$\sum_{m=0}^{n} \binom{n}{m} a^m b^{n-m} = 1,$$

which is important in the **binomial distribution** below where it is also assumed that $0 \le a, b \le 1$.

The coefficients of the factors in these expressions are easily generated by a method developed by **Blaise Pascal** (1623–1662) and known as **Pascal's triangle**. It is based on the iterative formula (see exercise 33)

$$\binom{n}{m} = \binom{n-1}{m-1} + \binom{n-1}{m}. \tag{7.16}$$

The associated "triangle" is developed row by row, with the nth row corresponding to the coefficients in the expansion of $(a+b)^n$. The coefficients up to $(a+b)^6$ are in (7.17), and these may be familiar from elementary algebra:

$$
\begin{array}{ccccccccccccc}
& & & & & & 1 & & & & & & \\
& & & & & 1 & & 1 & & & & & \\
& & & & 1 & & 2 & & 1 & & & & \\
& & & 1 & & 3 & & 3 & & 1 & & & \\
& & 1 & & 4 & & 6 & & 4 & & 1 & & \\
& 1 & & 5 & & 10 & & 10 & & 5 & & 1 & \\
1 & & 6 & & 15 & & 20 & & 15 & & 6 & & 1
\end{array}
\tag{7.17}
$$

$$\cdots$$

Notice that for any n, $\binom{n}{0} = \binom{n}{n} = 1$ and how, with clever spacing, each term of a row equals the sum of the terms right above it, implementing the iterative formula in (7.16).

r Subset Types

Now assume that there are r subsets of distinguishable objects, with n_j of type-j, $n_j \geq 0$, and with $\sum n_j - n$. Then the logic above carries forward identically, and we see that the number of such orderings is

$$
_{\bar{n}}C_n = \frac{n!}{n_1! n_2! \ldots n_r!},
\tag{7.18}
$$

where the nonstandard notation $_{\bar{n}}C_n$ is intended to connote that the choice made of the n objects is a vector $\bar{n} \equiv (n_1, n_2, \ldots, n_r)$. For a given n the collection of the number of such orderings

$$
\left\{ _{\bar{n}}C_n \,\middle|\, \bar{n} = (n_1, n_2, \ldots, n_r); \sum n_j = n \right\}
$$

are known as the **multinomial coefficients**.

The logic behind this formula is that there are $n!$ orderings of the n objects, momentarily considered to be distinct. For example, temporarily label the type-1 objects with numbers $1, 2, \ldots, n_1$, and so forth. Now select any one of these $n!$ orderings, and observe the positions of the type-1 objects. When this particular ordering was achieved, there were $n_1!$ possible orderings in which these type-1 objects could have been selected and placed into the given positions. Similarly, for any type-j, there would be $n_j!$ possible orderings in which these objects could have been selected and placed into the given positions of the selected ordering. In other words, the $n!$ orderings contain $n_1! n_2! \ldots n_r!$ copies of every distinct ordering, and hence one needs to divide by this factor to eliminate the redundancies.

Example 7.22   *Assume that we are given the* 10-*digit collection,*

$$\{1, 1, 2, 2, 2, 5, 5, 5, 5, 7\}.$$

*How many different base-*10 *numbers can be formed using all the digits? As before, there are* 10! *possible orderings, but with many multiple counts. Adjusting for these, we see that the total collection of distinct integers formed will be*

$$\frac{10!}{2!3!4!1!} = 12{,}600.$$

Multinomial Theorem

In the same way that the binomial coefficients can be found in the general expansion of the binomial $(a + b)^n$ so too can the multinomial coefficients in (7.18) be found in the general expansion of a multinomial $\left(\sum_{i=1}^{r} a_i\right)^n$. Specifically, we have that

$$\left(\sum_{i=1}^{r} a_i\right)^n = \sum_{n_1, n_2, \ldots n_r} \frac{n!}{n_1! n_2! \ldots n_r!} a_1^{n_1} a_2^{n_2} \ldots a_r^{n_r}, \tag{7.19}$$

where this summation is over all distinct r-tuples $(n_1, n_2, \ldots, n_r)$ so that $n_j \geq 0$ and $\sum_{j=1}^{r} n_j = n$.

As for the binomial theorem above, special identities are produced with simple applications of (7.19) in the special cases where $\sum_{i=1}^{r} a_i = 0$ or $\sum_{i=1}^{r} a_i = 1$. The latter case has an important application to the **multinomial distribution** below, where it is also assumed that $0 \leq a_i \leq 1$ for all i.

7.4   Random Variables

7.4.1   Quantifying Randomness

Notions of sample space, events, and probability measures are often introduced in the colorful and intuitive imagery of card hands dealt from one or more well-shuffled decks of cards, collections of colored balls drawn from an urn containing different numbers of colored balls with or without replacement, and sequences of flips of a fair or biased coin. While interesting, these models do not lend themselves to mathematical analysis very well because these contexts can obscure similarities or create misleading connections. If a problem is solved in the context of an urn problem, will it be apparent that the same procedure might be applied and the same result obtained in the very different context of dealt card hands? Or if a problem is solved in the context of flips of a biased coin, will it be apparent that the same procedure might be applied and result obtained in the very different context of the modeling of the prices of a common stock in discrete time steps?

The notion of a random variable was introduced for the purpose of stripping away the context of these problems, to reveal the common mathematical structures underlying them. In effect a random variable transfers the probabilities associated with these colorful events to probabilities associated with numerical values in $\mathbb{R}$. A few simple examples will illustrate the point.

Example 7.23

1. *Let's return to the sample space S of 10-flip sequences of a fair coin that, as we have seen, contains 2^{10} sample points and $2^{2^{10}}$ possible events, all with associated probabilities. We now define a function on the original sample space, as follows:*

$X(s) = n,$

where n is the number of Hs in $s \in S$. So X is a function, $X : S \to \{0, 1, 2, \ldots, 10\}$. Note that for any $n \in \{0, 1, 2, \ldots, 10\}$, the inverse $X^{-1}(n) \equiv A_n \in \mathcal{E}$ is a well-defined event of sample points with n Hs, and hence we can define implied probabilities on these integers by

$P(n) = \Pr[A_n].$

Of course, this particular random variable provides only one quantitative insight to this sample space, its events, and the associated probability structure, and there are many other insights that remain hidden. However, there are many more random variables that can be defined, each providing certain insights and hiding others. The particular definition of the random variable used is determined in such a way that the properties of S that are of interest to the analyst are revealed.

2. *As another example, one could imagine a game whereby after 10 flips of a fair coin, producing sample point s, the player receives a payoff of $Y(s) = \sum_{j=0}^{n} 10^j$, where n is the number of Hs in s. Now*

$Y : S \to \{1, 11, 111, \ldots, 11111111111\}.$

The range of Y here differs dramatically from the random variable X above, but the probabilities of the range values are the same in the sense that for any n,

$$\Pr\left[Y^{-1}\left(\sum_{j=0}^{n} 10^{j}\right)\right] = \Pr[X^{-1}(n)],$$

since in both cases these implied probabilities are defined by $\Pr[A_n]$, *the probability of the event in* S *defined by n Hs.*

3. *One can also change the probability structure by defining, for example,* $Z(s) = \sum_{j=1}^{10} s_j 10^j$, *where* s_j *denotes the jth flip, with* $s_j = 0$ *for a T, and* $s_j = 1$ *for a H. Now the range of Z differs significantly from that of Y, containing every integer that can be constructed with* 10 *digits, each of which is* 0 *or* 1. *There are consequently* 2^{10} *points in the range of Z, in contrast to* 11 *points in the range of X and Y. Also the probabilities on the range of Z depend not only on the total number of heads in a given sample point but also on the order of these heads in the sequence. So each event* A_n *above is split into* $\binom{10}{n}$ *events by Z. In essence, Z maps each sample point in* S *to a distinct integer and assigns a probability to this integer equal to the probability of the associated sample point.*

7.4.2   Random Variables and Probability Functions

Because this chapter addresses discrete probability theory, which is the theory as it applies to finite and countably infinite sample spaces, it is possible that the range of a random variable is any countable subset of $\mathbb{R}$ such as $\mathbb{N}$, $\mathbb{Z}$, or $\mathbb{Q}$, so we introduce a more economical way of demanding that $X^{-1}(r) \in \mathcal{E}$ for every r in the range of the random variable X. The idea is to use open intervals, (a, b), that are either bounded or unbounded. Then in every case, $X^{-1}[(a, b)]$ must be an event either because it is the finite or countable union of events of the form $X^{-1}(r)$ for $r \in (a, b)$, or because it is the null event, $\emptyset$, if this interval is disjoint from the range of X.

Use of open intervals in this definition is just a convention, of course, since $X^{-1}[(a, b)] \in \mathcal{E}$ for all open intervals if and only if $X^{-1}[[a, b]] \in \mathcal{E}$ for all closed intervals. To see this, first note that $X^{-1}[(a, b)] \in \mathcal{E}$ for all bounded or unbounded intervals implies that $X^{-1}[(-\infty, b)] \in \mathcal{E}$ and hence the complement in S, which is $X^{-1}[[b, \infty)] \in \mathcal{E}$. Similarly $X^{-1}[[a, \infty)] \in \mathcal{E}$. Also, if $X^{-1}[[b, \infty)] \in \mathcal{E}$ and $X^{-1}[[a, \infty)] \in \mathcal{E}$, then the intersection, $X^{-1}[[b, \infty)] \cap X^{-1}[[a, \infty)] \equiv X^{-1}[[a, b]] \in \mathcal{E}$. The reverse implication is demonstrated similarly.

Next we formalize the definition with this open set convention:

Definition 7.24   *Given a discrete sample space* S *and a complete collection of events* $\mathcal{E} = \{A \mid A \subset S\}$, *a **discrete random variable (r.v.)** is a function*

$$X : S \to \mathbb{R},$$

with $X[\mathcal{S}] = \{x_j\}_{j=1}^n$, *where possibly* $n = \infty$, *so that for any bounded or unbounded interval,* $(a, b) \subset \mathbb{R}$:

$$X^{-1}[(a, b)] \in \mathcal{E}.$$

The **probability density function (p.d.f.)** *or* **probability function** *associated with* X, *denoted* f *or* f_X, *is defined on the range of* X *by*

$$f(x_j) = \Pr[X^{-1}(x_j)]. \tag{7.20}$$

The **distribution function (d.f.)**, *or* **cumulative distribution function (c.d.f.)** *associated with* X, *denoted* F *or* F_X, *is defined on* $\mathbb{R}$ *by*

$$F(x) = \Pr[X^{-1}(-\infty, x]]. \tag{7.21}$$

Note that the c.d.f. is the sum of the p.d.f. values, since $\Pr[X^{-1}(-\infty, x]] = \sum_{x_j \le x} \Pr[X^{-1}(x_j)]$, and so

$$F(x) = \sum_{x_j \le x} f(x_j). \tag{7.22}$$

Graphically, when the sample space is finite, the c.d.f. has a "jump" at each value of x_j in the range of X, and the graph of $F(x)$ is horizontal otherwise. Such a function is often called a **step function** for apparent reasons. When the sample space is countably infinite, the c.d.f. will again look like a step function in the case of sparsely spaced range, $\{x_j\}$, such as the case for the positive integers. For a range with accumulation points, $\{x_j\}$, such as for the rationals in $[0, 1]$, the c.d.f. again would have jumps at each rational, but no flat spots or steps *per se*.

Remark 7.25   *Note that given any discrete random variable on* $\mathcal{S}$, *with* $X[\mathcal{S}] = \{x_j\}_{j=1}^n$, *where possibly* $n = \infty$, *the collection of events defined by* $\{X^{-1}[x_j]\}_{j=1}^n$ *are mutually exclusive, and hence for any collection of points,*

$$\Pr\left[\bigcup X^{-1}[x_j]\right] = \sum \Pr[X^{-1}[x_j]].$$

Example 7.26   *Let* $\mathcal{S}$ *be defined as the sample space of 3 flips of a fair coin, and* $X : \mathcal{S} \to \mathbb{R}$ *defined by* $X(s)$ *equals the number of Hs in s. So the range of* X, *as in definition 2.2.3,* $\mathrm{Rng}[X] = \{0, 1, 2, 3\}$. *The sample space* $\mathcal{S}$ *contains* $2^3 = 8$ *sample points, 1 each with 0 or 3 Hs, and 3 each with 1 or 2 Hs. This follows directly from the values of* $\binom{3}{j}$. *The probability of each sample point is* $\frac{1}{8}$. *Consequently the associated probability density function is defined by*

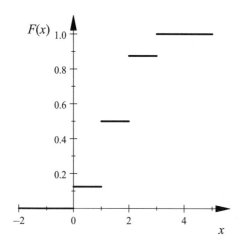

Figure 7.1
$F(x)$ for Hs in three flips

n: 0 1 2 3

$f(n)$: $\frac{1}{8}$ $\frac{3}{8}$ $\frac{3}{8}$ $\frac{1}{8}$

*The graph of the cumulative distribution function, $F(x)$, is seen in **figure 7.1**.*

7.4.3   Random Vectors and Joint Probability Functions

We begin with the simplest example and definition, and generalize later. Imagine that there are two random variables defined on the given sample space: $X, Y : \mathcal{S} \to \mathbb{R}$, which we think of as being combined into a **random vector** or a **vector-valued random variable**:

$$(X, Y) : \mathcal{S} \to \mathbb{R}^2.$$

Here, for a given sample point $s \in \mathcal{S}$, we define $(X, Y) : s \to (X(s), Y(s))$.

Generalizing the notion of open interval in the definition of random variable, we define the bounded or unbounded **open rectangle**, denoted $(\bar{a}, \bar{b})$, where $\bar{a} = (a_1, a_2)$, $\bar{b} = (b_1, b_2)$ and where $a_1 < b_1$ and $a_2 < b_2$, by

$$(\bar{a}, \bar{b}) = \{(x, y) \mid a_1 < x < a_2, b_1 < y < b_2. \tag{7.23}$$

A **closed rectangle**, $[\bar{a}, \bar{b}]$, or a semi-closed (or semi-open) rectangle, $[\bar{a}, \bar{b})$ or $(\bar{a}, \bar{b}]$, is defined similarly.

The requirement to qualify as a **random vector** is that the pre-image of all open rectangles be events, where for any point (x, y), the pre-image under (X, Y) is defined as

$$(X, Y)^{-1}[(x, y)] = X^{-1}(x) \cap Y^{-1}(y).$$

With this setup we can define the **joint probability density function** or **joint probability function**, $f(x_j, y_j)$, as the probability of the event $X^{-1}(x_j) \cap Y^{-1}(y_j)$, and correspondingly define the **joint cumulative distribution function** or **joint distribution function**, $F(x, y)$, as the probability of the event that is the pre-image of $(-\infty, \bar{b}]$, where $\bar{b} = (x, y)$. Then

$$F(x, y) = \sum_{(x_j, y_j) \leq (x, y)} f(x_j, y_j),$$

with the understanding that $(x_j, y_j) \leq (x, y)$ is shorthand for $x_j \leq x$ and $y_j \leq y$. This setup then easily generalizes to collections of 3 or more random variables, and we state the formal definition in this generality:

Definition 7.27   *Given a discrete sample space $\mathcal{S}$, a complete collection of events $\mathcal{E} = \{A \mid A \subset \mathcal{S}\}$, and a collection of random variables on $\mathcal{S}$, $\{X_k\}_{k=1}^n$, a **discrete random vector** is a function*

$$\overline{X} : \mathcal{S} \to \mathbb{R}^n,$$

where $\overline{X}(s) = (X_1(s), X_2(s), \ldots, X_n(s))$, with $X_k[\mathcal{S}] = \{x_{kj}\}_{j=1}^{n_k}$, and possibly $n_k = \infty$, for some or all k. For any bounded or unbounded open rectangle, $(\bar{a}, \bar{b}) \subset \mathbb{R}^n$, we require that

$$\overline{X}^{-1}((\bar{a}, \bar{b})) \equiv \bigcup_{\bar{x} \in (\bar{a}, \bar{b})} \overline{X}^{-1}(\bar{x}) \in \mathcal{E},$$

where $\overline{X}^{-1}(\bar{x})$ is defined for $\bar{x} = (x_1, x_2, \ldots, x_n)$ by

$$\overline{X}^{-1}(\bar{x}) = X_1^{-1}(x_1) \cap X_2^{-1}(x_2) \cap \cdots \cap X_n^{-1}(x_n).$$

*The **joint probability density function (p.d.f.)**, or **joint probability function**, associated with $\overline{X}$, denoted f or f_X, is defined on the range of $\overline{X}$ by*

$$f(x_1, x_2, \ldots, x_n) = \Pr[X_1^{-1}(x_1) \cap X_2^{-1}(x_2) \cap \cdots \cap X_n^{-1}(x_n)]. \tag{7.24}$$

*The **joint cumulative distribution function** (c.d.f.), or **joint distribution function** (d.f.)
associated with $\overline{X}$, denoted F or F_X, is defined on $\mathbb{R}^n$ by*

$$F(\bar{x}) = \Pr[\overline{X}^{-1}(-\infty, \bar{x}]].\tag{7.25}$$

As was the case for random variables above, because $\Pr[\overline{X}^{-1}(-\infty, \bar{x}]] = \sum_{\bar{x}' \leq \bar{x}} \Pr[\overline{X}^{-1}(\bar{x}')]$, where $\bar{x}' \leq \bar{x}$ is shorthand for $x_j' \leq x_j$ for all j, and $\bar{x}'$ is in the range of $\overline{X}$, the counterpart to (7.22) is

$$F(\bar{x}) = \sum_{\bar{x}' \leq \bar{x}} f(\bar{x}').\tag{7.26}$$

Example 7.28

1. On the sample space of 10-flip sequences of a fair coin, we could define random variables, $\{X_j\}_{j=1}^{10}$ on $s \in \mathcal{S}$ by

$$X_j(s) = \begin{cases} 1, & s_j = H, \\ -1, & s_j = T. \end{cases}$$

*In other words, each X_j is defined entirely in terms of the value of the jth flip. The
range of $\overline{X}$ is then the 2^{10} vectors in $\mathbb{R}^{10}$ defined by $\mathrm{Rng}(\overline{X}) = \{\bar{x} \in \mathbb{R}^{10} \mid x_j = \pm 1$
for all $j\}$. In this simple example the event $X_1^{-1}(x_1)$ contains all sequences with an H
for the first flip if $x_1 = 1$, and all sequences with a T for the first flip if $x_1 = -1$,
and similarly for other components. In addition $\overline{X}^{-1}(\bar{x}) = X_1^{-1}(x_1) \cap X_2^{-1}(x_2) \cap \cdots \cap
X_{10}^{-1}(x_{10})$ is a unique sample point for every $\bar{x} \in \mathrm{Rng}(\overline{X})$ and correspondingly,
$f(\bar{x}) = 2^{-10}$ for each such point.*

*2. Define $Y_1(s) = \sum_{j=1}^{5} X_j(s)$ and $Y_2(s) = \sum_{j=6}^{10} X_j(s)$, where $X_j(s)$ is defined in case
1. Now with $\overline{Y} \equiv (Y_1, Y_2)$, we have $\mathrm{Rng}(\overline{Y}) = \{\bar{y} \in \mathbb{R}^2 \mid y_1, y_2 = \pm 5, \pm 3, \pm 1\}$. The
number of sample points in $Y_j^{-1}(y_j)$ now varies by the value of y_j. For instance,
$Y_1^{-1}(5)$ is the event of all 2^5-flip sequences starting with HHHHH, whereas $Y_1^{-1}(1)$ is
the event of all flip sequences with 3-Hs and 2-Ts in the first five flips, of which there
are $\binom{5}{3}2^5 = 5 \cdot 2^6$ sample points. Correspondingly the value of $f(\bar{y}) = \Pr[Y_1^{-1}(y_1) \cap
Y_2^{-1}(y_2)]$ also varies over the range of $\overline{Y}$.*

7.4.4 Marginal and Conditional Probability Functions

Once a joint probability density function is defined on a sample space, it is natural
to consider additional probability functions. To set the stage, we start with an
example.

Example 7.29   *Consider the random variables $Y_1(s) = \sum_{j=1}^{3} X_j(s)$ and $Y_2(s) = \sum_{j=4}^{6} X_j(s)$ defined on the sample space of 6-flip sequences of a fair coin. As in case 1 in example 7.28 above, for $s \in S$, $X_j(s)$ is defined by*

$$X_j(s) = \begin{cases} 1, & s_j = H, \\ -1, & s_j = T. \end{cases}$$

The joint p.d.f. of the pair, $\overline{Y} \equiv (Y_1, Y_2)$, is defined on $\mathrm{Rng}(\overline{Y}) = \{\bar{y} \in \mathbb{R}^2 \mid y_1, y_2 = \pm 1, \pm 3\}$, which contains 16 points. The associated probabilities are given by

(y_1, y_2):	$(\pm 1, \pm 1)$	$(\pm 1, \pm 3)$	$(\pm 3, \pm 1)$	$(\pm 3, \pm 3)$
$f(y_1, y_2)$:	$\frac{9}{2^6}$	$\frac{3}{2^6}$	$\frac{3}{2^6}$	$\frac{1}{2^6}$

where there are 4 sample points represented in each numerical column. It is easy to see that the probabilities of the points in each column are the same by symmetry. For example, switching all $H \leftrightarrow T$ gives a 1:1 correspondence between the $(1,1)$ and $(-1,-1)$, while switching $H \leftrightarrow T$ only for the first 3 flips identifies $(1,1)$ and $(-1,1)$. Interchanging the first 3 and last 3 flips identifies $(1,3)$ and $(3,1)$, and so forth.

Since Y_1 and Y_2 are perfectly good random variables on their own, we can also define the p.d.f.s $f(y_1)$ and $f(y_2)$, which by symmetry will have the same values on the same 4 points:

y_j:	± 1	± 3
$f(y_j)$:	$\frac{3}{2^3}$	$\frac{1}{2^3}$

When calculating $f(y_j)$, intuition suggests that the original sample space was not necessary, and that it would have been easier to consider the sample space of 3-flip sequences of a fair coin. On the other hand, if the calculation was implemented in the original sample space S, every 3-flip outcome for the given y_1, say, would be counted 2^3 times, since in S, such an outcome would be associated with all 2^3 possible 3-flip sequences underlying y_2. Put another way, every 3-flip outcome for the given y_1 would be associated with all possible outcomes of y_2. Consequently we must have

$$f(y_1) = \sum_{y_2} f(y_1, y_2) \quad \text{and} \quad f(y_2) = \sum_{y_1} f(y_1, y_2).$$

A simple calculation relating these values to the defining probability measure on S, Pr, demonstrates that this is the case. In this context, $f(y_1)$ and $f(y_2)$ are called the **marginal probability density functions of the joint p.d.f.** $f(y_1, y_2)$

Another calculation of interest is for the so-called **conditional probability functions of the joint p.d.f.** $f(y_1, y_2)$, denoted $f(y_1 \mid y_2)$ and $f(y_2 \mid y_1)$. Focussing on $f(y_1 \mid y_2)$ for specificity, this p.d.f. is defined relative to the probability of the conditional event $A \mid B$, where $A = \{s \mid Y_1(s) = y_1\}$ and $B = \{s \mid Y_2(s) = y_2\}$. In other words, the conditional p.d.f. $f(y_1 \mid y_2)$ is defined as the probability of the conditional event $A \mid B$:

$$f(y_1 \mid y_2) = \Pr[A \mid B] = \Pr[Y_1^{-1}(y_1) \mid Y_2^{-1}(y_2)].$$

Once again, this conditional p.d.f. must be related to the joint p.d.f, $f(y_1, y_2)$, which provides probabilities for each event, $\Pr[Y_1^{-1}(y_1) \cap Y_2^{-1}(y_2)] = \Pr[A \cap B]$. Now in the preceding section on conditional events, we have from (7.3) that if $\Pr[B] \neq 0$, then $\Pr[A \mid B] = \frac{\Pr[A \cap B]}{\Pr[B]}$. Replacing this event notation with the corresponding p.d.f. notation, we conclude that

$$f(y_1 \mid y_2) = \frac{f(y_1, y_2)}{f(y_2)} \qquad \text{for } f(y_2) \neq 0,$$

with a corresponding formula for $f(y_2 \mid y_1)$.

Before formalizing these ideas in a definition, note that for a more general joint p.d.f., $f(y_1, y_2, \ldots, y_n)$, there are in fact $2^n - 2$ possible marginal p.d.f.s. Specifically, there are $\binom{n}{1}$ of the form $f(y_j)$, $\binom{n}{2}$ of the form $f(y_j, y_k)$ for $j \neq k$, and so forth. We get the -2 adjustment to the count because if no y_j is chosen, $\sum_{(y_1, y_2, \ldots, y_n)} f(y_1, y_2, \ldots, y_n) = 1$, which is not a probability function, whereas if all y_j are chosen, the original joint p.d.f. is produced.

In addition, for every such marginal p.d.f., one could define an associated conditional p.d.f., such as $f(y_1, y_2 \mid y_3, \ldots, y_n)$. However, the notation quickly becomes cumbersome, so the following definition will be presented both in the more limited generality of two random variables, a common framework for applications, and then for the general case:

Definition 7.30   *Given a random vector $\overline{Y} = (Y_1, Y_2)$ on a discrete sample space, S, and associated joint probability distribution function $f(y_1, y_2)$, the **marginal probability density functions**, denoted $f(y_1)$ and $f(y_2)$, are defined by*

$$f(y_1) = \sum_{y_2} f(y_1, y_2), \tag{7.27a}$$

$$f(y_2) = \sum_{y_1} f(y_1, y_2). \tag{7.27b}$$

The associated **conditional probability density functions**, denoted $f(y_1 \mid y_2)$ and $f(y_2 \mid y_1)$, are defined by

$$f(y_1 \mid y_2) = \frac{f(y_1, y_2)}{f(y_2)} \qquad \text{when } f(y_2) \neq 0, \tag{7.28a}$$

$$f(y_2 \mid y_1) = \frac{f(y_1, y_2)}{f(y_1)} \qquad \text{when } f(y_1) \neq 0. \tag{7.28b}$$

Note that the **law of total probability**, stated in the context of events in (7.4), can also be stated in terms of the joint, marginal, and conditional p.d.f. Specifically, we have from (7.28a) that $f(y_1, y_2) = f(y_1 \mid y_2)f(y_2)$, and also from (7.27a) that $f(y_1) = \sum_{y_2} f(y_1, y_2)$. Combining, we obtain the law of total probability:

$$f(y_1) = \sum_{y_2} f(y_1 \mid y_2)f(y_2), \tag{7.29}$$

and the analogous identity for $f(y_2)$.

For the more general definition, we introduce the notion of a **partition of the random vector** $\overline{Y} = (Y_1, Y_2, \ldots, Y_n)$ into two nonempty subsets of random variables $\overline{Y}_1 = (Y_{j_1}, Y_{j_2}, \ldots, Y_{j_m})$, and $\overline{Y}_2 = (Y_{i_1}, Y_{i_2}, \ldots, Y_{i_{n-m}})$, where this cumbersome notation is intended to imply that every Y_k is in one of $\overline{Y}_1$ and $\overline{Y}_2$ but not both.

Definition 7.31 *Given a random vector* $\overline{Y} = (Y_1, Y_2, \ldots, Y_n)$ *on a discrete sample space,* $\mathcal{S}$, *an associated joint probability distribution function* $f(y_1, y_2, \ldots, y_n)$, *and a partition,* $\overline{Y} = (\overline{Y}_1, \overline{Y}_2)$, *the **marginal probability density function**, denoted* $f(\overline{y}_1)$ *is defined by*

$$f(\overline{y}_1) = \sum_{\overline{y}_2} f(y_1, y_2, \ldots, y_n). \tag{7.30}$$

*The associated **conditional probability density function**, denoted* $f(\overline{y}_2 \mid \overline{y}_1)$, *is defined by*

$$f(\overline{y}_2 \mid \overline{y}_1) = \frac{f(y_1, y_2, \ldots, y_n)}{f(\overline{y}_1)} \qquad \text{when } f(\overline{y}_1) \neq 0. \tag{7.31}$$

We note that these general formulas also provide general versions of the law of total probability, but leave it to the reader to develop these formulas.

7.4.5 Independent Random Variables

Because a random variable X is defined so that the pre-image of open intervals $X^{-1}[(a, b)]$ are events in $\mathcal{E}$ with associated probabilities under the probability measure

Pr, it is natural to say that **two random variables are independent** if their pre-images of all intervals are stochastically independent as events in $\mathcal{E}$.

Definition 7.32   *Random variables X_1, X_2 on the discrete sample space $\mathcal{S}$ are **independent random variables** if for any intervals $(a_j, b_j) \subset \mathbb{R}$, bounded or unbounded, $X_1^{-1}[(a_1, b_1)]$ and $X_2^{-1}[(a_2, b_2)]$ are stochastically independent events in $\mathcal{E}$ as in (7.5). Equivalently, if $X_1 : \mathcal{S} \to \{x_{1j}\}$ and $X_2 : \mathcal{S} \to \{x_{2k}\}$, then X_1 and X_2 are independent if $X_1^{-1}[x_{1j}]$ and $X_2^{-1}[x_{2k}]$ are stochastically independent events for all x_{1j} and x_{2k}.*

*More generally, a collection of random variables $\{X_j\}_{j=1}^n$, where n may be ∞, are **mutually independent random variables** if every collection of events of the form $\{X_j^{-1}[(a_j, b_j)]\}_{j=1}^n$ are mutually independent events as in (7.6), or equivalently, $\{X_j^{-1}[x_{jk}]\}_{j=1}^n$ are mutually independent for any $x_{jk} \in \mathrm{Rng}[X_j]$.*

Example 7.33

1. *Define $\mathcal{S}$ as the sample space of all pairs of results achieved by rolling a fair die twice. Specifically, $\mathcal{S} = \{(d_1, d_2) \mid 1 \le d_j \le 6\}$, where d_1 denotes the result on the first roll, and d_2 the result on the second. By the assumption of fairness, each numerical value is equally likely and has probability of $\frac{1}{6}$ of occurrence, and consequently the probability function for $\mathcal{S}$ is defined as $\mathrm{Pr}[(d_1, d_2)] = \frac{1}{36}$ for every such sample point. Note that the values of this probability measure are influenced by the fact that the die throws were sequential, and hence order counts. On this ordered sample space, define first the random variables $X, Y : \mathcal{S} \to \mathbb{N}$ by*

$$X[(d_1, d_2)] = d_1,$$

$$Y[(d_1, d_2)] = d_2.$$

Intuition indicates that X and Y are independent random variables. To demonstrate this, note that for any $d_1, d_2 \in \{1, 2, \ldots, 6\}$, both $X^{-1}(d_1)$ and $Y^{-1}(d_2)$ are events in $\mathcal{S}$ of 6 points with measures under Pr of $\frac{1}{6}$. Also $X^{-1}(d_1) \cap Y^{-1}(d_2)$ contains a unique sample point, specifically, (d_1, d_2), which has measure $\frac{1}{36}$ under Pr. In other words, for all (d_1, d_2),

$$\mathrm{Pr}[X^{-1}(d_1) \cap Y^{-1}(d_2)] = \mathrm{Pr}[X^{-1}(d_1)]\,\mathrm{Pr}[Y^{-1}(d_2)].$$

2. *Now define a new random variable Z on $\mathcal{S}$ above as follows:*

$$Z[(d_1, d_2)] = d_1 + d_2.$$

Intuitively we expect X and Z not to be independent. That is because, if $Z[(d_1, d_2)] = 12$ (or 2), it must be the case that $X[(d_1, d_2)] = 6$ (or 1). More formally, Z assumes all integer values $2 \leq k \leq 12$, and the event defined by $Z^{-1}(k)$ has probabilities

k:	2	3	4	5	6	7	8	9	10	11	12
$\Pr[Z^{-1}(k)]$:	$\frac{1}{36}$	$\frac{2}{36}$	$\frac{3}{36}$	$\frac{4}{36}$	$\frac{5}{36}$	$\frac{6}{36}$	$\frac{5}{36}$	$\frac{4}{36}$	$\frac{3}{36}$	$\frac{2}{36}$	$\frac{1}{36}$

It is apparent that the numerator of $\Pr[Z^{-1}(k)]$ also represents the number of sample points in the associated event. As noted above, for each $1 \leq j \leq 6$, $X^{-1}(j)$ contains 6 sample points, and $\Pr[X^{-1}(j)] = \frac{1}{6}$ for all j. Now it is straightforward to justify that $X^{-1}(j) \cap Z^{-1}(k)$ contains one sample point or none. For example, $X^{-1}(1) \cap Z^{-1}(12) = \emptyset$, while $X^{-1}(4) \cap Z^{-1}(7) = (4,3)$. More generally, if $d_1 = j$ and $d_1 + d_2 = k$, there is a unique point provided that $1 \leq k - j \leq 6$, and no point otherwise. Hence $\Pr[X^{-1}(j) \cap Z^{-1}(k)]$ equals 0 or $\frac{1}{36}$, which can equal the product of probabilities of the respective events only when $k = 7$. Consequently X and Z are not independent.

3. *If instead of as in case 1, a pair of dice were thrown without keeping track of order, then the sample space, $\mathcal{S}'$, would contain only 21 rather than 36 sample points. One realization of this space is $\mathcal{S}' = \{(d_1, d_2) \mid 1 \leq d_1 \leq d_2 \leq 6\}$ where d_1 denotes the smaller result, d_2 the larger. The associated probability measure is then given by*

$$\Pr[(d_1, d_2)] = \begin{cases} \frac{1}{36} & d_1 = d_2, \\ \frac{1}{18} & d_1 < d_2. \end{cases}$$

Define the random variables $U, W : \mathcal{S}' \to \mathbb{N}$ by

$$U[(d_1, d_2)] = \min(d_1, d_2),$$

$$W[(d_1, d_2)] = \max(d_1, d_2).$$

Now U and W are not independent. For example, $\Pr[U^{-1}(1)] = \frac{11}{36}$, since this event contains the sample point

$$U^{-1}(1) = \{(1, d) \mid 1 \leq d \leq 6\},$$

which has measure $\frac{11}{36}$ by the above given probability measure on $\mathcal{S}'$. On the other hand, $\Pr[W^{-1}(1)] = \frac{1}{36}$, since $W^{-1}(1) = (1,1)$. Also $U^{-1}(1) \cap W^{-1}(1) = W^{-1}(1)$. Consequently

$$\Pr[U^{-1}(1) \cap W^{-1}(1)] \neq \Pr[U^{-1}(1)]\, \Pr[W^{-1}(1)].$$

The notion of independent random variables can also be defined in terms of the joint, conditional and marginal probability distribution functions.

Definition 7.34   *Given a random vector* $\overline{Y} = (Y_1, Y_2)$ *on a discrete sample space* $\mathcal{S}$ *and associated joint probability density function* $f(y_1, y_2)$, *the random variables* Y_1 *and* Y_2 *are* **independent random variables** *if*

$$f(y_1, y_2) = f(y_1)f(y_2),\tag{7.32a}$$

or equivalently if $f(y_2) \neq 0$,

$$f(y_1 \mid y_2) = f(y_1).\tag{7.32b}$$

More generally, given a random vector $\overline{Y} = (Y_1, Y_2, \ldots, Y_n)$ *on the discrete sample space* $\mathcal{S}$, *with associated joint probability density function* $f(y_1, y_2, \ldots, y_n)$, *the random variables* $\{Y_j\}$ *are* **mutually independent random variables** *if given any partition* $\overline{Y} = (\overline{Y}_1, \overline{Y}_2)$

$$f(y_1, y_2, \ldots, y_n) = f(\overline{Y}_1)f(\overline{Y}_2),\tag{7.33a}$$

or equivalently if $f(\overline{Y}_2) \neq 0$,

$$f(\overline{Y}_1 \mid \overline{Y}_2) = f(\overline{Y}_1).\tag{7.33b}$$

In particular, we then have

$$f(y_1, y_2, \ldots, y_n) = f(y_1)f(y_2)\ldots f(y_n).\tag{7.34}$$

7.5   Expectations of Discrete Distributions

7.5.1   Theoretical Moments

The definitions and notation for moments here closely parallel that given in section 3.3.2 for moments of sample data. This is no coincidence, as will be discussed below.

Expected Values

The general structure of the formulas below is seen repeatedly in probability theory. These calculations represent what are known as **expected value calculations**, and sometimes referred to as **taking expectations**. The general case is defined first, then specific examples are presented.

Definition 7.35   *Given a discrete random variable, $X : S \to \mathbb{R}$, and function $g(x)$ defined on the range of X, $\mathrm{Rng}[X] \subset \mathbb{R}$,* ***the expected value of*** *$g(X)$, denoted $\mathrm{E}[g(X)]$, is defined as*

$$\mathrm{E}[g(X)] = \sum_{s_j \in S} g(X(s_j)) \, \mathrm{Pr}(s_j). \tag{7.35}$$

If $\{x_j\} \subset \mathbb{R}$ denotes the range of X, and the p.d.f. of X is denoted by $f(x)$ so that $f(x_j) \equiv \mathrm{Pr}(s_j)$ with $x_j \equiv X(s_j)$, then this expected value can be defined by

$$\mathrm{E}[g(X)] = \sum_{j} g(x_j) f(x_j). \tag{7.36}$$

In either case, this ***expectation is only defined*** *when the associated summation is absolutely convergent, and so in the notation of (7.36), since $f(x_j) \geq 0$, it is required that*

$$\sum_{j} |g(x_j)| f(x_j) < \infty. \tag{7.37}$$

If (7.37) is not satisfied, we say that $\mathrm{E}[g(X)]$ does not exist.

Remark 7.36

1. *The condition in (7.37) is automatically satisfied if the $\{x_j\}$ is finite. The purpose of this restriction in the countably infinite case is to avoid the problem discussed in section 6.1.4, that if only conditionally convergent, the value of this summation is not well defined and depends on the order in which the summation is carried out.*

2. *All expectation formulas can be stated in terms of the random variable X, the sample space S, and its probability measure Pr, or directly in terms of the p.d.f. associated with X. In general, we will only provide the p.d.f. versions as in (7.36) and leave it as an exercise for the reader to formulate the sample space versions as in (7.35).*

3. *It is to be explicitly understood without further repetition that expectation definitions are valid only when the respective absolute convergence conditions as in (7.37) are satisfied.*

4. *When needed for clarity, a subscript is placed on the expectations symbol to identify what variable is involved in the expectation. For example, given p.d.f. $f(x)$, the meaning of $\mathrm{E}[X]$ is unambiguous, so expressing this as $\mathrm{E}_X[X]$ is redundant. On the other hand, the meaning of $\mathrm{E}[XY]$ is ambiguous, since it is not clear which variable is involved. So in this case one would clarify as $\mathrm{E}_X[XY]$ or $\mathrm{E}_Y[XY]$ or $\mathrm{E}_{XY}[XY]$.*

Of course, all expectations of random variables in finite sample spaces exist when $g(x)$ is defined and hence finitely valued on the range of X. However, for random variables on countably infinite sample spaces, expected values may not exist even when $g(x)$ is defined on the range of X.

Example 7.37 *If S is countably infinite, $X : S \to \mathbb{N}$ is defined by $X(s_j) = j$ with range equal to the positive integers, and $f(j)$ is given by $f(j) = \frac{c}{j^2}$, where c is chosen so that $\sum_j f(j) = 1$, then $E[X]$ does not exist, since $E[X] = \sum_j j \frac{c}{j^2} = \sum_j \frac{c}{j}$ is a multiple of the harmonic series and hence not finite. If instead X is defined by $X(s_j) = (-1)^j j$, then again $E[X]$ does not exist. This is because, although $E[X]$ is conditionally convergent, it is not absolutely convergent. Similarly it is easy to find p.d.f.s with finite expected values up to some exponent: $g(x) = x^n$, but with no finite expected values with larger exponents using power harmonic series from example 6.9 to define $f(j)$.*

On the assumption that expected values exist, they are easy to work with in terms of addition and scalar multiplication.

Proposition 7.38 *If $g(x)$ and $h(x)$ are functions for which $E[g(x)]$ and $E[h(x)]$ exist, and a, b, c are real numbers, then $E[ag(x) + bh(x) + c]$ exists and*

$$E[ag(x) + bh(x) + c] = aE[g(x)] + bE[h(x)] + c. \tag{7.38}$$

Proof This result is immediate from the definition, but we must first verify that $ag(x) + bh(x) + c$ satisfies (7.37). This, of course, follows from the triangle inequality

$$|ag(x) + bh(x) + c| \le |a|\,|g(x)| + |b|\,|h(x)| + |c|,$$

and the assumption that $E[g(x)]$ and $E[h(x)]$ exist.   ∎

On the other hand, expected values do not work well with multiplication and division, and as might be expected,

$$E[f(x)g(x)] \ne E[f(x)]E[g(x)],$$

$$E\left[\frac{f(x)}{g(x)}\right] \ne \frac{E[f(x)]}{E[g(x)]}.$$

Conditional and Joint Expectations

Expected value calculations can also be defined with respect to joint probability density functions, as well as conditional probability density functions. For example, if $\bar{X} = (X_1, X_2)$ is a random vector with joint p.d.f. $f(x_1, x_2)$, and $g(x_1, x_2)$ is defined on $\mathrm{Rng}[\bar{X}] \subset \mathbb{R}^2$, we define the **joint expectation of** $g(x_1, x_2)$ by

$$\mathrm{E}[g(X_1, X_2)] = \sum_{(x_1, x_2)} g(x_1, x_2) f(x_1, x_2). \tag{7.39}$$

Many such calculations are possible with differing values of $g(x_1, x_2)$. One important application of this type of formula is in the case where $\{X_j\}$ are independent trials from a given p.d.f. Another important example is for the **covariance** of two random variables. Both are addressed below.

If $f(x_1 \mid x_2)$ is one of the associated conditional p.d.f.s, and $g(x)$ is given, then the **conditional expected value** or **conditional expectation** is defined as

$$\mathrm{E}[g(X_1) \mid X_2 = x_2] = \sum_{x_1} g(x_1) f(x_1 \mid x_2). \tag{7.40}$$

Sometimes for clarity, though cumbersome, the conditional expectation symbol is written with a subscript of $X_1 \mid X_2$ as in $\mathrm{E}_{X_1 \mid X_2}[g(X_1) \mid X_2]$ or $\mathrm{E}[g(X_1) \mid X_2]$.

Remark 7.39   *Unlike most expected values, which provide numerical results, a conditional expectation can be interpreted as a function on the original sample space $\mathcal{S}$, defined by $s \to \mathrm{E}[g(X_1) \mid X_2(s)]$. It is in fact a random variable on $\mathcal{S}$, since the preimage of an open interval $(a, b) \subset \mathbb{R}$ is just the union of countably many events, which is an event in $\mathcal{E}$. It is then the case that the expectation of this random variable under the p.d.f. $f(x_2)$ equals the expectation of $g(x)$ using $f(x_1)$. In other words,*

$$\mathrm{E}_{X_2}[\mathrm{E}_{X_1 \mid X_2}[g(X_1) \mid X_2]] = \mathrm{E}_{X_1}[g(X_1)]. \tag{7.41}$$

The demonstration of this somewhat tediously notated formula is actually simple. By absolute convergence, we can reverse the order of the double summation and apply the law of total probability:

$$\mathrm{E}_{X_2}[\mathrm{E}_{X_1 \mid X_2}[g(X_1) \mid X_2]] = \sum_{x_2} \left[\sum_{x_1} g(x_1) f(x_1 \mid x_2) \right] f(x_2)$$

$$= \sum_{x_1} g(x_1) \left[\sum_{x_2} f(x_1 \mid x_2) f(x_2) \right]$$

$$= \sum_{x_1} g(x_1) f(x_1).$$

This interpretation of $\mathrm{E}[g(X_1) \mid X_2]$ as a random variable on $\mathcal{S}$ is critical in advanced probability theory.

Mean

The **mean** of X, denoted μ, is defined as $\mu = \mathrm{E}[X]$,

$$\mu = \sum_i x_i f(x_i). \tag{7.42}$$

In some applications, the random variable X may be defined in a complicated way, perhaps dependent on another random variable Y, and for which the conditional expectation, $\mathrm{E}[X \mid Y]$ is simpler to evaluate. An immediate application of (7.41) with $g(X) = X$ leads to the following identity between $\mathrm{E}[X]$ and the various conditional expectations $\mathrm{E}[X \mid Y]$, which is known as the **law of total expectation**:

$$\mathrm{E}[X] = \mathrm{E}[\mathrm{E}[X \mid Y]]. \tag{7.43}$$

While this formula may at first appear ambiguous, a moment of reflection justifies that it is well defined even without the subscript clutter of (7.41). The inner expectation can only be defined relative to the conditional p.d.f. $f(x \mid y)$ as $\mathrm{E}[X \mid Y] = \sum_i x_i f(x_i \mid Y)$. Once this expectation is performed, the remaining term is a function of Y alone, and hence the outer expectation must be calculated relative to the marginal p.d.f., $f(y)$. In other words,

$$\mathrm{E}[\mathrm{E}[X \mid Y]] = \sum_j \sum_i x_i f(x_i \mid y_j) f(y_j).$$

Variance

The **variance** of X, denoted σ^2, is defined as $\mathrm{E}[(X - \mu)^2]$:

$$\sigma^2 = \sum_i (x_i - \mu)^2 f(x_i), \tag{7.44}$$

and the **standard deviation**, denoted σ, is the positive square root of the variance. It is often more convenient to denote the variance by $\mathrm{Var}[X]$, and standard deviation by $\mathrm{s.d.}[X]$, as this notation has the advantage of making the random variable explicit. In addition one can also use the notation σ_X^2 and σ_X.

It is often easier to calculate variance by first expanding $(x_i - \mu)^2 = x_i^2 - 2\mu x_i + \mu^2$, and then using (7.38) to obtain

$$\sigma^2 = \mathrm{E}[X^2] - \mathrm{E}[X]^2. \tag{7.45}$$

As noted above in the discussion of the mean, it may be the case that the random variable X is defined in a complicated way, perhaps dependent on another random variable Y, and that $\text{Var}[X]$ is difficult to estimate directly, yet the conditional variance $\text{Var}[X \mid Y]$ is simpler. Of course, this conditional variance is well defined as the variance of X, utilizing the conditional p.d.f. $f(x \mid y)$. In other words,

$$\text{Var}[X \mid Y] = \sum_i (x_i - \mu_{X \mid Y})^2 f(x_i \mid Y),$$

where the conditional mean is defined, $\mu_{X \mid Y} = \text{E}[X \mid Y]$.

The question then becomes, can $\text{Var}[X]$ be recovered from the conditional variances $\text{Var}[X \mid Y]$ the same way that the mean can be recovered from the conditional means via (7.43)? The answer is "yes," but with a slightly more complicated formula, known as the **law of total variance**:

$$\text{Var}[X] = \text{E}[\text{Var}[X \mid Y]] + \text{Var}[\text{E}[X \mid Y]]. \tag{7.46}$$

Before addressing the derivation, note that the formula above is again well defined. As $\text{Var}[X \mid Y]$ and $\text{E}[X \mid Y]$ are functions only of Y, $\text{E}[\text{Var}[X \mid Y]]$ and $\text{Var}[\text{E}[X \mid Y]]$ must be calculated using the marginal p.d.f. $f(y)$, and the variance term is defined as in (7.44), with $\mu = \text{E}[\text{E}[X \mid Y]] = \text{E}[X]$. Summarizing, we have

$$\text{E}[\text{Var}[X \mid Y]] = \sum_i \text{Var}[X \mid y_i] f(y_i),$$

$$\text{Var}[\text{E}[X \mid Y]] = \sum_i (\text{E}[X \mid y_i] - \mu)^2 f(y_i).$$

To derive (7.46), we use the variance formula in (7.45), and substitute the law of total expectation in (7.41):

$$\text{Var}[X] = \text{E}[X^2] - (\text{E}[X]))^2$$

$$= \text{E}[\text{E}[X^2 \mid Y]] - (\text{E}[\text{E}[X \mid Y]])^2.$$

Now another application of (7.45) is

$$\text{E}[X^2 \mid Y] = \text{Var}[X \mid Y] + \text{E}[X \mid Y]^2,$$

which is inserted into the formula above to produce:

$$\text{Var}[X] = \text{E}[\text{Var}[X \mid Y]] + \text{E}[\text{E}[X \mid Y]^2] - (\text{E}[\text{E}[X \mid Y]])^2.$$

Finally, the last two terms are equal to $\mathrm{Var}[\mathrm{E}[X \mid Y]]$ by another application of (7.45), completing the derivation.

Because the laws of total probability, expectation, and variance are so important, the next proposition brings these results together:

Proposition 7.40 *Let X and Y be random variables on a discrete probability space S, with associated joint p.d.f. $f(x, y)$, marginal p.d.f.s $f(x)$ and $f(y)$, and conditional p.d.f. $f(x \mid y)$. Then:*

1. *Law of total probability*,

$$f(x) = \sum_y f(x \mid y) f(y). \tag{7.47}$$

2. *Law of total expectation*,

$$\mathrm{E}[X] = \mathrm{E}[\mathrm{E}[X \mid Y]]. \tag{7.48}$$

3. *Law of total variance*,

$$\mathrm{Var}[X] = \mathrm{E}[\mathrm{Var}[X \mid Y]] + \mathrm{Var}[\mathrm{E}[X \mid Y]]. \tag{7.49}$$

Example 7.41 *Let X denote the number of heads obtained in Y flips of a fair coin, where Y is the number of dots obtained in a roll of a fair die. The goal is to calculate $\mathrm{E}[X]$ and $\mathrm{Var}[X]$. To formalize a sample space, define S as the space of n-flips of a fair coin for $n = 1, 2, 3, \ldots, 6$. So $S = \{(F_1, F_2, \ldots, F_n) \mid 1 \le n \le 6\}$. Here $F_j = 1$ for an H on the jth flip, and 0 otherwise, so S contains $\sum_{n=1}^{6} 2^n = 2^7 - 1$ sample points. The probability measure is defined on each point by*

$$\mathrm{Pr}[(F_1, F_2, \ldots, F_n)] = \frac{1}{6} \frac{1}{2^n}.$$

Now X and Y are defined on S by

$$Y[(F_1, F_2, \ldots, F_n)] = n,$$

$$X[(F_1, F_2, \ldots, F_n)] = \sum_{j=1}^{n} F_j,$$

and so $\mathrm{Rng}[Y] = \{1 \le n \le 6\}$ and $\mathrm{Rng}[X] = \{0 \le m \le 6\}$. Also $f(n) = \frac{1}{6}$ for all n.

For $\mathrm{E}[X \mid Y = n]$ and $\mathrm{Var}[X \mid Y = n]$, we use formulas below in (7.99) from section 7.6.2 on the binomial distribution. Then $\mathrm{E}[X \mid Y = n] = \frac{n}{2}$, and from (7.48), $\mathrm{E}[X] = \mathrm{E}\left[\frac{n}{2}\right]$, so

$$E[X] = \frac{1}{12} \sum_{n=1}^{6} n = \frac{21}{12}.$$

Next, $\text{Var}[X \mid Y = n] = \frac{n}{4}$, *so* $E[\text{Var}[X \mid Y]] = \frac{21}{24}$. *Also from* $E[X \mid Y = n] = \frac{n}{2}$ *we obtain that*

$$\text{Var}[E[X \mid Y]] = E\left[\frac{n^2}{4}\right] - \left(E\left[\frac{n}{2}\right]\right)^2$$

$$= \frac{1}{24} \sum_{n=1}^{6} n^2 - \left(\frac{21}{12}\right)^2$$

$$= \frac{105}{144}.$$

Finally, using (7.49) obtains

$$\text{Var}[X] = \frac{21}{24} + \frac{105}{144} = \frac{231}{144}.$$

Covariance and Correlation

As noted above, there are many expected values that can be defined with a joint p.d.f. One common set of expectations, given $f(x_1, x_2, \ldots, x_n)$, is to evaluate the **covariance** between any two of these random variables. With the associated marginal densities $f(x_j)$, the respective means μ_j and variances σ_j^2 of each X_j can be calculated as discussed above. To calculate the covariance between X_i and X_j requires the joint p.d.f. $f(x_i, x_j)$. Although the notation is not standardized, we denote this expectation by σ_{ij}, and sometimes $\text{Cov}(X_i, X_j)$, the **covariance** is defined by $E[(X_i - \mu_i)(Y_j - \mu_j)]$:

$$\sigma_{ij} = \sum_{k,l} (x_k - \mu_i)(x_l - \mu_j) f(x_k, x_l). \tag{7.50}$$

With a slight abuse of notation, we can define $\sigma_{jj} \equiv \sigma_j^2 = \text{Var}[X_j]$.

Note that a calculation produces a result analogous to (7.45):

$$\sigma_{ij} = E[X_i Y_j] - E[X_i]E[Y_j]. \tag{7.51}$$

Also, if X_i and X_j are independent, then $f(x_i, x_j) = f(x_i)f(x_j)$, and it is apparent that $\sigma_{ij} = 0$, since

$$\sum_{kl}(x_k - \mu_i)(x_l - \mu_j)f(x_k, x_l) = \sum_{k}(x_k - \mu_i)f(x_k)\sum_{l}(x_l - \mu_j)f(x_l).$$

The **correlation** between x_i and x_j, denoted ρ_{ij}, and sometimes $\mathrm{Corr}(X_i, X_j)$, is defined as

$$\rho_{ij} = \frac{\sigma_{ij}}{\sigma_i \sigma_j}, \tag{7.52}$$

which is equivalently calculated as $\rho_{ij} = \sum_{k,l}\left(\frac{x_k - \mu_i}{\sigma_i}\right)\left(\frac{x_l - \mu_j}{\sigma_j}\right)f(x_k, x_l)$. The random variables are said to be **uncorrelated** if $\rho_{ij} = 0$, they are **positively correlated** if $\rho_{ij} > 0$, whereas they are said to be **negatively correlated** if $\rho_{ij} < 0$. As noted above, independent random variables are uncorrelated, and hence have $\rho_{ij} = 0$.

However, being uncorrelated is a weaker condition on two random variables than being independent.

Example 7.42   *Define $f(x, y)$ by*

$$f(x, y) = \begin{cases} \frac{1}{3}, & (x, y) = (-1, 1), \\ \frac{1}{3}, & (x, y) = (0, 0), \\ \frac{1}{3}, & (x, y) = (1, 1). \end{cases}$$

Then $f(x) = \frac{1}{3}$ for $x = -1, 0, 1$, and $f(y) = \frac{2}{3}$ for $y = 1$ and $f(y) = \frac{1}{3}$ for $y = 0$. Consequently X and Y are not independent, since $f(x, y) \neq f(x)f(y)$. On the other hand, X and Y are uncorrelated, since $\mathrm{E}[XY] = 0$, $\mathrm{E}[X] = 0$ and $\mathrm{E}[Y] = \frac{2}{3}$ imply that $\sigma_{XY} = \mathrm{E}[XY] - \mathrm{E}[X]\mathrm{E}[Y] = 0$, and so $\rho_{xy} = 0$.

An important application of the Cauchy–Schwarz inequality is as follows:

Proposition 7.43   *Given random variables X, Y with joint p.d.f. $f(x, y)$,*

$$|\sigma_{XY}| \leq \sigma_X \sigma_Y. \tag{7.53}$$

In other words,

$$-1 \leq \rho_{XY} \leq 1. \tag{7.54}$$

Proof   Since $f(x, y) \geq 0$, we have

$$\sigma_{XY} = \sum_{i,j}(x_i - \mu_X)(y_j - \mu_Y)f(x_i, y_j)$$

$$= \sum_{i,j}\left[(x_i - \mu_X)\sqrt{f(x_i, y_j)}\right]\left[(y_j - \mu_Y)\sqrt{f(x_i, y_j)}\right].$$

This second summation is seen to be an inner product, and by the Cauchy–Schwarz inequality, the square of this inner product is bounded by the product of the sums of squares:

$$\sigma_{XY}^2 \leq \sum_{i,j} \left[(x_i - \mu_X)\sqrt{f(x_i, y_j)} \right]^2 \sum_{i,j} \left[(y_j - \mu_Y)\sqrt{f(x_i, y_j)} \right]^2$$

$$= \sum_{i,j} (x_i - \mu_X)^2 f(x_i, y_j) \sum_{i,j} (y_j - \mu_Y)^2 f(x_i, y_j)$$

$$= \sum_i (x_i - \mu_X)^2 f(x_i) \sum_j (y_j - \mu_Y)^2 f(y_j) = \sigma_X^2 \sigma_Y^2. \qquad \blacksquare$$

The covariance also arises in the variance calculation of the sum of random variables, $X = \sum_{j=1}^n a_j X_j$ for constants $\{a_j\}$. The associated p.d.f. used in the expected value calculation is the joint p.d.f., $f(x_1, x_2, \ldots, x_n)$. With this we see that

$$\mathrm{E}\left[\sum_{j=1}^n a_j X_j \right] = \sum_{j=1}^n a_j \mathrm{E}[X_j]. \qquad (7.55)$$

Also

$$(X - \mathrm{E}[X])^2 = \left(\sum_{j=1}^n a_j [X_j - \mathrm{E}[X_j]] \right)^2$$

$$= \sum_{i=1}^n \sum_{j=1}^n a_i a_j [X_i - \mathrm{E}[X_i]][X_j - \mathrm{E}[X_j]].$$

After expectations are taken, this leads to

$$\mathrm{Var}\left[\sum_{j=1}^n a_j X_j \right] = \sum_{i=1}^n \sum_{j=1}^n a_i a_j \sigma_{ij} \qquad (7.56a)$$

$$= \sum_{j=1}^n a_j^2 \sigma_j^2 + 2 \sum_{i<j} a_i a_j \rho_{ij} \sigma_i \sigma_j. \qquad (7.56b)$$

Note that when the component random variables are independent, or simply uncorrelated:

$$\text{Var}\left[\sum_{j=1}^{n} a_j X_j\right] = \sum_{j=1}^{n} a_j^2 \sigma_j^2. \tag{7.57}$$

General Moments
Generalizing the definition of the mean of a random variable, the nth **moment**, denoted μ_n', is defined as $\text{E}[X^n]$ for $n \geq 0$:

$$\mu_n' = \sum_i x_i^n f(x_i), \tag{7.58}$$

so in particular, $\mu_0' = 1$, $\mu_1' = \mu$, and $\mu_2' = \sigma^2 + \mu^2$ as noted in (7.45).

Note that a direct application of (7.41) with $g(X) = X^n$ produces

$$\text{E}[X^n] = \text{E}[\text{E}[X^n \mid Y]],$$

as was used in the derivation of the law of total variance.

General Central Moments
Generalizing the definition of variance of a random variable, the nth **central moment**, denoted μ_n, is defined as $\text{E}[(X - \mu)^n]$ for $n \geq 0$:

$$\mu_n = \sum_i (x_i - \mu)^n f(x_i), \tag{7.59}$$

so in particular, $\mu_0 = 1$, $\mu_1 = 0$, and $\mu_2 = \sigma^2$.

Absolute Moments
When n is odd, the value of the moments $\text{E}[X^n]$ and/or $\text{E}[(X - \mu)^n]$ can reflect the cancellation of positive and negative terms. The notion of absolute moments is used to value the associated absolutely convergent series. The nth **absolute moment**, denoted $\mu_{|n|}'$, is defined as $\text{E}[|X|^n]$ for $n \geq 0$:

$$\mu_{|n|}' = \sum_i |x_i|^n f(x_i), \tag{7.60}$$

and the nth **absolute central moment**, denoted $\mu_{|n|}$, is defined as $\text{E}[|X - \mu|^n]$ for $n \geq 0$:

$$\mu_{|n|} = \sum_i |x_i - \mu|^n f(x_i). \tag{7.61}$$

This notation is descriptive but not standard.

Because of the condition in (7.37), absolute moments always exist when the corresponding moments exist. Of course, for n even, the absolute moments agree with the respective moments defined above. For n odd, the moments may agree with the absolute moments, for instance, if the range of X is positive, but the central moments and absolute central moments will not agree, since $\{x_i - \mu\}$ will always have both positive and negative terms.

Moment-Generating Function

The **moment-generating function (m.g.f.)**, as the name implies, reflects an expected value calculation that produces a function rather than a numerical constant. Denoted $M(t)$, or $M_X(t)$, it is defined as $\mathrm{E}[e^{Xt}]$:

$$M_X(t) = \sum_i e^{x_i t} f(x_i). \tag{7.62}$$

Of course, $M_X(0) = 1$, so the question of existence of the m.g.f. relates to existence for some interval, $|t| < T$. It is important to note at the outset that $M(t)$ **does not always exist**.

As we have seen before and will prove in chapter 9, the exponential function can be expanded into the power series:

$$e^x = \sum_{n=0}^{\infty} \frac{x^n}{n!}. \tag{7.63}$$

This series converges absolutely for all x by the ratio test,

$$\left| \frac{\frac{x^{n+1}}{(n+1)!}}{\frac{x^n}{n!}} \right| = \left| \frac{x}{n+1} \right| \to 0 \qquad \text{as } n \to \infty;$$

so what needs to be shown in chapter 9 is that the function of x defined by this series is indeed equal to e^x.

Substituting the corresponding expression for $e^{x_i t}$ into (7.62), and using the arithmetic properties of expected value noted above and the assumption of absolute convergence justified by the existence of $M_X(t)$, we derive

$$M_X(t) = \sum_i \sum_{n=0}^{\infty} \frac{(x_i t)^n}{n!} f(x_i) = \sum_{n=0}^{\infty} \frac{t^n}{n!} \sum_i x_i^n f(x_i),$$

and hence

$$M_X(t) = \sum_{n=0}^{\infty} \frac{\mu_n' t^n}{n!}. \tag{7.64}$$

Of course, since all terms in the summation are positive, all these manipulations require the assumption that $M_X(t)$ actually exists and so the series in (7.62) converges and hence converges absolutely. This is always the case for finite sample spaces, but not necessarily the case when the sample space is countably infinite. As seen in section 6.1.4, it is the absolute convergence of this series that justifies the manipulations in the double series and the reversal of the order of the summations.

In chapter 9 we will see that the moments $\{\mu_n'\}$ can in turn be recovered from the moment-generating function, or better said, "generated" from the m.g.f., if it converges in an interval containing 0. Specifically, with $M_X^{(n)}(t)$ denoting the nth **derivative** of the function $M_X(t)$ with respect to t, we will see that

$$\mu_n' = M_X^{(n)}(0). \tag{7.65}$$

A simple modification to the definition of the m.g.f. can be introduced that will generate the central moments. Specifically, since $X - \mu$ has the same p.d.f. as does X, its moment-generating function is defined by $M_{X-\mu}(t) = \sum_i e^{(x_i-\mu)t} f(x_i)$. Applying (7.63) obtains

$$M_{X-\mu}(t) = \sum_{n=0}^{\infty} \frac{\mu_n t^n}{n!}, \tag{7.66}$$

from which is produced

$$\mu_n = M_{X-\mu}^{(n)}(0). \tag{7.67}$$

For a joint probability density function, $f(x_1, x_2, \ldots, x_n)$, the moment-generating function is analogously defined. The definition above where $M_X(t) \equiv \mathrm{E}[e^{Xt}]$ is generalized so that the m.g.f. is now a function of $(t_1, t_2, \ldots, t_n)$, and defined with the aid of boldface vector notation as $M_{\mathbf{X}}(\mathbf{t}) \equiv \mathrm{E}[e^{\mathbf{X}\cdot\mathbf{t}}]$, where $\mathbf{X} \cdot \mathbf{t}$ denotes the inner product. In other words,

$$M_{\mathbf{X}}(\mathbf{t}) = \sum_{(x_1, x_2, \ldots, x_n)} e^{\sum x_i t_i} f(x_1, x_2, \ldots, x_n). \tag{7.68}$$

If the random variables in the definition of $f(x_1, x_2, \ldots, x_n)$ are independent, then (7.34) is satisfied, so

$$M_{\mathbf{X}}(\mathbf{t}) = \prod_{i=1}^{n} M_{X_i}(t_i).$$

If the random variables are independent and identically distributed, then with $Y = \sum_{i=1}^{n} X_i$ we derive from (7.34) and directly from $M_Y(t) = \mathrm{E}\left[e^{t\sum X_i}\right]$ that

$$M_Y(t) = [M_X(t)]^n. \tag{7.69}$$

Characteristic Function

The **characteristic function (c.f.)** is defined similarly to the m.g.f., and it will again be possible to generate moments from it, but it has the advantage that it always exists. The disadvantage to some is that while $M_X(t)$ is a function $M_X(t) : \mathbb{R} \to \mathbb{R}$, the characteristic function, denoted $C_X(t)$, is a function $C_X(t) : \mathbb{R} \to \mathbb{C}$. Specifically, $C_X(t) = \mathrm{E}[e^{iXt}]$, where i denotes the "imaginary unit" $i = \sqrt{-1}$, producing

$$C_X(t) = \sum_j e^{ix_j t} f(x_j). \tag{7.70}$$

It is straightforward to confirm that $C_X(t)$ exists for all $t \in \mathbb{R}$, since the summation converges absolutely. This is demonstrated using the triangle inequality and a consequence of **Euler's formula**: that $|e^{ix_j t}| = 1$ for all t and x_j. Specifically,

$$|C_X(t)| \le \sum_j |e^{ix_j t} f(x_j)| = \sum_j f(x_j) = 1.$$

Unlike the case of the m.g.f., which may not exist but is differentiable when it does exist, the characteristic function always exists, but it need not be differentiable. However, if all moments exist, then using the same manipulations above, justified by absolute converge, produces

$$C_X(t) = \sum_{n=0}^{\infty} \frac{\mu'_n (it)^n}{n!}, \tag{7.71}$$

and once again the moments can be recovered from this function as in (7.65). With analogous notation,

$$\mu'_n = \frac{1}{i^n} C_X^{(n)}(0). \tag{7.72}$$

Central moments can again be generated if they exist using $C_{X-\mu}(t) = \sum_j e^{i(x_j - \mu)t} f(x_j)$; then

$$C_{X-\mu}(t) = \sum_{n=0}^{\infty} \frac{\mu_n (it)^n}{n!}, \tag{7.73}$$

$$\mu_n = \frac{1}{i^n} C_{X-\mu}^{(n)}(0). \tag{7.74}$$

For a joint probability density function $f(x_1, x_2, \ldots, x_n)$, the characteristic function is analogously defined. The definition above where $C_X(t) = \mathrm{E}[e^{iXt}]$ is generalized so that the c.f. is a function of $(t_1, t_2, \ldots, t_n)$, and defined with the aid of boldface vector notation as $C_{\mathbf{X}}(\mathbf{t}) \equiv \mathrm{E}[e^{i\mathbf{X}\cdot\mathbf{t}}]$. In other words,

$$C_{\mathbf{X}}(\mathbf{t}) = \sum_{(x_1, x_2, \ldots, x_n)} e^{i \sum x_j t_j} f(x_1, x_2, \ldots, x_n). \tag{7.75}$$

Remark 7.44 *An important property of the moment-generating and characteristic functions is that they "characterize" the discrete probability density function (a property we will prove in chapter 8 but only in the case of finite discrete random variables). The proof in the more general cases requires the tools of real analysis and complex analysis. What "characterize" means is that if $C_X(t) = C_Y(t)$ or $M_X(t) = M_Y(t)$ for random variables X and Y, and for $t \in I$ where I is any open interval containing 0, then the discrete probability density functions are equal: $f(x) = g(y)$. In the finite discrete case this means that if $\{x_i\}_{i=1}^n$ and $\{y_j\}_{j=1}^m$ are the respective domains of these probability functions, arranged in increasing order, then $n = m$, $x_i = y_i$ and $f(x_i) = g(y_i)$ for all i. The m.g.f. and c.f. also characterize the p.d.f. of random variables in the more general cases to be developed later. Since the characteristic function always exists, this result can be applied to any p.d.f. and in any context.*

*7.5.2 Moments of Sample Data

An important application of the general random vector expectation formula (7.39) is to so-called **sample data expectations**. In this section we provide a theoretical framework for the sample statistics introduced in section 3.3.2.

Given a sample space S, we have the theoretical framework for a **random sample** or **independent trials** introduced in section 7.2.6 above. Specifically, recall that a random sample of size n was identified with a sample point in a new n-trial sample space, denoted S^n, that was given a probability structure defined in (7.7). In this section we apply this structure to random samples of a given random variable, and derive some important formulas related to the moments of these samples.

In the space S we assume that there is given a random variable X, and define a random vector $\overline{X} = (X_1, X_2, \ldots, X_n)$ on S^n. For $\bar{s} = (s_1, s_2, \ldots, s_n) \in S^n$, we define

$X_j(\bar{s}) = X(s_j)$. In the same way that $(s_1, s_2, \ldots, s_n)$ represents a random sample of n possible sample points, with probability in $\mathcal{S}^n$ defined so that $P_n(\bar{s}) = \prod_{j=1}^{n} \Pr(s_j)$, the random vector of values, $\bar{X}(\bar{s}) = (X_1(\bar{s}), X_2(\bar{s}), \ldots, X_n(\bar{s})) \in \mathbb{R}^n$ is a random sample of the values assumed by X on $\mathcal{S}$. In other words, the components of this random vector are **independent** in the formal meaning given in (7.34) above.

To see this, let $f(x_1, x_2, \ldots, x_n)$ be the joint p.d.f. defined on the $\text{Rng}[\bar{X}]$. That is,

$$f(x_1, x_2, \ldots, x_n) = P_n(X_1^{-1}(x_1), X_2^{-1}(x_2), \ldots, X_n^{-1}(x_n)).$$

Then by (7.7),

$$f(x_1, x_2, \ldots, x_n) = \prod_{j=1}^{n} \Pr(X_j^{-1}(x_j))$$

$$= \prod_{j=1}^{n} f(x_j),$$

where $f(x)$ is the p.d.f. of the random variable X.

In summary, we see that if a random variable X on $\mathcal{S}$ is generalized as above to a random vector $\bar{X}$ on the n-trial sample space $\mathcal{S}^n$, then the collection of component random variables $\{X_j(\bar{s})\}$ comprises independent random variables on $\mathcal{S}$ in the sense defined above, and

$$f(x_1, x_2, \ldots, x_n) = \prod_{j=1}^{n} f(x_j). \tag{7.76}$$

Initially this construction of a random sample may appear overly formal and unnecessary. In applications the random variable X is usually defined as the outcome of an experiment, or as an observation, and the notion of independent trials is understood as meaning that the experiment is repeated many times, or other observations are made. In such cases the truth of the identity in (7.76) would appear obvious to anyone that has flipped a coin, or rolled dice, and so forth. And in many applications this is a perfectly legitimate intuitive framework for what a random sample is, and perfectly legitimate justification for the meaning of independent sample.

But intuition does not always guarantee that a rigorous development is possible. So the construction above provides a rigorous construction, in a discrete sample space context, of what a random sample from a sample space represents, and also, what n independent trials of a random variable means. And better than our intuition, this formality will lead the way to the corresponding ideas in the less intuitive frameworks.

Definition 7.45

1. *Given a discrete sample space S and a random variable $X : S \to \mathbb{R}$, the terminology that $\{X_j\}_{j=1}^n$ are **n-independent and identically distributed (i.i.d.) random variables** will mean that $\bar{X} \equiv (X_1, X_2, \ldots, X_n)$ is a random vector on S^n, where for $\bar{s} = (s_1, s_2, \ldots, s_n) \in S^n$ the component random variables are defined by $X_j(\bar{s}) = X(s_j)$. In other words, the collection $\{X_j\}_{j=1}^n$ consists of independent random variables in that the joint p.d.f., $f(x_1, x_2, \ldots, x_n)$, satisfies (7.76), and each component random variable has the same probability density function as X. When $n = \infty$, the terminology that $\{X_j\}_{j=1}^\infty$ are **independent and identically distributed (i.i.d.) random variables** means that this is true of $\{X_j\}_{j=1}^n$ for any n.*

2. *The terminology that $\{x_j\}_{j=1}^n$ is a **random sample from X of size n** means that there is an $\bar{s} = (s_1, s_2, \ldots, s_n) \in S^n$, selected according to the probability measure P_n on S^n so that $(x_1, x_2, \ldots, x_n) = (X(s_1), X(s_2), \ldots, X(s_n))$. In practice, this sample can be generated iteratively by first selecting independent $\{s_j\}_{j=1}^n \subset S$ (see section 7.7 on generating random samples), and defining $(x_1, x_2, \ldots, x_n)$ as above.*

Remark 7.46   *It is standard notation in probability theory that a capital letter is used for the a random variable, such as X, while a lowercase letter, such as x, is used to represent a **realization**, or **sample point**, of the random variable selected according to the probabilities implied by the probability density function of X. Also note that the equivalence of the approaches to a random sample in 2 of definition 7.45 above is due to the probability measure P_n satisfying (7.7).*

Sample Mean
If $\{X_j\}_{j=1}^n$ are n independent and identically distributed (i.i.d.) random variables on S, the **sample mean**, denoted $\hat{X}$, is a random variable $\hat{X} : S^n \to \mathbb{R}$ defined by

$$\hat{X} \equiv \frac{1}{n} \sum_{j=1}^n X_j, \tag{7.77}$$

with probability density function given by $f(x_1, x_2, \ldots, x_n) = \prod_{j=1}^n f(x_j)$, where $f(x)$ is the p.d.f. of X.

When a specific sample is drawn or observed, that is, when $\{X_j\}_{j=1}^n = \{x_j\}_{j=1}^n$, the application of (7.77) to these data yields the numerical value denoted $\hat{\mu}$ or m in section 3.3.2.

The distinction here is the explicit recognition that any such observation $\{x_j\}_{j=1}^n$ is simply based on one sample point in the sample space $\mathcal{S}^n$, and that in more general terms, the calculation produces not a single and unique numerical value but only one of many possible values that the random variable $\hat{X}$ assumes on this sample space. Considered as a random variable, it is natural to inquire into its moments, as we do next.

Mean of the Sample Mean   By definition, we have that

$$E[\hat{X}] = \sum_{(x_1, x_2, \ldots, x_n)} \left(\frac{1}{n}\sum_{j=1}^n x_j\right) f(x_1, x_2, \ldots, x_n).$$

This formula simplifies using (7.76) and the observation that for any x_j,

$$\sum_{(x_1, x_2, \ldots, x_n)} x_j \prod_{k=1}^n f(x_k) = \sum_{x_j} x_j f(x_j) = E[X],$$

since $\sum_{x_k} f(x_k) = 1$ for $k \neq j$. Combining, we get that

$$E[\hat{X}] = E[X], \tag{7.78}$$

provided that $E[X]$ exists. In other words, the expected value of the sample mean is the expected value of the original random variable X.

Variance of the Sample Mean   Denoting $E[X]$ by μ, we have that

$$Var[\hat{X}] = E[(\hat{X} - \mu)^2]$$

$$= \sum_{(x_1, x_2, \ldots, x_n)} \left(\frac{1}{n}\sum_{j=1}^n (x_j - \mu)\right)^2 f(x_1, x_2, \ldots, x_n).$$

Again using (7.76), we get

$$Var[\hat{X}] = \frac{1}{n^2}\left[\sum_{j=1}^n (x_j - \mu)^2 f(x_j)\right] = \frac{\sigma^2}{n}.$$

This result is due to the fact that the mixed terms such as $(x_j - \mu)(x_k - \mu)f(x_j)f(x_k)$, with $j \neq k$, have expectation of 0, since the summations can be done sequentially.

Summarizing, we get

$$\mathrm{Var}[\hat{X}] = \frac{\sigma_X^2}{n}, \tag{7.79}$$

provided that σ_X^2 exists, and correspondingly

$$\mathrm{s.d.}[\hat{X}] = \frac{\sigma_X}{\sqrt{n}}. \tag{7.80}$$

m.g.f. of Sample Mean   Noting that $e^{t\hat{X}} = \prod_{j=1}^{n} e^{tX_j/n}$, and applying the same method as above, we get that

$$M_{\hat{X}}(t) = \left[M_X\left(\frac{t}{n}\right) \right]^n, \tag{7.81}$$

provided that $M_X\left(\frac{t}{n}\right)$ exists.

Sample Variance

If $\{X_j\}_{j=1}^{n}$ are n independent and identically distributed (i.i.d.) random variables on $\mathcal{S}$, the **unbiased sample variance** is defined as

$$\hat{V} = \frac{1}{n-1} \sum_{j=1}^{n} (X_j - \hat{X})^2, \tag{7.82}$$

where $\hat{X} = \frac{1}{n}\sum_{j=1}^{n} X_j$. $\hat{V}$ is again a random variable $\hat{V} : \mathcal{S}^n \to \mathbb{R}$ with a probability density function given by $f(x_1, x_2, \ldots, x_n) = \prod_{j=1}^{n} f(x_j)$, where $f(x)$ is the p.d.f. of X. Note that the sample variance is defined with the sample mean $\hat{X}$, and not the theoretical mean μ. As we will see, this is the reason that it is necessary to use $\frac{1}{n-1}$ in the formula above rather than the more natural value of $\frac{1}{n}$.

As was the case for the sample mean $\hat{X}$, when a specific sample is drawn or observed—that is, when $\{X_j\}_{j=1}^{n} = \{x_j\}_{j=1}^{n}$—the application of (7.82) to these data yields the numerical value denoted $\hat{\sigma}^2$ or s^2 in section 3.3.2, there defined with $n - 1$ rather than n. However, once again the perspective here is that any such observation $\{x_j\}_{j=1}^{n}$ is simply one sample point in the sample space $\mathcal{S}^n$, and that in more general terms, the calculation produces not a single and unique numerical value but only one of many possible values that the random variable $\hat{V}$ assumes on this sample space. Considered as a random variable, it is natural to inquire into its moments, as we do next.

Mean of Sample Variance   The calculation of $E[\hat{V}]$ is complicated by the fact that the random variables appear in two places in the squared terms, explicitly in the X_j terms, and implicitly in the $\hat{X}$ term. A simple trick is to write $X_j - \hat{X} = (X_j - \mu) - (\hat{X} - \mu)$ and $\hat{X} - \mu = \frac{1}{n}\sum_{k=1}^{n}(X_k - \mu)$, from which we get

$$(X_j - \hat{X})^2 = (X_j - \mu)^2 - 2(X_j - \mu)(\hat{X} - \mu) + (\hat{X} - \mu)^2$$

$$= (X_j - \mu)^2 - \frac{2}{n}\sum_{k=1}^{n}[(X_j - \mu)(X_k - \mu)] + \frac{1}{n^2}\sum_{i=1}^{n}\sum_{k=1}^{n}[(X_i - \mu)(X_k - \mu)].$$

Summed over j, the second and third term then combine, producing

$$\sum_{j=1}^{n}(X_j - \hat{X})^2 = \sum_{j=1}^{n}(X_j - \mu)^2 - \frac{1}{n}\sum_{i=1}^{n}\sum_{k=1}^{n}[(X_i - \mu)(X_k - \mu)].$$

Assuming that σ^2 exists, and taking expectations, we get

$$E\left[\sum_{j=1}^{n}(X_j - \hat{X})^2\right] = (n - 1)\sigma^2,$$

since the expectation of mixed terms in the double sum, when $i \neq k$, is 0 because of independence. This identity is equivalent to

$$E[\hat{V}] = \sigma^2. \tag{7.83}$$

It is this identity that motivates the use of the term "unbiased" for the sample variance formula given above. It is unbiased in the sense that the expected value of this statistic is the theoretical value of what is being estimated. In that sense, from (7.78) it is also the case that $\hat{X}$ is an unbiased estimator of the theoretical mean $\mu \equiv E[X]$, but this formula is never called the unbiased sample mean.

It is easy to check that if the theoretical mean, μ, is known and the sample variance defined as in (7.82) but with μ rather than $\hat{X}$, then the correct coefficient in (7.82) would be $\frac{1}{n}$, in that with this coefficient (7.83) would again be derived. But in most applications this is not relevant since sampling implies limited knowledge of the theoretical distribution and its theoretical moments, so it may be illogical to assume that μ is known.

Remark 7.47   *As it turns out, there is another calculation of sample variance that uses $\frac{1}{n}$ in its formulation rather than $\frac{1}{n-1}$, and yet also uses $\hat{X}$. This particular formulation is*

*known as the **maximum likelihood estimator of the sample variance**, and since the notation is not standardized, we use*

$$\hat{\sigma}^2_{MLE} = \frac{1}{n}\sum_{j=1}^{n}(x_j - \hat{X})^2, \tag{7.84}$$

*where $\hat{X} = \frac{1}{n}\sum_{j=1}^{n} x_j$. By the analysis above, if this version of a sample variance is
defined as a random variable on S^n analogously to $\hat{V}$, it is biased on the small side, in
that*

$$E[\hat{\sigma}^2_{MLE}] = \frac{n-1}{n}\sigma^2. \tag{7.85}$$

*The idea behind the MLE calculation is way ahead of our mathematical development,
but it can be presented in an intuitive way. Assume that a sample has been drawn or
observed, $\{x_j\}_{j=1}^{n}$, and for various reasons we believe a particular form for the p.d.f. of
the observed random variable, $f(x)$. In this case as in many, the assumed p.d.f. is that
of the **normal distribution**, which is formally introduced in chapter 8 and studied in
chapter 10. This distribution will be seen to be characterized by only two moments, μ
and σ^2. The question is then, given this assumed distribution, what estimates for μ and
σ^2 will maximize the probability of the observed sample? In other words, What estimates for μ and σ^2 will maximize the likelihood of observing the given sample?*

*Since the sample p.d.f. is $f(x_1, x_2, \ldots, x_n) = \prod_{j=1}^{n} f(x_j)$, as seen in (7.76), and $f(x)$
only depends on μ and σ^2, this question reduces to determining the values of these
parameters that maximize $\prod_{j=1}^{n} f(x_j)$, the probability of the sample point $(x_1, x_2, \ldots,
x_n)$ under this distributional assumption. This function to be maximized is actually a
function of the parameters μ and σ^2, since the sample point $(x_1, x_2, \ldots, x_n)$ is fixed
and known. Determining the maximum value of a function is an application of calculus
and will be seen in chapter 9 for one variable functions, while this particular application
with two variables requires multivariate calculus. As it turns out, the MLE estimators
for μ and σ^2 are $\hat{X}$ and $\hat{\sigma}^2_{MLE}$.*

Variance of Sample Variance   Because of (7.83) the needed calculation is that of
$\text{Var}[\hat{V}] = E[(\hat{\sigma}^2 - \sigma^2)^2]$, which involves some messy algebra and some determination
on the part of the analyst. To make this calculation reasonably tractable, we use the
approach in (7.45) that variance equals the second moment less the mean squared,
which becomes $\text{Var}[\hat{V}] = E[\hat{V}^2] - (E[\hat{V}])^2 = E[\hat{V}^2] - \sigma^4$. From the algebra in the
derivation of the mean of the sample variance, recall that

$$(n-1)\hat{V} = \sum_{j=1}^{n}(X_j - \hat{X})^2 = \sum_{j=1}^{n}Y_j^2 - \frac{1}{n}\sum_{i=1}^{n}\sum_{k=1}^{n}Y_iY_k,$$

where we simplify notation with $Y_j = X_j - \mu$. The key is that $\mathrm{E}[Y_j] = 0$, and so in any expression in which there is at least one Y_j term to the first power, the expectation will be zero and can be ignored.

This expression squared, which equals $(n-1)^2\hat{V}^2$, is then

$$\left[\sum_{j=1}^{n}(X_j-\hat{X})^2\right]^2 = \left[\sum_{j=1}^{n}Y_j^2\right]^2 - \frac{2}{n}\sum_{j=1}^{n}Y_j^2\sum_{i=1}^{n}\sum_{k=1}^{n}Y_iY_k + \frac{1}{n^2}\left[\sum_{i=1}^{n}\sum_{k=1}^{n}Y_iY_k\right]^2.$$

While initially ominous looking, we are only interested in determining how many terms of each "type" there are. For example, the squared first expression produces a sum of $Y_j^2Y_k^2$ terms, which fall into two types: one if $j = k$ and another if $j \neq k$. Any term of the first type has expectation μ_4, the fourth central moment of X, and any of the second type have expectation $\mu_2^2 = \sigma^4$.

Using the combinatorics discussed earlier, we have n terms of the first type and $n(n-1)$ of the second, since every j can be paired with $(n-1)$-ks. Hence the first expression becomes

$$\mathrm{E}\left[\sum_{j=1}^{n}Y_j^2\right]^2 = n\mu_4 + n(n-1)\sigma^4.$$

The second expression produces four types of terms:

$$Y_j^4, \quad Y_j^3Y_i, \quad Y_j^2Y_k^2, \quad Y_j^2Y_iY_k,$$

where the subscripts are meant to differ. These terms have expectations of μ_4, 0, σ^4, and 0, respectively, since $\mathrm{E}[Y_k] = 0$. The challenge is then counting the types, and all we are concerned with is the first and third type. Again, we draw on combinatorics and determine that there are n of the first type and $n(n-1)$ of the third. Combined, the second expression becomes

$$\mathrm{E}\left[-\frac{2}{n}\sum_{j=1}^{n}Y_j^2\sum_{i=1}^{n}\sum_{k=1}^{n}Y_iY_k\right] = -2\mu_4 - 2(n-1)\sigma^4.$$

The third expression produces five different types of terms, the four above and $Y_iY_jY_kY_l$. Of these five, we only need to evaluate the first and third, since all others have expectation of 0. Again, there are n of the first type, but for the third, the combinatorics for this expression are different. First off, $[\sum_{i=1}^{n} \sum_{k=1}^{n} Y_i Y_k]^2 = [\sum_{i=1}^{n} Y_i]^4$, and from the multinomial formula in (7.19), the coefficient of every $Y_j^2 Y_k^2$ term is $\frac{4!}{2!2!} = 6$, and as there are $\frac{n(n-1)}{2}$ different such terms with $j \neq k$, the total count is $3n(n-1)$. Combining produces the third expression:

$$\mathrm{E}\left[\frac{1}{n^2}\left[\sum_{i=1}^{n}\sum_{k=1}^{n} Y_i Y_k\right]^2\right] = \frac{1}{n}\mu_4 + \frac{3(n-1)}{n}\sigma^4.$$

Finally, the three expressions are combined to

$$\mathrm{E}\left[\left[\sum_{j=1}^{n}(X_j - \hat{X})^2\right]^2\right] = n\mu_4 + n(n-1)\sigma^4 - 2\mu_4 - 2(n-1)\sigma^4 + \frac{1}{n}\mu_4 + \frac{3(n-1)}{n}\sigma^4$$

$$= \left(n - 2 + \frac{1}{n}\right)\mu_4 + \left[(n-2)(n-1) + \frac{3(n-1)}{n}\right]\sigma^4.$$

Dividing by $(n-1)^2$ produces $\mathrm{E}[\hat{V}^2]$, and subtracting $(\mathrm{E}[\hat{V}])^2 = \sigma^4$ gives the final result:

$$\mathrm{Var}[\hat{V}] = \frac{1}{n}\mu_4 - \frac{n-3}{n(n-1)}\sigma^4, \tag{7.86}$$

as well as the associated result for $\hat{\sigma}_{MLE}^2$, interpreted as a random variable, by multiplying (7.86) by $\frac{(n-1)^2}{n^2}$:

$$\mathrm{Var}[\hat{\sigma}_{MLE}^2] = \frac{(n-1)^2}{n^3}\mu_4 - \frac{(n-1)(n-3)}{n^3}\sigma^4. \tag{7.87}$$

Other Sample Moments

Higher Order Moments   Due to the messiness of estimating the central moments, as observed for the estimates above related to the sample variance, we focus on estimates of the moments μ_k'. Given an independent and identically distributed sample $\{X_j\}_{j=1}^{n}$, the general **higher sample moment** estimation formula is

$$\hat{\mu}_k' = \frac{1}{n}\sum_{j=1}^{n} X_j^k. \tag{7.88}$$

The derivations of the following are assigned in exercise 13:

$$\mathrm{E}[\hat{\mu}_k'] = \mu_k', \tag{7.89a}$$

$$\mathrm{Var}[\hat{\mu}_k'] = \frac{1}{n}[\mu_{2k}' - (\mu_k')^2], \tag{7.89b}$$

provided that μ_k' and μ_{2k}' exist.

Remark 7.48   *The identities in exercise* 12 *between theoretical moments* $\{\mu_k'\}$ *and* $\{\mu_k\}$ *do not apply in the context of sample moments because, in that context,* μ_k *is typically defined relative to the sample mean* $\hat{X}$ *and not the theoretical mean* μ. *These formulas do apply if central moments are defined relative to the theoretical mean* μ, *but this is impractical in most circumstances, as noted above.*

Moment-Generating Function   Given an independent and identically distributed sample $\{X_j\}_{j=1}^n$, the **sample moment-generating function** estimation formula is

$$\hat{M}_X(t) = \frac{1}{n}\sum_{j=1}^n e^{tX_j}. \tag{7.90}$$

The function $\hat{M}_X(t)$ can be interpreted as a random variable on $\mathcal{S}^n$ for each t.

In exercise 34 are assigned the following:

$$\mathrm{E}[\hat{M}_X(t)] = M_X(t), \tag{7.91a}$$

$$\mathrm{Var}[\hat{M}_X(t)] = \frac{1}{n}[M_X(2t) - M_X^2(t)]. \tag{7.91b}$$

These functions are interpreted as valid for each t, provided that $M_X(t)$ and $M_X(2t)$ exist.

7.6   Discrete Probability Density Functions

Clearly, a random variable X conveys some but not all of the information about a sample space $\mathcal{S}$, its complete collection of events $\mathcal{E}$, and associated probability measure Pr, and it transfers this information to a collection of real numbers $\{x_j\}$ in the range of the random variable. In particular, a random variable allows us to think of the values in the range of X as "occurring" with certain probabilities. This is a good way to proceed for mathematical analysis because we can then study probability density functions and their properties objectively without having to reference the context of the original sample space or defining random variable. Indeed it is common to use a generic random variable X to define a given p.d.f., without reference to the defining sample space or events or to the functional form of X, in order to provide an objective language for investigating the properties of $f(x)$.

However, it is important to remember that the number x_j occurs with probability $f(x_j)$ not in isolation but because the event defined by $X^{-1}(x_j) \in \mathcal{E}$ occurs with probability $\Pr[X^{-1}(x_j)]$ in a given sample space $\mathcal{S}$.

In this section we list several of the most common examples of discrete p.d.f.s. Of course, there are infinitely many possible probability functions. In the finite case, if $\{x_j\}_{j=1}^n \subset \mathbb{R}$ and $\{f_j\}_{j=1}^n$ is any collection of real numbers, then one can define a p.d.f. by

$$f(x_j) = \frac{|f_j|}{\sum_{k=1}^n |f_k|}.$$

In the countably infinite case, if $\{x_j\}_{j=1}^\infty \subset \mathbb{R}$ and $\{f_j\}_{j=1}^\infty \in l_1$ as defined in chapter 6, then a p.d.f. can analogously be defined by

$$f(x_j) = \frac{|f_j|}{\sum_{k=1}^\infty |f_k|}.$$

Consequently any l_1-sequence can be used to define a p.d.f. in a countably infinite context. Of course, we use $|f_j|$ to ensure that $f(x_j) \geq 0$ for all j. In either the finite or countable case, the associated c.d.f.s are then defined by (7.22).

While these general constructions are useful to exemplify the range of potential p.d.f.s and some of their properties, there is a far more limited number of examples found in common practice.

7.6.1   Discrete Rectangular Distribution

The simplest probability density that can be imagined is one that assumes the same value on every sample point. The domain of this distribution is arbitrary but is conventionally taken as $\left\{\frac{j}{n}\right\}_{j=1}^n$ or $\left\{\frac{j}{n+1}\right\}_{j=0}^n$, so in either case $\mathrm{Dmn}[f(x)] \subset [0,1]$, where "Dmn" denotes the domain of the function as in definition 2.23. Rather than present two sets of formulas, we focus on the former definition and leave it as a general exercise to translate these to the latter setting if needed.

For a given n, the p.d.f. of the **discrete rectangular distribution**, sometimes called the **discrete uniform distribution**, is defined on $\left\{\frac{j}{n}\right\}_{j=1}^n$ by

$$f^R\left(\frac{j}{n}\right) = \frac{1}{n}, \qquad j = 1, 2, \ldots, n. \tag{7.92}$$

It is a relatively easy calculation to derive the mean and variance of this distribution, using the formulas

$$\sum_{j=1}^{n} j = \frac{n(n+1)}{2},$$

$$\sum_{j=1}^{n} j^2 = \frac{n(n+1)(2n+1)}{6},$$

which can be easily proved by mathematical induction. Implementing the necessary algebra, we derive

$$\mu_R = \frac{n+1}{2n},$$ (7.93a)

$$\sigma_R^2 = \frac{n^2-1}{12n^2}.$$ (7.93b)

Similarly the moment-generating function can be calculated as the sum of a geometric series, since $\frac{1}{n}\sum_{j=1}^{n} e^{jt/n} = \frac{1}{n}\sum_{j=1}^{n} (e^{t/n})^j$, producing

$$M_R(t) = \frac{e^{[1+(1/n)]t} - e^{t/n}}{n(e^{t/n}-1)}.$$ (7.94)

It is apparent from (7.93) that as $n \to \infty$, $\mu_R \to \frac{1}{2}$ and $\sigma_R^2 \to \frac{1}{12}$. Less apparent is what happens in the limit for the moment-generating function, due to the denominator, as it is clear that the numerator approaches $e^t - 1$. For the denominator we once again use the series expression for the exponential, to be proved in chapter 9, that $e^{t/n} = \sum_{j=0}^{\infty} \left(\frac{t}{n}\right)^j \frac{1}{j!}$. From this the denominator is seen to equal $t + \frac{h_n(t)}{n}$, where $h_n(t)$ is bounded as $n \to \infty$, so this denominator is seen to approach t. Hence as $n \to \infty$, $M_R(t) \to \frac{e^t-1}{t}$. As will be seen in chapter 10, these limiting values are the corresponding expressions for the continuous counterpart to the rectangular distribution defined on $[0,1]$.

One of the most important applications of this distribution will be to the problem of generating random samples from other distributions, a problem that is addressed in section 7.7 below.

This distribution can also be defined on an arbitrary closed interval $[a,b]$, generalizing the model above defined on $[0,1]$. The probability density is now defined on

$\left\{a + (b-a)\frac{j}{n}\right\}_{j=1}^{n}$ with values as in (7.92), and from (7.55) and (7.56) or directly we obtain

$$\mu_{R_{a,b}} = \frac{n-1}{2n}a + \frac{n+1}{2n}b, \tag{7.95a}$$

$$\sigma^2_{R_{a,b}} = (b-a)^2 \frac{n^2-1}{12n^2}. \tag{7.95b}$$

Limits as $n \to \infty$ are then $\mu_{R_{a,b}} = \frac{a+b}{2}$ and $\sigma^2_{R_{a,b}} = \frac{(b-a)^2}{12}$.

7.6.2   Binomial Distribution

For a given p, $0 < p < 1$, the **standard binomial** random variable is defined as $X_1^B : S \to \{0,1\}$, where the associated p.d.f. is defined on $\{0,1\}$ by $f(1) = p$, $f(0) = p' \equiv 1 - p$. This is often economically expressed as

$$X_1^B = \begin{cases} 1, & \Pr = p, \\ 0, & \Pr = p', \end{cases}$$

or to emphasize the associated p.d.f.,

$$f(X_1^B) = \begin{cases} p, & X_1^B = 1, \\ p', & X_1^B = 0. \end{cases} \tag{7.96}$$

A simple application for this random variable is the single coin flip so that $S = \{H, T\}$, and where a probability measure has been defined on S by $\Pr(H) = p$ and $\Pr(T) = p'$ so that $X_1^B(H) \equiv 1$ and $X_1^B(T) \equiv 0$. This random variable is sometimes referred to as a **Bernoulli trial**, and the associated c.d.f. as the **Bernoulli distribution**, after **Jakob Bernoulli** (1654–1705).

This standard formulation is then easily transformed to a **shifted standard binomial** random variable: $Y_1^B = b + (a - b)X_1^B$, which is defined as

$$Y_1^B = \begin{cases} a, & \Pr = p, \\ b, & \Pr = p', \end{cases}$$

where the example of $a = 1$, $b = -1$, is common in discrete stock price modeling.

Similarly this model can be extended to accommodate sample spaces of n-coin flips, producing the **general binomial** random variable, which now has two parameters, p and $n \in \mathbb{N}$. That is, $S = \{(F_1 F_2 \ldots F_n) \,|\, \text{all } F_j = H \text{ or } T\}$, and X_n^B is defined as the "head-counting" random variable:

$$X_n^B(F_1 F_2 \ldots F_n) = \sum_{j=1}^{n} X_1^B(F_j).$$

It is apparent that X_n^B assumes values $0, 1, 2, \ldots, n$, and that using the combinatorial analysis above, the associated probabilities are given by

$$X_n^B = \left\{ j; \mathrm{Pr} = \binom{n}{j} p^j (1 - p)^{n-j}, \qquad j = 0, 1, \ldots, n, \right.$$

or to emphasize the associated p.d.f.,

$$f^B(j) = \binom{n}{j} p^j (1 - p)^{n-j}, \qquad j = 0, 1, \ldots, n. \tag{7.97}$$

To derive these probabilities, we observe that if $(F_1 F_2 \ldots F_n) \in S$ is any sample point with j Hs, then $\mathrm{Pr}(F_1 F_2 \ldots F_n) = p^j (1 - p)^{n-j}$. Moreover, for any j, there are $\binom{n}{j}$ such sample points. Consequently the event $[X_n^B]^{-1}(j)$ in $\mathcal{E}$ has probability as given in (7.97). Of course, $\sum_{j=0}^{n} f^B(j) = 1$ by the binomial theorem in (7.15) with $a = p$ and $b = p'$, since then $a + b = 1$.

Finally, the mean, variance and moment-generating function of $f^B(j)$ are easier to handle using the fact, as was seen above, that $X_n^B = \sum_{j=1}^{n} X_{1j}^B$, where $\{X_{1j}^B\}$ are n independent, identically distributed standard binomials. For the **standard binomial** we readily obtain

$$\mu_B = p, \quad \sigma_B^2 = pq, \quad M_B(t) = pe^t + q. \tag{7.98}$$

Using the method of independent sums as seen in section 7.5.1 above on moments, also summarized in (7.38), (7.57), and (7.69), produces for any n the moments of the **general binomial**:

$$\mu_B = np, \quad \sigma_B^2 = npq, \quad M_B(t) = (pe^t + q)^n. \tag{7.99}$$

Note that the formulas in (7.99) at first appear inconsistent with those from the preceding section on the sample mean. This is because here we are working with a simple summation, while the earlier analysis was applied to the average of a summation.

It is sometimes necessary to be able to determine the **mode** of this distribution, defined as the value of j for which $f^B(j)$ is maximized, and which we denote by $\hat{j}$. We now show that the mode is any integer that satisfies

$$p(n + 1) - 1 < \hat{j} < p(n + 1), \tag{7.100}$$

so in general, it is possible to have two modes, and this occurs only when $p(n+1)$ is an integer. Otherwise, $\hat{j}$ is unique.

This result is derived from the identity

$$f_B(j+1) = \frac{p(n-j)}{(1-p)(j+1)} f_B(j).$$

From this formula it is apparent that $f_B(j+1) \geq f_B(j)$ if and only if $\frac{p(n-j)}{(1-p)(j+1)} \geq 1$. A bit of algebra produces that this occurs when $j \leq p(n+1) - 1$. In other words, the last j for which $f_B(j+1) \geq f_B(j)$, and the value of j that maximizes $f_B(j+1)$ satisfies $j \leq p(n+1) - 1$. From that point forward these probabilities begin to decrease. So the mode must satisfy $\hat{j} = j+1$, and from this analysis we conclude that $\hat{j} \leq p(n+1)$.

Now, if $\hat{j} - 1 = p(n+1) - 1$ is an integer, then this coefficient ratio is exactly 1. Hence $f_B(\hat{j} - 1) = f_B(\hat{j})$, and the binomial has two modes, one at each of $p(n+1) - 1$ and $p(n+1)$.

7.6.3   Geometric Distribution

For a given p, $0 < p < 1$, the geometric distribution is defined on the nonnegative integers, and its p.d.f. is given by

$$f^G(j) = p(1-p)^j, \qquad j = 0, 1, 2, \ldots. \tag{7.101}$$

This distribution is related to the standard binomial distribution in a natural way. The underlying sample space can be envisioned as the collection of all coin-flip sequences that terminate on the first H. So

$$S = \{H, TH, TTH, TTTH, \ldots\},$$

and the random variable X is defined as the number of flips before the first H. Consequently $f^G(j)$ above is the probability in S of the sequence of j-Ts and then 1-H, that is, the probability that the first H occurs after j-Ts.

Remark 7.49   *The geometric distribution is sometimes parametrized as*

$$f^{G'}(j) = p(1-p)^{j-1}, \qquad j = 1, 2, \ldots,$$

and then represents the probability of the first head in a coin flip sequence appearing on flip j. These representations are conceptually equivalent, but mathematically distinct due to the shift in domain. The result is that the moments for $f^{G'}(j)$ differ from those of $f^G(j)$ in that for $m \geq 1$,

$$\mu'_m(f^G) = (1 - p)\mu'_m(f^{G'}),$$

although coincidentally the variance remains the same.

Note that $\sum_{j=0}^{\infty} f^G(j) = 1$ as this geometric series can be summed with the methods of chapter 6. That is,

$$\sum_{j=0}^{\infty}(1 - p)^j = \frac{1}{p}.$$

The mean, variance, and moment-generating function of the geometric distribution can be calculated using various approaches, but the easiest of these to derive, surprisingly, is the m.g.f. as this is just another geometric series. Specifically:

$$M_G(t) = p\sum_{j=0}^{\infty}(1 - p)^j e^{jt} = p\sum_{j=0}^{\infty}[(1 - p)e^t]^j,$$

which is convergent by the ratio test if $(1 - p)e^t < 1$. Using the usual geometric series approach, we obtain

$$M_G(t) = \frac{p}{1 - (1 - p)e^t}. \tag{7.102}$$

The mean and variance can be derived from this expression with a bit of calculus from chapter 9 using (7.65), or directly (see exercise 15). This produces, with $p' \equiv 1 - p$,

$$\mu_G = \frac{p'}{p}, \quad \sigma_G^2 = \frac{p'}{p^2}. \tag{7.103}$$

7.6.4   Multinomial Distribution

The multinomial distribution reflects the combinatorial analysis in section 7.3.2 above for general orderings with r-subset types. For given $\{p_j\}_{j=1}^r$, $0 < p_j < 1$ with $\sum_{j=1}^r p_j = 1$, and fixed $n \in \mathbb{N}$, the multinomial p.d.f. is defined on every integer r-tuple, $(n_1, n_2, \ldots, n_r)$, with $0 \leq n_j$ and $\sum n_j = n$, by

$$f^M(n_1, n_2, \ldots, n_r) = \frac{n! p_1^{n_1} p_2^{n_2} \cdots p_r^{n_r}}{n_1! n_2! \ldots n_r!}. \tag{7.104}$$

There are several intuitive models for this distribution. One can imagine that at target practice, a girl scout with n arrows is shooting down-field at $r-1$ targets of different sizes, and has probability p_j of hitting the jth target, and probability $p_r = 1 - \sum_{j=1}^{r-1} p_j$ of missing them all. The sample space is the collection of all r-tuples of results, where each n_j denotes the number of arrows hitting the respective target (where $j = r$ denotes hitting the ground).

An alternative model can be achieved with the binomial model in (7.97). Now imagine that N sequences of n coin flips are to be generated. The question becomes, how many of these sequences will end up in each of the $r \equiv n+1$ "head-count buckets" implied by this model? Better said, for any nonnegative $(n+1)$-tuple: $(N_0, N_1, N_2, \ldots, N_n)$ with $\sum N_j = N$, what is the probability that exactly N_j sequences will have j-Hs for all j? In this application the probability of ending up in the j-heads bucket is given by $p_j = \binom{n}{j} p^j (1-p)^{n-j}, j = 0, 1, \ldots, n$.

In either sample space interpretation, we develop the p.d.f. formula in (7.104) using the same approach as for the binomial. For any r-tuple, $(n_1, n_2, \ldots, n_r)$, the probability of any specific such sequence is $p_1^{n_1} p_2^{n_2} \ldots p_r^{n_r}$. We now need to count how many of the sample points in the sample space have exactly this cell count. From section 7.3.2 on general orderings with r-subset types, the number is $\frac{n!}{n_1! n_2! \ldots n_r!}$, so the probability of the event defined by having exactly this many of each type is the product of this count factor with the probability above, which is formula (7.104).

Note that

$$\sum_{n_1, n_2, \ldots, n_r} f^M(n_1, n_2, \ldots, n_r) = 1,$$

by the multinomial theorem in (7.19), since $\sum_{j=1}^{r} p_j = 1$; hence $\left[\sum_{j=1}^{r} p_j \right]^n = 1$.

It is not difficult to show that if $(N_1, N_2, \ldots, N_r)$ is multinomial, with parameters $\{p_j\}_{j=1}^{r}$ and n, then each of the variables N_j has a binomial distribution. For example, calculating the marginal density of N_1, we have

$$f(n_1) = \sum_{n_2, \ldots, n_r} \frac{n! p_1^{n_1} p_2^{n_2} \ldots p_r^{n_r}}{n_1! n_2! \ldots n_r!}$$

$$= \frac{n! p_1^{n_1}}{(n - n_1)! n_1!} \sum_{n_2, \ldots, n_r} \frac{(n - n_1)! p_2^{n_2} \ldots p_r^{n_r}}{n_2! \ldots n_r!},$$

where this summation is over all $(r-1)$-tuples, $(n_2, \ldots, n_r)$, with $\sum_{j=2}^{r} n_j = n - n_1$. Now this summation has exactly the structure of a multinomial distribution, with parameters $\{p_j\}_{j=2}^r$ and $n - n_1 = \sum_{j=2}^r n_j$, but it cannot add up to 1 by the multinomial theorem because $\sum_{j=2}^r p_j = 1 - p_1 \neq 1$. This can be fixed by dividing each such p_j by $1 - p_1$, so the summation is identically 1. Fixing this division outside the summation proceeds as follows:

$$f(n_1) = \frac{n! p_1^{n_1} (1 - p_1)^{n - n_1}}{(n - n_1)! n_1!} \sum_{n_2, \ldots, n_r} \frac{(n - n_1)! \left(\frac{p_2}{1 - p_1}\right)^{n_2} \cdots \left(\frac{p_r}{1 - p_1}\right)^{n_r}}{n_2! \ldots n_r!}$$

$$= \binom{n}{n_1} p_1^{n_1} (1 - p_1)^{n - n_1},$$

which is the binomial density with parameters n and p_1. Consequently we know that every variable is similarly binomial, and by (7.99),

$$E[N_j] = n_j p_j, \quad \text{Var}[N_j] = n_j p_j (1 - p_j). \tag{7.105}$$

In the same way the marginal density of any group of distinct variables can be shown to be multinomial. For example, with two variables,

$$f(n_1, n_2) = \sum_{n_3, \ldots, n_r} \frac{n! p_1^{n_1} p_2^{n_2} \cdots p_r^{n_r}}{n_1! n_2! \ldots n_r!}$$

$$= \frac{n! p_1^{n_1} p_2^{n_2} (1 - p_1 - p_2)^{n - n_1 - n_2}}{n_1! n_2!}$$

$$\times \sum_{n_2, \ldots, n_r} \frac{(n - n_1 - n_2)! \left(\frac{p_3}{1 - p_1 - p_2}\right)^{n_3} \cdots \left(\frac{p_r}{1 - p_1 - p_2}\right)^{n_r}}{n_3! \ldots n_r!}$$

$$= \frac{n! p_1^{n_1} p_2^{n_2} (1 - p_1 - p_2)^{n - n_1 - n_2}}{n_1! n_2! (n - n_1 - n_2)!}.$$

This is a multinomial with parameters $\{p_1, p_2, 1 - p_1 - p_2\}$ and n. A similar formula is derived for $f(n_i, n_j)$, $i \neq j$. This joint p.d.f. is used in exercise 35 to derive the calculation that

$$\text{Cov}[N_i, N_j] = -n p_i p_j. \tag{7.106}$$

Finally, the moment-generating function of the multinomial distribution can be easily derived with the help of the multinomial theorem in (7.19). Using the definition in (7.68), where as above the summation is over all nonnegative r-tuples, $(n_1, n_2, \ldots, n_r)$, with $\sum n_j = n$, produces

$$M_{\mathbf{M}}(\mathbf{t}) = \sum_{(n_1, n_2, \ldots, n_r)} e^{\sum n_i t_i} \frac{n! p_1^{n_1} p_2^{n_2} \cdots p_r^{n_r}}{n_1! n_2! \ldots n_r!}$$

$$= \sum_{(n_1, n_2, \ldots, n_r)} \frac{n! (p_1 e^{t_1})^{n_1} \cdots (p_r e^{t_r})^{n_r}}{n_1! n_2! \ldots n_r!}.$$

Finally, applying the multinomial theorem, we obtain

$$M_{\mathbf{M}}(\mathbf{t}) = \left(\sum_{j=1}^{r} p_j e^{t_j} \right)^n. \tag{7.107}$$

The characteristic function is derived analogously, using (7.75).

7.6.5   Negative Binomial Distribution

The name of this distribution calls out yet another connection to the binomial distribution, and here we generalize the idea behind the geometric distribution. There $f^G(j)$ was defined as the probability of j-Ts before the first H. The negative binomial, $f^{NB}(j)$ introduces another parameter, k, and is defined as the probability of j-Ts before the kth-H. So when $k = 1$, the negative binomial is the same as the geometric. With that as an introduction, the p.d.f. is defined with parameters p, $0 < p < 1$, and $k \in \mathbb{N}$ as follows:

$$f^{NB}(j) = \binom{j+k-1}{k-1} p^k (1-p)^j, \qquad j = 0, 1, 2, \ldots. \tag{7.108}$$

This formula can be derived analogously to the geometric by considering in the sample space of all coin-flip sequences, those that are terminated on the occurrence of the kth-H. The probability of any such sequence with j-Ts and k-Hs is of course $p^k (1-p)^j$. Next we must determine the number of such sequences in the sample space. First off, since every such sequence terminates with an H, there are only the first $j + k - 1$ positions that need to be addressed. Each such sequence is then determined by the placement of the first $(k - 1)$-Hs, and so the total count of these sequences is $\binom{j+k-1}{k-1}$. Multiplying the probability and the count, we have (7.108).

As stated above, the negative binomial generalizes the geometric distribution and reduces to that distribution when the parameter $k = 1$. This is easily confirmed by comparing (7.108) with $k = 1$ to (7.101), recalling that $\binom{j}{0} = 1$ for all $j \geq 0$ since $0! \equiv 1$.

Finally, to demonstrate that $\sum_{j=0}^{\infty} f^{NB}(j) = 1$, which is equivalent to $\sum_{j=0}^{\infty} \binom{j+k-1}{k-1}(1-p)^j = p^{-k}$, we establish the following proposition. To simplify notation, we define $q = 1 - p$.

Proposition 7.50   *For $0 < q < 1$ and integer $k \geq 1$,*

$$(1-q)^{-k} = \sum_{j=0}^{\infty} \binom{j+k-1}{k-1} q^j. \tag{7.109}$$

Proof   We demonstrate this by induction, but first we must confirm that the series on the right of (7.109) actually converges. Defining $a_j = \binom{j+k-1}{k-1} q^j$, we derive the absolute value of the ratio of successive terms as

$$\left| \frac{a_{j+1}}{a_j} \right| = \frac{j+k}{j+1} |q|.$$

By the ratio test this series converges absolutely for $|q| < \lim \sup_{j \to \infty} \frac{j+1}{j+k} = 1$. As an absolutely convergent series, we are now able to manipulate the terms freely.

We use an induction proof, and first note that for $k = 1$, (7.109) reduces to $(1-q)^{-1} = \sum_{j=0}^{\infty} q^j$, which is easily derived as a geometric summation from chapter 6. Next assume that this formula is true for a given k, as well as for $k = 1$. Then we have that for $k + 1$,

$$(1-q)^{-k-1} = (1-q)^{-1}(1-q)^{-k}$$

$$= \sum_{i=0}^{\infty} q^i \sum_{j=0}^{\infty} \binom{j+k-1}{k-1} q^j$$

$$= \sum_{i=0}^{\infty} \sum_{j=0}^{\infty} \binom{j+k-1}{k-1} q^{j+i}$$

$$= \sum_{l=0}^{\infty} a_l q^l,$$

where the coefficient a_l is the sum of all the coefficients in the prior double sum for which $i + j = l$. So for given $l \geq 0$, $a_l = \sum_{j=0}^{l} \binom{j+k-1}{k-1}$, since for each such j there is a corresponding $i = l - j$. We finally need to show that $a_l = \binom{l+k}{k}$, as this is the appropriate coefficient in (7.109) for exponent $k + 1$. To do this, we apply (7.16) to each term with $j > 0$. We conclude that

$$\sum_{j=0}^{l}\binom{j+k-1}{k-1} = 1 + \sum_{j=1}^{l}\left[\binom{j+k}{k} - \binom{j+k-1}{k}\right]$$

$$= 1 + \sum_{j=1}^{l}\binom{j+k}{k} - \sum_{j=0}^{l-1}\binom{j+k}{k}$$

$$= \binom{l+k}{k},$$

as was to be proved.   ∎

Moments of the negative binomial are difficult to develop directly, as could be predicted from the length of the justification that $\sum_{j=0}^{\infty} f^{NB}(j) = 1$. However, like the geometric distribution, the moment-generating function is easily manageable using (7.109), as we now demonstrate.

By definition,

$$M^{NB}(t) = \sum_{j=0}^{\infty}\binom{j+k-1}{k-1}p^{k}(1-p)^{j}e^{jt}$$

$$= p^{k}\sum_{j=0}^{\infty}\binom{j+k-1}{k-1}[(1-p)e^{t}]^{j}.$$

Comparing the summation here with that in (7.109), we see that as long as $q \equiv (1-p)e^{t} < 1$, it must be the case that $\sum_{j=0}^{\infty}\binom{j+k-1}{k-1}[(1-p)e^{t}]^{j} = (1-q)^{-k}$. Combining, we obtain

$$M_{NB}(t) = \left(\frac{p}{1-(1-p)e^{t}}\right)^{k}. \tag{7.110}$$

Using this formula and (7.65) with the tools of chapter 9 produces the following results, with $q \equiv 1-p$:

$$\mu_{NB} = \frac{kq}{p}, \quad \sigma^{2}_{NB} = \frac{kq}{p^{2}}. \tag{7.111}$$

7.6.6   Poisson Distribution

The Poisson distribution is named for **Siméon-Denis Poisson** (1781–1840), who discovered its p.d.f. and its properties. This distribution is characterized by a single parameter $\lambda > 0$, and its p.d.f. is defined on the nonnegative integers by

$$f^P(j) = e^{-\lambda}\frac{\lambda^j}{j!}, \qquad j = 0, 1, 2, \ldots. \tag{7.112}$$

That $\sum_{j=0}^{\infty} f^P(j) = 1$ is an immediate application of (7.63), to be proved in chapter 9, since from that formula is produced $e^\lambda = \sum_{j=0}^{\infty} \frac{\lambda^j}{j!}$. Unfortunately, in order to develop other properties, we need to make an assumption of another result that will not be formally proved until chapter 9.

One important application of the Poisson distribution is that it provides a good approximation to the binomial distribution when the binomial parameter p is "small." Specifically, the binomial probabilities in (7.97) can be approximated by the Poisson probabilities above, with $\lambda = np$. Then for p small, and n large,

$$\binom{n}{j} p^j (1-p)^{n-j} \simeq e^{-np}\frac{(np)^j}{j!}. \tag{7.113}$$

This approximation was far more useful in pre-computer days, and comes from the result:

Proposition 7.51   *For $\lambda = np$ fixed, then as $n \to \infty$, binomial probabilities satisfy*

$$\binom{n}{j} p^j (1-p)^{n-j} \to e^{-\lambda}\frac{\lambda^j}{j!}. \tag{7.114}$$

In other words, as n increases and p decreases so that the product np is fixed and equal to λ, each of the probabilities of the binomial distribution will converge to the respective probabilities of the Poisson distribution.

Proof   First off,

$$\binom{n}{j} p^j (1-p)^{n-j} = \frac{n(n-1)\ldots(n-j+1)}{j!}\left(\frac{\lambda}{n}\right)^j\left(1-\frac{\lambda}{n}\right)^n\left(1-\frac{\lambda}{n}\right)^{-j}$$

$$= \frac{n(n-1)\ldots(n-j+1)}{n^j}\frac{\lambda^j}{j!}\left(1-\frac{\lambda}{n}\right)^n\left(1-\frac{\lambda}{n}\right)^{-j}.$$

Now the second term is fixed and independent of n, and the last is seen to converge to 1 as $n \to \infty$, as the exponent $-j$ is fixed. The first term equals the fixed product of j-terms $\prod_{k=0}^{j-1}\left(1 - \frac{k}{n}\right)$, and this product also converges to 1. The major subtlety here, and one we will not prove until chapter 9, is the result that for any real number λ, we have that $\left(1 - \frac{\lambda}{n}\right)^n \to e^{-\lambda}$ as $n \to \infty$. With that limit assumed, the proposition is proved.   ∎

Remark 7.52   *The requirement that p be small is typically understood as the condition that $p < 0.1$, or by symmetry, $p > 0.9$, while n large is understood as $n \geq 100$ or so.*

Another important property of the Poisson distribution is that it is the unique p.d.f. that characterizes **arrivals** during a given period of time under reasonable and frequently encountered assumptions. For example, the model might be one of auto-mobile arrivals at a stop light or toll booth, telephone calls to a switchboard, internet searches to a server, radio-active particles to a Geiger counter, insurance claims of any type (injuries, deaths, automobile accidents, etc.) from a large group of policy-holders, defaults from a large portfolio of loans or bonds, and so forth.

The required assumptions about such arrivals are that:

1. Arrivals in any interval of time are independent of arrivals in any other distinct interval of time.

2. For any interval of time of length $\frac{1}{n}$, measured in fixed units of time, the probability of one arrival is $\frac{\lambda}{n} + \frac{k_1}{n^2}$ as $n \to \infty$ for some constants λ and k_1.

3. The probability of two or more arrivals during any one of n intervals of time of length $\frac{1}{n}$ can be ignored as $n \to \infty$

We now show that under these conditions, if $f(j)$ denotes the probability of j arrivals during this unit interval of time, then with λ defined from assumption 2,

$$f(j) = f^P(j).$$

As will be seen below, the parameter λ in the Poisson p.d.f. equals μ_P and hence in this context the average number of arrivals during one unit of time.

The derivation begins by dividing the unit time interval into n-parts. Then $f(j) = f_1(j)$, where $f_1(j)$ denotes the probability of j-arrivals with at most one ar-rival in each subinterval, since by assumption 3 we can ignore in the limit the event that 2 or more arrivals occur in any subinterval. We then have that $f_1(j)$ is a general binomial probability because of the interval independence assumption in 1, and it equals the probability of one arrival in j-intervals, and none in $(n-j)$-intervals. This binomial probability is given in assumption 2. With the appropriate binomial coefficient we obtain

$$f_1(j) = \binom{n}{j} \left(\frac{\lambda}{n} + \frac{k_1}{n^2} \right)^j \left(1 - \frac{\lambda}{n} - \frac{k_1}{n^2} \right)^{n-j}.$$

Using the same approach as in proposition 7.51, we derive that $f_1(j) \to f^P(j)$ as $n \to \infty$. Here, however, we have that $p = \frac{\lambda}{n} + \frac{k_1}{n^2}$, so $np = \lambda + \frac{k_1}{n}$, and we require a generalized version of the above unproved fact that $\left(1 - \frac{\lambda}{n} - \frac{k_1}{n^2} \right)^n \to e^{-\lambda}$. In other words, the probability adjustment of $\frac{k_1}{n^2}$ is irrelevant in this limit as will be demonstrated in chapter 9.

Remark 7.53 *In many applications λ is defined as the average number of arrivals in a unit of time such as a minute, a month, or a year, depending on the application, and then the appropriate parameter for a period of length T-units of time is $\lambda' = \lambda T$ for any T.*

Turning next to expectations, we note that the moment-generating function is somewhat easier to derive than are the mean and variance. Specifically, $M_P(t) = e^{-\lambda} \sum_{j=0}^{\infty} \frac{\lambda^j}{j!} e^{jt} = e^{-\lambda} \sum_{j=0}^{\infty} \frac{(\lambda e^t)^j}{j!}$, where this summation is recognizable from (7.63) as $e^{\lambda e^t}$. Consequently we obtain

$$M_P(t) = e^{\lambda(e^t-1)}. \tag{7.115}$$

The mean and variance of the Poisson can then be derived from the m.g.f. or by a direct method assigned in exercise 16:

$$\mu_P = \lambda, \quad \sigma_P^2 = \lambda. \tag{7.116}$$

7.7 Generating Random Samples

In certain contexts random samples are observed, such as the daily market close prices, the periodic returns of a given security or investment index, the weekly rainfall in a given forest, the height measurements of girls upon their fourteenth birthday, or the number of hits on a Geiger counter in 30 seconds, or the number of bond defaults in a year, or the proportion of males just turning 65 years of age that will survive one year. Indeed the world is full of observations that can be construed by the observer as representing random sample points from an unknown probability distribution. The mathematical discipline of **statistics** concerns itself with the collection of such data, as well as the analysis and interpretation of these data.

On the other hand, past observations, often with a healthy dose of intuition and sometimes mathematical convenience, can lead one to assume that a given random variable of interest is in fact governed by a given probability density function. For example, an individual bond default or death could logically be assumed to be modeled by a standard binomial distribution, or the number of bond defaults or deaths perhaps modeled by a general binomial distribution or a Poisson approximation, while the average of many collections of observed random variables may be assumed to be normally distributed (see chapter 8). Such distributional assumptions can then be "calibrated" to observed data by choosing the distribution's parameters appropriately, or calibrated to characteristics assumed to hold in the future.

Once such a transition is made, from observing a random variable to assuming that the given random variable is governed by a given p.d.f., it is possible in theory to generate additional samples that can be studied. Such generated samples are used for insights that may not be possible based on observable data, often due to the sparseness of the observations or because one assumes that the p.d.f.s parameters in the future will differ from those underlying past observations.

For example, Chebyshev's inequality discussed in the next chapter assures that it is very unlikely to observe a random variable that is far from its mean when measured in units of standard deviations, but for many applications in finance, it is exactly the extreme events that are of most interest in the modeling. As another example, a market model calibrated during a bear market would need to have parameters modified to be applicable in a bull market.

So while the assumed p.d.f. has the potential to provide all the details on such extreme and other events neither observed nor perhaps observable, it does so with the inherent risk to the investigator that in most applications, such a p.d.f., is, after all, only an assumption. Nature almost never truly reveals underlying p.d.f.s nor promises to keep the parameters in any p.d.f.s constant. Nature doesn't even commit to using p.d.f.s, but in practice, it is convenient to assume such a commitment has been made, and to be mindful of the inherent risks of such an assumption.

That said, the purpose of this section is to present a very handy result with immediate application to the generation of random samples of the values of any random variable, given an assumed probability density function. First a definition.

Definition 7.54   *A collection $\{r_j\}_{j=1}^{n} \subset [0,1]$ is a **uniformly distributed random sample** if:*

1. *For any subinterval $\langle a, b \rangle \subset [0,1]$, where $\langle \ \rangle$ is intended to mean open, closed or mixed, $\Pr[r_j \in \langle a, b \rangle] = b - a$.*

2. *For any collection of subintervals $\{\langle a_j, b_j \rangle\}_{j=1}^{n}$, $\langle a_j, b_j \rangle \subset [0,1]$ for all j,*

$$\Pr[r_j \in \langle a_j, b_j \rangle \text{ for all } j] = \prod_{j=1}^{n} (b_j - a_j). \qquad (7.117)$$

It should be noted that part 1 of definition 7.54 implies that for any given $a \in [0, 1]$, $\Pr[r = a] = 0$. The term "uniform" means that the probabilities governing the location of each r_j value are proportional to the length of the interval in which such a value is sought. In addition the use of "random" is identical with that given in (7.9), where the probability of a joint event equals the product of the probabilities of the individual events. This is the essence of (7.117).

Remark 7.55 *This model can be imagined as the limiting situation for the discrete uniform p.d.f. as $n \to \infty$. This is because as $n \to \infty$, while the probabilities of individual points decrease to 0 under the discrete uniform p.d.f., the total probability of $r \in \langle a, b \rangle$ approaches $b - a$. In theory, however, the notion of uniformly distributed random sample is intended as a notion of continuous probability theory, as was the case noted above for the normal distribution. But, in practice, there is little difference between the uniform distribution above and the discrete uniform distribution for n large. Indeed all computers work in finite decimal (or binary) point precision, so in a given application, they are incapable of distinguishing x from $x + 10^{-m}$ for $m \geq M$, where M is generally about 16 or so. So with $n \geq 10^M$, the discrete uniform and continuous uniform are identical to your computer.*

The result in this section is simply that if $\{r_j\}_{j=1}^{n} \subset [0, 1]$ is a uniformly distributed random sample, then $\{F^{-1}(r_j)\}_{j=1}^{n} = \{X_j\}_{j=1}^{n}$ will be a random sample of the random variable X. In other words, $\{X_j\}_{j=1}^{n}$ are independent, identically distributed random variables in the sense of (7.34). So the problem of generating a random sample for any discrete random variable can be reduced to the problem of generating a uniformly distributed random sample from the interval $[0, 1]$, which is a problem that is solved in virtually any mathematical or calculation software.

The **inverse distribution function of a discrete random variable**, $F^{-1}(r)$, is defined:

Definition 7.56 *Let X be a random variable defined on a discrete sample space $\mathcal{S}$ with range $\{x_j\}$ and cumulative distribution function $F(x)$. Then for $r \in \mathbb{R}$,*

$$F^{-1}(r) = \min\{x_j \mid r \leq F(x_j)\}. \tag{7.118}$$

Example 7.57 *For simplicity, let X denote the binomial random variable,*

$$f(x) = \begin{cases} 0.25, & x = 0, \\ 0.75, & x = 1, \end{cases}$$

with distribution function

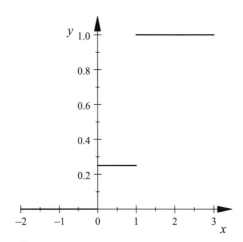

Figure 7.2
Binomial c.d.f.

$$F(x) = \begin{cases} 0, & x < 0, \\ 0.25, & 0 \le x < 1, \\ 1.0, & 1 \le x. \end{cases}$$

*The graph of $F(x)$ is seen in **figure 7.2**.*

From (7.118) the inverse distribution function is defined as

$$F^{-1}(r) = \begin{cases} 0, & 0 \le r \le 0.25, \\ 1, & 0.25 < r \le 1.0. \end{cases}$$

So, if $\{r_j\}_{j=1}^n$ is a uniformly distributed sample from the interval $[0,1]$, then for any r_j, $\Pr[r_j \in [0, 0.25]] = 0.25$, and hence $\Pr[F^{-1}(r_j) = 0] = 0.25$. Similarly $\Pr[r_j \in (0.25, 1.0]] = \Pr[r_j \in [0.25, 1.0]] = 0.75$. Hence $\Pr[F^{-1}(r_j) = 1] = 0.75$.

The proof of a simpler version of the general statement follows identically with this example, and is presented for completeness. By "simpler" is meant that we assume that the range of the random variable, which equals the domain of the probability density function, is **sparse**, meaning it has no accumulation points. The general statement and proof will then follow.

Proposition 7.58 *Let X be a discrete random variable on a sample space S with sparse range $\{x_j\}$ and distribution function $F(x)$. Then, if $\{r_j\}_{j=1}^n \subset [0,1]$ is a uniformly distributed random sample, $\{F^{-1}(r_j)\}_{j=1}^n$ is a random sample of X in the sense of (7.34).*

Proof If the collection $\{x_j\}$ is sparse and hence has no accumulation points, then enumerating in increasing order, we have that for any $r_j \neq 0$ there is a unique x_k so that $r_j \in (F(x_k), F(x_{k+1})]$. Since $\Pr[r_j = 0] = 0$, we ignore this case. Now, since $F^{-1}(r_j) = x_{k+1}$, we have that by the definition of uniformly distributed sample,

$$\Pr[F^{-1}(r_j) = x_{k+1}] = \Pr[r_j \in (F(x_k), F(x_{k+1})]] = F(x_{k+1}) - F(x_k) = f(x_{k+1}).$$

In other words, through F^{-1}, a uniformly distributed sample is transformed into a collection of outcomes of X with the correct probabilities. To demonstrate independence of $\{F^{-1}(r_j)\}_{j=1}^n$, let any collection $\{x_{k_j}\}_{j=1}^n$ be given; then

$$f(x_{k_1}, x_{k_2}, \ldots, x_{k_n}) = \Pr[F^{-1}(r_1) = x_{k_1}, F^{-1}(r_2) = x_{k_2}, \ldots, F^{-1}(r_n) = x_{k_n}]$$

$$= \Pr[r_1 \in (F(x_{k_1-1}), F(x_{k_1})], \ldots, r_n \in (F(x_{k_n-1}), F(x_{k_n})]]$$

$$= \prod_{j=1}^n [F(x_{k_j}) - F(x_{k_j-1})]$$

$$= \prod_{j=1}^n f(x_{k_j}),$$

where the third equality comes from the definition of $\{r_j\}_{j=1}^n$ as a uniformly distributed random sample. ∎

Example 7.59 *To generate a random sample of Poisson variables with $\lambda = 2$, we first calculate the appropriate half-open intervals for the r-values. Let $F(n) = \sum_{j=0}^n e^{-2} \frac{2^j}{j!}$ for $n = 0, 1, 2, \ldots$ and define the associated half-open intervals: $I_n = (F(n-1), F(n)]$, for $n = 0, 1, 2, \ldots$, where we note that $F(-1) = 0$ by definition. Then the length of I_n is given by $|I_n| = F(n) - F(n-1) = f(n) \equiv e^{-2} \frac{2^n}{n!}$, and it is clear that $\sum_{j=0}^\infty |I_j| = \sum_{j=0}^\infty f(n) = 1$. For any collection $\{r_j\}_{j=1}^n \subset [0, 1]$ generated using common software such as* Rand( ) *in Excel, the random sample of Poisson variables $\{F^{-1}(r_j)\}_{j=1}^n$ are defined by*

$$F^{-1}(r_j) = n \qquad if \ r_j \in I_n.$$

Note that if the range of the random variable $\{x_j\}$ has accumulation points, the proof above becomes compromised. For example, imagine a discrete random variable with range equal to the rational numbers in $[0, 1]$, ordered in some way. In this case $F(x)$ is well defined as in (7.22), but it is no longer true that $\{x_j\}$ can be enumerated in increasing order, nor is it true that for any $r_j \neq 0$ there is a unique x_k so that $r_j \in (F(x_k), F(x_{k+1})]$. The implication of this observation is not that the conclusion of the proposition above is false in this case, but that a somewhat more subtle argument is needed to demonstrate its truth.

Proposition 7.60   *Let X be a discrete random variable on a sample space S, with range $\{x_k\}$, and distribution function $F(x)$. Then, if $\{r_j\}_{j=1}^n \subset [0, 1]$ is a uniformly distributed random sample, $\{F^{-1}(r_j)\}_{j=1}^n$ is a random sample of X in the sense of (7.34).*

Proof   Let x_k be given. As above, our first goal is to show that $\Pr[F^{-1}(r_j) = x_k] = f(x_k)$. Consider the half-open interval about x_k, defined by $I_n = \left(x_k - \frac{1}{n}, x_k + \frac{1}{n}\right]$. Now, by (7.118), for $r \in [0, 1]$, $F^{-1}(r) \in I_n$ if and only if $x_k - \frac{1}{n} < \min\{x_j \mid r \le F(x_j)\} \le x_k + \frac{1}{n}$. That is, $F^{-1}(r) \in I_n$ if and only if,

$$r \in \left[F\left(x_k - \frac{1}{n}\right), F\left(x_k + \frac{1}{n}\right) \right].$$

So by the definition of uniformly distributed sample,

$$\Pr[F^{-1}(r) \in I_n] = F\left(x_k + \frac{1}{n}\right) - F\left(x_k - \frac{1}{n}\right)$$

$$= \sum_{x_k - 1/n < x_j \le x_k + 1/n} f(x_j).$$

Finally, as $n \to \infty$, $\Pr[F^{-1}(r) \in I_n] \to \Pr[F^{-1}(r) = x_k]$, and the summation above reduces to $f(x_k)$, demonstrating that $\Pr[F^{-1}(r_j) = x_k] = f(x_k)$. To demonstrate independence of $\{F^{-1}(r_j)\}_{j=1}^m$, let any collection $\{x_{k_j}\}_{j=1}^m$ be given, and define I_{n_j} as above for each x_{k_j}. Then

$$f(x_{k_1}, x_{k_2}, \ldots, x_{k_m}) = \Pr[F^{-1}(r_j) \in I_{n_j} \text{ for all } j]$$

$$= \Pr\left[r_j \in \left(F\left(x_{k_j} - \frac{1}{n}\right), F\left(x_{k_j} + \frac{1}{n}\right) \right) \text{ for all } j \right]$$

$$= \prod_{j=1}^m \left[F\left(x_{k_j} + \frac{1}{n}\right) - F\left(x_{k_j} - \frac{1}{n}\right) \right]$$

$$= \prod_{j=1}^m \left[\sum_{x_{k_j} - 1/n < x_l \le x_{k_j} + 1/n} f(x_l) \right],$$

where the second equality comes from the definition of uniformly distributed sample. Finally, letting $n \to \infty$, we obtain

$$f(x_{k_1}, x_{k_2}, \ldots, x_{k_m}) = \prod_{j=1}^{m} f(x_{k_j}).$$ ∎

7.8   Applications to Finance

7.8.1   Loan Portfolio Defaults and Losses

An example of a combination coin-flip and urn problem in finance that is typically implemented with computer algorithms is bond or loan default and loss modeling. Imagine a portfolio of n bonds, all with the same credit rating, say *Baa/BBB*. Assume that the event of a default in a given year is generated by an H-flip of a biased coin that produces heads, on average, in 75 of 1000 flips. We toss this coin n times, and record the number of heads; say it is n_H. We then go to the portfolio urn of bonds and select n_H bonds without replacement. These are the defaulted bonds in this trial, and the total par defaulted is denoted F_H. From this defaulted portfolio, losses can be modeled in terms of a fixed loss of λF_H, with $0 \leq \lambda \leq 1$ denoting a fixed loss ratio, or a **loss given default (LGD)** model can be implemented whereby losses vary according to some probability density function.

A simple example of how one might generate nonconstant losses is that with each defaulted bond, a die is rolled. So one dot represents a loss of $\frac{1}{6}$, or about 16.6% of par, and so forth to a roll of 6 dots, which represents a loss of 100%. More realistically one could use a variety of probability density functions calibrated to historic data, or create a sample space of losses constructed directly from all past defaults. In the former case, random losses are produced by first generating a uniform random sample on $[0, 1]$, then applying the approach of the last section. In the latter model this historic loss collection would be another urn containing the collection of LGDs experienced historically in percentage terms. Then with each bond selected from the original urn for default, an LGD is drawn from this second urn to determine the associated loss. This second urn would logically be sampled with replacement.

Individual Loss Model

More formally, let f_{jk} denote the loan amount of the jth bond or loan, in risk class k. Risk classes might be defined in terms of credit ratings for bonds or internal risk assessment criteria for other loans. For each risk class define the default random variable D_{jk} for this loan as a binomial with

$$D_{jk} = \begin{cases} 1, & \Pr = q_k, \\ 0, & \Pr = 1 - q_k. \end{cases}$$

Here the probability of default in the period is denoted q_k. The probability of continued payments during the period is $p_k = 1 - q_k$.

Note that the random variable D_{jk} will typically not depend on the loan other than through the risk class, and that the notational device of the subscript j is simply to confirm that each of the loans in the same risk class will have separate and independent coin flips. Notational use of D_k might suggest that all loans in a class default or do not default together, an unrealistic assumption.

Finally, the loss given default random variable, or **loss ratio**, for each loan, denoted L_{jk}, is again defined only by risk class as a random variable with range in the interval $[0, 1]$ and the same notational convention as for D_{jk}. Sometimes the **loan recovery** R_{jk} is modeled, representing the relative amount recovered from a borrower on default. Of course, $L_{jk} = 1 - R_{jk}$.

Total losses can now be notationally represented as the random variable that is given by the **individual loss model**:

$$L = \sum_{j,k} f_{jk} D_{jk} L_{jk}. \tag{7.119}$$

For each loan the random variable D_{jk} is generated, and for each loan for which $D_{jk} = 1$, the random variable L_{jk} is generated. Both random variables can be generated with the methodology of section 7.7 using uniformly distributed random samples from $[0, 1]$, since we want the collection $\{D_{jk}\}$ to be independent random variables, as well as the collection $\{L_{jk} \mid D_{jk} = 1\}$. Of course, the collections $\{D_{jk}\}$ and $\{L_{jk}\}$ cannot be independent, since $D_{jk} = 0$ implies that $L_{jk} = 0$, although there is no harm, other than with respect to wasted computational time, of generating $\{L_{jk}\}$ for all combinations of jk, and generating these as independent random variables that are also independent of the collection $\{D_{jk}\}$.

Specifically, for fixed k we generate $\{r_{jk}\} \subset [0, 1]$, one for each bond, and denoting by B_k the c.d.f. of the binomial with parameter q_k, we have

$$D_{jk} \equiv B_k^{-1}(r_{jk}) = \begin{cases} 1, & r_{jk} \leq q_k, \\ 0, & r_{jk} > q_k. \end{cases}$$

Similarly, from another collection, $\{r'_{jk}\} \subset [0, 1]$, one for each default, losses are generated using the assumed loss c.d.f. for each risk class. In other words, $L_{jk} = F_k^{-1}(r'_{jk})$, where F_k is the cumulative loss given default distribution function for risk class k.

Of course, such values need only be generated for those jk combinations for which $B_k^{-1}(r_{jk}) = 1$.

From this model one can calculate the mean and variance of losses using the conditional expectation formulas in (7.43) and (7.46). For example, conditioning on the random variable D_{jk} obtains

$$E[L] = \sum_{j,k} E[f_{jk} D_{jk} L_{jk}]$$

$$= \sum_{j,k} E[E[f_{jk} D_{jk} L_{jk} \mid D_{jk}]].$$

Now $E[f_{jk} D_{jk} L_{jk} \mid D_{jk} = 0] = 0$ and $E[f_{jk} D_{jk} L_{jk} \mid D_{jk} = 1] = f_{jk} E[L_{jk}] = f_{jk} E[L_k]$, since f_{jk} is a constant and the loss distribution depends only on the risk class. Said another way, for each risk class, $\{L_{jk}\}$ are independent and identically distributed with c.d.f. F_k. These "inner" conditional expectations can be expressed conveniently by

$$E[f_{jk} D_{jk} L_{jk} \mid D_{jk}] = f_{jk} E[L_k] D_{jk}.$$

Consequently, since $E[D_{jk}] = q_k$,

$$E[L] = \sum_{j,k} q_k f_{jk} E[L_k] \tag{7.120a}$$

$$= \sum_{k} q_k f_k E[L_k], \tag{7.120b}$$

where $f_k = \sum_j f_{jk}$, the total loan amount in this risk class.

Variance is similarly calculated with a conditioning approach. With the assumptions above regarding the generation of the collections $\{D_{jk}\}$ and $\{L_{jk}\}$, the losses on bonds are independent random variables. So the variance of the sum is the sum of the variances, and each of these variances is calculated by conditioning. In other words,

$$Var[L] = \sum_{j,k} Var[E[f_{jk} D_{jk} L_{jk} \mid D_{jk}]] + \sum_{j,k} E[Var[f_{jk} D_{jk} L_{jk} \mid D_{jk}]].$$

Now from the conditional expectation for $E[f_{jk} D_{jk} L_{jk} \mid D_{jk}]$ above, and $Var[D_{jk}] = q_k(1 - q_k)$, we get

$$Var[E[f_{jk} D_{jk} L_{jk} \mid D_{jk}]] = q_k(1 - q_k) f_{jk}^2 E[L_k]^2.$$

Next, from $\mathrm{Var}[f_{jk}D_{jk}L_{jk} \mid D_{jk}=0]=0$ and $\mathrm{Var}[f_{jk}D_{jk}L_{jk} \mid D_{jk}=1]=f_{jk}^2 \, \mathrm{Var}[L_k]$, we conclude that $\mathrm{Var}[f_{jk}D_{jk}L_{jk} \mid D_{jk}]=f_{jk}^2 \, \mathrm{Var}[L_k]D_{jk}$. Hence

$$\mathrm{E}[\mathrm{Var}[f_{jk}D_{jk}L_{jk} \mid D_{jk}]]=q_k f_{jk}^2 \, \mathrm{Var}[L_k].$$

Combining, we have

$$\mathrm{Var}[L]=\sum_{j,k} q_k(1-q_k)f_{jk}^2 \mathrm{E}[L_k]^2 + \sum_{j,k} q_k f_{jk}^2 \, \mathrm{Var}[L_k] \tag{7.121a}$$

$$=\sum_{k} q_k(1-q_k)f_k^{(2)}\mathrm{E}[L_k]^2 + \sum_{k} q_k f_k^{(2)} \, \mathrm{Var}[L_k]. \tag{7.121b}$$

Here we define $f_k^{(2)}=\sum_j f_{jk}^2$, which of course is not the same as f_k^2 for f_k defined above in (7.120b). These formulas can be rewritten if desired using $\mathrm{Var}[L_k]=\mathrm{E}[L_k^2]-\mathrm{E}[L_k]^2$.

Aggregate Loss Model

If the loan amounts in each risk class are similar and narrowly distributed, loan losses can also be modeled in what is called an **aggregate loss model**, or **collective loss model**. In each risk class, say class k, the collection of actual loan amounts $\{f_{jk}\}$, which contains n_k loans, is modeled as a portfolio of n_k loans of the same amount given by the average $\bar{f}_k \equiv \frac{1}{n_k}\sum_j f_{jk}$. Total losses can now be expressed as

$$L=\sum_{j,k} \bar{f}_k D_{jk}L_{jk}=\sum_k \bar{f}_k \sum_j D_{jk}L_{jk}.$$

Note that for each k, $N_k \equiv \sum_j D_{jk}$ is a random variable with a binomial distribution with parameters n_k and q_k, and by (7.99) we have $\mathrm{E}[N_k]=n_k q_k$ and $\mathrm{Var}[N_k]=n_k q_k(1-q_k)$. Also $\sum_j D_{jk}L_{jk}$ can be rewritten as

$$\sum_j D_{jk}L_{jk}=\sum_{D_{jk}\neq 0} L_{jk}$$

$$=N_k L_k'.$$

Here the random variable L_k' is given by

$$L_k'=\frac{1}{N_k}\sum_{j=1}^{N_k} L_{jk} \qquad \text{for } 1 \leq N_k \leq n_k,$$

and defined as the average loss ratio for class k, conditional on $N_k \geq 1$. In other words, L'_k is the average loss ratio conditional on there being a loss. This is consistent with the definition of L_{jk}, which is the loss ratio on loan j in class k given that a loss occurred, meaning that $D_{jk} = 1$.

Combining results, we have that in the case of loan amounts that are narrowly distributed by risk class, the individual loss model can be rewritten as an **aggregate loss model**:

$$L = \sum_k \bar{f}_k N_k L'_k. \tag{7.122}$$

Here N_k has binomial distribution with parameters n_k and q_k, $\bar{f}_k$ is the average loan amount in class k, and L'_k is a random variable equal to the average loss ratio in class k, conditional on $N_k \geq 1$.

We will now see that not surprisingly, L'_k has the same expected value as does the individual loss ratio random variable for class k, which is denoted L_k in the individual loss model above. On the other hand, L'_k will have a smaller variance than L_k, intuitively because L'_k is defined in terms of averages of the original $\{L_{jk}\}$, whereas L_k reflects no averaging.

First, $\mathrm{E}[L'_k]$ can be evaluated using the conditioning argument of (7.43), where subscripts are put on the expectation operators for clarity:

$$\mathrm{E}[L'_k] = \mathrm{E}_N\left[\mathrm{E}_L\left[\frac{1}{N_k}\sum_{j=1}^{N_k} L_{jk}\,\middle|\,N_k = n \geq 1\right]\right]$$

$$= \mathrm{E}_N\left[\frac{1}{n}\sum_{j=1}^{n} \mathrm{E}[L_k]\,\middle|\,n \geq 1\right]$$

$$= \mathrm{E}[L_k]\mathrm{E}_N[1\,|\,n \geq 1]$$

$$= \mathrm{E}[L_k]\sum_{n=1}^{n_k} \mathrm{Pr}[N_k = n\,|\,n \geq 1].$$

For the last step, it must be remembered that N_k is a binomial random variable, but conditional on the restriction that $N_k \geq 1$. So here $\mathrm{Pr}[N_k = n\,|\,n \geq 1] = \frac{\mathrm{Pr}[N_k=n]}{1-\mathrm{Pr}[N_k=0]}$, and so $\sum_{n=1}^{n_k} \mathrm{Pr}[N_k = n\,|\,n \geq 1] = 1$. Consequently

$$\mathrm{E}[L'_k] = \mathrm{E}[L_k]. \tag{7.123}$$

Next, while L_k' is a random variable with the same mean as L_k for each k fixed, it has a smaller variance. This is again derived from a conditioning argument in (7.46) as follows: From the mean calculation above, we have that $\mathrm{E}_L\left[\frac{1}{N_k}\sum_{j=1}^{N_k} L_{jk} \mid N_k = n \geq 1\right] = \mathrm{E}[L_k]$, a constant. Consequently the variance of this conditional expectation is 0.

On the other hand, for the conditional variance,

$$\mathrm{Var}\left[\frac{1}{N_k}\sum_{j=1}^{N_k} L_{jk} \middle| N_k = n \geq 1\right] = \mathrm{Var}\left[\frac{1}{n}\sum_{j=1}^{n} L_{jk}\right]$$

$$= \frac{1}{n^2}\sum_{j=1}^{n}\mathrm{Var}[L_{jk}]$$

$$= \frac{1}{n}\,\mathrm{Var}[L_k].$$

Combining and evaluating the expectation of this conditional variance obtains

$$\mathrm{Var}[L_k'] = \mathrm{Var}[L_k]\mathrm{E}_N\left[\frac{1}{N_k}\middle| N_k \geq 1\right], \tag{7.124}$$

where N_k has binomial distribution with parameters n_k and q_k, but conditional on $N_k \geq 1$. Apparently $\mathrm{E}\left[\frac{1}{N_k} \mid N_k \geq 1\right] < 1$, since by the conditional binomial probabilities, $\Pr[N_k = n \mid n \geq 1] = \frac{\Pr[N_k=n]}{1-\Pr[N_k=0]}$:

$$\mathrm{E}\left[\frac{1}{N_k}\middle| N_k \geq 1\right] = \sum_{n=1}^{n_k}\frac{1}{n}\binom{n_k}{n}\frac{q_k^n(1-q_k)^{n_k-n}}{1-(1-q_k)^{n_k}}.$$

So $\mathrm{Var}[L_k'] < \mathrm{Var}[L_k]$, since the summation is a weighted average $\sum_{n=1}^{n_k}\frac{1}{n}w_n$ where $\sum_{n=1}^{n_k} w_n = 1$.

The random variable N_k can also be modeled as Poisson for, in general, the associated q_k are quite small and satisfy in all but the most extreme cases the condition $q_k \leq 0.1$. In this case the Poisson parameter for risk class k is given by $\lambda_k = n_k q_k$, and the $\mathrm{E}\left[\frac{1}{N_k} \mid N_k \geq 1\right]$ calculated accordingly.

The mean and variance of L within the aggregate loss model can again be developed using the conditioning arguments. In exercise 17 is assigned the derivation of the following formulas, where $\mathrm{E}[L_k']$ is used for notational consistency, but recall from above that $\mathrm{E}[L_k'] = \mathrm{E}[L_k]$:

$$E[L] = \sum_k \bar{f}_k E[N_k] E[L'_k], \tag{7.125}$$

$$\mathrm{Var}[L] = \sum_k \bar{f}_k^2 E[L'_k]^2 \, \mathrm{Var}[N_k] + \sum_k \bar{f}_k^2 E[N_k^2] \, \mathrm{Var}[L'_k]. \tag{7.126}$$

Note that in these formulas, $E[N_k]$ and $\mathrm{Var}[N_k]$ reflect the whole distribution of N_k, and not the conditional distribution reflecting $N_k \geq 1$ as was used for the L'_k moments. Specifically, $E[N_k] = n_k q_k$ whether N_k is modeled as binomial or Poisson, since for the latter, $\lambda_k = n_k q_k$. However, $\mathrm{Var}[N_k]$ will equal $n_k q_k(1 - q_k)$ with the binomial, and $n_k q_k$ with the Poisson approximation. Also, while $E[L'_k]$ can be calculated directly from the assumed distribution for L_{jk} with k fixed, as was derived above, $\mathrm{Var}[L'_k]$ will be smaller than $\mathrm{Var}[L_k]$, due to the multiplicative factor of $E\left[\frac{1}{N_k} \mid N_k \geq 1\right]$.

As was the case for the individual loss model, the random variable L can be simulated using (7.122) and the approach in section 7.7 above to generating random samples from a distribution function. In this formula, for instance, N_k has binomial distribution with parameters n_k and q_k, or a Poisson distribution with $\lambda_k = n_k q_k$. In either case each simulation for class k involves first generating one uniformly distributed random variable $r \in [0,1]$ from which $N_k \equiv F_N^{-1}(r)$, with $F_N(x)$ denoting the cumulative distribution for N_k. Then, if $N_k > 0$, another N_k uniformly distributed variables are generated, $\{r_j\}_{j=1}^{N_k}$, from which loss ratios are defined by $\{L_{jk}\}_{j=1}^{N_k} = \{F_{L_k}^{-1}(r_j)\}_{j=1}^{N_k}$, with $F_{L_k}(x)$ the cumulative distribution function for L_k. The average loss ratio is then $L'_k = \frac{1}{N_k}\sum_{j=1}^{N_k} L_{jk}$. Each simulation then proceeds the same way.

7.8.2 Insurance Loss Models

With only a change in the definitions of the random variables, the individual and aggregate loss models can be used in a wide variety of insurance claims applications. For example, within a life insurance claims context, risk classes would typically be defined at least by age groups, with gender and/or insurance "ratings" classes not uncommon. Life insurance ratings are analogous to credit ratings on loans, only that here the goal is to identify individuals relative to mortality risk rather than the risk of default. Consequently, in this application, q_k is the probability of death in a period, often in a year, and f_{jk} denotes the life insurance policy "face amount" on a "net amount at risk" basis payable on death.

This so-called net amount at risk is an adjustment to the policy face amount that reflects the fact that for many insurance contracts, particularly those with level premiums paid by the insured, the insurer holds "reserves" backed by accumulated excess premiums. In many policies, while the net exposure amounts vary year to year, they are not random variables *per se*. In other words, there is no need for the random variable L_{jk} in a traditional life insurance model, since the "loss" on death is known in advance. When $D_{jk} = 1$, the entire net policy amount is paid, and hence $L_{jk} = 1$ as well.

It is also of interest to apply these models, and those below, over a multiple-year modeling horizon, changing model parameters each year. In such an application a random sequence is generated, and this is a special case of a stochastic process. Also note that with a multiple-year model, the present value of all losses could be modeled as a random variable, by introducing appropriate interest rates for discounting. These future interest rates could be modeled as fixed or as random variables.

For life insurance policies for which the death benefit is not fixed, such as is the case with variable life insurance, L_{jk} is once again a random variable. But in this application L_{jk} is a "multiplier" applied to the original policy face amount, and it usually reflects the performance in the financial markets, often with a minimum guarantee. It is also natural to allow $L_{jk} > 1$ in this model to accommodate favorable market environments.

These loss models also apply to various types of insurance policies for which the benefits are not fixed. For example, with a disability insurance policy, q_k would be the probability of disability in a period, again often a year, and the claim paid, symbolically $f_{jk}L_{jk}$, would be based on a probability distribution that reflects both past insurer claims patterns and trends, as well as amount limitations defined in the policy. It would be common practice to model the value of the claim as a present value of expected payments over the expected disability period. Specifically, f_{jk} could be modeled as the present value of the maximum claim allowed by the policy, and L_{jk} a loss ratio, $0 < L_{jk} \leq 1$, in the sense above.

Various types of health insurance benefits could be handled similarly, as could various benefits payable under property and casualty insurance policies, which include automobile insurance and home-owners or renters insurance.

7.8.3   Insurance Net Premium Calculations

Generalized Geometric and Related Distributions
Recall that the geometric density in (7.101) defined by $f^G(j) = p(1-p)^j$, $j = 0, 1, 2, \ldots$, provided the probability that j-Ts precede the first H in a sequence of binomial trials with $\Pr[H] = p$. The negative binomial distribution generalized this definition in that a new parameter k is introduced, and then $f^{NB}(j)$ represented the probability that j-Ts precede the kth-H in a sequence of binomial trials with $\Pr[H] = p$.

Another way of generalizing the geometric distribution is to allow the probability of a head to vary with the sequential number of the coin flip. Specifically, if $\Pr[H \mid j\text{th flip}] = p_j$, then with a simplifying change in notation to exclude the case $j = 0$, a **generalized geometric distribution** can be defined by the p.d.f.

$$f^{GG}(j) = p_j \prod_{k=1}^{j-1}(1 - p_k), \qquad j = 1, 2, 3, \ldots, \tag{7.127}$$

where $f^{GG}(j)$ is the probability of the first head appearing on flip j. By convention, when $j = 1$, $\prod_{k=1}^{0}(1 - p_k) = 1$.

Of course, as was demonstrated above, if $p_k = p > 0$ for all k, then $f^G(j)$ is indeed a p.d.f. in that $\sum_{j=0}^{\infty} p(1 - p)^j = 1$. With nonconstant probabilities, this conclusion is true but not obvious. Note, however, that if $0 < a \le p_k \le b < 1$ for all j, then the summation is finite, since $f^{GG}(j) < b(1 - a)^{j-1}$ and $\sum_{j=1}^{\infty} f^{GG}(j) < \frac{b}{a}$ by a geometric series summation.

In addition, letting $c_0 = 1$ and $c_j = \prod_{k=1}^{j}(1 - p_k)$ for $j \ge 1$, we have that

$$f^{GG}(j) = c_{j-1} - c_j,$$

and $\sum_{j=0}^{\infty} c_j \le \sum_{j=0}^{\infty}(1 - a)^j = \frac{1}{a}$. Consequently the alternating series version of this absolutely convergent series can be rearranged, producing

$$\sum_{j=1}^{\infty} f^{GG}(j) = \sum_{j=1}^{\infty}(c_{j-1} - c_j)$$

$$= c_0 + \sum_{j=1}^{\infty}(c_j - c_j)$$

$$= 1.$$

This probability density function is the essence of a **survival model**, although the notation switches from

$$p_k = \Pr[H \text{ on } k\text{th flip} \mid \text{all } T\text{s before } k\text{th flip}]$$

to

$$q_k = \Pr[\text{death in year } k \mid \text{survival for first } k - 1 \text{ years}],$$

where as expected, year 1 extends from time $t = 0$ to $t = 1$, and so forth. Consequently this conditional probability can also be expressed $q_k = \Pr[\text{death by time } k \mid$

alive at time $k - 1$]. In practice, for the group being modeled, $\{q_k\} \equiv \{q_{x+k-1}\}$, where $x =$ current age, and the standard actuarial notation: $q_{x+k-1} = \Pr\{\text{person age } x + k - 1 \text{ will die within one year}\}$. However, to simplify notation, we generally avoid the age-based notation unless it is needed for emphasis.

This definition of q_k may appear odd in that a natural reaction to the condition "alive at time $k - 1$" might well be "of course, they are alive at the beginning of the year, as that is what it means to die in year j!" But this obvious point is not the purpose of the condition.

Population census data as well as various insurance company and pension plan data usually present the probabilities of death on the basis of the q_k model definition. That is, among a group of individuals with comparable mortality risk that are alive at a point in time, say grouped by age x, what is the proportion that will die during the period? So $\hat{q}_x$, based on the sample, is just the ratio of those that die during the period to those alive and age x at the start of the period. From such studies various statistical methods are used to develop estimates of the underlying probabilities of death during the period by risk class, and this is denoted q_x for the various ages or otherwise defined risk classes.

The question of interest now, and one that a survival p.d.f. is intended to answer is, of an individual member of a group alive at time 0, say aged x, what is the probability of death in year k for $k = 1, 2, 3 \ldots$? The answer to this question is not a q_{x+k-1}-value, as this is only the probability of a death in a year k given survival to the beginning of that year. The necessary adjustment to q_{x+k-1} is to multiply by $\prod_{j=1}^{k-1}(1 - q_{x+j-1})$, since this now combines the probability of survival for the first $k - 1$ years, with the probability of death in year k.

For example, this model implies that two persons of age 25 and 30 can be expected to have different probabilities of death between ages 30 and 31, meaning between the 30th and 31st birthdays, and for the younger person the probability is smaller. This is not due to any projected favorable trends in the probabilities of death $\{q_x\}$ over time, but simply that for the younger individual there is some chance that life will terminate prior to reaching age 30. So the respective probabilities, using age-based notation, are $\prod_{k=0}^{4}(1 - q_{25+k})q_{30}$ and q_{30}, respectively. So the younger person has a lower chance of death in this year of age simply because they may not survive to the beginning of it!

The **mortality probability density**, $f^M(j)$, is therefore defined for $j = 1, 2, 3, \ldots$, and it denotes the probability that a person now alive will survive $j - 1$ years and die in year j, as is given by

$$f^M(j) - q_j \prod_{k=1}^{j-1}(1 - q_k), \qquad j = 1, 2, 3, \ldots. \tag{7.128}$$

Again, within an actuarial context, the notation for this probability would be

$$_{(j-1)|}q_x = q_{x+j-1} \prod_{k=1}^{j-1} (1 - q_{x+k-1}) \qquad \text{for } j \geq 1,$$

where $_{(j-1)|}q_x$ denotes the probability of a person now age x will survive $(j-1)$-years and die in year j, and so $_{0|}q_x = q_x$.

Associated with the p.d.f. $f^M(j)$ is the **mortality distribution function**, $F^M(j) = \sum_{k=1}^{j} f^M(k)$, and the **survival function**, $S^M(j) = 1 - F^M(j)$, which gives the probability that an individual survives j years.

Life Insurance Single Net Premium

One simple application of a survival model is to determine the **expected present value** of an insurance payment of \$1 to a person at the end of their year of death. This is a **whole life insurance** contract, meaning the coverage does not expire as of a specified point in time as is the case for **term life insurance**. Let I denote the random variable that equals the present value of this insurance payment. The expected value of I conditional on the death occurring in year j, denoted $E[I \mid j]$, equals $v^j \equiv (1+i)^{-j}$ for a constant annual rate of interest i. The expected value of I equals the expected value of these conditional expectations under $f^S(j)$ by (7.43), and hence

$$E[I] = \sum_{j=1}^{\infty} v^j q_j \prod_{k=1}^{j-1} (1 - q_k).$$

Here the use of ∞ is merely a notational convenience.

The calculated value of $E[I]$ is the expected value of this whole life insurance payment in units of present value at rate i. It is often denoted A_x in standard actuarial notation to identify the dependency on age x. It is a "single" premium, in that it reflects what would need to be received at $t = 0$ and invested at rate i to provide for the expected benefit. Put another way, if received from a large group of individuals of the same mortality risk and invested, all benefits would be payable with nothing left "in the end."

Also $E[I]$ is a "net" premium in that it provides only for the expected benefit; it does not provide for the various risks that would be assumed with such a contract (mortality, interest rate, etc.), nor does it provide for various levels of expenses associated with selling and maintaining this policy, nor the associated profits that the insurer requires as a return on risk capital invested.

To calculate the variance of I using conditioning and (7.46), let $Q_j = q_j \prod_{k=1}^{j-1} (1 - q_k)$. Then from the calculation above that $E[I \mid j] = v^j \equiv (1+i)^{-j}$, we have

$$\mathrm{Var}[\mathrm{E}[I\mid j]] = \mathrm{E}[(\mathrm{E}[I\mid j])^2] - (\mathrm{E}[\mathrm{E}[I\mid j]])^2$$

$$= \sum_{j=1}^{\infty} v^{2j} Q_j - \left[\sum_{j=1}^{\infty} v^j Q_j\right]^2.$$

Also the conditional variance is given by $\mathrm{Var}[I\mid j] = 0$, and so $\mathrm{E}[\mathrm{Var}[I\mid j]] = 0$. Combining, we get

$$\mathrm{Var}[I] = \sum_{j=1}^{\infty} v^{2j} Q_j - \left[\sum_{j=1}^{\infty} v^j Q_j\right]^2.$$

This basic life insurance benefit can be modified in various ways and handled similarly (see exercise 37).

Pension Benefit Single Net Premium

The survival function can also be used to evaluate, again on a single net premium basis, the cost to provide for an annual pension benefit to an individual, payable at the beginning of every year as long as the individual survives. This is an example of what is called a **life annuity** contract. Let B denote the random variable that equals the present value of these pension benefits or annuity payments. Then letting $\mathrm{E}[B\mid j]$ denote the expected value of this random variable conditional on death in year j we obtain $\mathrm{E}[B\mid j] = \sum_{k=1}^{j} v^{k-1} = (1+i)a_{j;i}$ in the notation of (2.11) of chapter 2. In other words, $a_{j;i} = \frac{1-(1+i)^{-j}}{i}$. Using (7.43), we have

$$\mathrm{E}[B] = (1+i) \sum_{j=1}^{\infty} a_{j;i} q_j \prod_{k=1}^{j-1} (1-q_k).$$

This value is a single net premium in the same sense as was $\mathrm{E}[I]$ above, providing for neither risks, expenses, nor profit. In exercise 19 is assigned the demonstration that $\mathrm{E}[B]$ can also be expressed in terms of the survival function $S^M(j)$. Using the same approach as for insurance benefits, we derive

$$\mathrm{Var}[B] = (1+i)^2 \sum_{j=1}^{\infty} a_{j;i}^2 Q_j - \left[(1+i) \sum_{j=1}^{\infty} a_{j;i} Q_j\right]^2.$$

Life annuity benefits can be guaranteed payable for a minimum of m years, and called an **m-year certain life annuity**, so that a_m is payable with probability $1 - \sum_{j=1}^{m+1} Q_j$, and thereafter for as long as life continues. Also annuity benefits need not be payable for the remainder of an individual's life; the annuity may be only payable for survival up through n years, and called an **n-year temporary life annuity** contract, or be guaranteed payable for a minimum of m years independent of survival to a maximum of n years, and called an **m-year certain, n-year temporary life annuity**, where logically $m < n$. Any of these annuity benefits can also be deferred k years, and called **k-year deferred** .... See exercises 20 and 21.

Life Insurance Periodic Net Premiums

It is common that whole life insurance is paid for not as a single premium but as a periodic premium, which we model as annually payable, although other payment frequencies are common. Denoting by π the net premium payable annually for the whole life insurance contract above, at the beginning of the year as long as the insured survives, we derive from $\pi E[B] = E[I]$,

$$\pi = \frac{\sum_{j=1}^{\infty} v^j Q_j}{(1+i) \sum_{j=1}^{\infty} a_{j;i} Q_j}.$$

These periodic payments can also be structured to be payable only several years.

Pension and annuity contracts can also be paid for with periodic payments when the annuity payments are deferred, as long as the payment period is less than or equal to the deferral period, and that for death during the deferral period there is a return of some fraction of payments with interest.

7.8.4 Asset Allocation Framework

The fundamental questions of asset allocation are:

1. Given a collection of risky assets, how does one model and evaluate the implications of allocating a given amount of wealth to each of these assets and a risk-free asset in different ways?

2. Can certain allocations be said to be "preferred" to others in the sense that their properties would be seen to be superior by any rational investor? Given allocations **W** and **V**, examples of this preference for **W** over **V** would be, where we informally define risk in terms of the investor failing to achieve the desired investment objectives:

W produces returns that are better than those of **V** no matter what happens in the market.

W and **V** have the same risk, but **W** has more expected return.

W and **V** have the same expected returns, but **W** has less risk.

3. Can certain allocations, **W**, be said to be "relatively optimally preferred" to others? For example:

W has a better expected return than any other allocation with the same risk.

W has less risk than any other allocation with the same expected return.

4. Can a certain allocation, **W**, be said to be "optimally preferred" in the sense that of all relatively optimally preferred allocations, **W** would be seen to be superior by any rational investor?

In this section we begin the analysis of asset allocation by addressing a framework for such investigations, which is the essence of question 1 above. We will return to this subject in later chapters with additional results as additional tools are developed. The most general analyses require the tools of multivariate calculus and linear algebra.

To this end, assume that a given finite collection of risky assets $\{A_j\}_{j=1}^n$ and a single risk-free asset T are given. By risky is meant that the return over the investor's horizon is uncertain, and risk-free means the return is certain over the investment horizon. Consequently the risk-free asset depends on the investor and the investment horizon. While a one-month T-bill in the United States is risk-free for a US dollar investor with a one-month investment horizon, it is neither risk-free for a US dollar "day trader" nor is it risk-free for euro investor with a one-month investment horizon.

Given this notion, it must be the case that for any investor group with a common investment horizon, there is effectively one risk-free investment vehicle, which is to say, any two such vehicles would share a common and unique return. This is because if there were two such investments with different returns, investors would sell the lower return investment and buy the higher return investment to create a **risk-free arbitrage**, or simply, **arbitrage**. Sales pressure on the former would lower its price and increase its return, while demand pressure on the latter would increase price and decrease return, until return equilibrium was achieved.

An **asset allocation** is a vector $\mathbf{W} = (w_0, w_1, \ldots, w_n)$, where w_0 represents the investment in T, and w_j the investment in A_j. This vector can be unitized in **relative terms**, where $\sum_{j=0}^n w_j = 1$, and correspondingly w_j denotes the proportion of total wealth invested in the given asset, or in **absolute terms**, where $\sum_{j=0}^n w_j = W_0$ and correspondingly w_j denotes the actual wealth invested in the given asset, with W_0 representing total initial wealth. The mathematical development in the two cases is similar, differing in predictable ways. To simplify notation, we assume that **W** is unitized in relative terms, and will explicitly acknowledge total wealth of W_0 when necessary.

To set notation, let R_j denote the random return from asset A_j over the investment horizon, which can be assumed to be discrete if for no other reason than investors' limited appetite for long decimal expressions in return reports, and let r_F denote the fixed risk-free return for the period. To simplify, assume that the investment horizon is one year, and that all rates are expressed as annual returns. Unless $w_j = 0$ for $j \geq 1$, it is apparent that the return on this portfolio allocation, R, is risky, and can be represented as

$$R = w_0 r_F + \sum_{j=1}^{n} w_j R_j. \tag{7.129}$$

The return R is a discrete random variable with probability function $f(R)$, the domain values of which depend on the given allocation as well as the p.d.f.s of the various R_j.

Given an asset allocation, the theoretical connection between $f(R)$ and $\{f(R_j)\}$ is, in general, complicated even when the latter are explicitly known. A counterexample where simplicity prevails is when the collection $\{R_j\}$ is assumed to have a multivariate normal distribution, which is not discrete, but then R will have a normal distribution. But this statement is way ahead of the tools developed so far.

Without additional tools it is also very difficult to even empirically simulate the implied p.d.f. for R, since this requires the simulation of collections $\{R_j\}$ of risky asset returns. While a random sample of returns on any one risky asset A_j can be simulated from its c.d.f. using the method described in section 7.7, the difficulty associated with generating the collection of returns for all assets is that it is virtually never the case that these returns are "independent," or the weaker statement, "uncorrelated." In other words, between virtually any two risky assets one evaluates historically, it is the case that the correlation between returns ρ is generally nonzero; in almost all nontrivial cases, it is positive, so $\rho > 0$. By a trivial case is meant that if one asset is a long position and the other a short position in a given security, then artificially one will have constructed a case with $\rho < 0$, and in fact $\rho = -1$. But most examples of long positions display positive correlations and more generally nonzero correlations, and consequently an empirical simulation of returns on the risky assets needs to reflect these correlations.

One popular approach to simulation is known as **historical simulation**, whereby one has access to contemporaneous return series for each of the assets in question: $\{(R_1^{(k)}, R_2^{(k)}, \ldots, R_n^{(k)}) \mid k = 1, 2, \ldots, N\}$. This notation implies that for each sequential time period k, which would be chosen in length to equal the investment horizon of interest, $(R_1^{(k)}, R_2^{(k)}, \ldots, R_n^{(k)})$ denotes the respective returns of the given assets during this period. For the same historical periods one would also identify the returns of the risk-free asset, denoted $\{r_F^{(k)}\}$. With these data series two simulations are possible:

1. Simulation of historical returns for the given allocation,

$$R^{(k)} = w_0 r_F^{(k)} + \sum_{j=1}^n w_j R_j^{(k)}.$$

2. Simulation of potential returns for the next period, where r_F is known,

$$R^{(k)} = w_0 r_F + \sum_{j=1}^n w_j R_j^{(k)},$$

which is in effect the model in (7.129).

From either model and a specified allocation $\{w_j\}_{j=0}^n$, a return data series is simulated, $\{R^{(k)}\}$, from which all moments of R can be calculated and $f(r)$ estimated. However, if it is desired to evaluate explicitly how these moments depend on the allocation parameters, an alternative approach is needed.

Specifically, sample moments from the historical return data can be used to estimate the various moments of the random variable R, without needing to fix the allocation parameters or explicitly calculate $f(R)$. For example, applying (7.38) to (7.129), we derive

$$E[R] = w_0 r_F + \sum_{j=1}^n w_j \mu_j, \quad \mu_j \equiv E[R_j], \tag{7.130}$$

and applying (7.56),

$$\text{Var}[R] = \sum_{i=1}^n \sum_{j=1}^n w_i w_j \sigma_i \sigma_j \rho_{ij}, \tag{7.131a}$$

$$\sigma_j^2 \equiv \text{Var}[R_j], \quad \rho_{ij} \equiv \text{Corr}[R_i, R_j]. \tag{7.131b}$$

Of course, if the goal is to calculate the mean and variance of end of period wealth, defined as $W_1 = W_0(1 + R)$, these would be calculated as

$$E[W_1] = W_0[1 + \mu], \quad \text{Var}[W_1] = W_0^2 \sigma^2, \tag{7.132}$$

where μ and σ^2 are commonly used notation for $E[R]$ and $\text{Var}[R]$, respectively.

Higher moments can similarly be estimated from the higher joint sample moments of the historical data. For example, the third central moment, $\mu_3 \equiv E[(R - \mu)^3]$, is developed from $R - \mu = \sum_{j=1}^{n} w_j(R_j - \mu_j)$, and hence

$$(R - \mu)^3 = \sum_{i=1}^{n} \sum_{j=1}^{n} \sum_{k=1}^{n} w_i w_j w_k (R_i - \mu_i)(R_j - \mu_j)(R_k - \mu_k).$$

This formula requires a bit of combinatorial manipulation, but the expectation will clearly involve terms as follows, where the subscripts are now intended to be distinct:

$$E[(R_i - \mu_i)(R_j - \mu_j)(R_k - \mu_k)], \quad E[(R_i - \mu_i)(R_j - \mu_j)^2], \quad E[(R_i - \mu_i)^3].$$

The analysis of these risk and return statistics, especially in terms of their behaviors for different allocation vectors, **W**, is now a question of evaluating these moments as functions of $(w_0, w_1, \ldots, w_n)$ considered as a point in $\mathbb{R}^{n+1}$. Such an analysis requires the more powerful tools of multivariate calculus and linear algebra to be complete. Still here we can appreciate what is to come with an informal analysis of the issue raised in question 2 above.

Given allocations **W** and **V**, there are many ways to define that **W** is "preferred" over **V**. For example, given the allocation $\mathbf{W} = (w_0, w_1, \ldots, w_n)$ define an **epsilon switch** allocation $\mathbf{W}_\epsilon^{ij}$ as equal to **W** except that w_i is increased by ϵ, and w_j is decreased by ϵ. Let R denote the random return under **W**, and R_ϵ^{ij} the return under $\mathbf{W}_\epsilon^{ij}$. An easy calculation produces

$$E[R_\epsilon^{ij}] - E[R] = \epsilon(\mu_i - \mu_j),$$

where for notational convenience we denote r_F by μ_0. Clearly, for $\epsilon > 0$ the expected return is increased or decreased according to whether $\mu_i > \mu_j$ or $\mu_i < \mu_j$.

For the variance analysis, the notation is simplified by noting that (7.131a) can be expressed as

$$\text{Var}[R] = \sum_{i=0}^{n} \sum_{j=0}^{n} w_i w_j \sigma_{ij}, \quad \sigma_{ij} \equiv \text{Cov}[R_i, R_j], \quad \sigma_{jj} \equiv \text{Var}[R_j], \qquad (7.133)$$

since for any $j \neq 0$, $\sigma_{0j} = \sigma_{j0} = 0$ and $\sigma_0^2 = 0$. With this formula the change in variance can be calculated, although in a more complicated way. The trick is to split the summation into terms that include i or j,

$$2w_i \sum_{k \neq i,j} w_k \sigma_{ik} + 2w_j \sum_{k \neq i,j} w_k \sigma_{kj} + 2w_i w_j \sigma_{ij} + w_i^2 \sigma_i^2 + w_j^2 \sigma_j^2,$$

and into terms that exclude both i and j,

$$\sum_{k \neq i,j} \sum_{l \neq i,j} w_k w_l \sigma_{kl}.$$

With this splitting, since only w_i and w_j are changed, we obtain with a bit of algebra

$$\mathrm{Var}[R_\epsilon^{ij}] - \mathrm{Var}[R] = 2\epsilon \sum_{k \neq i,j} w_k(\sigma_{ik} - \sigma_{kj}) + 2[\epsilon(w_i - w_j) - \epsilon^2]\sigma_{ij}$$

$$+ \epsilon^2(\sigma_i^2 + \sigma_j^2) + 2\epsilon(w_i\sigma_i^2 - w_j\sigma_j^2)$$

$$= \epsilon^2[\sigma_i^2 + \sigma_j^2 - 2\sigma_{ij}] + 2\epsilon\left[\sum_{k=0}^{n} w_k(\sigma_{ik} - \sigma_{kj}) + 2(w_i - w_j)\sigma_{ij}\right].$$

In other words, given any i and j, $\mathrm{Var}[R_\epsilon^{ij}] - \mathrm{Var}[R]$ is a quadratic function of ϵ that goes through the origin. So for fixed constants A and B that depend on i and j,

$$\mathrm{Var}[R_\epsilon^{ij}] - \mathrm{Var}[R] = A\epsilon^2 + 2B\epsilon.$$

Now in the proof of (7.54) it was shown by use of the Cauchy–Schwarz inequality, that $\sigma_{ij}^2 \leq \sigma_i^2\sigma_j^2$. From this we conclude that $-\sigma_i\sigma_j \leq \sigma_{ij} \leq \sigma_i\sigma_j$, and hence $A \geq 0$. Specifically,

$$0 \leq (\sigma_i - \sigma_j)^2 \leq A \leq (\sigma_i + \sigma_j)^2.$$

Now, if $B = 0$, then $\mathrm{Var}[R_\epsilon^{ij}] - \mathrm{Var}[R] \geq 0$ for all ϵ and the epsilon switch creates the same or more risk. If $B \neq 0$, this inequality for A implies that there is an interval for ϵ for which $\mathrm{Var}[R_\epsilon^{ij}] - \mathrm{Var}[R] < 0$, which is to say, that the variance has been decreased. Specifically, if $B > 0$, the variance reduction interval is $\epsilon \in \left(-\frac{2B}{A}, 0\right)$, whereas if $B < 0$, the variance reduction interval is $\left(0, \frac{2B}{A}\right)$. In both cases the point of maximal reduction is the interval midpoint.

This simple analysis can provide one answer to question 2 on an allocation being "preferred." Namely, if there is an i and j for which the expected return can be increased, $\mathrm{E}[R_\epsilon^{ij}] > \mathrm{E}[R]$, and variance of return decreased, $\mathrm{Var}[R_\epsilon^{ij}] < \mathrm{Var}[R]$, then this would appear to be a reasonable basis to claim that $\mathbf{W}_\epsilon^{ij}$ is preferred to $\mathbf{W}$. Of course, this is only a reasonable basis, since it ignores higher moments of these random variables with the two allocations.

7.8.5 Equity Price Models in Discrete Time

Stock Price Data Analysis

Let S_0 denote the price of an equity security at time zero. Many problems in finance relate to modeling the probability density functions and related characteristics of prices at a point in the future, or the evolution of such prices through time. Essential to this model is the notion that future stock prices, as well as the prices of futures contracts, currencies, interest rates, and so forth, are fundamentally random variables at time zero, even though their movements may well be fully or at least partially explainable after the fact. This is sometimes described by saying that future prices are random *ex ante*, but deterministic and possibly explainable *ex post*. These perspectives are not at odds.

Being explainable **ex post** means that one can develop certain cause and effect arguments that make the price effect understandable and even compelling, whereas being random **ex ante** means that one cannot predict what the future causes of price movements will be. In general, these causes evolve with the markets' **information processes**, which is the general model of how information emerges and travels through the markets. Randomness of price movements therefore reflects the randomness in the discovery, release, and dissemination of market relevant information.

Historical analysis also reinforces this view of randomness. If $\{S_j\}$ denotes a given stock price series evaluated at the market's close on a daily, weekly, or other regularly spaced basis over a reasonably long period of time, say 10 years or so, the collection of **period returns** $\{R_j\} \equiv \left\{ \frac{S_{j+1} - S_j}{S_j} \right\}$ can be plotted as a sequence, called a **time series**, and will generally appear to have many of the characteristics of a coin-flip sequence. Specifically, about 50:50 positive and negative results, with positive and negative runs of varying lengths.

Also, while one observes runs, a calculation of the correlation between successive returns, R_j and R_{j+1}, produces a so-called autocorrelation that is typically near 0. By **autocorrelation** is meant the correlation of a random variable with itself over time. An autocorrelation near 0 implies that on average, R_j provides little predictability to the value or even the sign of R_{j+1}, again like a series of coin flips. It is also the case that grouping ranges of returns, and plotting the associated approximate p.d.f. in a **histogram**, provides a familiar bell-shaped curve, seemingly almost normally distributed. But closer analysis proves that this distribution often has **fat tails** in the sense that the probabilities of normalized returns far from 0 exceed that allowed by the normal distribution.

These same characteristics are often observed in the **growth rate** series or **log-ratio return series**, $\{r_j\} \equiv \left\{ \ln\left(\frac{S_{j+1}}{S_j} \right) \right\} \equiv \{\ln(1 + R_j)\}$. The log-ratio returns tend to be the more popular for modeling, since in this case $S_{j+1} = S_j e^{r_j}$, whereas in terms of period returns, $S_{j+1} = S_j(1 + R_j)$. While this may appear of little mathematical consequence, the distinction comes from the modeling of prices n-periods forward:

Return model: $S_n = S_0 \prod_{j=0}^{n-1}(1 + R_j),$ (7.134)

Growth model: $S_n = S_0 e^{\sum_{j=0}^{n-1} r_j}.$ (7.135)

From these formulas it should be apparent that using $\{r_j\}$ as the collection of return variables requires the modeling of sums of random variables, whereas with $\{R_j\}$, we will be required to work with products. The log-ratio return parametrization is to be preferred simply because the mathematical analysis is more tractable in these terms.

Binomial Lattice Model
Now let μ and σ^2 denote the mean and variance of the log-ratio return series, where these parameters of necessity reflect some period of time, say $\Delta t = 1$, separating the data points. Knowing from history that $\{r_j\}$ has a bell-shaped distribution for small time intervals, one can approximate the log-ratio returns with binomial returns in anticipation of results of chapter 8:

$$S_{j+1} = S_j e^{B_j}.$$

Here $\{B_j\}$ are a random collection of i.i.d. binomials defined by

$$B = \begin{cases} u, & \Pr[u] = p, \\ d, & \Pr[d] = p', \end{cases}$$

where $p' \equiv 1 - p$ and p, u, and d are "calibrated" to achieve the desired moments from historical data as follows.

To derive all three model parameters from historical data will require three constraints. In practice, the analysis is often simplified by introducing one reasonable constraint judgementally. For example, by choosing $p = \frac{1}{2}$, $E[B] = \frac{1}{2}(u + d)$, $E[B^2] = \frac{1}{2}(u^2 + d^2)$, and $\text{Var}[B] = \frac{1}{4}(u - d)^2$. Consequently, in order to produce the two historical moments, it is required that

$$\frac{1}{2}(u + d) = \mu,$$

$$\frac{1}{4}(u-d)^2 = \sigma^2,$$

which is easily solved to produce the stock model

$$S_{j+1} = \begin{cases} S_j e^{\mu+\sigma}, & p = \frac{1}{2}, \\ S_j e^{\mu-\sigma}, & p' = \frac{1}{2}. \end{cases} \tag{7.136}$$

An alternative calibration is to constrain $d = \frac{1}{u}$; then using only mean and variance again, determine the parameters p and u.

From this $p = \frac{1}{2}$ model, stock prices in n time steps are seen to be binomially distributed with parameters n and p. This is because (7.135), with $r_j = \mu + b_j \sigma$ and

$$b_j = \begin{cases} 1, & \Pr = \frac{1}{2}, \\ -1, & \Pr = \frac{1}{2}, \end{cases}$$

produces

$$S_n = S_0 e^{n\mu + \sigma \sum_{j=0}^{n-1} b_j}, \tag{7.137}$$

where $\sum_{j=0}^{n-1} b_j$ assumes values of $\{-n+2k\}_{k=0}^n$ with probabilities $\{\binom{n}{k}\frac{1}{2^n}\}_{k=0}^n$.

This observation allows a notationally simpler parametrization of stock prices as follows:

$$S_n = S_0 e^{n(\mu-\sigma)+2\sigma B_n} = S_0 e^{nd+(u-d)B_n}; \tag{7.138a}$$

$$\Pr[B_n = j] = \binom{n}{j}\frac{1}{2^n} = \binom{n}{j}p^j(1-p)^{n-j}, \qquad j = 0, 1, \ldots, n. \tag{7.138b}$$

This formula is the basis of the **binomial lattice model** of stock prices whereby from an initial price of S_0 two prices are possible at $t = 1$, three prices are possible at $t = 2, \ldots$, and finally, $n + 1$ prices are possible at time n. Not uncommonly, these prices are represented in a positive integer lattice, with time plotted on the horizontal, and "state," or random stock price, along the vertical, as seen in **figure 7.3**.

The graph shown in the figure is usually oriented in the logical way, with lowest stock prices plotted at the bottom and associated with $B_n = 0$. From any "time-state" price, there are two possibilities in the next period, with the price directly to the right representing d, and the price to the northeast representing u, both with probability $\frac{1}{2}$ with this calibration. With the calibration assigned in exercise 23, the probability of the price directly to the right equals $1 - p$, while the price to the

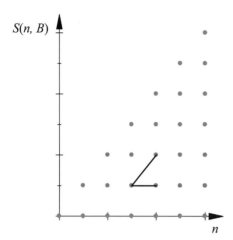

$S(n, B)$

n

Figure 7.3
Binomial stock price lattice

northeast has probability p. Looked at another way, the collection of $n + 1$ prices at time n are distributed as binomial variables with parameters n and $\frac{1}{2}$, as is indicated in (7.138), or more generally in exercise 23, distributed as binomial variables with parameters n and p. This provides a bell-shaped distribution of returns defined as $\ln[S_n/S_0]$, which is consistent with historical data. This will be formalized in chapter 8.

Binomial Scenario Model

An alternative and equally useful way to both conceptualize the evolution of stock prices, as well as to perform many types of calculations, is to generate **stock price paths**, or **stock price scenarios**. In contrast to the binomial lattice approach, which generates all possible prices up to time n under this model, the scenario approach generates one possible price path at a time. An example of a single price path is seen in **figure 7.4**.

Each such path requires the generation of n prices, since S_0 is given. In contrast, the generation of a complete lattice requires $\sum_{j=2}^{n+1} j = \frac{(n+1)(n+2)-2}{2}$ prices. The motivation for the scenario-based approach is often not combinatorial. Since there are 2^n possible paths, to generate them all requires $2^n n$ calculations when done methodically, and this materially exceeds $\frac{(n+1)(n+2)-2}{2}$ in total effort. In the typical situation the motivation for scenarios might be that the given problem cannot be solved within a lattice framework but can only be solved with generated paths.

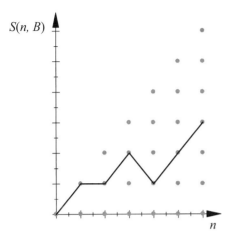

Figure 7.4
Binomial stock price path

For example, the price of a simple **European** or **American option** on a given common stock can be estimated on a lattice of stock prices. On the other hand, if the value of a European option at expiry reflects values of the stock's prices along whatever price path it followed, lattice-based methods do not work, and this calculation must be estimated with scenario-based calculations.

Scenario methods are also necessary in certain lattice models that are **nonrecombining**. The lattice model above is **recombining** in that from any given price the same price is produced two periods hence if the intervening returns were (u, d) or (d, u). Not all lattices have this property. A nonrecombining lattice is one for which $S^{(u,d)} \neq S^{(d,u)}$. In such a case generation of the entire lattice may be impossible, since the number of such prices is now $\sum_{j=1}^{n+1} 2^j = 2^{n+2} - 1$. For a nonrecombining model, even if lattice-based methods are theoretically possible, as in European option pricing, they are infeasible for large n, and scenario-based methods are required.

7.8.6   Discrete Time European Option Pricing: Lattice-Based

One-Period Pricing

Remark 7.61 *In this and other sections on option pricing, or more generally derivatives pricing, the underlying asset, denoted S, will be called a common stock. However, all of this theory applies to derivatives on any asset in which investors can take short positions. Of course, all assets allow investors to take long positions by simply* *acquiring them, so allowing a short position is somewhat restrictive. It is common language to call assets that can be shorted,* **investment assets**, *since these are assets commonly held in inventory by investors for their appreciation potential. Common stocks and stock indexes, fixed income investments and indexes, currencies, precious metals like gold and platinum, and all futures contracts are examples of investment assets; the general framework developed here is adaptable to derivatives on these assets. Other assets are called* **consumption assets**, *since these are assets that rarely are held in inventory except for consumption purposes; hence they are not available for lending and shorting. Examples include most commodities other than precious metals.*

Suppose that on a given stock with current value S_0, which will be assumed to pay no dividends, we seek to price a European option or other derivative security that expires in one period, and whose payoff is given by an arbitrary function of price at that time, denoted $\Lambda(S_1)$. Recall that the terminology "European" means that the option provides for no early exercise; it can only be exercised on the expiry date.

For example, if this option is a **European call** or a **put** with a **strike price of** K, the payoff function to the holder of the option is given:

Call option: $\Lambda(S_1) = \max(S_1 - K, 0),$ (7.139a)

Put option: $\Lambda(S_1) = \max(K - S_1, 0),$ (7.139b)

where the use of the "max" function is conventional and shorthand for the fact that the holder of the option, or the "long position," will either receive a positive payoff or nothing.

For the purposes here, the payoff function $\Lambda(S_1)$ can be arbitrary without affecting the mathematical development, but it is common in the market that $\Lambda(S_1) \geq 0$ for the long position, and $\Lambda(S_1) \leq 0$ for the short. Again, the mathematics does not require this, but the terminology is simplified in this case. An example of a derivatives security in the market that has both positive and negative payoffs is a **futures** contract, for which a long futures contract is equivalent to a long call and short put, and conversely, and either side of the contract can be paid or required to pay at expiry. To simplify the language below, we will assume that we are taking the perspective of the long position.

Assume that the stock price in one period is modeled,

$$S_1 = \begin{cases} S_0 e^u, & \Pr = p, \\ S_0 e^d, & \Pr = 1 - p, \end{cases}$$

for suitable p, u, d. Then the option payoff is either $\Lambda(S_0 e^u)$ or $\Lambda(S_0 e^d)$, which we denote by Λ^u and Λ^d, respectively. Naturally the price of this option at time 0, Λ_0, cannot equal or exceed the present value of the greater payoff, nor be equal to or less than the present value of the lesser payoff. In the former case, an investor would try to sell these options, and in the latter, buy them, thereby creating a chance (perhaps certain chance) of profit with no risk, which is an **arbitrage**, or a **risk-free arbitrage**. In theory, such a purchase would be financed by shorting T-bills, and the sale of options invested in T-bills, thereby insulating the trader from all risk.

Let r denote the continuous risk-free interest rate for this period, on a Treasury bill say. Note that it is nonstandard to quote r in other than annual units, and we correct this in chapter 8 where the appropriate time context is addressed. These bounds on Λ_0 can be expressed as

$$e^{-r} \min[\Lambda^u, \Lambda^d] < \Lambda_0 < e^{-r} \max[\Lambda^u, \Lambda^d].$$

Consequently there must be a unique real number q that can be called a probability, since $0 < q < 1$, so with $q' = 1 - q$,

$$\Lambda_0(S_0) = e^{-r}[q\Lambda^u + q'\Lambda^d]. \tag{7.140}$$

In other words, the market price must equal the **expected present value of the payoffs** at some as yet unspecified "probability" q.

It turns out that q can be derived because this option can be **replicated**. The idea of **replication** is that one can construct a portfolio of traded assets that has the same payoff as does the option. Hence the price of the option must equal the price of this portfolio, or else there will be an arbitrage opportunity. If the option was more expensive than the replicating portfolio, the savvy trader would sell the option, buy the portfolio for an immediate profit, and settle at expiry with no out of pocket cost. Similarly, if the option was cheaper than the replicating portfolio, the opposite trade would be implemented.

The replicating portfolio turns out to be a mix of stock and risk-free assets, usually referred to as T-bills. To see this, construct a portfolio of a shares of stock, and $\$b$ invested in T-bills, so the portfolio, denoted Π_0, is

$$\Pi_0 = \{aS_0, bT\},$$

where T denotes a $\$1$ investment in a T-bill. This portfolio costs $aS_0 + b$ at time 0, and at time 1 will have values

$$\Pi_1 = \begin{cases} aS_0e^u + be^r, & \Pr = p, \\ aS_0e^d + be^r, & \Pr = p'. \end{cases}$$

It is not difficult to determine the correct values of a, b so that $aS_0e^u + be^r = \Lambda^u$ and $aS_0e^d + be^r = \Lambda^d$. Specifically, we derive

$$a = \frac{\Lambda^u - \Lambda^d}{S_0(e^u - e^d)}, \quad b = e^{-r}\frac{e^u\Lambda^d - e^d\Lambda^u}{e^u - e^d}. \tag{7.141}$$

With these coefficients, a bit of algebra shows that the price of this portfolio at time 0, which is

$$\Lambda_0(S_0) = aS_0 + b, \tag{7.142}$$

can be expressed as in (7.140) with

$$q = \frac{e^r - e^d}{e^u - e^d}. \tag{7.143}$$

It must be the case that $0 < q < 1$, since the stock is a risky asset. Hence $e^d < e^r < e^u$, or else an arbitrage opportunity would exist.

We collect these results in a proposition:

Proposition 7.62 *Let $\Lambda(S)$ denote the payoff function for a one-period European derivatives contract on an investment asset with current price S_0, for which the end of period prices follow a binomial distribution as given in (7.138) with $n = 1$. Then the price of this derivatives contract $\Lambda_0(S_0)$ equals the price of the replicating portfolio given in (7.142), with coefficients given in (7.141). Alternatively, this price can be expressed as in (7.140) with probability q defined in (7.143).*

Remark 7.63 *This "probability" q is known as the **risk-neutral probability of an up-state**, since this is what the probability of an upstate must be in a risk-neutral world to justify the stock price of S_0. To better see this, first note that q is the unique probability that prices the current value of the common stock, S_0, to be equal to the risk-free present value of its expected future prices:*

$$S_0 = e^{-r}[qS_0e^u + q'S_0e^d]. \tag{7.144}$$

So why does this matter? We will see more on risk preference models in chapter 9, but the conclusion will be that risk-neutral investors do not charge for risk, as the term implies, and consequently they require the same return on all investments. Logically *this implies that the same return they require for all assets is the risk-free return. But what does "require the same return" mean if an asset has risk? The answer is that each investor can summarize risk through their own "utility function," and in general, will price assets in order to maximize the expected value of the utility of their wealth. This is usually called **maximizing expected utility**. For a risk-neutral investor, utility maximization turns out to be equivalent to pricing assets based on expected payoffs. Rewriting the stock-pricing formula above, we have*

$$S_0 e^r = [q e^u + q' e^d] S_0,$$

which shows that the expected payoff on S_0 under q provides the risk-free return.

Of course, no one believes investors to be risk neutral, but this is a good framework for describing how q can be interpreted. Indeed the model suggests that if investors expect a log-ratio return of μ, they likely believe that the probability of prices rising to $S_0 e^u$ is p, and not q. It is not difficult by example to show that $q \neq p$, and this can be proved with methods of chapter 9.

Multi-period Pricing

A two-period European option with payoff function $\Lambda(S_2)$ can be priced with the same methodology. If we know the prices of this option at time 1 in both stock price "states," $\Lambda(S_1^u)$ and $\Lambda(S_1^d)$, then the price at time 0 is given by (7.140) with risk-neutral probability q in (7.143):

$$\Lambda_0(S_0) = e^{-r}[q\Lambda(S_1^u) + q'\Lambda(S_1^d)]. \tag{7.145}$$

The argument is the same. This is the correct price because a replicating portfolio can be purchased for this amount that provides the correct future values whether the stock price rises or falls.

On the other hand, $\Lambda(S_1^u)$ can also be evaluated by this formula based on the payoffs at time 2,

$$\Lambda(S_1^u) = e^{-r}[q\Lambda(S_2^{2u}) + q'\Lambda(S_2^{u+d})],$$

and similarly for $\Lambda(S_1^d)$,

$$\Lambda(S_1^d) = e^{-r}[q\Lambda(S_2^{d+u}) + q'\Lambda(S_2^{2d})].$$

These formulas again follow, since these are the prices of the respective replicating portfolios. Note that the subscript in these formulas denotes time, and superscript denotes the stock state. For example, $S_2^{2u} = S_0 e^{2u}$, and so forth.

Inserting the second two formulas into the first, we see that $\Lambda_0(S_0)$ is again equal to the expected present value of the $t = 2$ payoffs, where the expectation is calculated with binomial probability q, and the present value at the risk-free rate r, producing

$$\Lambda_0(S_0) = e^{-2r}[q^2\Lambda(S_2^{2u}) + 2qq'\Lambda(S_2^{d+u}) + (q')^2\Lambda(S_2^{2d})]. \tag{7.146}$$

In exercise 39 is assigned the proof of the generalized version of this formula, for a European option with expiry in n time steps. The formula becomes

$$\Lambda_0(S_0) = e^{-nr}\sum_{j=0}^{n}\binom{n}{j}q^j(1-q)^{n-j}\Lambda(S_n^j), \qquad S_n^j = S_0e^{ju+(n-j)d}. \tag{7.147}$$

This price can also be expressed as the price of a replicating portfolio that replicates option prices at time 1-period, where the option prices are in turn given by an application of this same formula with $n-1$ periods to expiry:

$$\Lambda(S_1^u) = e^{-(n-1)r}\sum_{j=0}^{n-1}\binom{n-1}{j}q^j(1-q)^{n-1-j}\Lambda(S_{n-1}^{u_j}), \tag{7.148a}$$

$$\Lambda(S_1^d) = e^{-(n-1)r}\sum_{j=0}^{n-1}\binom{n-1}{j}q^j(1-q)^{n-1-j}\Lambda(S_{n-1}^{d_j}), \tag{7.148b}$$

where

$$S_{n-1}^{u_j} = S_1^u e^{ju+(n-1-j)d}; \quad S_{n-1}^{d_j} = S_1^d e^{ju+(n-1-j)d}.$$

In other words, $\Lambda_0(S_0)$ in (7.147) satisfies (7.145) with these values of $\Lambda(S_1^u)$ and $\Lambda(S_1^d)$, and from the preceding section we know that this is the same price as that of a replicating portfolio that replicates these option values. This result can be demonstrated directly by an application of (7.16).

By (7.147), the price of the option can be expressed as an **expected present value** under the assumption that the calculated value of q in (7.143) is the correct binomial probability of an upstate return of e^u. This will differ from the binomial probability of p that we started with, that reproduced the mean and variance of the stock's log-ratio returns.

Of course, the price in (7.147) is the theoretically correct price under the assumptions of this lattice. If two analysts calibrate their lattices to different assumptions of stock price behavior, or even the same assumptions but calibrated with different time steps of Δt, different prices will result, possibly materially different.

In finance, the p-model is referred to as the **real world model**, since it produces the statistical properties observed or believed to be valid in the real world, and the q-model is referred to as the **risk-neutral model**, since these probabilities correctly price the stock in a world where investors are risk neutral.

We collect these results in a proposition:

Proposition 7.64 *Let $\Lambda(S)$ denote the payoff function for an n-period European derivatives contract on an investment asset with current price S_0, for which the end of period prices follow a binomial distribution as given in (7.138). Then the price of this derivatives contract, $\Lambda_0(S_0)$, is given in (7.147) with probability q defined in (7.143). This price also equals that of the replicating portfolio given in (7.142), with coefficients given in (7.141), where the derivatives prices at time 1-period are given by (7.148).*

Remark 7.65 *It is important to recognize that while the pricing formula in (7.147) can be understood to provide a risk-neutral present value of option payoffs, this interpretation does not provide a compelling reason why the number produced is the theoretically correct market price. The logic that compels this conclusion is that $\Lambda_0(S_0)$ as given in that formula also satisfies the equation in (7.145):*

$$\Lambda_0(S_0) = e^{-r}[q\Lambda(S_1^u) + q'\Lambda(S_1^d)].$$

So by the analysis of the one-period model, this is the price of a portfolio that replicates the value of this option at the end of the first period. Each of these prices in turn equals what is needed to create a portfolio that replicates option prices in the next period, and so forth. In other words, by "rebalancing" the replicating portfolio each period after the first, and realizing this can be done with no additional costs, these replicating portfolios will track the emerging values of the option up to the final period in which the last replicating portfolio will replicate the actual option payoffs. That said, this argument ignores all real world market "friction" caused by trading costs and taxes, so the real world price will need to be adjusted somewhat for this.

Of course, the replication argument relies on the assumption that this option is on an investment asset as noted above. The reason for this is twofold. First off, the actual replicating portfolio will involve a short position in S when $a < 0$ in (7.141), which occurs when $\Lambda^u < \Lambda^d$. This is the case for a put option, for instance. Second, this argument does not in and of itself compel the conclusion that options **must** be sold at this price, it merely demonstrates that they could be sold near this price because the seller can hedge his risk with a replicating portfolio. In other words, selling the option creates a short position for the seller that can be hedged with a long position in the replicating portfolio. By "near this price" is meant adjusted for transactions costs and buyer convenience. Now, if the seller attempts to sell an option on an investment asset at a price materially different from the replicating portfolio cost, one of two things happen. Some investors will buy "cheap options" and hedge their position with short positions in the replicating portfolio. Some other investors will sell "dear options" and hedge with a long position in the replicating portfolio. In either case, the buying pressures would increase prices, and selling pressures would decrease prices. So in both cases investors move toward the price of the replicating portfolio, as adjusted for transactions costs.

7.8.7   Discrete Time European Option Pricing: Scenario Based

If N-paths are randomly generated, and $\{S_n^j\}_{j=0}^n$ denotes the $n+1$ possible stock prices in the recombining lattice above at time n, it is of interest to analyze the number of paths that arrive at each final state. In theory, we know from the lattice analysis above that the distribution of stock prices at time n is binomially distributed with parameters n, p in general, and hence $\Pr[S_n = S_n^j] = \binom{n}{j} p^j (1-p)^{n-j}$. Here p denotes the probability of a u-return, and stock prices are parametrized so that $j=0$ corresponds to the lowest price, $S_n^0 = e^{nd} S_0$, and $j = n$ corresponds to the highest price, $S_n^n = e^{nu} S_0$. On the other hand, we have shown that for the purposes of option pricing, we continue to use the stock price returns of e^u and e^d but switch the assumed probability of an upstate return from p to q given in (7.143).

In the lattice-based model, these q probabilities provide the likelihood of each final state for option pricing. Consequently, if from a sample of N-paths, N_j denotes the number that terminate at price S_n^j so that $\sum N_j = N$, then the $(n+1)$-tuple of integers $(N_0, N_1, \ldots, N_n)$ has a multinomial distribution with parameters N and $\{Q_j\}_{j=0}^n$, where $Q_j = \binom{n}{j} q^j (1-q)^{n-j}$. In other words, from (7.105) and (7.106) we conclude that

$$\mathrm{E}[N_j] = NQ_j, \tag{7.149a}$$

$$\mathrm{Var}[N_j] = NQ_j(1 - Q_j), \tag{7.149b}$$

$$\mathrm{Cov}[Q_j, Q_k] = -NQ_j Q_k. \tag{7.149c}$$

In a nonrecombining lattice, Q_j is again defined as the risk-neutral probability of terminating at price S_n^j, only in this case there are 2^n stock prices rather than $n+1$. The multinomial distribution is again applicable as are the moment formulas above.

As an application we develop the methodology for estimating the price of an n-period European option using the **scenario-based methodology**. For simplicity, we focus on the recombining lattice model, although the development is equally applicable in the more general case. To this end, let $\Lambda(S_n^j)$ denote the exercise value of the option at time n when the stock price, S_n^j, prevails. Given N-paths, define a random variable O_N, the **sample option price**,

$$O_N = \frac{e^{-nr}}{N} \sum_{j=0}^{n} N_j \Lambda(S_n^j). \tag{7.150}$$

Intuitively the random variable O_N is an estimate of the true option price based on a price scenario sample of size N. Although this formula at first looks completely different from the exact formula given in (7.147), the formulas are quite similar. Because of (7.149a), it is apparent that

$$\mathrm{E}\left[\frac{N_j}{N}\right] = Q_j \equiv \binom{n}{j} q^j (1-q)^{n-j},$$

and consequently the option price in (7.147) can be rewritten as

$$\Lambda_0(S_0) = e^{-nr} \sum_{j=0}^{n} \mathrm{E}\left[\frac{N_j}{N}\right] \Lambda(S_n^j).$$

In this form it is apparent that the difference between $\Lambda_0(S_0)$ and O_N is that for the former, the option exercise price of $\Lambda(S_n^j)$ is given the theoretically correct weight $\mathrm{E}\left[\frac{N_j}{N}\right]$, while for the latter, this weight is replaced by a sample-based estimate $\frac{N_j}{N}$. We should then expect that since the paths are generated in such a way as to arrive at each final stock price with the correct probability, the expected value of this random variable ought to equal $\Lambda_0(S_0)$, and this will be the case.

Even more important, it will turn out that as N increases, the probability that we are in error by any fixed amount goes to 0. These results are demonstrated in chapter 8. In addition the relationship of this pricing approach to the replication-based pricing above will be evaluated.

Exercises

Practice Exercises

1. Demonstrate that if $\mathcal{E}$ is a complete collection of events, and $A_j \in \mathcal{E}$ for $j = 1, 2, 3, \ldots$, then $\bigcap_j A_j \in \mathcal{E}$.

2. Confirm that in the sample space $\mathcal{S}$ of sequences of 10 flips of a fair coin, the event $A = \{x \mid x = HH \ldots\}$ contains exactly 25% of the total number of sequences, where this notation means that only the first two outcomes are fixed. In this demonstration, if you ignore what happens after the first 2 flips, justify explicitly that this is valid.

3. On the sample space of 10-flip sequences of a fair coin $\mathcal{S}$:

(a) Define three different random variables, $X : \mathcal{S} \to \mathbb{R}$. (*Hint*: For simplicity, identify an H with a 1, and a T with a 0.)

(b) Determine the associated ranges of these functions.

(c) Calculate $\Pr(a)$ for one a in the range of each X.

4. Generalize (7.2) and (7.1) in the following way:

(a) If a gambler is asked to bet m for a chance to win n, what is the probability of winning that will make this bet fair? Confirm that $m = 1$ gives (7.2).

(b) If the gambler knows that the probability of a win is p, what ratio of amount bet to amount won, $\frac{m}{n}$ from part (a), will make this a fair bet? Confirm that $m = 1$ gives (7.1).

(c) Show that if p is irrational in part (b), a fair bet requires an irrational value for $\frac{m}{n}$. Conclude that since bets and payoffs must be rational numbers, a bet with an irrational probability of winning can never be fair.

5. Show that if an event $B \in \mathcal{E}$ satisfies $\Pr(B) \neq 0$, then $\Pr(\cdot \mid B)$ satisfies all the definitional properties of a probability measure on the sample space $\mathcal{S}$.

6. Consider the sample space of 5 flips of a fair coin, where we identify an H with a 1, and a T with a 0. Define events A and B as

$$A = \left\{ s \in \mathcal{S} \,\middle|\, \sum_{i=1}^{3} s_i = 2 \right\} \quad \text{and} \quad B = \left\{ s \in \mathcal{S} \,\middle|\, \sum_{i=3}^{5} s_i = 1 \right\}.$$

(a) List the sample points in each event.

(b) Determine the probability of each event.

(c) What points are in the event $A \cap B$?

(d) Verify that $\Pr[A \cap B] = \Pr[A \mid B] \Pr[B]$.

7. Define the events $C_k, B_j \subset \mathcal{S}$, the sample space in exercise 6, by $C_k = \{s \in \mathcal{S} \mid \sum_{i=1}^{2} s_i = k\}$, and $B_j = \{s \in \mathcal{S} \mid \sum_{i=3}^{5} s_i = j\}$. Show for all j, k that C_k and B_j are independent events.

8. An urn contains 20 white, and 30 red balls.

(a) What is the probability of getting 8 or fewer red balls in a draw of 10 balls from this urn, with replacement? (*Hint*: $\Pr[A] = 1 - \Pr[\tilde{A}]$.)

(b) What is the probability of getting 8 or fewer red balls in a draw of 10 balls from this urn, without replacement? (*Hint*: In addition to the part (a) hint, note that while the individual probabilities associated with getting 9 red and 1 white ball reflect order, their product does not.)

9. Consider a simultaneous roll of two dice, and a sample space defined as $S = \{(n_1, n_2, \ldots, n_6)\}$, where n_j denotes the number of dice showing j dots.

(a) Let $X = \sum_{j=1}^{6} j n_j$, the total number of dots showing on a roll, and determine the range of X and associated p.d.f., $f(x_j)$.

(b) Develop a graph of the c.d.f. of X: $F(x)$

10. In the sample space in exercise 9, define $Y = \sum_{j=1}^{3} j n_j$, and consider the pair of random variables (X, Y).

(a) Determine the range of (X, Y) and associated p.d.f. $f(x, y)$.

(b) Calculate the marginal p.d.f.s $f(x)$ and $f(y)$.

(c) Calculate the conditional p.d.f.s $f(x \mid y)$ and $f(y \mid x)$, and confirm the law of total probability, that $f(x) = \sum_y f(x \mid y) f(y)$ and $f(y) = \sum_x f(y \mid x) f(x)$.

11. Demonstrate that the two definitions of independence of a collection of random variables are equivalent where one is framed in terms of independence of pre-image events in S as in definition 7.32, and the other in terms of joint and marginal probability distribution functions as in definition 7.34.

12. Given a random variable with moments up to order N, demonstrate that the collections of moments and central moments can each be derived from the other. Specifically, using properties of expectations, show that for $n \leq N$:

(a) $\mu_n = \sum_{j=0}^{n} (-1)^{n-j} \binom{n}{j} \mu'_j \mu^{n-j}$ (*Hint*: Use the binomial theorem.)

(b) $\mu'_n = \sum_{j=0}^{n} \binom{n}{j} \mu_j \mu^{n-j}$ (*Hint*: $X = [X - \mu] + \mu$.)

13. Given a sample, $\{x_j\}_{j=1}^{n}$, and $\hat{\mu}'_k$ defined in (7.88), show the following under the assumption of the existence of the stated moments:

(a) $E[\hat{\mu}'_k] = \mu'_k$

(b) $\text{Var}[\hat{\mu}'_k] = \frac{1}{n}[\mu'_{2k} - (\mu'_k)^2]$

14. Develop the details for deriving the formulas in (7.99) for the standard binomial X_n^B. Generalize this derivation to the analogously defined general binomial $Y_n^B = \sum_{j=1}^{n} Y_{1j}^B$, where

$$Y_1^B = \begin{cases} a, & \Pr = p, \\ b, & \Pr = p'. \end{cases}$$

15. For the geometric distribution, let $\mu_m' \equiv E[j^m] = p\sum_{j=0}^{\infty} j^m (1-p)^j$ for $m \in \mathbb{N}$ and $m \geq 1$, and analogously, $\mu_0' = 1$.

(a) Show that these moments can be produced iteratively by

$$\mu_m' = \frac{1-p}{p} \sum_{j=0}^{m-1} \binom{m}{j} \mu_j'.$$

(*Hint*: Show that for $m \geq 1$, $\sum_{j=1}^{\infty} j^m (1-p)^j = (1-p)[1 + \sum_{j=1}^{\infty} (j+1)^m (1-p)^j]$, and use the binomial theorem.)

(b) Derive the mean and variance formulas in (7.103) from part (a).

16. For the Poisson distribution with parameter λ, show that:

(a) $\mu_P = \lambda$ (*Hint*: $j\frac{\lambda^j}{j!} = \lambda \frac{\lambda^{j-1}}{(j-1)!}$.)

(b) $\sigma_P^2 = \lambda$ (*Hint*: $j^2 \frac{\lambda^j}{j!} = \lambda(j-1)\frac{\lambda^{j-1}}{(j-1)!} + \lambda\frac{\lambda^{j-1}}{(j-1)!}$.)

17. Derive the mean and variance formulas for the aggregate loss model in (7.125) and (7.126) using a conditioning argument. (*Hint*: Classes are independent, so derive for class k by conditioning on N_k. Recall that here, N_k is binomial or Poisson, but it is not conditional on $N_k \geq 1$.)

18. An automobile insurer wants to model claims for collision costs on 10,000 insured automobiles, 2000 "luxury class" and 8000 "standard class." It estimates that the annual probability of a collision on any given auto is 0.10 for standard class and 0.06 for luxury class. The average value of insured autos is $25,000 for luxury and $10,000 for standard. Experience dictates that when an accident occurs, the cost to repair is uniformly distributed as a percentage of car value, and is 25–75% for luxury, and 50–100% for standard. Total repair costs in the two classes are assumed to be independent.

(a) Create an individual loss model for the insurer, and with it determine the mean and variance of repair costs.

(b) Create an aggregate loss model for the insurer using the Poisson distribution, and with it determine the mean and variance of repair costs.

(*Hint*: The mean and variance of the uniform distribution equal the limits of these moments for the discrete rectangular distribution as $n \to \infty$. See (7.95) and also chapter 10.)

19. Demonstrate that the expected value of a life annuity can also be expressed in terms of the survival function by

$$E[B] = \sum_{j=0}^{\infty} v^j S^M(j).$$

20. Calculate $E[B]$ and $Var[B]$ using a conditioning argument in the following cases:

(a) Let B denote the random variable that equals the present value of annuity payments that are payable at the end of each year survived for life, but guaranteed payable for a minimum of m years, independent of survival. This is an "m-year certain life annuity."

(b) Let B denote the random variable that equals the present value of annuity payments that are only payable for survival up through the end of n years. This is an "n-year temporary life annuity."

(c) Let B denote the random variable that equals the present value of annuity payments that are only payable for survival up through the end of n years, but guaranteed payable for a minimum of m years, independent of survival, where $m < n$. This is an "m-year certain, n-year temporary life annuity."

21. Let B denote the random variable in each of parts (a) through (c) of exercise 20, but redefined to allow for a k year deferral of benefits. So each annuity is "k-year deferred" version of the annuity defined above. Consider the case where:

(a) No benefit is paid if death occurs during the first k years.

(b) A benefit of $1 is paid at the end of year of death if death occurs during the first k years.

(c) Show that the benefit in part (b) equals the benefit in part (a) plus a k year term life policy from exercise 37(a).

22. Assume that: $r_F = 0.05$, $\mu_1 = 0.065$, $\mu_2 = 0.09$, $\mu_3 = 0.15$, $\sigma_1^2 = (0.07)^2$, $\sigma_2^2 = (0.12)^2$, $\sigma_3^2 = (0.18)^2$, $\rho_{12} = 0.35$, $\rho_{23} = 0.4$, and $\rho_{13} = 0.25$:

(a) Develop formulas for the mean and variance of portfolio returns for an arbitrary allocation to three risky assets and the risk-free asset.

(b) Define $\mathbf{W} = (0.25, 0.25, 0.25, 0.25)$, and evaluate an epsilon shift between the risk-free and third risky asset. Graph both $E[R_\epsilon^{03}] - E[R]$ and $Var[R_\epsilon^{03}] - Var[R]$ as functions of ϵ for $-0.25 \le \epsilon \le 0.25$.

23. Generalize the calibration of the growth model for stock prices in (7.136) to develop formulas for u and d for arbitrary p, $0 < p < 1$, where $p = Pr[u]$, being explicit about the binomial probabilities that govern the associated price lattice in (7.138). (*Hint*: Proceed as before, showing that with the binomial B defined as in section 7.8.5, and $p' \equiv 1 - p$, $E[B] = pu + p'd$ and $Var[B] = pu^2 + p'd^2 - (pu + p'd)^2$.)

24. Price a 2-year European call, with strike price of 100, in the ways noted below. The stock has $S_0 = 100$, and based on time steps of $\Delta t = 0.25$ years, the quarterly log-ratios have been estimated to have $\mu_Q = 0.02$ and $\sigma_Q^2 = (0.07)^2$. The annual continuous risk-free rate is $r = 0.048$, and so for $\Delta t = 0.25$ years, you can assume that $r_Q = 0.012$.

(a) Develop a real world lattice of quarterly stock prices, with $p = \frac{1}{2}$, and price this option using (7.147).

(b) Evaluate the two prices of this option at time $t = 0.25$ from part (a), and construct a replicating portfolio at $t = 0$ for these prices. Demonstrate that the cost of this replicating portfolio equals the price obtained in part (a).

(c) Using exercise 23, price this option using (7.147) with the appropriate value of q based on a lattice for which $p = 0.25$.

(d) Generate one hundred 2-year paths in the risk-neutral world, each with quarterly time steps and using the model of part (a). Then estimate the price of this option using (7.150), by counting how many scenarios end in each stock price at time 2 years.

25. Demonstrate that the conclusion following (7.143), that $0 < q < 1$, follows from $e^d < e^r < e^u$, and that this latter conclusion is demanded by an arbitrage argument. (*Hint*: Show that if $e^r \leq e^d$ or $e^r \geq e^u$, then there would be a trade at time 0 that costs nothing, has no probability of a loss, and a positive probability of a profit over the period.)

Assignment Exercises

26. Prove the following properties of a probability measure based on the properties in definition 7.7:

(a) $\Pr(\emptyset) = 0$

(b) If $A, B \in \mathcal{E}$, $A \subset B$, then $\Pr(A) \leq \Pr(B)$. (*Hint*: Split B into disjoint sets.)

(c) If $A_j \in \mathcal{E}$ for $j = 1, 2, 3, \ldots$, then $\Pr(\bigcup_j A_j) \leq \sum \Pr(A_j)$. (*Hint*: Split $\bigcup_j A_j$ into disjoint sets.)

(d) If $A_j \in \mathcal{E}$ for $j = 1, 2, 3, \ldots$, then $\Pr(\bigcap_j A_j) \leq \min_j\{\Pr(A_j)\}$. (*Hint*: $\bigcap_j A_j \subset A_k$ for all k.)

27. Generalize exercise 2 and confirm that in the sample space $\mathcal{S}$ of sequences of n flips of a fair coin, the event A defined by specifying the values of any $m \leq n$ outcomes, contains exactly $\frac{100}{2^m}\%$ of the total number of sequences. As before, if you ignore what happens outside of these m flips, justify explicitly that this is valid.

28. Answer exercise 4 in the case of a lottery rather than a bet. (*Hint:* A lottery is the same as a bet with different payoffs.)

(a) If a gambler buys the lottery ticket for m, and either wins 0 or n, what is the probability of winning that will make this lottery fair?

(b) If the gambler knows that the probability of a win is p, what ratio of the cost of a ticket to amount won, $\frac{m}{n}$ from part (a), will make this a fair lottery?

(c) Show that if p is irrational in part (b), a fair lottery requires an irrational value for $\frac{m}{n}$. Conclude that since ticket prices and payoffs must be rational numbers, a lottery with an irrational probability of winning can never be fair.

29. Generalize the event B in exercise 6 to $B_j = \{s \in S \mid \sum_{i=3}^{5} s_i = j\}$.

(a) What points are in the event $A \cap B_j$ for all j?

(b) Show that $\bigcup B_j = S$.

(c) Confirm the law of total probability, that $\Pr[A] = \sum_j \Pr[A \mid B_j]\, \Pr[B_j]$.

30. Consider a simultaneous roll of 21 dice, and a sample space defined as $S = \{(n_1, n_2, \ldots, n_6)\}$, where n_j denotes the number of dice showing j dots. Develop formulaic or numerical solutions to the following:

(a) What is the probability of the sample point $s = (1, 2, 3, 4, 5, 6)$?

(b) What is the probability of the event $A = \{s \mid n_6 = 12 \text{ and } n_3 = 2\}$? (*Hint:* Can this event be defined in terms of (n_3, n_6, n_{other}) with adjusted probabilities?)

31. Consider a simultaneous flip of 5 unfair coins, $\Pr[H] = 0.3$, and a sample space defined as $S = \{(n_1, n_2) \mid n_1 \text{ denotes the number of } Hs, \text{ and } n_2 \text{ the number of } Ts\}$.

(a) Let $X = 0.01 \sum_{j=1}^{2} 10^j n_j$, and determine the range of X and associated p.d.f. $f(x_j)$.

(b) Develop a graph of the c.d.f. of X: $F(x)$.

32. In the sample space in exercise 31, define $Y = n_1$, and consider the pair of random variables (X, Y).

(a) Determine the range of (X, Y) and associated p.d.f. $f(x, y)$.

(b) Calculate the marginal p.d.f.s $f(x)$ and $f(y)$.

(c) Calculate the conditional p.d.f.s $f(x \mid y)$ and $f(y \mid x)$, and confirm the law of total probability, that $f(x) = \sum_y f(x \mid y) f(y)$ and $f(y) = \sum_x f(y \mid x) f(x)$.

33. Demonstrate algebraically the iterative formula in (7.16) underlying Pascal's triangle: $\binom{n}{m} = \binom{n-1}{m-1} + \binom{n-1}{m}$.

34. Given a sample $\{x_j\}_{j=1}^{n}$, and $\hat{M}_X(t)$ defined in (7.90), show the following under the assumption of the existence of the stated moments:

(a) $E[\hat{M}_X(t)] = M_X(t)$

(b) $\text{Var}[\hat{M}_X(t)] = \frac{1}{n}[M_X(2t) - M_X^2(t)]$

35. Using the 2-variable joint p.d.f. derived for the multinomial distribution and (7.50), show that for any two components with $i \neq j$: $\text{Cov}[N_i, N_j] = -np_i p_j$. (*Hint:* First justify:

$$E[N_1 N_2] = \sum_{n_1=1}^{n-1} \sum_{n_2=1}^{n-n_1} n_1 n_2 \frac{n! p_1^{n_1} p_2^{n_2} (1 - p_1 - p_2)^{n-n_1-n_2}}{n_1! n_2! (n - n_1 - n_2)!}.$$

Then split this summation as the product

$$\sum_{n_1=1}^{n-1} n_1 \frac{n! p_1^{n_1} (1 - p_1)^{n-n_1}}{n_1! (n - n_1)!} \times \sum_{n_2=1}^{n-n_1} n_2 \frac{(n - n_1)!}{n_2! (n - n_1 - n_2)!} \left(\frac{p_2}{1 - p_1}\right)^{n_2} \left(\frac{1 - p_1 - p_2}{1 - p_1}\right)^{n-n_1-n_2},$$

and note that this second summation is $E[n_2]$ with a certain binomial distribution. Alternatively, start with the double summation above, simplify $\frac{n_1 n_2}{n_1! n_2!}$, and look for the binomial theorem.)

36. A bond portfolio quantitative analyst wants to model credit losses on a $750 million portfolio, which includes three classes of credit risk: $250 million "low risk," $350 million "medium risk," and $150 million "high risk," where in each class the manager has maintained a $5 million average par investment exposure per credit. Annual default probabilities are 0.002, 0.009, and 0.025. Experience dictates that when a default occurs, the loss is uniformly distributed as a percentage of par value, and is 25–50% for low risk, 25–75% for medium risk, and 50–100% for high risk. Total credit losses in the three classes are assumed to be independent.

(a) Create an individual loss model for the analyst, and with it determine the mean and variance of credit losses.

(b) Create an aggregate loss model for the analyst using the Poisson distribution, and with it determine the mean and variance of credit losses.

(*Hint:* The mean and variance of the uniform distribution equal the limits of these moments for the discrete rectangular distribution as $n \to \infty$. See (7.95) and also chapter 10.)

37. Calculate $E[I_n]$ and $\text{Var}[I_n]$ using a conditioning argument in the following cases:

(a) Let I_n denote the random variable which equals the present value of a life insurance payment at the end of the year of death, but where a payment is made only if death occurs in the first n years. This is an "n-year term insurance" contract.

(b) Let I_n denote the random variable that equals the present value of a life insurance payment at the end of the year of death if death occurs in the first n years, or a payment of \$1 at time $t = n$ if the individual survives the n years. This is an "n-year endowment" contract.

38. Assuming that: $r_F = 0.03$, $\mu_1 = 0.095$, $\mu_2 = 0.19$, $\mu_3 = 0.15$, $\sigma_1^2 = (0.12)^2$, $\sigma_2^2 = (0.25)^2$, $\sigma_3^2 = (0.18)^2$, $\rho_{12} = 0.55$, $\rho_{23} = 0.4$, and $\rho_{13} = 0.20$:

(a) Develop formulas for the mean and variance of portfolio returns for an arbitrary allocation to three risky assets and the risk-free asset.

(b) Define $\mathbf{W} = (0.25, 0.25, 0.25, 0.25)$, and evaluate an epsilon shift between the second and third risky asset. Graph both $E[R_\epsilon^{23}] - E[R]$ and $\mathrm{Var}[R_\epsilon^{23}] - \mathrm{Var}[R]$ as functions of ϵ for $-0.25 \leq \epsilon \leq 0.25$.

39. Prove the formula in (7.147) using mathematical induction. (*Hint*: The formula is proved for $n = 1, 2$ already. Assume it to be true for n, and show it is true for $n + 1$ by applying the assumed formula to the two values of the option at time 1, $\Lambda(S_1^u)$ and $\Lambda(S_1^d)$. Recall exercise 33.)

40. Price a 2-year European put, with strike price of 100, in the ways noted below. The stock has $S_0 = 100$, and based on time steps of $\Delta t = 0.25$ years, the quarterly log-ratios have been estimated to have $\mu_Q = 0.025$, and $\sigma_Q^2 = (0.09)^2$. The annual continuous risk-free rate is $r = 0.06$, and so for $\Delta t = 0.25$ years you can assume that $r_Q = 0.015$.

(a) Develop a real world quarterly lattice of stock prices, with $p = \frac{1}{2}$, and price this option using (7.147).

(b) Evaluate the two prices of this option at time $t = 0.25$ from part (a), and construct a replicating portfolio at $t = 0$ for these prices. Demonstrate that the cost of this replicating portfolio equals the price obtained in part (a).

(c) Using exercise 23, price this option using (7.147) with the appropriate value of q based on a lattice for which $p = 0.35$.

(d) Generate one hundred 2-year paths in the risk-neutral world, each with quarterly time steps and using the model of part (a), and estimate the price of this option using (7.150), by counting how many scenarios end in each stock price at time 2 years.

41. Using (7.147), if Λ_0^C and Λ_0^P denote the $t = 0$ prices of European call and put options, respectively, both with a strike price of K and maturity of T, show that these prices satisfy **put-call parity**:

$$\Lambda_0^C + Ke^{-rT} = \Lambda_0^P + S_0, \tag{7.151}$$

where r denotes the risk-free rate in units of T.

8 Fundamental Probability Theorems

In this chapter is introduced several of the very important theorems from probability theory. Although a number of these results are somewhat challenging to demonstrate, they all have a great many applications. This is due to the great generality of the conclusions and the relatively minimal assumptions needed to produce them.

8.1 Uniqueness of the m.g.f. and c.f.

In this section we demonstrate a limited version of the result quoted in chapter 7, that if $C_X(t) = C_Y(t)$ or $M_X(t) = M_Y(t)$ for discrete random variables X and Y, and for some open interval I, containing 0, then the probability density functions are equal: $f_X(x) = g_Y(x)$. The narrower version of this result contemplated here assumes that these random variable have finite ranges. This result can be shown to be true in a more general context than finite discrete p.d.f.s, or even discrete p.d.f.s, but requires the tools of real analysis and complex analysis.

Proposition 8.1 *Let X and Y be finite discrete random variables with associated probability functions $f(x)$ and $g(y)$, and respective domains of $\{x_i\}_{i=1}^n$ and $\{y_j\}_{j=1}^m$, arranged in increasing order. If either $C_X(t) = C_Y(t)$ or $M_X(t) = M_Y(t)$ for $t \in I$, where I is an open interval containing 0, then $m = n$, $x_i = y_i$, and $f(x_i) = g(y_i)$ for all i.*

Proof If $M_X(t) = M_Y(t)$ for $t \in I$, then $\sum e^{tx_i} f(x_i) = \sum e^{ty_i} g(y_i)$. Consequently there are collections of real numbers $\{a_k\}$ and $\{b_k\}$, where the $\{b_k\}$ are all distinct, so that

$$\sum_{k=1}^{N} a_k e^{tb_k} = 0 \qquad \text{for } t \in I. \tag{8.1}$$

In other words, for cases where $x_i = y_j$ for some i and j, $a_k = f(x_i) - g(y_j)$ and $b_k = x_i = y_j$. In all other cases, a_k is either an $f(x_i)$ or a $g(y_j)$ term, and the associated b_k is x_i, respectively y_j. We now show that if (8.1) holds, then $a_k = 0$ for all k. This provides the result, since it means that for any $x_i = y_j$, it must be the case that $f(x_i) = g(y_j)$, whereas for any x_i or y_j with no "match," $f(x_i) = 0$ or $g(y_j) = 0$, respectively. The proof proceeds by induction on N. The result is apparently true for $N = 1$, since $a_1 e^{tb_1} = 0$ for $t \in I$ clearly implies that $a_1 = 0$. This result is also apparent for $N = 2$, since in this case it is concluded that $a_2 e^{t(b_2 - b_1)} = a_1$, but this is impossible unless $a_1 = a_2 = 0$, since $b_2 - b_1 \neq 0$. Assume next that the result holds for N, and that we seek to demonstrate the result for $N + 1$. Now $\sum_{k=1}^{N+1} a_k e^{tb_k} = 0$ implies that $\sum_{k=1}^{N} a_k e^{tc_k} = -a_{N+1}$ for $t \in I$, where $c_k = b_k - b_{N+1}$, and $\{c_k\}$ are all distinct and, importantly, all nonzero, since the $\{b_k\}$ are all distinct by assumption. Now, if

$s, t \in I$, this equation implies that $\sum_{k=1}^{N} a_k e^{tc_k} = \sum_{k=1}^{N} a_k e^{sc_k}$. This result can then be expressed as follows if $s \neq t$:

$$\sum_{k=1}^{N} a_k e^{sc_k} \left[\frac{e^{(t-s)c_k} - 1}{t - s} \right] = 0.$$

Now from (7.63) note that $\frac{e^{(t-s)c_k}-1}{t-s} = c_k + (t - s)\left[\frac{c_k^2}{2} + X_k\right]$, where X_k is an absolutely convergent summation of terms, all of which contain positive powers of $(t - s)$. Consequently, using the identities above obtains

$$\sum_{k=1}^{N} a_k c_k e^{sc_k} = \sum_{k=1}^{N} a_k \left[c_k - \frac{e^{(t-s)c_k} - 1}{t - s} \right] e^{sc_k}$$

$$= -(t - s) \sum_{k=1}^{N} a_k \left[\frac{c_k^2}{2} + X_k \right].$$

Now as $t \to s$, since each $X_k \to 0$ as noted above, we conclude that

$$\sum_{k=1}^{N} a_k c_k e^{sc_k} = 0.$$

From the induction step for N we conclude that $a_k c_k = 0$ for $1 \leq k \leq N$, and since $c_k \neq 0$, it must be the case that $a_k = 0$ for $1 \leq k \leq N$. Finally, this implies that $a_{N+1} = 0$ by substitution. To extend this proof to characteristic functions is immediate, with one subtlety, and that is the applicability of (7.63) to an exponential of the form e^{ix}, where $i = \sqrt{-1}$ and $x \in \mathbb{R}$. In this case the resulting power series is again seen to be absolutely convergent by the ratio test, and this series is equal to e^{ix} because that is how e^{ix} is defined!   ∎

Remark 8.2   *The proof above cannot be adapted to a countably infinite discrete probability function, and for that case an entirely different approach is needed, requiring a new and advanced set of tools. These tools will also handle this result for p.d.f.s that are not discrete. The problem is that while we could again conclude (8.1) with $N = \infty$, and the trick employed above adapted, this would only yield*

$$\sum_{k=2}^{\infty} a_k c_k e^{sc_k} = 0,$$

which provides no real simplification.

8.2 Chebyshev's Inequality

Chebyshev's inequality, sometimes spelled as Chebychev or Tchebysheff, applies to any probability density function that has a mean and variance, and hence it is quite generally applicable. It is named for its discoverer, **Pafnuty Chebyshev** (1821–1894). Chebyshev was a Russian mathematician, and hence the many transliterations of his name in English.

This inequality can be stated in many ways, and Chebyshev is actually a name now given to a family of inequalities as will be seen below. But this inequality is often applied as stated in the following proposition, when we are interested in an upper bound for the probability of the random variable being far from its mean, where "far" is measured in two common ways. Although the Chebyshev inequalities are stated here for discrete $f(x)$, it is an easy exercise to generalize these to continuous $f(x)$ using the tools of chapter 10.

Proposition 8.3 (*Chebyshev's inequality*) *If $f(x)$ is a discrete probability function with mean μ and variance σ^2, then for any real number $t > 0$,*

$$\Pr[|X - \mu| \geq t\sigma] \leq \frac{1}{t^2}. \tag{8.2}$$

Equivalently,

$$\Pr[|X - \mu| \geq s] \leq \frac{\sigma^2}{s^2}. \tag{8.3}$$

Proof By definition, $\sigma^2 = \sum_{x_i}(x_i - \mu)^2 f(x_i) \geq \sum_{|x_i-\mu|\geq t\sigma}(x_i - \mu)^2 f(x_i)$. In other words, in this last summation, only the x_i terms that satisfy $|x_i - \mu| \geq t\sigma$ are included. This second summation now satisfies $\sum_{|x_i-\mu|\geq t\sigma}(x_i - \mu)^2 f(x_i) \geq (t\sigma)^2 \sum_{|x_i-\mu|\geq t\sigma} f(x_i)$, and this last summation is seen to equal $\Pr[|X - \mu| \geq t\sigma]$. Combining the inequalities and dividing by σ^2 provides the first result. The second result is implied by the first with the substitution $t = \frac{s}{\sigma}$. ∎

Note that for any t with $t \leq 1$, this inequality provides no real limit on the associated probability, since in such a case, $\frac{1}{t^2} \geq 1$. However, using integral multiples of the standard deviation we obtain

$$\Pr[|X - \mu| \geq 2\sigma] \leq \frac{1}{4} = 0.25,$$

$$\Pr[|X - \mu| \geq 3\sigma] \leq \frac{1}{9} \approx 0.11,$$

$$\Pr[|X - \mu| \geq 4\sigma] \leq \frac{1}{16} \approx 0.06,$$

and so forth.

For example, if X^B has the binomial distribution with parameters n and p, then

$$\Pr[|X^B - np| \geq s] \leq \frac{np(1 - p)}{s^2}.$$

Similarly, for the negative binomial distribution with parameters p and k, we conclude that

$$\Pr\left[\left|X^P - \frac{k(1 - p)}{p}\right| \geq s\right] \leq \frac{k(1 - p)}{s^2 p^2}.$$

This inequality can be generalized in many ways. For example, an estimate of a probability of the form $\Pr[|X| \geq s]$ can be made with the same formula, except with μ_2' used instead of $\sigma^2 = \mu_2$. The proof above also readily applies to the case of μ_{2n} for any n, which then bounds the associated probabilities in terms of higher order central moments. In the case of odd central moments the proof only works when absolute values are introduced. We state the generalization in the form of absolute values, though the absolute value is redundant for even moments.

Proposition 8.4   *If $f(x)$ is a discrete probability function, with mean μ and absolute central moment $\mu_{|n|} \equiv \mathrm{E}[|X - \mu|^n]$ for $n \geq 1$, then for any real number $t > 0$,*

$$\Pr[|X - \mu| \geq t] \leq \frac{\mu_{|n|}}{t^n}. \tag{8.4}$$

Proof   By definition,

$$\mu_{|n|} = \sum_{x_i} |x_i - \mu|^n f(x_i) \geq \sum_{|x_i - \mu| \geq t} |x_i - \mu|^n f(x_i) \geq t^n \Pr[|X - \mu| \geq t],$$

and the result follows by division.   ∎

Once again, probabilities of the form $\Pr[|X| \geq t]$ can be bounded by the corresponding formula, with $\mu_{|n|}' \equiv \mathrm{E}[|X|^n]$. In this case, if the random variable has its range in the nonnegative real numbers, these estimates apply without the absolute values, that is, by using the moments μ_n' directly.

In exercise 1 is assigned the development of a probability estimate utilizing the moment-generating function $M_X(t)$.

Remark 8.5

1. *Note that when $n = 1$, the inequality in (8.4) restated in terms of $\mu'_{|1|} \equiv \mathrm{E}[|X|]$ is known as **Markov's inequality**, named for **Andrey Markov** (1856–1922), a student of Chebyshev. In other words,*

$$\Pr[|X| \geq t] \leq \frac{\mathrm{E}[|X|]}{t}. \tag{8.5}$$

2. *Note also that if $f(x)$ is a p.d.f. with $\mu_{|n|} = 0$ for some $n \geq 1$, then it must be the case that $\Pr[X = \mu] = 1$. In other words, the random variable X assumes only the value μ. This is because in (8.4) the inequality states that $\Pr[|X - \mu| \geq t] \leq 0$ for any $t > 0$, but since probabilities are nonnegative, we conclude that $\Pr[|X - \mu| \geq t] = 0$ for any $t > 0$ and therefore $\Pr[X = \mu] = 1$. Such a random variable is referred to as a **degenerate random variable**, and the associated p.d.f., a **degenerate probability density**, with no insult intended.*

There is also a one-sided version of the Chebyshev inequality that is useful when the focus of the investigation is on one and not both tails of the distribution. For instance, if we are modeling losses in a credit portfolio, we are interested in the probability of losses being large and positive relative to expected losses, and not so much interested in the probability that losses could be either large or small relative to this expected value. The following result gives a better bound than (8.3) in this case, and the amount of improvement grows with σ^2:

Proposition 8.6 (*Chebyshev's One-Sided Inequality*) *If $f(x)$ is a discrete probability function, with mean μ and variance σ^2, then for any real number $s > 0$,*

$$\Pr[X - \mu \geq s] \leq \frac{\sigma^2}{s^2 + \sigma^2}. \tag{8.6}$$

Proof For any value of t, we have

$$\Pr[X - \mu \geq s] = \Pr[X - \mu + t \geq s + t] \leq \Pr[(X - \mu + t)^2 \geq (s + t)^2].$$

This is because the last probability statement also encompasses $\Pr[-(X - \mu + t) \leq -(s + t)]$. Now, by the Markov inequality in (8.5) and a little algebra,

$$\Pr[(X - \mu + t)^2 \geq (s + t)^2] \leq \frac{\mathrm{E}[(X - \mu + t)^2]}{(s + t)^2} = \frac{\sigma^2 + t^2}{(s + t)^2}.$$

Summarizing we obtain

$$\Pr[X - \mu \geq s] \leq \frac{\sigma^2 + t^2}{(s+t)^2} \qquad \text{for any } t > 0.$$

Since t can be chosen arbitrarily, we do so to make the bound $\frac{\sigma^2+t^2}{(s+t)^2}$ as small as possible. Using the methods of calculus discussed in chapter 9, we find the value of t that minimizes this bound to be $t = \frac{\sigma^2}{s}$, and a substitution demonstrates that this produces the bound in (8.6). ∎

This one-sided inequality can also be expressed in units of the variance as in (8.2) as follows:

$$\Pr[X - \mu \geq t\sigma] \leq \frac{1}{t^2 + 1}. \tag{8.7}$$

8.3   Weak Law of Large Numbers

The so-called weak law of large numbers is actually a very powerful and general result with wide applicability but with the misfortune to be a relative of an even more general result, known as the strong law of large numbers. Like the Chebyshev inequality, it has the power of being applicable to virtually any probability distribution. Unlike the Chebyshev inequality, which requires that these distributions have both a mean and variance, the weak law requires only the existence of the first moment, but it is far easier to prove when the variance also exists.

Before giving its statement, recall that if a random variable X is defined on a discrete sample space S, then a random sample of size n of this random variable can be associated with a sample point in the n-**trial sample space**, denoted S^n, with probability structure defined in (7.7). The components of this sample point are then called **independent and identically distributed (i.i.d.) random variables**.

Proposition 8.7 (Weak Law of Large Numbers)   *For any n, let $\{X_i\}_{i=1}^n$ be independent and identically distributed random variables with common mean μ. Define the random variable $\hat{X}$ as the average, $\hat{X} = \frac{1}{n}\sum_{i=1}^n X_i$. Then for any $\epsilon > 0$:*

$$\Pr[|\hat{X} - \mu| > \epsilon] \to 0 \qquad \text{as } n \to \infty. \tag{8.8}$$

Remark 8.8   *Note that if $\{X_i\}_{i=1}^n$ are defined on the discrete sample space S, then $\hat{X}$ is a random variable defined on the n-trial sample space S^n. The formal meaning of the statement in (8.8) is that for any fixed $\epsilon > 0$, the events $V_\epsilon^n \subset S^n$ in the n-trial sample spaces S^n, defined by*

$$V_\epsilon^n = \{(X_1, \ldots, X_n) \mid |\hat{X} - \mu| > \epsilon\},$$

satisfy $\Pr[V_\epsilon^n] \to 0$ *as* $n \to \infty$.

The intuitive meaning of the statement in (8.8) can be described as follows: Suppose that for any n we can easily generate as many samples $\{X_i\}_{i=1}^n$ as desired, and for each sample calculate the associated sample average $\hat{X}$. On the real line we then plot the collection of averages and determine the proportion of these that are outside the interval $[\mu - \epsilon, \mu + \epsilon]$. The weak law asserts that for any $\epsilon > 0$, the proportion of sample averages outside this interval converges to 0 as $n \to \infty$. In general, the weak law provides no information on the speed at which this proportion converges, but see below the case where X also has a finite variance.

Proof   We prove this result in two cases. In applications the first case is often satisfied.

1. If the random variable X also has a variance σ^2, the weak law is an immediate consequence of Chebyshev's inequality and the formulas above for sample moments. As developed in (7.78) and (7.79), we have $\mathrm{E}[\hat{X}] = \mu$, and $\mathrm{Var}[\hat{X}] = \frac{\sigma^2}{n}$, which when substituted into (8.3) provides the result

$$\Pr[|\hat{X} - \mu| > \epsilon] \leq \frac{\sigma^2}{n\epsilon^2}. \tag{8.9}$$

This implies more than (8.8), and assures that this probability converges to 0 with a rate at least as fast as $\frac{c}{n}$ for $c = \frac{\sigma^2}{\epsilon^2}$.

2. In the general case we introduce the **method of truncation**, whereby, for each n and arbitrary but fixed $\lambda > 0$, the collection $\{X_i\}_{i=1}^n$ is truncated and split as

$$Y_i = \begin{cases} X_i - \mu, & |X_i - \mu| \leq \lambda n, \\ 0, & |X_i - \mu| > \lambda n; \end{cases}$$

$$Z_i = \begin{cases} 0, & |X_i - \mu| \leq \lambda n, \\ X_i - \mu, & |X_i - \mu| > \lambda n. \end{cases}$$

So $X_i - \mu = Y_i + Z_i$. Now with $\hat{Y}$ and $\hat{Z}$ defined as the associated averages, note that (see exercise 15)

$$\Pr[|\hat{X} - \mu| > \epsilon] \leq \Pr\left[|\hat{Y}| > \frac{\epsilon}{2}\right] + \Pr\left[|\hat{Z}| > \frac{\epsilon}{2}\right].$$

The weak law follows if it can be shown that for some $\lambda > 0$, the two probabilities on the right can be made as small as desired. For the first probability, note that since $|Y_1| \le \lambda n$,

$$E[(Y_1)^2] \le \lambda n E[|Y_1|] < \lambda n \mu_{|1|},$$

where $\mu_{|1|} = E[|X_1 - \mu|]$. Now $\{Y_i\}_{i=1}^n$ are independent because of the independence of $\{X_i\}_{i=1}^n$, and

$$\text{Var}[\hat{Y}] = \frac{1}{n}\,\text{Var}[Y_1] \le \frac{1}{n}E[(Y_1)^2] < \lambda \mu_{|1|}.$$

Then by Chebyshev's inequality,

$$\Pr\left[|\hat{Y} - E[\hat{Y}]| > \frac{\epsilon}{2}\right] \le \frac{4\lambda \mu_{|1|}}{\epsilon^2}.$$

But $E[\hat{Y}] \to E[\hat{X} - \mu] = 0$ as $n \to \infty$. So by choosing λ small, we can make $\Pr\left[|\hat{Y}| > \frac{\epsilon}{2}\right]$ as small as desired for any ϵ as $n \to \infty$.

For the second probability, we show that $\Pr[|\hat{Z}| > 0] \to 0$ as $n \to \infty$ for any λ. By a consideration of the associated events and the independence of $\{Z_i\}_{i=1}^n$, we write

$$\Pr[|\hat{Z}| > 0] \le \sum \Pr[|Z_i| > 0] = n\,\Pr[|Z_1| > 0].$$

But, by definition,

$$\Pr[|Z_1| > 0] = \Pr[|X_i - \mu| > \lambda n]$$

$$= \sum_{|x_i - \mu| > \lambda n} f(x_i)$$

$$\le \frac{1}{\lambda n} \sum_{|x_i - \mu| > \lambda n} |x_i - \mu| f(x_i).$$

Then, combining, we have

$$\Pr[|\hat{Z}| > 0] \le \frac{1}{\lambda} \sum_{|x_i - \mu| > \lambda n} |x_i - \mu| f(x_i),$$

which converges to 0 for any λ as $n \to \infty$. ∎

In the common application to a random variable with mean and variance, this law also provides a lower bound for the probability that the estimate will be close to the expected value. In other words, if μ and σ^2 exist, then

$$\Pr[|\hat{X} - \mu| \leq \epsilon] > 1 - \frac{\sigma^2}{n\epsilon^2}, \tag{8.10}$$

which is only useful, of course, when $\frac{\sigma^2}{n\epsilon^2} \leq 1$ or $\epsilon \geq \frac{\sigma}{\sqrt{n}}$. In the general case all that can be said is that

$$\Pr[|\hat{X} - \mu| \leq \epsilon] \to 1 \qquad \text{as } n \to \infty. \tag{8.11}$$

The formulation in (8.10) can then be understood in the context of providing a general **confidence interval** for the theoretical mean μ, which we may be interested in estimating using a sample mean $\hat{X}$. Specifically, define the closed interval I_ϵ by

$$I_\epsilon \equiv [\hat{X} - \epsilon, \hat{X} + \epsilon]. \tag{8.12}$$

Then the weak law of large numbers says that if $\{X_i\}_{i=1}^n$ are independent and identically distributed random variables with common mean μ and variance σ^2, then

$$\Pr[\mu \in I_\epsilon] > 1 - \frac{\sigma^2}{n\epsilon^2}. \tag{8.13}$$

To be clear, in any given application with sample statistic $\hat{X}$, it will be the case that either $\mu \in I_\epsilon$ or $\mu \notin I_\epsilon$. The probability statement in (8.13) needs to be interpreted in the context of n-trial sample space $\mathcal{S}^n$. Specifically, for $(X_1, X_2, \ldots, X_n) \in \mathcal{S}^n$, let $\hat{X} = \frac{1}{n}\sum_{i=1}^n X_i$, and define the event $\widetilde{V_\epsilon^n} \in \mathcal{E}$, the complement in $\mathcal{S}^n$ of the event in remark 8.8 above, by

$$\widetilde{V_\epsilon^n} \equiv \{(X_1, X_2, \ldots, X_n) \in \mathcal{S}^n \mid \mu \in [\hat{X} - \epsilon, \hat{X} + \epsilon]\},$$

where μ is the mean of the random variable X. Then (8.13) states that for any $\epsilon > 0$,

$$\Pr\left[\widetilde{V_\epsilon^n}\right] > 1 - \frac{\sigma^2}{n\epsilon^2}, \tag{8.14}$$

where σ^2 is the variance of X.

The weak law, with exactly the same proof and interpretations, applies to all of the sample moment estimates developed earlier, since all that was assumed in the proof above was that $\hat{X}$ is a random variable defined on n-trial sample space $\mathcal{S}^n$ and that the μ and σ^2 in (8.9) are, respectively, the mean and variance of this random variable.

Example 8.9   *With $\hat{\sigma}^2 = \frac{1}{n-1}\sum_{j=1}^{n}(X_j - \hat{X})^2$, the unbiased variance estimator, since $E[\hat{\sigma}^2] = \sigma^2$, we have that for any random sample of size n,*

$$\Pr[|\hat{\sigma}^2 - \sigma^2| > \epsilon] \leq \frac{(n-1)\mu_4 - (n-3)\sigma^4}{n(n-1)\epsilon^2},$$

where the upper bound for this probability reflects $\mathrm{Var}[\hat{\sigma}^2]$. For higher moments, with higher moment estimators defined by $\hat{\mu}'_k = \frac{1}{n}\sum_{j=1}^{n} X_j^k$, we have that for any random sample of size n,

$$\Pr[|\hat{\mu}'_k - \mu'_k| > \epsilon] \leq \frac{\mu'_{2k} - (\mu'_k)^2}{n\epsilon^2}.$$

Here again it is used that $E[\hat{\mu}'_k] = \mu'_k$, and the upper bound for this probability reflects $\mathrm{Var}[\hat{\mu}'_k]$.

The critical observation on all these probability estimates is that each probability is proportional to $\frac{1}{n}$, which is favorable as we can select $n \to \infty$, but is also proportional to $\frac{1}{\epsilon^2}$, which is unfavorable if we desire to have $\epsilon \to 0$. But for any desired margin of error ϵ, we can use these formulas to determine how large the sample size n needs to be so that the sample estimator will be within that margin of error with any probability that is desired.

Example 8.10   *To estimate the parameter $\lambda = E[X_P]$ for a Poisson distribution, the statement above produces*

$$\Pr[|\hat{X} - \lambda| > \epsilon] \leq \frac{\lambda}{n\epsilon^2},$$

which is initially a bit of a problem due to the presence of the unknown $\lambda = \mathrm{Var}(X_P)$ in the probability upper bound. However, it is commonly the case that a crude upper bound can be used successfully. For example, if a given sample produced $\hat{X} = 3$, we might be comfortable assuming $\lambda \leq 5$, and hence the probability statement above becomes

$$\Pr[|\hat{X} - \lambda| > \epsilon] \leq \frac{5}{n\epsilon^2}.$$

In order to have 1 *decimal point accuracy on the estimate for* λ, *we choose* $\epsilon = 0.05$ *and derive*

$$\Pr[|\hat{X} - \lambda| > 0.05] \leq \frac{2000}{n},$$

from which, with $n = 200{,}000$, *a random sample will have less than a* 1% *probability of producing an error in the first decimal place. Of course, if a smaller upper bound is assumed for* λ, *and/or a lower level of confidence desired, smaller samples will suffice.*

Remark 8.11   *This example reflects a practical constraint on the use of the weak law in empirical estimates. While this law provided a calculation of* $n = 200{,}000$ *to achieve the desired result, most statisticians would agree that this is an enormous sample, and almost certainly a sample size that is far bigger than what is truly needed. The problem is that the empirical weakness of this law is caused by its theoretical strength. Specifically, this law applies to every random variable that has a finite mean, or in the applications above, every random variable with finite mean and variance. Because of this generality, it would be unlikely that the formula provided would be efficient empirically when applied to any given random variable, which in many cases will have many more finite moments than the law requires. Consequently the weak law tends to be applied far more often in theoretical estimates than in empirical estimations.*

8.4   Strong Law of Large Numbers

The weak law of large numbers makes a statement about every n-trial sample space $\mathcal{S}^n$ associated with a random variable X with mean μ. Specifically, this law asserts that for any $\epsilon > 0$ the random variable $\hat{X} = \frac{1}{n}\sum_{i=1}^{n} X_i$ with i.i.d. $\{X_i\}_{i=1}^{n}$, "splits" this sample space into the event V_ϵ^n, of those sample points that are far from the mean in that $|\hat{X} - \mu| > \epsilon$, and the event $\widetilde{V}_\epsilon^n$, of those sample points that are close to the mean in that $|\hat{X} - \mu| \leq \epsilon$.

If we fix ϵ and assume that X has variance σ^2, the event V_ϵ^n has probability no more than $\frac{\sigma^2}{n\epsilon^2}$, which goes to 0, and event $\widetilde{V}_\epsilon^n$ has probability greater than $1 - \frac{\sigma^2}{n\epsilon^2}$, which goes to 1, both as $n \to \infty$. Without the assumption of the existence of σ^2, the same conclusions hold but without the information on rate of convergence.

Alternatively, for a fixed n, attempting to let $\epsilon \to 0$ in the case of finite variance provides ineffective probability bounds in that the event V_ϵ^n has probability bounded above by a quantity that goes to ∞ mathematically but to 1 logically. Likewise $\widetilde{V}_\epsilon^n$ has probability bounded below by a quantity that goes to $-\infty$ mathematically but to 0 logically.

On the other hand, if we choose $\epsilon \to 0$ carefully, say $\epsilon_n = n^{[a-(1/2)]}$ for $0 < a < \frac{1}{2}$, then we can simultaneously have that the probability of $V_{\epsilon_n}^n$ goes to zero as $n \to \infty$, and the error tolerance ϵ_n goes to zero. That is, with $\hat{X}_n$ denoting the sample mean random variable in S^n, and μ the corresponding theoretical mean, we obtain that as $n \to \infty$,

$$\Pr[|\hat{X}_n - \mu| > n^{[a-(1/2)]}] \le \frac{\sigma^2}{n^{2a}} \to 0.$$

We formalize this in a proposition.

Proposition 8.12   *Let S be a sample space and $\{X_i\}_{i=1}^n$ independent, identically distributed with mean μ and variance σ^2. If $\hat{X}_n = \frac{1}{n}\sum_{i=1}^n X_i$ denotes the average as a random variable in S^n, and $V_\epsilon^n \subset S^n$ is defined by*

$$V_\epsilon^n = \{(X_1, \ldots, X_n) \mid |\hat{X}_n - \mu| > \epsilon\},$$

then there is a sequence $\epsilon_n \to 0$ so that

$$\Pr[V_{\epsilon_n}^n] \to 0 \qquad \text{as } n \to \infty,$$

and correspondingly

$$\Pr[\widetilde{V_{\epsilon_n}^n}] \to 1 \qquad \text{as } n \to \infty.$$

Proof   Choose $\epsilon_n = n^{[a-(1/2)]}$ where $0 < a < \frac{1}{2}$, and apply the weak law of large numbers. ∎

Since this result gives that $\Pr[\widetilde{V_\epsilon^n}] \to 1$ as $n \to \infty$, it would be tempting to make the bold assertion that $\Pr[\hat{X}_n \to \mu] = 1$ as $n \to \infty$. But the proposition above is silent on the connection between the terms of any such sequence $\{\hat{X}_n\}$. Each sequential $\hat{X}_n$ term could be generated in at least one of two ways:

1. Model 1   Each sequential $\hat{X}_n$ term is generated and **independent** of the sample points that are chosen for $\hat{X}_j$ with $j < n$, meaning that for each n a new independent sample $(X_1, X_2, \ldots, X_n) \in S^n$ is produced.

2. Model 2   Each sequential $\hat{X}_n$ term is generated but **dependent** on the sample points that are chosen for $\hat{X}_j$ with $j < n$, so that $\hat{X}_{n+1}$ is defined with the same points as $\hat{X}_n$, which is $(X_1, X_2, \ldots, X_n)$, plus a new and independent sample point X_{n+1}.

The proposition above on the events V_ϵ^n gives no apparent statement on which model if either would allow the conclusion that $\Pr[\hat{X}_n \to \mu] = 1$ as $n \to \infty$. This proposition simply provides a statement about the probabilities of events defined in the sequential sample spaces S^n and confirms that these successive probabilities converge to 1. In either of these models of how $\{\hat{X}_n\}_{n=1}^\infty$ might be generated, we do not have a sample space with an associated probability structure, within which the collection $\{\hat{X}_n\}_{n=1}^\infty$ can be measured.

To better understand this point, we pursue these models in more detail. We will then see that model 2 is the model underlying the strong law of large numbers, and that this result is able to finesse a conclusion of $\Pr[\hat{X}_n \to \mu] = 1$ as $n \to \infty$, without the explicit construction of a probability space in which $\{\hat{X}_n\}_{n=1}^\infty$ can be measured.

8.4.1 Model 1: Independent $\{\hat{X}_n\}$

Intuitively for model 1 we need an "infinite product" sample space:

$$S^{(\infty)} \equiv S \times S^2 \times S^3 \times S^4 \times \cdots,$$

where each S^n denotes the n-trial sample space of sample points $\mathbf{X}_n \equiv (X_1, X_2, \ldots, X_n)$ and associated probability structure on which the random variable $\hat{X}_n = \frac{1}{n} \sum X_j$ is defined. The probability structures of the S^n would then need to be combined to a probability measure on this infinite product space in a way that is analogous to how the probability structure of $S^n \equiv S \times S \times S \times S \times \cdots \times S$ (n-times) was defined relative to the probability measure Pr on S. For any finite product $S^{(M)} = S \times S^2 \times S^3 \times S^4 \times \cdots \times S^M$, this sample space would be an example of a generalized M-trial sample space introduced in section 7.2.7, but for this model, this earlier construction must be generalized further to $M = \infty$.

The sequence $\{\hat{X}_n\}_{n=1}^\infty$ could then be defined in terms of a sample point in this product space $(\mathbf{X}_1, \mathbf{X}_2, \ldots, \mathbf{X}_n, \ldots)$, and the assertion $\Pr[\hat{X}_n \to \mu] = 1$ would have meaning. Namely $\Pr[\hat{X}_n \to \mu] = 1$ would mean that $\Pr[A] = 1$, where the event $A \subset S^{(\infty)}$ is defined as the collection of all sequences that so converge:

$$A \equiv \{(\mathbf{X}_1, \mathbf{X}_2, \ldots, \mathbf{X}_n, \ldots) \mid \hat{X}_n \to \mu\},$$

where each $\hat{X}_n$ is defined relative to the components of $\mathbf{X}_n$.

Alternatively, to attempt to avoid the construction of this sample space, let's recall the definition of limit. The statement $\hat{X}_n \to \mu$ means that for any $\epsilon > 0$ there is an integer N so that $|\hat{X}_n - \mu| < \epsilon$ for $n \geq N$. We could say that within this model, the expression $\Pr[\hat{X}_n \to \mu]$ is defined as the probability that for any $\epsilon > 0$ there is an integer N so that $|\hat{X}_n - \mu| < \epsilon$ for $n \geq N$.

Now by the weak law of large numbers, applied to the case where X has a finite variance, we know from (8.10) that for a given n this probability is greater than or equal to $1 - \frac{\sigma^2}{n\epsilon^2}$. In other words,

$$\Pr[|\hat{X}_n - \mu| < \epsilon] \geq \left(1 - \frac{\sigma^2}{n\epsilon^2}\right).$$

So by independence,

$$\Pr[|\hat{X}_n - \mu| < \epsilon \text{ for } n \geq N] = \prod_{n=N}^{\infty} [\Pr|\hat{X}_n - \mu| < \epsilon]$$

$$\geq \prod_{n=N}^{\infty} \left(1 - \frac{\sigma^2}{n\epsilon^2}\right).$$

Unfortunately, this leads to a dead end. Although beyond the tools we have developed so far the theory of **infinite products** is well developed in mathematics. As it turns out, the convergence of this infinite product to a number greater than 0 is related to the absolute convergence of the series $\left\{\frac{\sigma^2}{n\epsilon^2}\right\}$. Specifically, it will be shown in chapter 9 that given $\{x_n\}_{n=1}^{\infty}$ with $x_n > 0$ and $x_n \to 0$ as $n \to \infty$,

$$\prod_{n=1}^{\infty}(1 - x_n) = \begin{cases} 0, & \text{if } \sum x_n \text{ diverges,} \\ c > 0, & \text{if } \sum x_n \text{ converges.} \end{cases}$$

Of course here x_n is a multiple of the harmonic series, and we know from chapter 6 that $\sum x_n$ diverges. This implies that this infinite product has value 0 independent of N. In other words, we can only conclude what was obvious without any work, that in model 1, for any $\epsilon > 0$ and any N,

$$\Pr[|\hat{X}_n - \mu| < \epsilon \text{ for } n \geq N] \geq 0.$$

Equivalently, all that can be derived from the weak law is that

$$\Pr[\hat{X}_n \to \mu] \geq 0,$$

which is not a very deep insight.

8.4.2   Model 2: Dependent $\{\hat{X}_n\}$

In the second model for how $\{\hat{X}_n\}_{n=1}^{\infty}$ might be generated, we need a different sample space, one that is in effect the countably infinite version of $\mathcal{S}^n$,

$$\mathcal{S}^{\infty} \equiv \mathcal{S} \times \mathcal{S} \times \mathcal{S} \times \mathcal{S} \times \cdots,$$

with appropriate probability structure so that a sample point of the form $(X_1, X_2, \ldots, X_n, \ldots)$ can be selected, and associated sample mean sequence $\{\hat{X}_n\}_{n=1}^{\infty} \equiv \left\{\frac{1}{n}\sum_{j=1}^{n} X_j\right\}_{n=1}^{\infty}$ defined. Within such a space we could then define the event $A \subset \mathcal{S}^{\infty}$ as the collection of all sequences $(X_1, X_2, \ldots, X_n, \ldots) \in \mathcal{S}^{\infty}$ with associated mean sequences that satisfies $\hat{X}_n \to \mu$. Then the statement that $\Pr[\hat{X}_n \to \mu] = 1$ would mean that $\Pr[A] = 1$, where

$$A \equiv \{(X_1, X_2, \ldots, X_n, \ldots) \in \mathcal{S}^{\infty} \mid \hat{X}_n \to \mu\}.$$

The construction of this sample space would seem to be easy. We simply assert that

$$\mathcal{S}^{\infty} \equiv \{(X_1, X_2, \ldots, X_n, \ldots) \mid X_j \in \mathcal{S} \text{ for all } j\}.$$

The hard part, however, is the imposition of a probability measure. What is easy to demonstrate is that any attempt to generalize from (7.8) is hopeless. To attempt to define a probability function on $\mathcal{S}^{\infty}$ by $P_{\infty}[(s_1, s_2, \ldots)] = \prod_{j=1}^{\infty} \Pr(s_j)$ provides the immediate conclusion that $P_{\infty}[(s_1, s_2, \ldots)] = 0$ for all $(s_1, s_2, \ldots) \in \mathcal{S}^{\infty}$. Specifically, if $(s_1, s_2, \ldots)$ is any sample point, then in any nondegenerate space $\mathcal{S}$ it will be the case that $\Pr[s_j] \le p < 1$ for all j, and so $\prod_{j=1}^{N} \Pr(s_j) < p^N$, which converges to 0 as $N \to \infty$. The only counterexample to this conclusion is for a degenerate probability space $\mathcal{S} = \{s\}$ with one point in which $\Pr[s] = 1$. So another definitional approach is needed.

But any such approach will have to abandon the idea that sample points have nonzero probabilities since it can never be the case that such an $\mathcal{S}^{\infty}$ will be countable. Indeed, even for the simplest nondegenerate space, $\mathcal{S} \equiv \{0, 1\}$ underlying the standard binomial, $\mathcal{S}^{\infty}$ so defined contains the equivalent of the base-2 expansions of all real numbers in the interval $[0, 1]$ and hence is an uncountably infinite space. Assigning nonzero probabilities to an uncountable collection of sample points with the hope that these probabilities will add up to 1 is then doomed at the start. Why?

Because from the Cantor diagonalization approach in chapter 2, we know that every summation of the probabilities of sample points will of necessity omit many points, and hence any such sum must be unbounded and hence infinite. The only possible solution is to somehow identify a countable subcollection of points within $\mathcal{S}^{\infty}$, assign nonzero probabilities, and simply declare all other sample points to have probability 0. But since $\mathcal{S}^{\infty}$ is truly uncountable, it is clear that using such a construction to conclude that $\Pr[\hat{X}_n \to \mu] = 1$ would not answer the original question.

So another big idea is needed, but we do not have the necessary tools for such a product space with methods of this chapter. We will begin work on that big idea somewhat in chapter 10, which will address continuous probability theory, but the complete theory requires the tools of real analysis. It turns out that the strong law of large numbers addresses the desired result, produces a strong assertion, and avoids the construction of this infinite dimensional space. It addresses the sequence $\{\hat{X}_n\}_{n=1}^{\infty}$, which is defined in terms of a given sequence of independent X sample points $\{X_j\}_{j=1}^{\infty} \subset \mathcal{S}$, without constructing the sample space $\mathcal{S}^{\infty}$. But, if the strong law assures the conclusion that $\Pr[\hat{X}_n \to \mu] = 1$, without the space $\mathcal{S}^{\infty}$, what exactly does this conclusion mean?

8.4.3   The Strong Law Approach

The approach taken in the strong law of large numbers will be to strengthen the conclusion above, where it was shown that when σ^2 exists, there exists $\epsilon_n \to 0$ so that the events

$$V_{\epsilon_n}^n \equiv \{(X_1, \ldots, X_n) \,|\, |\hat{X}_n - \mu| > \epsilon_n\} \subset \mathcal{S}^n$$

satisfy $p_n \equiv \Pr[V_{\epsilon_n}^n] \to 0$ as $n \to \infty$. The idea was to choose $\epsilon_n = n^{a-(1/2)}$, $0 < a < \frac{1}{2}$.

While this result is meaningful, these probabilities do not converge to 0 very quickly. Indeed there is no N for which $\sum_{n=N}^{\infty} p_n < \infty$, since $p_n = \frac{\sigma^2}{n^{2a}}$, where $0 < 2a < 1$. In other words, the probabilities $p_n \to 0$ slower than the terms of the harmonic series, which we have seen does not converge. This is important because $\sum_{n=N}^{\infty} p_n = \sum_{n=N}^{\infty} \Pr[V_{\epsilon_n}^n]$. So if this summation could be made to converge, it would mean we could make this summation of probabilities as small as we want by choosing N big enough, and below we will see that this is enough to provide the desired conclusion in a logical way.

The problem of slow convergence is only partially caused by the goal of having the error tolerance, $\epsilon_n = n^{a-(1/2)}$, also converge to 0 as n increases. Even for fixed $\epsilon > 0$ we have seen from the weak law of large numbers that $p_n \equiv \Pr[V_{\epsilon}^n] = \frac{\sigma^2}{n\epsilon^2}$ by (8.9). While $p_n \to 0$ as $n \to \infty$, there again is no N so that $\sum_{n=N}^{\infty} p_n < \infty$. In other words, the best we can assert on the basis of the weak law is that for fixed $\epsilon > 0$, these probabilities decrease to 0 no faster than $\frac{1}{n}$ for a random collection $\{\hat{X}_n\}$.

The strong law of large numbers will apply to a collection of random variables $\{X_n\}_{n=1}^{\infty}$ defined on $\mathcal{S}$ and the associated sample mean sequence

$$\hat{X}_n \equiv \frac{1}{n}\sum_{j=1}^{n} X_j.$$

It will improve the results above in two ways:

1. The collection $\{X_n\}_{n=1}^{\infty}$ must be independent but need not be identically distributed. However, if not i.i.d., the collection of variances, $\{\sigma_i^2\}$ must not grow too fast with i.

2. It will be shown that with $\hat{\mu}_k \equiv \frac{1}{k}\sum_{j=1}^{k}\mu_j$, then for any $\epsilon > 0$,

$$\sum_{n=1}^{\infty} p_n < \infty,$$

where $p_n = p_n(\epsilon)$ is defined by

$$p_n = \Pr[|\hat{X}_k - \hat{\mu}_k| > \epsilon \text{ for at least one } k \text{ with } 2^{n-1} < k \le 2^n].$$

Hence for any $\delta > 0$ there is an N so that $\sum_{n=N}^{\infty} p_n < \delta$.

The strong law of large numbers then "finesses" the conclusion that $\Pr[\hat{X}_n \to \mu] = 1$ without the construction of $\mathcal{S}^{\infty}$ because of the critical statement in 2 which could not be derived from the weak law. First off, by definition,

$$\sum_{n=N}^{\infty} p_n = \Pr[|\hat{X}_k - \hat{\mu}_k| > \epsilon \text{ for at least one } k > 2^{N-1}].$$

So from 2 above we can state that for any $\delta > 0$ there is an $N = N(\epsilon)$ so that

$$\Pr[|\hat{X}_k - \hat{\mu}_k| > \epsilon \text{ for at least one } k > 2^{N-1}] < \delta.$$

In other words, for any $\delta > 0$ there is an N so that

$$\Pr[|\hat{X}_k - \hat{\mu}_k| \le \epsilon \text{ for all } k > 2^{N-1}] \ge 1 - \delta.$$

We return to this analysis after the statement and proof of the strong law of large numbers.

*8.4.4 Kolmogorov's Inequality

In order to prove the strong law, we need another and stronger inequality than Chebyshev's inequality, called **Kolmogorov's inequality**, named for **Andrey Kolmogorov** (1903–1987) who was also responsible for introducing an axiomatic framework for probability theory. Extending Chebyshev's inequality, Kolmogorov's inequality addresses a collection of random variables $\{X_i\}_{i=1}^n$ and provides a probability statement regarding their maximum summation.

Kolmogorov's inequality is stated for simplicity, under the assumption that $E[X_j] = 0$ for all j. However, this is not a true restriction. That is, if we are given $\{Y_j\}_{j=1}^n$ with $E[Y_j] = \mu_j$, we can apply the result to $X_j \equiv Y_j - \mu_j$, since it is clear that $\mathrm{Var}[X_j] = \mathrm{Var}[Y_j]$. And while this result requires that $\{X_j\}_{j=1}^n$ be independent random variables, it does not require that they be identically distributed, so it allows for differing variances.

Proposition 8.13 (*Kolmogorov's inequality*)   Let $\{X_i\}_{i=1}^n$ be independent random variables with $E[X_j] = 0$ and $\mathrm{Var}[X_j] = \sigma_j^2$. Then for $t > 0$,

$$\Pr\left\{ \max_{1 \le i \le n} \left| \sum_{j=1}^i X_j \right| > t \right\} \le \sum_{j=1}^n \frac{\sigma_j^2}{t^2}. \tag{8.15}$$

Remark 8.14   *Note that the event defined in (8.15) is an event in $\mathcal{S}^n$, where $\mathcal{S}$ is the common sample space on which $\{X_i\}_{i=1}^n$ are defined and independent. Note also that Kolmogorov's inequality is considerably stronger than is Chebyshev's inequality applied to this probability statement. The Chebyshev inequality would state that for any i, with $1 \le i \le n$,*

$$\Pr\left\{ \left| \sum_{j=1}^i X_j \right| > t \right\} \le \sum_{j=1}^i \frac{\sigma_j^2}{t^2},$$

since for independent random variables $\mathrm{Var}(\sum_{j=1}^i X_j) = \sum_{j=1}^i \sigma_j^2$. Of course, $\sum_{j=1}^i \frac{\sigma_j^2}{t^2} \le \sum_{j=1}^n \frac{\sigma_j^2}{t^2}$, so at first these inequalities appear similar. However, Chebyshev's inequality provides probability statements on n separate events, and it is silent on the question of the simultaneous occurrence of these n events. Kolmogorov's inequality says that the largest of the n Chebyshev probability bounds is sufficient to bound the probability of the worst case of these n events. Alternatively, Kolmogorov's inequality says that the largest of the n Chebyshev probability bounds is sufficient to bound the probability that all inequalities are satisfied simultaneously.

Proof   The idea of this proof is to eliminate the maximum function by introducing a new random variable that identifies the first summation for which $|\sum_{j=1}^i X_j| > t$, and then use a conditioning argument on this random variable. Consider the sequence $(\sum_{j=1}^i X_j)^2$, $i = 1, 2, \ldots, n$. For any collection of random variables $\{X_i\}_{i=1}^n$, define a new random variable $N = \min\{i \mid (\sum_{j=1}^i X_j)^2 > t^2\}$, but if $(\sum_{j=1}^i X_j)^2 \le t^2$ for all $i \le n$, define $N = n$. Then the events in $\mathcal{S}^n$ defined by

$$\left\{ \max_{1 \leq i \leq n} \left(\sum_{j=1}^{i} X_j \right)^2 > t^2 \right\},$$

$$\left\{ \left(\sum_{j=1}^{N} X_j \right)^2 > t^2 \right\},$$

are identical events with equal probabilities. Now by the Markov inequality applied to the second event, we get

$$\Pr\left\{ \left(\sum_{j=1}^{N} X_j \right)^2 > t^2 \right\} \leq \frac{\mathrm{E}[(\sum_{j=1}^{N} X_j)^2]}{t^2}.$$

Because of the assumption that $\mathrm{E}[X_j] = 0$, we have that $\mathrm{E}[(\sum_{j=1}^{N} X_j)^2] = \mathrm{Var}[\sum_{j=1}^{N} X_j]$, and so the proof will be complete if we can show that

$$\mathrm{Var}\left[\sum_{j=1}^{N} X_j \right] \leq \sum_{j=1}^{n} \sigma_j^2.$$

Note that this is a bit subtle because while $\{X_i\}_{i=1}^{N} \subset \{X_i\}_{i=1}^{n}$, N is a random variable, and hence we cannot simply assert that $\mathrm{Var}[\sum_{j=1}^{N} X_j] \leq \sum_{j=1}^{n} \sigma_j^2$. To demonstrate this upper bound, we use the law of total variance. First, for the conditional variance, $\mathrm{Var}[\sum_{j=1}^{N} X_j \mid N = k] = \mathrm{Var}[\sum_{j=1}^{k} X_j] = \sum_{j=1}^{k} \sigma_j^2$. Next, for the conditional mean, $\mathrm{E}[\sum_{j=1}^{N} X_j \mid N = k] = \mathrm{E}[\sum_{j=1}^{k} X_j] = 0$. We now have by (7.49),

$$\mathrm{Var}\left[\sum_{j=1}^{N} X_j \right] = \mathrm{E}\left[\sum_{j=1}^{k} \sigma_j^2 \right] + \mathrm{Var}[0].$$

For this last expectation, if $a_k = \Pr[N = k]$, then, since $\sum_{j=1}^{n} a_k = 1$,

$$\mathrm{E}\left[\sum_{j=1}^{k} \sigma_j^2 \right] = \sum_{k=1}^{n} a_k \left[\sum_{j=1}^{k} \sigma_j^2 \right] \leq \sum_{j=1}^{n} \sigma_j^2,$$

which follows by reversing the double summation: $\sum_{k=1}^{n} \sum_{j=1}^{k} = \sum_{j=1}^{n} \sum_{k=j}^{n}$. ∎

*8.4.5 Strong Law of Large Numbers

We next turn to the statement of the strong law of large numbers. The primary requirement is that while the collection of variances $\{\sigma_i^2\}$ do not need to be bounded, if unbounded, they cannot increase too fast. We provide this statement in both the simpler case of independent and identically distributed random variables, since that is the statement that is often sufficient for applications as well as in the more general case.

Proposition 8.15 (*Strong Law of Large Numbers 1*)   *Let $\{X_j\}_{j=1}^{\infty}$ be independent, identically distributed random variables with mean μ and variance σ^2, and define $\hat{X}_k = \frac{1}{k}\sum_{j=1}^{k} X_j$. For any $\epsilon > 0$ define the event $A_n \subset \mathcal{S}^{2^n}$,*

$$A_n = \{\overline{X} \mid |\hat{X}_k - \mu| > \epsilon \text{ for at least one } k \text{ with } 2^{n-1} < k \leq 2^n\},$$

where $\overline{X} \equiv (X_1, X_2, \ldots, X_{2^n}) \in \mathcal{S}^{2^n}$. Then

$$\sum_{n=1}^{\infty} \Pr[A_n] < \infty,$$

and hence for any $\delta > 0$ there is an N so that $\sum_{n=N}^{\infty} \Pr[A_n] < \delta$.

Proposition 8.16 (*Strong Law of Large Numbers 2*)   *Let $\{X_j\}_{j=1}^{\infty}$ be a sequence of mutually independent random variables with means $\{\mu_j\}_{j=1}^{\infty}$ and variances $\{\sigma_j^2\}_{j=1}^{\infty}$ with $\sum_{j=1}^{\infty} \frac{\sigma_j^2}{j^2} < \infty$. Define $\hat{X}_k = \frac{1}{k}\sum_{j=1}^{k} X_j$ and $\hat{\mu}_k = \frac{1}{k}\sum_{j=1}^{k} \mu_j$. For any $\epsilon > 0$ define the event $A_n \subset \mathcal{S}^{2^n}$,*

$$A_n = \{\overline{X} \mid |\hat{X}_k - \hat{\mu}_k| > \epsilon \text{ for at least one } k \text{ with } 2^{n-1} < k \leq 2^n\}, \tag{8.16}$$

where $\overline{X} \equiv (X_1, X_2, \ldots, X_{2^n}) \in \mathcal{S}^{2^n}$. Then

$$\sum_{n=1}^{\infty} \Pr[A_n] < \infty, \tag{8.17}$$

and hence for any $\delta > 0$ there is an N so that $\sum_{n=N}^{\infty} \Pr[A_n] < \delta$.

Proof   The event A_n can equivalently be defined as the event

$$A_n = \left[\max_{2^{n-1}<k\leq 2^n} \left| \sum_{j=1}^{k} Y_j \right| > k\epsilon \right],$$

where $Y_j = X_j - \mu_j$. In other words, $|\hat{X}_k - \hat{\mu}_k| > \epsilon$ for at least one k with $2^{n-1} < k \leq 2^n$ if and only if $\max_{(2^{n-1}<k\leq 2^n)}|\sum_{j=1}^{k} Y_j| > k\epsilon$. Note that $\Pr[A_n] < \Pr[A_n']$, where A_n' is defined in terms of $2^{n-1}\epsilon$ rahter than $k\epsilon$. By Kolmogorov's inequality, the probability of this latter event is given by

$$\Pr[A_n'] < \frac{1}{2^{2n-2}} \sum_{j=1}^{2^n} \frac{\sigma_j^2}{\epsilon^2}.$$

Hence

$$\sum_{n=1}^{\infty} \Pr[A_n] < \frac{4}{\epsilon^2} \sum_{n=1}^{\infty} \frac{1}{2^{2n}} \sum_{j=1}^{2^n} \sigma_j^2.$$

Note that in this double summation, each σ_j^2 is counted multiple times. In particular,

$$\sum_{n=1}^{\infty} \frac{1}{2^{2n}} \sum_{j=1}^{2^n} \sigma_j^2 = \sum_{j=1}^{\infty} \sigma_j^2 \sum_{2n \geq j}^{\infty} \frac{1}{2^{2n}}$$

$$\leq 2 \sum_{j=1}^{\infty} \frac{\sigma_j^2}{j^2},$$

since $\sum_{2n \geq j}^{\infty} \frac{1}{2^{2n}} \leq \sum_{n=j}^{\infty} \frac{1}{2^n} = \frac{1}{2^{j-1}} \leq \frac{2}{j^2}$ for $j \geq 4$. Hence $\sum_{n=1}^{\infty} \Pr[A_n] < \infty$. ∎

Remark 8.17 *The assumption in the general version of the strong law, that $\sum_{j=1}^{\infty} \frac{\sigma_j^2}{j^2} < \infty$, is certainly an assumption about the growth rate of σ_j^2 as $j \to \infty$. For example, if $\sigma_j^2 = \sigma^2$, which is the assumption of no growth, then, since $\sum_{j=1}^{\infty} \frac{1}{j^2} < \infty$ from chapter 6, the strong law applies. On the other hand, if $\sigma_j^2 = j\sigma^2$, so the standard deviation grows like $\sqrt{j}$, the strong law does not apply, since again from chapter 6, $\sum_{j=1}^{N} \frac{1}{j} \to \infty$ with N. Consequently linear variance growth, or equivalently, square root growth in standard deviation, is just a bit too fast for the strong law to apply. However, if $\sigma_j^2 = j^a \sigma^2$ for any $a < 1$, the strong law applies, since $\sum_{j=1}^{\infty} \frac{\sigma_j^2}{j^2} = \sigma^2 \sum_{j=1}^{\infty} \frac{1}{j^{2-a}} < \infty$ for $2 - a > 1$.*

Corollary 8.18 *Let $\{X_j\}_{j=1}^{\infty}$ be independent random variables with means $\{\mu_j\}_{j=1}^{\infty}$ and variances σ_j^2 with $\sum_{j=1}^{\infty} \frac{\sigma_j^2}{j^2} < \infty$, and for any k define $\hat{X}_k = \frac{1}{k} \sum_{j=1}^{k} X_j$ and $\hat{\mu}_k = \frac{1}{k} \sum_{j=1}^{k} \mu_j$. Then for any $\epsilon > 0$ and $\delta > 0$ there is an N so that*

$$\Pr[|\hat{X}_k - \hat{\mu}_k| > \epsilon \text{ for any } k > 2^N] < \delta.$$

Equivalently, for any $\epsilon > 0$ and $\delta > 0$ there is an N so that

$$\Pr[|\hat{X}_k - \hat{\mu}_k| \leq \epsilon \text{ for all } k > 2^N] > 1 - \delta. \tag{8.18}$$

Proof This follows from the observation that

$$[|\hat{X}_k - \hat{\mu}_k| > \epsilon \text{ for any } k > 2^N] = \bigcup_{n \geq N+1} A_n,$$

and the conclusion that $\sum_{n=1}^{\infty} \Pr[A_n] < \infty$. Hence for any $\delta > 0$ there is an N so that $\sum_{n=N+1}^{\infty} \Pr[A_n] < \delta$.   ∎

Remark 8.19 *Note that in this corollary,* $[|\hat{X}_k - \hat{\mu}_k| > \epsilon$ *for any* $k > 2^N]$ *is not an event in any of the n-trial sample spaces defined so far. Indeed, since this "event" is related to the entire collection of random variables, it would have to exist in* S^∞, *which we have not defined. In essence, with the strong law, we can avoid the construction of this event in* S^∞ *and finesse the result by defining this event as a union of the respective events in the* S^{2^n} *spaces for* $n \geq N+1$. *And this corollary estimates that the sum of the measures of all such events in all such sample spaces can be made as small as desired.*

And it is in this light that the strong law of large numbers provides the conclusion $\Pr[\hat{X}_n - \hat{\mu}_n \to 0] = 1$, *or in the case of identically distributed* X_n-*values, the conclusion* $\Pr[\hat{X}_n \to \mu] = 1$.

8.5   De Moivre–Laplace Theorem

The **De Moivre–Laplace theorem** is a special case of a very general result discussed below, known as the **central limit theorem**. The theorem of this section addresses the question of the "limiting distribution" of the binomial distribution as $n \to \infty$. Specifically, if $X^{(n)} \equiv \sum_{j=1}^{n} X_j^B$ is a binomially distributed random variable with parameters n and p, where X_j^B are i.i.d. standard binomial variables, we have from (7.97) the probability that for integers a and b,

$$\Pr[a \leq X^{(n)} \leq b] = \sum_{j=a'}^{b'} \binom{n}{j} p^j (1-p)^{n-j},$$

where $a' = \max(a, 0)$ and $b' = \min(b, n)$.

In this form it is difficult to specify what happens to this distribution as $n \to \infty$ because the range of the random variable is $[0, n]$ which varies with n. Put another way, we have from (7.99) that $\mathrm{E}[X^{(n)}] = np$ and $\mathrm{Var}[X^{(n)}] = np(1-p)$, so both the mean and variance of $X^{(n)}$ grow without bound as $n \to \infty$. In order to investigate quantitatively the probabilities under this distribution as $n \to \infty$, some form of scaling is necessary to stabilize results.

The approach used by **Abraham de Moivre** (1667–1754) in the special case of $p = \frac{1}{2}$, and many years later generalized to all p, $0 < p < 1$, by **Pierre-Simon Laplace** (1749–1827), was to consider what is now called the **normalized random variable**, $Y^{(n)}$, defined by

$$Y^{(n)} = \frac{X^{(n)} - \mathrm{E}[X^{(n)}]}{\sqrt{\mathrm{Var}[X^{(n)}]}}. \tag{8.19}$$

The random variable $Y^{(n)}$ has the same binomial probabilities as does $X^{(n)}$, of course, since for any n, $\mathrm{E}[X^{(n)}]$ and $\sqrt{\mathrm{Var}[X^{(n)}]}$ are constants. However, its range is now $\left\{ \frac{j - \mathrm{E}[X^{(n)}]}{\sqrt{\mathrm{Var}[X^{(n)}]}} \,\middle|\, 0 \le j \le n \right\}$, and a simple calculation using (7.38) yields that

$$\mathrm{E}[Y^{(n)}] = 0, \quad \mathrm{Var}[Y^{(n)}] = 1.$$

Consequently, with mean and variance both constant and independent of n, the question of investigating and potentially identifying the limiting distribution of $Y^{(n)}$ as $n \to \infty$ is better defined and its pursuit more compelling.

To this end, we first note two elementary but important results on $Y^{(n)}$:

Proposition 8.20 *Given $Y^{(n)}$ defined as in (8.19) where the binomial probability p satisfies $0 < p < 1$:*

1. *The range of $Y^{(n)}$ is unbounded both positively and negatively as $n \to \infty$.*

2. *If $y \in \mathbb{R}$, there is a sequence $\{y_n\}$ with $y_n \to y$, and each y_n is in the range of $Y^{(n)}$.*

Proof

1. Since $0 \le j \le n$, a simple calculation shows that with $q \equiv 1 - p$,

$$-\sqrt{n}\sqrt{\frac{p}{q}} \le \frac{j - \mathrm{E}[X^{(n)}]}{\sqrt{\mathrm{Var}[X^{(n)}]}} \le \sqrt{n}\sqrt{\frac{q}{p}}.$$

This result reduces to the unbounded symmetric interval $[-n, n]$ when $p = \frac{1}{2}$, and it is unbounded and asymmetrical otherwise as $n \to \infty$.

2. Let N denote the smallest integer so that $y \in \left(-\sqrt{N}\sqrt{\frac{p}{q}}, \sqrt{N}\sqrt{\frac{q}{p}} \right)$, where again $q = 1 - p$. This result is always possible, since these intervals grow without bound with N. Now it must be the case that there is a j, perhaps two such values, so that $y \in \left[\frac{j - Np}{\sqrt{Npq}}, \frac{j+1 - Np}{\sqrt{Npq}} \right]$, since the collection of these intervals covers $\left[-\sqrt{N}\sqrt{\frac{p}{q}}, \sqrt{N}\sqrt{\frac{q}{p}} \right]$. We then define y_0 as the left endpoint of this interval. For each value of $N + n$, where $n \ge 1$, now define y_n as the left endpoint of the interval for which $y \in$

$\left[\frac{j-(N+n)p}{\sqrt{(N+n)pq}}, \frac{j+1-(N+n)p}{\sqrt{(N+n)pq}} \right]$. There is again at least one such interval, since these intervals collectively cover $\left[-\sqrt{N+n}\sqrt{\frac{p}{q}}, \sqrt{N+n}\sqrt{\frac{q}{p}} \right]$. Since the length of the interval in this nth step is $\frac{1}{\sqrt{(N+n)pq}}$, which converges to 0 as $n \to \infty$, it is apparent by construction that $|y - y_n| \le \frac{1}{\sqrt{(N+n)pq}}$, and we can conclude that $y_n \to y$. ∎

Remark 8.21 *In this construction for the proof of 2, the right end points work equally well, as does a random selection from the two end points of each interval. In other words, there are infinitely many such sequences.*

Consequently for any $y \in \mathbb{R}$ we can investigate the existence of a probability density function $g(y)$ defined as

$$g(y) \equiv \lim_{n \to \infty} \Pr\{ Y^{(n)} = y_n \},$$

where $\{y_n\}$ is constructed so that $y_n \to y$. To be sure that such a pursuit is justified, one needs to ascertain that this limit makes sense and answers the original question: Is such a $g(y)$ the limiting density of the binomial p.d.f. for $Y^{(n)}$ as $n \to \infty$?

A moment of reflection demonstrates that this limit may well not answer this question, since it is the case that for any such sequence, $\{y_n\}$,

$$\Pr[Y^{(n)} = y_n] = \Pr[X^{(n)} = j_n],$$

where $j_n = y_n\sqrt{npq} + np$. So as $y_n \to y$, we see that $j_n \to \infty$, and hence it would appear logical that

$$\lim_{n \to \infty} \Pr[Y^{(n)} = y_n] = 0$$

for any y. In other words, as defined above, it would appear to be the case that $g(y) = 0$ for all y.

Before investigating this further, note that this conclusion is also compelled by the fact that if $g(y)$ is defined as above for every $y \in \mathbb{R}$, then it would not make sense to have $g(y) > 0$ for more than a countable subset of $\mathbb{R}$. This is because if $g(y) > 0$ for an uncountable set, then $\sum g(y)$ over all such values would have to be infinite and never equal 1 as is needed for a probability density. This follows from an argument analogous to the Cantor diagonalization process, that any attempt to enumerate and add up all such $g(y)$ values would of necessity omit all but a countable subcollection. Hence any such summation would of necessity be unbounded.

To formally show that $g(y) = 0$ for all y where $g(y)$ is defined above is somewhat difficult, but this conclusion will be an immediate consequence of the proof of the De Moivre–Laplace theorem that we now pursue. As will be seen, in order to get a true p.d.f. from the limit of the p.d.f.s of the associated $Y^{(n)}$ random variables, an adjustment factor is needed in the definition above of $g(y)$. Specifically, each probability $\Pr\{Y^{(n)} = y_n\}$ will be multiplied by $\sqrt{npq}$, and the product will then be shown to converge to the desired probability density function $h(y)$. In addition this proof will establish the speculation above that $\Pr\{Y^{(n)} = y_n\} \to 0$, since $\sqrt{npq} \to \infty$ and $\sqrt{npq}\, \Pr\{Y^{(n)} = y_n\} \to h(y)$ clearly implies this result.

The proof of this theorem depends on a famous approximation formula for $n!$, known as **Stirling's formula**, or **Stirling's approximation**, named for its discoverer, **James Stirling** (1692–1770), which is of interest in itself.

8.5.1   Stirling's Formula

To establish this approximation formula, we require another power series expansion from chapter 9 for the natural logarithm function $\ln(1 + x)$. The proof of this will depend on the same mathematical tools that will be used to prove the power series expansion of e^x noted in (7.63). The needed expansion here is

$$\ln(1 + x) = \sum_{n=1}^{\infty} (-1)^{n+1} \left(\frac{1}{n}\right) x^n \qquad \text{for } |x| < 1. \tag{8.20}$$

As was the case for the series expansion for e^x, the ratio test confirms absolute convergence of this series, since

$$\left| \frac{(-1)^{n+2} \left(\frac{1}{n+1}\right) x^{n+1}}{(-1)^{n+1} \left(\frac{1}{n}\right) x^n} \right| = \left| \frac{x}{\frac{n+1}{n}} \right| \to |x| \qquad \text{as } n \to \infty,$$

and consequently the restriction $|x| < 1$ assures absolute convergence. As $x \to -1$, this series approaches the negative of the harmonic series $-\sum_{n=1}^{\infty} \frac{1}{n}$, which diverges to $-\infty$. On the other hand, we will see in chapter 10 that as $x \to 1$, this series is well defined.

Note also that this formula can be written with $-x$, using $\ln(1 - x) = -\ln\left(\frac{1}{1-x}\right)$,

$$\ln\left(\frac{1}{1 - x}\right) = \sum_{n=1}^{\infty} \left(\frac{1}{n}\right) x^n \qquad \text{for } |x| < 1. \tag{8.21}$$

When combined with (8.20), this yields

$$\frac{1}{2}\ln\left(\frac{1+x}{1-x}\right) = \sum_{n=1}^{\infty}\left(\frac{1}{2n-1}\right)x^{2n-1} \qquad \text{for } |x| < 1, \tag{8.22}$$

since $\ln\left(\frac{1+x}{1-x}\right) = \ln(1+x) - \ln(1-x)$, and absolute convergence justifies rearranging the terms of these two series into a single series.

Proposition 8.22 (Stirling's Formula)   *As $n \to \infty$, we have the relative approximation $n! \sim \sqrt{2\pi}n^{n+(1/2)}e^{-n}$, in the sense that*

$$\frac{n!}{\sqrt{2\pi}n^{n+(1/2)}e^{-n}} \to 1 \qquad \text{as } n \to \infty. \tag{8.23}$$

Moreover the relative error in this approximation is given by

$$e^{1/(12n+1)} < \frac{n!}{\sqrt{2\pi}n^{n+(1/2)}e^{-n}} < e^{1/12n}. \tag{8.24}$$

Proof   We first show that there is a constant C so that $n! \sim e^C n^{n+(1/2)}e^{-n}$ has the noted properties. To this end, define $f_n = \ln\left(\frac{n!}{n^{n+(1/2)}e^{-n}}\right)$, which can be rewritten using properties of the logarithm

$$f_n = \ln n! - \left(n + \frac{1}{2}\right)\ln n + n.$$

We now show that there is a constant C so that $f_n \to C$. By exponentiation, this will then establish (8.23) with e^C in place of $\sqrt{2\pi}$. To do this, consider $f_n - f_{n+1}$. A calculation shows that

$$f_n - f_{n+1} = \left(n + \frac{1}{2}\right)\left(\ln\frac{n+1}{n}\right) - 1.$$

Expressing $\frac{n+1}{n} = \frac{1+x}{1-x}$, where $x = \frac{1}{2n+1}$, and using (8.22) with index m produces

$$f_n - f_{n+1} = \sum_{m=1}^{\infty}\left(\frac{1}{2m+1}\right)x^{2m},$$

which demonstrates that $f_n - f_{n+1} > 0$. Hence the sequence $\{f_n\}$ is decreasing. Further, since $\left(\frac{1}{2m+1}\right) < \frac{1}{3}$ except for $m = 1$, in which case we have equality

$$f_n - f_{n+1} < \frac{1}{3}\sum_{m=1}^{\infty}x^{2m}$$

$$= \frac{1}{3[(2n+1)^2 - 1]}$$

$$= \frac{1}{12n} - \frac{1}{12(n+1)}.$$

The last inequality shows that $f_n - \frac{1}{12n} < f_{n+1} - \frac{1}{12(n+1)}$, so $f_n - \frac{1}{12n}$ is increasing. Since $\frac{1}{12n} \to 0$, this implies that there is a constant C for which $f_n \to C$. The upper error bound in (8.24) also comes from this analysis. Because $f_n - \frac{1}{12n}$ is increasing with limit C, we have $f_n < C + \frac{1}{12n}$, and this can be exponentiated to the desired result. For the lower bound, the series expansion for $f_n - f_{n+1}$ above, using only the first term implies $f_n - f_{n+1} > \frac{1}{3}\left(\frac{1}{2n+1}\right)^2 > \frac{1}{12n+1} - \frac{1}{12(n+1)+1}$. As a result $f_n - \frac{1}{12n+1}$ is increasing and consequently $f_n > C + \frac{1}{12n+1}$. The final step is the demonstration that $e^C = \sqrt{2\pi}$, which we only sketch here and defer the details to chapter 10. This conclusion is a consequence of what is known as **Wallis' product formula for** $\frac{\pi}{2}$, named for its discoverer, **John Wallis** (1616–1703), which is

$$\frac{\pi}{2} = \prod_{n=1}^{\infty} \frac{(2n)^2}{(2n-1)(2n+1)}. \tag{8.25}$$

A calculation with much cancellation shows that

$$\prod_{n=1}^{m} \frac{(2n)^2}{(2n-1)(2n+1)} = \frac{2^{4m}(m!)^4}{(2m)!(2m+1)!}.$$

So this result can be written as

$$\frac{\pi}{2} = \lim_{m \to \infty} \frac{2^{4m}(m!)^4}{(2m)!(2m+1)!}.$$

Substituting the approximations for the factorial functions derived above, which are in the form $n! \sim e^C n^{n+(1/2)} e^{-n}$ completes the derivation that $e^C = \sqrt{2\pi}$. The proof of Wallis' formula involves mathematical tools of chapter 10 and an application of integration by parts.   ■

Remark 8.23

1. *Note that the approximation in Stirling's formula only converges in terms of relative error, and not in terms of absolute error. In fact from (8.24) we can conclude only that*

$$(e^{1/(12n+1)} - 1)\sqrt{2\pi n}\,n^{n+(1/2)}e^{-n} < n! - \sqrt{2\pi n}\,n^{n+(1/2)}e^{-n} < (e^{1/12n} - 1)\sqrt{2\pi n}\,n^{n+(1/2)}e^{-n},$$

which is an error interval that grows without bound.

2. *Also note that the convergence of the Wallis' product formula for $\frac{\pi}{2}$ is painfully slow. Indeed, defining $a_N = \prod_{n=1}^{N} \frac{(2n)^2}{(2n-1)(2n+1)}$, we have that $a_N = \frac{(2N)^2}{(2N-1)(2N+1)} a_{N-1}$, and the successive multiplicative factors $\frac{(2N)^2}{(2N-1)(2N+1)} = \frac{1}{1 - \frac{1}{4N^2}}$ converge to 1 very quickly.*

8.5.2   De Moivre–Laplace Theorem

With the aid of this approximation for $n!$, we can now address the primary result in this section.

Proposition 8.24 (*De Moivre–Laplace Theorem*)   *Let $X^{(n)}$ be a binomial random variable with parameters p and n, with $0 < p < 1$, and let $Y^{(n)}$ denote the normalized random variable in (8.19). For any $y \in \mathbb{R}$, and $\{y_n\}$ constructed so that $y_n \in \text{Rng}[Y^{(n)}]$ and $y_n \to y$, we have as $n \to \infty$,*

$$\sqrt{npq}\,\Pr\{Y^{(n)} = y_n\} \to \frac{1}{\sqrt{2\pi}}e^{-y^2/2}. \tag{8.26}$$

Proof   As noted above, with $j_n = y_n\sqrt{\text{Var}[X^{(n)}]} + \text{E}[X^{(n)}]$, we have that $\Pr\{Y^{(n)} = y_n\} = \Pr\{X^{(n)} = j_n\}$. Consequently

$$\sqrt{npq}\,\Pr\{Y^{(n)} = y_n\} = \sqrt{npq}\binom{n}{j_n}p^{j_n}(1-p)^{n-j_n}.$$

Using Stirling's formula applied to $\binom{n}{j}$, we write

$$\frac{n!}{j!(n-j)!} \sim \frac{\sqrt{2\pi n}\,n^{n+(1/2)}e^{-n}}{\sqrt{2\pi j}\,j^{j+(1/2)}e^{-j}\sqrt{2\pi(n-j)}\,(n-j)^{(n-j)+(1/2)}e^{-(n-j)}}$$

$$= \frac{1}{\sqrt{2\pi}}\sqrt{\frac{n}{j(n-j)}}\left(\frac{n}{n-j}\right)^{n-j}\left(\frac{n}{j}\right)^{j}.$$

In this analysis we shortcut with "$\sim$" the more technically accurate use of "$<$" and the necessary insertion of error terms in each of the Stirling approximations. We know from (8.24) that these approximations are collectively bounded above and below by exponential terms that converge to 1 as $n \to \infty$, since then $j_n \to \infty$.

With this restatement of the combinatorial term, the proof has two parts, since the $\frac{1}{\sqrt{2\pi}}$ term is apparently accounted for:

1. The first step is to show that

$$\sqrt{npq}\sqrt{\frac{n}{j_n(n-j_n)}} \to 1.$$

To this end, note that $\sqrt{\frac{j_n(n-j_n)}{n}} = \sqrt{n\left(\frac{j_n}{n}\right)\left(1-\frac{j_n}{n}\right)}$. But $\frac{j_n}{n} = p + y_n\sqrt{\frac{pq}{n}}$, and $1 - \frac{j_n}{n} = q - y_n\sqrt{\frac{pq}{n}}$, so

$$\sqrt{n\left(\frac{j_n}{n}\right)\left(1-\frac{j_n}{n}\right)} = \sqrt{npq + (q-p)y_n\sqrt{pqn} - y_n^2 pq}.$$

The ratio of $\sqrt{npq}$ to this term converges to 1 as $n \to \infty$, since $y_n \to y$, completing the first step.

2. The second step is to show that as $n \to \infty$,

$$\left(\frac{n}{n-j_n}\right)^{n-j_n}\left(\frac{n}{j_n}\right)^{j_n} p^{j_n}(1-p)^{n-j_n} = \left(\frac{n-j_n}{nq}\right)^{-(n-j_n)}\left(\frac{j_n}{np}\right)^{-j_n}$$

$$\to e^{-y^2/2}.$$

To do this, we first take -1 times the logarithm of the second expression and show that the resulting expression converges to $\frac{y^2}{2}$. From part 1 we have that $j_n = np + y_n\sqrt{pqn}$, and $n - j_n = nq - y_n\sqrt{pqn}$, from which $\frac{j_n}{np} = 1 + y_n\sqrt{\frac{q}{np}}$ and $\frac{n-j_n}{nq} = 1 - y_n\sqrt{\frac{p}{nq}}$. Hence

$$-\ln\left[\left(\frac{nq}{n-j_n}\right)^{n-j_n}\left(\frac{np}{j_n}\right)^{j_n}\right] = (nq - y_n\sqrt{pqn})\ln\left(1 - y_n\sqrt{\frac{p}{nq}}\right)$$

$$+ (np + y_n\sqrt{pqn})\ln\left(1 + y_n\sqrt{\frac{q}{np}}\right).$$

Next we apply the first three terms of the power series expansions for the logarithm in (8.20) to the expressions above, multiply, and collect terms. The "trick" in such a calculation is to not worry about any terms that will ultimately contain a factor of $n^{-1/2}$, or n^{-1}, and so forth, since these converge to 0 in the limit as $n \to \infty$. Since the terms in front of the logarithms contain a factor of n, the logarithm series is needed up to its third term, which is up to a factor of $n^{-3/2}$, and the product of this term with n will go to 0, as will all higher powers in the series.

Implementing this messy bit of algebra, and recalling that $y_n \to y$, produces

$$-\ln\left[\left(\frac{nq}{n-j_n}\right)^{n-j_n}\left(\frac{np}{j_n}\right)^{j_n}\right] = \frac{1}{2}y_n^2 + n^{-1/2}E(n)$$

$$\to \frac{1}{2}y^2,$$

with E denoting the remainder of the series' terms. This limit as $n \to \infty$ is justified by the observation that $E(n)$ is an absolutely convergent series with constant first term, and all other terms of the form $c_j n^{-a_j}$ for some $a_j > 0$.   ∎

8.5.3   Approximating Binomial Probabilities I

The De Moivre–Laplace theorem provides another handy way to approximate binomial probabilities, in addition to the Poisson distribution discussed in chapter 7. Rewriting (8.26) provides the approximation

$$\Pr\{Y^{(n)} = y_n\} \simeq \frac{1}{\sqrt{2\pi}\sqrt{npq}}e^{-y_n^2/2}. \tag{8.27}$$

In a given binomial application, a common calculation needed is one of the form $\Pr[a \le X^{(n)} \le a+b]$, where a and b are integers, and $X^{(n)}$ is binomially distributed with parameters n and p. Specifically,

$$\Pr[a \le X^{(n)} \le a+b] = \sum_{j=a}^{a+b}\frac{n!}{j!(n-j)!}p^j q^{n-j}.$$

This expression reflects the assumption that $0 \le a < a+b \le n$; otherwise, the summation begins at $j = 0$ and ends at $j = n$, as appropriate. While this is only an arithmetic calculation, for n large and the range $[a, a+b]$ wide, this calculation can be difficult even with advanced computing power.

To approximate this probability in such a case for n large, the Poisson p.d.f. can be used if p is small, say $p < 0.1$, as noted in chapter 7. In general, this approximation can also be implemented using (8.27) by converting this probability statement to a statement in the normalized variable $Y^{(n)} = \frac{X^{(n)} - np}{\sqrt{npq}}$. Specifically,

$$\Pr[a \le X^{(n)} \le a+b] = \Pr\left[\frac{a - np}{\sqrt{npq}} \le Y^{(n)} \le \frac{a+b-np}{\sqrt{npq}}\right].$$

Using the approximation in (8.27) above, with $y_0 = \frac{a-np}{\sqrt{npq}}$ and $y_k = y_{k-1} + \frac{1}{\sqrt{npq}}$, we get

$$\Pr[a \leq X^{(n)} \leq a+b] \simeq \frac{1}{\sqrt{2\pi}\sqrt{npq}} \sum_{k=0}^{b} e^{-y_k^2/2}, \tag{8.28}$$

which is a more manageable calculation. As noted above, this formula needs to be adjusted if either $a < 0$ and/or $a+b > n$ to ensure that the original summation includes at most the range $j = 0, 1, \ldots, n$.

Remark 8.25   *Note that while the De Moivre–Laplace theorem is stated in terms of sums of the standard binomial $X^{(n)} \equiv \sum_{j=1}^{n} X_j^B$, where $\{X_j^B\}$ are i.i.d. with $\Pr[X_j^B = 1] = p$ and $\Pr[X_j^B = 0] = 1 - p$, it is equally true for sums of shifted binomial random variables, where $\Pr[X_j^{B\prime} = c] = p$ and $\Pr[X_j^{B\prime} = d] = 1 - p$. This is because this variable can be expressed as*

$$X_j^{B\prime} = (c - d)X_j^B + d.$$

Consequently $\mathrm{E}[X_j^{B\prime}] = (c-d)\mathrm{E}[X_j^B] + d$, and $\mathrm{Var}[X_j^{B\prime}] = (c-d)^2\,\mathrm{Var}[X_j^B]$. Applying this to the normalized sums, we obtain

$$\frac{\sum_{j=1}^{n} X_j^{B\prime} - \mathrm{E}[\sum_{j=1}^{n} X_j^{B\prime}]}{\sqrt{\mathrm{Var}[\sum_{j=1}^{n} X_j^{B\prime}]}} = \frac{\sum_{j=1}^{n} X_j^{B} - \mathrm{E}[\sum_{j=1}^{n} X_j^{B}]}{\sqrt{\mathrm{Var}[\sum_{j=1}^{n} X_j^{B}]}} = Y^{(n)}.$$

In other words, the normalized summation of shifted binomial random variables equals the normalized summation of standard binomial random variables. Hence the De Moivre–Laplace theorem applies and (8.28) is adapted accordingly.

8.6   The Normal Distribution

8.6.1   Definition and Properties

The function

$$f(x) = \frac{1}{\sqrt{2\pi}} e^{-x^2/2}, \tag{8.29}$$

is in fact a continuous probability density function, although we will not have the mathematical tools to verify in what way this is true until chapter 10. This function

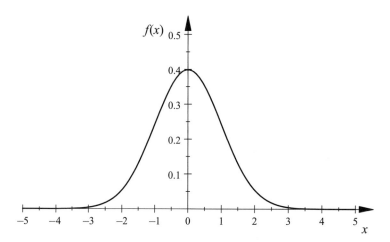

Figure 8.1
$f(x) = \frac{1}{\sqrt{2\pi}} e^{-x^2/2}$

is called the **normal density function**, and sometimes the **unit** or **standardized normal density function**. There is an associated distribution function, the **normal distribution function**, which again requires the tools of chapter 10 to formally define. When not specifically referring to either the density or distribution functions, it is common to simply refer to the **normal distribution**, particularly in reference to the graph of the density function in **figure 8.1**.

The normal distribution is also referred to as the **Gaussian distribution**, named for **Johann Carl Friedrich Gauss** (1777–1855), who used it as a model of measurement errors. The implied random variable, often denoted Z, is apparently not of the discrete type because it assumes all real values. In other words, Rng $Z = \mathbb{R}$. This distribution is of **continuous type**, and it may be the most celebrated example of a continuous probability distribution. The mathematics required for continuous distributions, and some more general distributions, will be developed in chapters 9 and 10, and we will return to study probability theory in these contexts.

It will be seen in chapter 10 that

$$E[Z] = 0, \quad \text{Var}[Z] = 1, \quad M_Z(t) = e^{t^2/2}, \quad C_Z(t) = e^{-t^2/2}, \tag{8.30}$$

and we express this p.d.f. relationship as $Z \sim N(0, 1)$. If X is a random variable so that $\frac{X-\mu}{\sigma} = Z$, then X is said to have a **general normal distribution**, denoted $X \sim N(\mu, \sigma^2)$, and using properties of expectations, one derives that

$$E[X] = \mu, \quad \text{Var}[X] = \sigma^2, \quad M_X(t) = e^{\mu t + \sigma^2 t^2/2}, \quad C_Z(t) = e^{i\mu t - \sigma^2 t^2/2}. \tag{8.31}$$

The graph of this density function is the familiar bell-shaped curve in **figure 8.1**.

As will be seen, associated with the normal p.d.f. is the normal distribution function $F(x)$, defined as in the discussion leading up to (7.22),

$$F(x) = \text{Pr}[Z^{-1}(-\infty, x]]$$

$$= \text{Pr}[Z \leq x].$$

The calculation of $F(x)$ from $f(x)$ used in (7.22) requires generalization here, whereby the summation of $f(x)$ values is replaced by the **integral** of $f(x)$ developed in chapter 10.

However, even with that mathematical insight and the needed tools, the normal distribution function $F(x)$ cannot be calculated exactly from the density function $f(x) = \frac{1}{\sqrt{2\pi}} e^{-x^2/2}$ and must be numerically approximated. Consequently it is common that many mathematical software packages supply this distribution function as a built-in formula, and also mandatory that every book in probability theory or statistics provides a table of $N(0,1)$ values at least for $x > 0$, often referred to as the standard normal table.

Such tables are easy to use because of the apparent symmetry of this function around the point $x = 0$. In other words, it is apparent that $\text{Pr}[Z \leq -a] = \text{Pr}[Z \geq a]$. Also $\text{Pr}[Z \leq -a] = 1 - \text{Pr}[Z < a] = 1 - \text{Pr}[Z \leq a]$, since $\text{Pr}[Z = a] = 0$. That is, $F(-a) = 1 - F(a)$. Consequently we calculate from the standard normal tables

$$\text{Pr}[a \leq Z \leq b] = \begin{cases} F(b) - F(a), & \text{if } 0 < a < b, \\ F(b) - [1 - F(-a)], & \text{if } a < 0 < b, \\ F(-a) - F(-b), & \text{if } a < b < 0. \end{cases} \tag{8.32}$$

Of course, if we have a table with positive and negative x values, or a computer built-in function, it is always the case that for $a < b$, $\text{Pr}[a \leq Z \leq b] = F(b) - F(a)$.

8.6.2   Approximating Binomial Probabilities II

Normal distribution tables can be used to approximate binomial probabilities as noted in (8.28), but a small adjustment is required. From that formula, it would be natural to assume that

$$\text{Pr}\left[\frac{a - np}{\sqrt{npq}} \leq Y^{(n)} \leq \frac{a + b - np}{\sqrt{npq}}\right] \simeq F\left(\frac{a + b - np}{\sqrt{npq}}\right) - F\left(\frac{a - np}{\sqrt{npq}}\right),$$

and of course, the presence of "$\simeq$" suggests this to be a "true" statement. However, it is not true that this approximation is as accurate as is possible. The problem with this approximation can be best observed by letting $b \to 0$, in which case the left-hand side becomes $\Pr\left[Y^{(n)} = \frac{a-np}{\sqrt{npq}}\right]$, and the right-hand side becomes $F\left(\frac{a-np}{\sqrt{npq}}\right) - F\left(\frac{a-np}{\sqrt{npq}}\right)$ $= 0$. This simple example highlights the error and illustrates the problem.

The binomial distribution for $X^{(n)}$ allocates a total probability of 1 among $n+1$ real values $0, 1, 2, \ldots, n$. In turn the binomial distribution for $Y^{(n)}$ allocates a total probability of 1 among $n+1$ real values

$$\frac{-np}{\sqrt{npq}}, \frac{1-np}{\sqrt{npq}}, \frac{2-np}{\sqrt{npq}}, \ldots, \frac{nq}{\sqrt{npq}}.$$

From (8.27) we have that $\Pr\left\{Y^{(n)} = \frac{a-np}{\sqrt{npq}}\right\} \simeq \frac{1}{\sqrt{2\pi}\sqrt{npq}} e^{-((a-np)/\sqrt{npq})^2/2}$, where we note that the multiplicative term $\frac{1}{\sqrt{npq}}$ is exactly equal to the distance between any two successive $Y^{(n)}$ values. In other words, this binomial probability is being approximated by the normal distribution, not at the point $\frac{a-np}{\sqrt{npq}}$ but over an interval around this point of length $\frac{1}{\sqrt{npq}}$.

Consequently one has for some $0 \le \lambda \le 1$,

$$\Pr\left[Y^{(n)} = \frac{a-np}{\sqrt{npq}}\right] \simeq \Pr\left[\frac{a-np-(1-\lambda)}{\sqrt{npq}} \le Z \le \frac{a-np+\lambda}{\sqrt{npq}}\right].$$

The usual convention is to take the symmetric value of $\lambda = \frac{1}{2}$, and hence

$$\Pr\left[Y^{(n)} = \frac{a-np}{\sqrt{npq}}\right] \simeq F\left(\frac{a-np+\frac{1}{2}}{\sqrt{npq}}\right) - F\left(\frac{a-np-\frac{1}{2}}{\sqrt{npq}}\right).$$

This is often referred to as the **half-interval adjustment**, or **half-integer adjustment**.

Extending this conventional approximation for a single probability, the binomial probability statement above, written in terms of the original binomial $X^{(n)}$, becomes

$$\Pr[a \le X^{(n)} \le a+b] \simeq F\left(\frac{a+b-np+\frac{1}{2}}{\sqrt{npq}}\right) - F\left(\frac{a-np-\frac{1}{2}}{\sqrt{npq}}\right). \tag{8.33}$$

Notation 8.26   *Because the normal distribution is so important in probability theory, it has inherited special notation that is almost universally recognized. As noted above, the standard normal random variable is usually denoted as Z, while the probability density function is denoted with the Greek letter phi, $\varphi(z) \equiv \frac{1}{\sqrt{2\pi}} e^{-z^2/2}$, and the distribution function either with the Greek capital phi, $\Phi(z)$, or as $N(z)$.*

*8.7 The Central Limit Theorem

There are many versions of the central limit theorem. All of them generalize the De Moivre–Laplace theorem in one remarkable way or another. In essence, what any version states and what makes any version indeed the "central" limit theorem is that under a wide variety of assumptions, the p.d.f. of the sum of n independent variables, after normalizing as in (8.19), converges to the normal distribution as $n \to \infty$.

Remarkably these random variables need not be identically distributed, just independent, although the need for normalization demands that these random variables have at least two moments: means and variances. When not identically distributed, there is a requirement that the sequence of variances does not grow too fast to preclude latter terms in the random variable series from increasingly dominating the summation, as well as a requirement that they do not converge to 0 so quickly that the average variance converges to 0.

These theorems can be equivalently stated in terms of the sum of these random variables, or their average. This is because from (7.38) we have

$$\frac{\sum_{j=1}^{n} X_j - \mathrm{E}[\sum_{j=1}^{n} X_j]}{\sqrt{\mathrm{Var}[\sum_{j=1}^{n} X_j]}} = \frac{\frac{1}{n}\sum_{j=1}^{n} X_j - \mathrm{E}\left[\frac{1}{n}\sum_{j=1}^{n} X_j\right]}{\sqrt{\mathrm{Var}\left[\frac{1}{n}\sum_{j=1}^{n} X_j\right]}},$$

since $\mathrm{E}\left[\frac{1}{n}\sum_{j=1}^{n} X_j\right] = \frac{1}{n}\mathrm{E}[\sum_{j=1}^{n} X_j]$ and $\mathrm{Var}\left[\frac{1}{n}\sum_{j=1}^{n} X_j\right] = \frac{1}{n^2}\mathrm{Var}[\sum_{j=1}^{n} X_j]$. So while the ranges of the sum and average of random variables are quite different, the associated normalized random variables are identical.

Consequently central limit theorems, in general, and the De Moivre–Laplace theorem, in particular, apply to the sums of random variables if and only if they apply to the averages of random variables. And similar to the result explored above for sums of general binomial random variables, the central limit theorem applies to $\sum_{j=1}^{n} X_j$ for i.i.d. $\{X_j\}$, and it applies to $\sum_{j=1}^{n} Y_j$ where $Y_j = aX_j + b$ for constants a and b.

Central limit theorems apply to all probability distributions that satisfy the given requirements, whether discrete, continuous, or mixed. Because of this generality there is no hope that a proof of such a result can proceed along the lines of the proof of the De Moivre–Laplace theorem, which relied heavily on the exact form of the binomial p.d.f. The tool used for these general proofs represents a sophisticated application of properties of the moment-generating function (m.g.f.), or more generally, the characteristic function (c.f.).

To set the stage, we provide a simplified proof of the central limit theorem in the case of independent, identically distributed discrete random variables that have moments of all orders and a convergent moment-generating function, and hence (7.66) applies. Mechanically, the proof works in settings other than discrete, but it requires manipulation properties of the moment-generating function that have only been proved in a discrete setting but are valid more generally. The proof can also be generalized to discrete distributions with only a few moments, and this will be discussed below.

That the conclusion of this theorem is consistent with the normal distribution depends on a fact that cannot be proved until chapter 10, that the moment-generating function of the unit normal distribution satisfies: $M_Z(t) = e^{t^2/2}$. In addition, as has been noted many times, and partially proved in section 8.1, the moment-generating function truly characterizes this and every distribution when it exists, so the standard normal distribution is the only distribution with the m.g.f. $M_Z(t) = e^{t^2/2}$.

Proposition 8.27 (*Central Limit Theorem*)   *Let X be a discrete random variable with moments of all orders and a convergent moment-generating function, and let $\{X_j\}_{j=1}^n$ be independent and identically distributed random variables. Denote by $X^{(n)}$ the average of this collection, $X^{(n)} = \frac{1}{n}\sum_{j=1}^n X_j$, and by $Y^{(n)}$ the normalized version, $Y^{(n)} = \frac{X^{(n)}-\mu}{\frac{\sigma}{\sqrt{n}}}$. If $M_{Y^{(n)}}(t)$ denotes the moment-generating function of $Y^{(n)}$, then*

$$M_{Y^{(n)}}(t) \to e^{t^2/2} \qquad as \ n \to \infty. \tag{8.34}$$

Proof   First note two properties of moment-generating functions that follow from the definition and properties of expectations (see exercise 8):

1. $M_{X/b}(t) = M_X\left(\frac{t}{b}\right)$.
2. $M_{\sum X_i}(t) = \prod M_{X_i}(t)$ if $\{X_i\}$ are independent.

From these it follows that with $Y^{(n)} = \sum_{j=1}^n \left(\frac{X_j-\mu}{\sqrt{n}\sigma}\right)$, we have

$$M_{Y^{(n)}}(t) = \prod_{j=1}^n M_{(X_j-\mu)}\left(\frac{t}{\sqrt{n}\sigma}\right)$$

$$= \left(M_{(X-\mu)}\left(\frac{t}{\sqrt{n}\sigma}\right)\right)^n,$$

where this last step follows from $\{X_j\}_{j=1}^n$ being i.i.d. Now, by (7.66),

$$M_{(X-\mu)}\left(\frac{t}{\sqrt{n}\sigma}\right) = \sum_{j=0}^{\infty} \frac{1}{j!}\mu_j\left(\frac{t}{\sqrt{n}\sigma}\right)^j.$$

Recalling that $\mu_0 = 1$, $\mu_1 = 0$ and $\mu_2 = \sigma^2$, we get

$$M_{(X-\mu)}\left(\frac{t}{\sqrt{n}\sigma}\right) = 1 + \frac{\sigma^2}{2}\left(\frac{t}{\sqrt{n}\sigma}\right)^2 + n^{-3/2}E(n)$$

$$= 1 + \frac{t^2}{2n} + n^{-3/2}E(n),$$

where $E(n) = \sum_{j=3}^{\infty}\frac{1}{j!}\mu_j\left(\frac{t}{\sigma}\right)^j n^{(3-j)/2}$. Now, since $M_X(t)$ is assumed convergent for $|t| < T$ say, $M_{X-\mu}(t) = e^{-\mu t}M_X(t)$ has the same interval of convergence, and hence $E(n)$ is convergent for $|t| < \sigma T$. As is true for $M_X(t)$, it is also true that $E(n)$ is a differentiable function of t, and hence a continuous function that attains its maximum and minimum on any closed interval $|t| \leq \sigma T - \epsilon$ (see proposition 9.39). Let K be defined so that $|E(n)| \leq K$ for all n on one such interval.

This expression can now be raised to the nth power, and a logarithm taken. The same trick is used here as in the proof of the De Moivre–Laplace theorem, in which we keep track of only the powers of n that are needed for the final limit, sometimes invoking a sample calculation to determine how many terms will be needed. This produces

$$\ln M_{Y^{(n)}}(t) = n\ln\left[1 + \frac{t^2}{2n} + n^{-3/2}E(n)\right]$$

$$= n\left[\left(\frac{t^2}{2n} + n^{-3/2}E(n)\right) - \frac{1}{2}\left(\frac{t^2}{2n} + n^{-3/2}E(n)\right)^2 + \cdots\right],$$

where in the second step is invoked the power series expansion for $\ln(1+x)$ from (8.20) with $x = \frac{t^2}{2n} + n^{-3/2}E(n)$. Now, since $|E(n)| \leq K$, we have that $|x| \leq \frac{t^2}{2n} + n^{-3/2}K < 1$ for n large, and $\ln(1+x)$ is absolutely convergent. Next the series above can be expanded and rearranged to produce

$$\ln M_{Y^{(n)}}(t) = \frac{1}{2}t^2 + F(n).$$

Now $F(n) = n^{-1/2}E(n) + n^{-1}\tilde{E}(n)$ is absolutely convergent for the same range of t, is continuous, and hence is bounded on closed subintervals. From this last step we conclude that $\ln M_{Y^{(n)}}(t) \to \frac{1}{2}t^2$ as $n \to \infty$, and hence $M_{Y^{(n)}}(t) \to e^{t^2/2}$.   ∎

The fact that this theorem allows a variety of generalizations can now be understood. For example, the assumption that X had "moments of all orders" was not really needed. What was needed was knowledge that $M_{(X-\mu)}\left(\frac{t}{\sqrt{n}\sigma}\right)$ could be approximated by

$$M_{(X-\mu)}\left(\frac{t}{\sqrt{n}\sigma}\right) = 1 + \frac{t^2}{2n} + n^{-3/2}E(n),$$

and where $E(n)$ is a bounded function of t on an interval $|t| \leq C$ as $n \to \infty$.

To reach a comparable conclusion in the case of a limited number of moments, one must work with the characteristic function, which always exists, and with that function it will be enough to assume that X has three moments using the tools of chapter 9 adapted to complex-valued functions such as $C_X(t)$.

Moreover, looking at the calculation above, we did not really need to have the error term, $E'(n) = n^{-3/2}E(n)$, with a factor of $n^{-3/2}$. If this coefficient was n^{-1-a} for any $a > 0$, this would be enough to again force the conclusion because then the leading coefficient of $F(n)$ would be $nE'(n) = n^{-a}E(n)$. It turns out that we can "almost" reach this conclusion if X has only two moments. The conclusion that can be reached, again with the adapted tools of chapter 9, is that this leading coefficient of $F(n)$ satisfies $nE'(n) \to 0$ as $n \to \infty$, and this again is enough for the conclusion.

As another example of a direction for generalization, suppose that $\{X_j\}_{j=1}^n$ are independent and have moments of all orders and convergent m.g.f.s but are not identically distributed. The normalized random variable $Y^{(n)}$ is defined as

$$Y^{(n)} = \frac{X^{(n)} - \mu^{(n)}}{\frac{\sigma^{(n)}}{\sqrt{n}}},$$

where $\mu^{(n)} = \frac{1}{n}\sum_{j=1}^n \mu_j$ and $[\sigma^{(n)}]^2 = \frac{1}{n}\sum_{j=1}^n \sigma_j^2$. Then all the steps up to $M_{Y^{(n)}}(t) = \prod_{j=1}^n M_{(X_j-\mu_j)}\left(\frac{t}{\sqrt{n}\sigma^{(n)}}\right)$ go through without any obstacle.

This approach produces, with the aid of (7.66) and taking of logarithms,

$$\ln M_{Y^{(n)}}(t) = \sum_{j=1}^n \ln\left[1 + \frac{\sigma_j^2}{2}\left(\frac{t}{\sqrt{n}\sigma^{(n)}}\right)^2 + n^{-3/2}E_j(n)\right]$$

$$= \sum_{j=1}^n \ln\left[1 + \frac{t^2}{2n}\left(\frac{\sigma_j}{\sigma^{(n)}}\right)^2 + n^{-3/2}E_j(n)\right]$$

$$= \frac{t^2}{2} \left[\frac{1}{n} \sum_{j=1}^{n} \left(\frac{\sigma_j}{\sigma^{(n)}} \right)^2 \right] + F(n)$$

$$= \frac{t^2}{2} + F(n),$$

where the last step is justified by the definition of $\sigma^{(n)}$.

Although everything looks harmless in this last expression, a closer examination of the new $F(n)$ expression reveals that this comes from summations and products of terms of the form $E_j(n) = \sum_{k=3}^{\infty} \frac{1}{k!} \mu_{jk} \left(\frac{t}{\sigma^{(n)}} \right)^k n^{(3-k)/2}$, where μ_{jk} denotes the kth central moment of X_j. As above, this can be reduced to essentially $\frac{1}{3!} \mu_{j3} \left(\frac{t}{\sigma^{(n)}} \right)^3 + n^{-1/2} \tilde{E}_j(n)$, which means that the first term in $F(n)$ is

$$n^{-3/2} \sum_{j=1}^{n} \frac{1}{3!} \mu_{j3} \left(\frac{t}{\sigma^{(n)}} \right)^3 = \frac{1}{3!} t^3 \frac{n^{-3/2} \sum_{j=1}^{n} \mu_{j3}}{\left(\frac{1}{n} \sum_{j=1}^{n} \sigma_j^2 \right)^{3/2}}$$

$$= \frac{1}{3!} t^3 \frac{\sum_{j=1}^{n} \mu_{j3}}{\left(\sum_{j=1}^{n} \sigma_j^2 \right)^{3/2}}.$$

So in order to be assured that $F(n)$ can be dismissed, it is necessary to assume that the absolute value of this ratio converges to 0. Now, by the triangle inequality applied twice,

$$\left| \sum_{j=1}^{n} \mu_{j3} \right| \le \sum_{j=1}^{n} |\mu_{j3}| \le \sum_{j=1}^{n} \mu_{|j3|},$$

where $\mu_{|j3|}$ denotes the third absolute central moment of X_j, which is $\mu_{|j3|} \equiv \mathrm{E}[|X_j - \mu_j|^3]$.

To assure this needed absolute convergence, it is common to define the condition in terms of the relative size of third absolute central moments to the variance terms:

$$\left(\frac{\sum_{j=1}^{n} \mu_{|j3|}}{(\sum_{j=1}^{n} \sigma_j^2)^{3/2}} \right)^{1/3} = \frac{(\sum_{j=1}^{n} \mu_{|j3|})^{1/3}}{(\sum_{j=1}^{n} \mu_{j2})^{1/2}} \to 0 \qquad \text{as } n \to \infty.$$

This assumption is a special case of what is known as **Lyapunov's condition**, after **Aleksandr Lyaponov** (1857–1918).

Note that in the case where $\{X_j\}_{j=1}^n$ are independent and identically distributed,

$$\frac{\sum_{j=1}^n \mu_{|j3|}}{\left(\sum_{j=1}^n \sigma_j^2\right)^{3/2}} = \frac{n\mu_{|3|}}{\left(n\sigma^2\right)^{3/2}} = \frac{\mu_{|3|}}{\sqrt{n}\sigma^3},$$

and then Lyapunov's condition is automatically satisfied.

8.8   Applications to Finance

8.8.1   Insurance Claim and Loan Loss Tail Events

For both the loan loss models and claims models of chapter 7, in which the mean and variance of the distributions were estimated, there is a natural interest in evaluating the **probability of severe loss events**, which in both cases is the probability $\Pr[L \geq A]$, or, $\Pr[L - \mathrm{E}[L] \geq C]$ for various values of assets A or capital C. In this notation one might envision A to be the assets allocated to cover all losses and insurance claims in a given period, or if $\mathrm{E}[L]$ has been placed on this balance sheet as a liability representing a provision for expected losses and claims, C then represents the capital allocated to cover excess losses. In this simple balance sheet framework, $A = \mathrm{E}[L] + C$ for each risk.

Of course, in the one-period model that we investigate the random loss variable L has two components in general:

- Insurance liability payments
- Credit losses on assets

So, if A denotes an asset portfolio at time 0, and L^A and L^I denote losses on assets and insurance payments respectively, then $\Pr[L \geq A]$ is shorthand for

$$\Pr[L \geq A] \equiv \Pr[L^I + L^A > A],$$

and $\Pr[L - \mathrm{E}[L] \geq C]$ is shorthand for

$$\Pr[L - \mathrm{E}[L] \geq C] \equiv \Pr[L^I - \mathrm{E}[L^I] + L^A - \mathrm{E}[L^A] > C],$$

where $C \equiv A - \mathrm{E}[L^I] - \mathrm{E}[L^A]$.

If assets are risk free, then adding more assets to A creates the same increase in C and this change has no effect on the volatility of losses.

However, when assets are risky, $\mathrm{E}[L^A]$ depends on A. Then adding assets to A creates a smaller increase in C, and also affects the volatility of losses. In this case it is simpler to think of $E[L^A]$ in terms of a **loss ratio random variable** R^A, in that $L^A = AR^A$. Hence we can define

$$\Pr[L \geq A] \equiv \Pr[L^I + AR^A > A], \tag{8.35}$$

and with $C \equiv A(1 - E[R^A]) - E[L^I]$,

$$\Pr[L - E[L] \geq C] \equiv \Pr[L^I - E[L^I] + A(R^A - E[R^A]) > C].$$

When such models are applied to a single business unit, the total entity is modeled as

$$\mathbf{A} = \mathbf{L} + \mathbf{C},$$

where $\mathbf{A}$ denotes total assets of the firm, $\mathbf{L}$ total liabilities representing provisions for all expected claims and losses, and $\mathbf{C}$ is total capital. Intuitively $\mathbf{A} = \sum A_j$, and similarly for $\mathbf{L}$ and $\mathbf{C}$, but the adequacy of corporate capital or assets cannot be assessed in terms of the capital or assets needs of each unit or risk separately.

Indeed, if C_j denotes the capital needed for the jth risk, in general, one has that $\mathbf{C} < \sum C_j$ because risks are not perfectly correlated. Hence tail events will not, in general, be realized together. To evaluate the entity in total, explicit assumptions are needed on the joint distribution of all risks. We ignore the broader question here and focus on the adequacy of assets or capital for the risks modeled in chapter 7, which were related to insurance claims or loan losses during a given fixed period.

We consider three approaches and introduce these in the model with risk-free assets so that $L^A = 0$. We then turn to the more general asset case.

Risk-Free Asset Portfolio

Chebyshev I   If insurance claims are modeled as in chapter 7, and $E[L]$ and $\text{Var}[L]$ calculated as in (7.120) and (7.121) for the individual loss model, or (7.125) and (7.126) for the aggregate loss model, the one-sided Chebyshev inequality in (8.6) can be used to deduce that for $A \geq E[L]$,

$$\Pr[L \geq A] \leq \frac{\text{Var}[L]}{(A - E[L])^2 + \text{Var}[L]}. \tag{8.36}$$

Since $A = E[L] + C$, this probability upper bound can also be expressed in terms of $C \geq 0$:

$$\Pr[L - E[L] \geq C] \leq \frac{\text{Var}[L]}{C^2 + \text{Var}[L]}. \tag{8.37}$$

Still this estimate can be considered crude because it is an estimate that applies to all distributions, and not necessarily one that specifically applies to the distribution at hand. In addition this estimate only reflects two moments of the given loss distribution, and no special information about the tail probabilities in this model.

Loss Simulation As noted in chapter 7, insurance claims under either the individual or aggregate loss model can be simulated using the approach in section 7.7 on generating random samples. These models are very general, and need to be adapted to a specific claims context as noted in chapter 7, but we discuss the general case.

Specifically, for the individual model in (7.119), losses are given by

$$L = \sum_{j,k} f_{jk} D_{jk} L_{jk},$$

where k denotes the risk class, j an enumeration of the individual exposures in this class, and f_{jk} the exposure on the jth exposure in class k.

To implement one simulation of L, one uniformly distributed random variable $r_{jk} \in [0,1]$ is first generated for each exposure of amount f_{jk} to determine if a loss occurred. If $r_{jk} < q_k$, with q_k the probability of a loss, then $D_{jk} = 1$ and there is a loss; otherwise, $D_{jk} = 0$. This procedure is equivalent to defining $D_{jk} = F_{B_k}^{-1}(r_{jk})$, where $F_{B_k}(x)$ is the distribution function for this binomial.

In addition for each exposure for which $D_{jk} = 1$, a new uniformly distributed random variable $r'_{jk} \in [0,1]$ is generated, and using the c.d.f. of the class k loss ratio random variable $F_k(x)$, we define the sampled loss ratio by $L_{jk} = F_k^{-1}(r'_{jk})$. This procedure then generates one simulation of the random variable L, and it can be repeated as many times as is desired.

Similarly, from (7.122) for the aggregate loss model,

$$L = \sum_k \bar{f}_k N_k L'_k,$$

each of the random variables N_k and L'_k needs to be generated. Here $\bar{f}_k$ denotes the average of the n_k exposures in class k. Since N_k denotes the total number of claims from this class, it can be modeled either as a binomial distribution with parameters n_k and q_k or as a Poisson distribution with $\lambda_k = n_k q_k$. In either case one simulation for class k requires first generating one uniformly distributed random variable $r \in [0,1]$, from which we define $N_k \equiv F_N^{-1}(r)$, where $F_N(x)$ denotes the assumed cumulative distribution for N_k. If $N_k > 0$, then another N_k uniformly distributed variables $\{r_j\}_{j=1}^{N_k}$ are generated from which are defined the loss ratios $\{L_{jk}\}_{j=1}^{N_k} =$

$\{F_k^{-1}(r_j)\}_{j=1}^{N_k}$, with $F_k(x)$ the cumulative distribution function for L_k. The average loss ratio is then $L_k' = \frac{1}{N_k}\sum_{j=1}^{N_k} L_{jk}$. Each additional simulation proceeds in the same way and is repeated as many times as is desired.

From these simulations one can now estimate $\Pr[L \geq A]$ directly from the generated data. Namely, if M denotes the total number of simulations, and M^A the number for which $L \geq A$, then

$$\Pr[L \geq A] \approx \frac{M^A}{M}. \tag{8.38}$$

If there is a shortcoming in this procedure, it is that for A large there may be very few sample points generated for which $L \geq A$. For example, if $\Pr[L \geq A] = p^A$, then given a simulation of M sample points for L,

$$E[M^A] = Mp^A,$$

$$\mathrm{Var}[M^A] = Mp^A(1 - p^A).$$

Consequently the mean and standard deviation of this probability estimate are

$$E\left[\frac{M^A}{M}\right] = p^A,$$

$$\mathrm{s.d.}\left[\frac{M^A}{M}\right] = \sqrt{\frac{p^A(1-p^A)}{M}},$$

and so by the De Moivre theorem, the $100(1-\alpha)\%$ confidence interval for $\frac{M^A}{M}$ is approximately

$$p^A\left(1 - z_{\alpha/2}\sqrt{\frac{(1-p^A)}{Mp^A}}\right) \leq \frac{M^A}{M} \leq p^A\left(1 + z_{1-(\alpha/2)}\sqrt{\frac{(1-p^A)}{Mp^A}}\right),$$

where $z_{\alpha/2}$ and $z_{1-(\alpha/2)}$ denote the respective percentiles on $N(0,1)$. This result can be better stated in terms of the relative error of the estimate:

$$1 - z_{\alpha/2}\sqrt{\frac{(1-p^A)}{Mp^A}} \leq \frac{\frac{M^A}{M}}{p^A} \leq 1 + z_{1-(\alpha/2)}\sqrt{\frac{(1-p^A)}{Mp^A}}.$$

Example 8.28   *If $p^A = 0.001$ and $\alpha = 0.05$, then the range of the ratio of the estimate $\frac{M^A}{M}$ to the actual value p^A, for a 95% confidence interval, is* $2z_{0.975}\sqrt{\frac{(1-p^A)}{Mp^A}} \simeq \frac{123.9}{\sqrt{M}}$, *since* $z_{0.975} \simeq 1.96$. *So to have this range equal to* p^A, *for a* 50% *relative estimate error "on average," requires,* $M \simeq 1.5 \times 10^{10}$ *simulations. If* $p^A \simeq 0.01$, *we have that* $2z_{0.975}\sqrt{\frac{(1-p^A)}{Mp^A}} \simeq \frac{39.0}{\sqrt{M}}$, *so to again have this range equal to* p^A, *for a* 50% *relative error requires* $M \simeq 15.2$ *million. Finally, for* $p^A \simeq 0.1$, *we require* $M \simeq 13,830$, *and for* $p^A \simeq 0.2$, *the number of simulations reduces to* $M \simeq 1537$.

Simulations and Chebyshev II   To avoid the estimation problem noted above when $\Pr[L \geq A]$ is small, which is the anticipated case for most problems of interest in assessing asset or capital adequacy, we use the simulation above to calibrate a new Chebyshev estimate. To this end, we first choose an initial asset level, A', so that $p^{A'} \equiv \Pr[L \geq A']$ is relatively large, say in the range: $0.10 \leq p^{A'} \leq 0.20$. Then approximately 10–20% of the simulations will produce losses in excess of this initial level.

Define L' to be the generated losses above this threshold. Specifically, L' is a conditional random variable:

$$L' = L \,|\, (L > A').$$

Formulaically, the distribution function of L' is given in terms of the distribution function of L by

$$F_{L'}(x) = \frac{F_L(x) - F_L(A')}{1 - F_L(A')}, \qquad x \geq A'.$$

From the simulated data, $\mathrm{E}[L']$ and $\mathrm{Var}[L']$ can be estimated, and from the one-sided Chebyshev inequality, we have for $A > \mathrm{E}[L']$,

$$\Pr[L' > A] \leq \frac{\mathrm{Var}[L']}{(A - \mathrm{E}[L'])^2 + \mathrm{Var}[L']}. \tag{8.39}$$

Note that $\Pr[L > A']$ is also estimated from the simulations as $\frac{M^{A'}}{M}$, and this is used next.

By the law of total probability, for any values of A and A',

$$\Pr[L > A] = \Pr[L > A \,|\, L < A'] \Pr[L < A'] + \Pr[L > A \,|\, L > A'] \Pr[L > A'].$$

For $A > A'$, we have that $\Pr[L > A \,|\, L < A'] = 0$. Also $\Pr[L > A \,|\, L > A'] = \Pr[L' > A]$, and therefore

$$\Pr[L > A] = \Pr[L' > A] \Pr[L > A'].$$

Finally, for $A > E[L']$, we have from (8.39) and (8.38) that

$$\Pr[L > A] \leq \frac{M^{A'}}{M} \frac{\text{Var}[L']}{(A - E[L'])^2 + \text{Var}[L']}. \tag{8.40}$$

Since $A = E[L] + C$, this probability upper bound can also be expressed in terms of C:

$$\Pr[L - E[L] \geq C] \leq \frac{M^{A'}}{M} \frac{\text{Var}[L']}{(C + E[L] - E[L'])^2 + \text{Var}[L']} \tag{8.41}$$

Risky Assets

Using (8.35), we write

$$\Pr[L \geq A] \equiv \Pr[L^I + AR^A > A];$$

the new challenge is the estimation of the moments of the random variable $L \equiv L^I + AR^A$ from two respective models. Of course, L^I is modeled as above in the risk-free asset case. For R^A the same models can be applied to a representative risky asset portfolio of amount A_0, and we then define the random variable R^A by

$$R^A = \frac{L^{A_0}}{A_0}.$$

We can then determine the mean and variance of R^A from the mean and variance of L^{A_0}, and simulate R^A from simulations of L^{A_0}.

The critical question in this context is the correlation between the random variables L^I and R^A. In some applications, such as for life insurance and credit losses, the assumption of independence seems justifiable. In others, for example, disability insurance and credit losses, or variable life insurance claims and stock portfolio losses, a nonzero correlation assumption is needed. This is because disability claims can be negatively correlated with the economy as are credit losses, so there is a positive correlation between L^I and R^A. Likewise variable life insurance minimum guarantees are more costly when equity markets are falling, so again there is a positive correlation between L^I and R^A.

We only investigate here the case of uncorrelated L^I and R^A and leave the more general development as an exercise. In this case,

$$E[L] = E[L^I] + AE[R^A],$$

$$\text{Var}[L] = \text{Var}[L^I] + A^2 \, \text{Var}[R^A].$$

Consequently the direct application of Chebyshev's inequality in (8.36) becomes for $A(1 - E[R^A]) > E[L^I]$, or $A > \frac{E[L^I]}{1-E[R^A]}$:

$$\Pr[L \geq A] \leq \frac{\text{Var}[L^I] + A^2 \text{Var}[R^A]}{(A(1 - E[R^A]) - E[L^I])^2 + \text{Var}[L^I] + A^2 \text{Var}[R^A]}. \tag{8.42}$$

For simulations, the random variables L^I and R^A are generated in pairs, and now (8.38) is applied directly, where M^A is again the number of paired scenarios for which $L \geq A$, which is equivalent to

$$L^I \geq A(1 - R^A).$$

Finally, the combined simulation and Chebyshev estimate works as above. First off, A' is defined so that $p^{A'} \equiv \Pr[L \geq A']$ is again in the range $0.1 \leq p^{A'} \leq 0.2$ where

$$\Pr[L \geq A'] = \Pr[L^I \geq A'(1 - R^A)].$$

Then L' is defined as the total loss random variable conditional on $L \geq A'$:

$$L' = L \,|\, (L > A'),$$

where $L = L^I + AR^A$.

The moments $E[L']$ and $\text{Var}[L']$ can be estimated from paired simulations, as can $\Pr[L > A'] = \frac{M^{A'}}{M}$. Note, however, that in general, there is no formulaic relationship between the conditional mean and variance of L' and the conditional means and variances of the components losses L^I and AR^A.

Finally, for $A > E[L']$, (8.40) again applies.

8.8.2   Binomial Lattice Equity Price Models as $\Delta t \to 0$

Let μ and σ^2 denote the mean and variance of the log-ratio return series as in chapter 7, where these parameters of necessity reflect the period of time separating the data points. By convention, and independent of the time period reflected in the data, these return statistics are always denominated in units of years. In other words,

$$\mu = E\left[\ln\left(\frac{S_{t+1}}{S_t}\right)\right], \quad \sigma^2 = \text{Var}\left[\ln\left(\frac{S_{t+1}}{S_t}\right)\right],$$

where the time parameter of these equity price observations, t, is denominated in years. Of course, if the raw data are spaced differently, say weekly or monthly, there may be a question as to how these estimates are defined if one chooses not to disregard most of the data. This question is addressed below.

Given this historical data series of annual log-ratio returns, which we now index with the natural numbers

$$R_j = \ln\left(\frac{S_{j+1}}{S_j}\right),$$

the density function usually appears bell-shaped, and tests confirm that this series appears reasonably uncorrelated. So one approximate model for projecting into the future assumes independent normally distributed returns. If $\{z_j\}$ denotes a random collection of standard normal variables, with $\mathrm{E}[z_j] = 0$, $\mathrm{Var}[z_j] = 1$, then $\{R_j\} \equiv \{\mu + z_j\sigma\}$ will be normally distributed and have the correct mean and variance, and the projection model becomes

$$S_{j+1} = S_j e^{\mu + z_j\sigma}.$$

While we have not proved this yet (see chapter 10), these standard normal variables are produced the same way as are discrete variables. That is, by starting with a uniformly distributed collection $\{x_j\} \subset [0,1]$, and defining $z_j = N^{-1}(x_j)$ with $N(x)$ the standard normal distribution function.

Alternatively, if the goal of the projection is to model prices in the distant future, we could approximate the log-ratio returns in this normal model with binomial returns, $R_j \simeq B_j$, defining

$$S_{j+1} = S_j e^{B_j}.$$

In this case $\{B_j\}$ are a random collection of binomials as in chapter 7,

$$B_j = \begin{cases} u, & \Pr[u] = p, \\ d, & \Pr[d] = 1 - p, \end{cases}$$

and here u and d are calibrated to achieve the desired moments of μ and σ^2.

The justification for these models being used as alternatives is that at a distant future point in time, $\sum_{j=1}^{n} B_j$ will be nearly normally distributed by the De Moivre–Laplace theorem, as long as n is large. Alternatively, if these models could be translated into models with small time steps of size Δt, the binomial approximation to the normal would be justified even for short-term projections, as long as Δt was small enough.

But how do the parameters μ and σ^2 depend on Δt?

Parameter Dependence on Δt

Since the modeling period is often fixed as $[0, T]$, say, n large is equivalent to $\Delta t \equiv \frac{T}{n}$ being small. But, of course, if Δt is taken as small, it may well be smaller than the original periods of time separating the data points on which μ and σ^2 were developed. Consequently in this section we first investigate a reasonable model for $\mu(\Delta t)$ and $\sigma^2(\Delta t)$, or the relationship between the log-ratio return mean and variance and the length of the time interval. For specificity, one may assume the intuitive model that μ and σ^2 are defined as annualized statistics so that the units of Δt are years, but all that is needed mathematically is that the statistics μ and σ^2 correspond to $\Delta t = 1$.

Specifically, assume that μ and σ^2 denote the mean and variance of the log-ratio return series $\{R_j\}$ for $\Delta t = 1$, and that B_j has been calibrated to the binomial model as in chapter 7. As derived in exercise 27 of that chapter, the general formulas for u and d, which define B_j for general p, $0 < p < 1$, equal

$$u = \mu + \left[\sqrt{\frac{p'}{p}}\right]\sigma, \quad d = \mu - \left[\sqrt{\frac{p}{p'}}\right]\sigma. \tag{8.43}$$

Now for $\Delta t = \frac{1}{m}$, so that there are m time steps in a given period, $[j, j+1]$, let $\{B_k(\Delta t)\}_{k=1}^m$ denote the associated subinterval random variables, defined by

$$S_{j+k/m} = S_{j+(k-1)/m}e^{B_k(\Delta t)}, \qquad k = 1, 2, \ldots, m.$$

If this model is applied iteratively to obtain $S_{j+1} = S_j e^{\sum B_k(\Delta t)}$, then it is apparent upon comparing it to the original model that

$$\sum_{k=1}^m B_k(\Delta t) = B_j.$$

In the same way that the collection $\{B_j\}$ were assumed in the model to be independent and identically distributed, it is logical to extend this assumption to $\{B_k(\Delta t)\}$. Namely we assume that for any Δt, the collection of subperiod log-ratio returns is independent and identically distributed.

Recall that the mean of a sum of random variables is the sum of the means, and the variance of an *independent* sum of random variables is the sum of the variances. Consequently we obtain $m\mu(\Delta t) = \mu$ and $m\sigma^2(\Delta t) = \sigma^2$ for the binomial model, and since $\Delta t = \frac{1}{m}$, this can be expressed as

$$\mu(\Delta t) = \mu\Delta t, \tag{8.44a}$$

$$\sigma^2(\Delta t) = \sigma^2\Delta t. \tag{8.44b}$$

For example, the binomial stock price model in time steps of $\Delta t \leq 1$ units for $p = \frac{1}{2}$ becomes

$$S_{t+\Delta t} = \begin{cases} S_t e^{\mu\Delta t + \sigma\sqrt{\Delta t}}, & \text{Pr} = \frac{1}{2}, \\ S_t e^{\mu\Delta t - \sigma\sqrt{\Delta t}}, & \text{Pr} = \frac{1}{2}, \end{cases} \tag{8.45}$$

with the analogous formula for general p using (8.43).

The normally distributed log-ratio return model can also be recalibrated to the new time interval with the same result based on the same calculation, that $\sum_{k=1}^{m} R_k(\Delta t) = R_j$, again producing (8.44).

Distributional Dependence on Δt

If $\{R_j\}$ are assumed to be independent and normally distributed, so too will be the subperiod returns $\{R_k(\Delta t)\}$. In other words,

$$S_{t+\Delta t} = S_t e^{R_t(\Delta t)},$$

where again, the collection $\{R_j(\Delta t)\}$ are i.i.d. and $N(\mu\Delta t, \sigma^2\Delta t)$. That is, for any time t,

$$R_t(\Delta t) = \mu\Delta t + z_t\sigma\sqrt{\Delta t},$$

where $\{z_t\}$ are i.i.d. and $N(0, 1)$.

This is demonstrated by the uniqueness of the moment-generating function or characteristic function as was introduced above. For example, if $\{R_j\}$ are normally distributed, $R \equiv R_j \sim N(\mu, \sigma^2)$, then from (8.31) we have $M_R(s) = e^{\mu s + \sigma^2 s^2/2}$. On the other hand, because of independence it must be the case that $M_{\sum R_k(\Delta t)}(s) = [M_{R_k(\Delta t)}(s)]^m$. Since $\sum_{k=1}^{m} R_k(\Delta t) = R$ and $\Delta t = \frac{1}{m}$, we derive

$$M_{R_k(\Delta t)}(s) = [e^{\mu s + \sigma^2 s^2/2}]^{1/m}$$

$$= e^{\mu\Delta t s + \sigma^2\Delta t s^2/2}.$$

This confirms both the mean and variance result in (8.44), as well as the result that $R_k(\Delta t) \sim N(\mu\Delta t, \sigma^2\Delta t)$.

In exercise 9 is assigned the demonstration that this result does not hold for binomially distributed R_j, despite the fact that we still have the moments result in

(8.44). In other words, there is a theoretical inconsistency in assuming for each Δt that log-ratio returns are independent and binomially distributed. However, we will now show that as $\Delta t \to 0$, this inconsistent binomial model converges and gives the same probability distribution of stock prices as does the assumption of normal log-ratio returns, which is consistent.

Real World Binomial Distribution as $\Delta t \to 0$
In this section we address the question of the limiting distribution of equity prices under the real world binomial model. Later, using the tools of chapter 9, we will be able to generalize this calculation to the question of the limiting distribution of equity prices under the risk-neutral binomial model. Such a derivation is of necessity more difficult, and hence the need for additional tools, since despite assuming the same values for future equity prices, the probabilities of the u and d returns change from numerically fixed values of p and p' to risk-neutral probabilities q and q' that depend on Δt.

For a fixed $T > 0$, where T is denominated in units of the time interval associated with μ and σ^2, we now investigate the limiting probability density function of S_T as $\Delta t \to 0$. For any given integer n, define $\Delta t = \frac{T}{n}$, and calibrate the n-step binomial lattice from $t = 0$ to $t = T$. Since $T = n\Delta t$, we have that for general p, as in (8.43),

$$S_T^{(n)} = S_0 e^{\sum B_j}, \tag{8.46a}$$

$$B_j = \begin{cases} \mu\Delta t + a\sigma\sqrt{\Delta t}, & \Pr = p, \\ \mu\Delta t - \frac{1}{a}\sigma\sqrt{\Delta t}, & \Pr = p', \end{cases} \quad j = 1, 2, \ldots, n, \tag{8.46b}$$

$$a = \sqrt{\frac{p'}{p}}. \tag{8.46c}$$

In other words, $\ln[S_T^{(n)}/S_0] = \sum_{j=1}^{n} B_j$ is a sum of n independent binomial random variables. Also, since $E[B_j] = \mu\Delta t$ and $\text{Var}[B_j] = \sigma^2\Delta t$, we obtain the following result, which is independent of n by construction:

$$E\left[\sum_{j=1}^{n} B_j\right] = \mu T, \quad \text{Var}\left[\sum_{j=1}^{n} B_j\right] = \sigma^2 T.$$

Now remark 8.25 following the proof of the de Moivre–Laplace theorem, here with $c = \mu\Delta t + a\sigma\sqrt{\Delta t}$ and $d = \mu\Delta t - \frac{1}{a}\sigma\sqrt{\Delta t}$ and general p, does not directly imply that the normalized summation of $\{B_j\}$ has a distribution that converges to the unit normal distribution as $n \to \infty$. The reason is that c and d are not constants here but change with n, since $\Delta t = \frac{T}{n}$. In other words, here we have $c = \frac{\mu T}{n} + \frac{a\sigma\sqrt{T}}{\sqrt{n}}$ and $d = \frac{\mu T}{n} - \frac{\sigma\sqrt{T}}{a\sqrt{n}}$.

So this summation of random variables is completely different from that accommodated by either the De Moivre–Laplace theorem or the central limit theorems, since here the basic random variables in the summation differ for each n, in that $B_j \equiv B_j^{(n)}$. Also there is no way to "freeze" these random variables to be independent of n. In the application at hand it is important for these random variables to change as $n \to \infty$ so that over the time interval $[0, T]$ the expected value of the sum is fixed at μT, and the variance of the sum is fixed at $\sigma^2 T$.

Still we can construct the normalized random variable $Y^{(n)}$ as in remark 8.25 and demonstrate that the unit normal is again produced in the limit. Specifically:

Proposition 8.29   *For B_j defined as in (8.46), let*

$$Y^{(n)} = \frac{\sum_{j=1}^{n} B_j - \mu T}{\sigma\sqrt{T}}. \tag{8.47}$$

Then as $n \to \infty$,

$$M_{Y^{(n)}}(s) \to e^{s^2/2}. \tag{8.48}$$

In other words, by (8.30), $Y^{(n)} \to N(0, 1)$.

Proof   Note that with $Y_j = \frac{B_j - \mu\Delta t}{\sigma\sqrt{T}}$, we have $Y^{(n)} = \sum_{j=1}^{n} Y_j$. Also, since

$$Y_j = \begin{cases} \frac{a}{\sqrt{n}}, & \Pr = p, \\ -\frac{1}{a\sqrt{n}}, & \Pr = p', \end{cases}$$

with $a = \sqrt{\frac{p'}{p}}$, we obtain with $\exp A \equiv e^A$,

$$M_{Y_j}(s) = p\exp\left(\frac{as}{\sqrt{n}}\right) + p'\exp\left(-\frac{s}{a\sqrt{n}}\right).$$

Using (7.63) and simplifying notation with $m_j \equiv pa^j + \frac{(-1)^j p'}{a^j}$ leads to

$$M_{Y_j}(s) = \sum_{j=0}^{\infty} m_j \frac{s^j}{j!} n^{-j/2}$$

$$= 1 + \frac{s^2 n^{-1}}{2} + n^{-3/2} E(n),$$

since $m_0 = 1$, $m_1 = 0$, and $m_2 = 1$. The rearrangement of these series is justified by their absolute convergence. The error term $E(n)$ is then also an absolutely convergent series for all n, and that as $n \to \infty$, we have that $E(n) \to m_3 \frac{s^3}{6}$. Consequently, since the $\{Y_j\}$ are independent, the m.g.f. of $Y^{(n)} = \sum_{j=1}^{n} Y_j$ is this expression raised to the nth power. Now, taking logarithms, we obtain

$$\ln M_{Y^{(n)}}(s) = n \ln \left[1 + \frac{s^2 n^{-1}}{2} + n^{-3/2} E(n) \right].$$

Next we apply (8.20) with $x = \frac{s^2 n^{-1}}{2} + n^{-3/2} E(n)$. This series is absolutely convergent for $x < 1$, which is to say, for n large enough. Then rearranging and keeping track of only the first few terms of the series, as the rest will converge to 0 as $n \to \infty$, we obtain

$$\ln M_{Y^{(n)}}(s) = n \sum_{j=1}^{\infty} (-1)^{j+1} \left(\frac{1}{j} \right) x^j$$

$$= n \left[\frac{s^2 n^{-1}}{2} + n^{-3/2} E(n) \right] + n^{-1} E'(n)$$

$$= \frac{s^2}{2} + n^{-1/2} [E(n) + n^{-1/2} E'(n)],$$

where $E'(n)$ is also absolutely convergent, and with $E'(n) \to \left[\frac{s^2}{2} \right]^2$ as $n \to \infty$. Finally, we see from this expression that as $n \to \infty$,

$$\ln M_{Y^{(n)}}(s) \to \frac{s^2}{2},$$

and from this we conclude (8.48) because of the continuity of the exponential function. So $Y^{(n)} \to N(0, 1)$, the standard normal variable by (8.30).   ■

Of course, since

$$Y^{(n)} = \frac{\ln[S_T^{(n)} / S_0] - \mu T}{\sigma \sqrt{T}},$$

we can apply the properties of the m.g.f. from exercise 8 to $\ln[S_T^{(n)} / S_0] = \sigma \sqrt{T} Y^{(n)} + \mu T$, to obtain

$$M_{\ln[S_T^{(n)} / S_0]}(s) = e^{\mu T s} M_{Y^{(n)}}(s \sigma \sqrt{T}). \tag{8.49}$$

The proposition above then asserts that as $n \to \infty$,

$$M_{\ln[S_T^{(n)}/S_0]}(s) \to e^{\mu Ts + \sigma^2 Ts^2/2},$$

and so

$$\ln\left[\frac{S_T^{(n)}}{S_0}\right] \to N(\mu T, \sigma^2 T).$$

This formula can be written as $\ln S_T^{(n)} \to \ln S_T$ as $n \to \infty$, where

$$\ln S_T \sim N(\ln S_0 + \mu T, \sigma^2 T). \tag{8.50}$$

In other words, in the limit of the real world binomial lattice model as $n \to \infty$, or equivalently as $\Delta t \to 0$, $\ln S_T$ will be normally distributed with a mean of $\ln S_0 + \mu T$ and variance of $\sigma^2 T$. This can equivalently be expressed as follows:

Corollary 8.30   *With $S_T^{(n)}$ defined as in (8.46), then $S_T^{(n)} \to S_T$ as $n \to \infty$ with*

$$S_T = S_0 e^X, \tag{8.51}$$

where $X \sim N(\mu T, \sigma^2 T)$.

Written in this form, S_T is said to have a **lognormal distribution**, which will be seen again in chapter 10.

Remark 8.31

1. *It was noted in section 7.8.5 and developed in exercise 23 of that chapter, that for any p with $0 < p < 1$, a binomial lattice with unit step-size can be calibrated with up and down state returns, u and d, so that $\mathrm{E}[S_{t+1}/S_t] = \mu$ and $\mathrm{Var}[S_{t+1}/S_t] = \sigma^2$ for arbitrary μ and σ^2. In section 8.8.2 this point was generalized to binomial lattices with step-size of Δt, so that now with $u(\Delta t)$ and $d(\Delta t)$, we obtain $\mathrm{E}[S_{t+\Delta t}/S_t] = \mu \Delta t$ and $\mathrm{Var}[S_{t+\Delta t}/S_t] = \sigma^2 \Delta t$. Further proposition 8.29 demonstrates that for any such choice of p and corresponding calibration, as $n \equiv \frac{T}{\Delta t} \to \infty$, the distribution of the binomial prices at time T, denoted $S_T^{(n)}$ satisfies*

$$\ln S_T^{(n)} \to N(\ln S_0 + \mu T, \sigma^2 T).$$

It is natural to wonder if the selection of p influences the speed of this convergence. A closer inspection of the proof of proposition 8.29 provides an insight. With the notation of that proof, we have

$$\ln M_{Y(n)}(s) = \frac{s^2}{2} + n^{-1/2} E(n) + n^{-1} E'(n),$$

where the $E(n)$ series equals $m_3 s^3/6 + O(n^{-1/2})$, and the $E'(n)$ series equals $[s^2/2]^2 + O(n^{-1/2})$. Consequently the speed of convergence could be improved from $O(n^{-1/2})$ to $O(n^{-1})$ if p could be selected to make $m_3 = 0$, and this is seen to occur when $p = 1/2$. In remark 9.158 we will return to this issue and there see that $p = 1/2$ also plays a partial role in improving the speed of convergence of the distribution of prices under the risk-neutral probability $q(\Delta t)$.

2. If returns are assumed to be normally distributed in each period, where $R_j = \mu \Delta t + z_j \sigma \sqrt{\Delta t}$, with $\Delta t = \frac{T}{m}$, then it is easy to see that at time T, independent of m,

$$S_T = S_0 e^{\sum_{j=1}^m R_j}$$

$$= S_0 e^{\sum_{j=1}^m [\mu \Delta t + z_j \sigma \sqrt{\Delta t}]}$$

$$= S_0 e^{\mu T + z \sigma \sqrt{T}}$$

$$= S_0 e^X,$$

where $X \sim N(\mu T, \sigma^2 T)$. In the third line of this calculation $\sum_{j=1}^m z_j \sim N(0, m)$ is used, and hence $\sum_{j=1}^m z_j = \sqrt{m} z$, where $z \sim N(0, 1)$, as can be verified by considering moment-generating functions. So the real world binomial lattice model converges as $\Delta t \to 0$ to exactly the same model of stock prices as does the normal return model. Interestingly this convergence occurs despite the fact that the assumption on subperiod returns having independent binomial distributions for all Δt is an inconsistent distributional assumption, as noted at the end of the last section.

Although providing the same equity price model in the limit, the advantage of the binomial model is that it provides a simpler framework within which to contemplate option pricing, which we address next.

8.8.3   Lattice-Based European Option Prices as $\Delta t \to 0$

The Model
In (7.147) was derived the lattice-based price of a European option, or other European-type derivative security with payoff function $\Lambda(S_T)$, by way of a replicating portfolio argument,

$$\Lambda_0(S_0) = e^{-nr} \sum_{j=0}^{n} \binom{n}{j} q^j (1-q)^{n-j} \Lambda(S_n^j),$$

$$S_n^j = S_0 e^{ju+(n-j)d}.$$

Here n denotes the number of time steps to the exercise date T, and the risk-neutral probability q is a function of the binomial stock returns u and d, as well as the period risk-free rate r. Recall from (7.143) that this relationship is given by

$$q = \frac{e^r - e^d}{e^u - e^d}.$$

Further recall the binomial stock returns calibrated in (8.43) to equal

$$u = \mu + \left[\sqrt{\frac{p'}{p}}\right]\sigma, \quad d = \mu - \left[\sqrt{\frac{p}{p'}}\right]\sigma,$$

where $0 < p < 1$, $p' \equiv 1 - p$, and μ and σ^2 denote the mean and variance of the log-ratio series for one time step. These formulas for u and d generalize those in (7.136), which were $u = \mu + \sigma$, $d = \mu - \sigma$, when $p = p' = \frac{1}{2}$.

Naturally, in this revised setting where T is fixed and time steps are defined by $\Delta t = \frac{T}{n}$, all these formulas are applicable with adjusted stock returns as in (8.44) and an adjusted risk-free rate. In other words, for the definition of q, we have

$$q(\Delta t) = \frac{e^{r(\Delta t)} - e^{d(\Delta t)}}{e^{u(\Delta t)} - e^{d(\Delta t)}}, \tag{8.52}$$

where

$$u(\Delta t) = \mu \Delta t + \left[\sqrt{\frac{p'}{p}}\right]\sigma\sqrt{\Delta t}, \tag{8.53a}$$

$$d(\Delta t) = \mu \Delta t - \left[\sqrt{\frac{p}{p'}}\right]\sigma\sqrt{\Delta t}. \tag{8.53b}$$

While not completely defensible, the common model for the risk-free rate is that with r denoting the rate for $\Delta t = 1$, which equals one year in practice,

$$r(\Delta t) = r\Delta t. \tag{8.54}$$

This model reflects the idea that the applicable continuous risk-free rate r is effectively fixed and that any investment for period $\Delta t \leq 1$ earns this same rate. This effectively ignores the term structure of risk-free investments, which can be observed historically to sometimes be a **normal term structure** for which $r(\Delta t) < r\Delta t$, sometimes an **inverted term structure** for which $r(\Delta t) > r\Delta t$, and sometimes a **flat term structure** for which $r(\Delta t) = r\Delta t$. That said, refinements to the assumption in (8.54) have little effect in practice, at least for common options with maturities within a few months.

European Call Option Illustration

To illustrate the behavior of the price of a European option as $\Delta t \to 0$, we assume that $\Lambda(S_n^j)$ is the exercise price of a call option: $\Lambda(S_n^j) = \max(S_n^j - K, 0)$. Inserting this exercise function into the formula above for $\Lambda_0(S_0)$, and recalling that $S_n^j = S_0 e^{ju+(n-j)d}$ and $n\Delta t = T$, we get

$$\Lambda_0^C(S_0) = e^{-nr\Delta t} \sum_{j=0}^{n} \binom{n}{j} q^j (1-q)^{n-j} \max(S_n^j - K, 0)$$

$$= e^{-rT} \left[\sum_{j=a}^{n} \binom{n}{j} q^j (1-q)^{n-j} S_n^j - K \sum_{j=a}^{n} \binom{n}{j} q^j (1-q)^{n-j} \right]$$

$$= S_0 \sum_{j=a}^{n} \binom{n}{j} (qe^u e^{-r\Delta t})^j [(1-q)e^d e^{-r\Delta t}]^{n-j} - e^{-rT} K \sum_{j=a}^{n} \binom{n}{j} q^j (1-q)^{n-j}.$$

Here a is defined by

$$a = \min\{j \mid S_n^j \geq K\}.$$

Note that if we define

$$\bar{q} = qe^u e^{-r\Delta t}, \tag{8.55}$$

then a calculation shows that $1 - \bar{q} = (1-q)e^d e^{-r\Delta t}$. In other words,

$$\Lambda_0^C(S_0) = S_0 \sum_{j=a}^{n} \binom{n}{j} \bar{q}^j (1-\bar{q})^{n-j} - e^{-rT} K \sum_{j=a}^{n} \binom{n}{j} q^j (1-q)^{n-j}$$

$$= S_0 \Pr[S_n \geq K \mid \text{Bin}(\bar{q}, n)] - e^{-rT} K \Pr[S_n \geq K \mid \text{Bin}(q, n)],$$

where $\text{Bin}(\bar{q}, n)$ is shorthand for the binomial distribution with parameters $\bar{q}$ and n, and similarly for $\text{Bin}(q, n)$. For both binomials, the subperiod stock returns are given by $u(\Delta t)$ and $d(\Delta t)$ above, where $\bar{q}$ and q, respectively, denote the probability of the return $u(\Delta t)$.

In more detail, the random variable S_n can be expressed with notation $\exp(A) \equiv e^A$:

$$S_n = S_0 \exp\left[\sum_{i=1}^{n} B_i\right],$$

where $\{B_i\}$ are independent and identically distributed binomial variables that assume values of $u(\Delta t)$ and $d(\Delta t)$. In the $\text{Bin}(\bar{q}, n)$ model, $\Pr[u(\Delta t)] = \bar{q}$, while in the $\text{Bin}(q, n)$ model, $\Pr[u(\Delta t)] = q$. With $\sum_{i=1}^{n} B_i$ denoted by $\bar{B}_{(n)}$ in the $\text{Bin}(\bar{q}, n)$ model, and by $B_{(n)}$ in the $\text{Bin}(q, n)$ model, the result above can be expressed as

$$\Lambda_0^C(S_0) = S_0 \Pr\left[\bar{B}_{(n)} \geq \ln\left[\frac{K}{S_0}\right]\right] - e^{-rT} K \Pr\left[B_{(n)} \geq \ln\left[\frac{K}{S_0}\right]\right].$$

Finally, we normalize the binomial random variables in the expression above for $\Lambda_0(S_0)$, subtracting the means of $\bar{\mu}_n$ and μ_n, respectively, and dividing by the standard deviations of σ_n and σ_n, respectively. Call these normalized binomials $\bar{B}'_{(n)}$ and $B'_{(n)}$, to produce

$$\Lambda_0^C(S_0) = S_0 \Pr\left[\bar{B}'_{(n)} \geq \frac{\ln\left[\frac{K}{S_0}\right] - \bar{\mu}_n}{\bar{\sigma}_n}\right]$$

$$- e^{-rT} K \Pr\left[B'_n \geq \frac{\ln\left[\frac{K}{S_0}\right] - \mu_n}{\sigma_n}\right]. \tag{8.56}$$

Remark 8.32   *As noted in chapter 7, q is called the **risk-neutral probability**. Utility functions will be discussed in chapter 9, but it will be seen there that $\bar{q}$ is **a risk-averter probability**. Unlike the risk-neutral probability, which is unique, any probability $\hat{q} > q$ is a risk-averter probability. So $\bar{q}$ is simply one example, since $u(\Delta t) > r\Delta t$ implies $\bar{q} > q$, and we will refer to it as the **special risk-averter probability**. However, despite the presence of a risk-averter probability in this option price, it is essential to understand that option pricing will be shown to be entirely independent of risk preferences, and the presence of $\bar{q}$ in the formula above is merely a mathematical artifact that simplifies the ultimate solution.*

To see this, note that the formula above for $\Lambda_0(S_0)$ can be expressed as

$$\Lambda_0^C(S_0) = e^{-rT} \sum_{j=0}^{n} \binom{n}{j} q^j (1-q)^{n-j} \max(S_n^j - K, 0)$$

$$= e^{-rT} \mathrm{E}[\max(S_n - K, 0) \mid \mathrm{Bin}(q, n)].$$

Clearly, in this formulation only the risk-neutral probability is needed for the option price. Restating this formula in terms of q and $\bar{q}$ just facilitates the study we discuss next and in chapter 9.

Black–Scholes–Merton Option-Pricing Formulas I

Because u, d, q, and $\bar{q}$, the parameters underlying B_n' and $\bar{B}_n'$, are all functions of $\Delta t = \frac{T}{n}$, there will be some work ahead to determine what are the limits of the two complicated probability expressions in (8.56) as $\Delta t \to 0$. We cannot, however, consider pursuing this analysis until we have some additional tools at our disposal from chapter 9, and even then the derivation will be seen to be subtle and somewhat challenging. We will also develop another approach using the chapter 10 tools, which circumvents the explicit analysis of u, d, q, and $\bar{q}$ as functions of Δt, or rather, studies this dependence from a different perspective using a new set of tools. This analysis will also be seen to be subtle and somewhat challenging. Both derivations will stand as testament to the depth and insight of the Black–Scholes–Merton results.

However, given the result above in section 8.8.2 on the limiting distribution of equity prices in the real world binomial lattice, it should not surprise the reader that both binomial random variables in (8.56) will be shown to converge in chapter 9 to normal variables:

$$\bar{B}_{(n)} \to N\left(\left[r + \frac{1}{2}\sigma^2\right]T, \sigma^2 T\right),$$

$$B_{(n)} \to N\left(\left[r - \frac{1}{2}\sigma^2\right]T, \sigma^2 T\right),$$

as $n \to \infty$, or equivalently, as $\Delta t \to 0$.

Remark 8.33 *Interestingly, within the real world binomial lattice analysis, the random variable that was normalized, $\sum_{j=1}^{n} B_j$, was a summation of binomials for which the probability p of u was fixed and independent of n but where the two values assumed by each B_j, u and d, changed with n. In the binomial models needed for option pricing, the random variable that is normalized is again of the form $\sum_{j=1}^{n} B_j$, with each B_j the* *same binomial as before, but where the probabilities that $B_j = u(\Delta t)$, which are q or $\bar{q}$, also now change with n.*

Assume for now this conclusion about the limiting distributions of the variables $\bar{B}_{(n)}$ and $B_{(n)}$. Then $\bar{B}'_{(n)}$ and $B'_{(n)}$ converge to the unit normal distribution. In other words,

$$\Pr\left[\bar{B}'_{(n)} \geq \frac{\ln\left[\frac{K}{S_0}\right] - \bar{\mu}_n}{\bar{\sigma}_n}\right] \rightarrow \Pr\left[Z \geq \frac{\ln\left[\frac{K}{S_0}\right] - \left(r + \frac{1}{2}\sigma^2\right)T}{\sigma\sqrt{T}}\right].$$

Because of the symmetry of the unit normal distribution, we have from (8.32) that $\Pr[Z \geq -d_1] = \Pr[Z \leq d_1] = \Phi(d_1)$, where Φ denotes the unit normal distribution function. Similarly the second probability statement can be expressed as $\Pr[Z \geq -d_2] = \Pr[Z \leq d_2] = \Phi(d_2)$.

Putting everything together, one arrives at the famous **Black–Scholes–Merton formula** for the price of a **European call option**, named for **Fischer Black** (1938–1995), **Myron S. Scholes** (b. 1941), and **Robert C. Merton** (b. 1944), for research published in papers by Black and Scholes, and Merton in the early 1970s, and for which Merton and Scholes received the 1997 Nobel Prize in Economics (sadly, such awards are not made posthumously).

The final result for a **European call option** is

$$\Lambda_0^C(S_0) = S_0\Phi(d_1) - e^{-rT}K\Phi(d_2), \tag{8.57a}$$

$$d_1 = \frac{\ln\frac{S_0}{K} + \left(r + \frac{1}{2}\sigma^2\right)T}{\sigma\sqrt{T}}, \tag{8.57b}$$

$$d_2 = \frac{\ln\frac{S_0}{K} + \left(r - \frac{1}{2}\sigma^2\right)T}{\sigma\sqrt{T}}. \tag{8.57c}$$

The related result for a **European put option** is

$$\Lambda_0^P(S_0) = e^{-rT}K\Phi(-d_2) - S_0\Phi(-d_1). \tag{8.58}$$

The approach used by Black–Scholes and Merton was close in spirit to that above, in the sense that they were able to replicate the option with a portfolio of stock and T-bills. They then concluded that the option must have a price equal to the price of this replicating portfolio. However, they used the advanced tools of stochastic calculus for this development (which will not be addressed until my next book, *Advanced*

Quantitative Finance, as mentioned in the Introduction). The approach taken here and in chapter 7, which used a binomial lattice approximation to stock price movements, and then replicated the option and evaluated the limit as $\Delta t \to 0$, is known as the **Cox–Ross–Rubinstein binomial lattice model** for option pricing. It was developed in a paper in the late 1970s by **John C. Cox, Stephen A. Ross, and Mark Rubinstein**.

Remark 8.34 *Using a binomial lattice with time step Δt to evaluate the price of a European option or other derivative security, which results in an application of (7.147), produces a price $\Lambda_0(S_0) \equiv \Lambda_0(S_0, \Delta t)$. This price reflects what is known as **discretization error**. In other words, the theoretically correct answer is obtained as $\Delta t \to 0$, and the lattice produces an error $\varepsilon^D(\Delta t) = \Lambda_0(S_0, 0) - \Lambda_0(S_0, \Delta t)$, which is caused by discretizing time and the p.d.f. of stock price movements. One consequence of this discretization is that for any Δt, the calculated value of $\Lambda_0(S_0, \Delta t)$ explicitly reflects the stock's mean log-ratio return μ as well as the real world probability used in the calibration, p, through the formulas for q, u, and d. For any Δt, the calculated value of the derivatives price will consequently vary somewhat as these parameters change. However, as one can explicitly appreciate in the Black–Scholes–Merton formulas, and will be seen to be true generally as $\Delta t \to 0$, these dependencies of option price on both μ and p disappear. Indeed in the formulas above there is no vestige of either parameter present, and in chapter 9 we will return to this point and observe this transition. In contrast, the variance of the stock's log-ratio return, σ^2, is quite evident in the final formulas, as is the risk-free rate, r.*

8.8.4   Scenario-Based European Option Prices as $N \to \infty$

The Model

If N-paths are randomly generated, and $\{S_n^j\}_{j=0}^n$ denotes the $n+1$ possible stock prices in the recombining lattice in section 8.8.3 above at time $n\Delta t = T$, it is of interest to analyze the number of paths that arrive at each final state. In theory, we know from the lattice analysis in section 8.8.2 that the distribution of stock prices at time n is binomially distributed in the real world with parameters n, p in general, and hence $\Pr[S_n = S_n^j] = \binom{n}{j} p^j (1-p)^{n-j}$. As in chapter 7, p denotes the probability of a u-return, $p' = 1 - p$ the probability of a d-return, and stock prices are parametrized so that $j = 0$ corresponds to the lowest price, $S_n^0 = e^{nd}S_0$, and $j = n$ corresponds to the highest price, $S_n^n = e^{nu}S_0$.

On the other hand, we have shown that for the purposes of option pricing, we continue to use the stock price returns of e^u and e^d but switch the assumed probability of an upstate return from the real world probability p to the risk-neutral probability q given in (8.52) above.

In the lattice-based model these q-probabilities determine the likelihood of each final equity price state that is relevant for option pricing. Consequently, if N_j denotes the number that terminate at price S_n^j from a sample of N paths so that $\sum N_j = N$, then the $(n+1)$-tuple of integers $(N_0, N_1, \ldots, N_n)$ has a multinomial distribution with parameters N and $\{Q_j\}_{j=0}^n$, where $Q_j = \binom{n}{j} q^j (1-q)^{n-j}$. From (7.105) and (7.106) we conclude that

$$\mathrm{E}[N_j] = NQ_j, \quad \mathrm{Var}[N_j] = NQ_j(1-Q_j), \quad \mathrm{Cov}[Q_j, Q_k] = -NQ_jQ_k.$$

In a nonrecombining lattice, Q_j is again defined as the risk-neutral probability of terminating at price S_n^j; only then there are 2^n stock prices rather than $n+1$. The multinomial distribution is again applicable in this case, as are the moment formulas above.

We now formalize the methodology for pricing an n-period European option using the scenario-based methodology introduced in section 7.8.7. For simplicity, we focus on the recombining lattice model, although the development is equally applicable in the more general case. To this end, let $\Lambda(S_n^j)$ denote the exercise value of the option or other derivative at time n when the stock price S_n^j prevails. Also assume that a time step of $\Delta t \equiv \frac{T}{n}$ has been chosen as in section 8.8.3, and that the binomial lattice is calibrated as in (8.52), (8.53), and (8.54).

Given N paths, define a random variable O_N, the **sample option price**, as in (7.150):

$$O_N = \frac{e^{-rT}}{N} \sum_{j=0}^n N_j \Lambda(S_n^j). \tag{8.59}$$

The random variable O_N is an estimate of the true option price based on a sample of size N. As was noted in section 7.8.7, the actual lattice-based price can be expressed

$$\Lambda_0(S_0) = e^{-rT} \sum_{j=0}^n \mathrm{E}\left[\frac{N_j}{N}\right] \Lambda(S_n^j),$$

and so the sample option price replaces the correct probability weight of $\mathrm{E}\left[\frac{N_j}{N}\right] = Q_j$ with the sample-based estimate of $\frac{N_j}{N}$.

Option Price Estimates as $N \to \infty$

We would expect that since the paths are generated in such a way as to arrive at each final stock price with the correct probability, the expected value of this random variable ought to equal $\Lambda_0(S_0)$, the value produced on the lattice with (7.147). Even more important, as N increases, we will prove that the probability that we are in error by any given amount goes to 0. The main result is as follows:

Proposition 8.35   *With O_N defined as in (8.59):*

1. *The expected value of O_N equals the lattice-based option price*

$$E[O_N] = \Lambda_0(S_0). \tag{8.60}$$

2. *If $\text{Var}[\Lambda(S_n^j)] < \infty$, where this variance is defined under $\{Q_j\}$, then for any $\epsilon > 0$,*

$$\Pr[|O_N - \Lambda_0(S_0)| > \epsilon] \to 0 \qquad \text{as } N \to \infty. \tag{8.61}$$

Proof   For property 1,

$$E[O_N] = e^{-rT} \sum_{j=0}^{n} E\left[\frac{N_j}{N}\right] \Lambda(S_n^j) = \Lambda_0(S_0),$$

since $E\left[\frac{N_j}{N}\right] = Q_j$ by (7.105). To demonstrate property 2, we use the Chebyshev inequality, which requires the variance of O_N. To this end, first note that using (7.56) obtains

$$\text{Var}[O_N] = \frac{e^{-2rT}}{N^2} \sum_{j=0}^{n} \text{Var}[N_j]\Lambda^2(S_n^j) + \frac{2e^{-2rT}}{N^2} \sum_{j<k} \text{Cov}[N_j, N_k]\Lambda(S_n^j)\Lambda(S_n^k)$$

$$= \frac{e^{-2rT}}{N^2} \left[\sum_{j=0}^{n} NQ_j(1 - Q_j)\Lambda^2(S_n^j) - 2\sum_{j<k} NQ_jQ_k\Lambda(S_n^j)\Lambda(S_n^k) \right]$$

$$= \frac{e^{-2rT}}{N} \left[\sum_{j=0}^{n} Q_j\Lambda^2(S_n^j) - \left(\sum_{j=0}^{n} Q_j\Lambda(S_n^j) \right)^2 \right]$$

$$= \frac{e^{-2rT}}{N} \, \text{Var}[\Lambda(S_n^j)].$$

Note that in the last step we used the identity that under $\{Q_j\}$, $\text{Var}[\Lambda(S_n^j)] = E[\Lambda^2(S_n^j)] - [E[\Lambda(S_n^j)]]^2$. From this derivation we conclude that as $N \to \infty$, we have $\text{Var}[O_N] \to 0$. Now, by Chebyshev's inequality, since $E[O_N] = \Lambda_0(S_0)$ by property 1,

$$\Pr[|O_N - \Lambda_0(S_0)| > \epsilon] < \frac{\text{Var}[O_N]}{\epsilon^2},$$

completing the proof.   ∎

Remark 8.36

1. *The price of a European derivative security obtained with the scenario-based model above will contain two types of error compared to the theoretically correct price. Denoting the price obtained with N-paths and time steps of Δt by $O_N(\Delta t)$, the errors are*

• **Discretization error**, *which is identical to that produced by the underlying lattice-based calculation and depends on Δt. This error is defined in remark 8.34 as*

$$\varepsilon^D(\Delta t) = \Lambda_0(S_0, 0) - \Lambda_0(S_0, \Delta t).$$

• **Estimation error**, *which is defined as*

$$\varepsilon^E(\Delta t) = \Lambda_0(S_0, \Delta t) - O_N(\Delta t),$$

is the error between the scenario-based option price estimate and the lattice-based value.

2. *As was seen in the proof above, the estimation error decreases with $\frac{1}{N}$ in the sense that*

$$\Pr[|\Lambda_0(S_0, \Delta t) - O_N(\Delta t)| > \epsilon] < \frac{e^{-2rT} \, \mathrm{Var}[\Lambda(S_n^j)]}{N\epsilon^2}.$$

Consequently, as was observed in proposition 8.12, we can choose $\epsilon_N \to 0$ in such a way that $N\epsilon_N^2 \to \infty$ and thereby ensure that as $N \to \infty$, all estimation error is theoretically eliminated. In practice, however, this elimination of error will be a slow and painful process, since in order for $N\epsilon_N^2 \to \infty$ it will be necessary to have $\epsilon_N \to 0$ slowly and/or have $N\epsilon_N^2 \to \infty$ slowly. For example, if $\epsilon = \frac{1}{N^a}$ for $0 < a < \frac{1}{2}$, both objectives are achieved, where $a \sim \frac{1}{2}$ provides faster $\epsilon_N \to 0$ and slower $N\epsilon_N^2 \to \infty$, and $a \sim 0$ does the opposite.

Scenario-Based Prices and Replication

As the last question for this section on scenario-based option pricing, we investigate the connection between option pricing based on sample scenarios and option pricing based on replication. First off, from (7.145), we know that replication-based prices can be rebalanced period to period. Rewriting that formula to reflect a period of length Δt produces

$$\Lambda_0(S_0) = e^{-r\Delta t}[q\Lambda(S_1^u) + q'\Lambda(S_1^d)].$$

Consequently, from the first conclusion of the proposition above, with the analogous notation

$$E[O_N] = e^{-r\Delta t}[qE[O_N^u] + q'E[O_N^d]],$$

the expected values of the scenario-based prices can also be rebalanced.

To investigate the one-period rebalancing of O_N to one of O_N^u and O_N^d, an assumption needs to be made about the collection of scenarios used for the latter calculations. We first assume that O_N^u is evaluated on the subset of the N original paths that start with a u, of which there are N^u, and similarly assume that O_N^d is evaluated on the subset of the N original paths that start with a d, of which there are N^d, and so $N^u + N^d = N$.

Next rewrite (8.59) as

$$O_N = \frac{e^{-rT}}{N}\left[\sum_{j=1}^{n} N_j^u \Lambda(S_n^j) + \sum_{j=0}^{n-1} N_j^d \Lambda(S_n^j)\right],$$

where $\{N_j^u\}$ is defined as the number of paths from the N^u subset that end at S_n^j, and similarly for $\{N_j^d\}$, where it is apparent by definition that $N_0^u = N_n^d = 0$.

Now to price O_N^u and O_N^d on these subsets of paths, we derive that

$$O_{N^u}^u = \frac{e^{-r(T-\Delta t)}}{N^u}\sum_{j=1}^{n} N_j^u \Lambda(S_n^j),$$

$$O_{N^d}^d = \frac{e^{-r(T-\Delta t)}}{N^d}\sum_{j=0}^{n-1} N_j^d \Lambda(S_n^j).$$

Finally, with a bit of algebra is obtained

$$O_N = e^{-r\Delta t}[q[a^u O_{N^u}^u] + q'[a^d O_{N^d}^d]], \tag{8.62}$$

where

$$a^u = \frac{N^u}{Nq}, \quad a^d = \frac{N^d}{Nq'}.$$

In summary, with O_N^u and O_N^d priced on the subsets of the original paths, O_N is the price of a replicating portfolio that will rebalance to $a^u O_N^u$ and $a^d O_N^d$ in the next period, and not O_N^u and O_N^d. So there is additional error in this rebalancing related to how far from 1 the a^u and a^d terms are. Of course,

$E[a^u] = E[a^d] = 1,$

$$Var[a^u] = \frac{q'}{Nq}, \quad Var[a^d] = \frac{q}{Nq'},$$

so for large N the rebalancing error over one period will be small. However, the process cannot be repeated to maturity because in each step the estimated prices are based on fewer and fewer paths.

Alternatively, if O_N^u and O_N^d are priced on new collections of N-paths each, there will be additional rebalancing error. Specifically, we obtain

$$O_N = e^{-r\Delta t}[q[b^u O_N^u] + q'[b^d O_N^d]], \tag{8.63}$$

where

$$b^u = \frac{a^u O_{N^u}^u}{O_N^u}, \quad b^d = \frac{a^d O_{N^u}^d}{O_N^d}.$$

In other words, O_N will equal the price of a portfolio that replicates values of $b^u O_N^u$ and $b^d O_N^d$.

Exercises

Practice Exercises

1. Show that if $f(x)$ is a discrete probability function, with m.g.f. $M(t)$, then for any real number $t > 0$,

$$Pr[X \geq t] \leq \frac{M(t)}{e^{t^2}}.$$

(*Hint:* $M(t) \geq \sum_{|x_i| \geq t} e^{tx_i} f(x_i)$.)

2. Market observers sometimes talk about 5-sigma or 10-sigma events, where sigma is the standard deviation. Such a statement is often used in the context of, "who could have possibly predicted this event?" as if all random variables were known to be normally distributed, and for which the probabilities of such events are indeed miniscule.

(a) Using the Chebyshev inequality, calculate the upper bound for the probability of a 5-sigma or worse event. A 10-sigma or worse event.

(b) Repeat part (a) using the one-sided Chebyshev inequality.

3. Apply the weak law of large numbers to determine the necessary sample size in the following cases to have 95% confidence:

(a) For the standard binomial distribution, estimate p to three decimal places ($\epsilon = 0.0005$) if it is known that $0.1 \le p \le 0.5$.

(b) For the negative binomial distribution with $k = 10$, estimate μ to two decimal places, where it is known that $p \le 0.1$.

4. Using the De Moivre–Laplace theorem (*Hint*: Recall the half-interval adjustment.):

(a) Approximate the probability that in one million flips of a biased coin with $\Pr[H] = 0.65$, the number of heads will be between 649,500 and 650,000.

(b) Approximate the probability that the number of tails will be 700,000 or more.

5. Using the central limit theorem (*Hint*: Recall the half-interval adjustment.):

(a) Approximate the probability of $\hat{X} \ge 79$ in a Poisson distribution with $\lambda = 75$, where $\hat{X}$ is a sample average of 50 independent trials.

(b) Approximate the probability of $76 \le \hat{X} \le 78$, with $\hat{X}$ based on a sample of 100.

6. Generalize the calibration of the growth model for stock prices in (8.45) to develop formulas for u and d for arbitrary p, $0 < p < 1$, and Δt.

7. Using the result of exercise 6, express $S_{m\Delta t}$ in terms of S_0 in two ways, paralleling the formulas in (7.137) and (7.138) but for general p and Δt, and being explicit about the binomial probabilities that govern the associated price lattice.

8. Demonstrate the following two properties of moment-generating functions, where X and X_i are discrete random variables, using the definition and properties of expectations:

(a) $M_{a+bX}(t) = e^{at}M_X(bt)$

(b) $M_{\sum X_i}(t) = \prod M_{X_i}(t)$ if $\{X_i\}$ are independent.

9. Using properties of the moment-generating function, show that if $\{B_j\}$ in the binomial lattice model are assumed to be independent and binomially distributed, then this will not imply that $\{B_k(\Delta t)\}$ are binomially distributed. (*Hint*: See exercise 8(b).)

10. Recall the claims model of exercise 18 of chapter 7:

(a) For both the individual and aggregate risk model estimates of the mean and variance of claims, apply the Chebyshev inequality in (8.36) to estimate the probability that claims exceed $8 million, $9.5 million, and $11 million.

(b) Estimate the probabilities from part (a) directly by a simulation method, with 1000 simulations, using (8.38).

(c) Using the simulations from part (b), and $C_0 = \$7.5$ million, estimate the conditional means and variances of the two models, and with these results estimate the probabilities in part (a) using (8.40).

11. (Compare with exercise 24 of chapter 7) Price a two-year European call, with strike price of 100, in the following ways. The stock price is $S_0 = 100$, and based on time steps of $\Delta t = 0.25$ years, the quarterly log-ratios have been estimated to have $\mu_Q = 0.02$, and $\sigma_Q^2 = (0.07)^2$. The annual continuous risk-free rate is $r = 0.048$.

(a) Develop a real world lattice of stock prices, with $p = \frac{1}{2}$ and time steps with $\Delta t = 0.05$, and price this option using (7.147) with the appropriate value of q.

(b) Evaluate the two prices of this option at time $t = 0.05$ from part (a), and construct a replicating portfolio at $t = 0$ for these prices. Demonstrate that the cost of this replicating portfolio equals the price obtained in part (a).

(c) Price this option using (7.147) with the appropriate value of q based on a lattice for which $p = 0.75$.

(d) Generate 500 two-year paths in the risk-neutral world using the same model as part (a), and estimate the price of this option using (7.150) by counting how many scenarios end in each stock price at time 2 years.

12. Generate another 99 prices for the exercise in 11(d) above, by generating another 99 batches of 500 two-year paths.

(a) Calculate the estimated price O_N using all $N = 50,000$ paths, and show that this is equivalent to simply averaging the 100 batch prices.

(b) Calculate the variance of the 100 batch prices, $\text{Var}[O_{500}]$, and use this to estimate the variance of the estimated price in part (a), $\text{Var}[O_N]$. (*Hint*: Recall that as a random variable O_N is the average of 100 prices.)

(c) With $\Lambda_0(S_0)$ defined as the lattice price obtained in exercise 11(a), and using $\text{Var}[O_{500}]$ from part (b), compare for various values of ϵ the proportion of the 100 prices that satisfy $|O_{500} - \Lambda_0(S_0)| > \epsilon$ to the upper bound for the probability of this event, $\frac{\text{Var}[O_{500}]}{\epsilon^2}$, developed in proposition 8.35.

Assignment Exercises

13. Let X be a discrete random variable.

(a) Prove that if $\mu_{|n|} \leq C$ for all n, then $\Pr[|X - \mu| \geq t] = 0$ for any $t > 1$. In other words, it must be the case that $\Pr[|X - \mu| \leq 1] = 1$. (*Hint*: Chebyshev.)

(b) Generalize part (a). Prove that if $\mu_{|n|} \leq C^n$ for all n, then $\Pr[|X - \mu| \geq t] = 0$ for any $t > C$.

(c) Conclude that if X has unbounded range, then it cannot be the case that $\mu_{|n|} \le C^n$ for any C.

14. Apply the weak law of large numbers to determine the necessary sample size in the following cases to have 95% confidence:

(a) For the geometric distribution, estimate the unbiased variance to one decimal place, where it is know that $p > 0.25$ (*Hint*: For the geometric, $\mu_4 = \frac{q}{p^2}\left(1 + \frac{9q}{p^2}\right)$.)

(b) For the Poisson distribution, estimate λ to two decimal places where it is known that $\lambda > 2$.

15. Demonstrate that in the proof of the weak law of large numbers:

$$\Pr[|\hat{X} - \mu| > \epsilon] \le \Pr\left[|\hat{Y}| > \frac{\epsilon}{2}\right] + \Pr\left[|\hat{Z}| > \frac{\epsilon}{2}\right].$$

(*Hint*: By the triangle inequality, $|\hat{X} - \mu| \le |\hat{Y}| + |\hat{Z}|$, and hence, if both $|\hat{Y}| \le \frac{\epsilon}{2}$ and $|\hat{Z}| \le \frac{\epsilon}{2}$, then $|\hat{X} - \mu| \le \epsilon$. Define events $A, B, C \subset \mathcal{S}$ by $A = \{(X_1, \ldots, X_n) \mid |\hat{X}_n - \mu| \le \epsilon\}$, $B = \{(X_1, \ldots, X_n) \mid |\hat{Y}| \le \frac{\epsilon}{2}\}$, and $C = \{(X_1, \ldots, X_n) \mid |\hat{Z}| \le \frac{\epsilon}{2}\}$. Then justify $B \cap C \subset A$, and use De Morgan's laws.)

16. Using the De Moivre–Laplace theorem (*Hint*: Recall the half-interval adjustment.):

(a) Approximate the probability that in one million flips of a biased coin with $\Pr[H] = 0.15$, the number of heads will be between 0 and 145,000 or between 149,500 and 150,000.

(b) Approximate the probability that the number of heads will be within 100 of the expected value.

17. Assuming that all the properties of expectations developed for discrete random variables apply to continuous random variables as well, derive (8.31) from (8.30).

18. Using the central limit theorem (*Hint*: Recall the half-interval adjustment.):

(a) Approximate the probability of $\hat{X} \ge 10$ in a geometric distribution with $p = 0.15$, where $\hat{X}$ represents an average from a sample of 40 trials.

(b) Approximate the probability of $4 \le \hat{X} \le 8$ where $\hat{X}$ is based on a sample of 60 from the same geometric distribution.

19. Demonstrate the following two properties of characteristic functions, where X and X_i are discrete random variables, using the definition and properties of expectations:

(a) $C_{a+bX}(t) = e^{iat} C_X(bt)$

(b) $C_{\sum X_i}(t) = \prod C_{X_i}(t)$ if $\{X_i\}$ are independent

20. Recall the credit model of exercise 36 of chapter 7:

(a) For both the individual and aggregate risk model estimates of the mean and variance of losses, apply the Chebyshev inequality in (8.36) to estimate the probability that losses exceed $8 million, $11 million, and $14 million.

(b) Estimate the probabilities from part (a) directly by a simulation method, with 1000 simulations, using (8.38).

(c) Using the simulations from part (b), and $C_0 = \$6$ million, estimate the conditional means and variances of the two models, and with these results estimate the probabilities in part (a) using (8.40).

21. (Compare with exercise 40 of chapter 7.) Price a two-year European put, with strike price of 100, in the following ways. The stock price is $S_0 = 100$, and based on time steps of $\Delta t = 0.25$ years, the quarterly log-ratios have been estimated to have: $\mu_Q = 0.025$, and $\sigma_Q^2 = (0.09)^2$. The annual continuous risk-free rate is $r = 0.06$.

(a) Develop a real world lattice of stock prices, with $p = \frac{1}{2}$ and time steps with $\Delta t = 0.05$, and price this option using (7.147) with the appropriate value of q.

(b) Evaluate the two prices of this option at time $t = 0.05$ using the same method as part (a), and construct a replicating portfolio at $t = 0$ for these prices. Demonstrate that the cost of this replicating portfolio equals the price obtained in part (a).

(c) Price this option using (7.147) with the appropriate value of q based on a lattice for which $p = 0.25$.

(d) Generate 500 two-year paths in the risk neutral world using the same model as part (a), and estimate the price of this option using (7.150) by counting how many scenarios end in each stock price at time 2 years.

22. Generate another 99 prices to the exercise in 21(d) above, by generating another 99 batches of 500 two-year paths.

(a) Calculate the estimated price O_N using all $N = 50,000$ paths, and show that this is equivalent to simply averaging the 100 batch prices.

(b) Calculate the variance of the batch prices, $\text{Var}[O_{500}]$, and use this to estimate the variance of the estimated price in part (a), $\text{Var}[O_N]$. (*Hint*: Recall that as a random variable O_N is the average of 100 prices.)

(c) With $\Lambda_0(S_0)$ defined as the lattice price obtained in exercise 21(a), and using $\text{Var}[O_{500}]$ from part (b), compare for various values of ϵ the proportion of the 100 prices that satisfy $|O_{500} - \Lambda_0(S_0)| > \epsilon$ to the upper bound for the probability of this event, $\frac{\text{Var}[O_{500}]}{\epsilon^2}$, developed in proposition 8.35.

9 Calculus I: Differentiation

9.1 Approximating Smooth Functions

Calculus is the mathematical discipline that studies properties of "smooth" functions. Intuitively a function is smooth if its values vary in a somewhat predictable way. So based on knowledge of its values and behavior at a given point, we can approximate its values "near" that given point. There are moreover various degrees of smoothness, and these in turn provide various degrees of accuracy in the approximation.

We begin by recalling the definition of a function introduced in chapter 2, and then introduce the simplest notion of smoothness, known as continuity, and some if its refinements. We will spend some time on these concepts because of their importance and subtlety. The next section then studies derivatives of a function, as well as Taylor series expansions, which are seen to both provide a formal basis for approximating function values, and for quantifying the notion of the accuracy of such an approximation.

In the process, we will finally be able to justify the earlier assumed power series expansions for e^x and $\ln x$, as well as demonstrate the validity of the limits needed in the development of the Poisson distribution, such as

$$\left(1 - \frac{\lambda}{n}\right)^n \to e^{-\lambda}, \qquad \text{as } n \to \infty.$$

Remark 9.1 *In general, the functions that appear to be addressed in calculus are real-valued functions of a real variable. In other words, functions*

$$f : X \to Y \quad where \quad X, Y \subset \mathbb{R}.$$

However, while the assumption that the domain of $f(x)$ is real is critical, and so $X = \mathrm{Dmn}(f) \subset \mathbb{R}$, there is often no essential difficulty in assuming f to be a complex-valued function of a real variable so that the range of $f(x)$, $Y = \mathrm{Rng}(f) \subset \mathbb{C}$. This is not often needed in finance, and the characteristic function is one of the few examples in finance where complex-valued functions are encountered.

One reason that $\mathrm{Dmn}(f) \subset \mathbb{R}$ is critical in the development of calculus is that we will often utilize the natural ordering of the real numbers. In other words, given $x, y \in \mathbb{R}$ with $x \neq y$, it must be the case that either $x < y$ or $y < x$. None of these proofs would generalize easily to functions of a complex variable where no such ordering exists. Indeed it turns out that the calculus of such functions is quite different and studied in what is called complex analysis. On the other hand, the only essential property of $\mathrm{Rng}(f)$ that is often assumed is that there is a metric with which one can define closeness and limits. Since $\mathbb{C}$ has a metric as noted in chapter 3, any proof that only *relies on the standard metric in* $\mathbb{R}$, *the absolute value, works equally well in* $\mathbb{C}$ *with its standard metric or any equivalent metric. In other words, the existence of an ordering in the range space doesn't matter for most results, and we simply need a metric structure.*

One counterexample to this statement on the range space is any result that addresses both $f(x)$ *and its inverse function,* $f^{-1}(y)$, *since in such a development,* $\mathrm{Dmn}(f^{-1}) = \mathrm{Rng}(f)$. *Another relates to statements about maximum or minimum values of* $f(x)$, *or intermediate values, which by definition implies an ordering. Such statements must be reviewed carefully to determine if only metric properties are needed, as may be the case for maximum or minimum values, or if the existence of an ordering is also needed, as is the case for an intermediate value.*

Because of the rarity of encountering complex-valued functions of a real variable in finance, all the statements in this chapter are either silent on the location of Y, *or explicitly assume* $Y \subset \mathbb{R}$. *In particular, no effort was made to explicitly frame all proofs in the general case* $Y \subset \mathbb{C}$, *since this overt generality seemed to have little purpose given the objectives of this book. However, any proof that is silent and relies only on a metric in* Y *will virtually always be seen to extend to the case where* $Y \subset \mathbb{C}$. *When a proof explicitly states that* $Y \subset \mathbb{R}$, *its generality must be thought through step by step, and in many cases it will be seen that again, only the metric in* Y *is used.*

The applicability of many results to a complex-valued function can also be justified by splitting the function values into real and imaginary parts. If $Y \subset \mathbb{C}$, *we write*

$$f(x) = g(x) + ih(x),$$

where both $g(x)$ *and* $h(x)$ *are real valued. The theory in this chapter can typically then be justifiably applied to* $f(x)$ *by applying it separately to* $g(x)$ *and* $h(x)$ *and combining results.*

9.2   Functions and Continuity

9.2.1   Functions

Definition 9.2   *A **function** is a rule, often represented notationally by* f, g, *and so forth, by which each element of one set of values, called the **domain** and denoted* $\mathrm{Dmn}(f)$, *is identified with a **unique** element of a second set of values, call the **range** and denoted* $\mathrm{Rng}(f)$.

The rule is often expressed by a formula such as

$$f(x) = x^2 + 3.$$

Here x is an element of the domain of the function f, while $f(x)$ is an element of the range of f. Functions are also thought of as "mappings" between their domain and range. The imagery of x being mapped to $f(x)$, is intuitively helpful at times. In this context, one might use the notation

$$f : X \to Y,$$

where X denotes the domain of f, and Y the range. It is also common to write $f(x)$ for both the function, which ought to be denoted only by f, and the value of the function at x. This bit of carelessness rarely causes confusion.

Note that while the definition of a function requires that $f(x)$ be unique for any x, it is not required that x be unique for any $f(x)$. For instance, the function above has $f(x) = f(-x)$ for any $x \neq 0$. Another way of expressing this is that a function can be a "many-to-one" rule, which includes one-to-one, but it cannot be a one-to-many rule.

An example of a one-to-many rule that is therefore not a function is

$$f(x) = \sqrt{x},$$

which assigns two values to every positive value of x, such as $f(4) = \pm 2$. In many applications one can transform such a rule into a function by simply defining its value to be one of the possible "branches" in the range. For example, the positive square root (or negative square root) are both functions.

A function that is in fact **one-to-one**, meaning that it satisfies $f(x) = f(x')$ iff $x = x'$, has the special property that it has an inverse that is also a function.

Definition 9.3   *Given a one-to-one function $f(x)$, $f : X \to Y$, the **inverse function**, denoted f^{-1}, is defined by*

$$f^{-1} : Y \to X,$$

$$f^{-1}(y) = x \quad \text{if } f(x) = y.$$

*In other words, $\mathrm{Dmn}(f^{-1}) = \mathrm{Rng}(f)$ and $\mathrm{Rng}(f^{-1}) = \mathrm{Dmn}(f)$. More generally, for an arbitrary function f and set A, the set $f^{-1}(A)$, the **pre-image of A under** f is defined by*

$$f^{-1}(A) = \{x \in \mathrm{Dmn}(f) \mid f(x) \in A\}.$$

Example 9.4   *The function $f(x) = x^2 + 3$ has no inverse if defined as a function with domain equal to all real numbers $\mathbb{R}$ because it is many-to-one on this domain, but it does have an inverse if the domain is restricted to any subset of the nonnegative or* *nonpositive real numbers. On the other hand, $f^{-1}(A)$ is defined for any set $A \subset \mathbb{R}$. For example, $f^{-1}([-1, 0]) = 0$ and $f^{-1}([1, 4]) = [-1, -2] \cup [1, 2]$.*

Functions can also be combined, or "composed," to produce so-called **composite functions**.

Definition 9.5   *If $g : X \to Y$ and $f : Y \to Z$, the composition of f and g, denoted $f \circ g$ or $f(g)$ is a function: $X \to Z$ defined by*

$$f \circ g(x) = f(g)(x) \equiv f(g(x)).$$

More generally, it is not necessary that $\mathrm{Dmn}(f) = \mathrm{Rng}(g)$, and $f(g)$ is well defined as long as $\mathrm{Rng}(g) \subset \mathrm{Dmn}(f)$.

Compositions of more than two functions are defined analogously, with the notational convention that functions are applied **right to left**. For instance,

$$f \circ g \circ h(x) \equiv f(g(h(x))),$$

which is evaluated as a mapping

$$x \to h(x) \to g(h(x)) \to f(g(h(x))).$$

Note finally that a composition of functions is not a "commutative" process, in that even when the domains and ranges of the functions allow the definition of both $f \circ g$ and $g \circ f$, in only the most trivial exceptional cases will these be equal. The rule is

$$f \circ g \neq g \circ f,$$

and so order matters!

9.2.2   The Notion of Continuity

Intuitively a function is said to be **continuous** at a given point x_0 if $f(x)$ **must** be close to $f(x_0)$ whenever x is close to x_0. In other words, $|f(x) - f(x_0)|$ will be "small" whenever $|x - x_0|$ is "small." Mathematicians formalize this notion with a logically complex statement that receives some discussion below.

Definition 9.6   *A function $f(x)$ is **continuous at a point** x_0 if for any value $\epsilon > 0$, one can find a $\delta > 0$, so that:*

- *$|f(x) - f(x_0)| < \epsilon$ whenever $|x - x_0| < \delta$, or equivalently,*
- *$|x - x_0| < \delta$ implies that $|f(x) - f(x_0)| < \epsilon$.*

The function $f(x)$ is **continuous on an interval** *if it is continuous at every point of that interval, and $f(x)$ is* **continuous** *if it is continuous at every point of its domain.*

Remark 9.7

1. *By convention, a function is defined to be* **continuous at the endpoint(s) of a closed interval** $[a, b]$ *if the definition applies with x restricted to that interval. The formal terminology is that $f(x)$ is* **continuous from the left at** b, *or* **continuous from the right at** a. *However, this formal language is often not used and a statement such as, $f(x)$* **is continuous on** $[a, b]$, *is universally understood in this sense.*

2. *Note that in this definition, the numerical value of δ depends on the value of ϵ. In a given application it is in fact required that this dependency can be formalized by a function so that $\delta \equiv \delta(\epsilon)$.*

Continuity at a point x_0 means that however small an open interval one constructs around $f(x_0)$, here the interval $(f(x_0) - \epsilon, f(x_0) + \epsilon)$, one can find an open interval around x_0, here the interval $(x_0 - \delta, x_0 + \delta)$, that gets mapped into it. In the case where x_0 is an endpoint of a closed interval $[a, b]$, this statement says that however small an open interval one constructs around $f(x_0)$, here the interval $(f(x_0) - \epsilon, f(x_0) + \epsilon)$, one can find a half-open interval, here the interval $(b - \delta, b]$ or $[a, a + \delta)$, that gets mapped into it.

Now the statement above about ϵ and δ is subtle, and even passive in tone. But this definition can be stated in a more active way.

Definition 9.8 $f(x)$ is **continuous at a point** x_0 if for any sequence $\epsilon_n \to 0$ we can find a sequence δ_n so that $|f(x) - f(x_0)| < \epsilon_n$ whenever $|x - x_0| < \delta_n$. In other words, by choosing x_n arbitrarily in the intervals $|x - x_0| < \delta_n$, we can be assured that $|f(x_n) - f(x_0)| < \epsilon_n$, and hence $|f(x_n) - f(x_0)| \to 0$.

In general, it will also be the case that $\delta_n \to 0$, but the example of $f(x) \equiv 1$ for all x shows that this need not be the case.

This ϵ-δ definition is one of many in mathematics, and it is close in structure to the ϵ-N definition used to define convergence of a sequence in chapter 5. This definition may seem stiff and formal. This is because continuity, which is intuitively a simple notion, is also quite subtle and somewhat difficult to define precisely. So this and other such definitions periodically fall in and out of favor among mathematics educators, and it is fair to say that at least some mathematicians have a love–hate relationship with this string of words that with practice rolls off their tongues like a religious chant.

In this book we pay homage to the tradition of such definitions, but at the same time acknowledge the pain and suffering they cause many students of the subject. So we do invest a bit more time in exploring their meaning. In point of fact, the traditional continuity chant is: "...for any $\epsilon > 0$, there is a $\delta > 0$ so that...," which we have adapted as above to make the point that determining if such a δ exists is typically an exercise in finding one that does work.

To explore this complicated notion, let's informally say that $f(x)$ is **continuous at** x_0 if *we can make $|f(x) - f(x_0)|$ as small as we want by choosing $|x - x_0|$ small enough*. We can also think of this as saying that the value of $f(x_0)$ can be predicted if we know the value of $f(x)$ for all x arbitrarily close to x_0. That is, we cannot be surprised at the value of $f(x_0)$ once we know the values of $f(x)$ for x near x_0.

The cause of the complexity in the definition is that continuity means more than simply that "we can find an x near x_0 so that $f(x)$ is near $f(x_0)$," or even "so that $f(x)$ is arbitrarily close to $f(x_0)$." Let's formalize these simpler statements and see what goes wrong.

Definition 9.9 (*Version 1*) $f(x)$ *is* **almost continuous at a point** x_0 *if for any* $\epsilon > 0$ *there is an x so that $|f(x) - f(x_0)| < \epsilon$.*

Well this version does not tell us very much, since it does not even ensure that x is anywhere near x_0.

Definition 9.10 (*Version 2*) $f(x)$ *is* **almost continuous at a point** x_0 *if for any* $\epsilon > 0$ *there is an x so that $|x - x_0| < \epsilon$ and $|f(x) - f(x_0)| < \epsilon$.*

This version 2 makes a bit more sense because at least we can be sure that as we require $f(x)$ to be nearer to $f(x_0)$, that there are x-values that work for which x becomes nearer to x_0. On the other hand, this definition allows there to be lots of x-values that are close to x_0 for which $f(x)$ is far, perhaps very far, from $f(x_0)$.

Example 9.11   *The classical example of this almost continuous situation is*

$$f(x) = \begin{cases} \sin \frac{1}{x}, & x \neq 0, \\ 0, & x = 0, \end{cases}$$

*as graphed in **figure 9.1**. This graph satisfies the definition of "almost continuous (version 2) at $x_0 = 0$," where $f(0) = 0$, since it is clear that "for any $\epsilon > 0$, there is an x so that $|x - x_0| < \epsilon$ and $|f(x) - f(x_0)| < \epsilon$." In fact "for any $\epsilon > 0$, there is an x so that $|x - x_0| < \epsilon$ and $f(x) = f(0)$".*

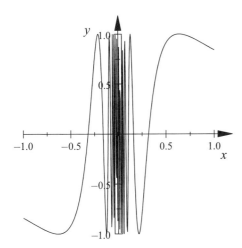

Figure 9.1

$$f(x) = \begin{cases} \sin\frac{1}{x}, & x \neq 0 \\ 0, & x = 0 \end{cases}$$

The inadequacy of this "almost continuous (version 2)" notion is further illustrated by the fact that if we arbitrarily define $f(0)$ as any number between -1 and 1, this definition is still satisfied. So the point is, what conclusions could be made about such a function at $x = 0$ if we can arbitrarily define its value there and still satisfy the definition? Obviously we cannot predict this value of $f(0)$ from knowing the value of $f(x)$ for x near 0.

Example 9.12 *The example above can be made even more compelling by considering*

$$g(x) = \begin{cases} \frac{1}{x}\sin\frac{1}{x}, & x \neq 0, \\ 0, & x = 0. \end{cases}$$

*We then have that $g(x)$ is "almost continuous (version 2) at $x = 0$," and and this will be true even if we define $g(0)$ as any real number! This is displayed in **figure 9.2**, where it is noted that $g(x)$ is unbounded both positively and negatively as $x \to 0$.*

The important detail that the definition of continuity adds to the definition of "almost continuous (version 2)," is that it demands that **the function f make all the values of $f(x)$ close to $f(x_0)$, for x near** x_0, not just some of them. In doing so, it allows the distance between x and x_0 to differ from the distance between $f(x)$ and $f(x_0)$, as long as we can choose the latter distance for any ϵ. So the final logic becomes the chant, "...for any value $\epsilon > 0$, one can find a $\delta > 0$ "

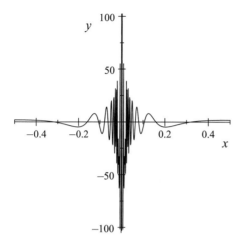

Figure 9.2

$$g(x) = \begin{cases} \frac{1}{x}\sin\frac{1}{x}, & x \neq 0 \\ 0, & x = 0 \end{cases}$$

Example 9.13   *The price of a 5-year zero-coupon bond per $1 par, in terms of an annual rate, is given by $P(r) = (1+r)^{-5}$. To see that this is a continuous function at $r_0 \in (0, \infty)$, the goal is to be able to make $|(1+r)^{-5} - (1+r_0)^{-5}|$ small by making $|r - r_0|$ small. To this end, note that*

$$|(1+r)^{-5} - (1+r_0)^{-5}| = \left| \frac{(1+r_0)^5 - (1+r)^5}{(1+r)^5(1+r_0)^5} \right|$$

$$< |(1+r_0)^5 - (1+r)^5|,$$

since $(1+r)^5(1+r_0)^5 \geq 1$ for $r \geq 0$, which we can assume by choosing $\epsilon < r_0$. Now, by the binomial theorem, $(1+r_0)^5 - (1+r)^5 = \sum_{j=1}^{5} \binom{5}{j}[r_0^j - r^j]$, since the $j = 0$ terms cancel. Each of the remaining terms $r_0^j - r^j$ for $j \geq 1$ can be factored:

$$r_0^j - r^j = (r_0 - r)\sum_{k=0}^{j-1} r_0^k r^{j-k-1}.$$

Combining, we get that $|(1+r)^{-5} - (1+r_0)^{-5}| < K|r_0 - r|$, where K is choosen as the largest numerical value of the $\sum_{j=1}^{5} \binom{5}{j} \sum_{k=0}^{j-1} r_0^k r^{j-k-1}$ factor. This bound would be determined by noting that $r < r_0 + \epsilon = ar_0$, for some $a > 1$, and then

$$K = \max_{r < r_0 + \epsilon} \sum_{j=1}^{5} \binom{5}{j} \sum_{k=0}^{j-1} r_0^k r^{j-k-1}$$

$$= \sum_{j=1}^{5} \binom{5}{j} r_0^{j-1} \sum_{k=0}^{j-1} a^{j-k-1} = \sum_{j=1}^{5} \binom{5}{j} \left(\frac{a^j - 1}{a - 1}\right) r_0^{j-1}.$$

Finally, from this we have that for any $\epsilon > 0$, $|(1+r)^{-5} - (1+r_0)^{-5}| < \epsilon$ if $|r_0 - r| < \frac{\epsilon}{K}$. That is, we define $\delta(\epsilon) = \frac{\epsilon}{K}$. In fact $P(r)$ is continuous on $r_0 \in (-1, \infty)$, but more care is needed for the numerical estimates, since there is an apparent problem at $r = -1$.

Note that in this example, no effort was made to determine the best value of K, for instance, by further restricting the range of allowable r-values in this maximum. To simply verify continuity, the analysis can be crude to simplify the derivation, or more refined. The conclusion of continuity does not depend on the size of this K, only that there was some function $\delta(\epsilon)$ that worked for any ϵ.

The Meaning of "Discontinuous"

Because of the logical complexity of the continuity definition, it makes sense to formalize the meaning of the notion that $f(x)$ is **not continuous** at x_0, that is, $f(x)$ **is discontinuous at a point** x_0. This idea could be needed for the proof of any statement of the form:

If property S, then $f(x)$ is continuous at x_0.

For example, we could choose to use a contrapositive proof, whereby we would attempt to prove

If $f(x)$ is discontinuous at x_0, then $\sim S$,

or a proof by contradiction, whereby we would attempt to prove

If property S and $f(x)$ is discontinuous at x_0, then $\sim S$.

In other words, for either of these approaches to a proof, a clear understanding is needed of the meaning of the statement that "$f(x)$ is discontinuous at x_0."

Using the ideas from chapter 1, we temporarily introduce statement notation:

$P \equiv f(x)$ is continuous at x_0,

$Q(\epsilon) \equiv |f(x) - f(x_0)| < \epsilon$,

$R(\delta) \equiv |x - x_0| < \delta$.

Then we have that P is defined by

$$P \Leftrightarrow \forall \epsilon \exists \delta \forall x (R(\delta) \Rightarrow Q(\epsilon)).$$

The logical development of $\sim P$ proceeds as follows, recalling that the universal quantifiers are negations of each other:

$$\sim P \leftrightarrow \sim [\forall \epsilon \exists \delta \forall x (R(\delta) \Rightarrow Q(\epsilon))]$$

$$\leftrightarrow \exists \epsilon \sim [\exists \delta \forall x (R(\delta) \Rightarrow Q(\epsilon))]$$

$$\leftrightarrow \exists \epsilon \forall \delta \sim [\forall x (R(\delta) \Rightarrow Q(\epsilon))]$$

$$\leftrightarrow \exists \epsilon \forall \delta \exists x \sim (R(\delta) \Rightarrow Q(\epsilon))$$

$$\leftrightarrow \exists \epsilon \forall \delta \exists x (R(\delta) \wedge \sim Q(\epsilon)).$$

Summarizing, we obtain:

Definition 9.14 $f(x)$ *is **discontinuous at a point** x_0 if there is an $\epsilon > 0$ so that for any $\delta > 0$ we can find an x with $|x - x_0| < \delta$ and yet $|f(x) - f(x_0)| \geq \epsilon$. More generally given this ϵ, for any sequence $\delta_n \to 0$, we can find x_n so that $|x_n - x_0| < \delta_n$ and $|f(x_n) - f(x_0)| \geq \epsilon$. So $x_n \to x_0$ but $f(x_n) \nrightarrow f(x_0)$.*

As will be seen below, every continuous function has the useful property that it preserves convergence of sequences. To set the stage for this, first recall the ϵ-N definition of convergence from chapter 5, which is generalized here to functions. To obtain this generalization, note that if $\{x_n\}$ is a sequence, we can define a function $f : \mathbb{N} \to \mathbb{R}$ by

$$f(n) = x_n.$$

Definition 9.15 *A sequence $\{x_n\}$ **converges to** $x < \infty$ **as** $n \to \infty$, denoted $x_n \to x$, if, given any $\epsilon > 0$, one can find an $N \in \mathbb{N}$ so that*

$$|x_n - x| < \epsilon \quad \text{whenever} \quad n \geq N.$$

*Analogously, a function $f(x)$ **converges to a limit** $L < \infty$ **as** $x \to \infty$, denoted $\lim_{x \to \infty} f(x) = L$, if, given any $\epsilon > 0$, one can find an N so that*

$$|f(x) - L| < \epsilon \quad \text{whenever} \quad x \geq N.$$

*More generally, a function $f(x)$ **converges to a limit** $L < \infty$ **as** $x \to x_0 < \infty$, denoted $\lim_{x \to x_0} f(x) = L$, if, given any $\epsilon > 0$, one can find a $\delta > 0$ so that*

$|f(x) - L| < \epsilon$ *whenever* $|x - x_0| < \delta$.

In other words, convergence of a sequence implies that eventually all the terms of the sequence get arbitrarily close to the limiting value. For convergence of a function we require that $f(x)$ can be made arbitrarily close to L by choosing x close enough to x_0, or in the case of $x_0 = \infty$, the definition is adapted to ensure that $f(x)$ can be made arbitrarily close to L by choosing x large enough.

Remark 9.16   *It is important to understand that the notion of a limit of a function in the definition above is two sided. That is to say, because of the absolute values in the convergence criterion, the statement $\lim_{x \to x_0} f(x) = L$ means that a limiting value for $f(x)$ exists whether $x \to x_0$ "from the right," so that $x > x_0$, or "from the left," so that $x < x_0$, and that these limits are equal. "One-sided" limits can also be defined:*

Definition 9.17   *A function $f(x)$ **converges to a limit** $L < \infty$ **from the left** as $x \to x_0 < \infty$, denoted $\lim_{x \to x_0^-} f(x) = L$, if, given any $\epsilon > 0$, one can find a $\delta > 0$ so that*

$|f(x) - L| < \epsilon$ *whenever* $x_0 - \delta < x < x_0$.

*A function $f(x)$ **converges to a limit** $L < \infty$ **from the right** as $x \to x_0 < \infty$, denoted $\lim_{x \to x_0^+} f(x) = L$, if, given any $\epsilon > 0$, one can find a $\delta > 0$ so that*

$|f(x) - L| < \epsilon$ *whenever* $x_0 < x < x_0 + \delta$.

Notation 9.18   *To economize on language, it is common to say that $\lim_{x \to x_0} f(x)$ exists for all $x_0 \in [a, b]$, as a brief way of stating that $\lim_{x \to x_0} f(x)$ exists for all $x_0 \in (a, b)$, and also that $\lim_{x \to a^+} f(x)$ and $\lim_{x \to b^-} f(x)$ exist.*

Example 9.19   *It is instructive to demonstrate by the definitions above that if $f(x) = \frac{x^2}{1-x^2}$, then $\lim_{x \to 0} f(x) = 0$ and $\lim_{x \to \infty} f(x) = -1$.*

1. *For the limit $\lim_{x \to 0} f(x)$, we can arbitrarily restrict attention to $|x| < 0.1$ say, since we only care about the limit at $x = 0$. To make $\left| \frac{x^2}{1-x^2} \right|$ small for $|x|$ small, note that*

$$\left| \frac{x^2}{1-x^2} \right| < \frac{100}{99} x^2 < \frac{100}{99} |x|,$$

since $|x| < 0.1$ implies that $\frac{1}{1-x^2} < \frac{100}{99}$ and $x^2 < x$. So to make $\left| \frac{x^2}{1-x^2} \right| < \epsilon$, we can choose $|x| < \delta(\epsilon) \equiv \frac{99}{100} \epsilon$.

2. *For the limit* $\lim_{x \to \infty} f(x)$, *to make* $\left| \frac{x^2}{1-x^2} - (-1) \right|$ *small for x large, note that*

$$\left| \frac{x^2}{1-x^2} + 1 \right| = \left| \frac{1}{x^2 - 1} \right| < \frac{1}{x},$$

since $x^2 - 1 > x$ *for* $x > 3$, *say. So to make* $\left| \frac{x^2}{1-x^2} - (-1) \right| < \epsilon$, *we can choose* $N \equiv \frac{1}{\epsilon}$.

From the definitions above it should also be apparent that the statement: $f(x)$ is **continuous at** x_0, is equivalent to the statement that $\lim_{x \to x_0} f(x) = f(x_0)$. To say that $f(x)$ is **continuous on** (a, b) is equivalent to the statement that $\lim_{x \to x_0} f(x) = f(x_0)$ for all $x_0 \in (a, b)$. Finally, the notion of one-sided limits implies that the statement, $f(x)$ is **continuous on** $[a, b]$ is equivalent to the statement that $\lim_{x \to x_0} f(x) = f(x_0)$ for all $x_0 \in (a, b)$, and also that $\lim_{x \to a^+} f(x) = f(a)$, and $\lim_{x \to b^-} f(x) = f(b)$.

This observation provides another simple way to think about functions that are discontinuous at a point x_0.

Definition 9.20   *The function $f(x)$ is **discontinuous at** x_0 if either:*

1. $\lim_{x \to x_0} f(x)$ *does not exist, or*

2. $\lim_{x \to x_0} f(x)$ *does exist and equals L, but $f(x_0) \neq L$.*

For example, $f(x) = \frac{1}{x}$ is discontinuous at $x = 0$ both because $\lim_{x \to 0} \frac{1}{x}$ does not exist, and because $f(0)$ is not defined. On the other hand,

$$g(x) = \begin{cases} x, & x \neq 0 \\ 1, & x = 0 \end{cases}$$

is discontinuous at $x = 0$ not because $\lim_{x \to 0} g(x)$ does not exist but because $g(0) \neq \lim_{x \to 0} g(x) = 0$.

*The Metric Notion of Continuity
Stated as in the definition above, continuity is seen to be a fundamentally "metric" notion. Recall from chapter 3 that $|x|$ is a norm on $\mathbb{R}$ that gives rise to a metric or distance function, defined by

$$d(a, b) = |a - b|.$$

Consequently the definition of continuity explicitly utilizes this notion of distance, and with this notion it requires that we can make $|f(x) - f(x_0)|$ as small as we want by choosing $|x - x_0|$ small enough. In other words, for any value of $\epsilon > 0$, one can find a $\delta > 0$ so that $d(f(x), f(x_0)) < \epsilon$ whenever $d(x, x_0) < \delta$.

The importance of this observation is that all of the development below for real-valued functions of a real variable, $f(x)$, carries over with only a notational change to functions defined between any two metric spaces. For example, the notion of a **continuous complex-valued function of a complex variable**, as well as other examples, can be framed directly in terms of the respective metrics. We leave this general point here for now, and continue to develop the theory in the more familiar setting of $\mathrm{Dmn}(f) \subset \mathbb{R}$.

Also more generally, one can develop additional intuition for continuity by introducing a more geometric interpretation. Recall the open ball constructions from chapter 4 in (4.1):

$$B_r(x) = \{ y \in \mathbb{R} \,|\, |x - y| < r \}.$$

The definition of continuity can then be restated in two ways, each of which has apparent application to the more general framework of later chapters and more advanced mathematical treatments.

Definition 9.21

1. $f(x)$ is **continuous at a point** x_0 if for any value $\epsilon > 0$ one can find a $\delta > 0$ so that

$$f(B_\delta(x_0)) \subset B_\epsilon(f(x_0)).$$

2. $f(x)$ is **continuous at a point** x_0 if for any integer $n > 0$ one can find an integer $m > 0$ so that

$$f(B_{1/m}(x_0)) \subset B_{1/n}(f(x_0)).$$

In other words, continuity at x_0 means that however small an open ball one constructs around $f(x_0)$, one can find an open ball around x_0 that gets mapped into it.

Interpreted this way, it is again apparent that the notion of continuity is very generally applicable to all metric spaces. Below it will be seen to be applicable even beyond metric spaces.

Sequential Continuity

Another notion of continuity that is equivalent to that above is the notion of **sequential continuity**, which we define next.

Definition 9.22 $f(x)$ **is sequentially continuous at** x_0 if, given any sequence $\{x_n\}$ such that $x_n \to x_0$, then $f(x_n) \to f(x_0)$. Similarly $f(x)$ is **sequentially continuous on an interval** if it has this property at every point of the interval.

Proposition 9.23 $f(x)$ *is continuous at* x_0 *if and only if it is sequentially continuous at* x_0.

Proof   See exercise 28.   ∎

9.2.3   Basic Properties of Continuous Functions

While providing various intuitive frameworks for continuity, none of the preceding definitions provide an accessible approach to demonstrating that a given function is continuous in any but the simplest cases. For example, how might one prove that $f(x) = x^5 + x^4 + x^3\sqrt{(x^6+4)} + x^{(x^2+x)}$ is continuous for all $x > 0$? Certainly the prospect of determining δ for a given $\epsilon > 0$ is not appealing, and determining the general formula for $\delta(\epsilon)$ is even less so.

The following propositions state that the notion of continuity combines well arithmetically, and in a variety of other ways.

Proposition 9.24   *If* $f(x)$ *and* $g(x)$ *are continuous at* x_0, *then the following are also continuous at* x_0:

1. $af(x) + b$, *for* $a, b \in \mathbb{R}$
2. $f(x) + g(x)$
3. $f(x) - g(x)$
4. $f(x)g(x)$
5. $\frac{f(x)}{g(x)}$ *if* $g(x_0) \neq 0$

Proof   In each case the objective is to show that if we can make both $|f(x) - f(x_0)|$ and $|g(x) - g(x_0)|$ arbitrarily small by choosing $|x - x_0|$ small, that this property transfers to the given combinations. Denoting by $\delta(\epsilon)$ the value that works for both f and g given ϵ, which is defined as the smaller of the respective values, we find $\delta'(\epsilon)$, the value that is needed for the given combination.

1. $|[af(x) + b] - [af(x_0) + b]| = |a|\,|f(x) - f(x_0)|$, so we can choose $\delta'(\epsilon) = \delta\left(\frac{\epsilon}{|a|}\right)$.

2. $|[f(x) + g(x)] - [f(x_0) + g(x_0)]| \leq |f(x) - f(x_0)| + |g(x) - g(x_0)|$ by the triangle inequality, so we choose $\delta'(\epsilon) = \delta\left(\frac{\epsilon}{2}\right)$.

3. This follows from part 1, with $a = -1$ and $b = 0$, and then part 2 applied to the continuous $f(x)$ and $-g(x)$.

4. By the triangle inequality,

$$|f(x)g(x) - f(x_0)g(x_0)| = |[f(x)g(x) - f(x_0)g(x)] + [f(x_0)g(x) - f(x_0)g(x_0)]|$$

$$\leq M|f(x) - f(x_0)| + |f(x_0)|\,|g(x) - g(x_0)|,$$

where M denotes any upper bound for $|g(x)|$ on $|x - x_0| < \delta$. Such an upper bound must exist; that is, if we are given that $|g(x) - g(x_0)| < \epsilon$, then since $g(x) = g(x_0) + (g(x) - g(x_0))$, we have by the triangle inequality that $|g(x)| < |g(x_0)| + \epsilon$. We can hence choose $\delta'(\epsilon) = \min\left[\delta\left(\frac{\epsilon}{2M}\right), \delta\left(\frac{\epsilon}{2|f(x_0)|}\right)\right]$ if $f(x_0) \neq 0$. Otherwise, if $f(x_0) = 0$, then $|f(x)g(x) - f(x_0)g(x_0)| \leq M|f(x) - f(x_0)|$, and we take $\delta'(\epsilon) = \delta\left(\frac{\epsilon}{M}\right)$.

5. First off, since $g(x_0) \neq 0$, and $g(x)$ is continuous at x_0, for $\epsilon = \frac{g(x_0)}{2}$ there is a δ'' so that $|g(x) - g(x_0)| < \frac{g(x_0)}{2}$ for $|x - x_0| < \delta''$. Consequently $g(x) \neq 0$ for $|x - x_0| < \delta''$. Next

$$\left| \frac{f(x)}{g(x)} - \frac{f(x_0)}{g(x_0)} \right| = \left| \frac{f(x)g(x_0) - f(x_0)g(x)}{g(x)g(x_0)} \right|$$

$$= \left| \frac{f(x)g(x_0) - f(x_0)g(x_0) + f(x_0)g(x_0) - f(x_0)g(x)}{g(x)g(x_0)} \right|$$

$$\leq \left| \frac{f(x) - f(x_0)}{g(x)} \right| + \left| \frac{f(x_0)}{g(x_0)} \frac{g(x_0) - g(x)}{g(x)} \right|$$

$$\leq m|f(x) - f(x_0)| + cm|g(x) - g(x_0)|,$$

where m is the maximum value of $\frac{1}{g(x)}$ for $|x - x_0| < \delta''$ and $c = \left| \frac{f(x_0)}{g(x_0)} \right|$. We can now choose $\delta'(\epsilon) = \min\left[\delta\left(\frac{\epsilon}{2m}\right), \delta\left(\frac{\epsilon}{2cm}\right), \delta''\right]$.   ∎

Example 9.25   *Returning to the question on verification of the continuity of complicated functions, the proposition above provides useful tools. Since $f(x) = x$ is obviously continuous with $\delta(\epsilon) = \epsilon$, it follows that every integer power of x is also continuous, since these are products of $f(x)$, as is any polynomial in x, since this equals sums and scalar multiples of these continuous integer powers of x. Similarly every **rational function**, defined as a ratio of polynomials, is continuous everywhere the denominator polynomial is nonzero.*

The final building blocks for confirming the continuity of complicated functions follow in a series of propositions below:

1. *The first proposition addresses inverses of one-to-one functions, which will imply that $f(x) = x^{1/n}$ is continuous on $x \geq 0$ for all integer $n \in \mathbb{N}$, as is $f(x) = x^{-1/n}$ for $x > 0$.*

2. *The second proposition addresses compositions of continuous functions, from which one derives the continuity of many common functions, for instance, $f(x) = x^{m/n}$ for all integers $m, n \neq 0$, as well as various linear combinations of such functions and ratios of these combinations with nonzero denominators.*

3. *Last, the common exponential functions $f(x) = a^x$ for some real number $a > 0$ require direct verification of continuity, from which the associated logarithms $g(x) = \log_a x$ will be continuous for $x > 0$ as these are inverse functions to the exponentials. Then for irrational exponents q the continuity of $f(x) = x^q$ follows for $x > 0$ by noting that $f(x) = e^{q \ln x}$, which is a composition of continuous functions.*

Proposition 9.26   *If $f(x)$ is continuous at x_0 and one-to-one in an open interval about x_0, then f^{-1} is continuous at $f(x_0)$.*

Proof   Assume that $f(x)$ is continuous at x_0 and one-to-one on an open interval I about x_0, and let $\bar{J} \subset I$ be the closure of a bounded open subinterval, with $x_0 \in J$. We restrict f to $\bar{J}$ and show that f^{-1} is then continuous at $f(x_0)$ by a proof by contradiction. If f^{-1} is discontinuous at $f(x_0)$, then there exists an $\epsilon' > 0$ and a sequence $\{y_n\} \subset f(\bar{J})$ so that $|y_n - f(x_0)| < \frac{1}{n}$ yet $|f^{-1}(y_n) - f^{-1}(f(x_0))| = |x_n - x_0| > \epsilon'$ for all n. Now, since $\bar{J}$ is compact and $\{x_n\} \subset \bar{J}$, there is an accumulation point $x' \in \bar{J}$ and a subsequence $\{x_n'\} \subset \{x_n\}$ so that $x_n' \to x'$. Hence, since $|x_n - x_0| > \epsilon'$ for all n, it follows that $|x_n' - x_0| > \epsilon'$, and so $|x' - x_0| \geq \epsilon'$. However, $|y_n - f(x_0)| = |f(x_n) - f(x_0)| < \frac{1}{n}$ implies that $|f(x_n') - f(x_0)| \to 0$. But $x_n' \to x'$ and continuity of $f(x)$ then implies $|f(x_n') - f(x')| \to 0$ and so $f(x') = f(x_0)$. We now have a contradiction. Namely $|x' - x_0| > \epsilon'$ and $f(x') = f(x_0)$ contradicts that f is one-to-one. ∎

The following proposition applies to the composition of any collection of continuous functions, by iteration:

Proposition 9.27   *If $g(x)$ is continuous at x_0, and $f(x)$ is continuous at $g(x_0)$, then $f(g(x))$ is continuous at x_0.*

Proof   Given $\epsilon > 0$, the goal is to find $\delta(\epsilon)$ so that $|f(g(x)) - f(g(x_0))| < \epsilon$ when $|x - x_0| < \delta(\epsilon)$. By continuity of $f(x)$, we conclude for any $\epsilon < 0$ that $|f(g(x)) - f(g(x_0))| < \epsilon$ if $|g(x) - g(x_0)| < \delta'(\epsilon)$, where δ' denotes the associated function for $f(x)$. Next, by the continuity of $g(x)$, we conclude that $|g(x) - g(x_0)| < \delta'(\epsilon)$ when $|x - x_0| < \delta''(\delta'(\epsilon))$, where δ'' denotes the associated function for $g(x)$. Hence we choose $\delta(\epsilon) = \delta''(\delta'(\epsilon))$. ∎

Finally, we address the exponential and logarithmic functions.

Proposition 9.28   *The function $f(x) = e^x$ is continuous for all $x \in \mathbb{R}$.*

Proof   Given x_0, $e^x - e^{x_0} = e^{x_0}[e^{x-x_0} - 1]$ so that e^x is continuous at x_0 if, for any ϵ, we can find a δ so that $e^{x_0}|e^{x-x_0} - 1| < \epsilon$ whenever $|x - x_0| < \delta$. Since e^{x_0} is just a number, this result will follow if e^y is continuous at $y = 0$. Then for any ϵ' we can find a δ' so that $|e^y - 1| < \epsilon'$ whenever $|y| < \delta'$, and so given ϵ, we define $\epsilon' = \frac{\epsilon}{e^{x_0}}$

and $\delta = \delta'$. In summary, if e^y is continuous at $y = 0$, it is continuous everywhere. Now by section 9.3.3, $e > 1$, and we have that $e^y > 1$ and $e^{-y} < 1$ for $y > 0$. Hence e^x is a monotonically increasing function on $\mathbb{R}$, meaning, if $x' < x$, then $e^{x'} < e^x$. This is because if $x = x' + x''$ for some $x'' > 0$, then $e^x = e^{x'}e^{x''} > e^{x'}$. Also, since $(e^y - 1)^2 \geq 0$, we derive by expansion that

$$e^y - 1 \geq 1 - e^{-y} \geq 0.$$

So, if for any $\epsilon > 0$, there is a δ so that $0 \leq e^y - 1 < \epsilon$ whenever $0 \leq y < \delta$; then also $0 < 1 - e^{-y} < \epsilon$, and hence for $|y| < \delta$ it follows that $|e^y - 1| < \epsilon$ and the proof of continuity at $y = 0$ will be complete. To this end, let $\epsilon > 0$ be given, and consider the sequence $x_n = e^{y_n}$ where $y_n > 0$ and $y_n \to 0$ monotonically. Consequently $x_n > 1$ for all n. Also the monotonicity of e^x implies that x_n is a monotonically decreasing sequence. It is also bounded from below by 1, and hence it has a unique accumulation point x_0. If $x_0 = 1$, we are done. But assume that $x_0 > 1$. Then $x_n \to x_0$, and is monotonically decreasing. Therefore

$$e = x_n^{1/y_n} > x_0^{1/y_n},$$

but this is a contradiction, since $x_0 > 1$ and $y_n \to 0$ implies $x_0^{1/y_n} \to \infty$. Consequently $x_0 = 1$. ∎

Example 9.29 *The continuity of e^x implies the continuity of its inverse function, $\ln x$ for $x > 0$, since e^x is one-to-one. For $a > 0$ the function $f(x) = a^x$ is then continuous as a composite function, since $a^x = e^{x \ln a}$. Similarly the continuity of $\log_a x$ follows for $x > 0$ and $a > 0$, since $\log_a x = \frac{\ln x}{\ln a}$, and also of $x^x = e^{x \ln x}$ for $x > 0$.*

9.2.4 Uniform Continuity

As noted in the preceding section, a formal demonstration of continuity requires an explicit expression for δ as a function of ϵ, $\delta \equiv \delta(\epsilon)$. It should also be noted that such a demonstration can be complicated by the fact that while the value of δ in the definition of continuity apparently depends on ϵ, it can also in general depend on x_0, so $\delta \equiv \delta(\epsilon, x_0)$.

Example 9.30 *The function $f(x) = 1/x$ is continuous throughout its domain: $\mathrm{Dmn}(f) = \{x \mid x \neq 0\}$. However, it is not difficult to verify that for a given ϵ and positive x_0, that the associated δ is also a function of ϵ and x_0:*

$$\delta(\epsilon, x_0) = \frac{\epsilon x_0^2}{1 + \epsilon |x_0|}.$$

This is justified for $x_0 > 0$ say, by noting that if $|x - x_0| < \delta$, with $\delta < \frac{x_0}{2}$ to keep $x > 0$;
then

$$\left| \frac{1}{x} - \frac{1}{x_0} \right| < \frac{\delta}{xx_0} < \frac{\delta}{(x_0 - \delta)x_0}.$$

To have $\left| \frac{1}{x} - \frac{1}{x_0} \right| < \epsilon$, we solve for δ producing the formula above. Consequently, for a given ϵ, δ can be arbitrarily large if $|x_0|$ is large, yet it must be choosen increasingly small as $|x_0|$ approaches 0. This is, of course, also apparent from the graph of $f(x)$.

An important notion is that of **uniform continuity**, whereby it is possible to choose δ to be independent of x_0.

Definition 9.31   *$f(x)$ is **uniformly continuous on an interval** if for any value $\epsilon > 0$ one can find a $\delta > 0$ so that for all x and y in the interval, $|f(x) - f(y)| < \epsilon$ whenever $|x - y| < \delta$. Similarly $f(x)$ is **uniformly continuous** if it satisfies this property for all x and y in its domain.*

Example 9.32   *$f(x) = 1/x$ is uniformly continuous on any closed interval $[a, b]$ not containing the origin. This is easily demonstrated using example 9.30 in that one chooses δ to equal the minimum value of $\delta(x_0)$ for x_0 in the interval, which is apparently the value of $\delta(x)$ at the endpoint of the interval closest to 0.*

This example is generalized below. But note that the idea of uniform continuity is that for any $\epsilon > 0$ the associated δ in the definition of continuity, which in general is a function of both x and ϵ, $\delta(x, \epsilon)$, satisfies $\delta(x, \epsilon) > \delta(\epsilon) > 0$ for all x for some other function, $\delta(\epsilon)$. So what keeps a continuous function from being uniformly continuous is that for a given ϵ, the $\delta(x, \epsilon)$ values get arbitrarily close to 0 as x varies. This was seen in example 9.30, where $f(x) = 1/x$.

We return to this point after the next result. Its proof relies on a simple but important property of closed and bounded intervals which we have encountered before in chapter 4 in proposition 4.17. We prove this simpler version directly.

Proposition 9.33   *If $\{r_j\}$ is a bounded infinite sequence of reals $\{r_j\} \subset [a, b]$, then there is a subsequence $\{r'_j\}$ and a point $r \in [a, b]$ so that $r'_j \to r$ as $j \to \infty$.*

Proof   Divide the interval into halves: $\left[a, \frac{a+b}{2}\right]$ and $\left[\frac{a+b}{2}, b\right]$. Then one or both of these subintervals contains an infinite subsequence of $\{r_j\}$, and we choose that subinterval if unique, or an arbitrary subinterval otherwise. We also choose r'_1 to be any point in the choosen interval. We then divide that subinterval in half, and once again observe that one or both of the new subintervals contains an infinite subsequence. So we choose one, as well as r_2' in that subinterval. Continuing in this manner, we obtain a sequence of nested intervals of length $\frac{a+b}{2^j}$, each of which contains one member of the desired sequence $\{r_j'\}$. It is clear that the intersection of all choosen subintervals is a single point r, since, if it contained more than one point, it would also contain the interval spanning the two points, in contradiction to the fact that the lengths of these subintervals converge to 0 by the halving property of the construction. Finally, by construction $|r_j' - r| < \frac{a+b}{2^j}$, so $r_n' \to r$ as required.   ∎

By the Heine–Borel theorem, the closed and bounded interval $[a, b]$ is compact. This result is then a special case of the general chapter 4 result noted above that if a compact set K contains an infinite sequence $\{r_j\}$, then there is a subsequence $\{r_j'\}$ and a point $r \in K$ so that $r_j' \to r$ as $j \to \infty$. However, this proof was supplied rather than simply quoting the proposition 4.17 result because in this application, as in many, the construction in the special case is revealing and too simple to avoid.

Note that this proposition addresses the existence of such a point r, and it cannot be improved to assert the uniqueness of this point. Indeed it is possible that in every subinterval of the construction above there is an infinite subsequence of the original sequence $\{r_j\}$.

Example 9.34   *Let $\{r_j\}$ denote an arbitrary enumeration of the rational numbers in $[a, b]$. Then the construction above shows that for every real number $r \in [a, b]$ there is a subsequence $\{r_j'\} \subset [a, b]$ so that $r_j' \to r$ as $j \to \infty$. One simply chooses, at each step, the subinterval that contains the given point, r.*

Proposition 9.35 (*Version 1*)   *If $f(x)$ is continuous on a closed and bounded interval $[a, b]$, then it is uniformly continuous on this interval.*

Proof   Assume that $\epsilon > 0$ is given. For each number $r \in [a, b]$, let $\delta(r) \equiv \delta(r, \epsilon)$ denote the associated delta for this ϵ. We claim that $\{\delta(r)\}$ is bounded away from 0, and that we can take δ in the definition of uniform continuity to be equal to any nonzero lower bound for this collection. To show this boundedness, assume that it is not, and a contradiction will be revealed. That is, assume that there is a sequence of real numbers r_j with $\delta(r_j) \to 0$. Then for each positive integer k there is an associated r_k and x_k so that $|f(r_k) - f(x_k)| \geq \epsilon$ and $|r_k - x_k| < \frac{1}{k}$. If such points did not exist for $k \geq K$, say, then $f(x)$ would be uniformly continuous with $\delta = \frac{1}{K}$. Now we demonstrate a contradiction to the continuity of $f(x)$. The sequences $\{r_k\}$ and $\{x_k\}$ have subsequences that converge by the proposition above, and must converge to the same point in $[a, b]$, since $|r_k - x_k| < \frac{1}{k}$. But since $|f(r_k) - f(x_k)| \geq \epsilon$, we cannot have $\{f(r_k)\}$ and $\{f(x_k)\}$ convergent to the same point, contradicting the sequential continuity, and hence continuity of $f(x)$. Hence $\{\delta(r)\}$ is bounded away from 0 and the proof is complete.   ∎

From the comments above on compactness, one would have to think that this result is somehow related to the compactness of the interval $[a, b]$, and that on this basis the result will generalize. Recall that by compactness is meant that every collection of open intervals that cover $[a, b]$ contains a finite subcover, which is to say, a finite subcollection that also covers this interval. We demonstrate this general case with an alternative proof.

Proposition 9.36 (*Version 2*)   *If $f(x)$ is continuous on a compact set, $K \subset \mathbb{R}$, then it is uniformly continuous on K.*

Proof   Assume that $\epsilon > 0$ is given. For each number $r \in K$, let $\delta(r)$ denote the associated delta for $\frac{\epsilon}{2}$. Next consider the interval defined by $\delta(r)$ for a given r: $I_r = \left\{ r' \,|\, |r - r'| < \frac{\delta(r)}{2} \right\}$. The reason for this sleight of hand of dividing ϵ by 2 will be apparent in a moment. Now consider $\{I_r\}$ for all $r \in K$. Clearly, this is an open cover for K, which due to compactness, has a finite subcover, $\{I_{r_j}\}_{j=1}^n$. Define $\delta(\epsilon) = \frac{1}{2} \min\{\delta(r_j)\}$ and let $r', r'' \in K$ with $|r' - r''| < \delta(\epsilon)$. Then, since $r' \in I_{r_j}$ for some j, $|r' - r_j| < \frac{\delta(r_j)}{2}$, and so $|f(r') - f(r_j)| < \frac{\epsilon}{2}$. Also, by the triangle inequality,

$$|r'' - r_j| \leq |r'' - r'| + |r' - r_j|$$

$$< \delta(\epsilon) + \frac{\delta(r_j)}{2} \leq \delta(r_j),$$

and hence $|f(r_j) - f(r'')| < \frac{\epsilon}{2}$. Finally, by another application of the triangle inequality,

$$|f(r') - f(r'')| \leq |f(r') - f(r_j)| + |f(r_j) - f(r'')|$$

$$< \epsilon.$$ ∎

Remark 9.37   *A few comments are in order:*

1. *There are basically two approaches to the kind of proof just given:*

• *Reverse engineer all the intermediate steps so that one gets the desired conclusion that $|f(r') - f(r'')| < \epsilon$ in the last line of the proof. This is the approach used above. It fits right into the definition that "for any $\epsilon > 0$ one can find a δ so that . . . ." The advantage* *is that the continuity definition is produced verbatim; the disadvantage, which the reader undoubtedly encountered, is the temporary mystery associated with the $\frac{1}{2}$ factors, which in other proofs may be $\frac{1}{3}, \frac{1}{4}$, and so forth.*

• Ignore the reverse engineering and ultimately derive something like $|f(r') - f(r'')| < 4\epsilon$. Then we prove a statement like "given $\epsilon > 0$ there is a δ so that if $|r' - r''| < \delta$, then $|f(r') - f(r'')| < 4\epsilon$." Of course, this is logically equivalent to the original idea, but some find the presence of the 4 in the conclusion to be aesthetically unpleasant.

The present author alternates between these approaches, and generally prefers the second approach in personal research, and the first approach in communications. However, the reverse engineering required to produce a clean conclusion can at times add unjustifiable complexity to the derivation, and so will sometimes be abandoned.

2. *One can easily imagine going through the proof above almost verbatim if K is a compact subset of any metric space (X, d) and $f(x)$ is a continuous function from X to $\mathbb{R}$, or from X to another metric space (Y, d'). See exercises 5 and 30.*

9.2.5   Other Properties of Continuous Functions

A few other fundamental results on continuous functions are addressed next. The first is simple but powerful. Namely the sign of a continuous function at a point must be preserved in some open interval about that point.

Proposition 9.38   *If $f(x)$ is continuous at x_0, and $f(x_0) \neq 0$, then there is an interval about x_0, say $I = (x_0 - a, x_0 + a)$ for some $a > 0$, so that*

$$f(x_0) > 0 \Rightarrow f(x) > 0 \qquad \text{for all } x \in I,$$

$$f(x_0) < 0 \Rightarrow f(x) < 0 \qquad \text{for all } x \in I.$$

Proof   We demonstrate the result for $f(x_0) > 0$. By continuity, for $\epsilon = \frac{1}{2} f(x_0)$ say, there is a δ so that

$$|f(x) - f(x_0)| < \frac{1}{2} f(x_0) \quad \text{when} \quad |x - x_0| < \delta.$$

If this inequality is rewritten without the absolute values, it implies that

$$\frac{1}{2} f(x_0) < f(x) < \frac{3}{2} f(x_0) \qquad \text{for } x_0 - \delta < x < x_0 + \delta,$$

and this completes the proof with $a = \delta$.   ∎

The next result is that a continuous function is bounded on a compact interval, but far more important, such a function actually achieves these bounds at points **within** the interval.

Proposition 9.39 *If $f(x)$ is continuous on a closed and bounded (i.e., compact) interval $[a, b]$, then $f(x)$ attains its maximum and minimum values within this interval. That is, there are points $x^{\min}, x^{\max} \in [a, b]$, so that for all $x \in [a, b]$,*

$$f(x^{\min}) \le f(x) \le f(x^{\max}).$$

Proof Because $f(x)$ is uniformly continuous on $[a, b]$, it must be bounded from above and below. Indeed, for an arbitrary value of ϵ, there is an associated δ so that $|f(x) - f(y)| < \epsilon$ whenever $|x - y| < \delta$. This implies that the range of $f(x)$ must be contained in an interval of length $N\epsilon$, where integer $N > \frac{(b-a)}{\delta}$, since we can then cover $[a, b]$ with N intervals of length δ. Now, because $f(x)$ is bounded, there must be a greatest lower bound, and least upper bound, which we denote by L and U. By definition, we can construct two sequences $\{x_n^L\}$ and $\{x_n^U\}$, both in $[a, b]$ and so that $f(x_n^L) \to L$, and $f(x_n^U) \to U$. By the proposition above, these sequences must each have subsequences that converge to points in $[a, b]$, $x_n^L \to x^{\min}$ and $x_n^U \to x^{\max}$, and by the continuity of $f(x)$, this convergence is preserved by f so that $f(x_n^L) \to f(x^{\min})$ and $f(x_n^U) \to f(x^{\max})$. Hence, again by continuity, $L = f(x^{\min})$ and $U = f(x^{\max})$. ∎

Remark 9.40 *Note that the idea of a maximum or minimum in mathematics is different from what one may understand of these terms informally. Outside mathematics, the notion of a maximum is one of biggest, while the notion of a minimum is one of smallest. In mathematics, the term maximum simply means that there is no value of x with $f(x) > f(x^{\max})$; it does not preclude the possibility that there are many values of x with $f(x) = f(x^{\max})$, and likewise for the term minimum. While in the real world, such an interpretation is not excluded by the language, it tends to be excluded in practice. For example, the statement, "I got the maximum grade in my class on the math final" would generally not be expected to include the possibility that everyone got the same grade. In mathematics the possibility that $f(x) = f(x^{\max})$ for all x is explicitly allowed and encompassed by the notion that "$f(x)$ attains its maximum at x_0."*

The final result reinforces the intuitive notion that the graph of a continuous function must be drawn without the pencil leaving the paper, or in updated imagery, on your computer without your finger leaving the mouse button. In other words, with no holes or gaps in the graph.

Proposition 9.41 (*Intermediate Value Theorem*) *If $f(x)$ is continuous on a closed and bounded (i.e., compact) interval $[a,b]$, then $f(x)$ attains every value between its maximum and minimum values. That is, for any point y so that $f(x^{\min}) \leq y \leq f(x^{\max})$, there is a point $c \in [a,b]$ with*

$$f(c) = y. \tag{9.1}$$

Proof   Let y be given. We define $A = \{x \in [a,b] \mid f(x) \leq y\}$. Let x^A denote the least upper bound of the set A, and let $\{x_n\} \subset A$ be a sequence so that $x_n \to x^A$. Then, by continuity, $f(x_n) \to f(x^A) \leq y$. Because x^A is a least upper bound for A, it must also be the case that there is sequence $\{x'_n\} \subset \tilde{A} \equiv \{x \mid f(x) \geq y\}$ with $x'_n \to x^A$. By continuity, we have that $f(x'_n) \to f(x^A)$, and hence that $f(x^A) \geq y$. Combining, we see that $f(x^A) = y$, and the conclusion follows.   ∎

While the notion of continuity assures us what the value of $f(x_0)$ will be based on values of $f(x)$ for x "near" x_0, it provides no insight as to how quickly the value of $f(x)$ approaches this value. The notions of **Lipschitz** and **Hölder continuity** address this question next.

9.2.6   Hölder and Lipschitz Continuity

Definition 9.42   *$f(x)$ is **Hölder continuous at a given point** x_0 **of order** $\alpha > 0$ if there is a constant $C \equiv C(x_0)$ so that*

$$|f(x) - f(x_0)| \leq C|x - x_0|^\alpha. \tag{9.2}$$

*More generally, we say that $f(x)$ is **Hölder continuous of order** $\alpha > 0$ **on an interval** or simply **Hölder continuous of order** $\alpha > 0$, if it is Hölder continuous at every point of the interval or of its domain. In the special case when $\alpha = 1$, $f(x)$ is called **Lipschitz continuous** instead of **Hölder continuous of order 1**.*

Notation 9.43   *To simplify terminology, the statement that "$f(x)$ is **Hölder continuous of order** $\alpha > 0$" will be be intended to include the $\alpha = 1$ Lipschitz case.*

Lipschitz continuity is named for **Rudolf Lipschitz** (1832–1903), and Hölder continuity is named for **Otto Hölder** (1859–1937). In practice, one only considers Hölder continuity of order $\alpha \leq 1$, since the only functions that can be continuous of higher order, except at isolated points, are the constant functions: $f(x) = c$. The demonstration of this follows in the next section in two steps:

1. Once derivatives are defined and studied, we will see that such a function has a derivative that is identically 0 everywhere.

2. With the help of the mean value theorem, we will then see that the only continuous functions with an identically 0 derivative are the constant functions.

These notions of continuity can also be thought of as providing an explicit functional relationship between the ϵ and δ in the definition of continuity. Specifically, a Hölder continuous function can be defined as a continuous function for which given ϵ one can choose $\delta(\epsilon)$ by

$$\delta(\epsilon) = \left(\frac{\epsilon}{C}\right)^{1/\alpha}.$$

Knowing that a function is Hölder continuous is valuable, since this knowledge provides an explicit estimate of exactly how fast $f(x)$ converges to $f(x_0)$ in terms of the distance between x and x_0. For instance, a Lipschitz continuous function converges with speed $|\Delta x| \equiv |x - x_0|$, whereas a Hölder continuous function of order $\frac{1}{2}$ converges with speed $\sqrt{|\Delta x|}$. In general, this speed of convergence implies an approximation formula:

$$f(x_0) - C|x - x_0|^\alpha \le f(x) \le f(x_0) + C|x - x_0|^\alpha. \tag{9.3}$$

This notion of speed of convergence is formalized in mathematics in terms of "Big O" and "Little o" notation as follows.

Big O and Little o Convergence

Definition 9.44   *A function $f(x)$ is **Big O** of $g(x)$ as $x \to a$, denoted*

$$f(x) = O(g(x)) \qquad as\ x \to a,$$

if there is a $C \ne 0$ and $\delta > 0$ so that

$$\frac{|f(x)|}{|g(x)|} \le C \qquad for\ |x - a| < \delta.$$

*Similarly a function $f(x)$ is **Little o** of $g(x)$ as $x \to a$, denoted*

$$f(x) = o(g(x)) \qquad as\ x \to a,$$

if

$$\frac{|f(x)|}{|g(x)|} \to 0 \qquad as\ x \to a.$$

Remark 9.45 *In most applications in this book, we will be interested in expressing* $|\Delta f| \equiv |f(x + \Delta x) - f(x)|$ *in terms of* $g(x) = |\Delta x|^{\alpha}$. *The common language we use is,* "Δf *is* **Big O of order** α" *or* "**Little o of order** α." *Also of interest in this context is* $O(1)$, *which means* $|f(x)| \leq C$ *as* $x \to a$, *and especially* $o(1)$, *which means* $f(x) \to 0$ *as* $x \to a$.

Example 9.46 *If* $f(x)$ *is Hölder continuous at* x *of order* α, *then*

$$|\Delta f| = O(|\Delta x|^{\alpha}),$$

where $\Delta f \equiv f(x + \Delta x) - f(x)$, *but if* $f(x)$ *is simply continuous at* x, *then*

$$|\Delta f| = o(1)$$

as $\Delta x \to 0$.

Because the definition of continuity can be informally summarized by

$$|\Delta f| \to 0 \qquad \text{as } |\Delta x| \to 0,$$

it is tempting to think that every continuous function must be Hölder continuous of some order α, perhaps a value of α quite close to 0, In other words:

Question: If $|\Delta f| \to 0$, as $|\Delta x| \to 0$, must it be the case that $|\Delta f| = O(|\Delta x|^{\alpha})$ for some $\alpha > 0$?

Answer: "No." A continuous function's speed of convergence can be slower than Hölder at any order.

Example 9.47 *Consider:*

$$f(x) = \begin{cases} \frac{1}{\ln|x|}, & x \neq 0, \\ 0, & x = 0. \end{cases}$$

First off, this function is continuous at $x = 0$, *as can be seen by considering* $x_n = e^{-n}$, *for example, and evaluating* $f(x)$. *But it is not Hölder continuous of any order. This is demonstrated by considering* $x_n = e^{-n/\alpha}$ *for an arbitrary value of* $\alpha > 0$. *Then since* $f(x_n) = -\frac{\alpha}{n}$, *and* $x_n^{\alpha} = e^{-n}$, *if* $f(x)$ *was Hölder continuous of order* α, *then there would exist* $C > 0$ *so that*

$$|f(x_n)| \leq C|x_n^{\alpha}| \qquad \text{as } n \to \infty,$$

which in turn implies that

$\alpha \leq Cne^{-n} \qquad \text{as } n \to \infty.$

But since $ne^{-n} \to 0$, no such $\alpha > 0$ can exist.

It is also tempting to think that because Little o convergence is faster than Big O convergence, it must be the case that Little o implies Big O convergence at a higher order. In other words:

Question: If $|\Delta f| = o(|\Delta x|^{\alpha})$, must $|\Delta f| = O(|\Delta x|^{\alpha+\epsilon})$ for some $\epsilon > 0$?

Answer: "No." While $o(|\Delta x|^{\alpha})$ is faster than $O(|\Delta x|^{\alpha})$, it can be slower than $O(|\Delta x|^{\alpha+\epsilon})$ for any $\epsilon > 0$.

Example 9.48 *Take $g(x) = x^{\alpha} f(x)$, with $f(x)$ defined in example 9.27 above. Then the same analysis shows that at $x = 0$, $|\Delta g| = o(|\Delta x|^{\alpha})$, but that we do not have $|\Delta g| = O(|\Delta x|^{\alpha+\epsilon})$ for any $\epsilon > 0$.*

9.2.7   Convergence of a Sequence of Continuous Functions

There is another important notion related to continuity which we introduce with the following question:

Question: If $f_n(x)$ is a sequence of continuous functions, and there is a function $f(x)$ so that for every x, $f_n(x) \to f(x)$ as $n \to \infty$, must $f(x)$ be continuous?

Answer: In general, the answer is no, and this conclusion is easy to exemplify.

Example 9.49 *Define*

$$f(x) = \begin{cases} 1, & x \leq 0, \\ 0, & x > 0, \end{cases}$$

and

$$f_n(x) = \begin{cases} 1, & x \leq 0, \\ 1 - nx, & 0 < x \leq \frac{1}{n}, \\ 0, & x > \frac{1}{n}. \end{cases} \tag{9.4}$$

It is clear that $f_n(x)$ is continuous for all n, and that $f(x)$ is not continuous at $x = 0$. Also, for every x, $f_n(x) \to f(x)$ as $n \to \infty$. To understand why $f(x) \nrightarrow f(0) = 1$ as $x \to 0$, we expand for any given n,

$$f(x) - f(0) = [f(x) - f_n(x)] + [f_n(x) - f_n(0)] + [f_n(0) - f(0)].$$

As $x \to 0$, only the first term in brackets requires analysis, since by continuity of each $f_n(x)$, the second term goes to 0 for any n and the third term is identically 0. Now note that

$$f(x) - f_n(x) = \begin{cases} 0, & x \le 0, \\ nx - 1, & 0 < x \le \frac{1}{n}, \\ 0, & x > \frac{1}{n}. \end{cases}$$

In other words, although $f_n(x) \to f(x)$ for each x as $n \to \infty$, it does so increasingly slowly as $x \to 0$. That is, for any $x > 0$ we have $f_n(x) \to f(x)$ because $f_n(x) = f(x) = 0$ for $n > \frac{1}{x}$. But for $0 < x \le \frac{1}{n}$ we have

$$f(x) - f(0) = [f(x) - f_n(x)] - nx$$

$$= -1.$$

The following definition introduces an important notion of convergence that proves to give the affirmative conclusion to the question above. It will be seen that this definition eliminates the problem observed in this example, whereby the speed of convergence varies greatly with n.

Definition 9.50   *A function sequence $f_n(x)$ is said to **converge pointwise** to $f(x)$ on an interval I if for every $x \in I$, $f_n(x) \to f(x)$ as $n \to \infty$. That is, for any $\epsilon > 0$ there is an integer $N = N(x)$ so that $|f_n(x) - f(x)| < \epsilon$ for $n > N(x)$. Pointwise convergence on an arbitrary set $K \subset \mathbb{R}$ is defined similarly. A function sequence $f_n(x)$ is said to **converge uniformly** to $f(x)$ on an interval I if for any $\epsilon > 0$ there is an integer N, independent of x, so that for $x \in I$: $|f_n(x) - f(x)| < \epsilon$ for $n > N$. Uniform convergence on an arbitrary set $K \subset \mathbb{R}$ is defined similarly.*

It should be clear from the definition that uniform convergence implies pointwise convergence. Also example 9.49 provides an illustration that this implication cannot be reversed in general. In that example $f_n(x) \to f(x)$ pointwise for every $x \in \mathbb{R}$, but this convergence is not uniform. For example, with $\epsilon = \frac{1}{2}$, since $|f_n(x) - f(x)| = 1 - nx$ for $0 < x \le \frac{1}{n}$, we have that for any n, $|f_n(x) - f(x)| > \frac{1}{2}$ for $0 < x \le \frac{1}{2n}$. In other words, we cannot have $|f_n(x) - f(x)| < \epsilon$ for all n and all $|x| < \delta$ independent of how small a value of δ is chosen since for any n with $\frac{1}{2n} < \delta$, the calculations above show that $|f_n(x) - f(x)| > \frac{1}{2}$ for $0 < x < \frac{1}{2n}$.

The next result demonstrates that unlike what was seen to be the case for pointwise convergence, uniform convergence preserves continuity.

Proposition 9.51 *If $f_n(x)$ is a sequence of continuous functions that converge uniformly to $f(x)$ on an interval I, then $f(x)$ is continuous on I.*

Remark 9.52 *Note that by the proposition 9.33, if I is a closed and bounded (i.e., compact) interval, $[a,b]$, then each $f_n(x)$ is in fact uniformly continuous on $[a,b]$, and the same will be true for $f(x)$ once it is shown to be continuous.*

Proof Let $x_0 \in [a,b]$ and $\epsilon > 0$ be given. To prove that $f(x)$ is continuous at x_0, we show that there exists δ so that $|f(x) - f(x_0)| < \epsilon$ when $|x - x_0| < \delta$. To this end, let N be given as in the definition of uniform continuity to ensure that $|f_n(x) - f(x)| < \frac{\epsilon}{3}$ for all x provided that $n > N$. For any such n, let δ be the value associated with $f_n(x)$ to ensure that $|f_n(x) - f(x_0)| < \frac{\epsilon}{3}$ for $|x - x_0| < \delta$. We write

$$f(x) - f(x_0) = [f(x) - f_n(x)] + [f_n(x) - f_n(x_0)] + [f_n(x_0) - f(x_0)],$$

and by the triangle inequality, for $|x - x_0| < \delta$, we have

$$|f(x) - f(x_0)| \leq |f(x) - f_n(x)| + |f_n(x) - f_n(x_0)| + |f_n(x_0) - f(x_0)|$$

$$< \frac{\epsilon}{3} + \frac{\epsilon}{3} + \frac{\epsilon}{3} = \epsilon. \qquad \blacksquare$$

Remark 9.53 *The term "uniform" in the context of convergence is conceptually identical to the use of that term in the context of continuity.*

1. *For continuity, the general requirement is the existence of a δ for every ϵ, but in general, this δ can depend on both ϵ and the point x. What is required for uniform continuity is that δ may depend on ϵ but not the point x.*

2. *For pointwise convergence, the general requirement is the existence of an N for every ϵ, but again this N can depend on both ϵ and the point x. What is required for uniform convergence is that N may depend on ϵ but not the point x.*

The notion of "uniformity" in both contexts removes the dependency on x.

The property of uniform convergence can also be stated in terms of the Cauchy criterion, as was the case for convergence of numerical sequences in chapter 5.

Proposition 9.54 *A function sequence $f_n(x)$ converges uniformly to a function $f(x)$ on $K \subset \mathbb{R}$ if and only if for any $\epsilon > 0$ there is an integer N so that if $n, m > N$, then $|f_n(x) - f_m(x)| < \epsilon$ for all $x \in K$.*

Proof   If $f_n(x)$ converges uniformly to $f(x)$, then for $\epsilon > 0$ there is an integer N so that $|f_n(x) - f(x)| < \frac{\epsilon}{2}$ for all x provided that $n > N$. Now if $n, m > N$, we have by the triangle inequality,

$$|f_n(x) - f_m(x)| \leq |f_n(x) - f(x)| + |f(x) - f_m(x)|$$

$$< \epsilon,$$

which is the Cauchy criterion. Conversely, given the Cauchy criterion, the numerical sequence $f_n(x)$ is a Cauchy sequence by chapter 4 for every x, and hence it converges to some number for every x, which we denote by $f(x)$. Now given $\epsilon > 0$, the Cauchy criterion states that for all x, $|f_n(x) - f_m(x)| < \epsilon$ if $n, m > N$. Letting $m \to \infty$, we conclude that for all x, $|f_n(x) - f(x)| < \epsilon$ if $n > N$, and so $f_n(x) \to f(x)$ uniformly. $\blacksquare$

*Series of Functions
An important corollary to proposition 9.51 above relates to series of functions, $\sum_{j=1}^{\infty} g_j(x)$. First a definition.

Definition 9.55   *Given a sequence of functions $g_j(x)$ defined on a common interval I, and a function $g(x)$ also defined on I, the function series $\sum_{j=1}^{\infty} g_j(x)$ is said to **converge pointwise** to $g(x)$ if with $f_n(x) \equiv \sum_{j=1}^{n} g_j(x)$ for any $\epsilon > 0$ there is an integer $N = N(x)$ so that $|f_n(x) - g(x)| < \epsilon$ for $n > N(x)$. A function series $\sum_{j=1}^{\infty} g_j(x)$ is said to **converge uniformly** to $g(x)$ on an interval $J \subset I$ if for any $\epsilon > 0$ there is an integer N, independent of x, so that for $x \in J$: $|f_n(x) - g(x)| < \epsilon$ for $n > N$. Pointwise and uniform convergence of a series of functions on an arbitrary set $K \subset \mathbb{R}$ are defined analogously.*

There is an immediate application of proposition 9.51 to series of continuous functions that converge uniformly.

Proposition 9.56   *If $g_j(x)$ is a sequence of continuous functions defined on an interval I, and $\sum_{j=1}^{\infty} g_j(x)$ converges uniformly to a function $g(x)$, then $g(x)$ is continuous on I.*

Proof   Define the function sequence $f_n(x) \equiv \sum_{j=1}^{n} g_j(x)$. Then each $f_n(x)$ is continuous on I, as a finite sum of continuous functions, and $f_n(x) \to g(x)$ uniformly by assumption. Consequently the continuity of $g(x)$ follows from proposition 9.51 above. $\blacksquare$

*Interchanging Limits
There is another important consequence to proposition 9.51 that is useful in practice and relates to interchanging the order of limits. This is a manipulation that is always

dangerous in mathematics and one that needs to be approached with caution. Specifically, the question here is:

Question: If $f_n(x) \to f(x)$ for each x as $n \to \infty$, when is

$$\lim_{x \to y} \lim_{n \to \infty} f_n(x) = \lim_{n \to \infty} \lim_{x \to y} f_n(x)?$$

Partial Answer: The functions in (9.4) of example 9.49 show that pointwise convergence $f_n(x) \to f(x)$ is not enough to allow this interchange.

Example 9.57   *With $y = 0$ in example 9.49, we have that $\lim_{n \to \infty} \lim_{x \to 0} f_n(x) = 1$, while $\lim_{x \to 0} \lim_{n \to \infty} f_n(x) = \lim_{x \to 0} f(x)$ is not even defined, since this limit is 0 if approached from the right and 1 if approached from the left. In the notation introduced in definition 9.17, $\lim_{x \to 0^+} f(x) = 0$, and $\lim_{x \to 0^-} f(x) = 1$. So it appears that this example fails because $f(x)$ is not continuous at y.*

The affirmative result for interchanging limits is again provided by uniform convergence. We provide a simple result first that is often adequate in practice, and a more general result in proposition 9.59. In section 9.4 on convergence of a sequence of derivatives we will return to this question. The simple result follows immediately from the proposition above.

Proposition 9.58   *If $f_n(x)$ is a sequence of continuous functions that converge uniformly to $f(x)$ on a closed and bounded (i.e., compact) interval $[a, b]$, then for any $y \in [a, b]$,*

$$\lim_{x \to y} \lim_{n \to \infty} f_n(x) = \lim_{n \to \infty} \lim_{x \to y} f_n(x). \qquad (9.5)$$

Proof   This result is immediate from proposition 9.51 by the restatement of (9.5) which is justified by the sequential convergence and continuity assumptions, as:

$$\lim_{x \to y} f(x) = \lim_{n \to \infty} f_n(y).$$

Since $f(x)$ is continuous on $[a, b]$, $\lim_{x \to y} f(x) = f(y)$. Also, since $y \in [a, b]$, we have $\lim_{n \to \infty} f_n(y) = f(y)$.   ∎

Surprisingly, it turns out that the property of uniform convergence is so strong that it allows the interchange of limits even when the point y is outside the interval of uniform convergence as long as it is a limit point of this interval, and $\lim_{x \to y} f_n(x)$ exists for all n.

Proposition 9.59 *Let $f_n(x)$ be a sequence of continuous functions that converge uniformly to $f(x)$ on an interval I, and let $y \in \bar{I}$, the closure of I. If $\lim_{x \to y} f_n(x)$ exists for all n, then (9.5) holds.*

Proof Since this limit is assumed to exist, we define $f_n(y) \equiv \lim_{x \to y} f_n(x)$. Of course, if $y \in I$, then this definition reproduces the original value of $f_n(y)$ by continuity but otherwise extends the domain and range of $f_n(x)$ when $y \in \bar{I} \sim I$. By the Cauchy criterion for uniform convergence, we conclude that for any $\epsilon > 0$ there is an N so that for all $x \in I$,

$$|f_n(x) - f_m(x)| < \epsilon, \qquad n, m > N.$$

Also the assumption that $\lim_{x \to y} f_n(x)$ exists for all n justifies letting $x \to y$ in this inequality and yields that

$$|f_n(y) - f_m(y)| \leq \epsilon, \qquad n, m > N.$$

So $f_n(y)$ is a Cauchy numerical sequence as $n \to \infty$, and hence converges to a number by chapter 5, which is labeled $f(y)$. Note that by construction, $f(y) = \lim_{n \to \infty} \lim_{x \to y} f_n(x)$. The goal is to now show that $f(y) = \lim_{x \to y} \lim_{n \to \infty} f_n(x) = \lim_{x \to y} f(x)$. To do this, note that for $x \in I$, by the triangle inequality,

$$|f(x) - f(y)| \leq |f(x) - f_n(x)| + |f_n(x) - f_n(y)| + |f_n(y) - f(y)|.$$

This summation can be made small for n large enough, since for $\epsilon > 0$ given above and the various definitions of convergence:

1. $f_n(x) \to f(x)$ uniformly for $x \in I$, means there is an N_1 so that $|f(x) - f_n(x)| < \epsilon$ for all x for $n > N_1$.

2. $f_n(y) \to f(y)$, means there is an N_2 so that $|f_n(y) - f(y)| < \epsilon$ for $n > N_2$.

3. $f_n(x) \to f_n(y)$ for any n, means there is δ_n so that $|x - y| < \delta_n$ implies that $|f_n(x) - f_n(y)| < \epsilon$.

Combining, we conclude that for $N' = \max(N_1, N_2)$, and $|x - y| < \delta_{N'}$ that

$$|f(x) - f(y)| < 3\epsilon.$$

In other words, $f(y) = \lim_{x \to y} f(x)$. ∎

Remark 9.60 *In the example of $I = (a, b)$, the proposition 9.59 result states that if $\lim_{x \to a} f_n(x) \equiv f_n(a)$ exists for all n, then uniform convergence on I gives more information about what happens at a. This result assures that it must be the case that:*

1. $\lim_{n \to \infty} f_n(a)$ exists,

2. $\lim_{x \to a} f(x)$ exists, and

3. $\lim_{n \to \infty} f_n(a) = \lim_{x \to a} f(x)$.

*9.2.8   Continuity and Topology

In addition to the interpretation that the continuity of a function implies metric properties—that $f(x)$ can be made arbitrarily close to $f(x_0)$ by choosing x close to x_0—continuity also has topological implications. That is, continuous functions have predictable behaviors on open, closed, connected, and compact sets.

Remark 9.61   *In the statement and proof below, recall that $f^{-1}(A)$, the pre-image of a set A under f, is defined even if f is not one-to-one, which is to say, even when f^{-1} is not defined as a function. Specifically, $f^{-1}(A) = \{x \mid f(x) \in A\}$.*

Proposition 9.62   *If $f(x)$ is a continuous function, $f : \mathbb{R} \to \mathbb{R}$, then:*

1. *$f^{-1}(G)$ is open for every open set $G \subset \mathbb{R}$*

2. *$f^{-1}(F)$ is closed for every closed set $F \subset \mathbb{R}$*

3. *$f(C)$ is connected for every connected set $C \subset \mathbb{R}$*

4. *$f(K)$ is compact for every compact set $K \subset \mathbb{R}$*

Proof

1. Given $G \subset \mathbb{R}$ open, to show that $f^{-1}(G)$ is open is to show that for any $x_0 \in f^{-1}(G)$, there is an open ball about x_0, $B_r(x_0)$, with $B_r(x_0) \subset f^{-1}(G)$. Now since G is open, there is a ball about $f(x_0)$ contained in G. That is, for some $\epsilon > 0$ we have $B_\epsilon(f(x_0)) \subset G$. Given ϵ, by the continuity of f there is a $\delta > 0$ so that $|x - x_0| < \delta$ implies that $|f(x) - f(x_0)| < \epsilon$. That is, $f(B_\delta(x_0)) \subset B_\epsilon(f(x_0))$. Consequently $B_\delta(x_0) \subset f^{-1}(G)$, and hence $f^{-1}(G)$ is open.

2. Given $F \subset \mathbb{R}$ closed, the complement of F: $\tilde{F} \equiv \mathbb{R} \sim F$, is open, so by 1, $f^{-1}(\tilde{F})$ is also open. Consequently $\widetilde{f^{-1}(\tilde{F})}$ is closed. The final step is to show that $\widetilde{f^{-1}(\tilde{F})} = f^{-1}(F)$. The proof of the equivalent statement that $\widetilde{f^{-1}(\tilde{F})} = \widetilde{f^{-1}(F)}$, for an arbitrary set F, is left as exercise 31.

3. We argue by contradiction. Suppose that $C \subset \mathbb{R}$ is connected but that $f(C)$ is not. Then there are open sets G_1 and G_2 so that $f(C) \subset G_1 \cup G_2$ yet $G_1 \cap G_2 = \emptyset$. Now, by definition, $C \subset f^{-1}(G_1 \cup G_2)$, but also $f^{-1}(G_1 \cup G_2) = f^{-1}(G_1) \cup f^{-1}(G_2)$ as is easily demonstrated. However, $G_1 \cap G_2 = \emptyset$, implies that $f^{-1}(G_1) \cap f^{-1}(G_2) = \emptyset$, and by part 1, both $f^{-1}(G_1)$ and $f^{-1}(G_2)$ are open, contradicting the assumption that C is connected.

4. Assume that $K \subset \mathbb{R}$ is compact, and let $\{G_\alpha\}$ be an open cover of $f(K)$. That is, $f(K) \subset \bigcup G_\alpha$. We need to show that there is a finite subcollection $\{G_j\}_{j=1}^n \subset \{G_\alpha\}$ so that $f(K) \subset \bigcup_{j=1}^n G_j$. Now by part 1, $\{f^{-1}(G_\alpha)\}$ is an open cover of K, and since K is compact, there is a finite subcover $K \subset \bigcup_{j=1}^n f^{-1}(G_j)$. Hence $f(K) \subset \bigcup_{j=1}^n G_j$, demonstrating that $f(K)$ is compact. ∎

Remark 9.63 *Note that parts 1 and 2 in this proposition can be stated in terms of "if and only if," and not just as an implication. In other words, a function is continuous if and only if $f^{-1}(G)$ is open for every open set G, or equivalently $f^{-1}(F)$ is closed for every closed set F. For example, if $f^{-1}(G)$ is open for every open set G, then for $f(x_0) \in G$ there is an open ball $B_\epsilon(f(x_0)) \subset G$, and by assumption, $f^{-1}(B_\epsilon(f(x_0)))$ is an open set that contains x_0. So, by definition, there is an open ball $B_\delta(x_0) \subset f^{-1}(B_\epsilon(f(x_0)))$, which means that $f(B_\delta(x_0)) \subset B_\epsilon(f(x_0))$, and these are the ϵ and δ needed for the definition of continuity.*

The importance of this observation is that it motivates the definition of continuous function on, or between, general topological spaces.

Definition 9.64 *If $f : X \to Y$ is a function defined on a topological space X, and taking values in a topological space Y, then we define f to be **continuous** if $f^{-1}(G)$ is open in X for all G open in Y.*

The proposition 9.62 result on preserving openness is explicitly related to the **inverse** of a continuous function, as it is not true in general that a continuous function itself will preserve openness. As an example of G open but $f(G)$ closed:

Example 9.65 *Consider the function: $f(x) = x^2(x^2 - 2)$ in **figure 9.3**. It is clear from the graph that $f(G)$ need not be open when G is open. For instance, if $G = (-a, a)$ for any $1 < a \leq \sqrt{2}$, then $f(G) = [-1, 0]$.*

It is also the case that in general, F closed does not imply that $f(F)$ is closed. However, from part 4 of the proposition above, such an example would have to be one for which the set F is closed and unbounded. This is because if F is closed and bounded it is compact by the Heine–Borel theorem, and hence so too is $f(F)$ by part 4. But in a metric space, compact means closed and bounded, and so $f(F)$ must then also be closed and bounded.

Example 9.66 *The classic example of F closed and unbounded and $f(F)$ not closed is $F = \{-n\pi \mid n = 0, 1, 2, 3, \ldots\}$ and the continuous function $f(x) = e^x \cos x$. Of course, since the complement of F is the union of open intervals, F is clearly closed. However, $f(F)$ is seen to equal $\{(-1)^n e^{-n\pi} \mid n = 0, 1, 2, 3, \ldots\}$, since $\cos(-n\pi) = (-1)^n$. The set $f(F)$ is not closed because a closed set must contain all of its limit points. However, $x = 0$, is apparently a limit point of this set but not an element of this set.*

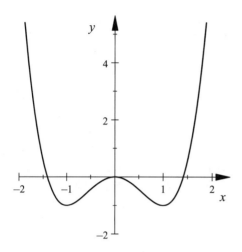

Figure 9.3
$f(x) = x^2(x^2 - 2)$

Note that in each of these counterexamples the given function $f(x)$ was seen to be a many-to-one function. This was necessary because, for one-to-one continuous functions, all the statements of the proposition above generalize:

Proposition 9.67 *If $f(x)$ is a continuous one-to-one function, $f : \mathbb{R} \to \mathbb{R}$, then:*

1. $f(G)$ *is open for every open set $G \subset \mathbb{R}$.*
2. $f(F)$ *is closed for every closed set $F \subset \mathbb{R}$.*
3. $f^{-1}(C)$ *is connected for every connected set $C \subset \mathbb{R}$.*
4. $f^{-1}(K)$ *is compact for every compact set $K \subset \mathbb{R}$.*

Proof The proof follows from the fact that because $f(x)$ is a continuous one-to-one function, $f^{-1}(x)$ is also continuous by proposition 9.26, and hence we can apply proposition 9.62. ∎

9.3   Derivatives and Taylor Series

9.3.1   Improving an Approximation I

In the preceding section various notions of continuity were reviewed and their properties discussed. To motivate the discussion of this section, we begin with an informal attempt to improve upon the definition of continuity in terms of its implication for approximating a function's values. Recall that if $f(x)$ is continuous at x_0, then $f(x)$ can be approximated by $f(x_0)$ for x "near" x_0. In the case of Hölder continuity, we can even determine the order of magnitude of this error as seen in (9.3).

Furthering this investigation, it is natural to inquire into the approximation of $f(x)$ near x_0, not simply by a constant $f(x_0)$ but instead by a "linear" term that varies proportionally with $\Delta x = x - x_0$:

$$f(x) \approx f(x_0) + a\Delta x,$$

where a is a constant.

To be effective as an approximation tool, we require that the error in this approximation goes to 0 as $\Delta x \to 0$. That is, at the minimum, we require that

$$f(x) - [f(x_0) + a\Delta x] \to 0 \qquad \text{as } \Delta x \to 0,$$

or equivalently

$$f(x) - f(x_0) - a\Delta x = o(1) \qquad \text{as } \Delta x \to 0.$$

Here we recall definition 9.44 that $o(1)$ means that this expression converges to 0 as $\Delta x \to 0$.

However, a moment of thought reveals the weakness in this idea. Namely, if $f(x)$ is continuous at x_0, the minimal requirement above is satisfied for any constant a, so we have gained nothing with the addition of the extra term of $a\Delta x$ in the approximation. This approximation would be an improvement, however, if the error term could somehow be changed from $o(1)$ to $o(\Delta x)$.

To this end, we rewrite:

$$f(x) - f(x_0) - a\Delta x \equiv \left(\frac{f(x) - f(x_0)}{\Delta x} - a \right)\Delta x.$$

In order for this expression to go to 0 in a way that supports better approximations, and provides a method of determining the appropriate value of a, we require that

$$\frac{f(x) - f(x_0)}{\Delta x} - a = o(1) \qquad \text{as } \Delta x \to 0. \qquad (9.6)$$

Then, by recognizing the extra Δx term above and recalling that $o(1)\Delta x = o(\Delta x)$, we see that we can improve the approximation of $f(x)$ for x near x_0 by the resulting value of a, and that for this value

$$f(x) - f(x_0) - a\Delta x = o(\Delta x). \qquad (9.7)$$

In other words, if the limit in (9.6) exists, we can dramatically improve our ability to approximate from the case of general continuity,

$$f(x) - f(x_0) \to 0 \qquad \text{as } x \to x_0,$$

with no information on speed of convergence, to (9.7). This tells us that the convergence $f(x) \to f(x_0)$ is $O(\Delta x)$, and once we account for the linear term $a\Delta x$, we achieve an approximation and convergence that is in fact $o(\Delta x)$. This discussion motivates the following development.

9.3.2   The First Derivative

We formalize in a definition the condition required in (9.6).

Definition 9.68 $f(x)$ is **differentiable at** x_0, or has a **first derivative at** x_0, denoted $f'(x_0)$, or $\frac{df}{dx}\big|_{x=x_0}$, if the following limit exists:

$$f'(x_0) = \lim_{\Delta x \to 0} \frac{f(x_0 + \Delta x) - f(x_0)}{\Delta x}. \qquad (9.8)$$

Similarly $f(x)$ is **differentiable on an open interval** $(a,b) \equiv \{x \,|\, a < x < b\}$, or has a **first derivative everywhere on** (a,b), if the limit in (9.8) exists for all $x_0 \in (a,b)$.

Remark 9.69

1. The ratio $\frac{f(x_0 + \Delta x) - f(x_0)}{\Delta x}$ represents the slope of the **secant line** between the points $(x_0, f(x_0))$ and $(x_0 + \Delta x, f(x_0 + \Delta x))$, on the graph of $y = f(x)$. Consequently, as $\Delta x \to 0$, the derivative can be interpreted as the slope of the **tangent line** to the graph of $y = f(x)$ at the point $(x_0, f(x_0))$. The equation of this tangent line, which can be used to approximate $f(x)$ for x near x_0, is then

$$y = f(x_0) + f'(x_0)(x - x_0). \qquad (9.9)$$

2. One can introduce the notion of a **one-sided derivative** at the endpoints of a closed interval $[a,b]$, by restricting the limit in (9.8) to $\lim_{\Delta x \to 0^+}$ for $f'(a)$, or $\lim_{\Delta x \to 0^-}$ for $f'(b)$. In general, however, most of our applications will relate to the standard two-sided limit.

From the earlier discussion in section 9.3.1, it should be clear that there is an alternative way to define the notion that $f(x)$ is **differentiable at** x_0 that avoids the sometimes troublesome division by Δx and can be easier to apply in derivations to come. Specifically:

Definition 9.70   *$f(x)$ is **differentiable at** x_0 if there is number $f'(x_0)$ and an "error" function $\varepsilon_f(x_0 + \Delta x)$ with $\varepsilon_f(x_0 + \Delta x) \to 0$ as $\Delta x \to 0$, and for which*

$$f(x_0 + \Delta x) - f(x_0) = \Delta x(f'(x_0) + \varepsilon_f(x_0 + \Delta x)). \tag{9.10}$$

That this definition is equivalent to the former follows from the observation that the limit in (9.8) means that for any given $\Delta x \neq 0$, we have that $\frac{f(x_0 + \Delta x) - f(x_0)}{\Delta x} = f'(x_0) + error$. This error term, denoted $\varepsilon_f(x_0 + \Delta x)$ in (9.10), must converge to 0 as $\Delta x \to 0$.

Example 9.71

1. *If $f(x) = c$, a constant, then trivially, $f'(x) = 0$. Not so obviously, but as was noted in section 9.2.6, constant functions are the only continuous functions with this property.*

2. *One easily derives that for any positive integer n, $f(x) = x^n$ is differentiable, and*

$$\left. \frac{dx^n}{dx} \right|_{x=x_0} = nx_0^{n-1}. \tag{9.11}$$

This result is immediate for $n = 1$ by the definition, while for $n \geq 2$ one derives this from the binomial formula:

$$(x + \Delta x)^n = x^n + nx^{n-1}\Delta x + O(\Delta x^2).$$

3. *The absolute value function $f(x) = |x|$ is differentiable for $x \neq 0$. We obtain, by definition,*

$$f'(x) = \begin{cases} 1, & x > 0, \\ -1, & x < 0. \end{cases}$$

The absolute value function is not differentiable at $x = 0$ because the limit in (9.8) produces $+1$ when $\Delta x > 0$, and -1 when $\Delta x < 0$.

From (9.8) we derive the following:

Proposition 9.72   *If $f(x)$ is differentiable at x_0, then it is continuous there. Moreover $f(x)$ is Lipschitz continuous at x_0.*

Proof   From (9.8), as $\Delta x \to 0$,

$$f(x_0 + \Delta x) - f(x_0) = \Delta x \frac{f(x_0 + \Delta x) - f(x_0)}{\Delta x}$$

$$\to 0 \cdot f'(x_0) = 0,$$

so $f(x)$ is continuous. This derivation also shows that

$$f(x_0 + \Delta x) - f(x_0) = O(\Delta x) \qquad \text{as } \Delta x \to 0,$$

so $f(x)$ is Lipschitz continuous. ∎

Remark 9.73 *The converse of this proposition is false because Lipschitz continuity simply requires that with $x \equiv x_0 + \Delta x$,*

$$\left| \frac{f(x) - f(x_0)}{\Delta x} \right| \leq C \qquad \text{as } \Delta x \to 0.$$

Lipschitz continuity does not require that this ratio converge to a limit. The simplest example of this is the next:

Example 9.74 $f(x) = |x|$ *is Lipschitz continuous at $x = 0$ but not differentiable there, since the left- and right-sided limits produced by (9.8) are -1 and $+1$, respectively as noted in example 9.71.*

9.3.3 Calculating Derivatives

Demonstrating that complicated functions are differentiable, and finding their derivatives, can be difficult and tedious based on the definitions above. The following three results provide a systematic approach to verifying differentiability and determining derivatives of many common functions.

Proposition 9.75 *If $f(x)$ and $g(x)$ are differentiable at x_0, then so too is:*

1. $h(x) = af(x) \pm bg(x)$, with $h'(x_0) = af'(x_0) \pm bg'(x_0)$
2. $h(x) = f(x)g(x)$, with $h'(x_0) = f'(x_0)g(x_0) + f(x_0)g'(x_0)$
3. $h(x) = \frac{1}{g(x)}$ if $g(x_0) \neq 0$, with $h'(x_0) = \frac{-g'(x_0)}{g^2(x_0)}$
4. $h(x) = \frac{f(x)}{g(x)}$ if $g(x_0) \neq 0$, with $h'(x_0) = \frac{f'(x_0)g(x_0) - f(x_0)g'(x_0)}{g^2(x_0)}$

Proof See exercises 6 and 32. See also exercise 34 for a generalization of 2 known as the **Leibniz rule**, which is reminiscent of the binomial theorem. ∎

The next two results are more subtle, so we provide details of the proofs.

Proposition 9.76 *If $g(x)$ is differentiable at x_0 and $f(x)$ is differentiable at $g(x_0)$, then so too is*

5. $h(x) = f(g(x))$ at x_0, with $h'(x_0) = f'(g(x_0))g'(x_0)$

Proof Note that if $g(x)$ is differentiable at x_0 and $f(x)$ is differentiable at $y_0 = g(x_0)$, then from (9.10),

$$g(x_0 + \Delta x) - g(x_0) = \Delta x(g'(x_0) + \varepsilon_g(x_0 + \Delta x)),$$

$$f(y_0 + \Delta y) - f(y_0) = \Delta y(f'(y_0) + \varepsilon_f(y_0 + \Delta y)).$$

Consequently, noting that $y_0 + \Delta y = g(x_0 + \Delta x)$, we write

$$h(x_0 + \Delta x) - h(x_0) = f(g(x_0 + \Delta x)) - f(g(x_0))$$

$$= [g(x_0 + \Delta x) - g(x_0)][f'(g(x_0)) + \varepsilon_f(g(x_0 + \Delta x))]$$

$$= \Delta x[g'(x_0) + \varepsilon_g(x_0 + \Delta x)][f'(g(x_0)) + \varepsilon_f(g(x_0 + \Delta x))].$$

By definition that $g(x)$ is differentiable at x_0, $\varepsilon_g(x_0 + \Delta x) \to 0$ as $\Delta x \to 0$. Also $\varepsilon_f(y_0 + \Delta y) \to 0$ as $\Delta y \to 0$, but since $\Delta y = g(x_0 + \Delta x) - g(x_0)$, we have by the continuity of $g(x)$ that $\Delta y \to 0$ as $\Delta x \to 0$. Multiplying out the final expression, we derive with a notational change

$$h(x_0 + \Delta x) - h(x_0) = \Delta x[f'(g(x_0))g'(x_0) + \varepsilon_h(x_0 + \Delta x)],$$

where $\varepsilon_h(x_0 + \Delta x) \to 0$ as $\Delta x \to 0$, with the error term given by

$$\varepsilon_h(x_0 + \Delta x) = g'(x_0)\varepsilon_f(g(x_0 + \Delta x)) + f'(g(x_0))\varepsilon_g(x_0 + \Delta x)$$

$$+ \varepsilon_g(x_0 + \Delta x)\varepsilon_f(g(x_0 + \Delta x)).$$

Hence $h(x)$ is differentiable by (9.10).   ∎

Proposition 9.77 *If $g(x)$ is differentiable at x_0, $g'(x_0) \neq 0$, and $g'(x)$ is continuous on an interval about x_0, then*

6. *$h(y) = g^{-1}(y)$ is differentiable at $y_0 = g(x_0)$, with $h'(y_0) = \frac{1}{g'(x_0)}$*

Remark 9.78 *Note that we do not explicitly assume that $g(x)$ is one-to-one, or even one-to-one "near" x_0. While this result may appear odd, since we require the existence of $g^{-1}(y)$ "near" y_0 so that its derivative there is well defined, this requirement on $g(x)$ is assured by the assumption that $g'(x_0) \neq 0$ and the continuity of $g'(x)$ (see exercise 7).*

Proof From (9.10), we need to show that if $g'(x_0) \neq 0$,

$$g^{-1}(y_0 + \Delta y) - g^{-1}(y_0) = \Delta y\left(\frac{1}{g'(x_0)} + \varepsilon_{g^{-1}}(y_0 + \Delta y)\right)$$

for some error function with $\varepsilon_{g^{-1}}(y_0 + \Delta y) \to 0$ as $\Delta y \to 0$. Now, if $g^{-1}(y_0) \equiv x_0$, and $g^{-1}(y_0 + \Delta y) \equiv x_0 + \Delta x$, then $\Delta y = g(x_0 + \Delta x) - g(x_0)$, and the equation above is notationally equivalent to showing that

$$\Delta x = [g(x_0 + \Delta x) - g(x_0)]\left(\frac{1}{g'(x_0)} + \varepsilon_{g^{-1}}(g(x_0 + \Delta x))\right).$$

This in turn is equivalent to

$$g(x_0 + \Delta x) - g(x_0) = \frac{\Delta x g'(x_0)}{1 + g'(x_0)\varepsilon_{g^{-1}}(g(x_0 + \Delta x))}$$

$$= \Delta x(g'(x_0) + \tilde{\varepsilon}_{g^{-1}}(g(x_0 + \Delta x))),$$

where with some algebra, we can derive

$$\tilde{\varepsilon}_{g^{-1}}(g(x_0 + \Delta x)) \equiv -\frac{[g'(x_0)]^2 \varepsilon_{g^{-1}}(g(x_0 + \Delta x))}{1 + g'(x_0)\varepsilon_{g^{-1}}(g(x_0 + \Delta x))}.$$

Now, by the differentiability of $g(x)$ at x_0, we have that there is an $\varepsilon_g(x_0 + \Delta x)$ so that

$$g(x_0 + \Delta x) - g(x_0) = \Delta x(g'(x_0) + \varepsilon_g(x_0 + \Delta x)).$$

Comparing expressions, we will be done if we can solve

$$\varepsilon_g(x_0 + \Delta x) = -\frac{[g'(x_0)]^2 \varepsilon_{g^{-1}}(g(x_0 + \Delta x))}{1 + g'(x_0)\varepsilon_{g^{-1}}(g(x_0 + \Delta x))}$$

for the needed error function, $\varepsilon_{g^{-1}}(g(x_0 + \Delta x))$, and demonstrate that it has the right properties. A bit of algebra yields

$$\varepsilon_{g^{-1}}(g(x_0 + \Delta x)) = \frac{-\varepsilon_g(x_0 + \Delta x)}{[g'(x_0)]^2 + g'(x_0)\varepsilon_g(x_0 + \Delta x)}.$$

Finally, as $\Delta y \equiv g(x_0 + \Delta x) - g(x_0) \to 0$, we can conclude that $\Delta x \to 0$ because of the one-to-oneness assured by exercise 7. Hence as $\Delta y \to 0$, we have that $\varepsilon_g(x_0 + \Delta x) \to 0$ and also $\varepsilon_{g^{-1}}(g(x_0 + \Delta x)) = \varepsilon_{g^{-1}}(y_0 + \Delta y) \to 0$, and the proof is complete.   ∎

Remark 9.79   *After the somewhat detailed proof of the derivative of the inverse function, here is a really easy proof—provided that $h(y) = g^{-1}(y)$ is explicitly assumed to* *be one-to-one near y_0 and differentiable at y_0. Since the composition $g(h(y))$ is the simple function $g(h(y)) = y$, we can take the derivative of both sides using the composition formula in property 5 of proposition 9.76 above, evaluated at y_0, to obtain*

$$g'(h(y_0))h'(y_0) = 1.$$

The conclusion follows with $h(y_0) = x_0$. Now exercise 7 demonstrates that $g^{-1}(y)$ is one-to-one near y_0, but there is no easy way to demonstrate that $g^{-1}(y)$ is differentiable at y_0 without the added details of the proof above.

Example 9.80   *Some examples of the wide applicability of these propositions are:*

1. *From (9.11) and property 1 above, one easily finds the derivative of any polynomial function, while with property 4, one finds the derivative of any rational function, which is a ratio of polynomials, at points for which the denominator polynomial is nonzero. Similarly one finds the derivative of various composites of polynomial and rational functions using property 5. In addition, 6 is useful in generalizing (9.11) from positive integers to rationals of the form $\frac{1}{m}$, since $g(y) = y^{1/m}$ is inverse to $f(x) = x^m$, which with properties 5 and 3 can be further generalized to all rational number exponents (positive or negative) of the form $\frac{n}{m}$. For these non-integer rational exponents, the domains of the functions are restricted to $x \geq 0$ for positive exponents, and $x > 0$ for negative exponents.*

As a specific case,

If $f(x) = \displaystyle\sum_{i=0}^{n} a_i x^i$, then $f'(x) = \displaystyle\sum_{i=1}^{n} i a_i x^{i-1}$,

since the derivative of the constant a_0 is zero. Similarly, with $f(x)$ as above, and $g(x) = x^q$ for q rational, define the function $h(x) \equiv g(f(x))$:

If $h(x) \equiv \left[\displaystyle\sum_{i=0}^{n} a_i x^i \right]^q$, then $h'(x) = q \left[\displaystyle\sum_{i=0}^{n} a_i x^i \right]^{q-1} \displaystyle\sum_{i=1}^{n} i a_i x^{i-1}$.

2. *However, these formulas do not confirm differentiability, nor provide the derivative of the exponential functions $f(x) = a^x$ for $a > 1$. In exercise 8 it is noted that*

$$\frac{da^x}{dx} = a^x \ln a, \qquad a > 1, \tag{9.12}$$

is a corollary of the formula for the natural exponential:

$$\frac{de^x}{dx} = e^x. \tag{9.13}$$

For this latter formula it is easy to see that

$$\frac{f(x + \Delta x) - f(x)}{\Delta x} = a^x \frac{a^{\Delta x} - 1}{\Delta x},$$

and the base of the "natural exponential," e, can be defined as the real number that satisfies

$$\lim_{\Delta x \to 0} \frac{e^{\Delta x} - 1}{\Delta x} = 1, \tag{9.14}$$

from which (9.13) follows immediately. That there exists such a number e that satisfies the limit in (9.14) is not apparent, but this numerical value can be expressed in an equivalent way as in (9.19) below, and shown to exist by direct arguments (see case 7 below and the following section).

For a derivative example with $f(x)$ as in case 1, and $g(x) = e^x$, define the function $h(x) \equiv g(f(x))$:

If $h(x) = e^{\sum_{i=0}^{n} a_i x^i}$, then $h'(x) = \left(e^{\sum_{i=0}^{n} a_i x^i}\right) \sum_{i=1}^{n} i a_i x^{i-1}$.

3. *The natural exponential provides a basis for extending (9.11) to any real number exponent. That is, for any real number r, $g(x) \equiv x^r$ can be defined by $g(x) = e^{r \ln x}$ on the domain $x > 0$. Applying (9.13) and property 5 in the proposition, we get $g'(x) = \frac{r}{x} e^{r \ln x} = \frac{r}{x} x^r = r x^{r-1}$. In other words,*

If $g(x) = x^r$, $x > 0$, $r \in \mathbb{R}$, then $g'(x) = r x^{r-1}$. \tag{9.15}

4. *Let $f(x) = e^{ix}$, where $i = \sqrt{-1}$. We have from **Euler's formula** in (2.5) that*

$$e^{ix} = \cos x + i \sin x.$$

Now, if $b \in \mathbb{R}$ and $g(x) = e^{bx} = (e^b)^x$, then from (9.12) we derive $g'(x) = b e^{bx}$. This formula also turns out to be true for $b \in \mathbb{C}$, but we do not prove this since it is not essential to this book's goals. But this fact allows an easy derivation of the derivatives of $\sin x$ and $\cos x$. Namely

$$i e^{ix} = \frac{de^{ix}}{dx} = \frac{d \cos x}{dx} + i \frac{d \sin x}{dx},$$

but also

$$ie^{ix} = -\sin x + i \cos x.$$

Comparing, we derive (with a bit of cheating that the derivative formula above is valid for $b \in \mathbb{C}$):

$$\frac{d \sin x}{dx} = \cos x, \quad \frac{d \cos x}{dx} = -\sin x. \tag{9.16}$$

Remark 9.81   *To make these ideas rigorous, we must first derive (9.16) directly from the definition of $f'(x)$ using trigonometric identities. These formulas imply each function is infinitely differentiable (see definition 9.91). From the methods of Taylor series used below, it turns out that e^x, $\sin x$, and $\cos x$ are each analytic and have convergent series representations. The function e^{ix}, or generally e^{ibx} for $b \in \mathbb{R}$, can then be defined in terms of the Taylor series expansion for e^x by substitution, and shown to be absolutely convergent. Moreover, if $c \in \mathbb{C}$, $c = a + bi$, then define $e^{cx} = e^{ax}e^{ibx}$. Finally, the associated Taylor series for e^{ibx}, $\sin bx$ and $\cos bx$, can be shown to satisfy*

$$e^{ibx} = \cos bx + i \sin bx,$$

which for $b = 1$ is Euler's formula.

5. *Because $f(y) = \ln y$, defined on $y > 0$, is the inverse function of $g(x) = e^x$ defined on $\mathbb{R}$, we can apply property 6 in the proposition above to conclude that*

$$\frac{d \ln y}{dy} = \frac{1}{y}. \tag{9.17}$$

Also, since $\log_a y = \frac{1}{\ln a} \ln y$ for $a > 1$, we obtain from property 1 of the proposition, since $\frac{1}{\ln a}$ is a constant,

$$\frac{d \log_a y}{dy} = \frac{1}{y \ln a}. \tag{9.18}$$

6. *With the formula for the derivative of $\ln x$, we are now in the position to clarify a couple of limits that were used in the chapter 7 development of the Poisson distribution. Specifically, we need to show that for any real number λ and constant k,*

$$\left(1 - \frac{\lambda}{n} + \frac{k}{n^2}\right)^n \to e^{-\lambda} \qquad as \; n \to \infty.$$

Taking natural logarithms, this is equivalent to showing that

$$n \ln\left(1 - \frac{\lambda}{n} + \frac{k}{n^2}\right) \to -\lambda \qquad \text{as } n \to \infty.$$

Consider the function $f(x) = \ln(1 - \lambda x + kx^2)$, which is differentiable for $1 - \lambda x + kx^2 > 0$, and this in turn is valid for any choice of constants for x close enough to 0. In particular, $f(x)$ is differentiable at $x = 0$, and from the development above we have that $f'(x) = \frac{-\lambda + 2kx}{1 - \lambda x + kx^2}$, and so $f'(0) = -\lambda$. Applying the formula for the derivative $f'(0)$, and observing that $f(0) = 0$, we have

$$-\lambda = \lim_{\Delta x \to 0} \frac{f(\Delta x)}{\Delta x}.$$

Finally, substituting $\Delta x = \frac{1}{n}$ and letting $n \to \infty$ completes the derivation.

7. *A simple yet elegant corollary to case 6 is the following definition of e, obtained with $k = 0$ and $\lambda = -1$:*

$$e = \lim_{n \to \infty} \left(1 + \frac{1}{n}\right)^n, \tag{9.19}$$

which also follows from (9.14) by setting $\Delta x = \frac{1}{n}$ and letting $n \to \infty$.

Remark 9.82   *Obviously, to avoid circular logic, one of cases 2, 5, 6, and 7 of example 9.80 must be independently derived, and the others then follow. The usual approach, as noted above, is to first establish the limit in (9.19) directly by analysis of the sequence $a_n = \left(1 + \frac{1}{n}\right)^n$ (see the following section). From this the limit in (9.14) and differentiability of e^x and a^x follow, as then does the differentiability of $\ln x$ and $\log_a x$, and then finally the limits in case 6 above.*

8. *As noted above, $f(x) = |x|$ is differentiable everywhere except for $x = 0$. However, if $p > 1$, the function $g(x) = |x|^p$ is differentiable everywhere. This follows from noting that since*

$$g(x) = \begin{cases} x^p, & x \geq 0, \\ (-x)^p, & x \leq 0, \end{cases}$$

we can apply (9.15) in example 9.80 to produce for $x \neq 0$,

$$g'(x) = \begin{cases} px^{p-1}, & x > 0, \\ -p(-x)^{p-1}, & x < 0. \end{cases}$$

For x = 0,

$$\frac{g(\Delta x) - g(0)}{\Delta x} = \begin{cases} (\Delta x)^{p-1}, & \Delta x > 0, \\ -|\Delta x|^{p-1}, & \Delta x < 0, \end{cases}$$

and hence $g'(0) = 0$. Combining, we obtain the result:
  If $g(x) = |x|^p$, $p > 1$, then

$$g'(x) = \begin{cases} p|x|^{p-1}, & x \geq 0, \\ -p|x|^{p-1}, & x \leq 0. \end{cases} \tag{9.20}$$

A Discussion of e

The simplest approach to deriving the numerical value of e involves two steps:

Step 1. Define e by

$$e = \sum_{n=0}^{\infty} \frac{1}{n!}.$$

That this summation converges follows directly from chapter 6 and the ratio test. Since $b_n = \frac{1}{n!}$, we see that as $n \to \infty$,

$$\frac{b_{n+1}}{b_n} = \frac{1}{n+1} \to 0.$$

It is also apparent that $\frac{1}{n!} \leq \frac{1}{2^{n-1}}$ for $n \geq 1$, so by evaluation of the geometric series,

$$e = 1 + \sum_{n=1}^{\infty} \frac{1}{n!}$$

$$\leq 1 + \sum_{n=0}^{\infty} \frac{1}{2^n} = 3.$$

In fact

$$e \approx 2.718281828459\ldots. \tag{9.21}$$

Step 2. Define $a_n = \left(1 + \frac{1}{n}\right)^n$ as in (9.19) of case 7 of example 9.80 above. We now show that $a_n \to e$. By the binomial theorem,

$$a_n = \sum_{j=0}^{n} \binom{n}{j} \frac{1}{n^j}$$

$$= 1 + \sum_{j=1}^{n} \left[\prod_{k=0}^{j-1} \left(1 - \frac{k}{n} \right) \right] \frac{1}{j!}.$$

From this result we conclude that since $\prod_{k=0}^{j-1} \left(1 - \frac{k}{n} \right) \leq 1$,

$$a_n \leq e_n < e,$$

where $e_n = \sum_{j=0}^{n} \frac{1}{j!}$ is the partial sum that converges to e above. It is also apparent that

$$a_n < a_{n+1},$$

since a_{n+1} has one more positive term in the summation above, and for the other terms, the coefficients of $\left\{ \frac{1}{j!} \right\}_{j=1}^{n}$ increase from $\prod_{k=0}^{j-1} \left(1 - \frac{k}{n} \right)$ to $\prod_{k=0}^{j-1} \left(1 - \frac{k}{n+1} \right)$. Because a_n is an increasing sequence and is bounded above by e, this sequence converges by chapter 5 to a say, where $a \leq e$. To see that $a = e$, note that for $m > n$,

$$a_m = 1 + \sum_{j=1}^{m} \left[\prod_{k=0}^{j-1} \left(1 - \frac{k}{m} \right) \right] \frac{1}{j!}$$

$$> 1 + \sum_{j=1}^{n} \left[\prod_{k=0}^{j-1} \left(1 - \frac{k}{m} \right) \right] \frac{1}{j!}.$$

Letting $m \to \infty$, we conclude that since $a_m \to a$ and $\prod_{k=0}^{j-1} \left(1 - \frac{k}{m} \right) \to 1$,

$$a \geq e_n.$$

Combining, we have

$$a_n \leq e_n \leq a,$$

and hence $a_n \to e$ as desired.

9.3.4   Properties of Derivatives

One important and well-known result for differentiable functions is the following mean value theorem, which often goes under the moniker of the MVT. Graphically, recalling (9.9), the MVT states that if $f(x)$ satisfies the given properties on $[a, b]$, then there is a point $c \in (a, b)$ so that the slope of the tangent line to $y = f(x)$ at c, or $f'(c)$, equals the slope between the endpoints of the graph of $f(x)$ on $[a, b]$. The endpoints are, of course, $(a, f(a))$ and $(b, f(b))$.

Proposition 9.83 (*Mean Value Theorem*) *If $f(x)$ is differentiable on (a, b) and continuous on $[a, b]$, then there is a number $c \in (a, b)$, so that*

$$f'(c) = \frac{f(b) - f(a)}{b - a}. \tag{9.22}$$

Proof Define a new function

$$g(x) = f(x) - \frac{f(b) - f(a)}{b - a}(x - a).$$

Then $g(a) = g(b) = f(a)$, and $g'(x) = f'(x) - \frac{f(b)-f(a)}{b-a}$, so the proof follows if we can show that there is a $c \in (a, b)$ with $g'(c) = 0$. The next proposition provides this conclusion. ∎

Proposition 9.84 (*Rolle's Theorem*) *If $g(x)$ is differentiable on (a, b) and continuous on $[a, b]$, with $g(a) = g(b)$, then there is a number $c \in (a, b)$, so that $g'(c) = 0$.*

Proof If $g(x)$ is constant on $[a, b]$, then the conclusion follows for all $c \in (a, b)$. If not constant, then as a continuous function on $[a, b]$, $g(x)$ must achieve both its maximum and minimum value on this interval. Since $g(x)$ is assumed to be nonconstant and $g(a) = g(b)$, at least one of these must occur within (a, b), and we denote this value by c. Now, if $g(c)$ is a maximum, we conclude that

$$\frac{g(x) - g(c)}{x - c} \begin{cases} \leq 0, & x \geq c, \\ \geq 0, & x \leq c, \end{cases}$$

and with the opposite inequalities at a minimum. Since the limit must exist as $x \to c$, and equal $g'(c)$, we conclude that the only possible value for this limit is 0. ∎

Remark 9.85

1. *With the aid of the mean value theorem, we return to the point made in section 9.2.6 on Hölder continuity, that being, if $f(x)$ is Hölder continuous of order $\alpha > 1$ on an interval (a, b), then $f(x) = c$, a constant on this interval. To see this, first note that if $f(x)$ has this order of continuity at x_0, then*

$$\left| \frac{f(x) - f(x_0)}{\Delta x} \right| = O(\Delta x^{\alpha - 1}),$$

and hence $f'(x_0) = 0$. Consequently, if $f(x)$ has this order of continuity throughout an interval (a, b), then $f'(x) = 0$ for all $x \in (a, b)$. By the MVT, for any interval $[c, d] \subset (a, b)$ there is $e \in [c, d]$ with $\frac{f(d) - f(c)}{d - c} = f'(e)$, and we conclude from $f'(e) = 0$ that $f(d) = f(c)$, so $f(x)$ is constant. Of course, there is no such conclusion if $f(x)$ satisfies this Hölder condition at an isolated point, as the functions $f(x) = x^\alpha$ for $\alpha > 1$ demonstrate at $x = 0$.

2. *Another consequence of (9.22) noted in item 1 is that if $f'(x) \equiv 0$ on an interval (a, b), then for any $c, d \in (a, b)$, we must have that $f(c) = f(d)$. In other words, the only functions with identically 0 first derivatives are the constant functions.*

The proof of Rolle's theorem produces a **necessary** condition on a point $c \in (a, b)$ to be a **relative maximum** or a **relative minimum** of $f(x)$ on $[a, b]$, but first a definition.

Definition 9.86   *A point c is a **relative minimum** of a function $f(x)$ if there is an open interval I, with $c \in I$, so that for all $x \in I$, $f(c) \leq f(x)$. The point c is a **relative maximum** of $f(x)$ if there is an open interval I containing c so that for all $x \in I$, $f(c) \geq f(x)$.*

When $f(x)$ is a differentiable function, it is often easy to find all possible candidates for relative minimums and relative maximums. Specifically, at any such point, $f'(c) = 0$.

Proposition 9.87   *If c is a relative maximum or relative minimum of $f(x)$, and $f(x)$ is differentiable at c, then $f'(c) = 0$.*

Proof   As in the proof of Rolle's theorem, at a relative minimum,

$$\frac{f(x) - f(c)}{x - c} \begin{cases} \geq 0, & x \geq c, \\ \leq 0, & x \leq c, \end{cases}$$

and the inequalities reverse at a relative maximum. As $x \to c$, the existence of $f'(c)$ implies that these ratios converge to the same value, which must therefore be 0.   ■

Example 9.88

1. *Note that a differentiable function does not necessarily have $f'(x) = 0$ at a **global maximum** or **global minimum** on $[a, b]$, since such extreme values may occur at an interval endpoint. For example, $f(x) = x$ is a simple function that achieves its global maximum and minimum on the endpoints of every closed interval $[a, b]$, and yet $f'(x) \equiv 1$.*

2. *Also $f'(c) = 0$ is **only** a necessary condition for a relative maximum or mini is not sufficient as the function $f(x) = x^3$ exemplifies at $c = 0$.*

Because of the importance of the points at which the derivative of a function is zero, these points warrant a special name.

Definition 9.89 *Given a differentiable function $f(x)$, the points for which $f'(c) = 0$ are known as the **critical points** of $f(x)$.*

Critical points are the first place one looks to find relative maximums or minimums of a differentiable function. Because such an analysis will only reveal a function's relative maximums and minimums, for global maximums and minimums on a closed and bounded interval, the second place to be evaluated are the interval's endpoints. For global maximums and minimums on an open interval, (a, b), bounded or unbounded, one needs to consider the function's values as $x \to a$ and $x \to b$, and in such cases the function may be **unbounded**, meaning the global maximum (respectively, minimum) is ∞ (respectively, $-\infty$).

A final simple property, but a useful one to highlight, was noted in the proof of the derivative formula for the inverse function in proposition 9.77. Its proof is assigned as exercise 7, and will be omitted.

Proposition 9.90 *If $f(x)$ is differentiable at x_0, $f'(x_0) \neq 0$, and $f'(x)$ is continuous in an open interval containing x_0 then there is an open interval about x_0, say $I = (x_0 - a, x_0 + a)$ for some $a > 0$, so that on I, $f(x)$ is one-to-one and monotonic. Specifically, if $x, y \in I$ and $x < y$, then*

$$f'(x_0) > 0 \Rightarrow f(x) < f(y),$$

$$f'(x_0) < 0 \Rightarrow f(x) > f(y).$$

9.3.5 Improving an Approximation II

Another significant conclusion that can be drawn from the mean value theorem is a numerical refinement of the rate of convergence of $f(x)$ to $f(x_0)$ in the case where $f(x)$ has a **bounded** derivative. Specifically, if $M = \max\{f'(x) \mid x \in (a, b)\}$, then for any $x, x_0 \in (a, b)$, we have from (9.10) and the triangle inequality that

$$|f(x) - f(x_0)| \leq M|x - x_0|. \tag{9.23}$$

While this bound is in theory less powerful than (9.7), which we rewrite here for comparability,

$$|f(x) - f(x_0)| \leq f'(x_0)|x - x_0| + o(|x - x_0|),$$

in practice, it can be more valuable when M is easily estimated, since this inequality works uniformly for any x and x_0 in the interval, rather than only at a point, x_0. This estimate also avoids the extra Little o term that, while useful when we have $\Delta x \to 0$, is not useful for numerical estimates when Δx is fixed and finite, since its exact formula is unknown.

Also note that by rewriting (9.7) with $a = f'(x_0)$, we achieve the following approximation:

$$f(x) = f(x_0) + f'(x_0)\Delta x + o(\Delta x), \tag{9.24}$$

where as usual, $x = x_0 + \Delta x$. We will see below that this is a special case of a Taylor series expansion of $f(x)$.

Comparing (9.24) with (9.9), we identify the error between the tangent line approximation and the graph of the function to be $o(\Delta x)$.

9.3.6   Higher Order Derivatives

In order to pursue higher order approximations to $f(x)$ near x_0, we define the following notion:

Definition 9.91   *For each integer $n > 1$, the **nth derivative of** $f(x)$ **at** x_0, denoted $f^{(n)}(x_0)$, or, $\left.\frac{d^n f}{dx^n}\right|_{x=x_0}$, is defined iteratively by*

$$f^{(n)}(x_0) \equiv \lim_{\Delta x \to 0} \frac{f^{(n-1)}(x) - f^{(n-1)}(x_0)}{\Delta x}, \tag{9.25}$$

*when this limit exists. One then says that $f(x)$ **is n-times differentiable at** x_0, or on an interval (a, b), and so forth. If $f^{(n)}(x_0)$ exists for all n, we say that $f(x)$ is **infinitely differentiable at** x_0, or infinitely differentiable on an interval, and so forth. The existence of the nth **derivative of** $f(x)$ can also be expressed in a way that is analogous to (9.10):*

$$f^{(n-1)}(x_0 + \Delta x) - f^{(n-1)}(x_0) = \Delta x(f^{(n)}(x_0) + \varepsilon_{f^{(n-1)}}(x_0 + \Delta x)).$$

Note that if $f(x)$ is n-times differentiable at x_0, then by proposition 9.72 each of the first $(n-1)$ derivatives must be continuous at x_0. Also note above that a function's nth derivative is calculated sequentially, by calculating in turn the function's derivatives, first, then second, and so on. Below we investigate numerical estimation of derivatives that are developed directly from values of the function.

Example 9.92 *Let $f(x) = x^N$, where N is a positive integer. Then as was shown in example 9.71 above, $f'(x) = Nx^{N-1}$. By iteration, we derive that*

$$\frac{dx^N}{dx^n} = \begin{cases} \frac{N!}{(N-n)!} x^{N-n}, & n \le N, \\ 0, & n > N, \end{cases}$$

*where we recall below the **factorial notation** and related **binomial coefficients**.*

Definition 9.93

1. *If N is a positive integer, then $N!$, or N **factorial** is defined as*

$$N! = N(N-1)(N-2)\ldots 2 \cdot 1,$$

and also $0! = 1$ (see chapter 10, the gamma distribution, for a compelling motivation for the definition of $0!$).

2. *If N and M are nonnegative integers, $0 \le M \le N$, the **binomial coefficient**, $\binom{N}{M}$ is defined as*

$$\binom{N}{M} = \frac{N!}{M!(N-M)!}.$$

9.3.7 Improving an Approximation III: Taylor Series Approximations

Generalizing the analysis above that led to (9.24), we introduce next the general **Taylor series**. To this end, assume that we want to approximate $f(x)$ with an nth order polynomial, generalizing the first order approximation in (9.24). In other words, the goal is to approximate $f(x)$ by

$$f(x) \approx \sum_{j=0}^{n} a_j (x - x_0)^j,$$

where here we express Δx as $x - x_0$ for specificity below.

If we assume that $f(x)$ is n-times differentiable, we can differentiate this expression using example 9.92 above, and substitute $x = x_0$ to solve for the coefficients a_j. For example,

$$f(x_0) = \sum_{j=0}^{n} a_j (x_0 - x_0)^j = a_0,$$

$$f'(x_0) = \sum_{j=1}^{n} j a_j (x_0 - x_0)^{j-1} = a_1,$$

$$f^{(2)}(x_0) = \sum_{j=2}^{n} j(j-1)a_j(x_0 - x_0)^{j-2} = 2a_2,$$

$$\vdots$$

$$f^{(m)}(x_0) = \sum_{j=m}^{n} \frac{j!}{(j-m)!} a_j(x_0 - x_0)^{j-m} = m!a_m \qquad \text{for } m \leq n.$$

From this calculation we derive the nth-**order Taylor polynomial for** $f(x)$ **centered at** x_0:

$$f(x) \approx \sum_{j=0}^{n} \frac{1}{j!} f^{(j)}(x_0)(x - x_0)^j. \tag{9.26}$$

This expansion is named for **Brook Taylor** (1685–1731) who published the approximation result in (9.26) in the early 1700s, although it was apparently discovered some time earlier by **James Gregory** (1638–1675). When $x_0 = 0$, this series approximation is sometimes referred to as a **Maclaurin series**, named for **Colin Maclaurin** (1698–1746), who applied this idea to trigonometric functions.

As a first application we derive the nth-order Taylor polynomial for e^x, first referenced in chapter 6 and applied in chapter 7.

Example 9.94   *With* $f(x) = e^x$, *and* $x_0 = 0$, *we have that* $f^{(n)}(x_0) = e^{x_0} = 1$ *for all n, and so*

$$e^x \approx \sum_{j=0}^{n} \frac{1}{j!} x^j.$$

We next investigate the error in the approximation in (9.26). Of course, if $f(x)$ is a polynomial of degree n, the nth-order Taylor polynomial will exactly reproduce $f(x)$. In fact from (9.26) it is apparent that for any such polynomial, the coefficient of x^j equals the jth-derivative of the polynomial divided by $j!$, where these derivatives are evaluated at $x = 0$. In general, however, there will be a remainder, also called the **error term**.

We now investigate one property of this remainder.

Proposition 9.95   *If* $f(x)$ *is n-times differentiable on an interval* (a, b), *with* $f^{(j)}(x)$ *continuous on* $[a, b]$ *for* $j \leq n - 1$, *then for* $x, x_0 \in [a, b]$,

$$f(x) = \sum_{j=0}^{n} \frac{1}{j!} f^{(j)}(x_0)(x - x_0)^j + O(\Delta x^n), \qquad (9.27)$$

where $\Delta x = x - x_0$. *In addition, if $f^{(n)}(x)$ is continuous on $[a, b]$, then the error improves slightly to*

$$f(x) = \sum_{j=0}^{n} \frac{1}{j!} f^{(j)}(x_0)(x - x_0)^j + o(\Delta x^n). \qquad (9.28)$$

Proof   For $x, x_0 \in [a, b]$ given, with $x_0 < x$ for specificity, define the constant $A \equiv A(x, x_0)$, so that

$$f(x) = \sum_{j=0}^{n-1} \frac{1}{j!} f^{(j)}(x_0)(x - x_0)^j + A \frac{(x - x_0)^n}{n!},$$

and define the residual function

$$g(y) = f(x) - \sum_{j=0}^{n-1} \frac{1}{j!} f^{(j)}(y)(x - y)^j - A \frac{(x - y)^n}{n!}.$$

Now by the assumptions of the proposition, $g(y)$ is continuous on $[a, b]$ and differentiable on (a, b). Also $g(x) = g(x_0) = 0$. So by Rolle's theorem, there is a value $c \in (x_0, x)$ so that $g'(c) = 0$. A calculation, using the product rule for derivatives, produces $g'(y)$:

$$g'(y) = -\sum_{j=0}^{n-1} \frac{1}{j!} f^{(j+1)}(y)(x - y)^j + \sum_{j=1}^{n-1} \frac{1}{(j-1)!} f^{(j)}(y)(x - y)^{j-1}$$

$$+ A \frac{(x - y)^{n-1}}{(n-1)!}.$$

A careful look at the two summations reveals that the first $n - 1$ terms of the first sum cancel with the $n - 1$ terms of the second, leaving

$$g'(y) = -\frac{1}{(n-1)!} f^{(n)}(y)(x - y)^{n-1} + A \frac{(x - y)^{n-1}}{(n-1)!}.$$

The conclusion of Rolle's theorem, that there is a $c \in (x_0, x)$ so that $g'(c) = 0$, can be rewritten as

$$f^{(n)}(c) = A.$$

Hence we have that for some $c \in (x_0, x)$,

$$f(x) = \sum_{j=0}^{n-1} \frac{1}{j!} f^{(j)}(x_0)(x - x_0)^j + f^{(n)}(c) \frac{(x - x_0)^n}{n!}. \tag{9.29}$$

The same conclusion follows if $x_0 > x$. From (9.29) we have then that for some $c \in (x_0, x)$ or $c \in (x, x_0)$,

$$f(x) = \sum_{j=0}^{n} \frac{1}{j!} f^{(j)}(x_0)(x - x_0)^j + [f^{(n)}(c) - f^{(n)}(x_0)] \frac{(x - x_0)^n}{n!}.$$

The error term is seen to be $O(\Delta x^n)$ if we only know that $f^{(n)}(x)$ exists. However, if $f^{(n)}(x)$ is also continuous so that $f^{(n)}(c) - f^{(n)}(x_0) \to 0$ as $\Delta x \to 0$, then this error is seen to be $o(\Delta x^n)$.   ∎

Notation 9.96   *Given $x, x_0 \in (a, b)$, there is a convenient notational devise for identifying a point c that is "between" x and x_0, which is to say that $c \in (x_0, x)$ if $x_0 < x$, and $c \in (x, x_0)$ if $x < x_0$. Stated more succinctly, there exists θ, with $0 < \theta < 1$, so that $c = x_0 + \theta \Delta x$, where $\Delta x = x - x_0$, and this is used below.*

Example 9.97   *From example 9.94, we have that since e^x is infinitely differentiable, and hence has continuous derivatives of all orders, then for any n,*

$$e^x - \sum_{j=0}^{n} \frac{1}{j!} x^j = o(x^n) \qquad as\ x \to 0.$$

Analytic Functions
It turns out that in many applications, the Taylor polynomials not only provide high-order approximations to the given function at x_0 as $\Delta x \to 0$, but also these polynomials approximate the function everywhere as $n \to \infty$. Such functions are called **analytic functions**.

Definition 9.98   *A function $f(x)$ is called **analytic in a neighborhood of** x_0 if it can be expanded in a convergent Taylor series:*

$$f(x) = \sum_{j=0}^{\infty} \frac{1}{j!} f^{(j)}(x_0)(x - x_0)^j, \tag{9.30}$$

for x in an open interval centered on x_0. In other words, for every x in this interval,

$$f(x) = \lim_{n \to \infty} \sum_{j=0}^{n} \frac{1}{j!} f^{(j)}(x_0)(x - x_0)^j.$$

It is apparent that every polynomial is analytic, since all but a finite number of derivatives satisfy $f^{(j)}(x_0) = 0$, as are many familiar functions such as e^x, $\ln x$, $\sin x$, and $\cos x$. Each is analytic everywhere in their respective domains of definition. Proving analyticity, however, requires some new tools, as developed in (9.34) below. The formula (9.27) does not help, even if we know that $f(x)$ is infinitely differentiable and this formula holds for all n. The reason is that this expression only provides information about the behavior of the Taylor polynomial as $\Delta x \to 0$. To be analytic at x_0 requires that $f_n(x) \to f(x)$ as $n \to \infty$ for x in a neighborhood of x_0, where $f_n(x)$ denotes the nth-degree Taylor polynomial in (9.26).

While analyticity requires the existence of infinitely many derivatives, the following classical example demonstrates that it requires more than just this. In other words, infinite differentiability is a necessary condition for a function to be analytic, but it is not a sufficient condition.

Example 9.99 *Define $f(x)$ by*

$$f(x) = \begin{cases} e^{-1/x^2}, & x \neq 0, \\ 0, & x = 0. \end{cases}$$

Then every derivative of $f(x)$ is a finite sum of terms of the form

$$c \frac{e^{-1/x^2}}{x^j}.$$

So $f^{(n)}(x)$ exists for all $x \neq 0$, but also it is possible to justify that for all n, $f^{(n)}(x) \to 0$ as $x \to 0$. To see this, substitute $y = \frac{1}{x}$, obtaining sums of terms of the form $cy^j e^{-y^2}$, and let $y \to \infty$. Then, as $y \to \infty$, since $y^j < e^y$ for any j, we conclude that

$$cy^j e^{-y^2} < ce^{-y(y-1)} \to 0 \qquad as \ y \to \infty.$$

In other words, $f^{(n)}(0) = 0$ for all n, and hence the Taylor polynomials evaluated at $x_0 = 0$ satisfy $f_n(x) \equiv 0$ for all n. Consequently we cannot have that $f_n(x) \to f(x)$ as $n \to \infty$ for x in a neighborhood of 0, and we conclude that $f(x)$ is infinitely differentiable but not analytic at 0.

Note that the definition of analytic above does not require that the Taylor series converge absolutely, only that it converges. This is in contrast to the definition of a power series in chapter 6 for which the interval of convergence and radius of convergence are defined in a way to ensure that these series converge absolutely. However, many analytic functions do indeed converge absolutely, and using chapter 6 methods, we can readily identify two conditions that assure absolute convergence. Both conditions relate to the growth of $\{f^{(n)}(x_0)\}$ as $n \to \infty$.

Proposition 9.100   *Let $f(x)$ be an analytic function given by (9.30) in the interval $|x - x_0| < R$.*

1. *If*

$$\limsup_{n} \left| \frac{f^{(n+1)}(x_0)}{(n+1)f^{(n)}(x_0)} \right| = L < \infty, \tag{9.31}$$

then the Taylor series is absolutely convergent for $|x - x_0| < R'$, where $R' = \frac{1}{L}$.

2. *If there is an x' so that*

$$\left| \frac{f^{(n)}(x_0)}{n!} (x' - x_0)^n \right| \le C \qquad \text{for all } n, \tag{9.32}$$

then the Taylor series is absolutely convergent for $|x - x_0| < R''$, where $R'' = |x' - x_0|$.

Proof   Statement 1 follows from the ratio test in chapter 6, which assures absolute convergence if the limit superior of the ratios of successive terms is less than 1. Letting $c_n \equiv \frac{f^{(n)}(x_0)}{n!} (x - x_0)^n$, we write

$$\limsup_{n} \left| \frac{c_{n+1}}{c_n} \right| = \limsup_{n} \left| \frac{f^{(n+1)}(x_0)}{(n+1)f^{(n)}(x_0)} \right| |x - x_0|,$$

$$= L|x - x_0|,$$

so absolute convergence is assured if $L|x - x_0| < 1$. Statement 2 follows from the comparison test. Specifically, (9.32) implies that

$$\left| \frac{f^{(n)}(x_0)}{n!} (x - x_0)^n \right| \le C \left| \frac{x - x_0}{x' - x_0} \right|^n$$

$$< Cr^n,$$

where $r < 1$ if $|x - x_0| < |x' - x_0|$, and this Taylor series is therefore bounded by a convergent geometric series.   ∎

A useful corollary of this result is as follows:

Proposition 9.101 *If*

$$f(x) = \sum_{j=0}^{\infty} \frac{1}{j!} f^{(j)}(x_0)(x - x_0)^j, \quad g(x) = \sum_{j=0}^{\infty} \frac{1}{j!} g^{(j)}(x_0)(x - x_0)^j,$$

are analytic functions that are absolutely convergent for $|x - x_0| < R$, then for any $a, b \in \mathbb{R}$, $h(x) \equiv af(x) + bg(x)$ is analytic, absolutely convergent for $|x - x_0| < R$, and $h(x) = \sum_{j=0}^{\infty} \frac{1}{j!} h^{(j)}(x_0)(x - x_0)^j$.

Proof That $h(x)$ is absolutely convergent follows from the triangle inequality and the absolute convergence of $f(x)$ and $g(x)$:

$$\left| a \sum_{j=0}^{\infty} \frac{1}{j!} f^{(j)}(x_0)(x - x_0)^j + b \sum_{j=0}^{\infty} \frac{1}{j!} g^{(j)}(x_0)(x - x_0)^j \right|$$

$$\leq |a| \sum_{j=0}^{\infty} \frac{1}{j!} |f^{(j)}(x_0)(x - x_0)^j| + |b| \sum_{j=0}^{\infty} \frac{1}{j!} |g^{(j)}(x_0)(x - x_0)^j|.$$

That the Taylor series for $h(x)$ is given in terms of the derivatives of $h(x)$ also follows by the absolute convergence of the series

$$a \sum_{j=0}^{\infty} \frac{1}{j!} f^{(j)}(x_0)(x - x_0)^j + b \sum_{j=0}^{\infty} \frac{1}{j!} g^{(j)}(x_0)(x - x_0)^j,$$

which justifies the rearrangement of these terms to

$$\sum_{j=0}^{\infty} \frac{1}{j!} [af^{(j)}(x_0) + bg^{(j)}(x_0)](x - x_0)^j. \qquad \blacksquare$$

Remark 9.102 *While the Taylor series of an analytic function need not be absolutely convergent, the partial sums of these series are pointwise convergent. Hence these partial sums will be uniformly convergent on any compact set inside the interval of convergence.*

9.3.8 Taylor Series Remainder

In this section we present a useful and explicit expression for the remainder term implicit in (9.26) and seen in the development of (9.27) and (9.28). Another expression for this remainder will be seen in section 10.8.

Defining $f_n(x)$ as the nth-order Taylor polynomial in (9.26), we write

$$f_n(x) = \sum_{j=0}^{n} \frac{1}{j!} f^{(j)}(x_0)(x - x_0)^j.$$

Proposition 9.95 provides qualitative information on the error term:

$$f(x) = f_n(x) + R_n(x).$$

Summarizing, we have from (9.27) and (9.28) that:

1. $R_n(x) = O(\Delta x^n)$ in all cases, requiring only that $f^{(n)}(x)$ exists on (a, b).

2. $R_n(x) = o(\Delta x^n)$ if $f^{(n)}(x)$ is also continuous on this interval.

Now, what if $f^{(n)}(x)$ is also differentiable on this interval? Then proposition 9.95 states that $f(x)$ can be approximated by $f_{n+1}(x)$ with an error of $R_{n+1}(x) = O(\Delta x^{n+1})$. Alternatively, the last term in $f_{n+1}(x)$ can be moved to the error term so that $f(x)$ can be approximated by $f_n(x)$, with an error of

$$R'_n(x) = R_{n+1}(x) + \frac{1}{(n+1)!} f^{(n+1)}(x_0)(x - x_0)^{n+1} = O(\Delta x^{n+1}).$$

However, it turns out that in this case where we assume that $f(x)$ has one additional derivative $f^{(n+1)}(x)$, an explicit expression for this remainder can also be derived. If this additional derivative is continuous, this explicit expression provides a useful upper bound for this error everywhere in the given interval. This remainder is often used for proving convergence of a Taylor series, as well as providing numerical estimates for given x_0 and Δx, while the upper bound is used for proving analyticity on a given interval.

Proposition 9.103 *If $f(x)$ is $(n+1)$-times differentiable on an interval (a, b), with $f^{(j)}(x)$ continuous on $[a, b]$ for $j \leq n$, and $x, x_0 \in (a, b)$, then there exists θ, with $0 < \theta < 1$, so that*

$$f(x) = \sum_{j=0}^{n} \frac{1}{j!} f^{(j)}(x_0)(x - x_0)^j + \frac{1}{(n+1)!} f^{(n+1)}(c)(x - x_0)^{n+1}, \tag{9.33}$$

where $c = x_0 + \theta \Delta x$. In other words, c is between x and x_0, and so $c \in (x_0, x)$ if $x_0 < x$, and $c \in (x, x_0)$ if $x < x_0$. In addition, if $f^{(n+1)}(x)$ is continuous on $[a, b]$, then there exists $M > 0$ so that for all $x, x_0 \in (a, b)$,

$$\left| f(x) - \sum_{j=0}^{n} \frac{1}{j!} f^{(j)}(x_0)(x - x_0)^j \right| \le \frac{M}{(n+1)!} |x - x_0|^{n+1}. \tag{9.34}$$

Proof   The expression in (9.33) follows immediately from (9.29) in the proof of proposition 9.95. In addition, if $f^{(n+1)}(x)$ is continuous on $[a, b]$, then from proposition 9.39 this function attains its upper and lower bounds in this interval. Here M denotes the larger of the absolute values of these bounds.   ■

Remark 9.104   *The remainder term in the Taylor series expansion in (9.33) is known as the **Lagrange form of the remainder**, after **Joseph-Louis Lagrange** (1736–1813), who proved the mean value theorem and derived this remainder term from this result. Another form of this remainder, named for **Augustin Louis Cauchy**, will be developed in section 10.8.*

Example 9.105

1. *We can apply this proposition to the **infinite product** encountered in section 8.4.1, in the discussion preceding the strong law of large numbers. Given $\{x_n\}_{n=1}^{\infty}$ with $x_n > 0$ and $x_n \to 0$ as $n \to \infty$, we show that*

$$\prod_{n=1}^{\infty} (1 - x_n) = \begin{cases} 0, & \text{if } \sum x_n \text{ diverges}, \\ c > 0, & \text{if } \sum x_n \text{ converges}. \end{cases}$$

Applying (9.33) with $n = 1$ to $f(x) = \ln(1 - x)$, and recalling that $f'(x) = \frac{-1}{1-x}$ and $f''(x) = \frac{-1}{(1-x)^2}$, we obtain the following with $x_0 = 0$, where it is also assumed that $x_n < 1$:

$$\ln(1 - x_n) = -x_n - \frac{1}{2}(\theta_n x_n)^2, \qquad 0 < \theta_n < 1.$$

Consequently, since all but a finite number of x_n satisfy $x_n < 1$, we can ignore these exceptions since they do not influence the conclusion, and obtain

$$\ln \prod_{n=1}^{N} (1 - x_n) = -\sum_{n=1}^{N} x_n - \frac{1}{2} \sum_{n=1}^{N} (\theta_n x_n)^2.$$

Now, if $\sum_{n=1}^{\infty} x_n = \infty$, then we have that $\ln \prod_{n=1}^{\infty}(1 - x_n) = -\infty$, and hence $\prod_{n=1}^{\infty}(1 - x_n) = 0$. On the other hand, if $\sum_{n=1}^{\infty} x_n = s < \infty$, then since $\theta_n, x_n < 1$, it is apparent that $\sum_{n=1}^{\infty}(\theta_n x_n)^2 = s' < s$. So $\ln \prod_{n=1}^{\infty}(1 - x_n) = -s - \frac{1}{2}s'$, and $\prod_{n=1}^{\infty}(1 - x_n) = e^{-s-(s'/2)}.$

2. *If* $\{x_n\}_{n=1}^{\infty}$ *satisfies* $x_n \to 0$ *as* $n \to \infty$ *without the restriction* $x_n > 0$, *the same second convergence conclusion follows provided that* $\sum |x_n|$ *converges. This condition assures that* $\sum_{n=1}^{\infty}(\theta_n x_n)^2$ *converges, since for* $|x_n| < 1$,

$$\sum_{n=1}^{\infty}(\theta_n x_n)^2 < \sum_{n=1}^{\infty}x_n^2 < \sum |x_n|.$$

Note that $x_n = \frac{(-1)^n}{\sqrt{n}}$ *demonstrates the necessity of this condition of absolute convergence, since although* $\sum_{n=1}^{\infty}x_n = s < \infty$, *all that can be said about this second summation is that* $\sum_{n=1}^{\infty}(\theta_n x_n)^2 < \sum_{n=1}^{\infty}\frac{1}{n}$, *which is divergent.*

The upper bound for the remainder term is often useful in proving analyticity of a given function. However, we will see that this estimate sometimes fails to provide a proof of convergence of a Taylor series because it is somewhat crude, reflecting the maximum of $f^{(n+1)}(x)$ on $[a,b]$ in general, or more specifically for given x, x_0, the maximum of $f^{(n+1)}(x)$ on $[x, x_0]$ or $[x_0, x]$. According to (9.29) in the proof of proposition 9.95, we really only need an estimate of the absolute value of $f^{(n+1)}(x)$ at an intermediate point c, which is generally unknown. This crudeness can be a problem when this interval maximum is large.

As noted above, there are other forms of the remainder, and the Cauchy form, which reflects an average of $f^{(n+1)}(t)$ between x and x_0, will be developed in section 10.8, and seen to succeed in proving analyticity in cases where the Lagrange remainder fails.

Example 9.106   *We now address three Taylor series quoted and used in prior chapters. The series for* $f(x) = \frac{1}{1-x}$ *was used in example 6.47 in chapter 6, the* **exponential series** e^x *was used in chapter 7 in the development of moment relationships, and the* **natural logarithm series** $\ln(1 + x)$ *was needed for the chapter 8 development of Stirling's formula and other results.*

1. *With* $f(x) = \frac{1}{1-x} = (1-x)^{-1}$ *and* $x_0 = 0$ *it is easy to derive that* $f^{(n)}(x) = n!(1-x)^{-n-1}$ *and so* $f^{(n)}(0) = n!$. *Noting that* $(1-x)^{-n-1}$ *is an increasing function on* $(-\infty, 1)$, *and hence*

$$\max_{[0,x]}(1-y)^{-n-1} = \begin{cases} (1-x)^{-n-1}, & 0 < x < 1, \\ 1, & x \leq 0, \end{cases}$$

we obtain from (9.34) that

$$\left| \frac{1}{1-x} - \sum_{j=0}^{n}x^j \right| \leq \begin{cases} \left|\frac{x}{1-x}\right|^{n+1}, & 0 < x < 1, \\ |x|^{n+1}, & x \leq 0. \end{cases}$$

In chapter 6 it was shown that $\sum_{j=0}^{\infty} x^j$ converges for $|x| < 1$. Consequently the Lagrange remainder only proves that $\frac{1}{1-x} = \sum_{j=0}^{\infty} x^j$ in the second case where $-1 < x \le 0$, since then $|x|^{n+1} \to 0$ as $n \to \infty$. For $0 < x < 1$, $\frac{x}{1-x} > 1$, and hence $\left|\frac{x}{1-x}\right|^{n+1} \to \infty$ as $n \to \infty$. We will return to this example in chapter 10 with a different remainder estimate and proof of convergence to $f(x)$ in this case.

2. *With $f(x) = e^x$ in (9.34) and $x_0 = 0$, recall that $f^{(j)}(x) = e^x$ and $f^{(j)}(0) = 1$ for all j, and so*

$$\left| e^x - \sum_{j=0}^{n} \frac{1}{j!} x^j \right| \le \begin{cases} \frac{e^x}{(n+1)!} |x|^{n+1}, & x > 0, \\ \frac{1}{(n+1)!} |x|^{n+1}, & x \le 0, \end{cases}$$

since the maximum value of $f^{(n+1)}(y)$ over $[0, x]$ is e^x when $x > 0$ and is 1 when $x \le 0$. Now in chapter 6 it was shown that $\sum_{j=0}^{\infty} \frac{1}{j!} x^j$ converges for all $x \in \mathbb{R}$. By Simpson's rule applied to $(n+1)!$,

$$\frac{(n+1)!}{\sqrt{2\pi}(n+1)^{n+(3/2)} e^{-(n+1)}} = \frac{(n+1)!}{\sqrt{2\pi n}\left(\frac{n+1}{e}\right)^{n+1}} \to 1,$$

and so $(n+1)!$ grows faster than $\left(\frac{n+1}{e}\right)^{n+1}$ and hence much faster than x^{n+1} for any x. This shows that for any value of x, this error goes to zero, and the Taylor series converges to e^x as $n \to \infty$. In other words, e^x is an analytic function, and as was noted in (7.63),

$$e^x = \sum_{j=0}^{\infty} \frac{1}{j!} x^j \qquad \text{for all } x \in \mathbb{R}. \tag{9.35}$$

3. *With $f(x) = \ln(1 + x)$, we obtain*

$$f'(x) = \frac{1}{1+x}, f^{(2)}(x) = \frac{-1}{(1+x)^2}, \ldots, f^{(n)}(x) = \frac{(-1)^{n+1}(n-1)!}{(1+x)^n}.$$

Consequently with $x_0 = 0$, $f(0) = 0$ and $f^{(n)}(0) = (-1)^{n-1}(n-1)!$ for $n \ge 1$. Also, to find M, note that since $\frac{1}{(1+y)^{n+1}}$ is a decreasing function for $y > -1$,

$$\max_{[0,x]} \frac{1}{(1+y)^{n+1}} = \begin{cases} 1, & 0 \le x, \\ \frac{1}{(1+x)^{n+1}}, & -1 < x < 0. \end{cases}$$

By (9.34) we obtain

$$\left| \ln(1+x) - \sum_{j=1}^{n} \frac{(-1)^{j+1}}{j} x^j \right| \le \begin{cases} \frac{1}{n+1} |x|^{n+1}, & y \in [0,x], x \ge 0, \\ \frac{1}{n+1} \left| \frac{x}{1+x} \right|^{n+1}, & y \in [x,0], -1 < x \le 0. \end{cases}$$

It was shown in chapter 6 that $\sum_{j=1}^{\infty} \frac{(-1)^{j+1}}{j} x^j$ converges absolutely for $|x| < 1$ and conditionally for $x = 1$, and diverges for $x = -1$. So as in case 1 of this example, the Lagrange remainder only yields a partial result. That is, $\ln(1+x) = \sum_{j=1}^{\infty} \frac{(-1)^{j+1}}{j} x^j$ in the first case where $0 \le x \le 1$, since then $\frac{|x|^{n+1}}{(n+1)} \to 0$ as $n \to \infty$, and in part of the second case where $-1 < x \le 0$. Specifically, for this latter range of x, we have that $\left| \frac{x}{1+x} \right| \le 1$ if $-\frac{1}{2} \le x \le 0$ and hence $\frac{1}{n+1} \left| \frac{x}{1-x} \right|^{n+1} \to 0$ as $n \to \infty$. But for $-1 < x < -\frac{1}{2}$, we see that $\left| \frac{x}{1+x} \right| > 1$, so $\frac{1}{n+1} \left| \frac{x}{1-x} \right|^{n+1} \to \infty$ as $n \to \infty$. We will return to this example in chapter 10 with a different remainder estimate and proof of convergence in this case.

With this analysis applied to $x = 1$, we can conclude that

$$\ln 2 = \sum_{j=1}^{\infty} \frac{(-1)^{j+1}}{j}, \tag{9.36}$$

deriving the numerical value of the alternating harmonic series as was noted in example 6.10.

9.4   Convergence of a Sequence of Derivatives

Expanding the discussion in section 9.2.7 on convergence of a sequence of continuous functions, there is an analogous discussion related to derivatives which we introduce with the following questions:

Question 1:   If $f_n(x)$ is a sequence of differentiable functions, and there is a function $f(x)$ so that $f_n(x) \to f(x)$ pointwise as $n \to \infty$, must $f(x)$ be differentiable?

Question 2:   If $f(x)$ in question 1 is differentiable, must $f_n'(x) \to f'(x)$ for every x as $n \to \infty$?

Question 3:   If $f_n(x) \to f(x)$ uniformly rather than pointwise, do the answers to questions 1 and 2 change?

Answer: The answer to all three questions is, in general, "no," and this is easy to exemplify.

Example 9.107

1. *Define*

$$f_n(x) = \begin{cases} x^{1+(1/n)}, & x \ge 0, \\ (-x)^{1+(1/n)}, & x < 0. \end{cases}$$

Then each $f_n(x)$ *is differentiable, with*

$$f_n'(x) = \begin{cases} \left(1 + \frac{1}{n}\right) x^{1/n}, & x \geq 0, \\ -\left(1 + \frac{1}{n}\right)(-x)^{1/n}, & x < 0. \end{cases}$$

Now $f_n(x) \rightarrow f(x) \equiv |x|$, *which is not differentiable at* $x = 0$, *and for* $x \neq 0$, $f'(x) = 1$ *for* $x > 0$ *and* $f'(x) = -1$ *for* $x < 0$. *Also, it is the case that* $f_n'(x) \rightarrow f'(x)$ *for* $x \neq 0$, *since* $|x|^{1/n} \rightarrow 1$ *as* $n \rightarrow \infty$ *for all* $x \neq 0$. *This observation provides hope, albeit temporary, that the answer to the second question might be "yes."*

2. *Define*

$$f_n(x) = \frac{\sin nx}{\sqrt{n}}.$$

Then each $f_n(x)$ *is differentiable, with*

$$f_n'(x) = \sqrt{n} \cos nx.$$

Now $f_n(x) \rightarrow f(x) \equiv 0$ *for all* x *since* $|\sin nx| \leq 1$, *and* $f(x)$ *is differentiable everywhere with* $f'(x) \equiv 0$. *However,* $f_n'(0) = \sqrt{n} \rightarrow \infty$, *while* $f_n'(\pi)$ *alternates between* $\pm \sqrt{n}$, *and* $f_n'\left(\frac{\pi}{2}\right)$ *cycles through the sequence* $\{0, -\sqrt{n}, 0, \sqrt{n}\}$, *and so forth.*

3. *Finally, although uniform convergence provided a positive result in section 9.2.7 above in terms of preserving continuity it does not help here. Case 1 converges uniformly on compact sets and case 2 converges uniformly, so the same negative conclusions follow. Note, however, that in case 1 the sequence of derivatives,* $f_n'(x)$, *does not converge uniformly by the Cauchy criterion on any interval that contains* 0, *since as* $n \rightarrow \infty$:

$$f_n'(x) \rightarrow \begin{cases} 1, & x > 0, \\ -1, & x < 0. \end{cases}$$

For case 2, the sequence of derivatives $f_n'(x)$ *does not converge uniformly on any interval.*

Although not the most general statement, the following positive result is adequate in most applications.

Proposition 9.108 *If* $f_n(x)$ *is a sequence of continuously differentiable functions and there is a function* $f(x)$ *so that on some interval I,* $f_n(x) \rightarrow f(x)$ *uniformly and* $f_n'(x)$

converge uniformly by the Cauchy criterion, then $f(x)$ is differentiable and $f_n'(x) \to$
$f'(x)$.

Proof From propositions 9.51 and 9.54 on uniform convergence in section 9.2.7,
the assumption that $f_n'(x)$ are continuous and converge uniformly by the Cauchy criterion implies that there is a continuous function, $g(x)$ say, so that $f_n'(x) \to g(x)$ uniformly. What is left to prove is that $f(x)$ is differentiable and $f'(x) = g(x)$. To this
end, fix $x_0 \in I$, and define the "finite difference functions" for $x \neq x_0$:

$$D_n(x) = \frac{f_n(x) - f_n(x_0)}{x - x_0}, \quad D(x) = \frac{f(x) - f(x_0)}{x - x_0}.$$

The assumption that $f_n(x) \to f(x)$ uniformly implies that for $x \neq x_0$,

$$D_n(x) \to D(x) \qquad \text{as } n \to \infty.$$

Since $f_n(x)$ is differentiable,

$$\lim_{x \to x_0} D_n(x) = f_n'(x_0) \qquad \text{for all } n.$$

We now show that for fixed $x \neq x_0$, $D_n(x)$ converges uniformly as $n \to \infty$. This follows in two steps. First off, the mean value theorem applied to $f_n(x) - f_m(x)$ yields
that for some y between x and x_0,

$$|f_n(x) - f_m(x) - f_n(x_0) + f_m(x_0)| = |f_n'(y) - f_m'(y)| \, |x - x_0|.$$

Second, the uniform convergence of $f_n'(x)$ means that for any $\epsilon > 0$ there is an N so
that for $n, m > N$ and any $y \in I$,

$$|f_n'(y) - f_m'(y)| < \epsilon.$$

Combining these steps, we derive that for $n, m > N$ and $x \neq x_0$,

$$|D_n(x) - D_m(x)| < \epsilon,$$

and so $D_n(x)$ converges uniformly as $n \to \infty$ for $x \neq x_0$. Combining these pieces, and
noting that since x_0 is a limit point of the set $I - x_0$, the limits below can be reversed
because of proposition 9.60. This produces

$$f'(x_0) \equiv \lim_{x \to x_0} D(x)$$

$$= \lim_{x \to x_0} \lim_{n \to \infty} D_n(x)$$

$$= \lim_{n \to \infty} \lim_{x \to x_0} D_n(x)$$

$$= \lim_{n \to \infty} f_n'(x_0).$$ ∎

9.4.1   Series of Functions

The preceding proposition 9.108 generalizes easily to a series of functions.

Proposition 9.109   *If $g_j(x)$ is a sequence of continuously differentiable functions, and there is a function $g(x)$ so that on some interval I, $\sum_{j=1}^{n} g_j(x)$ converges uniformly to $g(x)$ as $n \to \infty$ and $\sum_{j=1}^{n} g_j'(x)$ converges uniformly by the Cauchy criterion, then $g(x)$ is differentiable and $\sum_{j=1}^{n} g_j'(x) \to g'(x)$. In other words,*

$$g'(x) = \lim_{n \to \infty} \sum_{j=1}^{n} g_j'(x).$$

Remark 9.110   *In plain language, the uniform convergence of a series of continuously differentiable functions yields a differentiable function when the series of derivatives also converge uniformly, and the derivative of this limit function equals the sum of the derivatives of terms in the series. That is, uniform convergence of the series and its derivatives justifies differentiating term by term, which means that*

$$\left(\sum_{j=1}^{\infty} g_j(x) \right)' = \sum_{j=1}^{\infty} g_j'(x).$$

Proof   Define $f_n(x) = \sum_{j=1}^{n} g(x)$. Then $f_n(x)$ is continuously differentiable for all n as a finite sum of continuously differentiable functions, and by assumption, $f_n(x) \to g(x)$ uniformly. Also $f_n'(x) \equiv \sum_{j=1}^{n} g_j'(x)$, and so $f_n'(x)$ converges uniformly by the Cauchy criterion. The result follows from proposition 9.108 above.   ∎

9.4.2   Differentiability of Power Series

We have seen that in order to have any hope of expanding a given function as a Taylor series, such a function must be infinitely differentiable. However, not all infinitely differentiable functions can be represented as convergent Taylor series, as

$$f(x) = \begin{cases} e^{-1/x^2}, & x \neq 0, \\ 0, & x = 0, \end{cases}$$

analyzed in example 9.99 above illustrates. Here $f^{(n)}(0) = 0$ for all n, so the Taylor series centered at $x_0 = 0$ satisfies

$$\sum_{j=0}^{\infty} \frac{1}{j!} f^{(j)}(0) x^j \equiv 0,$$

and so cannot possibly represent this function in any neighborhood of this point.

The property of $f(x)$ called **analytic** above, or more precisely, analytic in a neighborhood of x_0, means more than just that this function is infinitely differentiable at x_0. It means that the function can be represented by a Taylor series centered at $x = x_0$, and that this series is convergent *to the function values* in some neighborhood of this point. The emphasis on "to the function values" is deliberate, since the function above has a Taylor series centered on $x_0 = 0$ that is convergent everywhere, but it does not converge to $f(x)$ for any $x \neq 0$.

Now a Taylor series is a special case of a power series introduced in chapter 6, and it is natural to ask:

Question: If a function $f(x)$ is defined as the power series $f(x) = \sum_{j=0}^{\infty} c_j (x - x_0)^j$ that is convergent for $|x - x_0| < R$ for some $R > 0$:

1. Is $f(x)$ infinitely differentiable, and if so, how is $f^{(n)}(x)$ evaluated?

2. If infinitely differentiable, is $f(x)$ an analytic function in the sense of (9.30)?

3. If an analytic function, and $f(x)$ is expanded in a Taylor series about x_0, must it be the case that

$$c_n = \frac{f^{(n)}(x_0)}{n!}?$$

The following proposition addresses these questions, and provides affirmative responses. It is largely a corollary to proposition 9.109 above on series of functions, but it is stated here to clarify that a small amount of thought needs to be applied to assure that the uniformity of convergence needed for the result above applies.

Proposition 9.111   *If a function $f(x)$ is defined by the power series*

$$f(x) = \sum_{j=0}^{\infty} c_j (x - x_0)^j \tag{9.37}$$

and has an interval of convergence given by $|x - x_0| < R$ for some $R > 0$, then:

1. $f(x)$ is infinitely differentiable, and

$$f^{(n)}(x) = \sum_{j=n}^{\infty} c_j \frac{j!}{(j-n)!} (x-x_0)^j \qquad (9.38)$$

is absolutely convergent for $|x - x_0| < R$. In other words, power series are infinitely differentiable and can be differentiated term by term.

2. $f(x)$ is analytic in the sense of (9.30), so

$$f(x) = \sum_{n=0}^{\infty} \frac{f^{(n)}(x_0)}{n!} (x-x_0)^n,$$

and this series is absolutely convergent on $|x - x_0| < R$. Further

$$\frac{f^{(n)}(x_0)}{n!} = c_n. \qquad (9.39)$$

In other words, power series expansions are unique.

Proof   Define $f_n(x)$ as the partial summation associated with $f(x)$:

$$f_n(x) = \sum_{j=0}^{n} c_j (x-x_0)^j.$$

For the moment, assume the radius of convergence, $R < \infty$, where we recall that R is defined in chapter 6 by $R = \frac{1}{L}$, where L is given in (6.20):

$$L = \limsup_{j \to \infty} \left\{ \frac{|c_{j+1}|}{|c_j|} \right\}.$$

Then it is apparent that $f_n(x)$ is continuous, $f_n(x) \to f(x)$ pointwise on $|x - x_0| < R$, and hence by exercise 30(b) converges uniformly on the compact $|x - x_0| \le R - \epsilon$, for any $\epsilon > 0$. Also $f_n(x)$ is differentiable,

$$f_n'(x) = \sum_{j=1}^{n} j c_j (x-x_0)^{j-1},$$

and we now show that $f_n'(x)$ converges pointwise on $|x - x_0| < R$ by demonstrating that the series $\sum_{j=1}^{\infty} j c_j (x - x_0)^{j-1}$ has the same interval of convergence as the series for $f(x)$. By the ratio test,

$$\limsup_{j \to \infty} \left\{ \frac{|(j+1)c_{j+1}(x-x_0)^j|}{|jc_j(x-x_0)^{j-1}|} \right\} = \limsup_{j \to \infty} \left\{ \frac{j+1}{j} \frac{|c_{j+1}|}{|c_j|} |x-x_0| \right\}$$

$$= \limsup_{j \to \infty} \left\{ \frac{|c_{j+1}|}{|c_j|} \right\} |x-x_0|$$

$$= L|x-x_0|.$$

So the series $f_n'(x)$ converges on $|x-x_0| < R$ and hence also converges uniformly by the Cauchy criterion on $|x-x_0| \le R - \epsilon$. By proposition 9.108, it follows that $f(x)$ is differentiable, and $f'(x) = \lim_{n \to \infty} f_n'(x)$ for all $|x-x_0| \le R - \epsilon$. Since this is true for all $\epsilon > 0$, the result in (9.38) follows for $n = 1$. However, $f'(x) = \sum_{j=1}^{\infty} jc_j(x-x_0)^{j-1}$ is now a power series to which the same argument applies, and by iteration, (9.38) follows for all n. If $R = \infty$, the same argument applies except that compact sets needed for uniform convergence are defined, $|x-x_0| \le R'$ for any $R' < \infty$. This proves part 1 of the proposition.

For part 2, it is apparent from (9.38) by substitution that $f^{(n)}(x_0) = n!c_n$, and so the Taylor series centered on x_0 converges absolutely for $|x-x_0| < R$ because it is identical to the power series.   ∎

Remark 9.112   *Of course, the notion that power series representations are unique, as stated in part 2 of proposition 9.111, is meant in the sense that if for some x_0,*

$$f(x) = \sum_{j=0}^{\infty} c_j(x-x_0)^j = \sum_{j=0}^{\infty} d_j(x-x_0)^j$$

for $|x-x_0| < R$ with $R > 0$, then $c_j = d_j = \frac{f^{(j)}(x_0)}{j!}$ for all j. A given analytic function has many Taylor series expansions for different values of x_0, of course. For example, expanding about $x = 0$ and $x = 1$, we have

$$e^x = \sum_{j=0}^{\infty} \frac{x^j}{j!} = \sum_{j=0}^{\infty} \frac{e(x-1)^j}{j!}.$$

By the proposition above, every power series is an analytic function in its interval of convergence in the sense of definition 9.98 in section 9.3.7.

Example 9.113   *In section 7.5.1 formulas were introduced for the moment-generating function and characteristic function of a discrete random variable, and it was claimed that each was equal to a power series reflecting the moments of the given random vari-*

able. For example, if $f(x)$ is the probability density function of a given discrete random variable $X : S \rightarrow \{x_i\}_{i=1}^{\infty} \subset \mathbb{R}$, the moment-generating function is defined by

$$M_X(t) = \sum_{i=1}^{\infty} e^{tx_i} f(x_i),$$

when this series converges, and also converges absolutely, for t in an interval I about $t = 0$. Now e^{tx_i} is an analytic function for all t, and expressing it as a Taylor series, we have

$$M_X(t) = \sum_{i=1}^{\infty} \sum_{j=0}^{\infty} \frac{(tx_i)^j}{j!} f(x_i).$$

Since this series is absolutely convergent on I, we can interchange the order of summation by the analysis in section 6.1.4 to produce (7.64):

$$M_X(t) = \sum_{j=0}^{\infty} \frac{t^j}{j!} \sum_{i=1}^{\infty} x_i^j f(x_i)$$

$$= \sum_{j=0}^{\infty} \frac{t^j \mu_j'}{j!}.$$

As a convergent power series on I, we now have that $M_X(t)$ is infinitely differentiable on I, and (9.38) can be applied to produce

$$M_X^{(n)}(t) = \sum_{j=n}^{\infty} \frac{t^{j-n} \mu_j'}{(j-n)!},$$

which produces (7.65) when $t = 0$ is substituted:

$$\mu_n' = M_X^{(n)}(0).$$

The same analysis works for the characteristic function $C_X(t)$, when all moments exist, and demonstrates the analogous properties of this function. As noted before, this requires the use of the power series expansion for e^{itx_j}, with a complex exponent, and this series is seen to be absolutely convergent by the triangle inequality. However, $C_X(t)$ need not be infinitely differentiable at $t = 0$, and will have the same number of derivatives there as $f(x)$ has moments.

Product of Taylor Series

The next discussion in this section relates to the product of two analytic functions. Obviously, if $f(x)$ and $g(x)$ are any two analytic functions, the function $h(x) \equiv f(x)g(x)$ is well defined. The question here is, if $f(x)$ and $g(x)$ are given as absolutely convergent Taylor series centered on x_0, with respective radii of convergence of R and R', is $h(x)$ analytic? If so, what is the power series representation of $h(x)$ and what is its radius of convergence?

The following proposition addresses this question, and expands the result in proposition 9.101, which addressed the analyticity of $af(x) + bg(x)$ for $a, b \in \mathbb{R}$, when $f(x)$ and $g(x)$ are analytic.

Proposition 9.114 *Let $f(x)$ and $g(x)$ be analytic functions and given as convergent power series centered on x_0:*

$$f(x) = \sum_{n=0}^{\infty} \frac{f^{(n)}(x_0)}{n!}(x - x_0)^n, \quad g(x) = \sum_{n=0}^{\infty} \frac{g^{(n)}(x_0)}{n!}(x - x_0)^n,$$

which are absolutely convergent for $|x - x_0| < R$. Then $h(x) \equiv f(x)g(x)$ is an analytic function, absolutely convergent for $|x - x_0| < R$:

$$h(x) = \sum_{n=0}^{\infty} d_n(x - x_0)^n, \tag{9.40}$$

where

$$d_n = \sum_{k=0}^{n} \frac{f^{(k)}(x_0)g^{(n-k)}(x_0)}{k!(n - k)!}. \tag{9.41}$$

Proof Because $f(x)$ and $g(x)$ are absolutely convergent, the conclusion follows directly from proposition 6.52. Specifically, (9.41) follows from (6.22). ∎

We now have an immediate corollary from this proposition, known as the **Leibniz rule** for the nth-derivative of the product of two n-times differentiable functions, named for **Gottfried Wilhelm Leibniz** (1646–1716). This corollary applies to the product of analytic functions, but is true under the weaker assumption that the functions each are simply n-times differentiable. Exercise 34 assigns the proof of this formula in this general case, using mathematical induction.

Proposition 9.115 *If $f(x)$ and $g(x)$ are analytic functions, absolutely convergent for $|x - x_0| < R$, then for $h(x) = f(x)g(x)$,*

$$h^{(n)}(x) = \sum_{k=0}^{n} \binom{n}{k} f^{(k)}(x) g^{(n-k)}(x) \qquad \text{for } |x - x_0| < R. \tag{9.42}$$

Proof This formula for $h^{(n)}(x)$ is true for $x = x_0$ because $h(x)$ is analytic, and hence

$$h(x) = \sum_{n=0}^{\infty} \frac{h^{(n)}(x_0)}{n!} (x - x_0)^n.$$

Comparing this expansion with (6.22), produces $h^{(n)}(x_0) = n! d_n$ and the result follows since $\binom{n}{k} = \frac{n!}{k!(n-k)!}$. For any other x with $|x - x_0| < R$, a Taylor series can be centered on x, and will be absolutely convergent on any interval $(x - R', x + R') \subset (x_0 - R, x_0 + R)$. With this Taylor series, and the above derivation, (9.42) follows for all such x. ∎

*Division of Taylor Series
The last discussion in this section relates to the division of two analytic functions, or the reciprocal of an analytic function. Obviously, if $f(x)$ and $h(x)$ are any two analytic functions, the function $g(x) \equiv \frac{h(x)}{f(x)}$ is well defined if $f(x) \neq 0$. When $h(x) \equiv 1$, the function $g(x)$ is the reciprocal of $f(x)$. The question here is, if $f(x)$ and $h(x)$ are given as absolutely convergent Taylor series centered on x_0, with $f(x_0) \neq 0$, and common radius of convergence of R, is $g(x)$ analytic? If so, what is the power series representation of $h(x)$ and what is its radius of convergence?

The following proposition addresses this question:

Proposition 9.116 *Let $f(x)$ and $h(x)$ be analytic functions and given as convergent power series centered on x_0,*

$$f(x) = \sum_{n=0}^{\infty} \frac{f^{(n)}(x_0)}{n!} (x - x_0)^n, \quad h(x) = \sum_{n=0}^{\infty} \frac{h^{(n)}(x_0)}{n!} (x - x_0)^n,$$

which are absolutely convergent for $|x - x_0| < R$ and where $f(x_0) \neq 0$. Then $g(x) \equiv \frac{h(x)}{f(x)}$ is an analytic function,

$$g(x) = \sum_{n=0}^{\infty} c_n (x - x_0)^n, \tag{9.43}$$

where

$$c_0 = \frac{h(x_0)}{f(x_0)} \tag{9.44a}$$

$$c_n = \frac{1}{f(x_0)} \left[\frac{h^{(n)}(x_0)}{n!} - \sum_{k=0}^{n-1} \frac{f^{(n-k)}(x_0)c_k}{(n-k)!} \right], \tag{9.44b}$$

which is absolutely convergent for $|x - x_0| < R'$ for some $R' > 0$.

Proof   Because $f(x)$ and $h(x)$ are absolutely convergent, the conclusion follows directly from proposition 6.53, which also showed that $\frac{1}{f(x)}$ is absolutely convergent. Specifically, (9.44) follows from (6.25).   ■

Remark 9.117   *In section 9.8.10 below on the risk-neutral probability $q(\Delta t)$ will be an analysis of the ratio of analytic functions and an application of this result, or equivalently, an application of formulas (6.25). However, it is often the case that the power series for the ratio can be derived directly and more easily by a "long division" of the power series of $h(x)$ by the power series of $f(x)$ rather than by generating these coefficients iteratively through formulas such as in (9.44) or (6.25). The importance of the proposition above is that it assures that this ratio function is analytic in a neighborhood of x_0, so we can generate only a few of the terms and still be sure that the remainder will converge to 0 with the order of magnitude implied by the number of terms generated. Without such a result, we could be generating and using a partial sum of a series for which the remainder did not converge.*

Because $c_n = \frac{g^{(n)}(x_0)}{n!}$, we have an immediate corollary from this proposition for the nth-derivative of the ratio of two analytic functions within the interval of convergence. This corollary applies to the ratio of analytic functions because we use the result above, but is true under the general assumption that the functions are each n-times differentiable, as can be proved using mathematical induction.

Proposition 9.118   *If $g(x) \equiv \frac{h(x)}{f(x)}$, with $h(x)$ and $f(x)$ given in proposition 9.116, then*

$$g^{(n)}(x) = \frac{1}{f(x)} \left[h^{(n)}(x) - \sum_{k=0}^{n-1} \binom{n}{k} f^{(n-k)}(x)g^{(k)}(x) \right], \qquad n \geq 1.$$

Proof   This result follows from (9.44), and also follows from the Leibniz rule in (9.42) by writing $h(x) = f(x)g(x)$ and iteratively solving for $g^{(n)}(x)$.   ■

9.5   Critical Point Analysis

9.5.1   Second-Derivative Test

With the help of section 9.3.8 on Taylor series, it is now possible to classify the critical points of a differentiable function. Proposition 9.87 above provided a necessary condition in order that x_0 be a relative maximum or relative minimum of $f(x)$, namely that $f'(x_0) = 0$. In other words, a necessary condition is that x_0 be a **critical point** of $f(x)$. The second and higher derivatives now provide a sorting of these cases.

Proposition 9.119   *If $f(x)$ is a twice differentiable function with $f'(x_0) = 0$, and $f''(x)$ is continuous in a neighborhood of x_0, then:*

1. x_0 *is a relative minimum of $f(x)$ if $f''(x_0) > 0$*
2. x_0 *is a relative maximum of $f(x)$ if $f''(x_0) < 0$*
3. x_0 *can be either or neither if $f''(x_0) = 0$*

Proof   First off, in cases 1 and 2, as was demonstrated in proposition 9.38, if $f''(x)$ is continuous at x_0, then there is an interval about x_0, say $I = (x_0 - a, x_0 + a)$, within which $f''(x)$ has the same sign as it does at x_0. The result in these cases then follows immediately from the Taylor series representation in (9.33) with $n = 1$. Since $f'(x_0) = 0$,

$$f(x) = f(x_0) + \frac{1}{2} f''(y)(x - x_0)^2,$$

where $y = x_0 + \theta \Delta x$ with $0 < \theta < 1$. Choosing x in the interval I within which the sign of $f''(y)$ equals the sign of $f''(x_0)$, the result follows. Case 3 is easily handled by examples below.   ∎

Example 9.120

1. *Simple examples of a relative maximum and minimum in cases 1 and 2 are given by $f(x) = \pm x^2$ with $x_0 = 0$. We then have $f'(x_0) = 0$ and $f''(x_0) = \pm 2$.*
2. *For case 3, we use $f(x) = \pm x^4$ with $x_0 = 0$ for examples of a maximum and minimum when $f'(x_0) = 0$ and $f''(x_0) = 0$; and $f(x) = x^3$ provides a simple example of $f'(0) = 0$ and $f''(x_0) = 0$ but with $x_0 = 0$ being neither a maximum or minimum. A critical point that is not a maximum or minimum is a **point of inflection** or **inflection point** of $f(x)$, although inflection points need not be critical points. See also definition 9.137.*

Definition 9.121   *Given twice differentiable $f(x)$, the point x_0 is a **point of inflection** or **inflection point** of $f(x)$ if $f''(x)$ changes sign between $x < x_0$ and $x > x_0$.*

Example 9.122   *For continuous $f''(x)$, we note that $f''(x_0) = 0$ is therefore a necessary condition for a point of inflection by proposition 9.41, but not a sufficient condition, as $f(x) = x^4$ exemplifies at $x_0 = 0$. Also a point of inflection need not be a critical point, as $f(x) = x^3 = x$ exemplifies at $x_0 = 0$.*

Remark 9.123   *In the case where $f'(x_0) = 0$ and $f''(x_0) = 0$, we can resolve the nature of $f(x)$ at x_0 if $f(x)$ has enough derivatives by determining the first value of n for which $f^{(n)}(x_0) \neq 0$. Again based on (9.24), as long as $f^{(n)}(x)$ is continuous in a neighborhood of x_0, we can conclude that:*

1. *If n is even, x_0 will be a relative minimum if $f^{(n)}(x_0) > 0$ and a relative maximum if $f^{(n)}(x_0) < 0$.*

2. *If n is odd, x_0 will be an inflection point, independent of the sign of $f^{(n)}(x_0)$.*

Functions of the form $f(x) = \pm x^m$ provide simple examples of this generalization with $x_0 = 0$. See section 9.6 on concave and convex functions for more details on points of inflection, and especially example 9.140.

As will be seen in an l_p-norm example in example 9.129 in the next section, it is not always convenient or even possible to evaluate $f''(x_0)$ to determine if x_0 is a maximum or a minimum. In such cases there is an alternative first derivative test that is sometimes more convenient to apply.

Proposition 9.124   *Let $f(x)$ be a differentiable function, with $f'(x_0) = 0$, and assume that there is an open interval I, with $x_0 \in I$, on which $f'(x)$ is continuous. Then:*

1. *If $f'(x)$ is a strictly increasing function on I, then x_0 is a relative minimum of $f(x)$.*

2. *If $f'(x)$ is a strictly decreasing function on I, then x_0 is a relative maximum of $f(x)$.*

Proof   By (9.33) and $n = 0$, for $x \in I$ there is $y = x_0 + \theta \Delta x$, with $0 < \theta < 1$, so that

$$f(x) = f(x_0) + f'(y)\Delta x.$$

Now, if $f'(x)$ is a strictly increasing function on I, then since $f'(x_0) = 0$, we conclude that $f'(x) < 0$ for $x < x_0$ and $f'(x) > 0$ for $x > x_0$. But for $x \in I$, by definition of $y \in I$, it must be the case that $f'(y)\Delta x > 0$. So $f(x) > f(x_0)$ and x_0 is a relative minimum of $f(x)$. When $f'(x)$ is a strictly decreasing function on I, then $f'(y)\Delta x < 0$, so $f(x) < f(x_0)$ and x_0 is a relative maximum of $f(x)$.   ∎

*9.5.2   Critical Points of Transformed Functions

When pursuing a critical point analysis of a given function $f(x)$, it is often convenient to first transform the function by taking a composite of $f(x)$ with another function $g(x)$ and consider the critical points of $g(f(x))$. For example, if $f(x)$ is given as an exponential function $f(x) = e^{j(x)}$, it would be natural to prefer to evaluate the derivatives of $\ln f(x) = j(x)$ rather than derivatives of $f(x)$. This same idea applies when $f(x)$ is the ratio of functions, $f(x) = \frac{j(x)}{k(x)}$, or the product, $f(x) = j(x)k(x)$, where again $\ln f(x)$ would be simpler to differentiate than $f(x)$ as long as the various functions are positive so that the logarithm is well defined. In these examples the function used in the composition is given by $g(x) = \ln x$.

Similar considerations apply if $f(x)$ is given as the natural logarithm of a positive function $f(x) = \ln j(x)$, where composing with $g(x) = e^x$ gives $e^{f(x)} = j(x)$, or if $f(x)$ is a power of a function $f(x) = j(x)^a$, where forming the composition with $g(x) = x^{1/a}$ produces a simpler function. In each case composition produces a simpler function to differentiate.

In all such cases the question is: What is the relationship between the critical points of $f(x)$ and those of $g(f(x))$? The next proposition summarizes the result.

Proposition 9.125 *Let $f(x)$ be a differentiable function, and $g(x)$ a differentiable function that is well defined on* $\mathrm{Rng}(f)$. *Then, if x_0 is a critical point of $f(x)$, it will also be a critical point of $g(f(x))$.*

Proof The function $h(x) \equiv g(f(x))$ is differentiable on $\mathrm{Dmn}(f)$, and using the results from proposition 9.76 produces

$$h'(x) = g'(f(x))f'(x).$$

Consequently, if $f'(x_0) - 0$, then $h'(x_0) - 0$.   ∎

In other words, the critical points of $f(x)$ will be a subset of the critical points of $h(x)$. However, we see from the formula above for $h'(x)$ that critical points of the transformed function $h(x)$ need not be critical points of $f(x)$ unless one knows that $g'(f(x_0)) \neq 0$.

Example 9.126 *Take $g(x) = e^x$, $\ln x$, or $x^{1/a}$. Then $g'(x) = e^x$, $\frac{1}{x}$, and $\frac{1}{a}x^{(1-a)/a}$, respectively. In the first two cases, since $g'(x)$ has no zero values, the critical points of $f(x)$ and those of $h(x)$ agree. In the third case of $g(x) = x^{1/a}$, it appears possible for $h(x)$ to inherit extra critical points at any value of x for which $f(x_0) = 0$, since then*

$$h'(x_0) = g'(f(x_0))f'(x_0) = \frac{1}{a}(f(x_0))^{(1-a)/a}f'(x_0) = 0.$$

But this conclusion requires that $\frac{1-a}{a} > 0$, which is equivalent to $\frac{1}{a} > 1$ or $0 < a < 1$, since otherwise, $(0)^{(1-a)/a}$ is meaningless. On the other hand, this transformation would typically only be considered when $f(x) = j(x)^a$, which equals 0 only when $j(x_0) = 0$. But, if $0 < a < 1$, such an $f(x)$ is not differentiable when $j(x_0) = 0$, so the differentiability of $f(x)$ assures that $j(x_0) \neq 0$ and no additional critical points are inherited in this case as well.

In summary, the three simple transformations illustrated above will exactly preserve the critical points of $f(x)$, as long as the differentiability assumptions of the
proposition are satisfied. For more general transformations, for which $g'(f(x_0)) = 0$
for some x_0, the critical points of $f(x)$ will be augmented by the critical points of
$g(x)$ on $\mathrm{Rng}(f)$.

We turn next to the second derivative test:

Proposition 9.127 *Let $f(x)$ be a twice differentiable function, and $g(x)$ a twice differentiable function that is well defined on $\mathrm{Rng}(f)$. Then, if x_0 is a critical point of $f(x)$
that is a relative maximum or relative minimum of $f(x)$, x_0 will have the same property
for $g(f(x))$ if $g'(f(x_0)) > 0$, and the opposite property if $g'(f(x_0)) < 0$.*

Proof The function $h(x) \equiv g(f(x))$ is twice differentiable on $\mathrm{Dmn}(f)$, and

$$h''(x) = g''(f(x))[f'(x)]^2 + g'(f(x))f''(x).$$

Consequently, if $f'(x_0) = 0$, then

$$h''(x_0) = g'(f(x_0))f''(x_0),$$

and $f''(x_0)$ and $h''(x_0)$ will have the same sign if $g'(f(x_0)) > 0$, and opposite signs if
$g'(f(x_0)) < 0$.   ∎

Example 9.128

1. *If the transforming function, $g(x)$, is an increasing function so that $g'(x) > 0$ for all
x, proposition 9.125 ensures that the critical points of $f(x)$ and $h(x)$ coincide, while
proposition 9.127 ensures that maximums will coincide with maximums, and minimums
with minimums. As examples, $g(x) = e^x$ is an increasing function for all x, while $\ln x$ is
an increasing function for $x > 0$, as is $x^{1/a}$ as long as $a > 0$.*

2. *If the transforming function is a decreasing function so that $g'(x) < 0$ for all x, the
critical points of $f(x)$ and $h(x)$ will again coincide, but maximums and minimums will
be reversed. In such a case is is easier to work with the transforming function: $\tilde{g}(x) \equiv
-g(x)$, which is increasing, to avoid the necessity of remembering that maximums and
minimums will reverse under $g(x)$.*

Recall the following problem from section 3.3.2 on tractability of the l_p-norms:
Suppose that we are given a collection of data points $\{x_i\}_{i=1}^n$ that we envision either
as distributed on the real line $\mathbb{R}$ or as a point $\mathbf{x} = (x_1, x_2, \dots, x_n) \in \mathbb{R}^n$. Assume that
for notational simplicity we arrange the data points in increasing order $x_1 \leq x_2 \leq \cdots$

$\leq x_n$. The goal is to find a single number x_p that best approximates these points in the l_p-norm, where $p \geq 1$. That is, find x_p so that

$\|(x_1 - x_p, x_2 - x_p, \ldots, x_n - x_p)\|_p$ is minimized.

This problem can be envisioned as a problem in $\mathbb{R}$ or as a problem in $\mathbb{R}^n$, but we choose the former to apply the tools of this chapter. The problem then becomes

$$\text{Minimize:} \quad f(x) = \left(\sum_{i=1}^{n} |x_i - x|^p \right)^{1/p}.$$

This problem was solved in chapter 3 by direct methods in the cases of $p = 1, 2, \infty$, where we recall that for $p = \infty$ the l_p-norm problem is defined as

$$\text{Minimize:} \quad f(x) = \max_{i}\{|x_i - x|\}.$$

We now return to this example for other values of p.

Example 9.129 *To apply the tools of this chapter, we require $f(x)$ to be differentiable, and we have seen in example 9.80 (case 8) that this requires that $1 < p < \infty$. Because $g(x) = x^p$ is an increasing function, the maximums and minimums of $f(x)$ and $f(x)^p$ agree as noted in example 9.128 above, with $\frac{1}{a} = p$. This follows because $f(x) > 0$ for all x except in the trivial case where all $x_j = c$, and so $f(c) = 0$. We ignore this case, since then $x_p = c$ apparently.*

Suppose that $\{x_j\}_{j=1}^{n}$ contain m different values, which we denote by $\{y_j\}_{j=1}^{m}$ in increasing order, and that the original set contains n_j of each y_j. The goal is then to find the minimum of $h(x) = g(f(x))$:

$$h(x) = \sum_{j=1}^{m} n_j |y_j - x|^p, \qquad 1 < p < \infty.$$

From (9.20), we have that

$$h'(x) = \begin{cases} -p \sum_{j=1}^{m} n_j (y_j - x)^{p-1}, & x \leq y_1, \\ p \sum_{j=1}^{k} n_j (x - y_j)^{p-1} - p \sum_{j=k+1}^{m} n_j (y_j - x)^{p-1}, & y_k \leq x \leq y_{k+1}, \\ p \sum_{j=1}^{m} n_j (x - y_j)^{p-1}, & y_m \leq x. \end{cases}$$

Note that $h'(x)$ is continuous, since its values at the interval endpoints $\{y_j\}_{j=1}^{m}$ are well defined even though they are defined piecewise continuously on intervals. Also $h'(x)$ is *negative for $x \leq y_1$ and positive for $y_m \leq x$. So by the intermediate value theorem, there is at least one point x', with*

$$y_1 < x' < y_m \quad and \quad h'(x') = 0.$$

Specifically, if $y_k \leq x' \leq y_{k+1}$, then

$$\sum_{j=1}^{k} n_j p(x' - y_j)^{p-1} = \sum_{j=k+1}^{m} n_j p(y_j - x')^{p-1}.$$

When $p = 2$, this equation can be explicitly solved, producing

$$x' = \frac{\sum_{j=1}^{m} n_j y_j}{\sum_{j=1}^{m} n_j} = \frac{1}{n} \sum_{j=1}^{n} x_j,$$

as derived in chapter 3.

We can confirm that x' is always unique for $1 < p < \infty$, since $h'(x)$ is a strictly increasing function. This is apparent for $x \leq y_1$ and $y_m \leq x$ but also for $y_k \leq x \leq y_{k+1}$, since as x increases, the positive summation increases and the negative summation decreases. This analysis also confirms that x' is a minimum of $h(x)$, as noted in proposition 9.124, so $x' = x_p$.

Note that in general, we can not use a second derivative test to confirm that x' is a minimum, since $h'(x)$ is differentiable only if $p \geq 2$. The differentiability problem for $1 < p < 2$ occurs for $x = y_j$ for any j, as $h'(x)$ is differentiable otherwise for any x. If we assume that $p \geq 2$, or if $1 < p < 2$ and $x' \neq y_j$ for any j, then the second derivative test can be used:

$$h''(x) = p(p-1) \sum_{j=1}^{m} n_j |y_j - x|^{p-2}.$$

From this we can conclude that $h''(x') > 0$, even though x' is not explicitly known, and hence x' is a minimum.

9.6   Concave and Convex Functions

9.6.1   Definitions

In previous chapters the notions of convexity and concavity have been encountered. First we recall the definitions:

Definition 9.130   *A function $f(x)$ is **concave** on an interval I, which can be open, closed or semi-closed, finite or infinite, if for any $x, y \in I$,*

$$f(tx + (1 - t)y) \geq tf(x) + (1 - t)f(y) \qquad for\ t \in [0, 1]. \tag{9.45}$$

*A function $f(x)$ is **convex** on I if, for any $x, y \in I$,*

$$f(tx + (1 - t)y) \leq tf(x) + (1 - t)f(y) \qquad for\ t \in [0, 1]. \tag{9.46}$$

*When the inequalities are strict for $t \in (0, 1)$, such functions are referred to as **strictly concave** and **strictly convex**, respectively.*

Remark 9.131   *Note that $f(x)$ is concave if and only if $-f(x)$ is convex, and conversely. Consequently most propositions need only be proved in one case, and the other case will follow once the effect of the minus sign on the result is reflected.*

Interestingly, the properties of concavity and convexity are quite strong. As it turns out, concave and convex functions are always continuous on open intervals and are in fact Lipschitz continuous.

Proposition 9.132   *If $f(x)$ is concave or convex on an open interval I, then it is Lipschitz continuous on I.*

Proof   We demonstrate this for a convex function $f(x)$. Then, if $g(x)$ is concave, the result follows from the continuity of the convex $-g(x)$. To this end, let $y \in I$ be given, and let $J = (y - a, y + a)$ be defined so that $[y - a, y + a] \subset I$. Since I is open, there is an open interval about y contained in I by definition, and we simply choose a smaller open interval J whose closure is also in I. Let $M = \max(f(y - a), f(y + a))$. For any $x \in J$, we conclude that $f(x) \leq M$, since any such point can be expressed $x = (1 - t)(y - a) + t(y + a)$ for some $t \in (0, 1)$, and since $f(x)$ is convex, (9.46) provides this conclusion. Now let $x \in J$ be given and assume for the moment that $x \geq y$. To standardize notation, let $x = y + ta$ for some $t \in [0, 1]$, then we have that

$$y - a < y \leq x \equiv y + ta < y + a.$$

Now, by construction,

$$x = (1 - t)y + t(y + a).$$

In order to write x as a linear combination of $y - a$ and this same y, an algebraic exercise produces

$$y = \frac{t}{1+t}(y-a) + \frac{1}{1+t}x,$$

where both $\frac{t}{1+t}, \frac{1}{1+t} \in [0,1]$. Now, from the convexity of $f(x)$, and the definition of M, we conclude that

$$f(x) \le (1-t)f(y) + tM,$$

$$f(y) \le \frac{t}{1+t}M + \frac{1}{1+t}f(x).$$

Using the first inequality for an upper bound, the second for the lower bound, provides

$$-t[M - f(y)] \le f(x) - f(y) \le t[M - f(y)].$$

That is,

$$|f(x) - f(y)| \le t|M - f(y)|.$$

Since $t = \frac{x-y}{a}$, we have the final result for Lipschitz continuity:

$$|f(x) - f(y)| \le \frac{|M - f(y)|}{a}(x-y) \qquad \text{for } x \ge y.$$

An identical construction applies when $x \le y$, by expressing $x = y - ta$, so $y - a < x \le y < y + a$. Combining the resulting inequalities, we get

$$|f(x) - f(y)| \le C|x - y|. \qquad\qquad\qquad \blacksquare$$

Example 9.133   *It is important to note that this proposition does not extend to the result that a convex/concave function on a closed interval is continuous. For example, on the interval* $[0,1]$, *define*

$$f(x) = \begin{cases} x(x-1), & 0 < x \le 1, \\ -1, & x = 0. \end{cases}$$

Then $f(x)$ *is apparently concave, and equally apparently, not continuous.*

When a function is differentiable, it is relatively easy to confirm when it is either concave or convex.

Proposition 9.134   *There are two derivatives-based tests that characterize convexity and concavity:*

1. *If $f(x)$ is differentiable, then:*

(a) *$f(x)$ is concave on an interval if and only if $f'(x)$ is a decreasing function on that interval.*

(b) *$f(x)$ is convex on an interval if and only if $f'(x)$ is an increasing function on that interval.*

(c) *$f(x)$ is strictly concave iff $f'(x)$ is strictly decreasing, and strictly convex iff $f'(x)$ is strictly increasing.*

2. *If $f(x)$ is twice differentiable, then:*

(a) *$f(x)$ is concave on an interval if and only if $f''(x) \leq 0$ on that interval.*

(b) *$f(x)$ is convex on an interval if and only if $f''(x) \geq 0$ on that interval.*

(c) *Strict concavity and strict convexity follow from $f''(x) < 0$, or $f''(x) > 0$, respectively.*

Remark 9.135

1. *We use the term "decreasing" in case 1 when we could have used the more complicated notion of "nonincreasing." The point is that "decreasing" here means that if $x < y$, then $f(x) \geq f(y)$. When we want to specify that $x < y \Rightarrow f(x) > f(y)$, we use the terminology "strictly decreasing Similar remarks apply to the term "increasing."*

2. *It may be apparent that the first five statements in this proposition were stated in terms of "$f(x)$ is concave/convex if and only if...." For part 2(c), the second derivative statement is not a characterization of strict concavity or convexity but is a sufficient condition. That this second derivative restriction is not necessary is easily exemplified by $f(x) = \pm x^4$ on the interval $[-1, 1]$, say. It is apparent that these functions are strictly convex $(+)$ and concave $(-)$ on the interval, yet $f''(0) = 0$.*

Proof Treating these statements in turn:

1. Given differentiable $f(x)$, and $y < x$, define the function:

$$g(t) = f(tx + (1 - t)y),$$

for $t \in [0, 1]$. Note that $g'(t) = f'(tx + (1 - t)y)(x - y)$. Applying (9.33) with $n = 0$, and $t_0 = 0, 1$, we get

$$g(t) = g(0) + tg'(\theta_1), \qquad 0 < \theta_1 < t,$$

$$g(t) = g(1) + (t - 1)g'(\theta_2), \qquad t < \theta_2 < 1.$$

Substituting back the original functions produces

$$f(tx + (1 - t)y) = f(y) + t(x - y)f'(y + \theta_1(x - y)),$$

$$f(tx + (1 - t)y) = f(x) + (t - 1)(x - y)f'(y + \theta_2(x - y)).$$

Next, multiplying the first equation by $1 - t$ and the second by t and adding produces

$$f(tx + (1 - t)y) = (1 - t)f(y) + tf(x) + E(t)$$

where the error function is defined as

$$E(t) = (x - y)t(1 - t)[f'(y + \theta_1(x - y)) - f'(y + \theta_2(x - y))].$$

To investigate the sign of $E(t)$, recall $y < x$. So the sign of $E(t)$ is the same as the sign of the term in square brackets. Now since $\theta_1 < \theta_2$ by construction, $y + \theta_1(x - y) < y + \theta_2(x - y)$ and we conclude that:

$E(t) \geq 0$ iff $f'(x)$ is decreasing, and then $f(x)$ is concave,

$E(t) \leq 0$ iff $f'(x)$ is increasing, and then $f(x)$ is convex.

If $f'(x)$ is strictly monotonic, then $f(x)$ is either strictly concave or strictly convex, since then $E(t) > 0$ or $E(t) < 0$, respectively.

2. Turning next to twice differentiable $f(x)$, let $y < x$ be given. Applying (9.33) to $f'(x)$ with $n = 0$, and $x_0 = y$, we get

$$f'(x) = f'(y) + (x - y)f''(\theta), \qquad y < \theta < x.$$

Now, if $f''(\theta) \leq 0$, for all θ, it is apparent that $f'(x) \leq f'(y)$, and hence $f'(x)$ is a decreasing function and $f(x)$ is concave by part 1. Similarly, if $f''(\theta) \geq 0$, we conclude that $f(x)$ is convex.

So the restrictions on $f''(x)$ in parts 2(a) and 2(b) assure concavity and convexity. To demonstrate that these restrictions on $f''(x)$ are assured by the assumptions of convavity or convexity, we argue the concavity result by contradiction, and the convexity result is identical. Assume that $f(x)$ is concave on an interval and that there is some x in the interval with $f''(x) > 0$. Then

$$\lim_{t \to 0} \frac{f'(x + t) - f'(x)}{t} > 0.$$

By definition of limit, we conclude that there exists $\epsilon > 0$ so that $\frac{f'(x+t) - f'(x)}{t} > 0$ for $|t| < \epsilon$. Hence, taking $0 < t < \epsilon$, we conclude that

$f'(x + t) > f'(x)$.

So $f'(x)$ is a strictly increasing function on $[x, x + \epsilon)$, contradicting the concavity of $f(x)$ by part 1(a).

Finally, for part 2(c) if $f''(\theta) < 0$ or $f''(\theta) > 0$ for all θ, then strict concavity (respectively, strict convexity) is assured by the identity above between $f'(x)$ and $f'(y)$ for $y < x$. ∎

Example 9.136

1. *As noted in section 3.1.5 for the proof of Young's inequality, $f(x) = \ln x$ is concave, in fact strictly concave, on $(0, \infty)$. This function has derivatives $f'(x) = \frac{1}{x}$ and $f''(x) = -\frac{1}{x^2}$. Observing that $f'(x)$ is strictly decreasing, or that $f''(x) < 0$, on $(0, \infty)$, the conclusion follows.*

2. *As noted in section 3.2.2 and in the proof of proposition 6.33, $f(x) = x^p$ is strictly convex on $(0, \infty)$ for $p > 1$. Here $f'(x) = px^{p-1}$ and $f''(x) = p(p - 1)x^{p-2}$. Observing that $f'(x)$ is strictly increasing, or that $f''(x) > 0$, on $(0, \infty)$, the conclusion follows.*

3. *As a third example, $f(x) = e^x$ is strictly convex on $\mathbb{R}$, since $f'(x) = e^x$ is strictly increasing. Alternatively, $f''(x) = e^x > 0$ for all x.*

Returning to the discussion on **points of inflection**, we begin with a definition.

Definition 9.137 *A point x_0 is a **point of inflection of** $f(x)$ if there is an interval (a, b) containing x_0 so that $f(x)$ is concave on (a, x_0) and convex on (x_0, b), or conversely.*

Example 9.138 *The point $x = 0$ is a point of inflection of $f(x) = x^3$, since $f''(x) = 6x$, which is positive for $x > 0$, and hence $f(x)$ is convex on $(0, \infty)$. Also $f''(x)$ is negative for $x < 0$, and so $f(x)$ is concave on $(-\infty, 0)$. For this example, $f'(x) = 0$, so $x = 0$ is also a critical point. But inflection points need not be critical points. For example, $g(x) = x^3 + bx$ satisfies $g''(x) = 6x$, so $x = 0$ is again an inflection point, and yet $g'(0) = b$ can be any value we choose.*

In the same way that potential relative maximums and minimums can be identified by inspecting the critical points of a function where $f'(x) = 0$, there is a necessary condition in order for a point to be a point of inflection.

Proposition 9.139 *If x_0 is a point of inflection of a twice differentiable function $f(x)$ with $f''(x)$ continuous, then $f''(x_0) = 0$.*

Proof This follows immediately from proposition 9.134 above, since a twice differentiable function satisfies $f''(x) \leq 0$ when concave and $f''(x) \geq 0$ when convex.

Since $f''(x)$ is continuous, $f''(x_0) = \lim_{x \to x_0} f(x)$, and this common value must therefore be 0.   ∎

Example 9.140   *As noted in section 9.5.1, functions of the form, $f(x) = ax^n$, for integer $n > 2$, and $a \in \mathbb{R}$, provide a variety of possible behaviors when $f''(0) = 0$. For n even, it is apparent that $x_0 = 0$ is a relative minimum if $a > 0$, and a relative maximum if $a < 0$. For n odd, it is also apparent that for $a > 0$, the second derivative satisfies $f''(x) > 0$ for $x > 0$ and conversely for $x < 0$. Hence $x_0 = 0$ is a point in inflection. The same conclusion is reached for $a < 0$.*

More generally, as noted in remark 9.123, if $f(x)$ is a function with $f^{(j)}(x_0) = 0$ for $j = 1, \ldots, n-1$, and $f^{(n)}(x_0) \neq 0$, with $f^{(n)}(x)$ continuous, then if n is even, x_0 will be a relative minimum if $f^{(n)}(x_0) > 0$ and a relative maximum if $f^{(n)}(x_0) < 0$. This follows from (9.24):

$$f(x) = f(x_0) + \frac{1}{n!} f^{(n)}(y)(x - x_0)^n,$$

where y is between x and x_0. Since $f^{(n)}(x)$ is continuous, there is an interval about x_0, I, within which $f^{(n)}(y)$ has the same sign as $f^{(n)}(x_0)$, as noted in proposition 9.38. Consequently, if $f^{(n)}(y) > 0$ for $y \in I$, then since n is even, $f(x) \geq f(x_0)$ and x_0 is a relative minimum, and the same argument applies if $f^{(n)}(y) < 0$.

It was also noted that if n is odd, x_0 will be a point of inflection independent of the sign of $f^{(n)}(x_0)$. To see this, note that (9.24) can also be applied to the function $g(x) = f''(x)$, for which $g(x_0) = 0$ and $g^{(j)}(x_0) = 0$ for $j = 1, \ldots, n-3$:

$$g(x) = \frac{1}{(n-2)!} g^{(n-2)}(y)(x - x_0)^{n-2}.$$

In other words,

$$f''(x) = \frac{1}{(n-2)!} f^{(n)}(y)(x - x_0)^{n-2}.$$

Now, if $f^{(n)}(y) > 0$ for $y \in I$, then since n is odd, $f''(x) < 0$ for $x < x_0$ and conversely for $x > x_0$. If $f^{(n)}(y) < 0$ for $y \in I$, the same argument applies and produces $f''(x) > 0$ for $x < x_0$, and conversely for $x > x_0$. So since $f''(x)$ changes sign at $x = x_0$, this point is a point of inflection by proposition 9.134.

9.6.2   Jensen's Inequality

An important consequence of a function $f(x)$ being concave or convex is that it allows the prediction of the relationship between

$E[f(X)]$ and $f(E[X])$,

where X is a random variable with a given probability density function $g(x)$, and E denotes the expectation of the given quantity as defined in chapter 7. The result that will be developed here will apply only to discrete p.d.f.s at this time, but once the necessary tools are developed, it can be shown to be true in a far more general context.

To this end, first note that the definition of convexity and concavity, while given in the context of two points, is true for any finite number.

Proposition 9.141   *If $f(x)$ is concave on an interval I, and $\{x_i\}_{i=1}^n \subset I$ and $\{t_i\}_{i=1}^n \subset \mathbb{R}$ with $t_i \geq 0$ for all i and $\sum t_i = 1$, then*

$$f\left(\sum_{i=1}^n t_i x_i\right) \geq \sum_{i=1}^n t_i f(x_i). \tag{9.47}$$

Similarly, if $f(x)$ is convex, then

$$f\left(\sum_{i=1}^n t_i x_i\right) \leq \sum_{i=1}^n t_i f(x_i). \tag{9.48}$$

Proof   The proof is by induction. The result is true for $n = 2$ by definition. Assuming it is true for n, let $\{x_i\}_{i=1}^{n+1} \subset I$, and $\{t_i\}_{i=1}^{n=1} \subset \mathbb{R}$ be given. Define

$$t = t_1, \quad x = x_1, \quad 1 - t = \sum_{i=2}^{n+1} t_i, \quad y = \frac{\sum_{i=2}^{n+1} t_i x_i}{\sum_{i=2}^{n+1} t_i},$$

and apply the definition to $f(tx + (1-t)y)$, obtaining in the convex case

$$f\left(\sum_{i=1}^n t_i x_i\right) \leq t_1 f(x_1) + \left(\sum_{i=2}^{n+1} t_i\right) f\left(\sum_{i=2}^{n+1} s_i x_i\right),$$

where $s_i = \frac{t_i}{\sum_{i=2}^{n+1} t_i}$. Now since $\sum_{i=2}^{n+1} s_i = 1$, apply the assumption that the result holds for n to this last term, obtaining

$$f\left(\sum_{i=2}^{n+1} s_i x_i\right) \leq \sum_{i=2}^{n+1} s_i f(x_i),$$

and the proof is complete after substitution for s_i and multiplication by $\left(\sum_{i=2}^{n+1} t_i\right)$. ∎

This result has two immediate applications. The first is to the proof of the **arithmetic-geometric mean inequality**.

Proposition 9.142   *If* $\{x_i\}_{i=1}^{n} \subset \mathbb{R}$, *and* $x_i \geq 0$ *for all i, then*

$$\frac{1}{n}\sum_{i=1}^{n} x_i \geq \left(\prod_{i=1}^{n} x_i\right)^{1/n}.$$ (9.49)

Proof   See exercise 12.   ∎

Now consider the earlier question on the relationship between $E[f(X)]$ and $f(E[X])$. If X is a finite discrete random variable, with p.d.f. $g(x)$ and range $\{x_i\}_{i=1}^{n}$, then since $g(x_i) > 0$ for all i, and $\sum_{i=1}^{n} g(x_i) = 1$, proposition 9.141 assures that

$E[f(X)] \leq f(E[X])$ if $f(x)$ is concave,

$E[f(X)] \geq f(E[X])$ if $f(x)$ is convex.

Both results follow from

$$E[f(X)] = \sum_{i=1}^{n} f(x_i)g(x_i),$$

$$f(E[X]) = f\left(\sum_{i=1}^{n} x_i g(x_i)\right).$$

Rather than formalize this limited result, we generalize it to the case of an arbitrary discrete p.d.f., for which we need a new approach.

Proposition 9.143   *If* $f(x)$ *is differentiable, then for any a,*

$f(x) \leq f(a) + f'(a)(x - a)$ *if* $f(x)$ *is concave,*

$f(x) \geq f(a) + f'(a)(x - a)$ *if* $f(x)$ *is convex.*

In addition, if $f(x)$ *is strictly concave or strictly convex, then the inequalities are strict.*

Remark 9.144   *This result is true without the assumption of differentiability, but where* $f'(a)$ *is replaced by a different function of a. This function is closely related to the "derivative," and in fact is defined as a one-sided derivative whereby in the definition in (9.8),* Δx *is restricted to be only positive or only negative. It then turns out that* *concave and convex functions have both of these one-sided derivatives at every point, and that they agree, except perhaps on a countable collection of points. In other words, concave or convex $f(x)$ is not only Lipschitz continuous as proved in proposition 9.132, but also differentiable, except perhaps on a countable collection of points. However, we have no further use for this generalization, so we will not develop it. We will instead simply assume differentiability.*

Proof   By the mean value theorem, we have that for any a,

$$f(x) = f(a) + f'(\theta)(x - a),$$

where θ is between x and a. For example, if $x > a$, then $x > \theta > a$. Now, if $f(x)$ is concave, $f'(x)$ is a decreasing function, and hence $f'(\theta) \leq f(a)$ if $x > a$, and $f'(\theta) \geq f(a)$ if $x < a$. In both cases $f'(\theta)(x - a) \leq f'(a)(x - a)$. If $f(x)$ is convex, the inequalities reverse. When strictly concave or strictly convex, the first derivative inequalities are sharp and so too are the inequalities in the conclusion.   ∎

We now turn to an important result related to concave and convex functions, known as **Jensen's inequality**, and named for its discoverer, **Johan Jensen**, (1859–1925).

Proposition 9.145 (*Jensen's Inequality*)   *Let $f(x)$ be a differentiable function, and X a discrete random variable with range contained in the domain of f, namely* $\mathrm{Rng}(X) \subset \mathrm{Dmn}(f)$. *Then*

$$\mathrm{E}[f(X)] \leq f(\mathrm{E}[X]) \quad \textit{if } f(x) \textit{ is concave,} \tag{9.50a}$$

$$\mathrm{E}[f(X)] \geq f(\mathrm{E}[X]) \quad \textit{if } f(x) \textit{ is convex.} \tag{9.50b}$$

If strictly concave or strictly convex, the inequalities are strict.

Proof   Let $a = \mathrm{E}[X]$ in the proposition 9.143. Since $\mathrm{E}[f'(a)(x - a)] = f'(a)\mathrm{E}[(x - a)] = 0$, the result follows.   ∎

Remark 9.146

1. *Continuous probability distributions will be studied in chapter 10, but it is noted here that once introduced and the notion of $\mathrm{E}[f(X)]$ is defined, the simplicity of the proof above will carry over to this case without modification.*

2. *Note that an easy calculation directly demonstrates that if $f(x)$ is an affine function, $f(x) = ax + b$ for constants a and b, which is both concave and convex, then*

$$\mathrm{E}[f(X)] = f(\mathrm{E}[X]).$$

9.7   Approximating Derivatives

As it turns out, the Taylor series approximations in the earlier sections can be used not only to approximate a given function but also in developing approximation formulas for its various derivatives.

9.7.1   Approximating $f'(x)$

For a function with only one derivative, we have directly from the definition in (9.8) that $f'(x_0)$ can be approximated by $\frac{f(x)-f(x_0)}{\Delta x}$, but this provides no information on the rate of convergence. Using (9.27) with $n = 1$ does not help, as even in the case of continuous $f'(x)$ the error is seen to be $o(\Delta x)/\Delta x = o(1)$, which just means the error goes to 0 at some rate of speed, a fact already known from the existence of $f'(x)$.

If we assume that $f(x)$ has two derivatives, we can use (9.27) with $n = 2$. Specifically, from $f(x) = \sum_{j=0}^{2} \frac{1}{j!} f^{(j)}(x_0)(x - x_0)^j + O(\Delta x^2)$ we obtain by subtracting $f(x_0)$, dividing by Δx, and solving for $f'(x_0)$,

$$f'(x_0) \approx \frac{f(x_0 + \Delta x) - f(x_0)}{\Delta x} + O(\Delta x). \tag{9.51}$$

This approximation formula is known as the **forward difference approximation**, and the error of $O(\Delta x)$ comes from the second derivative term in (9.27) divided by Δx.

This approximation can be improved if there are three derivatives, by applying (9.27) with $n = 3$ to both $f(x_0 + \Delta x)$ and $f(x_0 - \Delta x)$ and subtracting. Then the second derivative terms cancel out, and we obtain

$$f'(x_0) \approx \frac{f(x_0 + \Delta x) - f(x_0 - \Delta x)}{2\Delta x} + O(\Delta x^2). \tag{9.52}$$

This approximation formula is known as the **central difference approximation**, and the error term comes from the $O(\Delta x^3)$ term in (9.27) divided by Δx.

The formula in (9.52) can also be applied if $f(x)$ has only two derivatives, but then the error is again $O(\Delta x)$ as in (9.51).

9.7.2   Approximating $f''(x)$

Once again applying (9.27) with $n = 3$ to both $f(x_0 + \Delta x)$ and $f(x_0 - \Delta x)$ and adding, then subtracting $2f(x_0)$, we obtain in the case of three derivatives:

$$f''(x) \approx \frac{f(x_0 + \Delta x) + f(x_0 - \Delta x) - 2f(x_0)}{(\Delta x)^2} + O(\Delta x). \tag{9.53}$$

This approximation formula is also known as the **central difference approximation**, and the error term comes from the $O(\Delta x^3)$ term in (9.27) divided by Δx^2. If $f(x)$ has four derivatives at x_0, we can apply (9.27) with $n = 4$. The resulting error term will be $O(\Delta x^2)$, since then the third derivatives will cancel.

9.7.3 Approximating $f^{(n)}(x)$, $n > 2$

Methods similar to those above can be applied but are somewhat more complex. The reason is that one needs to determine collections of increments, $\{\Delta x_j\}_{j=1}^n$, and numerical coefficients, $\{a_j\}_{j=1}^n$, so that the Taylor polynomial for $\sum_{j=1}^n a_j f(x_0 + \Delta x_j)$ will have all derivative terms cancel, except for the last, which we wish to approximate. One then solves for this last derivative, producing the desired approximation formula and associated error term. This problem is readily solvable with the tools of linear algebra.

9.8 Applications to Finance

9.8.1 Continuity of Price Functions

Continuity is a pervasive notion in many applications, including those in finance, and one that tends to be assumed in virtually every situation without question, or even explicit recognition. For example, the value at time t of a \$100 investment at time 0 with an annual rate of interest of r is given by

$$f(r, t) = 100(1 + r)^t.$$

Fixing r for the moment, it would be nearly universally assumed that f is a continuous function of t, in that for any t_0,

$$\lim_{t \to t_0} f(r, t) = f(r, t_0).$$

In other words, the value of the investment grows smoothly with time; there are no unexpected jumps in the account value. One similarly assumes that for t fixed, if r is close to r_0, it would be expected to be the case that $f(r, t)$ will be close to $f(r_0, t)$, and $\lim_{r \to r_0} f(r, t) = f(r_0, t)$.

Of course, this function is not uniformly continuous in either r or t unless we restrict the range of allowable values to a closed and bounded interval. A 25 basis point change in r has a much smaller absolute effect on f when r is large than when r is small. In other words, given ϵ, the value of δ needed so that $|r - r_0| < \delta$ implies that $|f(r,t) - f(r_0,t)| < \epsilon$ increases as r_0 increases. For $t = 15$ and $r_0 = 0.05$, a value of $\epsilon = \$1$ can be achieved with $\delta \approx 0.0007$ or about 7 basis points, whereas for $r_0 = 0.25$, the associated $\delta \approx 0.00082$ or about 8.2 basis points.

This lack of uniform continuity is fairly mild and uneventful over the typical range of market rates, and is also mild compared to that observed when one considers f as a function of t. In this case, δ decreases as t_0 increases. Again starting with $t_0 = 15$ and $r = 0.05$, a value of $\epsilon = \$1$ can be achieved with $\delta \approx 0.0007$, whereas for $t_0 = 30$, the associated $\delta \approx 0.00035$ or about 3.5 basis points.

Similar remarks apply to the host of fixed income type pricing formulas. For example, a general discounted present value of a series of cash flows

$$f(r) = \sum_{t=0}^{n} c_t(1+r)^{-t},$$

as well as the counterpart formula for an n-year semiannual coupon bond in (2.15),

$$P(i,r) = F\frac{r}{2}a_{2n;i/2} + Fv_{i/2}^{2n},$$

are given by continuous functions.

A similar conclusion applies to the price of a preferred stock in (2.21),

$$P(i,r) = \frac{Fr}{i}, \qquad i > 0,$$

or the valuation of common stock using the discounted dividend model with growth in (2.22),

$$V(D,g,r) = D\frac{1+g}{r-g}, \qquad r > g,$$

as well as to forward prices on a given traded security in (2.24),

$$F_0(S_0, r_T, T) = S_0(1 + r_T)^T.$$

Within the domains of these functions, identified with the tools in this chapter as functions of a single variable by holding the others constant, intuition compels that each will produce continuous pricing results, although typically not uniformly continuous. Based on the theory above, we easily confirm that such intuition if formally verifiable on the respective price function domains.

9.8.2 Constrained Optimization

The notion of continuity is important for **constrained optimization** problems. As seen in chapters 3 and 4, a general problem can be framed as

Maximize (minimize): $g(\mathbf{x})$,

Given: $\mathbf{x} \in A \equiv \{\mathbf{x} \in \mathbb{R}^n \mid f(\mathbf{x}) = c\}$.

Since $A = f^{-1}(c)$, if f^{-1} is continuous, then the topological result in proposition 9.67 generalizes and A will be a compact set because c is compact. In addition, generalizing the proposition 9.39 result for continuous functions on closed and bounded intervals, if $g(\mathbf{x})$ is continuous, it must attain its maximum and minimum on every compact set. So continuity provides a theoretical assurance of the existence of at least one solution to such optimization problems. A similar analysis applies if $A = \{\mathbf{x} \mid f(\mathbf{x}) \in C\}$ where C is any compact set, or if the problem has a finite number of constraints, and $A = \bigcap_j \{\mathbf{x} \mid f_j(\mathbf{x}) \in C_j\}$ where C_j is compact for all j.

9.8.3 Interval Bisection

Another example comes from chapters 4 and 5 where **interval bisection** was introduced as a method to solve equations of the form

$$f(x) = c.$$

In those chapters this method was illustrated with $f(x)$ denoting the price of a bond with yield x and c denoting the bond's current price. In other words, the goal was to find the bond's yield to maturity.

This method involves constructing two sequences of values: $\{x_n^+\}$ and $\{x_n^-\}$ with the property that:

1. $x_n^+ \leq x_n^-$,
2. $f(x_n^-) \leq c \leq f(x_n^+)$,
3. $|x_n^+ - x_n^-| \leq \frac{|x_0^+ - x_0^-|}{2^n}$; that is, $|x_n^+ - x_n^-| = O(2^{-n})$.

In chapter 5 it was shown that $x_n^+ - x_n^- \to 0$ implies that there is an $\bar{x}$ to which both sequences converge. Then, if $f(x)$ is a continuous function, as is the case for the price function of a bond, it is also sequentially continuous. Consequently, $x_n^{+/-} \to \bar{x}$ assures that $f(x_n^{+/-}) \to f(\bar{x})$. Finally, because $f(x_n^-) \leq c \leq f(x_n^+)$, we conclude that $f(\bar{x}) = c$.

Of course, if $f(x)$ is a continuous function, the intermediate value theorem assures the existence of $\bar{x}$ with $f(\bar{x}) = c$ as soon as the first two terms of the sequences are found with $f(x_0^-) \le c \le f(x_0^+)$. The method of interval bisection simply provides a numerical procedure for estimating this value.

9.8.4   Minimal Risk Asset Allocation

Say that two risky assets are given, A_1 and A_2, to which we desire to allocate a given dollar investment with weights w_1 and $w_2 = 1 - w_1$. Let the return random variables be denoted R_j, $j = 1, 2$, and analogously, the mean returns and standard deviation of returns denoted μ_j and σ_j, $j = 1, 2$; let the correlation between these returns be ρ.

The portfolio random return is given as a function of weight $w \equiv w_1$:

$$R = wR_1 + (1 - w)R_2.$$

Using the results from chapter 7, we derive

$$E[R] = w\mu_1 + (1 - w)\mu_2, \tag{9.54a}$$

$$\text{Var}[R] = w^2\sigma_1^2 + (1 - w)^2\sigma_2^2 + 2w(1 - w)\rho\sigma_1\sigma_2. \tag{9.54b}$$

Considered as a function of w, it is apparent that $E[R] = \mu_2 + (\mu_1 - \mu_2)w$ achieves its maximum and minimum only at the endpoints of any allowable interval for w, such as $[0, 1]$ if no short positions are allowed. In other words, $E[R]$ has no critical points. On the other hand,

$$\text{Var}[R] = (\sigma_1^2 + \sigma_2^2 - 2\rho\sigma_1\sigma_2)w^2 + 2\sigma_2(\rho\sigma_1 - \sigma_2)w + \sigma_2^2,$$

is a quadratic function of w, and hence it has a minimum or maximum depending on the sign of the coefficient of w^2.

This coefficient of w^2 is evidently positive when $\sigma_1 \ne \sigma_2$, since $-1 \le \rho \le 1$ by proposition 7.43, and

$$\sigma_1^2 + \sigma_2^2 - 2\rho\sigma_1\sigma_2 = (\sigma_1 - \sigma_2)^2 + 2(1 - \rho)\sigma_1\sigma_2.$$

Hence there is a minimal risk allocation. If $\sigma_1 = \sigma_2$, the same conclusion applies unless $\rho = 1$, in which case $\text{Var}[R]$ is constant and $E[R]$ is linear, and acheives its maximum and minimum at the endpoints of any allowable interval for w.

Denoting $\text{Var}[R]$ as $V(w)$, we have that

$$V'(w) = 2(\sigma_1^2 + \sigma_2^2 - 2\rho\sigma_1\sigma_2)w + 2\sigma_2(\rho\sigma_1 - \sigma_2).$$

Hence the risk-minimizing critical point, where $V'(w^{min}) = 0$, is given by

$$w^{min} = \frac{\sigma_2(\sigma_2 - \rho\sigma_1)}{\sigma_1^2 + \sigma_2^2 - 2\rho\sigma_1\sigma_2}. \tag{9.55}$$

Since $V''(w) = 2(\sigma_1^2 + \sigma_2^2 - 2\rho\sigma_1\sigma_2) > 0$ except in the trivial case of $\sigma_1 = \sigma_2$ and $\rho = 1$, the second derivative test confirms what we already knew, that w^{min} is a relative minimum of this variance function.

Since the denominator of w^{min} is, with one exception, always positive, the sign of w^{min} is determined by the sign of the numerator, $\sigma_2(\sigma_2 - \rho\sigma_1)$, which is determined by the sign of $\sigma_2 - \rho\sigma_1$. Specifically, the risk-minimizing allocation to A_1 satisfies

$$w^{min} > 0 \quad \text{if } \rho < \frac{\sigma_2}{\sigma_1}, \tag{9.56a}$$

$$w^{min} = 0 \quad \text{if } \rho = \frac{\sigma_1}{\sigma_2}, \tag{9.56b}$$

$$w^{min} < 0 \quad \text{if } \rho > \frac{\sigma_2}{\sigma_1}. \tag{9.56c}$$

It is easy to verify that if one of these assets is the risk-free asset, this analysis yields the obvious conclusion that the minimal risk allocation is $w_j = 1$ in the risk-free asset. (See exercise 39.)

9.8.5 Duration and Convexity Approximations

The same way that many of the most common pricing functions above can be shown to be continuous, they are easily shown to be differentiable on their domains of definition. For instance, the price of an n-year bond with annual cash flows and annual yield, $f(r) = \sum_{t=0}^{n} c_t(1 + r)^{-t}$, is easily differentiated to produce

$$f'(r) = -\sum_{t=1}^{n} tc_t(1 + r)^{-t-1},$$

$$f''(r) = \sum_{t=1}^{n} t(t + 1)c_t(1 + r)^{-t-2}.$$

For the price of a preferred stock, with $P(i) = \frac{Fr}{i}$, we have $P'(i) = -\frac{Fr}{i^2}$ and $P''(i) = \frac{2Fr}{i^3}$.

With such derivatives, one can then approximate the bond price at r based on information on the bond price function at r_0, and similarly for the preferred stock, using (9.26) and error estimates based on (9.33) or (9.27). In general, for fixed income applications, such approximations are restated in terms of relative derivatives, defined as follows:

Definition 9.147   *If $f(r)$ denotes the price of a fixed income security as a function of its yield r, the* **(modified) duration of** *$f(r)$* **at** *r_0, denoted $D(r_0)$, and the* **convexity of** *$f(r)$* **at** *r_0, denoted $C(r_0)$, are defined when $f(r_0) \neq 0$ by*

$$D(r_0) = -\frac{f'(r_0)}{f(r_0)} \tag{9.57a}$$

$$= \frac{\sum_{t=1}^{n} tc_t(1+r_0)^{-t-1}}{\sum_{t=0}^{n} c_t(1+r_0)^{-t}}, \tag{9.57b}$$

$$C(r_0) = \frac{f''(r_0)}{f(r_0)} \tag{9.58a}$$

$$= \frac{\sum_{t=1}^{n} t(t+1)c_t(1+r_0)^{-t-2}}{\sum_{t=0}^{n} c_t(1+r_0)^{-t}}. \tag{9.58b}$$

Of course, duration and convexity are functions of r as is the original price function, but in practice, one is often focused on the value of these functions at the current yield level of r_0 rather than in their functional attributes. The formulas above reflect the assumption of annual cash flows and an annual yield rate r and are easily generalized. For instance, with semiannual yields and cash flows we have for an n-year security: $f(r) = \sum_{t=0}^{2n} c_{t/2}\left(1+\frac{r}{2}\right)^{-t}$; and duration and convexity are again defined as relative derivatives of this function. For instance,

$$D(r_0) = -\frac{f'(r_0)}{f(r_0)} = \frac{\sum_{t=1}^{2n} \frac{1}{2} tc_{t/2}\left(1+\frac{r_0}{2}\right)^{-t-1}}{\sum_{t=0}^{2n} c_{t/2}\left(1+\frac{r_0}{2}\right)^{-t}}.$$

For the preferred stock, one has $D(i_0) = \frac{1}{i_0}$.

Also note that the definition of duration above is often labeled **modified duration** to distinguish it from an earlier notion of **Macaulay duration**, named for **Frederick Macaulay (1882–1970)**. Macaulay introduced this calculation in 1938, which in the annual yield case is

$$D^{Mac}(r_0) = \frac{\sum_{t=1}^{n} tc_t(1 + r_0)^{-t}}{\sum_{t=0}^{n} c_t(1 + r_0)^{-t}}, \tag{9.59}$$

with analogous definitions for other yield nominal bases. Modified duration is then easily seen to equal Macaulay duration divided by $(1 + r)$, or in the semiannual case by $\left(1 + \frac{r}{2}\right)$, and so forth.

Note that this Macaulay duration formula can be interpreted as a weighted "time to cash receipt" measure:

$$D^{Mac}(r_0) = \sum_{t=1}^{n} tw_t,$$

$$w_t = \frac{c_t(1 + r)^{-t}}{\sum_{t=0}^{n} c_t(1 + r)^{-t}}.$$

Using the values in (9.57) and (9.58), one then has the following approximations from (9.26):

$$f(r) \approx f(r_0)[1 - D(r_0)(r - r_0)], \tag{9.60}$$

known as the **duration approximation**, as well as

$$f(r) \approx f(r_0)\left[1 - D(r_0)(r - r_0) + \frac{1}{2}C(r_0)(r - r_0)^2\right], \tag{9.61}$$

known as the **duration approximation with a convexity adjustment**. The second formula provides one way to understand and quantify the price sensitivity "benefit" of a large, positive convexity value. Whether rates increase or decrease, a large positive convexity value will improve the benefit of duration when this duration effect is positive, and it will mitigate somewhat the harm of duration when this duration effect is negative. This convexity benefit is offset, of course, by the price one predictably pays for this extra convexity in terms of a lower yield.

Note that the historical justification for labeling the measure in (9.57) as "modified duration" was that it was recognized that Macaulay duration could be used to approximate the price change of a bond, as in (9.60), if this measure was first modified by dividing by a factor $(1 + r)$, or in the semiannual case, $\left(1 + \frac{r}{2}\right)$, and so forth, thereby producing a modified duration measure.

Dollar-Based Measures

In the case where $f(r_0) = 0$, which can easily happen when $f(r)$ denotes the price of a net portfolio such as a long/short bond portfolio, or a hedged bond portfolio, or when $f(r)$ is the price of a derivatives contract such as an interest rate swap or futures contract, duration and convexity are not defined. In this case one works with **dollar duration**, $D^{\$}(r_0)$, and **dollar convexity**, $C^{\$}(r_0)$. In general, these measures are defined in one of two ways as follows:

$$D^{\$}(r_0) \equiv D(r_0)f(r_0) = -f'(r_0), \tag{9.62a}$$

$$C^{\$}(r_0) \equiv C(r_0)f(r_0) = f''(r_0). \tag{9.62b}$$

When $f(r_0) = 0$ and duration and convexity are not defined, these dollar measures are defined directly in terms of the price functions derivatives.

In this case of $f(r_0) = 0$, the approximation formulas in (9.60) and (9.61) more closely resemble standard Taylor series polynomials in (9.34), except for the conventional use of $D^{\$}(r_0) = -f'(r_0)$. So the formulas become

$$f(r) \approx f(r_0) - D^{\$}(r_0)(r - r_0), \tag{9.63}$$

$$f(r) \approx f(r_0) - D^{\$}(r_0)(r - r_0) + \frac{1}{2}C^{\$}(r_0)(r - r_0)^2. \tag{9.64}$$

From (9.27) we see that in all cases the error in the duration approximation is $O(\Delta r)$, while with a convexity adjustment it is $O(\Delta r)^2$. Using (9.34), one can also express the maximum error in the duration approximation for $\frac{f(r)}{f(r_0)}$ in terms of the maximum of the convexity function between r and r_0:

$$\left| \frac{f(r)}{f(r_0)} - [1 - D(r_0)(r - r_0)] \right| \le \frac{M}{2}(r - r_0)^2, \qquad M = \max_{\tilde{r} \in \{r, r_0\}} |C(\tilde{r})|.$$

Similarly the formula with a convexity adjustment involves the maximum of $\left| \frac{f^{(3)}(r)}{f(r)} \right|$ on $\{r, r_0\}$ where this notation is intended to denote the interval $[r, r_0]$ or $[r_0, r]$, depending on which of r_0 and r is larger.

When $f(r_0) = 0$, these error bounds follow directly from (9.34). So M reflects the maximum of $|f''(r)|$ on $\{r, r_0\}$ for the duration approximation and the maximum of $|f^{(3)}(r)|$ on $\{r, r_0\}$ for the approximation with a convexity adjustment.

Embedded Options

For more complicated fixed income price functions, such as those associated with securities with embedded options, the approximations above are again used. However, because there is no formulaic approach to calculating derivatives in this case, such derivatives are approximated using formulas such as in (9.51), (9.52), and (9.53) for an appropriately choosen value of Δr. In such cases one often calls the associated duration and convexity measures **effective duration** and **effective convexity**, in part to highlight the fact that embedded options have been accounted for and in part to highlight the dependency on an assumed Δr value used in the estimate. More important, this terminology is intended to distance such calculations from those for fixed cash flow securities for which these measures also have interpretations in terms of the time distribution of the cash flows. When embedded options are present, all such connections may cease to exist.

For example, a security such as an **interest only (IO) strip** of a collateralized mortgage obligation (CMO) can have a negative effective duration, despite the fact that all payments are made in the future. This is because such securities have the property that they increase in value when rates rise. On the other hand, a **principal only (PO) strip** of a CMO, because of the extreme sensitivities of the price function, can have an effective duration measure significantly in excess of the maximum time to receipt of the last projected cash flow. In both cases this is because of the embedded prepayment option in the underlying mortgages.

Naturally duration approximations apply equally well to price functions of common and preferred stock, and one sometimes even sees notions of duration and convexity applied to such securities calculated as above. For example, the price of a common stock with fixed growth rate dividends is given as $V(r) = D\frac{1+g}{r-g}$, where here D denotes the dollar value of the last dividend. This function is clearly differentiable for $r > g$, the logical domain of definition. The modified duration of this price function is then calculated as

$$D(r_0) = \frac{1}{r_0 - g}.$$

Rate Sensitivity of Duration
In addition to providing a second-order adjustment to the duration approximation in (9.61), convexity is relevant for determining the sensitivity of the duration measure to changes in interest rates, and this is in turn relevant in terms of suggesting how often duration rebalancing may be necessary for the applications of the next section. Defining the duration and convexity functions, $D(r)$ and $C(r)$, as in (9.57) and (9.58) on the assumption that $P(r) \neq 0$, we have

$$D(r) = -\frac{P'(r)}{P(r)}, \quad C(r) = \frac{P''(r)}{P(r)}.$$

It is straightforward to evaluate $D'(r)$ and obtain

$$D'(r) = D^2(r) - C(r).\tag{9.65}$$

Consequently, from the first-order Taylor series for $D(r)$,

$$D(r) = D(r_0) + [D^2(r_0) - C(r_0)](r - r_0),\tag{9.66}$$

it is apparent that as yields increase, duration will decrease if $D^2(r_0) < C(r_0)$, and conversely, and the opposite is true as yields decrease.

This provides another way to understand the price sensitivity benefit associated with large positive convexity. Specifically, when $C(r_0)$ exceeds $D^2(r_0)$, the duration of the security decreases as rates rise, and increases as rates fall. Consequently the duration effect on price is enhanced when positive, and mitigated when negative. Of course, small, and especially negative, convexity works oppositely, enhancing the duration effect on price when negative, and mitigating this effect when positive.

But again it is important to note that convexity in a security is not a "free good" when positive, nor a "free bad" when negative. Convexity attributes of a security influence its desirability, and hence price, so there is an expected price and yield offset to the effect of the convexity adjustment.

9.8.6   Asset–Liability Management

The most important application of the notions of duration and convexity may be to hedging interest rate risk in a portfolio, which is a major component of **asset–liability management**, also called **asset–liability risk management**, and to the cognoscenti, **ALM**. The general setup is that one has an asset portfolio $A(i)$ whose value is modeled as a function which depends on the single interest rate i, as well as a liability portfolio, $L(i)$, which depends on the same rate. The focus of asset–liability management is then on the surplus, net worth, or capital of this entity:

$$S(i) = A(i) - L(i).$$

In particular, the focus is on managing the interest rate risk of this net position or some function of this net position. In this sense, asset–liability risk management is in fact **surplus risk management** or **capital risk management**. As will be seen below, neither label for this endeavor adequately describes the broad range and applicability of this theory.

That A, L, and hence S depend on a single interest rate is of course an oversimplifying assumption in the real world, where both assets and liabilities are likely to be multivariate functions dependent on many interest rates and, in general, different interest rates. However, in one application of this general theory, A and L are evaluated on their respective collections of interest rates, and the parameter i denotes the common change in all rates. In other words, in this application, while the initial interest rate structures are realistic, the simplifying assumption is that all structures move in parallel. In this application the model is often referred to as the **parallel shift model**.

To address these general multivariate price function models requires additional tools from multivariate calculus. That said, even in this simplistic context, important notions can be introduced and understood which underlie the generalizations possible in that framework.

To ground the reader in specific applications of this theory, consider the following:

Example 9.148

1. *Assets, liabilities, and surplus for a **financial intermediary** such as a life insurer, property and casualty insurer, commercial bank, or pension fund correspond to the respective portfolios on the entities' balance sheets. However, in these applications it is important to recognize that the function values $A(i)$, $L(i)$, and $S(i)$ are not intended to denote the firms' carrying values on their balance sheets. Carrying values are reflective of various accounting conventions prescribed by **generally accepted accounting principles (GAAP)** for publicly traded firms, which can vary from country to country, although these principles are now in the process of converging to an **international accounting standard (IAS)**. In the case of US insurance companies, there is also an accounting framework known as **statutory accounting**, promulgated by the state insurance regulators, the focus of which is on a conservative estimation of the firms' capital adequacy. For a pension plan, **valuation accounting** is the common basis which is reflective of both regulatory and market valuation principles.*

*Instead of carrying values, the values implied by $A(i)$, $L(i)$, and $S(i)$ are intended to be market values, or in the case of illiquid or nontradable positions, **fair values** defined as the market price "between a willing seller and willing buyer in a competitive market." Of course, in many accounting frameworks, it is the market value that determines the carrying value. The point here is that whether or not it is defined that way, the focus of asset–liability management is on market value, broadly defined. That said, one important responsibility of an ALM manager is to ensure that strategies formulated in this environment will have well-understood, and favorable, or at least acceptably adverse, effects in the respective accounting regime(s).*

2. *For a **fixed income hedge fund** or trading desk of an investment bank, $A(i)$ and $L(i)$ could denote the market values of the long and short positions, respectively.*

3. *In a general* **asset-hedging** *application, $A(i)$ is a portfolio of assets, and $L(i)$, which may not exist at the moment, is the intended hedging portfolio which intuitively will represent a "short" position in the market, or a financial derivatives overlay. In such an application, defining $S(i) = A(i) - L(i)$ as the net position is a notational convenience, and one must be careful about "signs." If $L(i)$ denotes the market value of securities, and if these securities are shorted, then the net risk position is $A(i) - L(i)$. On the other hand, if $L(i)$ denotes the market value of the hedging position, then the net position is $A(i) + L(i)$. To avoid confusion, hedges are often set up within the former framework, where $L(i)$ denotes the value of a position, which is then shorted, and hence the math works out with a "$-$" sign.*

4. *For a general* **liability-hedging** *application, such as that related to the issuance of debt, it is the $L(i)$ that is the given, and one might be interested in establishing a hedging position $A(i)$. Again, it is important to be mindful of the signs used in the analysis.*

5. *Finally, in* **fixed income portfolio management** *such as for a mutual fund, $A(i)$ would naturally denote the value of the portfolio, and one can notionally define $L(i)$ as a position in the portfolio's benchmark index of the same initial dollar value. Then $A(i) - L(i)$ can be evaluated by the portfolio manager to identify interest rate risk positions vis-à-vis the benchmark, and trades evaluated in the asset portfolio to manage this exposure.*

To develop some results and unambiguously address the sign problem, imagine that we wish to quantify the risk profile of $S(i) = A(i) - L(i)$ for a firm as in the first example above. We assume that initially the interest rate variable has value i_0, and hence the initial value of surplus is $S(i_0) = A(i_0) - L(i_0)$. In the parallel shift model, $i_0 = 0$, reflecting valuation on today's interest rate structures, and the general shift $i_0 \to i$ is really $0 \to i$.

To calculate duration and convexity of any portfolio is easy, since the portfolio values reflect simple weighted averages of the individual securities' values. For example, assume that the asset portfolio value is a sum of security values:

$$A(i) = \sum_{j=1}^{n} A_j(i),$$

where to avoid definitional problems we assume that $A_j(i_0) \neq 0$ for all j and $A(i_0) \neq 0$. The second condition is not superfluous, since $\{A_j(i_0)\}$ values may be both positive and negative.

Then because derivatives of sums equal sums of derivatives by proposition 9.75, it is straightforward to derive (see exercise 19) with $w_j = \frac{A_j(i_0)}{A(i_0)}$:

$$D^A(i_0) = \sum_{j=1}^{n} w_j D_j(i_0), \tag{9.67a}$$

$$C^A(i_0) = \sum_{j=1}^{n} w_j C_j(i_0), \tag{9.67b}$$

Of course, $\sum_{j=1}^{n} w_j = 1$, although $\{w_j\}$ may contain both positive and negative values.

Depending on the goals of the ALM program, the risk in $S(i_0)$ associated with a change in interest rates may be defined in one of several ways. If $T(i)$ denotes the **target risk measure**, three of which are illustrated below, the first step is to calculate the second-order Taylor series expansion of $T(i)$ as in (9.61):

$$T(i) \approx T(i_0)\left[1 - D^T(i_0)(i - i_0) + \frac{1}{2}C^T(i_0)(i - i_0)^2\right].$$

The error in this approximation is $O(\Delta i^3)$ by (9.27) if $T^{(3)}(i)$ exists, and $o(\Delta i^3)$ by (9.28) if $T^{(3)}(i)$ is continuous.

The risk to this function from the shift $i_0 \to i$ comes from duration risk $D^T(i_0)$, which presents a **signed risk of order** $O(\Delta i)$, and from convexity risk $C^T(i_0)$, which presents an **unsigned risk of order** $O(\Delta i^2)$. By a signed risk is meant that the effect on $T(i)$ by the shift $i_0 \to i \equiv i_0 + \Delta i$, depends on the sign of Δi, as in $\pm$, and on the magnitude of Δi, whereas for an unsigned risk the effect does not depend on sign but only the magnitude of Δi.

The Holy Grail of ALM is then to seek to achieve the following structure:

$$D^T(i_0) = 0, \tag{9.68a}$$

$$C^T(i_0) > 0. \tag{9.68b}$$

This then results in a target risk measure with the classical **immunized risk profile** as graphed in **figure 9.4**.

To some practitioners, the goal of **risk immunization** is considered unrealistic, since the resulting portfolio would appear to represent a risk-free arbitrage in the market. No matter what becomes of interest rates, a profit is made. This criticism has some merit as a cautionary statement about what is and is not possible, but the simple notion that "immunization is impossible because to do so would be to create a risk-free arbitrage, a free lunch, and this is impossible," overstates the case.

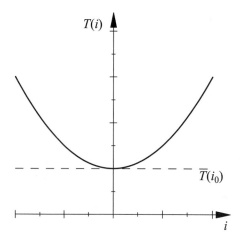

Figure 9.4
$$T(i) \approx T(i_0)\left[1 + \tfrac{1}{2}C^T(i_0)(i - i_0)^2\right]$$

In order to be a true risk-free arbitrage, all of the following would need to be true, and in practice, they never are:

1. The trade from the original target portfolio, to the immunized portfolio, can be done in a cost free way.

2. The resulting immunized portfolio earns more than the risk-free rate at all times.

3. The risk associated with $i_0 \rightarrow i$ summarizes all the risks of the portfolio; no other risks exist and no new risks are added.

So, in practice, the pursuit of immunization will not create a risk-free arbitrage but will create a framework within which many of the risks of the portfolio can be summarized, and various hedging trades evaluated from a cost/benefit perspective.

Three approaches to $T(i)$ are developed next. The goal here is not to present the only, or even the best, approaches but to illustrate the broad applicability of this general methodology.

Surplus Immunization, Time $t = 0$
The target measure is simply the current value of **surplus**:

$$T(i) = S(i).$$

Because $S'(i) = A'(i) - L'(i)$, and similarly for $S''(i)$, a simple calculation produces the following as long as $S(i_0) \neq 0$, and these should be understood as special cases of (9.67):

$$D^S(i_0) = \frac{A(i_0)}{S(i_0)} D^A(i_0) - \frac{L(i_0)}{S(i_0)} D^L(i_0), \tag{9.69}$$

$$C^S(i_0) = \frac{A(i_0)}{S(i_0)} C^A(i_0) - \frac{L(i_0)}{S(i_0)} C^L(i_0). \tag{9.70}$$

To achieve the objectives in (9.68) then requires that

$$D^A(i_0) = \frac{L(i_0)}{A(i_0)} D^L(i_0), \tag{9.71}$$

$$C^A(i_0) > \frac{L(i_0)}{A(i_0)} C^L(i_0). \tag{9.72}$$

In the case of $A(i_0) = L(i_0)$, and hence $S(i_0) = 0$, these conditions formally reduce to

$$D^A(i_0) = D^L(i_0), \tag{9.73}$$

$$C^A(i_0) > C^L(i_0). \tag{9.74}$$

But note that (9.73) is **not a legitimate deduction** from (9.71), since this latter formula was developed under the assumption that $S(i_0) \neq 0$, which is to say, $A(i_0) \neq L(i_0)$. Still, in the case where $S(i_0) = 0$, one can work directly with the original Taylor series expansions of $S(i)$ in (9.27), which is to say, the dollar duration and dollar convexity approach, and it will be seen that the immunizing conditions in (9.73) and (9.74) are produced, and legitimately so (see exercise 42).

Surplus Immunization, Time $t > 0$

If $Z_t(i)$ denotes the market price of a t-period, risk-free zero-coupon bond that matures for \$1 at time t, the **forward value of surplus**, denoted $S_t(i)$, is defined by

$$S_t(i) \equiv \frac{S(i)}{Z_t(i)}.$$

The intuition for this definition is that if surplus was now liquidated and invested in zeros, this would be the value produced at time t with certainty. In that sense, $S_t(i)$ is the value achievable at time t with the current portfolio and interest rates at level i if liquidated and reinvested.

Immunizing the forward value of surplus means that

$$T(i) = S_t(i),$$

and this requires conditions that depend on t that reduce to those above when $t = 0$. To this end, we first calculate $S_t'(i)$ and $S_t''(i)$. Although a bit messy, the following is produced if $S(i_0) \neq 0$ (see exercise 46):

$$D^{S_t}(i_0) = D^S(i_0) - D^{Z_t}(i_0),$$ (9.75a)

$$C^{S_t}(i_0) = C^S(i_0) - C^{Z_t}(i_0) - 2D^{Z_t}(i_0)[D^S(i_0) - D^{Z_t}(i_0)].$$ (9.75b)

Applying (9.68), the immunizing conditions are

$$D^S(i_0) = D^{Z_t}(i_0),$$ (9.76a)

$$C^S(i_0) > C^{Z_t}(i_0).$$ (9.76b)

Note that as $t \to 0$, it is apparent that $D^{Z_t}(i_0) \to 0$ and $C^{Z_t}(i_0) \to 0$, and so the conditions in (9.76) reduce to those in (9.71) and (9.72). Also, in the case of $S(i_0) = 0$, one can work directly with the Taylor series for $S_t(i) = (A(i) - L(i))/Z_t(i)$ to produce the conditions in (9.73) and (9.74), and hence the result is then independent of t.

Surplus Ratio Immunization
The **surplus ratio**, denoted $R(i)$, is defined by

$$R(i) = \frac{S(i)}{A(i)}.$$

It is unnecessary to specify whether this is the time 0 surplus ratio or the time $t > 0$ ratio, since it is easy to see that

$$R_t(i) \equiv \frac{S_t(i)}{A_t(i)} = \frac{S(i)}{A(i)}.$$

To immunize the surplus ratio is to set

$$T(i) = R(i).$$

As a ratio, the duration and convexity formulas for $R(i)$ are identical to those of the ratio function $S_t(i)$ in (9.75), with only a change in notation, which we record here, when $S(i_0) \neq 0$:

$$D^R(i_0) = D^S(i_0) - D^A(i_0),$$ (9.77a)

$$C^R(i_0) = C^S(i_0) - C^A(i_0) - 2D^A(i_0)[D^S(i_0) - D^A(i_0)].$$ (9.77b)

Applying (9.68) to these formulas produces

$$D^S(i_0) = D^A(i_0), \tag{9.78a}$$

$$C^S(i_0) > C^A(i_0). \tag{9.78b}$$

Note that (9.78) reduces to (9.73) and (9.74) when (9.69) and (9.70) are used to eliminate the dependence on $S(i)$.

It is also the case that (9.73) and (9.74) present the correct immunizing conditions for the surplus ratio when $S(i_0) = 0$, as can be derived by working directly with $R'(i)$ and $R''(i)$, or simply recognizing that immunizing $S(i)$ when $S(i_0) = 0$ is identical to immunizing $R(i)$ in this case, and hence, (9.73) and (9.74) follow immediately.

9.8.7 The "Greeks"

Although duration and convexity, which are relative derivative measures, are the conventional way to measure and quote the sensitivities of fixed income instruments and associated interest rate based derivative securities, for most other financial instruments, sensitivities are expressed directly in terms of the derivatives of the price functions. For example, the price of a put or call option based on the Black–Scholes–Merton formulas in chapter 8 is clearly a function of:

S_0: stock price

σ: stock price volatility

r: risk-free rate

t or T: time to expiry

The name "Greeks" is given to the various derivatives of this price function, and further applied to other financial derivative securities on currencies, commodities, common stock indexes, futures contracts, and so forth. With O used to denote the price of the given security, which is a function of these variables, the derivatives of O are labeled with Greek letters, and sometimes with a fictional "Greek" letter:

$$\text{delta:} \quad \Delta = \frac{dO}{dS}, \tag{9.79a}$$

$$\text{gamma:} \quad \Gamma = \frac{d^2O}{dS^2}, \tag{9.79b}$$

$$\text{rho:} \quad \rho = \frac{dO}{dr}, \tag{9.79c}$$

"vega": $v = \dfrac{dO}{d\sigma}$,   (9.79d)

theta: $\theta = \dfrac{dO}{dt}$.   (9.79e)

Note that the Greek symbol for "vega" is actually the lowercase Greek letter nu.

While we will not formally address multivariate functions, for the purposes of the definitions above, derivatives can be defined as if the price function in question is a function of only the variable of interest. Also with this convention the Taylor series results above can be applied. For instance,

$$O(S) \approx O(S_0) + \Delta(S - S_0) + \frac{1}{2}\Gamma(S - S_0)^2,$$

and we know that the error is $O(\Delta S^2)$. However, to approximate this price function simultaneously in all variables will require some new tools from multivariate calculus.

From the formulas above, the Greeks allow the risk evaluation of general equity-based and financial derivatives-based portfolios, and with this model, hedging strategies can be formulated that are parallel to those discussed in section 9.8.6 on asset-liability management.

9.8.8 Utility Theory

An important application of the notions of concavity and convexity in finance and economics is within the subject of utility theory, which provides a mathematical framework and model for understanding a given person's choices among various risky alternatives. Such **risk preferences** are expressed all the time, of course, such as when an individual chooses among various risky investments, or between risky and risk-free assets, as well as when that individual decides what kind of insurance to buy, or how much, or even whether or not to buy. Indeed it is also expressed in terms of an individual's propensity to gamble, as well as in the particular games of chance that attract more versus attract less.

While this subject can be studied within a formal axiomatic framework, we instead take an informal approach but note its origins. The key result is called the **von Neumann–Morgenstern theorem**, named for its discoverers: **John von Neumann** (1903–1957) and **Oskar Morgenstern** (1902–1977). This theorem states that if an individual has risk preferences that are consistent and satisfy certain other logical relationships, then there is a function $u(x)$, the **utility function**, so that "preference" can be predicted by the expected value of $u(W(X))$, where $W(X)$ denotes the value of the individual's wealth as a function of the realization of the risky variable X. The calibration of $u(x)$ as an increasing function is done so that "more is better than less," or "greater utility is preferred to less," and hence the objective of a decision maker is to maximize the "expected utility" of wealth, $E[u(W(X))]$.

In this setting, W_0 is often used to denote the initial wealth of the decision maker at the time of the decision.

Investment Choices

Within this risk-preference framework an investment of $I \leq W_0$ over time period $[0, T]$ with risky returns defined by a random variable Y will be deemed attractive compared to a risk-free investment if and only if

$$E[u(W(Y))] > u(W_0(1 + r)^T).$$

Here r denotes the annual risk-free rate for the period, and

$$W(Y) = I(1 + Y) + (W_0 - I)(1 + r)^T.$$

This framework also works for $I > W_0$, in which case the investment involves a short position in the risk-free asset.

More generally, this investment will be preferred to another investment with risky returns defined by a random variable Y', for an investment of I, if and only if

$$E[u(W(Y))] > E[u(W(Y'))],$$

where the wealth functions, $W(Y)$ and $W(Y')$ are defined as above.

Of course, the decision of how much to invest can also be addressed in this framework, since the optimum I, given Y, is the value that maximizes $E[u(W(Y))]$, for a given investment. This maximum might well be at $I < 0$, $I = 0$, or $I > W_0$.

Insurance Choices

Insurance decisions can also be posed in this framework, where now X denotes a risky loss that an individual confronts and is contemplating insuring. If insurance costs P, then the individual will insure if

$$u(W_0 - P) > E[u(W_0 - X)].$$

For partial versus complete insurance, the choice would be to completely insure if

$$u(W_0 - P) > E[u(W_0 - P_\lambda - (1 - \lambda)X)],$$

where P_λ is the cost to insure $100\lambda\%$ of the loss. One could also determine the value of λ which maximizes $\mathrm{E}[u(W_0 - P_\lambda - (1 - \lambda)X)]$.

Gambling Choices
For a gambling choice, say the purchase of a lottery ticket with a cost of L, the decision will be to gamble if

$$\mathrm{E}[u(W_0 - L + Y)] > u(W_0),$$

where Y is the random pay-off from the gamble.

Utility and Risk Aversion
As noted above, utility functions are calibrated as increasing functions, and hence given the assumption of differentiability it is always the case that $u'(x) > 0$. The essence of the risk preference, however, is defined by the sign of the second derivative, $u''(x)$. Specifically, we have the terminology:

Risk averse: $u''(x) < 0$, and so $u(x)$ is strictly concave.   (9.80a)

Risk neutral: $u''(x) \equiv 0$, and so $u(x)$ is linear (affine).   (9.80b)

Risk seeking: $u''(x) > 0$, and so $u(x)$ is strictly convex.   (9.80c)

   The motivation for this terminology comes from an application of Jensen's inequality to specific risk preference questions, as will be seen below. Note that by (9.33) with $n = 1$, we have $u''(x) \equiv 0$ if and only if $u(x) = ax + b$, and hence justifying the terminology that this is a linear utility function (the formal term is "affine" unless $b = 0$).
   To evaluate an investment over a fixed horizon, it must be recognized that the decision to not invest in a risky asset should not be modeled as if the funds will remain dormant. The more logical alternative would be to assume that the choice is between a risky and a risk-free investment. Assume that over the investment horizon in question, the risk-free rate per period, a year say, can be expressed as r. To invest over $[0, T]$, measured in an integer number of periods, with X denoting the risky period return, the choice is between

Risk-free investment: $u(W_0 + I((1 + r)^T - 1))$,

Risky investment: $\mathrm{E}\left[u\left(W_0 + I\left(\prod_{j=1}^{T}(1 + X_j) - 1\right)\right)\right].$

The following proposition summarizes the result for an investment choice, and exercises 22 and 47 assign the task of developing the conclusions as they apply to an insurance choice or a gamble.

Proposition 9.149   *Given a planning horizon of $[0, T]$, a decision maker will be indifferent between the risky investment and the risk-free investment, depending on the relationship between $\mathrm{E}[\prod_{j=1}^{T}(1 + X_j)]$ and $(1 + r)^T$, as follows:*

1. *If risk averse, indifference requires $\mathrm{E}[\prod_{j=1}^{T}(1 + X_j)] = (1 + r + a)^T$; for some $a > 0$.*

2. *If risk neutral, indifference requires $\mathrm{E}[\prod_{j=1}^{T}(1 + X_j)] = (1 + r)^T$.*

3. *If risk seeking, indifference requires $\mathrm{E}[\prod_{j=1}^{T}(1 + X_j)] = (1 + r - a)^T$ for some $a > 0$.*

Remark 9.150

1. *Note the intuitive justification for the risk preference terminology. For a risk-averse investor, in order to be indifferent between the risky and risk-free investment, the risky investment must have an expected return in excess of the risk-free rate. In other words, a risk-averse investor requires a "positive risk premium" on the expected return in order to be willing to take the risk of a possible lower return. A risk seeker will be indifferent even with an expected return below the risk-free rate. In essence, such an investor is willing to give up expected return for the opportunity to do better with a risky return. Finally, a risk-neutral investor is "neutral" to risk, and is willing to take risk with no associated adjustment to the expected return versus the risk-free rate.*

2. *The proposition above is stated in terms of an annual or period nominal interest rate r, but it can be equivalently stated in terms of a continuously compounded risk-free rate r'. For example, for a risk-neutral investor the condition becomes*

$$\mathrm{E}\left[\prod_{j=1}^{T}(1 + X_j)\right] = e^{r'T}.$$

Proof   The decision maker will be indifferent if

$$u(W_0 + I((1 + r)^T - 1)) = \mathrm{E}\left[u\left(W_0 + I\left(\prod_{j=1}^{T}(1 + X_j) - 1\right)\right)\right].$$

Now, if the investor is risk averse, and hence with a strictly concave utility function, we have from Jensen's inequality in (9.50a) that

$$\mathrm{E}\left[u\left(W_0 + I\left(\prod_{j=1}^{T}(1 + X_j) - 1\right)\right)\right] < u\left(W_0 + I\left(\mathrm{E}\left[\prod_{j=1}^{T}(1 + X_j)\right] - 1\right)\right).$$

Comparing, we see that for a risk-averse investor to be indifferent requires that

$$u(W_0 + I((1 + r)^T - 1)) < u\left(W_0 + I\left(\mathrm{E}\left[\prod_{j=1}^{T}(1 + X_j)\right] - 1\right)\right),$$

and recalling that $u(x)$ is an increasing function, we obtain the first result. That is, for some $a > 0$,

$$W_0 + I((1 + r + a)^T - 1) = W_0 + I\left(\mathrm{E}\left[\prod_{j=1}^{T}(1 + X_j)\right] - 1\right).$$

For the risk-neutral investor, this second last equation is

$$u(W_0 + I((1 + r)^T - 1)) = u\left(W_0 + I\left(\mathrm{E}\left[\prod_{j=1}^{T}(1 + X_j)\right] - 1\right)\right),$$

and hence the second result. Finally, for a risk seeker with strictly convex utility function, by (9.50b),

$$\mathrm{E}\left[u\left(W_0 + I\left(\prod_{j=1}^{T}(1 + X_j) - 1\right)\right)\right] > u\left(W_0 + I\left(\mathrm{E}\left[\prod_{j=1}^{T}(1 + X_j)\right] - 1\right)\right),$$

and the final equation to solve is

$$u(W_0 + I((1 + r)^T - 1)) > u\left(W_0 + I\left(\mathrm{E}\left[\prod_{j=1}^{T}(1 + X_j)\right] - 1\right)\right).$$

Since $u(x)$ is increasing, we obtain the third conclusion.   ∎

Example 9.151

1. *The **risk-neutral probability** was introduced in section 7.8.6, and generalized in section 8.8.3 to an arbitrary period of length Δt, and is defined by*

$$q(\Delta t) = \frac{e^{r\Delta t} - e^{d(\Delta t)}}{e^{u(\Delta t)} - e^{d(\Delta t)}}.$$

Here r denotes the annualized risk-free rate, assumed constant, $u(\Delta t)$ and $d(\Delta t)$ the assumed upstate and downstate returns of the stock in the period, and $q(\Delta t)$ the probability of an upstate. As was shown in chapter 7, and easily generalized to a period of length Δt, the expected value of the stock at time Δt under q satisfies

$$E_q[S_{\Delta t}] = e^{r\Delta t}S_0.$$

In other words, with $1 + X = \frac{S_{\Delta t}}{S_0}$ equal to the random period return,

$$E[1 + X] = e^{r\Delta t},$$

justifying by the proposition above that $q(\Delta t)$ is the probability of an upstate for a risk neutral investor willing to pay S_0 for this security.

2. *A special **risk-averter probability** was also introduced in chapter 8 in connection with the Black–Scholes–Merton pricing formulas and defined in (8.55) by*

$$\bar{q}(\Delta t) = q(\Delta t)e^{u(\Delta t)}e^{-r\Delta t}.$$

A simple calculation now produces, dropping the Δt for notational simplicity, that $1 - \bar{q} = (1 - q)e^{d}e^{-r\Delta t}$ and

$$E_{\bar{q}}[S_{\Delta t}] = \bar{q}(S_0 e^{u}) + (1 - \bar{q})(S_0 e^{d})$$

$$= [e^{u} - e^{u+d-r\Delta t} + e^{d}]S_0.$$

Although not immediately apparent, $E_{\bar{q}}[1 + X] > e^{r\Delta t}$, and so $\bar{q}(\Delta t)$ is the probability of an upstate for a risk-averse investor willing to pay S_0 for this security. This conclusion follows from the algebraic steps:

$$e^{u} - e^{u+d-r\Delta t} + e^{d} > e^{r\Delta t} \quad iff:$$

$$e^{u-r\Delta t} - e^{u+d-2r\Delta t} + e^{d-r\Delta t} - 1 > 0 \quad iff:$$

$$(e^{u-r\Delta t} - 1)(1 - e^{d-r\Delta t}) > 0.$$

The validity of this last inequality follows from $d(\Delta t) < r\Delta t < u(\Delta t)$.

Examples of Utility Functions

Remark 9.152 *Note that by the definition above of risk preference, we can conclude that:*

1. *If $u(x)$ is any utility function, then $\tilde{u}(x) \equiv au(x) + b$ has the same properties for any $a, b \in \mathbb{R}$ and $a > 0$ in terms of risk aversion, risk neutrality or risk seeking because $\tilde{u}''(x) = au''(x)$.*

2. *In addition, for $a, b \in \mathbb{R}$ and $a > 0$, a decision maker with utility function $\tilde{u}(x)$ will make identical decisions as one with $u(x)$. This conclusion follows from the fact that $E[\tilde{u}(W(x))] = aE[u(W(x))] + b$, and hence in any of the preceding decision inequalities between an expected utility and a fixed utility, or between two expected utilities, the a and b play no role.*

3. *Because of 1 and 2, utility functions are sometimes calibrated so that $u(W_0) = 0$ and/or $u(0) = -1$.*

4. *If $u(x)$ is a risk-averse utility function, then $au(x) + b$ is risk-seeking for $a < 0$ and any b, and conversely.*

A few common examples of risk-averse utility functions defined on $x \geq 0$ follow. Each can be made to represent risk seeking preference by multiplying by -1 by remark 4 above.

Example 9.153

1. *Exponential Utility*:

$$u(x) = 1 - e^{-kx}, \qquad k > 0.$$

2. *Quadratic Utility*:

$$u(x) = ax - bx^2, \qquad a, b > 0.$$

Note that this utility function violates the $u'(x) > 0$ assumption, at least for $x > \frac{a}{2b}$.

3. *Power Utility*:

$$u(x) = \frac{1}{\lambda}x^\lambda, \qquad \lambda > 0.$$

4. *Logarithmic Utility*:

$$u(x) = \ln\left(1 + \frac{x}{c}\right), \qquad c > 0.$$

9.8.9   Optimal Risky Asset Allocation

Assume that an investor with utility function $u(x)$ and initial wealth W_0 wants to make an optimal allocation between a risky asset, with period return random variable X, and the risk-free asset, with period return r. If I denotes the investment in the risky asset, this investor's risky utility after investment for T periods is

$$u\left(W_0(1+r)^T + I\left(\prod_{j=1}^{T}(1+X_j) - (1+r)^T \right) \right).$$

For notational ease, we assume the planning horizon $T = 1$, so the risky utility value is

$$u(W_0(1+r) + I(X-r)).$$

Here r can denote the fixed risk-free return for the period, or the variable com-pounded risk-free returns over subperiods.

Now, if we temporarily assume that $u(x)$ is an analytic function, this risky utility can be expanded about $W_0(1+r)$ to produce

$$u(W_0(1+r) + I(X-r)) = \sum_{k=0}^{\infty} \frac{1}{k!} u^{(k)}(W_0(1+r))(I(X-r))^k.$$

If only differentiable to order m, this expansion holds up to the mth derivative as a Taylor series, with error no worse than $O(\Delta x^m)$ with $\Delta x = I(X - r)$. For notational simplicity, we maintain the upper summation limit of ∞.

To simplify the analysis, and because of the second point made in remark 9.152, this utility function can be transformed to the form: $\tilde{u}(x) = au(x) + b$ with $a > 0$, without changing any conclusions that we may draw. Since we assume that $u'(W_0(1+r)) > 0$, we define

$$\tilde{u}(x) = \frac{u(x) - u(W_0(1+r))}{u'(W_0(1+r))}.$$

This then produces

$$\tilde{u}(W_0(1+r) + I(X-r)) = \sum_{k=1}^{\infty} \frac{1}{k!} \frac{u^{(k)}(W_0(1+r))}{u'(W_0(1+r))} I^k (X-r)^k$$

$$= \sum_{k=1}^{\infty} \frac{1}{k!} \tilde{u}_k I^k (X-r)^k,$$

with

$$\tilde{u}_k = \frac{u^{(k)}(W_0(1+r))}{u'(W_0(1+r))},$$

and so $\tilde{u}_1 \equiv 1$.

The **Arrow–Pratt measure of absolute risk aversion**, r_{AP}, is defined by

$$r_{AP} = -\tilde{u}_2 = -\frac{u''(W_0(1+r))}{u'(W_0(1+r))}, \tag{9.81}$$

and named for **Kenneth J. Arrow** (b. 1921) and **John W. Pratt** (b. 1931). Since $u'(W_0(1+r)) > 0$, this measure of risk aversion is positive for a risk-averter, negative for a risk-seeker, and identically zero for a risk-neutral investor. Moreover a larger positive r_{AP} implies greater risk aversion, and a more negative r_{AP} implies greater risk seeking, as will be seen below.

Taking expected values, we derive

$$E[\tilde{u}(W_0(1+r) + I(X-r))] = \sum_{k=1}^{\infty} \frac{1}{k!} \tilde{u}_k I^k E[(X-r)^k]. \tag{9.82}$$

Using only the first two terms of this series,

$$E[\tilde{u}(W_0(1+r) + I(X-r))] \approx I E[(X-r)] - \frac{I^2 r_{AP}}{2} E[(X-r)^2], \tag{9.83}$$

and an optimum value of I can be found for a risk-averse investor, where by optimum is meant utility maximizing.

Letting $f_2(I)$ denote the right-hand side of (9.83) as a function of I, we derive

$$f_2'(I) = E[(X-r)] - I r_{AP} E[(X-r)^2],$$

$$f_2''(I) = -r_{AP} E[(X-r)^2].$$

So this expected utility function has a critical point at

$$I_0 = \frac{E[(X-r)]}{r_{AP} E[(X-r)^2]}, \tag{9.84}$$

which will be a relative maximum if $f_2''(I_0) < 0$.

For a risk-averse investor, with $r_{AP} > 0$, or equivalently $u''(W_0(1+r)) < 0$, I_0 is always a relative maximum of the expected utility. If $E[(X-r)] > 0$, the typical case for risky assets, then $I_0 > 0$, and such an investor will go long to maximize expected utility. If $E[(X - r)] < 0$, this investor will short the risky asset, since then $I_0 < 0$. Note that in either case, as the Arrow–Pratt measure increases, this investor will go long less (respectively, short less) to obtain the optimum utility.

For a risk-seeker, with $r_{AP} < 0$, or equivalently $u''(W_0(1 + r)) > 0$, I_0 is always a relative minimum of the expected utility. This is logical since in this case, considered as a function of I, the expression in (9.83) is of the form $g(I) = aI + bI^2$, with $b > 0$, and so utility is only maximized at the endpoints of whatever interval for I is allowed. In other words, a risk-seeker will maximize utility by comparing a long position with maximal leverage, to the maximal short position in the risky asset, and choose the option with greater utility.

The value of the expected utility function at I_0 is given by

$$E[\tilde{u}(W_0(1 + r) + I_0(X - r))] = \frac{E[(X - r)]^2}{2r_{AP}E[(X - r)^2]}.$$

This maximum expected utility can be equivalently expressed in terms of the **Sharpe ratio** developed by **William F. Sharpe** (b. 1934):

$$E[\tilde{u}(W_0(1 + r) + I_0(X - r))] = \frac{s^2}{2r_{AP}}, \tag{9.85}$$

where the Sharpe ratio is defined by

$$s = \frac{E[(X - r)]}{\sqrt{E[(X - r)^2]}}. \tag{9.86}$$

Remark 9.154 *The significance of the Sharpe ratio is that for every risk-averse investor, optimal utility in (9.85) can be increased by choosing the risky asset with the largest Sharpe ratio.*

When r is assumed constant, maximizing s is equivalent to maximizing

$$s' = \frac{\mu - r}{\sigma}, \tag{9.87}$$

where μ and σ are the mean and standard deviation of X. This follows since $X - r = (X - \mu) + (\mu - r)$ and a calculation produces $s = \frac{s'}{\sqrt{1+(s')^2}}$. Consequently s is maximized when s' is maximized. The formula in (9.87) is also called the Sharpe ratio.

9.8.10   Risk-Neutral Binomial Distribution as $\Delta t \to 0$

In section 8.8.2 was shown that the real world binomial model for equity prices converged to the lognormal distribution. Specifically, as defined in (8.46) and repeated here, let

$$S_T^{(n)} = S_0 e^{\sum B_j},$$

where

$$B_j = \begin{cases} \mu\Delta t + a\sigma\sqrt{\Delta t}, & \Pr = p, \\ \mu\Delta t - \frac{1}{a}\sigma\sqrt{\Delta t}, & \Pr = p', \end{cases} \qquad j = 1, 2, \ldots, n,$$

with $a = \sqrt{\frac{p'}{p}} = \frac{p'}{\sqrt{pp'}}$, and $-\frac{1}{a} = \frac{-p}{\sqrt{pp'}}$.

Then as $\Delta t \to 0$, we have as in (8.50),

$$\ln S_T^{(n)} \to_p N(\ln S_0 + \mu T, \sigma^2 T),$$

where we emphasize the real world probability with the notation "$\to_p$". With S_T denoting the limiting random variable, this can be equivalently written as in (8.51):

$$S_T = S_0 e^X,$$

where $X \sim N(\mu T, \sigma^2 T)$. This is the definition of a lognormal random variable (see chapter 10 for more details on this distribution).

In this section we investigate the limiting distribution of the same equity prices, but rather than using the binomial probability p appropriate for real world modeling, we use the risk-neutral probability q, as is implicitly assumed in the option pricing formulas in chapters 7 and 8. This limiting distribution is needed for the Black–Scholes–Merton pricing formulas for European put and call options introduced in section 8.8.3.

In the next section we investigate the limiting distribution under the special risk averter probability $\bar{q}$, also needed for the Black–Scholes–Merton pricing formulas, defined in (8.55) by $\bar{q} = q e^u e^{-r\Delta t}$, where $u = \mu\Delta t + a\sigma\sqrt{\Delta t}$.

The added complexity in these investigations is the fact that unlike p, the probability q, and hence also $\bar{q}$, is a function of Δt as noted in (8.52):

$$q(\Delta t) = \frac{e^{r\Delta t} - e^{d(\Delta t)}}{e^{u(\Delta t)} - e^{d(\Delta t)}}.$$

Here r is the assumed constant risk-free interest rate, and $r(\Delta t) = r\Delta t$ is assumed linear in Δt, while the upstate and downstate equity returns for B_j are again given as

$$u(\Delta t) = \mu\Delta t + a\sigma\sqrt{\Delta t},$$

$$d(\Delta t) = \mu\Delta t - \frac{1}{a}\sigma\sqrt{\Delta t}.$$

To facilitate this first inquiry, we require a more accessible formula for $q(\Delta t)$ that makes the functional dependence on Δt more manageable.

Analysis of the Risk-Neutral Probability: $q(\Delta t)$
The goal of this section is to derive the following expansion:

Proposition 9.155 *With $q(\Delta t)$ defined as above, we have that*

$$q(\Delta t) = p + \frac{[r - \mu - \frac{1}{2}\sigma^2]}{\frac{\sigma}{\sqrt{pp'}}}\sqrt{\Delta t} + \left(p - \frac{1}{2}\right)\left(r - \mu - \frac{\sigma^2}{6}\right)\Delta t$$

$$+ \left[\frac{(r-\mu)^2 + (r-\mu)\sigma^2\left(\frac{1}{6pp'} - 1\right) + \frac{\sigma^4}{12}}{\frac{2\sigma}{\sqrt{pp'}}}\right]\Delta t^{3/2} + O[\Delta t^2]. \tag{9.88}$$

First off, to investigate the behavior of $q(\Delta t)$ as $\Delta t \to 0$, we need to do some analysis, since direct substitution of $\Delta t = 0$ leads to $\frac{0}{0}$. Dividing out the common term $e^{d(\Delta t)}$, and then applying (9.35) to each exponential term in this expression produces

$$q(\Delta t) = \frac{\exp\left(\frac{1}{a}\sigma\sqrt{\Delta t} + (r-\mu)\Delta t\right) - 1}{\exp\left((a + \frac{1}{a})\sigma\sqrt{\Delta t}\right) - 1}$$

$$= \frac{\frac{1}{a}\sigma\sqrt{\Delta t} + \left[\frac{1}{2}\left(\frac{1}{a}\sigma\right)^2 + (r-\mu)\right]\Delta t + O(\Delta t^{3/2})}{(a + \frac{1}{a})\sigma\sqrt{\Delta t} + \frac{1}{2}\left[(a + \frac{1}{a})\sigma\right]^2\Delta t + O(\Delta t^{3/2})}.$$

In this format, we can divide numerator and denominator by the common factor $\sqrt{\Delta t}$, substitute $\Delta t = 0$ and obtain

$$q(0) = \frac{\frac{1}{a}}{a + \frac{1}{a}} = p.$$

Perhaps surprisingly, as $\Delta t \to 0$ the risk-neutral probability converges to p, the real world probability:

$q(\Delta t) \to p$ as $\Delta t \to 0$.

It would be entirely justified at this point to expect that this conclusion should imply that the limiting distribution under this risk-neutral probability $q(\Delta t)$ is the same as that derived in chapter 8 for the real world probability p. Quite remarkably, this expectation will be proved to be false, and we will see that although $q(\Delta t) \to p$, it does so slowly enough that the limiting distribution of prices is changed from what was earlier derived.

To see this, we need to complete the analysis of $q(\Delta t)$, in effect by deriving more terms in its Taylor series than the constant term p. To do this, we could just start taking derivatives of $q(\Delta t)$, but a moment of reflection will prove it a painful pursuit, so we explore another approach. An approach that is appealing is based on proposition 9.116 of section 9.4.2. To this end, let us assume that

$$q(\Delta t) = p + \sum_{n=1}^{\infty} q_n (\sqrt{\Delta t})^n. \tag{9.89}$$

Then since both numerator and denominator of the function $q(\Delta t)$ are analytic functions of the variable $\sqrt{\Delta t}$ about $\sqrt{\Delta t} = 0$, so too is this ratio function $q(\Delta t)$, as proved in that section.

Remark 9.156 *Note that we do not claim that the numerator or denominator of $q(\Delta t)$, or $q(\Delta t)$ itself, are analytic functions of Δt about $\Delta t = 0$, which they cannot be since $\sqrt{\Delta t}$ is not even differentiable at $\Delta t = 0$. For example, while $f(x) = e^x$ is an everywhere analytic function of x, $g(x) = e^{\sqrt{x}}$ is not even differentiable at $x = 0$, since $g'(x) = \frac{1}{2\sqrt{x}} e^{\sqrt{x}}$. On the other hand, while not analytic in Δt, all three functions have absolutely convergent series as functions of $\sqrt{\Delta t}$. For example, since the numerator of $q(\Delta t)$ is an analytic function of $\sqrt{\Delta t}$, it is absolutely convergent for $|\sqrt{\Delta t}| < R$, for some R, which in this case we know to be $R = \infty$. The same is true for the denominator of $q(x)$, and hence for $q(x)$ itself for $0 \le \sqrt{\Delta t} < R'$ for some $R' > 0$.*

To simplify notation, it is appealing to substitute $x = \sqrt{\Delta t}$, and express $q(x)$ as

$$q(x) = \frac{\exp(pdx + cx^2) - 1}{\exp(dx) - 1},$$

$$c = r - \mu, \quad d = \frac{\sigma}{\sqrt{pp'}}.$$

The Taylor series for numerator and denominator then become

$$q(x) = \frac{\sum_{j=1}^{\infty} \frac{1}{j!}(pdx + cx^2)^j}{\sum_{k=1}^{\infty} \frac{1}{k!}(dx)^k}.$$

Expanding these expressions to $O(x^4)$ to put in the format of a ratio of power series, with $r(x)$ and $s(x)$ denoting the numerator and denominator, respectively, we obtain

$$r(x) = (pd)x + \left(c + \frac{1}{2}d^2p^2\right)x^2 + \left(cdp + \frac{1}{6}d^3p^3\right)x^3 + O(x^4),$$

$$s(x) = dx + \frac{1}{2}d^2x^2 + \frac{1}{6}d^3x^3 + O(x^4).$$

The goal is to determine $\{q_n\}$ in (9.89) so that

$$\left(p + \sum_{n=1}^{\infty} q_n x^n\right) s(x) = r(x), \tag{9.90}$$

which we can implement using (6.24). Although algebraically tedious, and prone to initial missteps, this approach is significantly easier than evaluating the derivatives of $q(x)$ directly as a ratio function.

Alternatively, since $q_n = \frac{q^{(n)}(0)}{n!}$, we could evaluate the derivatives of $q(x)$ indirectly by differentiating the identity

$$r(x) = q(x)s(x), \tag{9.91}$$

and solving. Specifically, we have by the Leibniz formula in (9.42), that for x in the interval about 0 for which $q(x)$ is analytic and hence infinitely differentiable,

$$r^{(n)}(x) = \sum_{k=0}^{n} \binom{n}{k} q^{(k)}(x) s^{(n-k)}(x).$$

This can be solved iteratively at $x = 0$. Then recalling that $s(0) = 0$, we obtain

$$q(0) = \frac{r'(0)}{s'(0)}, \tag{9.92a}$$

$$q^{(n-1)}(0) = \frac{1}{ns'(0)} \left[r^{(n)}(0) - \sum_{k=0}^{n-2} \binom{n}{k} q^{(k)}(0) s^{(n-k)}(0) \right], \qquad n \geq 2; \tag{9.92b}$$

and substituting $q_n = \frac{q^{(n)}(0)}{n!}$ into (9.89) will produce the desired result.

This is a different approach only methodologically from what was developed in (6.24) and not a new approach in theory. Here we developed an iteration for derivative values of $q(x)$ from those of $r(x)$ and $s(x)$, and constructed $q(x)$ as a Taylor series. To use (6.24), we would first construct the Taylor series for $r(x)$ and $s(x)$, which reflect these derivatives, and then iteratively generate the coefficients of the series for $q(x)$.

An easy calculation using the definition that $s(x) = \exp(dx) - 1$ produces

$$s^{(k)}(0) = d^k, \qquad k \geq 1.$$

The function $r(x) = \exp(pdx + cx^2) - 1$ is a bit more complicated because of the quadratic in the exponent, but to four derivatives we have

$$r'(x) = (pd + 2cx)\exp(pdx + cx^2),$$

$$r''(x) = [2c + (pd + 2cx)^2]\exp(pdx + cx^2),$$

$$r^{(3)}(x) = [6c(pd + 2cx) + (pd + 2cx)^3]\exp(pdx + cx^2),$$

$$r^{(4)}(x) = [12c^2 + 12c(pd + 2cx)^2 + (pd + 2cx)^4]\exp(pdx + cx^2).$$

Correspondingly,

$$r'(0) = pd,$$

$$r''(0) = 2c + (pd)^2,$$

$$r^{(3)}(0) = 6cdp + (pd)^3,$$

$$r^{(4)}(0) = 12c^2 + 12c(pd)^2 + (pd)^4.$$

Substituting into (9.92), we get

$$q(0) = p,$$

$$q'(0) = \frac{1}{2s'(0)}[r^{(2)}(0) - q^{(0)}(0)s^{(2)}(0)]$$

$$= \frac{c}{d} - \frac{dpp'}{2},$$

$$q''(0) = \frac{1}{3s'(0)}\left[r^{(3)}(0) - \sum_{k=0}^{1}\binom{3}{k}q^{(k)}(0)s^{(3-k)}(0)\right]$$

$$= 2\left(p - \frac{1}{2}\right)\left(c - \frac{d^2 pp'}{6}\right),$$

$$q^{(3)}(0) = \frac{1}{4s'(0)}\left[r^{(4)}(0) - \sum_{k=0}^{2}\binom{4}{k}q^{(k)}(0)s^{(4-k)}(0)\right]$$

$$= \frac{3c^2}{d} + 3cd\left[\frac{1}{6} - pp'\right] + \frac{1}{4}(pp')^2 d^3.$$

Recalling that $q_n = \frac{q^{(n)}(0)}{n!}$, $c = r - \mu$, and $d = \frac{\sigma}{\sqrt{pp'}}$, we obtain the final result in (9.88) after a bit more algebra.

Of course, $\tilde{q}(\Delta t) \equiv 1 - q(\Delta t)$, needed below, is easily developed from this expression by replacing p with p' and changing the sign of all other coefficients from positive to negative.

Notation 9.157   *Note that we use $\tilde{q}(\Delta t)$ to denote the complementary probability of $q(\Delta t)$, whereas in other applications the complement of p was denoted p'. The notation $q'(\Delta t)$ will be avoided for this purpose because of the confusion it would cause with the standard notation for the derivative of $q(\Delta t)$.*

Remark 9.158   *In remark 8.31 was discussed the relationship between the choice of the real world probability of an upstate, denoted p, and the speed of convergence of the distribution of binomial lattice prices to the normal distribution in (8.50). There it was concluded that $p = 1/2$ provided faster convergence by changing the error term in the development from $O(n^{-1/2})$ to $O(n^{-1})$. Because the risk neutral probabilities are also functions of $\Delta t = T/n$, it is natural to expect that speed of convergence of the binomial lattice under the risk-neutral probability depends not only on p but also on other parameters used in the lattice calibration. Indeed (9.88) indicates that $q(\Delta t)$ converges to p relatively slowly, with order of magnitude $O(\sqrt{\Delta t}) = O(n^{-1/2})$. But it is also apparent that if a lattice is to be developed only for option pricing, then choosing $\mu = r - \sigma^2/2$ causes $q(\Delta t)$ to converge to p with order of magnitude $O(\Delta t) = O(n^{-1})$. If additionally we select $p = 1/2$, the convergence improves to $O((\Delta t)^{3/2}) = O(n^{-3/2})$. Of course, choosing $p = 1/2$ is harmless, but choosing $\mu = r - \sigma^2/2$ does not provide a lattice that will, in general, be useful for real world stock price modeling. But this* *calibration is often used in practice for option pricing because it accelerates option price convergence as a function of Δt. And this choice is further justified by the observation that in the limit of the Black–Scholes–Merton option-pricing formulas, the real world parameter μ plays no role in any case, as noted in remark 8.34 of section 8.8.3, so we are justified to choose this value at will. Of course, if the goal is to produce a realistic stock price lattice for real world modeling and option pricing, one must choose a realistic μ and tolerate the fact that option prices will converge more slowly as $\Delta t \to 0$.*

Risk-Neutral Binomial Distribution as $\Delta t \to 0$

We are now in a position to investigate the limiting distribution of the binomial model under the risk-neutral probabilities. First off, with the analogous setup from chapter 8 in (8.46), we define

$$S_T^{(n)} = S_0 e^{\sum B_j},$$

where for $j = 1, 2, \ldots, n$,

$$B_j = \begin{cases} u(\Delta t) \equiv \mu \Delta t + a\sigma\sqrt{\Delta t}, & \Pr = q(\Delta t), \\ d(\Delta t) \equiv \mu \Delta t - \frac{1}{a}\sigma\sqrt{\Delta t}, & \Pr = 1 - q(\Delta t), \end{cases}$$

with $a = \sqrt{\frac{p'}{p}} = \frac{p'}{\sqrt{pp'}}$, $-\frac{1}{a} = \frac{-p}{\sqrt{pp'}}$ and $q(\Delta t) = \frac{e^{r\Delta t} - e^{d(\Delta t)}}{e^{u(\Delta t)} - e^{d(\Delta t)}}$.

The goal of this section is to prove the following:

Proposition 9.159   *With $S_T^{(n)}$ and $q(\Delta t)$ defined as above, then as $\Delta t \to 0$, in contrast to (8.50),*

$$\ln\left[\frac{S_T^{(n)}}{S_0}\right] \to_q \ln\left[\frac{S_T}{S_0}\right] \sim N\left(\left(r - \frac{1}{2}\sigma^2\right)T, \sigma^2 T\right), \tag{9.93a}$$

or

$$\ln S_T^{(n)} \to_q \ln S_T \sim N\left(\ln S_0 + \left(r - \frac{1}{2}\sigma^2\right)T, \sigma^2 T\right), \tag{9.93b}$$

where the limit symbol "$\to_q$" is used to emphasize the dependence of this result on the risk-neutral probability structure.

With S_T denoting the limiting random variable, this can be equivalently written as

$$S_T = S_0 e^X, \tag{9.94}$$

where $X \sim N\big((r - \frac{1}{2}\sigma^2)T, \sigma^2 T\big)$. So S_T satisfies the definition of a lognormal random variable (see chapter 10 for more details on this distribution).

This is truly a remarkable result when contrasted with the limits under the real world probability p stated in proposition 8.30. Of course, it may not seem remarkable that changing the binomial probability from p to $q(\Delta t)$ changes the moments of the limiting distribution of $\ln[S_T/S_0]$, here from $N(\mu T, \sigma^2 T)$ to $N(r - \frac{1}{2}\sigma^2)T, \sigma^2 T)$. What is remarkable is that as seen above, this change occurs despite the fact that $q(\Delta t) \to p$ as $\Delta t \to 0$.

As a first step in the investigation, we first note that under $q(\Delta t)$, using (9.88),

$$\mathrm{E}\left[\ln\left[\frac{S_{t+\Delta t}}{S_t}\right]\right] = \left(r - \frac{1}{2}\sigma^2\right)\Delta t + O[\Delta t^{3/2}], \tag{9.95a}$$

$$\mathrm{Var}\left[\ln\left[\frac{S_{t+\Delta t}}{S_t}\right]\right] = \sigma^2 \Delta t + O[\Delta t^{3/2}]. \tag{9.95b}$$

This derivation is assigned in exercise 24 below. So even with this relatively simple calculation it is apparent that despite the fact that $q(\Delta t) \to p$ as $\Delta t \to 0$, this convergence occurs in a way that introduces a permanent shift in the mean of this distribution compared to the earlier result.

To now demonstrate the result on the limiting distribution, we again resort to a moment-generating function argument. Because of the effect $q(\Delta t)$ has on the mean of the distribution, there is no benefit in attempting to parallel the development in section 8.8.2 in which we worked with the normalized random variable $Y^{(n)}$ rather than the actual random variable $B^{(n)} = \sum_{j=1}^{n} B_j \equiv \ln[S_T^{(n)}/S_0]$. There, with $Y^{(n)}$ we could eliminate the Δt-terms and only work with simplified $\sqrt{\Delta t}$-terms of B_j. Here, the normalized variable is actually more difficult to work with than the original random variable, so we work directly with $B^{(n)}$.

For the moment-generating function of $B^{(n)}$, first note that with $a = \sqrt{\frac{p'}{p}} = \frac{p'}{\sqrt{pp'}}$, $-\frac{1}{a} = \frac{-p}{\sqrt{pp'}}$ and $d = \frac{\sigma}{\sqrt{pp'}}$ as in the $q(\Delta t)$ analysis above,

$$M_{B_j}(s) = e^{\mu s \Delta t}\left(q(\Delta t)e^{as\sigma\sqrt{\Delta t}} + \tilde{q}(\Delta t)e^{-(s\sigma\sqrt{\Delta t})/a}\right)$$

$$= e^{\mu s \Delta t}\left(q(\Delta t)e^{dsp'\sqrt{\Delta t}} + \tilde{q}(\Delta t)e^{-dsp\sqrt{\Delta t}}\right),$$

where $\tilde{q}(\Delta t) \equiv 1 - q(\Delta t)$. Because the $\{B_j\}$ are independent and identically distributed, $M_{B^{(n)}}(s) = \prod_{j=1}^{n} M_{B_j}(s)$, and so since $n\Delta t = T$,

$$M_{B^{(n)}}(s) = e^{\mu Ts}\left(q(\Delta t)e^{dsp'\sqrt{\Delta t}} + \tilde{q}(\Delta t)e^{-dsp\sqrt{\Delta t}}\right)^{T/\Delta t}.$$

The goal is to show that

$$M_{B^{(n)}}(s) \to e^{(r-\sigma^2/2)Ts+(\sigma^2 Ts^2)/2}.$$

The challenge here is to evaluate

$$\lim_{\Delta t \to 0}\left(q(\Delta t)e^{dsp'\sqrt{\Delta t}} + \tilde{q}(\Delta t)e^{-dsp\sqrt{\Delta t}}\right)^{1/\Delta t}.$$

Since $f(y) = y^T$ is a continuous function for $T \geq 0$, if it is shown that $y(\Delta t) \to y_0$ as $\Delta t \to 0$, where

$$y(\Delta t) \equiv \left(q(\Delta t)e^{dsp'\sqrt{\Delta t}} + \tilde{q}(\Delta t)e^{-dsp\sqrt{\Delta t}}\right)^{1/\Delta t},$$

then $f(y(\Delta t)) \to f(y_0)$, so we can exponentiate this limit after it is evaluated. This limit of $y(\Delta t)$ can in turn be evaluated by working with $z(\Delta t) \equiv \ln y(\Delta t)$, since $g(y) = e^y$ is continuous, and hence, if $z(\Delta t) \to z_0$, then $y(\Delta t) = e^{z(\Delta t)} \to e^{z_0} = y_0$.

Working with $z(\Delta t)$, which we express for notational simplicity as $z(x)$, we have

$$z(x) = \frac{\ln\left(q(x)e^{dsp'\sqrt{x}} + \tilde{q}(x)e^{-dsp\sqrt{x}}\right)}{x},$$

and the goal is to determine $\lim_{x\to 0} z(x)$. Note that by reversing the above sequence of steps, we have

$$M_{B^{(n)}}(s) = e^{\mu Ts}\left[e^{z(0)}\right]^T.$$

So once $z_0 \equiv \lim_{x\to 0} z(x)$ is determined, we will conclude from the continuity of the exponential and power functions that

$$M_{B^{(n)}}(S) \to e^{\mu Ts+z_0 T}. \tag{9.96}$$

Of course, in order for the claim above in (9.93) to be validated by this derivation, we must show that

$$z_0 = \left(r - \mu - \frac{1}{2}\sigma^2\right)s + \frac{1}{2}\sigma^2 s^2. \tag{9.97}$$

The details are a bit messy, and provided below for completeness.

*Details of the Limiting Result   To derive (9.97), note that with

$$A(x) \equiv q(x)e^{dsp'\sqrt{x}} + \tilde{q}(x)e^{-dsp\sqrt{x}},$$

where $d = \frac{\sigma}{\sqrt{pp'}}$:

1. $A(x)$ is continuous on $x \geq 0$ and $A(0) = 1$.

2. The series expansions of the 4 functions in the definition of $A(x)$ are absolutely convergent for some interval, $0 \leq x < R'$ for $s = 1$ as noted in the remark 9.156 following (9.89), and hence this remains true for $0 \leq s \leq 1$. Consequently the series expansion for $A(x)$ can be developed by manipulating these series, and rearranging as desired (recall the section 6.1.4 discussion on rearrangements of absolutely convergent series).

3. Because of item 1, for any $\epsilon > 0$ there is a δ so that if $0 \leq x < \delta$, we have that $|A(x) - 1| < \epsilon$. So we let $\epsilon = \frac{1}{2}$, say, and conclude that $A(x) = 1 + B(x)$, where $|B(x)| < \frac{1}{2}$ for $0 \leq x < \delta$. As a small technicality, we only consider $0 \leq x < R'$, with R' defined in item 2 above if $R' < \delta$.

4. By item 2, the series expansion for $B(x)$ is also absolutely convergent for $0 \leq x < \min(\delta, R')$.

We now complete the derivation of this section's result by the proof of two claims.

Claim 9.160   *If* $A(x) = 1 + x[z_0 + C(x)]$, *where* $C(x)$ *has an absolutely convergent series expansion on* $0 \leq x < \min(\delta, R')$, *with* $C(0) = 0$, *then*

$$\lim_{x \to 0} z(x) = z_0.$$

Proof   Because $z(x) = \frac{1}{x} \ln[A(x)] = \frac{1}{x} \ln[1 + x(z_0 + C(x))]$, and $|x(z_0 + C(x))| = |B(x)| < \frac{1}{2}$ for $0 \leq x < \min(\delta, R')$ by item 3 above, the power series for $\ln(1 + y)$ can be utilized, and this is an absolutely convergent series:

$$\ln[1 + x(z_0 + C(x))] = \sum_{j=1}^{\infty} \frac{(-1)^{j+1} x^j (z_0 + C(x))^j}{j}$$

$$= x(z_0 + C(x)) + x^2 \sum_{j=2}^{\infty} \frac{(-1)^{j+1} x^{j-2} (z_0 + C(x))^j}{j}.$$

Consequently

$$z(x) = (z_0 + C(x)) + x \sum_{j=2}^{\infty} \frac{(-1)^{j+1} x^{j-2} (z_0 + C(x))^j}{j}.$$

Since $C(0) = 0$, we conclude that $z(x) \to z_0$ as $x \to 0$ as claimed. ∎

We now show that $A(x)$ has the required properties with z_0 as given in (9.97) and, hence by (9.96), will complete the proof of (9.93).

Claim 9.161 $A(x) = 1 + x\left[(r - \mu - \frac{1}{2}\sigma^2)s + \frac{1}{2}\sigma^2 s^2 + C(x)\right]$, *where $C(x)$ has an absolutely convergent series expansion on $0 \le x < \min(\delta, R')$, with $C(0) = 0$.*

Proof   With $A(x) \equiv q(x)e^{dsp'\sqrt{x}} + \tilde{q}(x)e^{-dsp\sqrt{x}}$, we have, since $q(x) + \tilde{q}(x) = 1$,

$$A(x) = 1 + q(x)(e^{dsp'\sqrt{x}} - 1) + \tilde{q}(x)(e^{-dsp\sqrt{x}} - 1)$$

$$= 1 + \sum_{i=0}^{\infty} q_i x^{i/2} \sum_{j=1}^{\infty} \frac{(dsp')^j x^{j/2}}{j!} + \sum_{i=0}^{\infty} \tilde{q}_i x^{i/2} \sum_{j=1}^{\infty} \frac{(-dsp)^j x^{j/2}}{j!},$$

where all series are absolutely convergent for $0 \le x < \min(\delta, R')$ as noted above. Here $\{q_i\}$ are defined as in (9.89) using (9.88) and $\{\tilde{q}_i\}$ are defined as the corresponding coefficients for $\tilde{q}(x)$. Consequently

$$\tilde{q}_0 = 1 - q_0 = p', \quad \tilde{q}_i = -q_i, \quad i \ge 1.$$

Each of these two series products in the expansion of $A(x)$ can be expanded as in (6.22) and (6.23), and combined to produce

$$A(x) = 1 + \sum_{n=1}^{\infty} (d_n + \tilde{d}_n) x^{n/2},$$

with

$$d_n = \sum_{k=1}^{n} q_{n-k} \frac{(dsp')^k}{k!}, \quad \tilde{d}_n = \sum_{k=1}^{n} \tilde{q}_{n-k} \frac{(-dsp)^k}{k!}.$$

The claim will be complete by now showing that $d_1 + \tilde{d}_1 = 0$ and $d_2 + \tilde{d}_2 = (r - \mu - \frac{1}{2}\sigma^2)s + \frac{1}{2}\sigma^2 s^2$. To this end, recall that $d = \frac{\sigma}{\sqrt{pp'}}$,

$$d_1 + \tilde{d}_1 = q_0(dsp') + \tilde{q}_0(-dsp)$$

$$= p(dsp') - p'(dsp) - 0.$$

Also, since $q_1 = \frac{[r-\mu-(\sigma^2/2)]}{\sigma/\sqrt{pp'}}$ by (9.88),

$$d_2 + \tilde{d}_2 = q_1(dsp') + \tilde{q}_1(-dsp) + q_0 \frac{(dsp')^2}{2} + \tilde{q}_0 \frac{(-dsp)^2}{2}$$

$$= q_1 ds + \frac{1}{2} d^2 pp' s^2$$

$$= \left(r - \mu - \frac{1}{2}\sigma^2\right)s + \frac{1}{2}\sigma^2 s^2. \qquad \blacksquare$$

Putting this all together, we have from (9.96) and the above claims that

$$M_{B^{(n)}}(S) \rightarrow e^{\mu Ts + (r-\mu-(1/2)\sigma^2)Ts + (1/2)\sigma^2 Ts^2}$$

$$= e^{(r-(1/2)\sigma^2)Ts + (1/2)\sigma^2 Ts^2}.$$

In other words, as in (9.93),

$$B^{(n)} \equiv \ln\left[\frac{S_T^{(n)}}{S_0}\right] \rightarrow_q N\left(\left(r - \frac{1}{2}\sigma^2\right)T, \sigma^2 T\right).$$

*9.8.11 Special Risk-Averter Binomial Distribution as $\Delta t \rightarrow 0$

Fortunately, we do not need to repeat the long section above to determine the other limiting distribution needed for the Black–Scholes–Merton pricing formulas for European put and call options as noted in section 8.8.3. We simply need to adapt the work above to this modified situation.

Analysis of the Special Risk-Averter Probability: $\bar{q}(\Delta t)$
Because $\bar{q}(\Delta t) = q(\Delta t)e^{u(\Delta t)}e^{-r\Delta t}$, we can relatively easily determine the series expansion for $\bar{q}(\Delta t)$ from the series expansion for $q(\Delta t)$ given in (9.88), and the series expansion for $e^{u(\Delta t)-r\Delta t}$. This derivation is possible because each of these series is absolutely convergent for $0 \leq \Delta t < R$ for some $R > 0$. So we can multiply, using the section 6.3.1 results on multiplying series in (6.22) and (6.23), and rearrange summations at will. Consequently, as will be needed below, the series for $\bar{q}(\Delta t)$ is also absolutely convergent.

The goal of this section is to derive the following expansion:

Proposition 9.162 *With $\bar{q}(\Delta t)$ defined as above, we have that*

$$\bar{q}(\Delta t) = p + \frac{[r - \mu + \frac{1}{2}\sigma^2]}{\sigma/\sqrt{pp'}}\sqrt{\Delta t}$$

$$+ \left[\left(p - \frac{1}{2}\right)\left(r - \mu - \frac{7\sigma^2}{6}\right) - p^2\left(r - \mu + \frac{1}{2}\sigma^2\right)\right]\Delta t$$

$$+ O[\Delta t^{3/2}]. \tag{9.98}$$

Denoting the coefficients of the $q(\Delta t)$ series as $\{q_i\}$ as above, and the corresponding coefficients of the $\bar{q}(\Delta t)$ series by $\{\bar{q}_i\}$, we have from $\bar{q}(\Delta t) = q(\Delta t)e^{u(\Delta t)}e^{-r\Delta t}$ that

$$\sum_{n=0}^{\infty}\bar{q}_n(\Delta t)^{n/2} = \sum_{k=0}^{\infty}q_k(\Delta t)^{k/2}\sum_{j=0}^{\infty}\frac{[-c\Delta t + dp'\sqrt{\Delta t}]^j}{j!}.$$

Here, as in the development of (9.88), we use the simplifying notation $c = r - \mu$ and $d = \frac{\sigma}{\sqrt{pp'}}$. If each of these series is then expanded, (6.23) can be applied to derive the needed $\bar{q}_i$-terms.

Knowing from the proof of the second claim for the $q(\Delta t)$ analysis that we only require this expansion up to the $\sqrt{\Delta t}$, but calculating the Δt term for good measure, we derive

$$\sum_{k=0}^{\infty}q_k(\Delta t)^{k/2} = q_0 + q_1\sqrt{\Delta t} + q_2\Delta t + \cdots,$$

$$\sum_{j=0}^{\infty}\frac{[-c\Delta t + dp'\sqrt{\Delta t}]^j}{j!} = 1 + dp'\sqrt{\Delta t} + \left(-c + \frac{1}{2}(dp')^2\right)\Delta t + \cdots,$$

and so

$$\bar{q}_0 = q_0,$$

$$\bar{q}_1 = q_1 + q_0 dp',$$

$$\bar{q}_2 = q_2 + q_1 dp' + q_0\left(-c + \frac{1}{2}(dp')^2\right).$$

Implementing the necessary algebra with the coefficients from (9.88), recalling $c = r - \mu$ and $d = \frac{\sigma}{\sqrt{pp'}}$, produces (9.98).

Special Risk-Averter Binomial Distribution as $\Delta t \to 0$

We are now in a position to derive the limiting distribution of the binomial model under the special risk-averter probabilities. Specifically, we begin with the analogous setup to that above for the risk-neutral analysis:

$$S_T^{(n)} = S_0 e^{\sum B_j},$$

where for $j = 1, 2, \ldots, n$,

$$B_j = \begin{cases} u(\Delta t) \equiv \mu \Delta t + a\sigma\sqrt{\Delta t}, & \Pr = \bar{q}(\Delta t), \\ d(\Delta t) \equiv \mu \Delta t - \frac{1}{a}\sigma\sqrt{\Delta t}, & \Pr = 1 - \bar{q}(\Delta t), \end{cases}$$

with $\bar{q}(\Delta t) = q(\Delta t)e^{u(\Delta t)}e^{-r\Delta t}$, $a = \sqrt{\frac{p'}{p}} = \frac{p'}{\sqrt{pp'}}$, and $-\frac{1}{a} = \frac{-p}{\sqrt{pp'}}$.

The goal of this section is to prove the following:

Proposition 9.163 *With $S_T^{(n)}$ and $\bar{q}(\Delta t)$ defined as above, then as $\Delta t \to 0$, in contrast to both (8.50) and (9.93):*

$$\ln\left[\frac{S_T^{(n)}}{S_0}\right] \to_{\tilde{q}} \ln\left[\frac{S_T}{S_0}\right] \sim N\left(\left(r + \frac{1}{2}\sigma^2\right)T, \sigma^2 T\right), \tag{9.99a}$$

or

$$\ln S_T^{(n)} \to_{\tilde{q}} \ln S_T \sim N\left(\ln S_0 + \left(r + \frac{1}{2}\sigma^2\right)T, \sigma^2 T\right), \tag{9.99b}$$

where the limit symbol "$\to_{\tilde{q}}$" is used to emphasize the dependence of this result on the special risk-averter probability structure.

With S_T denoting the limiting random variable, this can be equivalently written as

$$S_T = S_0 e^X, \tag{9.100}$$

where $X \sim N\left((r + \frac{1}{2}\sigma^2)T, \sigma^2 T\right)$. So once again, S_T has a lognormal distribution, as defined and studied in chapter 10.

As a first step in the investigation, we note that under $\bar{q}(\Delta t)$, using (9.98),

$$E\left[\ln\left[\frac{S_{t+\Delta t}}{S_t}\right]\right] = \left(r + \frac{1}{2}\sigma^2\right)\Delta t + O[\Delta t^{3/2}], \tag{9.101a}$$

$$\text{Var}\left[\ln\left[\frac{S_{t+\Delta t}}{S_t}\right]\right] = \sigma^2 \Delta t + O[\Delta t^{3/2}]. \tag{9.101b}$$

This derivation is assigned in exercise 48 below. So even with this relatively simple calculation, it is apparent that even though $\bar{q}(\Delta t) \to p$ and $\bar{q}(\Delta t) - q(\Delta t) \to 0$ as $\Delta t \to 0$, this convergence occurs slowly enough to cause a different permanent shift in the mean of this distribution compared to the earlier results in sections 8.8.2 and 9.8.10.

Details of the Limiting Result

For the limiting result, a moment of review in the risk-neutral case will confirm that there was only one step in that long derivation where the series for $q(\Delta t)$ actually mattered, and that was in the derivation of the second claim at the end of the section, in which the z_0 needed in (9.97) was derived. We state the modified second claim here, with all notation the same as before.

Claim 9.164   With $\bar{A}(x)$ defined by $\bar{A}(x) \equiv \bar{q}(x)e^{dsp'\sqrt{x}} + \tilde{\bar{q}}(x)e^{-dsp\sqrt{x}}$, where $\tilde{\bar{q}}(x) = 1 - \bar{q}(x)$, then $\bar{A}(x) = 1 + x[(r \quad \mu + \frac{1}{2}\sigma^2)s + \frac{1}{2}\sigma^2 s^2 + \bar{C}(x)]$, where $\bar{C}(x)$ has an absolutely convergent series expansion on $0 \le x < \min(\delta, R')$, with $\bar{C}(0) = 0$.

Proof   The derivation that $\bar{A}(x) = 1 + \sum_{n=1}^{\infty}(\bar{d}_n + \tilde{\bar{d}}_n)x^{n/2}$ is identical to that above, with the series coefficients in (9.98), $\bar{q}_i$, replacing those from (9.88), q_i. That is,

$$\bar{A}(x) = 1 + \bar{q}(x)(e^{dsp'\sqrt{x}} - 1) + \tilde{\bar{q}}(x)(e^{-dsp\sqrt{x}} - 1)$$

$$= 1 + \sum_{i=0}^{\infty}\bar{q}_i x^{i/2}\sum_{j=1}^{\infty}\frac{(dsp')^j x^{j/2}}{j!} + \sum_{i=0}^{\infty}\tilde{\bar{q}}_i x^{i/2}\sum_{j=1}^{\infty}\frac{(-dsp)^j x^{j/2}}{j!}$$

$$= 1 + \sum_{n=1}^{\infty}(\bar{d}_n + \tilde{\bar{d}}_n)x^{n/2}.$$

Here

$$\bar{d}_n = \sum_{k=1}^{n}\bar{q}_{n-k}\frac{(dsp')^k}{k!}, \quad \tilde{\bar{d}}_n = \sum_{k=1}^{n}\tilde{\bar{q}}_{n-k}\frac{(-dsp)^k}{k!}.$$

The only steps of the proof that differ and need to be checked relate to the first 2 terms of the series. For example,

$$\bar{d}_1 + \tilde{\bar{d}}_1 = 0$$

because $\bar{q}_0 = q_0 = p$, and $d = \frac{\sigma}{\sqrt{pp'}}$. Also

$$\bar{d}_2 + \tilde{\bar{d}}_2 = \bar{q}_1 ds + \frac{1}{2}d^2 pp's^2$$

$$= \left(r - \mu + \frac{1}{2}\sigma^2\right)s + \frac{1}{2}\sigma^2 s^2,$$

which follows from $\bar{q}_1 = \frac{\left[r - \mu + \frac{1}{2}\sigma^2\right]}{\sigma/\sqrt{pp'}}$. ∎

9.8.12 Black–Scholes–Merton Option-Pricing Formulas II

We began the derivation of the famous Black–Scholes–Merton pricing formulas for European put and call options in section 8.8.3. For a T-period European call on an equity S, with a strike price of K, it was derived that the price at time 0, defined as the price of a replicating portfolio on a binomial lattice with $\Delta t = \frac{T}{n}$, is given in the equation preceding (8.56) by

$$\Lambda_0(S_0) = S_0 \Pr\left[\bar{B}_{(n)} \geq \ln\left[\frac{K}{S_0}\right]\right] - e^{-rT}K \Pr\left[B_{(n)} \geq \ln\left[\frac{K}{S_0}\right]\right].$$

Recall than $\bar{B}_{(n)} = \sum_{i=1}^{n} B_i$ in the $\text{Bin}(\bar{q}, n)$ model, where $\{B_i\}$ are i.i.d. binomials and have upstate and downstate values of $u(\Delta t)$ and $d(\Delta t)$ with special risk-averter probabilities $\bar{q}(\Delta t)$ and $1 - \bar{q}(\Delta t)$, respectively, and $B_{(n)}$ is identically defined in the $\text{Bin}(q, n)$ model, but with the risk-neutral probability $q \equiv q(\Delta t)$.

The proofs in the prior two sections show that $\bar{B}_{(n)} \to N\left((r + \frac{1}{2}\sigma^2)T, \sigma^2 T\right)$ and that $B_{(n)} \to N\left((r - \frac{1}{2}\sigma^2)T, \sigma^2 T\right)$. Consequently, with Z_1 and Z_2 denoting these normal variates, and $\Phi(z)$ the unit normal cumulative distribution function,

$$\Pr\left[\bar{B}_{(n)} \geq \ln\left[\frac{K}{S_0}\right]\right] \to \Pr\left[Z_1 \geq \ln\left[\frac{K}{S_0}\right]\right]$$

$$= 1 - \Phi\left(\frac{\ln\left[\frac{K}{S_0}\right] - (r + \frac{1}{2}\sigma^2)T}{\sigma\sqrt{T}}\right)$$

$$= \Phi\left(\frac{\ln\left[\frac{S_0}{K}\right] + (r + \frac{1}{2}\sigma^2)T}{\sigma\sqrt{T}}\right),$$

where the last step follows from the symmetry of the normal distribution, which implies that $1 - \Phi(z) = \Phi(-z)$.

Similarly

$$\Pr\left[B_{(n)} \geq \ln\left[\frac{K}{S_0}\right]\right] \rightarrow \Pr\left[Z_2 \geq \ln\left[\frac{K}{S_0}\right]\right]$$

$$= \Phi\left(\frac{\ln\left[\frac{S_0}{K}\right] + \left(r - \frac{1}{2}\sigma^2\right)T}{\sigma\sqrt{T}}\right).$$

Combining results, we have derived the Black–Scholes–Merton pricing formula for a European call option:

$$\Lambda_0^C(S_0) = S_0\Phi(d_1) - e^{-rT}K\Phi(d_2), \tag{9.102a}$$

$$d_1 = \frac{\ln\frac{S_0}{K} + \left(r + \frac{1}{2}\sigma^2\right)T}{\sigma\sqrt{T}}, \tag{9.102b}$$

$$d_2 = \frac{\ln\frac{S_0}{K} + \left(r - \frac{1}{2}\sigma^2\right)T}{\sigma\sqrt{T}}. \tag{9.102c}$$

A European put option is now easy to price. While the payoff function at expiry for a call is

$$\Lambda^C(S_T) = \max(S_T - K, 0), \tag{9.103}$$

for a put option we have

$$\Lambda^P(S_T) = \max(K - S_T, 0). \tag{9.104}$$

Consequently the payoff function for a portfolio that includes a short put and a long call is

$$\Lambda^C(S_T) - \Lambda^P(S_T) = S_T - K.$$

In other words, this portfolio has value equal to $S_T - K$ at time T, which means is can be replicated by a portfolio of one long share, and a short position in a T-bill that matures for K. Consequently the price of this options portfolio at $t = 0$ equals the price of this replicating portfolio and therefore satisfies

$$\Lambda_0^C(S_0) - \Lambda_0^P(S_0) = S_0 - Ke^{-rT}. \tag{9.105}$$

This famous identity in prices, forced by this replication argument, is known as **put-call parity**.

Exercise 23 assigns the task of deriving the Black–Scholes–Merton pricing formula for a European put option using put-call parity, the price above of a European call option, and symmetry properties of the unit normal distribution. The formula with the same notation as for a call is

$$\Lambda_0^P(S_0) = e^{-rT}K\Phi(-d_2) - S_0\Phi(-d_1). \tag{9.106}$$

Exercises

Practice Exercises

1. For each of the following collections of functions, determine the given composite functions:

(a) $f(x) = x^{-n}$ and $g(i) = 1 + \frac{i}{2}$: $f(g(i))$ and $g(f(x))$

(b) $f(x) = \sum_{j=1}^{n} x^{-j}$ and $g(i) = 1 + \frac{i}{2}$: $f(g(i))$ and $g(f(x))$

(c) $f(x) = e^{rx}$, $g(y) = \ln y$, $h(z) = \sum_{j=1}^{n} z^j$: $f \circ g \circ h(z)$, $g \circ f \circ h(z)$ and $f \circ h \circ g(y)$

2. Demonstrate that the following functions are continuous at the given points. (*Hint*: Demonstrate directly or make use of the propositions on combining known continuous functions.)

(a) $r(i) = (1+i)^2$ for all $i \in \mathbb{R}$.

(b) $s(i) = (1+i)^n$ for all $i \in \mathbb{R}$, where $n \in \mathbb{N}$.

(c) $f(x) = (1+x)^{-n}$ for all $x > -1$, where $n \in \mathbb{N}$.

(d) $g(z) = \sum_{j=0}^{N} b_j z^j$ for $z \in \mathbb{R}$, where $b_j \in \mathbb{R}$, $N \in \mathbb{N}$.

(e) $a(i) = \begin{cases} \frac{1-(1+i)^{-n}}{i}, & i > -1, i \neq 0 \\ n, & i = 0 \end{cases}$ where $n \in \mathbb{N}$. (*Hint*: Consider $(1+i)^n a(i)$ and recall the binomial theorem.)

3. Demonstrate that the following functions are not continuous as indicated:

(a) $f(x) = \begin{cases} \sin \frac{1}{x}, & x \neq 0, \\ 0, & x = 0, \end{cases}$ is not continuous at $x = 0$.

(b) $g(y) = \begin{cases} 1, & x \geq 3, \\ -1, & x < 3, \end{cases}$ is not continuous at $y = 3$.

4. Of the functions in exercise 2, demonstrate that 2(a), (b), and (d) are uniformly continuous on $(-1, 1]$, and that 2(c) and (e) are not.

5. Explicitly write out the definitions of continuous, sequentially continuous, and uniformly continuous for a function $f(x)$ defined on a metric space (X, d), and with range in:

(a) $\mathbb{R}$, under the standard metric

(b) a general metric space (Y, d')

6. Show that if $f(x)$ and $g(x)$ are differentiable at x_0, then so is $h(x)$. (*Hint*: The goal is to express $h(x) - h(x_0)$ in terms of $f(x) - f(x_0)$, $g(x) - g(x_0)$ and other terms that are easy to work with. Consider (9.10).)

(a) $h(x) = af(x) \pm bg(x)$, and $h'(x_0) = af'(x_0) \pm bg'(x_0)$

(b) $h(x) = f(x)g(x)$ and $h'(x_0) = f'(x_0)g(x_0) + f(x_0)g'(x_0)$

7. Show that if $g(x)$ is differentiable and $g'(x)$ continuous in an open interval containing x_0 and $g'(x_0) \neq 0$, then there is an interval about x_0, say $(x_0 - a, x_0 + a)$, for some $a > 0$, where $g(x)$ is one-to-one. (*Hint*: Assume $g'(x_0) > 0$, and note that if $\lim_{\Delta x \to 0} \frac{g(x_0 + \Delta x) - g(x_0)}{\Delta x} = g'(x_0) > 0$, then for $\epsilon = \frac{1}{2}g'(x_0)$ there is a δ so that

$$\left| \frac{g(x_0 + \Delta x) - g(x_0)}{\Delta x} - g'(x_0) \right| < \frac{1}{2}g'(x_0),$$

for $|\Delta x| < \delta$. What does this say about $g(x_0 + \Delta x) - g(x_0)$? Consider also $g'(x_0) < 0$.)

8. Show that $\frac{da^x}{dx} = a^x \ln a$, for $a > 0$ follows from the identity: $a^x = e^{x \ln a}$. (*Hint*: $a^x = f(g(x))$ with $g(x) = x \ln a$ and $f(y) = e^y$.)

9. Calculate the derivative of the functions in exercise 2, and determine if any restrictions are needed on the domains given there.

10. Find the Taylor series expansions for the following functions, and determine when they converge.

(a) $f(x) = (1 + x)^{-1}$ with $x_0 = 0$

(b) $g(y) = (1 - y)^{-n}$ with $y_0 = 0$

(c) $h(z) = e^{-rz}$ with $z_0 = 0$

11. Confirm where each of the following functions is concave or convex on their respective domains:

(a) $f(x) = e^{-x^2}$, $x \in \mathbb{R}$

(b) $h(y) = (1 + y)^{-n}$, n a positive integer, $y > -1$

(c) $l(z) = \ln(1 + z)$, $z > -1$

12. Prove the arithmetic-geometric means inequality. If $x_i \geq 0$ for all i,

$$\frac{1}{n}\sum_{i=1}^{n} x_i \geq \left(\prod_{i=1}^{n} x_i\right)^{1/n}.$$

(*Hint*: The result is apparently true if some $x_i = 0$, so assume all $x_i > 0$. Take logarithms and consider if $\ln x$ is a concave or convex function.)

Remark 9.165 *When $\{x_i\}$ are both positive and negative, this inequality is satisfied with the collection, $\{|x_i|\}$.*

13. Show by considering the product of Taylor series, that for $a, b \in \mathbb{R}$: $e^{ax}e^{bx} = e^{(a+b)x}$. Justify the reordering of these summations to get the intended result. (*Hint*: Use the binomial theorem and (9.41).)

14. Show, using a Taylor series expansion, that if $f(x) = \ln(1 + x)$, for $x > -1$, then $f'(x) = \frac{1}{1+x}$. Justify differentiating term by term as well as the convergence of the final series to the desired answer.

15. Derive the risk-minimizing allocation between two assets, as well as the resulting portfolio's mean return and standard deviation of return:

(a) If $\mu_1 - 0.05$, $\sigma_1 - 0.09$, $\mu_1 - 0.08$, $\sigma_1 = 0.15$, $\rho = 0.4$

(b) If $\mu_1 = 0.05$, $\sigma_1 = 0.09$, $\mu_1 = 0.08$, $\sigma_1 = 0.15$, $\rho = 0.6$

(c) If $\mu_1 = 0.05$, $\sigma_1 = 0.09$, $\mu_1 = 0.08$, $\sigma_1 = 0.15$, $\rho = 0.8$

16. For the exponential ($k = 9 \cdot 10^{-5}$), quadratic ($a = 1$, $b = 4 \cdot 10^{-6}$), power ($\lambda = 0.01$), and logarithmic ($c = 10{,}000$) utility functions, determine the optimal risky asset allocation between the risk-free asset with $r = 0.03$ and a risky asset with $\mu = 0.10$ and $\sigma = 0.18$, where $W_0 = 100{,}000$. (*Hint*: See exercise 38.)

17. Calculate the duration and convexity of the following price functions exactly, and using the approximation formulas with both $\Delta i = 0.01$, and $\Delta i = 0.001$. For duration, compare the results of (9.52) with (9.51). Assume 100 par.

(a) 10-year zero coupon bond with a yield of 8% semiannual

(b) 3-year, 6% semiannual coupon bond, with a yield of 7% semiannual.

18. For each of the price functions in exercise 17, compare the prices predicted by the forward difference duration approximation with $\Delta i = 0.01$ to those predicted with the convexity adjustment, again using the convexity approximation with $\Delta i = 0.01$, and then to the exact prices. Do this exercise shifting the original pricing yields $\pm 3\%$, $\pm 2\%$, $\pm 1\%$, $\pm 0.5\%$, $\pm 0.1\%$.

19. Prove for a portfolio of fixed income securities with price function given by

$$P(i) = \sum_{j=1}^{n} P_j(i)$$

that the duration and convexity of the portfolio, assuming $P(i) \neq 0$, and $P_j(i) \neq 0$ for all j, is given by

$$D(i) = \sum_{j=1}^{n} w_j D_j(i), \quad C(i) = \sum_{j=1}^{n} w_j C_j(i),$$

where $w_j = \frac{P_j(i)}{P(i)}$, and hence $\sum_{j=1}^{n} w_j = 1$.

Remark 9.166 *It is important to note that you will not need to make an assumption about the signs of $\{P_j(i)\}$ to prove this result. So this result applies equally well to long positions, $P_j(i) > 0$, short positions, $P_j(i) < 0$, or a mixed portfolio of longs and shorts.*

20. Given an asset portfolio of \$250 million of duration 6 bonds, and \$225 million of liabilities of duration 4.5, determine the necessary "target" duration for assets to achieve immunization of surplus in the following cases, as well as the necessary asset trade. Assume that bonds are homogeneous and can be sold in any amount, and that cash is to be reinvested in duration 1 assets. (*Hint*: Surplus is a long portfolio of assets and a short portfolio of liabilities. See exercise 19.)

(a) Surplus immunization at $t = 0$

(b) Surplus immunization at $t = 2$, where $Z_2(i)$ is priced at $i = 0.03$ semiannual

(c) Surplus ratio immunization

21. Using the Black–Scholes–Merton formula for a call option, from (9.102), derive the Delta of a call option as

$$\Delta^C = \Phi(d_1).$$

(*Hint*: This is a challenging calculation. It is seductive to think that because the first part of the BSM formula is $S_0 \Phi(d_1)$, that this derivative, $\frac{d\Delta}{dS_0}$ is obvious, but it is not, since both d_1 and d_2 are functions of S_0 also. Once you have the derivative expression, see what is needed to achieve the desired answer.)

22. Develop the relationship between an individual's risk preference and their willingness to insure a given risk, where the indifference equation is

$$u(W_0 - P) = E[u(W_0 - X)],$$

with P as the insurance premium and X the risk insured against. In other words, how is the resulting relationship between P and $E[X]$ determined by $u''(x)$? (*Hint*: Use Jensen's inequality.)

23. Derive the Black–Scholes–Merton pricing formula for a European put option in (9.106) using put-call parity, and the Black–Scholes–Merton price of a European call option in (9.102).

24. Investigate the moments of $\ln[S_{t+\Delta t}/S_t]$ under the risk-neutral probability.

(a) Derive (9.95) using the expansion of $q(\Delta t)$ in (9.88). (*Hint*: Only keep track of the terms in $q(\Delta t)$ and $1 - q(\Delta t)$ up to $O(\sqrt{\Delta t})$, since the higher order terms will be part of the error, $O[\Delta t^{3/2}]$, as will be confirmed next.)

(b) Demonstrate that this shift in the mean is caused only by the coefficient of $\sqrt{\Delta t}$ in the expansion of $q(\Delta t)$, and that the higher order terms have no effect on these moments larger than $O[\Delta t^{3/2}]$.

Assignment Exercises

25. For each of the following collections of functions, determine the given composite functions:

(a) $f(x) = e^{-rx}$ and $g(z) = \sum_{j=1}^{n} z^j$: $f(g(z))$ and $g(f(x))$

(b) $f(x) = \frac{1}{x}$ and $g(y) = \sum_{j=1}^{n} y^{-j}$: $f(g(y))$ and $g(f(x))$

(c) $f(x) = \left(1 + \frac{i}{12}\right)^x$, $g(y) = \ln y$, $h(z) = \sum_{j=1}^{n} \frac{z}{j}$: $f \circ g \circ h(z)$, $g \circ f \circ h(z)$ and $f \circ h \circ g(y)$

26. Demonstrate that the following functions are continuous at the given points. (*Hint*: demonstrate directly or make use of the propositions on combining continuous functions.)

(a) $h(x) = e^{rx}$ for all $x \in \mathbb{R}$, for any $r \in \mathbb{R}$.

(b) $g(z) = \frac{1}{\sqrt{2\pi}} e^{-z^2}$ for all $z \in \mathbb{R}$.

(c) $h(z) = \sum_{j=0}^{N} \frac{b_j}{z^j}$, where $b_j \in \mathbb{R}$, $N \in \mathbb{N}$, for $z > 0$.

(d) $r(i) = m \ln\left(1 + \frac{i}{m}\right)$ for $m \in \mathbb{N}$ and all $i > -1$. (*Note*: $r(i)$ is the continuous rate equivalent to the *mthly* nominal rate i, as will be studied in chapter 10.)

(e) $f(x) = \frac{1}{x^2}$ for $x \neq 0$.

27. Demonstrate that the following functions are not continuous as indicated:

(a) $i(z) = \begin{cases} 1, & z \text{ rational}, \\ -1, & z \text{ irrational}, \end{cases}$ is not continuous at any $z \in \mathbb{R}$.

(b) $f(x) = \begin{cases} n, & x = n \in \mathbb{Z}, \\ \frac{1}{x^2}, & x \notin \mathbb{Z}, \end{cases}$ is not continuous at any $n \in \mathbb{Z}$ except $n = 1$.

28. Prove that $f(x)$ is continuous at x_0 if and only if it is sequentially continuous at x_0. (*Hint*: If continuous, consider the definition in conjunction with definition of $x_n \to x_0$. Prove the reverse implication by contradiction, if $f(x)$ is not continuous. ....)

29. Of the functions in exercise 26, demonstrate that the functions in (a), and (b) are uniformly continuous on $(-1, 1]$, that the function in (d) is uniformly continuous on $(-1, 1]$ only when $m > 1$, and that the functions in (c) and (e) are not uniformly continuous on $(0, 1]$. (*Note*: The function in (c) is constant and hence uniformly continuous in the trivial case when $N = 0$, so assume $N > 0$ for this exercise.)

30. (a) Prove that if $f(x)$ is continuous on a compact set $K \subset X$, where (X, d) is a metric space, then it is uniformly continuous on K. Assume that the range of $f(x)$ is a general metric space (Y, d'), or if easier, first consider the case where $f : X \to \mathbb{R}$. (*Hint*: First review the chapter proof when $X = \mathbb{R}$.)

(b) Show that if $f(x) = \sum_{j=0}^{\infty} a_j(x - x_0)^j$ is a power series that converges on

$$I = \{x \mid |x - x_0| < R\},$$

and if $f_n(x)$ denotes the partial sum of this series, then $f_n(x) \to f(x)$ uniformly on any compact set $K \subset I$.

31. Show that if $f(x)$ is an arbitrary function, $f : \mathbb{R} \to \mathbb{R}$, then $f^{-1}(\tilde{F}) = \widetilde{f^{-1}(F)}$ for any set $F \subset \mathbb{R}$.

32. Show that if $f(x)$ and $g(x)$ are differentiable at x_0, then so is $h(x)$. (*Hint*: The goal is to express $h(x) - h(x_0)$ in terms of $f(x) - f(x_0)$, $g(x) - g(x_0)$, and other terms that are easy to work with. Consider (9.10).)

(a) $h(x) = \frac{1}{g(x)}$ if $g(x_0) \neq 0$, and $h'(x_0) = \frac{-g'(x_0)}{g^2(x_0)}$

(b) $h(x) = \frac{f(x)}{g(x)}$ if $g(x_0) \neq 0$, and $h'(x_0) = \frac{f'(x_0)g(x_0) - f(x_0)g'(x_0)}{g^2(x_0)}$

33. Calculate the derivative of the functions in exercise 26, and determine if any restrictions are needed on the domains given there.

34. Prove the **Leibniz rule** for the nth-derivative of the product of two n-times differentiable functions as given in (9.42). Namely, if $h(x) = f(x)g(x)$, then

$$h^{(n)}(x) = \sum_{k=0}^{n} \binom{n}{k} f^{(k)}(x) g^{(n-k)}(x),$$

where $f^{(0)}(x) \equiv f(x)$, and similarly $g^{(0)}(x) \equiv g(x)$. (*Hint*: Use mathematical induction.)

35. Find the Taylor series expansions for the following functions, and determine when they converge:

(a) $P(r) = \frac{D}{r}$ with $r_0 = 0.05$.

(b) $f(x) = \sin x$ with $x_0 = 0$. (*Hint*: Use (9.16).)

(c) $g(x) = \cos x$ with $x_0 = 0$.

(d) Confirm using parts (b) and (c), that in terms of the resulting Taylor series,

$$e^{ix} = \cos x + i \sin x,$$

which is Euler's formula from (2.5) in chapter 2.

36. Confirm where each of the following functions is concave or convex on their respective domains:

(a) $j(w) = w^r$, for $r > 0$, $w \geq 0$

(b) $a(u) = \frac{1}{u}$, $u \neq 0$

(c) $z(v) = e^v$, $v \in \mathbb{R}$

37. Show, using a Taylor series expansion, that if $f(x) = e^{-rx}$ for $r > 0$, that $f'(x) = -rf(x)$. Justify differentiating term by term.

38. Derive the Arrow–Pratt measure of absolute risk aversion, r_{AP}, for the exponential ($k = 9 \cdot 10^{-5}$), quadratic ($a = 1$, $b = 4 \cdot 10^{-6}$), power ($\lambda = 0.01$), and logarithmic ($c = 10,000$) utility functions where $r = 0.03$ and $W_0 = 100,000$.

39. Using the general formula for the risk of a portfolio in (9.54b), derive the obvious result that the risk-minimizing allocation between a risky asset and a risk-free asset is $w_j = 1$ in the risk-free asset.

40. Calculate the duration and convexity of the following price functions exactly, and using the approximation formulas with both $\Delta i = 0.01$, and $\Delta i = 0.001$. For the duration, compare the results of (9.52) with (9.51). Assume 100 par for part (a), and a loan of 100 in part (b).

(a) 8% annual dividend preferred stock, with an annual yield of 10%

(b) A 5-year, monthly repayment schedule loan made with a monthly loan rate of 10%, priced with a yield of 12% monthly

41. For each of the price functions in exercise 40, compare the prices predicted by the forward difference duration approximation with $\Delta i = 0.01$, to those predicted with the convexity adjustment, again using the convexity approximation with $\Delta i =$

0.01, and then to the exact prices. Do this exercise shifting the original pricing yields $\pm 3\%$, $\pm 2\%$, $\pm 1\%$, $\pm 0.5\%$, $\pm 0.1\%$.

42. Derive the immunizing conditions in (9.73) where $S(i_0) = 0$. (*Hint*: Determine what conditions ensure that $S'(i_0) = 0$ and $S''(i_0) > 0$.)

43. Given a fixed income hedge fund with asset portfolio of \$900 million of duration 4.5 bonds, and \$850 million of debt of duration 2.5, determine the necessary "target" duration for assets to achieve immunization of the hedge fund equity in the following cases, as well as the necessary asset trade. Assume that bonds are homogeneous and can be sold in any amount, and reinvested in duration 0.25 assets. (*Hint*: Equity is a long portfolio of assets and a short portfolio of liabilities. See exercise 19.)

(a) Equity immunization at $t = 0$

(b) Equity immunization at $t = 1$, where $Z_1(i)$ is priced at $i = 0.025$ semiannual

(c) Equity ratio immunization

44. Derive the Delta of a put option as priced by the Black–Scholes–Merton formula from (9.106):

$$\Delta^P = \Phi(d_1) - 1.$$

(*Hint*: Consider exercise 21 and put-call parity from (9.105).)

45. Using exercises 21 and 44, calculate the gamma of a put and call option as priced by the Black–Scholes–Merton formulas, and show that they are the same:

$$\Gamma^{P/C} = \frac{\Phi'(d_1)}{S_0 \sigma \sqrt{T}},$$

where Φ' is the derivative of the normal distribution function, which is the normal density function: $\Phi'(d_1) = \phi(d_1)$. (See section 10.5.2.)

46. With the forward value of surplus, $S_t(i)$, defined as

$$S_t(i) \equiv \frac{S(i)}{Z_t(i)},$$

calculate $S_t'(i)$ and $S_t''(i)$, as well as the duration and convexity formulas:

$$D^{S_t}(i_0) = D^S(i_0) - D^{Z_t}(i_0),$$

$$C^{S_t}(i_0) = C^S(i_0) - C^{Z_t}(i_0) - 2D^{Z_t}(i_0)[D^S(i_0) - D^{Z_t}(i_0)].$$

47. Develop the relationship between an individual's risk preference and their willingness to engage in a given bet, where the indifference equation is

$$E[u(W_0 - L + Y)] = u(W_0),$$

with L = cost of gamble, and Y = potential payoff. In other words, how is the resulting relationship between L and $E[Y]$ determined by $u''(x)$? (*Hint*: Use Jensen's inequality.)

48. Repeat exercise 24 for the moments of the special risk-averter distribution in (9.101).

10 Calculus II: Integration

10.1 Summing Smooth Functions

In this chapter we study the earliest conception of the integral, or generalized summation, of a function as it applies to continuous and certain generalizations of continuous functions. This approach to integration was first introduced on a rigorous basis by **Bernhard Riemann** (1826–1866), who despite his short life was responsible for a remarkable number of acclaimed mathematical discoveries, many of which bear his name. Here we also develop the relationship between this integral and derivative, and explore some of the consequences of this relationship. In the final section, we explore the strengths and limitations of the Riemann integral. This will serve as background for the more general integration theories of real analysis.

Remark 10.1 *In general, the functions that appear to be addressed in calculus are real-valued functions of a real variable. In other words, these are functions*

$$f : X \to Y,$$

where $X, Y \subset \mathbb{R}$. However, while the assumption that the domain of $f(x)$ is real is critical, $X = \mathrm{Dmn}(f) \subset \mathbb{R}$, there is often no essential difficulty in assuming f to be a complex-valued function of a real variable so that the range of $f(x)$, $Y = \mathrm{Rng}(f) \subset \mathbb{C}$. This generalization is not often needed in finance, and the characteristic function is one of the few examples in finance where complex-valued functions are encountered.

One reason that $\mathrm{Dmn}(f) \subset \mathbb{R}$ is critical in the development of calculus is that we will often utilize the natural ordering of the real numbers. In other words, given $x, y \in \mathbb{R}$ with $x \neq y$, then it must be the case that either $x < y$ or $y < x$. None of these proofs would generalize easily to functions of a complex variable where no such ordering exists. Indeed it turns out that the calculus of such functions is quite different and studied in what is called complex analysis.

Because of the rarity of encountering complex-valued functions of a real variable in finance, all the statements in this chapter are either silent on the location of Y, or explicitly assume $Y \subset \mathbb{R}$. In particular, no effort was made to explicitly frame all proofs in the general case $Y \subset \mathbb{C}$, since this overt generality seemed to have little purpose given the objectives of this book.

The applicability of many of the results of calculus to a complex-valued function can often be justified by splitting the function values into "real" and "imaginary" parts. If $Y \subset \mathbb{C}$, we write

$$f(x) = g(x) + ih(x),$$

where both $g(x)$ and $h(x)$ are real valued. For an integration theory, ordering in the range space matters as will be immediately observed, and so splitting $f(x)$ into "real" and "imaginary" parts, where both $g(x)$ and $h(x)$ are real valued, is how one must proceed. The integration theory in this chapter can usually then be applied to $f(x)$ by applying it separately to $g(x)$ and $h(x)$ and combining results.

10.2   Riemann Integration of Functions

10.2.1   Riemann Integral of a Continuous Function

The intuitive idea behind the definition of a Riemann integral is that of finding the "signed area" between the graph of a given continuous function $f(x)$ and the x-axis over the interval $[a, b]$, where $a < b$. By "signed" is meant that area above the x-axis is counted as "positive" area and that below is "negative" area. This is done by first approximating this area with a collection of non-overlapping rectangles.

For example, splitting the interval $[a, b]$ into n-subintervals of length $\Delta x = \frac{b-a}{n}$, and choosing one point in each subinterval, $\tilde{x}_i \in [a + (i-1)\Delta x, a + i\Delta x]$ for $i = 1,$ $2, \ldots, n$, we can produce an approximation

$$\text{Signed area} \approx \sum_{i=1}^{n} f(\tilde{x}_i)\Delta x.$$

Of course, the goal is then to determine conditions on $f(x)$ that assure that this approximation converges as $\Delta x \to 0$, or equivalently as $n \to \infty$, and that it converges independently of how one chooses the $\tilde{x}_i$ values in the subintervals.

When $f(x)$ is a nonnegative function $f(x) \geq 0$, this signed area corresponds with the usual notion of area. However, for general $f(x)$, it is important to note that for functions that are both positive and negative, the integral provides the "net" area between the function's graph and x-axis, whereby area above the axis is counted as positive area and that below as negative. The integral then provides a "netting" of the two values, which could be positive, negative, or zero.

If we assume that $f(x)$ is a continuous function, then on every closed subinterval, $[a + (i-1)\Delta x, a + i\Delta x]$, it attains its maximum value, M_i, and minimum value, m_i, and we can conclude that for any choice of the $\tilde{x}_i$ values,

$$\sum_{i=1}^{n} m_i\Delta x \leq \sum_{i=1}^{n} f(\tilde{x}_i)\Delta x \leq \sum_{i=1}^{n} M_i\Delta x.$$

The smaller summation is referred to as a **lower Riemann sum**, while the larger sum is correspondingly referred to as an **upper Riemann sum**. All other summations of this type are simply called **Riemann sums**.

More generally, one can define these summations with respect to an arbitrary **partition** of the interval $[a, b]$ into subintervals $[x_{i-1}, x_i]$:

$$a = x_0 < x_1 < \cdots < x_{n-1} < x_n = b,$$

where we again choose $\widetilde{x}_i \in [x_{i-1}, x_i]$ and define $\Delta x_i = x_i - x_{i-1}$. We obtain

$$m(b - a) \leq \sum_{i=1}^{n} m_i \Delta x_i \leq \sum_{i=1}^{n} f(\widetilde{x}_i) \Delta x_i \leq \sum_{i=1}^{n} M_i \Delta x_i \leq M(b - a), \tag{10.1}$$

where M_i and m_i denote the maximum and minimum values of the continuous $f(x)$ on the subinterval $[x_{i-1}, x_i]$, while M and m denote these defined on the full interval $[a, b]$.

Even more generally, if $f(x)$ is not continuous on $[a, b]$ but is bounded, we can achieve the same set of inequalities by defining M_i, and m_i, as the **least upper bound**, or **l.u.b.**, and **greatest lower bound**, or **g.l.b., respectively**, of $f(x)$ on each subinterval. Specifically, for $i = 1, \ldots, n$,

$$M_i = \text{l.u.b.}\{f(x) \mid x \in [x_{i-1}, x_i]\}$$

$$= \min\{y \mid y \geq f(x) \quad \text{for } x \in [x_{i-1}, x_i]\}, \tag{10.2}$$

$$m_i = \text{g.l.b.}\{f(x) \mid x \in [x_{i-1}, x_i]\}$$

$$= \max\{y \mid y \leq f(x) \quad \text{for } x \in [x_{i-1}, x_i]\}.$$

The question of convergence of Riemann sums in the context of a general partition is now defined in terms of the partition becoming increasingly fine. Specifically, with

$$\mu \equiv \max_{1 \leq i \leq n} \{x_i - x_{i-1}\}, \tag{10.3}$$

convergence is investigated as $\mu \to 0$. The measure μ is often referred to as the **mesh size of the partition**.

From (10.1) it is clear that both the question of convergence of the Riemann sums, as well as the independence of these limits from the choice of the $\widetilde{x}_i$ values can be addressed together. Namely both questions can be answered in the affirmative if we can show that the upper and lower Riemann sums converge to the same value as $\mu \to 0$. With this in mind, we have the following definition.

Definition 10.2 $f(x)$ is ***Riemann integrable*** on an interval $[a, b]$ if as $\mu \to 0$ we have that

$$\left[\sum_{i=1}^{n} M_i\Delta x_i - \sum_{i=1}^{n} m_i\Delta x_i\right] \to 0, \tag{10.4}$$

where M_i and m_i are defined in (10.2). In this case we define the **Riemann integral of** $f(x)$ **over** $[a,b]$, by

$$\int_a^b f(x)\,dx = \lim_{\mu\to 0}\sum_{i=1}^{n} f(\tilde x_i)\Delta x_i, \tag{10.5}$$

which exists and is independent of the choice of $\tilde x_i \in [x_{i-1}, x_i]$ by (10.1). The function $f(x)$ is then called the **integrand**, and the constants a and b the **limits of integration** of the integral.

Remark 10.3   Sometimes, for added clarity, the above integral is called a **definite integral**, in contrast to an **indefinite integral** introduced in section 10.5.2 on the derivative of an integral.

The following result is central to the theory, but it is not the most general result. It requires both that $f(x)$ be continuous and that the interval $[a,b]$ be bounded.

Proposition 10.4   If $f(x)$ is continuous on bounded $[a,b]$, then $f(x)$ is Riemann integrable.

Proof   Since $f(x)$ must be uniformly continuous on closed and bounded $[a,b]$ by proposition 9.35, we have that for any $\epsilon > 0$ there is a δ so that

$$|f(x) - f(x')| < \epsilon \qquad \text{if } |x - x'| < \delta.$$

Hence, if the mesh size of a given partition of $[a,b]$ satisfies $\mu \le \delta$, then on any subinterval

$$|M_i - m_i| < \epsilon.$$

The triangle inequality then produces

$$\left|\sum_{i=1}^{n} M_i\Delta x_i - \sum_{i=1}^{n} m_i\Delta x_i\right| \le \sum_{i=1}^{n} |M_i - m_i|\Delta x_i$$

$$< \epsilon(b-a),$$

so the difference between upper and lower summations converges to 0 as $\epsilon \to 0$.   ■

Next a Riemann integral over an interval can in fact be calculated in pieces.

Proposition 10.5   *If $f(x)$ is continuous on bounded $[a,b]$ and $a < c < b$, then*

$$\int_a^b f(x)\,dx = \int_a^c f(x)\,dx + \int_c^b f(x)\,dx. \qquad (10.6)$$

Proof   Clearly, if we choose partitions of the interval $[a,b]$ so that one of the partition points $x_i = c$, this result is immediate as we simply split the upper and lower Riemann sums into those applicable to $[a,c]$ and those applicable to $[c,b]$. More generally, assume that the point c is within one of the subintervals of a partition. That is, assume that $c \in [x_{i-1}, x_i]$. Denoting by M_i^1 the l.u.b. of $f(x)$ on $[x_{i-1}, c]$, and M_i^2 the l.u.b. of $f(x)$ on $[c, x_i]$, it is clear that $M_i^k \leq M_i$, the l.u.b. of $f(x)$ on $[x_{i-1}, x_i]$ where $k = 1, 2$. With analogous notation, $m_i^k \geq m_i$. Hence, with Δx_i^1 used as notation for $c - x_{i-1}$, and Δx_i^2 as notation for $x_i - c$,

$$[M_i^1 - m_i^1]\Delta x_i^1 + [M_i^2 - m_i^2]\Delta x_i^2 \leq [M_i - m_i]\Delta x_i,$$

and hence, as $\Delta x_i \to 0$, the terms in the Riemann sums that reflect intervals that contain c converge to 0.   ∎

Remark 10.6   *It should be noted that the above proof demonstrated that the terms in the Riemann sums that reflect intervals that contained c could be discarded since they converged to 0. In other words, it was demonstrated that for this function, as $\epsilon \to 0$,*

$$\int_a^{c-\epsilon} f(x)\,dx \to \int_a^c f(x)\,dx, \qquad (10.7a)$$

$$\int_{c+\epsilon}^b f(x)\,dx \to \int_c^b f(x)\,dx. \qquad (10.7b)$$

This observation provides an easy generalization to the proposition above in the case where $f(x)$ is only continuous on the bounded open interval (a,b), as long as it is also bounded there.

Proposition 10.7   *If $f(x)$ is continuous and bounded on bounded (a,b), then $f(x)$ is Riemann integrable on $[a,b]$. Further, for any $\epsilon_1, \epsilon_2 \to 0$,*

$$\int_a^b f(x)\,dx = \lim_{\epsilon_1, \epsilon_2 \to 0} \int_{a-\epsilon_1}^{b-\epsilon_2} f(x)\,dx. \qquad (10.8)$$

Proof Given any partition of the interval $[a, b]$, say

$$a = x_0 < x_1 < \cdots < x_{n-1} < x_n = b,$$

we must prove (10.4). Now, since $[x_1, x_{n-1}] \subset (a, b)$, we conclude that $f(x)$ is continuous and hence Riemann integrable on this interval. Also, since it is bounded, we can assume that on $(a, x_1] \cup [x_{n-1}, b)$ the function $f(x)$ satisfies $m \leq f(x) \leq M$. Finally, with $\Delta x_i = x_i - x_{i-1}$, then as $\mu \to 0$,

$$\left| \sum_{i=1}^{n} M_i \Delta x_i - \sum_{i=1}^{n} m_i \Delta x_i \right|$$

$$\leq \sum_{i=1}^{n} |M_i - m_i| \Delta x_i$$

$$= |M - m| [\Delta x_1 + \Delta x_n] + \sum_{i=2}^{n-1} |M_i - m_i| \Delta x_i$$

$$\to 0.$$

So $f(x)$ is Riemann integrable on $[a, b]$. Also, since $|f(x)| \leq M'$ on $(a, a - \epsilon_1] \cup [b - \epsilon_2, b)$, we have

$$\left| \int_a^b f(x)\, dx - \int_{a-\epsilon_1}^{b-\epsilon_2} f(x)\, dx \right| \leq M'(\epsilon_1 + \epsilon_2),$$

proving (10.8).   ∎

A useful result in applications is that the Riemann integral of a linear combination of functions can be easily simplified to integrals of the components summands. In its simplest form, and one that has an obvious generalization, we have:

Proposition 10.8 *If $f(x)$ and $g(x)$ are Riemann integrable on $[a, b]$, then so too is $cf(x) + dg(x)$ for any $c, d \in \mathbb{R}$, and*

$$\int_a^b [cf(x) + dg(x)]\, dx = c \int_a^b f(x)\, dx + d \int_a^b g(x)\, dx. \tag{10.9}$$

Proof That $f(x)$ and $g(x)$ are Riemann integrable on $[a, b]$ implies that each can be expressed as

$$\int_a^b f(x)\,dx = \lim_{\mu \to 0} \sum_{i=1}^n f(\tilde{x}_i)\Delta x_i,$$

$$\int_a^b g(x)\,dx = \lim_{\mu \to 0} \sum_{i=1}^n g(\tilde{x}_i)\Delta x_i,$$

where μ denotes the mesh size of the partition, and $\{\tilde{x}_i\}$ are arbitrary points in the subintervals of each partition. Now, for any partition and collection of subinterval points,

$$\sum_{i=1}^n [cf(\tilde{x}_i) + dg(\tilde{x}_i)]\Delta x_i = c\sum_{i=1}^n f(\tilde{x}_i)\Delta x_i + d\sum_{i=1}^n g(\tilde{x}_i)\Delta x_i.$$

Consequently, by taking the limit as $\mu \to 0$, we conclude both the integrability of $cf(x) + dg(x)$ as well as the formula in (10.9).   ∎

Finally, there is a **triangle inequality** for Riemann integrals that is useful in many estimation problems.

Proposition 10.9   *If $f(x)$ is continuous on bounded $[a,b]$, then*

$$\left| \int_a^b f(x)\,dx \right| \le \int_a^b |f(x)|\,dx. \tag{10.10}$$

Proof   First off, note that if $f(x)$ is continuous on bounded $[a,b]$, so too is $|f(x)|$, and hence the second integral is well defined (see exercise 23). Also, if $\{x_n\}$ is any convergent numerical sequence, then

$$\left| \lim_{n\to\infty} x_n \right| = \lim_{n\to\infty} |x_n|,$$

since if $x_n \to x$, then by (10.139) in exercise 23, $|x_n| \to |x|$. Using these facts and the definition of this integral in (10.5), we have by the triangle inequality,

$$\left| \int_a^b f(x)\,dx \right| = \left| \lim_{\mu \to 0} \sum_{i=1}^n f(\tilde{x}_i)\Delta x_i \right|$$

$$= \lim_{\mu \to 0} \left| \sum_{i=1}^n f(\tilde{x}_i)\Delta x_i \right|$$

$$\leq \lim_{\mu \to 0} \sum_{i=1}^{n} |f(\tilde{x}_i)| \Delta x_i$$

$$= \int_a^b |f(x)| \, dx. \qquad\qquad\qquad \blacksquare$$

Remark 10.10 *It is important to note that while $f(x)$ being continuous implies that $|f(x)|$ is continuous, the reverse implication is patently false. A simple example defined on $[0,1]$ is*

$$f(x) = \begin{cases} 1, & x \text{ rational}, \\ -1, & x \text{ irrational}. \end{cases}$$

Then $|f(x)| \equiv 1$ and is therefore continuous, but $f(x)$ is not continuous at any point.

10.2.2   Riemann Integral without Continuity

The result that continuous functions are Riemann integrable on closed and bounded intervals is a good example of mathematical overkill. Just the brevity of the proof indicates that continuity is a very powerful assumption, and probably far more than is actually needed to make the Riemann sums converge. The case of continuous functions on infinite intervals will be addressed below as so-called improper integrals. Here we address the issue of continuity on the bounded interval $[a, b]$.

Finitely Many Discontinuities

Example 10.11 *Define the function*

$$f(x) = \begin{cases} x^2, & 0 \leq x < 1, \\ x^2 + 5, & 1 \leq x \leq 2, \end{cases}$$

*with graph in **figure 10.1**. Based on the proof of (10.6) above, one could hardly be surprised that $f(x)$ is Riemann integrable, and that*

$$\int_0^2 f(x) \, dx = \int_0^1 x^2 \, dx + \int_1^2 (x^2 + 5) \, dx,$$

where the first integral is defined by (10.8). As we will see below, this integral sum has value $\frac{23}{3}$. The formal verification of this splitting reflects the proofs of (10.6) and (10.8). The central idea was the fact that the terms in the Riemann sums that reflect the subintervals that contain any given point c could be shown to converge to 0. In point

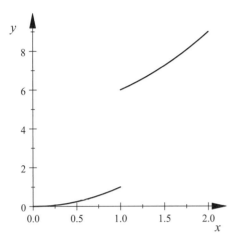

Figure 10.1

$$f(x) = \begin{cases} x^2, & 0 \le x < 1 \\ x^2 + 5, & 1 \le x \le 2 \end{cases}$$

of fact, the proof of (10.6) did not utilize the assumption that $f(x)$ was continuous at c, but only that the function was bounded in each of the partitions' subintervals that contained c. This boundedness assumption was explicit in the proof of (10.8). The value of $f(c)$ is entirely irrelevant as long as the function is bounded in an interval about c.

This example easily generalizes to the case of a bounded function $f(x)$, continuous on an interval $[a, b]$ except at a finite collection of points $\{\hat{x}_j\}_{j=1}^n$, that may contain one or both of the interval endpoints. Such a function is called **piecewise continuous on** $[a, b]$. The proof, as in the example above, simply notes that the terms of the Riemann sums that reflect these points of discontinuity add nothing to the value of the integral in the limit as $\mu \to 0$.

Formalizing this notion:

Definition 10.12 *A function $f(x)$ is **piecewise continuous** on $[a, b]$ if there exists points*

$$a \le \hat{x}_0 < \hat{x}_1 < \hat{x}_2 < \cdots < \hat{x}_n \le b$$

so that on each open interval, $(\hat{x}_{j-1}, \hat{x}_j)$, $f(x)$ is bounded and continuous.

Remark 10.13 *Depending on the application, one might be distressed that this definition does not require that $f(\hat{x}_j)$ is even defined. For the existence of the Riemann integral we do not need these values to be defined, but only that $f(x)$ is bounded as noted in* *the proof of (10.8). However, if one wishes to define values for $f(\hat{x}_j)$ in this definition, it would typically be required that $f(\hat{x}_j)$ is defined as one of the limits:* $\lim_{x \to \hat{x}_j^+} f(x)$ *or* $\lim_{x \to \hat{x}_j^-} f(x)$.

Of course, the boundedness assumption on $f(x)$ is critical, since this is what limits the values of M_i and m_i in each such interval of the Riemann sum, and is necessary to support the conclusion that these exceptional terms decrease to 0 as $\mu \to 0$. That is, if $[x_{i_j-1}, x_{i_j}]$ is any such interval in the partition containing the point of discontinuity $\hat{x}_j$, the associated term of the Riemann sum, for any $\tilde{x}_{i_j} \in [x_{i_j-1}, x_{i_j}]$ satisfies

$$m_{i_j} \Delta x_{i_j} \leq f(\tilde{x}_{i_j}) \Delta x_{i_j} \leq M_{i_j} \Delta x_{i_j}.$$

Consequently, as $\Delta x_{i_j} \to 0$, so too does $f(\tilde{x}_{i_j}) \Delta x_{i_j} \to 0$, since M_{i_j} cannot increase and m_{i_j} cannot decrease, as the intervals about the given point of discontinuity decrease.

Notation 10.14   *The notation in the above paragraph and in some of what follows is a bit cumbersome, but necessary. The problem is that each of the exceptional points $\{\hat{x}_j\}$ will be found in one subinterval of every partition which defines a Riemann sum, but not the same subinterval. So it is inaccurate to claim that $\hat{x}_j \in [x_{j-1}, x_j]$, for instance, since each $\hat{x}_j$ is fixed, yet the number of subintervals in the partition increases with n. So each $\hat{x}_j$ will be in a different subinterval in each partition. So the notation used is $\hat{x}_j \in [x_{i_j-1}, x_{i_j}]$, indicating that $[x_{i_j-1}, x_{i_j}]$ is one of the partition's $[x_{i-1}, x_i]$ subintervals, and in particular the subinterval that contains $\hat{x}_j$.*

In addition to boundedness, another critical assumption in this demonstration of integrability is that the collection of points of discontinuity $\{\hat{x}_j\}_{j=1}^n$ was finite, so that this collection of points could be contained in a collection of partition subintervals $\{[x_{i_j-1}, x_{i_j}]\}_{j=1}^n$, the total lengths of which could be made as small as desired. Then, despite the fact that $M_{i_j} - m_{i_j} \nrightarrow 0$ on these subintervals, as in the example above and **figure 10.1**, one still has the desired result that these terms will add nothing to the Riemann sum in the limit. This is because the total length of these intervals, $\sum (M_{i_j} - m_{i_j}) \Delta x_{i_j}$, can then be made arbitrarily small as $\mu \to 0$ even if $M_{i_j} - m_{i_j}$ do not decrease to 0.

This discussion leads to the following proposition, which we state without separate proof, relying on the discussion above and proofs that the reader can formalize. Also in the next section this result will be further generalized with proof.

Proposition 10.15   *Let $f(x)$ be a bounded function, continuous on bounded $[a,b]$ except at points $\{\hat{x}_j\}_{j=1}^n \subset [a,b]$ written in increasing order. Then $f(x)$ is Riemann integrable on $[a,b]$. Generalizing (10.6), we have*

$$\int_a^b f(x)\,dx = \int_a^{\hat{x}_1} f(x)\,dx + \sum_{j=1}^{n-1} \int_{\hat{x}_j}^{\hat{x}_{j+1}} f(x)\,dx + \int_{\hat{x}_n}^b f(x)\,dx, \tag{10.11}$$

where the first integral is 0 if $a = \hat{x}_1$, and the last is 0 if $\hat{x}_n = b$. Each integral is to be interpreted in the sense of (10.8).

*Infinitely Many Discontinuities

The proposition above relied on an important "covering property" of a finite collection of points that is referred to as the property of being a **set of measure 0**. The Cantor ternary set in section 4.2 was a set of measure 0. This property means that this collection of points $\{\hat{x}_j\}_{j=1}^n$ can be contained in a collection of intervals $\{[x_{i_j-1}, x_{i_j}]\}_{j=1}^n$, the total lengths of which, $\sum \Delta x_{i_j}$, can be made as small as desired. This allows the conclusion that despite the fact that $M_{i_j} - m_{i_j} \nrightarrow 0$ on $[x_{i_j-1}, x_{i_j}]$, the total contribution to the Riemann sum satisfies

$$\sum (M_{i_j} - m_{i_j})\Delta x_{i_j} \to 0.$$

This property of being a set of measure 0 is in fact shared by any countable collection of points. For example, given $\{\hat{x}_j\}_{j=1}^\infty$ and any $\epsilon > 0$, the closed intervals

$$\left\{ \left[\hat{x}_j - \frac{\epsilon}{2^{j+1}}, \hat{x}_j + \frac{\epsilon}{2^{j+1}} \right] \right\}_{j=1}^\infty$$

have lengths $\left\{ \frac{\epsilon}{2^j} \right\}_{j=1}^\infty$ and total length $\sum_{j=1}^\infty \frac{\epsilon}{2^j} = \epsilon$. In other words, $\{\hat{x}_j\}_{j=1}^\infty$ is a set of measure 0.

This generalizes to:

Proposition 10.16   If $\{E_j\}_{j=1}^\infty$ is a countable collection of sets of measure 0, then $\bigcup E_j$ has measure 0.

Proof   First, we cover each set E_j with intervals of total length $\frac{\epsilon}{2^j}$, which is possible since E_j has measure 0. Then $\bigcup E_j$ can be covered by the unions of these covering intervals, and their total length will be no greater than ϵ as noted above.   ∎

We now pursue a proposition that identifies how far the arguments above on the continuity of $f(x)$ can be pushed and still maintain the conclusion of Riemann integrability. This result was proved by **Bernhard Riemann**. The critical observation is that if M_i and m_i are defined as in (10.2) for a collection of intervals: $\{(x_{i-1}, x_i)\}$, where all such intervals contain a given point, x', then $M_i - m_i \to 0$ as $\Delta x_i \to 0$ if and only if $f(x)$ is continuous on $x' \equiv \bigcap(x_{i-1}, x_i)$. This result follows from the definition of continuity (see exercise 3).

Generalizing this idea, we introduce a convenient notation which measures the variability of a function on a given interval, as well as its continuity or discontinuity at a given point.

Definition 10.17   *Given an open interval, $I = (x_{i-1}, x_i)$, denote by $\omega(x; I)$, the **oscillation of $f(x)$ on I:***

$$\omega(x; I) = [M_i - m_i],$$

*where M_i and m_i are defined as in (10.2) but applied to the open interval I. In addition, denote by $\omega(x)$, the **oscillation of $f(x)$ at x:***

$$\omega(x) = g.l.b.\{\omega(x; I)\} \qquad \text{for all } I \text{ with } x \in I.$$

We also define E_N by

$$E_N = \left\{ x \mid \omega(x) \geq \frac{1}{N} \right\},$$

and $E \equiv \bigcup_{N \geq 1} E_N = \{x \mid \omega(x) > 0\}$.

By the discussion preceding this definition and exercise 3:

- $\omega(x) = 0$ if and only if $f(x)$ is continuous at x, and equivalently,
- $\omega(x) > 0$ if and only if $f(x)$ is discontinuous at x.

Consequently E is the collection of discontinuities if $f(x)$.

Example 10.18   *The function graphed in figure 10.1 has $\omega(1) = 5$, and $\omega(x) = 0$ for all $x \in (0, 1) \cup (1, 2)$.*

We next demonstrate two facts that will be necessary for the proposition below.

Proposition 10.19   *The set E_N is a closed set for every N. Hence the set of discontinuities of any function is equal to a countable union of closed sets.*

Proof   Because a set is closed if and only if it contains all of its limit points, we demonstrate that if x is a limit point of E_N, then $\omega(x) \geq \frac{1}{N}$ and so $x \in E_N$. To this end, if I is any open interval containing x, I also contains a point $x' \in E_N$ by definition of limit point. Hence, with M and m defined on I by (10.2), we have that $M - m \geq \omega(x')$ since $\omega(x')$ is the g.l.b. of all such values over all such intervals I. But also

$\omega(x') \geq \frac{1}{N}$, since $x' \in E_N$. Since $M - m \geq \frac{1}{N}$ for any open interval containing x, the g.l.b. of such values also satisfies this inequality, and hence $\omega(x) \geq \frac{1}{N}$ and $x \in E_N$. $\blacksquare$

Remark 10.20

1. *A set that is the countable union of closed sets is sometimes referred to as an* **F_σ-set**, *pronounced "F-sigma set." The F represents the standard notation for a closed set, as this notion apparently originated in France with the word "fermé," while the "sigma" denotes the French word for summation or "union" of closed sets, "somme." An F_σ-set can be open, closed or neither as demonstrated by the examples of $\left\{ \left[\frac{1}{n}, 1 - \frac{1}{n} \right] \right\}$, $\left\{ \left[-\frac{1}{n}, 1 + \frac{1}{n} \right] \right\}$, and $\left\{ \left[\frac{1}{n}, 1 + \frac{1}{n} \right] \right\}$, with respective unions of $(0,1)$, $[-1,2]$, and $(0,2]$. The rational numbers are also an F_σ-set and another example of one that is neither open nor closed.*

2. *The complement of the sets E_N, defined by*

$$\tilde{E}_N = \left\{ x \mid \omega(x) < \frac{1}{N} \right\},$$

are consequently open sets. So the set of continuity points of a given function is the countable intersection of these open sets. Such a set is sometimes referred to as a **G_δ-set**, *pronounced "G-delta set." The G represents the standard notation for an open set, as this notion apparently originated in Germany with the word for area, "Gebiet," while the "delta" denotes the German word for "intersection" of these closed sets, or "Durchschnitt." A G_δ-set can be open, closed, or neither and can be exemplified as above. The irrational numbers are also a G_δ-set that is neither open nor closed, since this set equals the intersection of the open sets:*

$$G_q = (-\infty, q) \cup (q, \infty)$$

for all $q \in \mathbb{Q}$.

3. *By De Morgan's laws, the complement of a G_δ-set is an F_σ-set, and conversely. For example, the complement of a countable union of closed sets is a countable intersection of open sets, and conversely.*

The oscillation function is also important in that knowing its values sheds light on the maximum potential difference between a function's upper and lower Riemann sums, as the next proposition formalizes.

Proposition 10.21   *If $\omega(x) < c$ for all $x \in [a,b]$, then there is a partition of this interval so that*

$$\sum_{i=1}^{n} M_i \Delta x_i - \sum_{i=1}^{n} m_i \Delta x_i < c(b - a).$$

Proof   Since $\omega(x) = \text{g.l.b.}\{\omega(x; I)\}$ for all I with $x \in I$, for every x we can choose an open interval I with $\omega(x; I) < c$, and by shrinking each such I as necessary, we can find an open interval J with closure $\bar{J} \subset I$, and $\omega(x; J) < c$. The collection of all such J is an open cover of the compact interval $[a, b]$, so there is a finite subcover $\{J_k\}_{k=1}^{m}$. The desired partition is now defined by the collection of endpoints of this family of intervals that are within $[a, b]$, as well as the points a and b. On every such partition interval $\{J_k'\}_{k=1}^{n}$ we have $\omega(x; J_k') < c$, and so

$$\sum_{i=1}^{n} [M_i - m_i] \Delta x_i < c \sum_{i=1}^{n} \Delta x_i = c(b - a). \qquad \blacksquare$$

We now present the main result, which provides a necessary and sufficient condition on a bounded function $f(x)$ in order to ensure Riemann integrability on any bounded interval $[a, b]$. It was proved by **Bernhard Riemann**.

Proposition 10.22 (*Riemann Existence Theorem*)   *If $f(x)$ is a bounded function on the finite interval $[a, b]$, then $\int_a^b f(x)\, dx$ exists if and only if $f(x)$ is continuous except on a collection of points $E \equiv \{x_\alpha\}$ of measure 0. That is, for any $\epsilon > 0$, there is a countable collection of intervals $\{I_\alpha\}$ so that $x_\alpha \in I_\alpha$ for all α, and $\sum |I_\alpha| < \epsilon$, where $|I_\alpha|$ denotes the length of the interval I_α.*

Proof   We first assume that $\int_a^b f(x)\, dx$ exists, which means that $\sum_{i=1}^{n} [M_i - m_i] \Delta x_i \to 0$ for any partition with $\mu \equiv \max\{\Delta x_i\} \to 0$. For a given ϵ and integer N, choose a partition with

$$\sum_{i=1}^{n} [M_i - m_i] \Delta x_i < \frac{c}{N}.$$

We now show that E_N has measure 0, and hence the countable union $E = \bigcup E_N$ that equals the set of all discontinuities also has measure 0 by proposition 10.16. Any points of discontinuity of $f(x)$ in E_N that happen to be among the endpoints of this partition's intervals clearly have measure 0, since there are at most $n + 1$ such points. So we consider only such discontinuity points within these subintervals. Let $\{I_j\}_{j=1}^{m}$ denote the subset of partition intervals that have at least one point of discontinuity from E_N in their interior. Then on any such interval $\frac{1}{N} \le M_j - m_j$, since $\frac{1}{N}$

is defined as the g.l.b. of such values among all intervals which contain points of E_N. Consequently, as a subset of the original partition,

$$\frac{1}{N}\sum_{j=1}^{m}|I_j| \le \sum_{i=1}^{n}[M_i - m_i]\Delta x_i < \frac{\epsilon}{N},$$

and hence $\sum|I_j| < \epsilon$ as was to be proved.

Next assume that bounded $f(x)$ is continuous except on a collection of points $E \equiv \{x_\alpha\}$ of measure 0. For any N, $E_N \subset E$ and must also have measure 0, and hence for any $\epsilon > 0$ there is a family of open intervals $\{I_\alpha\}$ so that $E_N \subset \bigcup I_\alpha$ and $\sum|I_\alpha| < \epsilon$. Now, since E_N is closed and a subset of the compact set $[a,b]$, it must also be compact and there is a finite subcollection $\{I_j\}_{j=1}^{n}$ with the same properties: $E_N \subset \bigcup_{j \le n} I_j$ and $\sum_{j \le n}|I_j| < \epsilon$. Also note that since $f(x)$ is bounded on $[a,b]$, there is an M and m so that for any partition of $[a,b]$, the associated M_i and m_i satisfy:

$$m \le m_i \le M_i \le M.$$

Now $[a,b] - \bigcup_{j \le n} I_j$ equals a finite collection of closed intervals, say $\{K_j\}_{j=1}^{m}$, and $\omega(x) < \frac{1}{N}$ for any $x \in K_j$, since each K_j is in the complement of E_N. By proposition 10.21, there is then a partition of each closed interval K_j so that

$$\sum_{i=1}^{m'}M_i\Delta x_i - \sum_{i=1}^{m'}m_i\Delta x_i < \sum \frac{|K_j|}{N},$$

where m' denotes the total number of subintervals in these partitions. With these partitions for $\{K_j\}_{j=1}^{m}$, and the I_j intervals as their own partitions, we have that the associated Riemann upper and lower sums can be split between the two groups of intervals:

$$\sum_{i=1}^{n}M_i\Delta x_i - \sum_{i=1}^{n}m_i\Delta x_i < \frac{\sum|K_j|}{N} + (M-m)\sum|I_j|$$

$$< \frac{(b-a)}{N} + (M-m)\epsilon,$$

where M and m are the bounds for $f(x)$ throughout $[a,b]$. Since N and ϵ were arbitrary, we see that there exist partitions which make the upper and lower Riemann sums differ by an arbitrarily small amount. Now given arbitrary partitions with $\mu \to 0$, these will eventually become finer than the partitions constructed, and hence will satisfy the same bounds. Consequently $f(x)$ is Riemann integrable on $[a, b]$.   ∎

Remark 10.23

1. *Sets of "measure 0," play a central role in real analysis. There, an integration theory is introduced which is more general than Riemann integration, and for which sets of measure 0 again do not matter. However, unlike the Riemann integral, which requires continuity outside this set, this generalized integral requires less. It is known as the **Lebesgue integral** and named for **Henri Léon Lebesgue** (1875–1941). This generalization eliminates the counterintuitive properties of the Riemann integral that are discussed in section 10.3 below on examples of the Riemann integral.*

2. *As a point on terminology, when a function $f(x)$ has a certain property, "except on a set of measure 0," it is often said that $f(x)$ has the certain property **almost everywhere**, and this is often shortened to (**a.e.**). For example, proposition 10.22 states that a bounded function $f(x)$ is Riemann integrable on a bounded interval $[a, b]$ if and only if $f(x)$ is continuous (a.e.).*

10.3   Examples of the Riemann Integral

In this section we illustrate the range of applicability of the notion of the Riemann integral to functions that are continuous except on a set of measure 0, and then use one example to illustrate when the integral fails to exist.

The first example provides the classic case of how one often thinks about Riemann integration as it applies to functions that are continuous except on a set of measure 0. This classic example is for a function $s(x)$ that is **piecewise continuous**, which is to say that $s(x)$ is defined as a bounded and continuous function on each of a collection of non-overlapping intervals. These functions were introduced in section 10.2.2.

The piecewise continuous terminology is descriptive because it literally means "continuous in pieces." When this continuous function is constant on each interval, it is typically called a **step function**, for apparent reasons. A Riemann sum can then be thought of as an approximation to the integral with a step function defined so that on each subinterval, the step function assumes some value of $f(x)$ in that interval. For the upper and lower Riemann sums, these values of $f(x)$ are chosen as the maximum and minimum values of the function on each subinterval.

Example 10.24

1. *Define a function on the interval* $[0, 2]$ *as follows: First split the interval into*

$$[0, 2] = [0, 1) \cup \left[1, 1\frac{1}{2}\right) \cup \left[1\frac{1}{2}, 1\frac{3}{4}\right) \cup \left[1\frac{3}{4}, 1\frac{7}{8}\right) \cup \cdots \cup [2].$$

In other words, split $[0, 2] = \bigcup_{n=0}^{\infty} I_n \cup [2]$, *where* $I_0 \equiv [0, 1)$ *and*

$$I_n = \left[\sum_{j=0}^{n-1} \frac{1}{2^j}, \sum_{j=0}^{n} \frac{1}{2^j}\right) \qquad for\ n = 1, 2, 3, \ldots.$$

Next, define a function by

$$s(x) = \begin{cases} \frac{1}{2^n}, & x \in I_n, \\ 1, & x = 2, \end{cases}$$

*which is graphed in **figure 10.2**. Since this bounded function is continuous except on the countable collection of points* $\left\{\sum_{j=0}^{n} \frac{1}{2^j}\right\}_{n=0}^{\infty} \cup \{2\}$, *it must be Riemann integrable by proposition 10.22. As the length of* I_n *is* $\frac{1}{2^n}$ *for all n, and the Riemann sums containing the points of discontinuity add nothing in the limit, we have that*

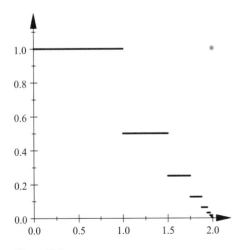

Figure 10.2
Piecewise continuous $s(x)$

$$\int_0^2 s(x)\,dx = \sum_{n=0}^{\infty} \int_{I_n} s(x)\,dx$$

$$= \sum_{n=0}^{\infty} \frac{1}{2^n}\frac{1}{2^n}$$

$$= \sum_{n=0}^{\infty} \frac{1}{4^n} = \frac{4}{3},$$

using the methods of chapter 6 on geometric series.

The next examples generalize this idea, in that there are no longer intervals on which $f(x)$ is piecewise continuous.

Example 10.25

2. *Define a function on the interval $[0,1]$ by*

$$f(x) = \begin{cases} 1, & x = 0, \\ \frac{1}{n}, & x \text{ rational, } x = \frac{m}{n} \text{ in lowest terms,} \\ 0, & x \text{ irrational,} \end{cases}$$

*which is graphed for n up to $n = 13$ in **figure 10.3**. This function is not continuous at any rational number. For example, if $x = \frac{m}{n}$ in lowest terms, $f(x) = \frac{1}{n}$ and yet any interval that contains x also contains irrational numbers for which $f(x) = 0$. So, if $0 < \epsilon < \frac{1}{n}$, there can be no δ for which $\left| f\left(\frac{m}{n}\right) - f(x) \right| < \epsilon$ for all x with $\left| \frac{m}{n} - x \right| < \delta$, since there are always irrational x for which $\left| f\left(\frac{m}{n}\right) - f(x) \right| = \frac{1}{n}$. What may be surprising is that $f(x)$ is continuous at every irrational number. To see this, let x be an irrational number and $\epsilon > 0$ be given. Choose N so that $\frac{1}{N} < \epsilon$. From the finite collection of rationals $\{\frac{m}{n} \mid n \le N, m \le n\}$, there is one that is closest to x; choose δ to be smaller than this closest distance. That is, define*

$$\delta < \min\left\{ \left| x - \frac{m}{n} \right| \mid n \le N, m \le n \right\}.$$

By construction, any rational in the interval $(x - \delta, x + \delta)$ must be of the form $\frac{m}{M}$ for $M > N$, and so $\left| f(x) - f\left(\frac{m}{M}\right) \right| = \left| \frac{1}{M} \right| < \frac{1}{N} < \epsilon$. Consequently $f(x)$ is continuous at irrational x.

Since the points of discontinuity are the rational numbers that are a set of measure 0, this function is Riemann integrable by proposition 10.22. It is apparent that

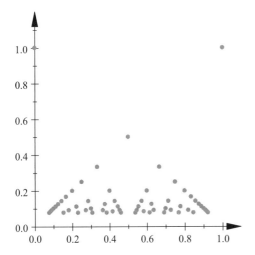

Figure 10.3

$$f(x) = \begin{cases} 1, & x = 0 \\ \frac{1}{n}, & x = \frac{m}{n} \text{ in lowest terms} \\ 0, & x \text{ irrational} \end{cases}$$

$$\int_0^1 f(x)\, dx = 0,$$

since $f(x) = 0$ on the points of continuity. This result can also be justified directly by Riemann sums. For any N, we can construct non-overlapping intervals of total length ϵ that cover $\left\{ \frac{m}{n} \mid n \le N, m \le n \right\}$. Since $f\left(\frac{m}{n}\right) \le 1$ at every point of this set, these Riemann sums add up to no more than ϵ. Moreover, since $f(x) \le \frac{1}{N+1}$ outside these intervals by construction, Riemann sums associated with these points can contribute no more than $\frac{1}{N+1}$. In other words, for any ϵ and N we can find a Riemann sum so that

$$0 \le \sum_{i=1}^n f(\tilde{x}_i)\Delta x_i \le \epsilon + \frac{1}{N+1},$$

and these Riemann sums can be made to converge to 0 by choosing $\epsilon \to 0$ and $N \to \infty$.

3. *In the preceding case 2, $f(x)$ can be redefined in many ways in terms of the assignment of values for $f\left(\frac{m}{n}\right)$. All that is needed is that the sequence $f\left(\frac{m}{n}\right) \to 0$ as $n \to \infty$. Such an assignment of values is critical for continuity on the irrationals, and hence critical for the existence of the Riemann integral. For example, if $g(n) \to 0$ monotonically as $n \to \infty$, redefining $f\left(\frac{m}{n}\right) = g(n)$ provides a comparable result.*

4. *Note that with $f(x)$ defined above in case 2, if we define*

$$g(x) = 1 - f(x),$$

then $g(x)$ is Riemann integrable with integral equal to 1, and $g(x)$ is again only continuous on the irrational numbers where it has value identically equal to 1. For the rational numbers, $g(x)$ assumes values $g\left(\frac{m}{n}\right) = 1 - \frac{1}{n}$, where $\frac{m}{n}$ is given in lowest terms.

The next example nudges cases 2 through 4 of example 10.25 one more step and Riemann integrability now fails.

Example 10.26

5. *Define $h(x)$ on $[0,1]$ by*

$$h(x) = \begin{cases} 1, & x \text{ irrational,} \\ 0, & x \text{ rational.} \end{cases}$$

Note that except on the rationals, a set of measure 0, $h(x) = g(x)$ in case 4 of example 10.25. And yet $g(x)$ is Riemann integrable with integral 1 and $h(x)$ is not integrable. That $h(x)$ is not Riemann integrable is easy to see, since for any partition of $[0,1]$ the upper Riemann sum will equal 1 while the lower sum equals 0.

This series of examples presents much of the range of applicability of Riemann integration, from its beauty and power as applied to continuous functions and certain generalizations to its inherent shortcomings, which can be summarized as a conflicted relationship between this integral and sets of measure zero:

1. If $f(x)$ is continuous except on a set of measure 0, it is Riemann integrable. This implies that in this sense, sets of measure 0 are irrelevant.

2. On the other hand, if one starts with a continuous function, say $t(x) \equiv 1$ on $[0,1]$, and redefines it on a set of measure 0, sometimes integrability is preserved as was the case for $g(x)$ in example 4 above, and sometimes it is not, as is the case for $h(x)$ in example 5 above.

3. For some sets of measure zero, such as finite sets, one can redefine the function at these points arbitrarily without influencing integrability. This is the case for simple step functions, with finitely many steps.

4. On the other hand, infinite sets of measure zero can create different outcomes:

• If the set has accumulation points, like the rationals, an integrable function must be redefined carefully on this set to maintain Riemann integrability.

• If the set is sparse with no accumulation points, such as the integers, Riemann integrability will be independent of the definition of values of the function at these points. In other words, the integral will exist or not, independent of these values.

As it turns out, this conflicted relationship has more to do with the Riemann approach to integration than it has to do with some intrinsic property of continuous functions. In real analysis a different approach to the integral will eliminate all the confusion about sets of measure 0 in favor of the simplest answer. Namely, sets of measure 0 will not matter in that they will not influence integrability, and this conclusion will be independent of whether the set is finite or infinite, and in the latter case, whether the set has accumulation points, like the rationals, or is sparsely distributed like the integers.

10.4   Mean Value Theorem for Integrals

The next result is an immediate consequence of the intermediate value theorem for continuous functions introduced in proposition 9.41. It is known as the "first" mean value theorem for integrals for a reason that will be seen below in proposition 10.52.

Proposition 10.27 (*First Mean Value Theorem for Integrals*)   *Let $f(x)$ be continuous on bounded $[a, b]$. Then there is a $c \in [a, b]$ so that*

$$\int_a^b f(x)\, dx = f(c)(b - a). \tag{10.12}$$

Proof   Because $f(x)$ is continuous, it achieves its maximum M and minimum m on this interval, and hence

$$m(b - a) \le \int_a^b f(x)\, dx \le M(b - a).$$

Consequently $\frac{1}{b-a} \int_a^b f(x)\, dx$ also has value within the same bounds as $f(x)$. By the intermediate value theorem of chapter 9, there must therefore be a $c \in [a, b]$ with $f(c) = \frac{1}{b-a} \int_a^b f(x)\, dx$.   ∎

Remark 10.28   *Rewrite (10.12) as*

$$f(c) = \frac{1}{b - a} \int_a^b f(x)\, dx,$$

and consider the integral as the net area under $f(x)$ on the interval $[a,b]$. Then the value $f(c)$ can be interpreted as the average value of $f(x)$ on this interval in that this signed area also equals the signed area of a rectangle with base length $b - a$, and height $f(c)$.

One application of this result that is more interesting than it is applicable in practice is the following. For any partition of an interval, there is a set of intermediate points for which the Riemann sum is exact. This is another classic example of an "existence" theorem in mathematics. It confirms existence but provides no insight to how one identifies or constructs these points.

Proposition 10.29   *Let $f(x)$ be continuous on bounded $[a,b]$. Then given any partition*

$$a = x_0 < x_1 < \cdots < x_{n-1} < x_n = b,$$

there exists $\{\tilde{x}_i\}_{i=1}^n$, with $\tilde{x}_i \in [x_{i-1}, x_i]$ so that with $\Delta x_i = x_i - x_{i-1}$,

$$\int_a^b f(x)\,dx = \sum_{i=1}^n f(\tilde{x}_i)\Delta x_i. \tag{10.13}$$

Proof   By (10.6), we can conclude that

$$\int_a^b f(x)\,dx = \sum_{i=1}^n \int_{x_{i-1}}^{x_i} f(x)\,dx,$$

and the proposition above and (10.12) assure that for each integral on the right, there is a point $\tilde{x}_i \in [x_{i-1}, x_i]$ that gives the stated result. ∎

Remark 10.30   *This proof demonstrates that for a continuous function and any partition, one can in theory choose the intermediate $\tilde{x}_i$ values so that the Riemann sum exactly reproduces the limiting integral value. Of course, this proof and conclusion rely on the first mean value theorem for integrals, which sheds no light on how such intermediates values can be determined; it only confirms their existence. Consequently this result is not useful in practice in terms of obtaining exact integrals with Riemann sums.*

Two other important corollaries to this Proposition provide simple and also useful results:

Proposition 10.31   *Let $f(x)$ be continuous on bounded $[a,b]$, and assume that for every $[c,d] \subset [a,b]$,*

$$\int_c^d f(x)\,dx = 0.$$

Then $f(x) \equiv 0$ on $[a,b]$.

Proof By proposition 10.27, we conclude that for every $[c,d] \subset [a,b]$, there is a point $c' \in [c,d]$ with $\int_c^d f(x)\,dx = f(c')(d-c)$, and hence $f(c') = 0$ for every such c'. Letting $d = c + \Delta c$, we conclude that for every Δc there is a $c' \in [c, c + \Delta c]$ with $f(c') = 0$. Letting $\Delta c \to 0$, we have $c' \to c$ and the continuity of $f(x)$ provides $f(c) = 0$. This proves $f(c) = 0$ for all $c \in [a,b)$, and hence $f(b) = 0$ by continuity. ∎

Proposition 10.32 *Let $f(x)$ and $g(x)$ be continuous on bounded $[a,b]$, and assume that for every $[c,d] \subset [a,b]$,*

$$\int_c^d f(x)\,dx = \int_c^d g(x)\,dx.$$

Then $f(x) \equiv g(x)$ on $[a,b]$.

Proof The conclusion is immediate, since $f(x) - g(x)$ satisfies the hypothesis of proposition 10.31, and so $\int_c^d [f(x) - g(x)]\,dx = 0$. ∎

10.5   Integrals and Derivatives

There are two related results which connect the notions of derivative, as developed in chapter 9, and that of Riemann integral as developed above. The first is the result obtained when a derivative is integrated.

10.5.1   The Integral of a Derivative

Proposition 10.33 (*Fundamental Theorem of Calculus, Version I*) *Let $f(x)$ be a differentiable function so that $f'(x)$ is continuous on bounded $[a,b]$. Then*

$$\int_a^b f'(x)\,dx = f(b) - f(a). \tag{10.14}$$

Proof We know that the integral exists since $f'(x)$ is assumed to be continuous. Consequently we can define it in terms of any Riemann sums, which is to say, any partition with $\mu \to 0$. Given any partition, we have by the mean value theorem of proposition 9.83 that in every subinterval, $[x_{i-1}, x_i]$ there is an $\tilde{x}_i$ so that $f'(\tilde{x}_i)\Delta x_i$ $= f(x_i) - f(x_{i-1})$. Choosing these $\tilde{x}_i$, we have

$$\sum_{i=1}^{n} f'(\tilde{x}_i)\Delta x_i = \sum_{i=1}^{n} [f(x_i) - f(x_{i-1})] = f(b) - f(a),$$

since this middle summation "telescopes" by cancellation to only the first and last terms.   ∎

Notation 10.34   *It is common in calculus to use the notation $f(x)|_a^b$ for the right-hand side of (10.14). In other words,*

$$f(x)|_a^b \equiv f(b) - f(a).$$

Example 10.35

1. *With $f(x) = e^x$, since $f'(x) = f(x)$, we have that for any closed interval $[a,b]$,*

$$\int_a^b e^x \, dx = e^b - e^a.$$

2. *With*

$$f(x) = \begin{cases} x^2, & 0 \le x < 1, \\ x^2 + 5, & 1 \le x \le 2, \end{cases}$$

the example in **figure 10.1**, *then on $(0,1) \cup (1,2)$, we have that $f(x) = F'(x)$, where*

$$F(x) = \begin{cases} \frac{x^3}{3}, & 0 \le x < 1, \\ \frac{x^3}{3} + 5x, & 1 \le x \le 2. \end{cases}$$

So

$$\int_0^2 f(x)\,dx = \lim_{\epsilon \to 0} \int_0^{1-\epsilon} x^2 \, dx + \int_1^2 (x^2 + 5)\,dx$$

$$= \lim_{\epsilon \to 0} \frac{x^3}{3}\bigg|_0^{1-\epsilon} + \left(\frac{x^3}{3} + 5x\right)\bigg|_1^2$$

$$= \frac{23}{3}.$$

The applicability of the Fundamental Theorem of Calculus I is apparent. If one is attempting to integrate a continuous function, $f(x)$, and this function is recognized as the derivative of another function $F(x)$, so that $F'(x) = f(x)$, then the Riemann summation and limiting process to evaluating the integral can be circumvented, since

$$\int_a^b f(x)\,dx = F(b) - F(a). \tag{10.15}$$

Because of the Fundamental Theorem, it is the case that in many calculus texts, integration of a function $f(x)$ is transformed into techniques for determining the associated function $F(x)$, the so-called **antiderivative** of $f(x)$, since then integration is simplified. Unfortunately, not all continuous functions are the derivatives of other recognizable functions, and even for those that are, finding the functional form of the correct $F(x)$ can be difficult at best. Consequently efficient numerical techniques are often useful and will be discussed in section 10.10.

Definition 10.36   *Given a continuous function $f(x)$, the **antiderivative** of $f(x)$, sometimes denoted $\int f(x)\,dx$ and sometimes $F(x)$ when $f(x)$ is clear from the context, is any function such that*

$$F'(x) = f(x). \tag{10.16}$$

It is important to note that the antiderivative of a function is not unique. In particular, if $F(x)$ is an antiderivative, so too is $F(x) + C$ for any $C \in \mathbb{R}$. Of course, as is easily seen, for the purpose of evaluating an integral by (10.15), any antiderivative works equally well, and in practice, one typically uses $C = 0$.

Also note that the notational convention $F(x) = \int f(x)\,dx$ is a bit careless and yet not uncommon. On the left, the variable x denotes the label for the domain variable of the function F, while on the right, the variable x is a so-called **dummy variable**, which could be denoted y, z, l, α, or any other letter. It is the same as the dummy variable in a summation, in the sense that $\sum_{j=a}^b j^2$ is identical to $\sum_{k=a}^b k^2$. What is meant by this notation is that $F(x)$ is a function with derivative $f(x)$. Although notationally careless, to instead define $F(x) = \int f(y)\,dy$, say, is notationally ambiguous. This notation will be made precise in section 10.5.2, where the second statement of the Fundamental Theorem is developed.

Remark 10.37   *The intuitive framework for the Riemann integral, $\int_a^b f(x)\,dx$, is one of "net area" between the curve, $y = f(x)$, and the x-axis, over the interval $[a, b]$, as* *noted above. For this purpose it was assumed that a < b. However, in this case the meaning of $\int_b^a f(x)\,dx$ can also be introduced in a consistent way by proposition 10.33 above. Namely from (10.14) we can conclude that the definition*

$$\int_b^a f(x)\,dx \equiv -\int_a^b f(x)\,dx \tag{10.17}$$

provides a consistent generalization of the definition of the Riemann integral.

The version above of the Fundamental Theorem in (10.14) also provides another simple and often utilized conclusion. First recall:

1. It is apparent from the definition of $f'(x)$, that if $f(x)$ is increasing, in that $f(x) \le f(x')$ for all $x \le x'$, then $f'(x) \ge 0$ for all x.

2. If $f(x)$ is strictly increasing, in that $f(x) < f(x')$ for all $x < x'$, then it is not true that $f'(x) > 0$ for all x, and the conclusion remains only that $f'(x) \ge 0$ for all x. For example, consider $f(x) = x^3$, which is strictly increasing but $f'(0) = 0$.

3. Similarly, if $f(x)$ is decreasing or strictly decreasing, then $f'(x) \le 0$ for all x.

The question addressed next is the reversal of these implications. Namely, what if anything does the "sign" of $f'(x)$ predict about the behavior of $f(x)$? The conclusions below will be seen to be a bit weaker than those drawn with the tools of chapter 9. Specifically, using (9.33) with $n = 0$ provides the same conclusions without the assumed continuity of $f'(x)$. However, there is value in comparing results with different approaches.

Proposition 10.38 *Let $f(x)$ be continuous with continuous $f'(x)$ on an interval $[a,b]$. Then for any $[c,d] \subset [a,b]$,*

1. $f'(x) \ge 0$ on $[c,d] \Rightarrow f(c) \le f(d)$, and $f(x)$ is increasing
2. $f'(x) > 0$ on $[c,d] \Rightarrow f(c) < f(d)$, and $f(x)$ is strictly increasing
3. $f'(x) \le 0$ on $[c,d] \Rightarrow f(c) \ge f(d)$, and $f(x)$ is decreasing
4. $f'(x) < 0$ on $[c,d] \Rightarrow f(c) > f(d)$, and $f(x)$ is strictly decreasing

Proof Each of these statements easily follows from (10.14), which in the notation here states that:

$$\int_c^d f'(x)\,dx = f(d) - f(c).$$

Now by the definition of the Riemann integral, the sign of $\int_c^d f'(x)\,dx$ follows from the sign of $f'(x)$ on $[c,d]$ if this sign is consistent, and the proposition's four statements follow from this observation. ∎

10.5.2   The Derivative of an Integral

The second result on the relationship between derivative and Riemann integral looks somewhat different, but is equivalent. To this end, we introduce the notion of an **indefinite integral**, whereby the variable x is used as one of the limits of integration. This terminology is complemented by sometimes calling $\int_b^a f(x)\,dx$ a **definite integral** as noted in remark 10.3 following the definition of this Riemann integral.

Definition 10.39   *The indefinite integral of a continuous function $f(x)$ is defined by*

$$F(x) = \int_a^x f(y)\,dy. \tag{10.18}$$

If $a < x$, the value is given directly by the definition of Riemann integral, whereas if $x < a$, this function is defined as in (10.17) of remark 10.37 above as $-\int_x^a f(y)\,dy$.

The next result provides an alternative view of the connection between integral and derivative.

Proposition 10.40 (*Fundamental Theorem of Calculus, Version II*)   *Let $f(x)$ be a continuous function on bounded $[a,b]$, and define*

$$F(x) = F(a) + \int_a^x f(y)\,dy, \qquad a \le x \le b, \tag{10.19}$$

where $F(a)$ is arbitrarily defined. Then $F(x)$ is differentiable on (a,b), and

$$F'(x) = f(x). \tag{10.20}$$

Proof   First off, it may not be obvious that $F(x)$ is even continuous. However, the MVT for integrals in proposition 10.27 assures us that for $x' < x$,

$$F(x) - F(x') = \int_{x'}^x f(y)\,dy$$

$$= f(c)(x - x') \quad \text{for some } c \in [x', x].$$

Now since $f(x)$ is assumed continuous on $[x', x]$, we conclude that $F(x') \to F(x)$ as $x' \to x$, and so $F(x)$ is continuous. This same equation also shows that

$$\frac{F(x) - F(x')}{x - x'} = f(c) \quad \text{for some } c \in [x', x].$$

Hence, as $x' \to x$, $\frac{F(x)-F(x')}{(x-x')} \to f(x)$, again due to the continuity of $f(x)$. ∎

Remark 10.41

1. *This proposition is also called the **Fundamental Theorem of Calculus** because it is equivalent to the statement in proposition 10.33 above. Letting $x = b$, and using the conclusion that $F'(x) = f(x)$, we derive from the FTC II proposition that*

$$F(b) = F(a) + \int_a^b F'(y)\,dy,$$

which is the statement of the earlier Fundamental Theorem of Calculus I in (10.14) with a small change in notation.

On the other hand, FTC I clearly holds with b replaced by x, and can be rearranged to produce

$$f(x) = f(a) + \int_a^x f'(y)\,dy,$$

where $f(x)$ is given and assumed differentiable with continuous derivative. If this same $f(x)$ could be achieved with a different continuous function $g(x)$,

$$f(x) = f(a) + \int_a^x g(y)\,dy,$$

we would conclude that $\int_a^x [g(y) - f'(y)]\,dy = 0$ for all x. Therefore, by subtraction, $\int_x^{x+\Delta x} [g(y) - f'(y)]\,dy = 0$ for any x and Δx, and by the proposition 10.32 result from the MVT we would conclude that $g(y) = f'(y)$ for all y.

2. *This statement of the Fundamental Theorem also makes precise the notational carelessness of the expression often used for antiderivatives: $F(x) = \int f(x)\,dx$. This is shorthand for $F(x) = F(a) + \int_a^x f(y)\,dy$, in (10.19), utilizing the fact that the values of both a and $F(a)$ are irrelevant.*

Example 10.42

1. *A simple application of FTCII is that it provides an interesting new definition of $\ln x$. Specifically, since $f'(x) = \frac{1}{x}$ for $f(x) = \ln x$, and $\ln 1 = 0$, we can conclude that for $x \geq 1$,*

$$\ln x = \int_1^x \frac{dy}{y}. \tag{10.21}$$

2. *As another application with $f(x) = e^x$, since $f'(x) = e^x$, we have for any a and $x \geq a$,*

$$e^x = e^a + \int_a^x e^y \, dy.$$

Letting $a \to -\infty$, we obtain an example of an "improper integral" discussed below:

$$e^x = \int_{-\infty}^x e^y \, dy.$$

10.6 Improper Integrals

10.6.1 Definitions

The preceding sections provide insight as to how one can attempt to extend the integral of some functions in one of two ways:

1. Define $\int_a^b f(x) \, dx$ for continuous $f(x)$ where $b = \infty$ and/or $a = -\infty$.

2. Define $\int_a^b f(x) \, dx$ for $f(x)$ that is continuous on every closed interval $[c, d] \subset [a, b]$ but is unbounded at a and/or b.

In both cases the resulting integrals are called **improper integrals** if they exist because they are defined outside the general framework of the Riemann existence theorem in proposition 10.22 above. Recall that this theorem applied to bounded functions on a bounded interval $[a, b]$ that are continuous almost everywhere, which is to say, everywhere except on a set of measure 0.

Each of these extensions can potentially be defined as a limit of well-defined Riemann integrals. For example, if $F'(x) = f(x)$ is continuous on $[-M, N]$ for all M, N, then define

$$\int_{-\infty}^{\infty} f(x) \, dx \equiv \lim_{N, M \to \infty} \int_{-M}^{N} f(x) \, dx = \lim_{N, M \to \infty} [F(N) - F(-M)]. \tag{10.22}$$

Similarly, if $F'(x) = f(x)$ is continuous on $[a + \delta, b - \epsilon]$ for all $\epsilon, \delta > 0$, then define

$$\int_a^b f(x)\,dx \equiv \lim_{\epsilon,\delta \to 0} \int_{a+\delta}^{b-\epsilon} f(x)\,dx = \lim_{\epsilon,\delta \to 0} [F(b-\epsilon) - F(a+\delta)].$$ (10.23)

The only remaining question is the existence of these limits. Since such limits involve two variates, we formalize the definition in the predictable way:

Definition 10.43 $\lim_{x,y \to a} f(x,y) = L$ *for* $a < \infty$ *and* $L < \infty$, *if for any* $\epsilon > 0$ *there is a* δ *so that* $|f(x,y) - L| < \epsilon$ *when* $|x-a| < \delta$ *and* $|y-a| < \delta$. *Also,* $\lim_{x,y \to \infty} f(x,y) = L$ *if for any* $\epsilon > 0$ *there is an* N *so that* $|f(x,y) - L| < \epsilon$ *when* $x, y > N$.

Example 10.44

1. $\int_1^\infty x^{-a}\,dx$ *is well defined iff* $a > 1$, *since by* (10.14), *for* $a \neq -1$,

$$\int_1^\infty x^{-a}\,dx = \lim_{x \to \infty} \left[\frac{x^{1-a}}{1-a} - \frac{1}{1-a} \right]$$

$$= \begin{cases} \frac{1}{a-1}, & \text{if } a > 1, \\ \infty, & \text{if } a < 1. \end{cases}$$

Also, if $a = -1$,

$$\int_1^\infty x^{-1}\,dx = \lim_{N \to \infty} \ln N = \infty.$$

2. $\int_0^1 x^a\,dx$ *is well defined iff* $a > -1$ *since for* $a \neq -1$,

$$\int_0^1 x^a\,dx = \frac{1}{\alpha+1} - \lim_{\epsilon \to 0} \frac{\epsilon^{\alpha+1}}{\alpha+1}$$

$$= \begin{cases} \frac{1}{\alpha+1}, & \text{if } a > -1, \\ \infty, & \text{if } a < -1. \end{cases}$$

Also, if $a = -1$,

$$\int_0^1 x^{-1}\,dx = -\lim_{\epsilon \to 0} \ln \epsilon = \infty.$$

10.6.2   Integral Test for Series Convergence

In this section is introduced another test for the convergence of a numerical series, the so-called **integral test**, as noted in chapter 6. At first it may seem odd that the

Riemann integral can be used to determine the convergence of a series. But since these integrals are limits of Riemann sums, and each Riemann sum is a finite series, and for an improper integral, an infinite series, the connection is not so surprising.

To motivate the method, we provide yet another proof of the divergence of the **harmonic series**:

Example 10.45 *Consider the series:* $\sum_{n=1}^{\infty} \frac{1}{n}$. *For* $f(x) = \ln x$, *we have from (10.21) that* $\ln x = \int_1^x \frac{dy}{y}$, *and so by definition,* $\int_1^\infty \frac{dy}{y} = \infty$. *Now splitting the integral into unit intervals, we conclude that*

$$\sum_{n=1}^{\infty} \int_n^{n+1} \frac{dy}{y} = \infty.$$

Note that on each unit interval, by using a single upper and lower Riemann sum, we have

$$\frac{1}{n+1} < \int_n^{n+1} \frac{dy}{y} < \frac{1}{n},$$

and hence

$$\sum_{n=1}^{\infty} \int_n^{n+1} \frac{dy}{y} < \sum_{n=1}^{\infty} \frac{1}{n},$$

proving divergence.

With this example as motivation, we now present the integral test for a series.

Proposition 10.46 (*Integral Test*) *Let* $\sum_{n=1}^{\infty} a_n$ *be a given series, and* $f(x)$ *a continuous function on* $[1, \infty)$ *with*

$$a_{n+1} \leq f(x) \leq a_n \qquad \text{for } x \in [n, n+1]. \tag{10.24}$$

Then $\sum_{n=1}^{\infty} a_n$ *and* $\int_1^\infty f(x)\,dx$ *both converge or both diverge.*

Proof By the given assumptions, for all $n \geq 1$,

$$a_{n+1} \leq \int_n^{n+1} f(x)\,dx \leq a_n,$$

and by addition,

$$\left(\sum_{n=1}^{\infty} a_n\right) - a_1 \leq \int_1^{\infty} f(x)\, dx \leq \sum_{n=1}^{\infty} a_n.$$

The result follows by comparison.   ∎

It is important to note that this test not only gives insight to the convergence or divergence of a given series, but it is also useful for numerical estimates in the case of convergence, and a growth rate analysis in the case of divergence. To this end, consider the partial sum version of the inequalities above from $n = 1$ to $N - 1$. It produces

$$\left(\sum_{n=1}^{N} a_n\right) - a_1 \leq \int_1^{N} f(x)\, dx \leq \left(\sum_{n=1}^{N} a_n\right) - a_N.$$

Rearranging, we obtain

$$\int_1^{N} f(x)\, dx + a_N \leq \sum_{n=1}^{N} a_n \leq \int_1^{N} f(x)\, dx + a_1. \tag{10.25}$$

In the case of series divergence the integral provides an estimate of the rate of divergence.

In the case of convergence, and so $a_N \to 0$ by letting $N \to \infty$, the integral produces an estimate of the summation:

$$\int_1^{\infty} f(x)\, dx \leq \sum_{n=1}^{\infty} a_n \leq \int_1^{\infty} f(x)\, dx + a_1 \tag{10.26}$$

We next consider an example of each case.

Example 10.47

1. *As an application in the case of convergence, recall the power harmonic series* $\sum_{n=1}^{\infty} \frac{1}{n^p}$ *for* $p > 1$, *which converges by the analysis in example 6.9. With* $f(x) = x^{-p}$, *a calculation gives that*

$$\int_1^{\infty} f(x)\, dx = \left(\frac{x^{1-p}}{1-p}\right)\Bigg|_{x=1}^{\infty} = \frac{1}{p-1},$$

and so

$$\frac{1}{p-1} \le \sum_{n=1}^{\infty} \frac{1}{n^p} \le \frac{p}{p-1}.$$

For the partial sums of this series, the same approach provides for all N,

$$\frac{1-N^{1-p}}{p-1} + N^{-p} \le \sum_{n=1}^{N} \frac{1}{n^p} \le \frac{p-N^{1-p}}{p-1}.$$

2. *As an application in the case of divergence, we return to the harmonic series and* $f(x) = \frac{1}{x}$. *Since* $\int_1^N \frac{dx}{x} = \ln N$,

$$\ln N + \frac{1}{N} \le \sum_{n=1}^{N} \frac{1}{n} \le \ln N + 1,$$

and so the harmonic series partial sums are larger than, but within 1 *unit of, the value of* $\ln N$. *A more detailed analysis below demonstrates that the following limit exists:*

$$\lim_{N \to \infty} \left[\sum_{n=1}^{N} \frac{1}{n} - \ln N \right] = \gamma \approx 0.577215664902\ldots. \tag{10.27}$$

This constant, γ, *is known as the* **Euler constant**, *after its discoverer,* **Leonhard Euler** *(1707–1783).*

Note that by expressing $N = \prod_{n=1}^{N-1} \left(\frac{n+1}{n} \right)$, *we derive that*

$$\sum_{n=1}^{N} \frac{1}{n} - \ln N = \frac{1}{N} + \sum_{n=1}^{N-1} \left[\frac{1}{n} - \ln \left(1 + \frac{1}{n} \right) \right].$$

Consequently, applying (9.33) and n $= 1$ *to* $\ln(1+x)$, *we obtain that there is* $\{c_n\}$ *with* $0 < c_n < 1$, *so that*

$$\sum_{n=1}^{N-1} \left[\frac{1}{n} - \ln \left(1 + \frac{1}{n} \right) \right] = \sum_{n=1}^{N-1} \left[\frac{1}{2} \left(\frac{n}{n + c_n} \right)^2 \frac{1}{n^2} \right]$$

$$\le \sum_{n=1}^{N-1} \frac{1}{2n^2} < \infty.$$

So the sequence in (10.27), minus the inconsequential term $\frac{1}{N}$, *is increasing and bounded, and hence converges as* $N \to \infty$ *by proposition 5.18.*

10.7   Formulaic Integration Tricks

The most important trick for formulaically evaluating an integral, $\int_a^b f(x)\,dx$, is given by the Fundamental Theorem of Calculus in (10.15). Namely one attempts to find the antiderivative of $f(x)$, which means any function $F(x)$ so that $F'(x) = f(x)$. We use the term "any" function $F(x)$, since we know that if $F(x)$ is one such function, then $F(x) + C$ is another for any constant C. Of course, this constant is eliminated in the application of (10.15).

There are many tricks that one can use to assist in the identification of such an antiderivative. In general, these methods allow one to simplify the problem, in one or many steps, and thereby reveal the antiderivative in pieces. In this section we consider two approaches. The emphasis will be on definite integrals. However, any process that allows the general evaluation of a definite integral provides the formula for an antiderivative by (10.15).

Specifically, any formula of the sort

$$\int_a^b f(x)\,dx = F(b) - F(a),$$

can be rewritten with $b = y$, say, to produce

$$F(y) = F(a) + \int_a^y f(x)\,dx.$$

This is an antiderivative of the integrand, $f(y)$, for any value of a, and any assignment of the value of $F(a)$.

Remark 10.48   *In practice, a simple implementation of this idea of finding an antiderivative is to convert b to x in the expression F(b) − F(a), discard any terms in the formula that are constant and independent of x, and add the arbitrary constant C.*

10.7.1   Method of Substitution

The method of substitution is more akin to *trompe-l'œil*, the art form of tricking the eye, than it is a new mathematical method. But sometimes tricking the eye into seeing a simpler problem is exactly what is needed to get an integration problem started.

This method is an application of the formula for differentiation of a composite function:

$$[F(G(x))]' = f(G(x))g(x),$$

where, of course, $F'(x) = f(x)$ and $G'(x) = g(x)$. By the Fundamental Theorem, the integral of the derivative is easy to evaluate, and

$$\int_a^b f(G(x))g(x)\,dx \equiv \int_a^b [F(G(x))]'\,dx$$

$$= F(G(b)) - F(G(a)).$$

So the difficulty in applying this result is recognizing when a given integrand is in fact of this form, and that is where *trompe-l'œil* helps. We illustrate first with an example.

Example 10.49   *Let's evaluate $\int_a^b e^x(e^x+4)^{20}\,dx$. One elementary but tedious approach is to expand the integrand into a summation of terms of the form $\sum c_i e^{d_i x}$, which is easy to integrate term by term by the Fundamental Theorem since $\int c_i e^{d_i x}\,dx = \frac{c_i}{d_i}e^{d_i x} + e_i$ for arbitrary constant e_i. Alternatively, we may observe that this is a composite function, where $f(y) = y^{20}$, $G(x) = e^x + 4$, and $e^x(e^x+4)^{20} = f(G(x))g(x)$, and hence since $F(y) = \frac{y^{21}}{21}$,*

$$\int_a^b e^x(e^x+4)^{20}\,dx = \frac{1}{21}[(e^b+4)^{21} - (e^a+4)^{21}].$$

Written as an antiderivative,

$$\int e^x(e^x+4)^{20}\,dx = \frac{1}{21}(e^x+4)^{21} + C.$$

Admittedly, this calculation required us to keep track of the various components of the composite function, and in many cases this mental tracking can be complex. The method of substitution is intended to simplify the tracking with a neat notational device.

For this example, the **method of substitution** is to define a new "variable" u, which is indeed a function of x by $u = e^x + 4$, and correspondingly define the "differential" $du = \frac{du}{dx}dx = e^x\,dx$. We then have the eye fooled into seeing:

$$\int e^x(e^x+4)^{20}\,dx = \int u^{20}\,du,$$

where this second antiderivative is elementary and equals $\frac{u^{21}}{21}$. One can then substitute back to get $\int e^x(e^x+4)^{20}\,dx = \frac{(e^x+4)^{21}}{21}$, and with the antiderivative in hand, the Fundamental Theorem provides the integration result above.

Note that we in fact could have modified this process to apply directly to the integral with limits, by introducing the u limits that correspond to the x limits. That is,

$$\int_a^b e^x(e^x+4)^{20}\,dx = \int_{e^a+4}^{e^b+4} u^{20}\,du.$$

In this example, the variable u is taking the place of the function $G(x)$ in the composite function above, while du is accounting for the $g(x)\,dx$ term. So the **method of substitution** can be described as converting something complex looking into something simple looking. In other words, tricking the eye,

$$\int_a^b f(G(x))g(x)\,dx = \int_{G(a)}^{G(b)} f(u)\,du, \tag{10.28}$$

where the substitution is: $u = G(x)$, $du = g(x)\,dx$, and the x-limits of integration are converted into u-limits.

In applications, one sometimes guesses as to an appropriate definition for the variable u and sees how the integral is transformed. For substitution to succeed two things must occur:

1. There is some substitution $u = G(x)$ into the initial integral that converts it into an integral of the form $\int f(u)\,du$ in (10.28).

2. The integral produced can be handled directly, or with the further application of this or another technique.

10.7.2   Integration by Parts

As the name suggests, integration by parts provides an algorithm for reducing an integrand to a new integrand that is hopefully simper to deal with. It gives "part" of the final result, and is derived from the formula for the derivative of a product of two functions.

Let $F(x)$ and $G(x)$ be two differentiable functions with derivatives $f(x)$ and $g(x)$, respectively. In other words, $F'(x) = f(x)$ and $G'(x) = g(x)$. Then the derivative of $F(x)G(x)$ is given by

$$[F(x)G(x)]' = f(x)G(x) + F(x)g(x).$$

By the Fundamental Theorem, it is easy to integrate $[F(x)G(x)]'$, namely

$$\int_a^b [F(x)G(x)]'\,dx = F(b)G(b) - F(a)G(a).$$

So the integration by parts idea is to convert the integral of something "hard," say $f(x)G(x)$, into an integral of something that is hopefully easier, namely $F(x)g(x)$, and a second very easy integral of $[F(x)G(x)]'$. We then get the **integration by parts formula**:

$$\int_a^b f(x)G(x)\,dx = F(b)G(b) - F(a)G(a) - \int_a^b F(x)g(x)\,dx. \qquad (10.29)$$

Note that the application of (10.29) requires two things in order to succeed in solving the problem:

1. The integrand can be split into a product $f(x)G(x)$ of a function $f(x)$ for which we can find the antiderivative $F(x)$, and a function $G(x)$ that we can differentiate to obtain $g(x)$;

2. This splitting produces a final integrand $F(x)g(x)$, which is easier to work with than the initial integrand $f(x)G(x)$.

In some applications, this process is implemented repeatedly. In other applications, trial and error and/or creative thinking is required, as the next few examples illustrate.

Example 10.50

1. *Consider the evaluation of $\int_a^b x^3 e^{x^2}\,dx$. Since we do not know the antiderivative of e^{x^2}, it is natural to guess that we ought to define $f(x) = x^3$ and $G(x) = e^{x^2}$, producing $F(x) = \frac{x^4}{4}$ and $g(x) = 2xe^{x^2}$. Unfortunately, the final integral in (10.29) is then $\frac{1}{2}\int_a^b x^5 e^{x^2}\,dx$, which is worse than what we started with. So, if this method is to work, and in many cases it does not, we must find a way to move e^{x^2} into the definition of $f(x)$. A little thought reveals that while finding the antiderivative of e^{x^2} appears impossible, the antiderivative of $f(x) = xe^{x^2}$ is $F(x) = \frac{1}{2}e^{x^2}$. So we define $G(x) = x^2$ with $g(x) = 2x$ and obtain*

$$\int_a^b x^3 e^{x^2}\,dx = \frac{1}{2}[b^2 e^{b^2} - a^2 e^{a^2}] - \int_a^b xe^{x^2}\,dx$$

$$= \frac{1}{2}[b^2 e^{b^2} - a^2 e^{a^2}] - \frac{1}{2}[e^{b^2} - e^{a^2}].$$

Written as an antiderivative,

$$\int x^3 e^{x^2}\,dx = \frac{1}{2}[x^2 e^{x^2} - e^{x^2}] + C.$$

2. *Let's next evaluate $\int_a^b \ln x \, dx$ where we assume $b > a > 0$, to ensure that the integrand is well defined and continuous. The case $a > b > 0$ is similarly handled by remark 10.37. In this case we have only one function visible, implying that this method has no chance of success. Certainly we would likely choose $G(x) = \ln x$, with $g(x) = \frac{1}{x}$, since to assign $\ln x$ to $f(x)$ is to require the calculation of its antiderivative $F(x)$, which we do not know or else we would just apply the Fundamental Theorem. With this definition of $G(x)$, there is no other choice than to try $f(x) = 1$, with $F(x) = x$, and hope for the best. We then get*

$$\int_a^b \ln x \, dx = b \ln b - a \ln a - \int_a^b 1 \, dx$$

$$= b \ln b - a \ln a - (b - a).$$

Again, written as an antiderivative,

$$\int \ln x \, dx = x \ln x - x + C.$$

3. *Consider $\int_a^b x^n e^x \, dx$ for integer n. Since both x^n and e^x are easily differentiated as well as easily integrated, it appears that there is some choice in their assignment to $f(x)$ and $G(x)$. However, if we assign $G(x) = e^x$ and $f(x) = x^n$, it is clear that we are moving in the wrong direction and that the final integral is $\frac{-1}{n+1} \int_a^b x^{n+1} e^x \, dx$. Reversing the assignment, we obtain a final integral of $-n \int_a^b x^{n-1} e^x \, dx$, and the process can be repeated until the final integral is $K \int_a^b e^x \, dx$, for constant $K = \pm n!$, at which point it is easily completed. See exercises 8 and 27.*

*10.7.3 Wallis' Product Formula

As a final application of the method underlying integration by parts, we return to the development of **Wallis' product formula**, introduced in section 8.5.1 in the derivation of Stirling's formula. Recall that this product formula, as stated in (8.25), is

$$\frac{\pi}{2} = \prod_{n=1}^{\infty} \frac{(2n)^2}{(2n-1)(2n+1)}. \tag{10.30}$$

To this end, first note that if $h(x) = \sin^{n-1} x \cos x$, then by (9.16) is derived that

$$h'(x) = (n-1) \sin^{n-2} x \cos^2 x - \sin^n x$$

$$= (n-1) \sin^{n-2} x (1 - \sin^2 x) - \sin^n x$$

$$= (n-1) \sin^{n-2} x - n \sin^n x,$$

where this derivation also used

$$\sin^2 x + \cos^2 x = 1. \tag{10.31}$$

Now, for $n > 1$, $\int_0^{\pi/2} h'(x)\, dx = h\left(\frac{\pi}{2}\right) - h(0) = 0$, and hence

$$\int_0^{\pi/2} \sin^n x\, dx = \frac{n-1}{n} \int_0^{\pi/2} \sin^{n-2} x\, dx, \qquad n > 1. \tag{10.32}$$

This identity can then be applied to even, $n = 2m$, and odd, $n = 2m + 1$, integers and iterated to produce

$$\int_0^{\pi/2} \sin^{2m} x\, dx = \frac{2m-1}{2m} \frac{2m-3}{2m-2} \cdots \frac{1}{2} \int_0^{\pi/2} dx$$

$$= \frac{\pi}{2} \prod_{j=0}^{m-1} \left(\frac{2m - 2j - 1}{2m - 2j} \right),$$

and similarly since $\int_0^{\pi/2} \sin x\, dx = 1$,

$$\int_0^{\pi/2} \sin^{2m+1} x\, dx = \frac{2m}{2m+1} \frac{2m-2}{2m-1} \cdots \frac{2}{3} \int_0^{\pi/2} \sin x\, dx$$

$$= \prod_{j=0}^{m-1} \left(\frac{2m - 2j}{2m - 2j + 1} \right).$$

For the next step, we must divide these expressions and solve for $\frac{\pi}{2}$, producing

$$\frac{\pi}{2} = \prod_{j=0}^{m-1} \left(\frac{2m - 2j}{2m - 2j + 1} \right) \prod_{j=0}^{m-1} \left(\frac{2m - 2j}{2m - 2j - 1} \right) \frac{\int_0^{\pi/2} \sin^{2m} x\, dx}{\int_0^{\pi/2} \sin^{2m+1} x\, dx}$$

$$= \prod_{j=0}^{m-1} \frac{(2m - 2j)^2}{(2m - 2j + 1)(2m - 2j - 1)} \frac{\int_0^{\pi/2} \sin^{2m} x\, dx}{\int_0^{\pi/2} \sin^{2m+1} x\, dx}.$$

This formula is then rewritten by defining $n = m - j$ and changing the j product to an n product from $n = 1$ to $n = m$:

$$\frac{\pi}{2} = \prod_{n=1}^{m} \frac{(2n)^2}{(2n+1)(2n-1)} \frac{\int_0^{\pi/2} \sin^{2m} x \, dx}{\int_0^{\pi/2} \sin^{2m+1} x \, dx}.$$

The final step is to let $m \to \infty$, but to do so requires a demonstration that the ratio of integrals converges to 1. Because $0 < \sin x < 1$ for $0 < x < \frac{\pi}{2}$, it follows that for any m,

$$\int_0^{\pi/2} \sin^{2m+1} x \, dx < \int_0^{\pi/2} \sin^{2m} x \, dx < \int_0^{\pi/2} \sin^{2m-1} x \, dx.$$

If this set of inequalities is divided by $\int_0^{\pi/2} \sin^{2m+1} x \, dx$, and (10.32) applied, we obtain

$$1 < \frac{\int_0^{\pi/2} \sin^{2m} x \, dx}{\int_0^{\pi/2} \sin^{2m+1} x \, dx} < 1 + \frac{1}{2m},$$

and so this ratio of integrals converges to 1, and (10.30) is demonstrated.

10.8   Taylor Series with Integral Remainder

In sections 9.3.7 and 9.3.8 Taylor series were introduced and some properties studied. In this section we revisit this idea, and with the aid of integration by parts, develop a new form for the remainder term which can be contrasted with the representation in (9.33).

To this end, we first observe that given x_0, we have by the second form of the Fundamental Theorem in (10.19) that

$$h(x) = h(x_0) + \int_{x_0}^{x} h'(z) \, dz,$$

where we use the general notation $h(x)$ to avoid confusion with the functions in (10.29). We can now apply integration by parts, expressing $h'(z) = f(z)G(z)$ with $f(z) = 1$ and $G(z) = h'(z)$, and using the arbitrary yet convenient constant term to express $F(z) = -(x - z)$, and we obtain

$$h(x) = h(x_0) + h'(x_0)(x - x_0) + \int_{x_0}^{x} h''(z)(x - z) \, dz.$$

We then express $h''(z)(x - z) = f(z)G(z)$ with $f(z) = (x - z)$, $G(z) = h''(z)$, and $F(z) = -\frac{1}{2}(x - z)^2$, and so on, producing the following:

Proposition 10.51 *Let $h(x)$ be $(n + 1)$-times differentiable on (a, b), with all derivatives continuous on $[a, b]$. Then for $x, x_0 \in (a, b)$,*

$$h(x) = \sum_{j=0}^{n} \frac{1}{j!} h^{(j)}(x_0)(x - x_0)^j + \frac{1}{n!} \int_{x_0}^{x} h^{(n+1)}(z)(x - z)^n \, dz. \tag{10.33}$$

Proof This follows by mathematical induction, using integration by parts as noted above.   ∎

The remainder term in the Taylor series expansion in (10.33) is known as the **Cauchy form** of the remainder after **Augustin Louis Cauchy** (1789–1857), who is credited with the first rigorous proof of the Taylor theorem. Another form of this remainder, named for **Joseph-Louis Lagrange** was developed in section 9.3.8.

If we compare the remainder terms of Cauchy and Lagrange, adjusting for notation, we obtain

$$\frac{1}{n!} \int_{x_0}^{x} f^{(n+1)}(z)(x - z)^n \, dz = \frac{1}{(n + 1)!} f^{(n+1)}(y)(x - x_0)^{n+1}, \tag{10.34}$$

where the point y depends on x and satisfies $x_0 < y < x$ or $x < y < x_0$. It turns out that this relationship between these remainders is a special case of what is known as the **second mean value theorem** for integrals:

Proposition 10.52 (*Second Mean Value Theorem for Integrals*) *Let $f(x)$ and $g(x)$ be continuous on bounded $[a, b]$, and $g(x) \geq 0$. Then there is a point $c \in [a, b]$ so that*

$$\int_{a}^{b} f(x)g(x) \, dx = f(c) \int_{a}^{b} g(x) \, dx. \tag{10.35}$$

Proof Since this result is obviously true if $g(x) \equiv 0$, we can assume that $g(x) > 0$ somewhere on this interval and hence that $\int_{a}^{b} g(x) \, dx > 0$. Now as $f(x)$ is continuous on $[a, b]$, it attains its maximum, M, and minimum, m, on this interval. From the definition of Riemann integral, we have that $m \leq f(x) \leq M$ implies that

$$m \int_{a}^{b} g(x) \, dx \leq \int_{a}^{b} f(x)g(x) \, dx \leq M \int_{a}^{b} g(x) \, dx,$$

which in turn implies that

$$m \le \frac{\int_a^b f(x)g(x)\,dx}{\int_a^b g(x)\,dx} \le M.$$

Since this ratio is between the minimum and maximum values attained by $f(x)$, we conclude from the intermediate value theorem in (9.1) of proposition 9.41 that there is a $c \in [a, b]$ so that $f(c)$ equals this ratio.   ■

Of course, the expression in (10.12), the first mean value theorem for integrals, is a special case of this result for which $g(x) \equiv 1$. Also the Taylor remainders in (10.34) are another special case, since is is easy to evaluate the remaining integral:

$$\int_{x_0}^x (x - z)^n\,dz = \frac{(x - x_0)^{n+1}}{n + 1}.$$

One advantage of the Cauchy form of the remainder is that it reflects an averaging of the $f^{(n+1)}(z)$ values on the given interval, whereas the Lagrange remainder is a point estimate. Consequently, to prove convergence of a Taylor series, and hence the analyticity of a given function, the Cauchy remainder can give more useful estimates than those based on the maximum value of $f^{(n+1)}(z)$ as is required in proposition 9.103 with the Lagrange remainder.

Recall example 9.106 in the cases for which the Lagrange remainder provided only partial results.

Example 10.53

1. *With $f(x) = \frac{1}{1-x} = (1 - x)^{-1}$ and $x_0 = 0$, it is easy to justify that $f^{(n)}(x) = n!(1 - x)^{-n-1}$ and so $f^{(n)}(0) = n!$. Although in chapter 6 it was shown that $\sum_{j=0}^{\infty} x^j$ converges for $|x| < 1$, it was seen in example 9.106 that the Lagrange remainder only proved that $\frac{1}{1-x} = \sum_{j=0}^{\infty} x^j$ in the case $-1 < x \le 0$, since then the Lagrange remainder converged to 0 as $n \to \infty$. For $0 < x < 1$ this remainder diverged. Using the Cauchy form above, we have*

$$\frac{1}{n!}\int_0^x f^{(n+1)}(z)(x - z)^n\,dz = (n + 1)\int_0^x (1 - z)^{-n-2}(x - z)^n\,dz$$

$$= (n + 1)\int_0^x \left(\frac{x - z}{1 - z}\right)^n (1 - z)^{-2}\,dz.$$

Now on the interval $0 \le z \le x$, where $0 < x < 1$, the function $g(z) = \frac{x-z}{1-z}$ is positive and deceasing and $g(z) \le x$. Also on this same interval $h(z) = (1-z)^{-2}$ is positive and increasing and $h(z) \le (1-x)^{-2}$. Consequently we obtain that

$$\left| \frac{1}{1-x} - \sum_{j=0}^{n} x^j \right| \le (n+1) \frac{x^n}{(1-x)^2} \int_0^x dz = (n+1) \frac{x^{n+1}}{(1-x)^2},$$

and so for $0 < x < 1$, this remainder converges to 0 as $n \to \infty$. This completes the demonstration that $f(x) = \frac{1}{1-x}$ is an analytic function on $(-1, 1)$ and given by the series expansion

$$\frac{1}{1-x} = \sum_{j=0}^{\infty} x^j, \qquad -1 < x < 1.$$

2. *With $f(x) = \ln(1+x)$ we obtain*

$$f'(x) = \frac{1}{1+x}, f^{(2)}(x) = \frac{-1}{(1+x)^2}, \dots, f^{(n)}(x) = \frac{(-1)^{n+1}(n-1)!}{(1+x)^n}.$$

Consequently $f(0) = 0$ and $f^{(n)}(0) = (-1)^{n-1}(n-1)!$ for $n \ge 1$. It was shown in example 9.106 that $\sum_{j=1}^{\infty} \frac{(-1)^{j+1}}{j} x^j$ converges for $|x| < 1$ and $x = 1$. But as in case 1, the Lagrange remainder only yields a partial result, that $\ln(1+x) = \sum_{j=1}^{\infty} \frac{(-1)^{j+1}}{j} x^j$ in the case of $-\frac{1}{2} \le x \le 1$, since then the Lagrange remainder converged to 0 as $n \to \infty$. For $-1 < x < -\frac{1}{2}$ this remainder diverged. Using the Cauchy form above obtains

$$\frac{1}{n!} \int_0^x f^{(n+1)}(z)(x-z)^n \, dz = (-1)^{n+2} \int_0^x (x-z)^n (1+z)^{-n-1} \, dz$$

$$= (-1)^{n+3} \int_x^0 \left(\frac{x-z}{1+z} \right)^n (1+z)^{-1} \, dz,$$

where it is noted that the limits of integration were reversed and offset by multiplication by -1 to better accommodate the x-values contemplated. Repeating the analysis above on this integrand, we have that on the interval $x \le z \le 0$ for $-1 < x < -\frac{1}{2}$, the function $g(z) = \left| \frac{x-z}{1+z} \right|$ is positive and increasing and $g(z) \le |x|$, while the function $h(z) = (1+z)^{-1}$ is positive and decreasing and $h(z) \le (1+x)^{-1}$. We now obtain, using (10.10),

$$\left| \ln(1+x) - \sum_{j=1}^{n} \frac{(-1)^{j+1}}{j} x^j \right| \leq \int_x^0 \left| \frac{x-z}{1+z} \right|^n (1+z)^{-1} \, dz$$

$$\leq \frac{|x|^{n+1}}{1+x},$$

and so for $-1 < x < -\frac{1}{2}$, *this remainder converges to* 0 *as* $n \to \infty$. *This completes the demonstration that* $f(x) = \ln(1+x)$ *is an analytic function on* $(-1, 1]$ *and given by the series expansion*

$$\ln(1+x) = \sum_{j=1}^{\infty} \frac{(-1)^{j+1}}{j} x^j, \qquad -1 < x \leq 1.$$

10.9   Convergence of a Sequence of Integrals

10.9.1   Review of Earlier Convergence Results

An important situation that often arises in mathematics is related to a sequence of functions $\{f_n(x)\}$ that is known to converge in some sense to a function $f(x)$. If each function in the sequence is known to have a certain property, can it be concluded that $f(x)$ will also have this property? The typical application, of course, is where the functions are simple in some way and have a desirable property that is easy to establish, and the question pursued is whether we can infer that this desirable property is also shared by $f(x)$.

For example, it was shown in chapter 9 in section 9.2.7 on convergence of a sequence of continuous functions that continuity is a property that does not in general transfer well from the functions $f_n(x)$ to the function $f(x)$ if convergence is defined pointwise. In other words, if for each x the numerical sequence $f_n(x)$ converges to the point $f(x)$, it is possible that each function in the sequence is continuous, yet $f(x)$ is not. The simple example given there was

$$f(x) = \begin{cases} 1, & x \leq 0, \\ 0, & x > 0, \end{cases}$$

and

$$f_n(x) = \begin{cases} 1, & x \leq 0, \\ 1 - nx, & 0 < x \leq \frac{1}{n}, \\ 0, & x > \frac{1}{n}. \end{cases}$$

Although $f_n(x) \to f(x)$ for all x, the continuity of $f_n(x)$ is lost at $x = 0$ because the convergence becomes increasingly slow, the closer x is to 0. This insight also produces the solution to the problem, and that is, if $f_n(x) \to f(x)$ uniformly, continuity is preserved, where by "uniformly" is meant that $|f_n(x) - f(x)|$ can be made arbitrarily small **for all** x by making n large enough.

For the property of differentiability, it was shown in section 9.4 on convergence of a sequence of derivatives that neither pointwise nor uniform convergence of $f_n(x) \to f(x)$ was enough to ensure that the differentiability of $f_n(x)$ would either imply the differentiability of $f(x)$ or, in the case where $f'(x)$ existed, imply the convergence $f_n'(x) \to f'(x)$. An example for nonexistence of $f'(x)$ was given by

$$f_n(x) = \begin{cases} x^{1+(1/n)}, & x \ge 0, \\ (-x)^{1+(1/n)}, & x \le 0, \end{cases}$$

$$f(x) = |x|,$$

since here $f'(0)$ does not exist.

The example for when $f'(x)$ exists for all x but $f_n'(x) \nrightarrow f'(x)$ was given by

$$f_n(x) = \frac{\sin nx}{\sqrt{n}},$$

$$f(x) \equiv 0.$$

10.9.2   Sequence of Continuous Functions

The general questions of this section are:

Question 1: If $f_n(x)$ is Riemann integrable over $[a,b]$ for all n and $f_n(x) \to f(x)$ pointwise, will it be the case that $\int_a^b f(x)\,dx$ exists and $\int_a^b f_n(x)\,dx \to \int_a^b f(x)\,dx$?

Question 2: In general, what kind of convergence of integrable functions $f_n(x)$ will ensure integrability of $f(x)$ and the convergence of integral values, and what if any bearing do properties of the interval of integration have on these results?

As it turns out, question 1 is relatively easy, but question 2 is far more subtle and difficult than is the related investigation on continuity or differentiability. We address question 1 and an important portion of question 2 here. This discussion will be greatly expanded within the framework of real analysis.

Answer 1: Pointwise convergence of $f_n(x)$ that are Riemann integrable over bounded $[a,b]$ assures neither the integrability of $f(x)$ nor, in the case where $f(x)$ is integrable, the convergence of integral values. Examples of these behaviors follow.

Example 10.54

1. *For any ordering of the rational numbers in* $[0, 1]$, $\{r_j\}_{j=1}^{\infty}$, *define*

$$f_n(x) = \begin{cases} 1, & x = r_j, \ 1 \le j \le n, \\ \frac{1}{n}, & elsewhere. \end{cases}$$

Then $f_n(x)$ *is continuous except at* n *points and hence is integrable, and* $\int_0^1 f_n(x)\,dx = \frac{1}{n}$. *However,* $f_n(x) \to f(x)$ *pointwise, where*

$$f(x) = \begin{cases} 1, & x \ rational, \\ 0, & x \ irrational, \end{cases}$$

which is nowhere continuous and hence not Riemann integrable.

2. *Define for* $n \ge 1$,

$$f_n(x) = \begin{cases} 2^n, & \frac{1}{2^n} \le n \le \frac{1}{2^{n-1}}, \\ 0, & elsewhere. \end{cases}$$

Then $f_n(x)$ *converges pointwise on* $[0, 1]$, *but not uniformly, to the continuous and hence integrable function* $f(x) \equiv 0$. *Also, for all* n, *a simple calculation gives that* $\int_0^1 f_n(x)\,dx = 1$, *but obviously,* $\int_0^1 f(x)\,dx = 0$.

Answer 2:   The next two propositions in this and the next section provide two cases where the desired conclusions follow. The first result calls for the uniform convergence of continuous functions on a bounded interval, the second, found in the next section, generalizes this result.

Proposition 10.55   *If* $\{f_n(x)\}$ *is a sequence of continuous functions on a closed and bounded interval* $[a, b]$, *and there is a function* $f(x)$ *so that* $f_n(x) \to f(x)$ *uniformly, then* $f(x)$ *is Riemann integrable and*

$$\int_a^b f_n(x)\,dx \to \int_a^b f(x)\,dx. \tag{10.36}$$

In other words,

$$\int_a^b f(x)\,dx = \lim_{n \to \infty} \int_a^b f_n(x)\,dx. \tag{10.37}$$

Proof   First off, $\int_a^b f_n(x)\,dx$ exists for all n, since these functions are continuous and the interval is bounded. Also uniform convergence assures the continuity of $f(x)$ by proposition 9.51, and hence the existence of $\int_a^b f(x)\,dx$, so the only question is one of convergence of the values of the integrals in (10.36), that $\int_a^b [f_n(x) - f(x)]\,dx \to 0$. To this end, uniform convergence implies that for any $\epsilon > 0$ there is an $N(\epsilon)$ so that $|f_n(x) - f(x)| < \epsilon$ for all $x \in [a, b]$ for $n > N$. Hence for any partition $a = x_0 < x_1 < \cdots < x_n = b$, where $y_j \in [x_{j+1} - x_j]$, the Riemann sum is bounded:

$$\left| \sum_j [f_n(y_j) - f(y_j)][x_{j+1} - x_j] \right| \leq \sum |f_n(y_j) - f(y_j)||[x_{j+1} - x_j]$$

$$< \epsilon \sum [x_{j+1} - x_j]$$

$$= \epsilon(b - a), \qquad n > N.$$

Consequently $\int_a^b [f_n(x) - f(x)]\,dx \to 0$, and because of the linearity of the integral in (10.9) this result is equivalent to demonstrating (10.36). ∎

Remark 10.56 *Note that (10.37) can be rewritten to emphasize that this is another example of reversing the order of two limiting operations as in proposition 9.58. Recall that the integral is defined as the limit of Riemann sums, and (10.37) becomes*

$$\int_a^b \left[\lim_{n \to \infty} f_n(x) \right] dx = \lim_{n \to \infty} \left[\int_a^b f_n(x)\,dx \right].$$

10.9.3 Sequence of Integrable Functions

The preceding result can be generalized, in that the assumption of the continuity of $f_n(x)$ can be relaxed to just the assumptions of boundedness and Riemann integrable.

Proposition 10.57 *If $\{f_n(x)\}$ is a sequence of bounded, Riemann integrable functions on a closed and bounded interval $[a, b]$, and there is a function $f(x)$ so that $f_n(x) \to f(x)$ uniformly, then $f(x)$ is Riemann integrable and (10.36) holds.*

Proof First off, we show $f(x)$ is indeed Riemann integrable. By the characterization of integrability on bounded intervals in the Riemann existence theorem of proposition 10.22, it is enough to prove that $f(x)$ is bounded and continuous except on a set of measure zero.

To this end, let E_n denote the set of discontinuity points of $f_n(x)$ that has measure 0 because of the integrability assumption, and let $E = \bigcup E_n$. Then E also has measure 0 by proposition 10.16, and we will show that $f(x)$ is continuous outside E. It is important to note that $f(x)$ will also in general be continuous on many, even all, of the points in E, but we cannot be assured of this and in any case do not need this for the desired result.

By uniform convergence we have that for any $\epsilon > 0$ there is an $N = N(\epsilon)$ so that $|f(y) - f_n(y)| < \epsilon$ for all $y \in [a, b]$ and all $n \geq N$. Let $x \in [a, b] - E$, and since $f_N(x)$ is continuous at x, there is a δ_N for this same ϵ so that $|f_N(x) - f_N(y)| < \epsilon$ if $|x - y| < \delta_N$. By the triangle inequality, if $|x - y| < \delta_N$, then

$$|f(x) - f(y)| \leq |f(x) - f_N(x)| + |f_N(x) - f_N(y)| + |f_N(y) - f(y)|$$

$$< 3\epsilon,$$

and so $f(x)$ is continuous outside E, a set of measure 0.

Boundedness of $f(x)$ also follows from the uniform convergence and the boundedness of $f_n(x)$. For $n \geq N$ above,

$$|f(x)| \leq |f(x) - f_n(x)| + |f_n(x)|$$

$$< \epsilon + C_n,$$

where C_n denotes the maximum of bounded $|f_n(x)|$ on $[a, b]$.

To next show the convergence of integrals in (10.36), uniform continuity implies that for all $x \in [a, b]$ and $n \geq N$,

$$-\epsilon < f(x) - f_n(x) < \epsilon,$$

which implies that for $n \geq N$,

$$-\epsilon(b - a) < \int_a^b [f(x) - f_n(x)] \, dx < \epsilon(b - a).$$

As ϵ is arbitrary, this demonstrates (10.36).   ∎

10.9.4 Series of Functions

As was the case in section 9.27 on sequences of continuous functions and section 9.4 on sequences of differentiable functions, the propositions above on sequences of integrable functions easily yield comparable results on series of functions which converge uniformly. We state only the more general case.

Proposition 10.58 *If $g_j(x)$ is a sequence of bounded, Riemann integrable functions, and there is a function $g(x)$ so that on some interval $[a, b]$ the series $\sum_{j=1}^{\infty} g_j(x)$ converges uniformly to $g(x)$, then $g(x)$ is Riemann integrable, and $\sum_{j=1}^{n} \int_a^b g_j(x) \, dx \to \int_a^b g(x) \, dx$ as $n \to \infty$.*

Remark 10.59   *The uniform convergence of a series of integrable functions y...* *integrable function whose integral equals the sum of the integrals of terms in the series. That is, uniform convergence justifies integrating term by term, then summing, which means*

$$\int_a^b g(x)\, dx = \lim_{n\to\infty} \sum_{j=1}^n \int_a^b g_j(x)\, dx.$$

Proof   Define $f_n(x) = \sum_{j=1}^n g(x)$. Then $f_n(x)$ is bounded and Riemann integrable for all n as a finite sum of bounded integrable functions, and by assumption, $f_n(x) \to g(x)$ uniformly. Also $\int_a^b f_n(x)\, dx \equiv \sum_{j=1}^n \int_a^b g_j(x)\, dx$. So the result follows from proposition 10.57.   ∎

10.9.5   Integrability of Power Series

We next apply the above result on series of functions to the special case of a power series. It is largely a corollary to the proposition above on series of functions, but it is stated here to clarify that a small amount of thought needs to be applied to ensure the uniformity of convergence as the above result requires.

Proposition 10.60   *Assume that a function $f(x)$ is defined by the power series*

$$f(x) = \sum_{j=0}^\infty c_j(x - x_0)^j \tag{10.38}$$

and has an interval of convergence given by $I = \{x \mid |x - x_0| < R\}$ for some $R > 0$. Then $f(x)$ is Riemann integrable on any bounded interval $[a, b] \subset I$, and

$$\int_a^b f(x)\, dx = \sum_{j=0}^\infty \frac{c_j}{j+1} [(b - x_0)^{j+1} - (a - x_0)^{j+1}]. \tag{10.39}$$

In other words, a power series can be integrated term by term within its interval of convergence.

Proof   Of course, $f(x)$ is infinitely differentiable as was demonstrated in section 9.42, and hence it is continuous on I and Riemann integrable on any bounded interval within I. Define $f_n(x)$ as the partial summation associated with $f(x)$,

$$f_n(x) = \sum_{j=0}^n c_j(x - x_0)^j.$$

The function $f_n(x)$ is continuous and hence Riemann integrable for all n. As a finite summation the integral is given as

$$\int_a^b f_n(x)\,dx = \sum_{j=0}^n \int_a^b c_j(x - x_0)^j\,dx$$

$$= \sum_{j=0}^n \frac{c_j}{j+1}[(b - x_0)^{j+1} - (a - x_0)^{j+1}].$$

Because $f_n(x) \to f(x)$ pointwise on $|x - x_0| < R$, this convergence is uniform on the compact $[a, b] \subset I$ by exercise 30(b) of chapter 9. So by proposition 10.55:

$$\int_a^b f(x)\,dx = \lim_{n\to\infty} \int_a^b f_n(x)\,dx$$

$$= \sum_{j=0}^\infty \frac{c_j}{j+1}[(b - x_0)^{j+1} - (a - x_0)^{j+1}]. \qquad \blacksquare$$

Remark 10.61

1. *Note that in this result on the integral of a series of functions, it is apparent that the series of integrals is convergent, in fact absolutely convergent. First off, by the triangle inequality,*

$$\left| \sum_{j=0}^\infty \frac{c_j}{j+1}[(b - x_0)^{j+1} - (a - x_0)^{j+1}] \right|$$

$$\leq \sum_{j=0}^\infty \frac{|c_j|}{j+1}|(b - x_0)|^{j+1} + \sum_{j=0}^\infty \frac{|c_j|}{j+1}|(a - x_0)|^{j+1}.$$

Now, by the ratio test, for any $x \in I$,

$$\limsup_{j\to\infty} \frac{\frac{|c_{j+1}|}{j+2}|(x - x_0)|^{j+2}}{\frac{|c_j|}{j+1}|(x - x_0)|^{j+1}} = \limsup_{j\to\infty} \frac{|c_{j+1}|}{|c_j|}|(x - x_0)|,$$

and this limit is less than 1 exactly when $|(x - x_0)| < R$, since by definition, $\frac{1}{R} = \limsup_{j\to\infty} \frac{|c_{j+1}|}{|c_j|}$.

2. *This proposition applies to absolutely convergent Taylor series of analytic functions, of course, since the partial sums of these converge pointwise and hence uniformly on any bounded interval,* $[a, b]$.

10.10   Numerical Integration

When an integral of a function $f(x)$ is required, there may be no apparent way to apply the result of the Fundamental Theorem of Calculus version I because the given function is not the derivative of a recognized function. In such a case a numerical algorithm is required, and there are many to choose from.

The most basic approach comes from the definition of the Riemann integral itself, in (10.5). We simply partition the interval into a finite collection of subintervals, of equal or unequal size, choose a point from each subinterval, and use the approximation:

$$\int_a^b f(x)\,dx \approx \sum_{i=1}^n f(\tilde{x}_i) \Delta x_i,$$

where

$$\tilde{x}_i \in \left[a + \sum_{j=1}^{i-1} \Delta x_j, a + \sum_{j=1}^i \Delta x_j \right].$$

This is an approximation to the result because, by definition, the exact value of the integral is produced by this procedure as the mesh size of the partition μ, defined in (10.3), converges to 0. What is unknown, of course, is the quality of this approximation for a given partition, or the rate at which the error of the approximation goes to 0 as $\mu \to 0$.

In the following, we consider only the case of equal partitions, where $\mu = \Delta x \equiv \frac{b-a}{n}$.

10.10.1   Trapezoidal Rule

One useful way of defining the "quality of an approximation methodology" is to determine the class of functions for which the methodology produces an exact result. The bigger the class, the better is the quality of the approximation. For example, upper and lower Riemann sums will, in general, only provide an exact answer for a constant function $f(x) = d$, or more generally, a piecewise constant function $f(x) = d_i$ for $x \in [x_i, x_{i+1}]$, where $x_0 = a$ and $x_n = b$. This piecewise constant function is also known as a **step function**.

Using a slight modification of this technique, we can expand this class of functions to include all linear or affine functions $f(x) = cx + d$, as well as piecewise linear functions $f(x) = c_i x + d_i$ for $x \in [x_i, x_{i+1}]$. The simple modification involves defining $f(\tilde{x}_i)$ in the Riemann summation above at the midpoint of the interval or, more generally, for other applications, replacing $f(\tilde{x}_i)$ with the average of the value of the function at the endpoints of each subinterval:

$$\int_a^b f(x)\,dx \approx \sum_{i=1}^n \left[\frac{f(x_{i-1}) + f(x_i)}{2} \right] \Delta x.$$

This methodology produces what is known as the **trapezoidal rule** and can be rewritten as

$$\int_a^b f(x)\,dx \approx \frac{1}{2} \left[f(x_0) + 2 \sum_{i=1}^{n-1} f(x_i) + f(x_n) \right] \Delta x. \tag{10.40}$$

It is apparent from geometric considerations that this approximation is exact for affine functions, and for properly chosen partitions, piecewise affine functions.

To evaluate the error in this approximation, recall from (10.6) that we can evaluate this integral over each subinterval separately, and then simply add up the results. Similarly we can investigate the quality of a proposed approximation over each subinterval, and the overall error of the approximation is simply the sum of the subinterval errors.

For notational simplicity we evaluate the trapezoidal approximation over the first subinterval, $[a, a + \Delta x]$. From the Taylor series expansion in (9.33) with $n = 1$ and $x_0 = a$, and assuming continuous $f^{(2)}(x)$, we get

$$f(x) = f(a) + f'(a)(x - a) + \frac{1}{2} f^{(2)}(y)(x - a)^2,$$

where $y \equiv y(x)$ and $a < y(x) < x$. Since $f(x)$ is continuous, we infer that $f^{(2)}(y)$ is also a continuous function of x.

Integrating this formula over the interval $[a, a + \Delta x]$ produces for $I \equiv \int_a^{a+\Delta x} f(x)\,dx$,

$$I = f(a)\Delta x + \frac{1}{2} f'(a)\Delta x^2 + \frac{1}{2} \int_a^{a+\Delta x} f^{(2)}(y(x))(x - a)^2\,dx$$

$$= f(a)\Delta x + \frac{1}{2} f'(a)\Delta x^2 + \frac{1}{3!} f^{(2)}(z)\Delta x^3.$$

Note that the last step is justified by the second MVT for integrals in (10.35) applied to the function $f^{(2)}(y(x))$. In other words, $f^{(2)}(z)$ is defined as $f^{(2)}(y(c))$ for some $c \in [a, a + \Delta x]$, and consequently since $a < y(x) < x$, we conclude that $z \in [a, a + \Delta x]$.

The trapezoidal approximation over this interval, using the Taylor expansion above, is:

$$I^T = \frac{1}{2}[f(a) + f(a + \Delta x)]\Delta x$$

$$= \frac{1}{2}\left[2f(a)\Delta x + f'(a)\Delta x^2 + \frac{1}{2}f^{(2)}(y)\Delta x^3\right],$$

where $y \equiv y(\Delta x)$ and $a < y(\Delta x) < a + \Delta x$. Subtracting these expressions produces

$$I - I^T = -\left[\frac{1}{4}f^{(2)}(z) - \frac{1}{6}f^{(2)}(y)\right]\Delta x^3 \qquad \text{for } a < y, z < a + \Delta x.$$

Finally, for $d > b > 0$, consider the expression

$$\frac{df^{(2)}(z) - bf^{(2)}(y)}{d - b} = f^{(2)}(z) + \frac{b}{d - b}[f^{(2)}(z) - f^{(2)}(y)].$$

It is apparent that this expression is strictly between $f^{(2)}(z)$ and $f^{(2)}(y)$, and hence by the continuity of $f^{(2)}(x)$ and the intermediate value theorem in (9.1), there is a w between y and z so that $f^{(2)}(w) = \frac{df^{(2)}(z) - bf^{(2)}(y)}{d - b}$. Applying this to the trapezoidal approximation above, where $d = \frac{1}{4}$ and $b = \frac{1}{6}$, we conclude that

$$I - I^T = -\frac{1}{12}f^{(2)}(w)\Delta x^3 \qquad \text{for } a < w < a + \Delta x.$$

Summarizing, we have derived the following result:

Proposition 10.62 *If $f(x)$ is a twice differentiable function with continuous $f^{(2)}(x)$ on the bounded interval $[a, b]$, with partition given by $\{x_i\}_{i=0}^n = \{a + i\Delta x\}_{i=0}^n$ and $\Delta x = \frac{b-a}{n}$, then the error in the trapezoidal approximation defined in (10.40) is given by*

$$I - I^T = -\frac{1}{12}f^{(2)}(w)\frac{(b - a)^3}{n^2} \tag{10.41}$$

for some $w \in [a, b]$. If $|f^{(2)}(x)| \leq M_2$ on $[a, b]$, the absolute error bound is given by

$$|I - I^T| \leq \frac{M_2(b - a)^3}{12n^2}. \tag{10.42}$$

Proof Applying the analysis above to each subinterval and adding, we derive

$$I - I^T = -\frac{1}{12}\sum_{i=1}^{n} f^{(2)}(w_i)\Delta x^3$$

for $a + (i - 1)\Delta x < w_i < a + i\Delta x$. Now, since $\frac{1}{n}\sum_{i=1}^{n} f^{(2)}(w_i)$ is bounded by the maximum and minimum values of $f^{(2)}(x)$ on $[a,b]$, by the intermediate value theorem there is a $w \in [a,b]$ that equals this value. Substituting $\Delta x = \frac{b-a}{n}$ completes the proof of (10.41). From this, (10.42) follows by taking absolute values and bounding $f^{(2)}(w)$ by its maximum, M_2. ∎

Remark 10.63 *Note that the error estimate of the trapezoidal approximation is* $O(\Delta x^2)$. *Specifically, since* $\Delta x = \frac{b-a}{n}$, *we have from (10.42) that*

$$|I - I^T| \leq \frac{M_2(b-a)}{12}\Delta x^2. \tag{10.43}$$

10.10.2 Simpson's Rule

With a bit more effort, Simpson's rule improves the error in the trapezoidal rule approximation from $O(\Delta x^2)$ to $O(\Delta x^4)$. The additional effort required is to utilize the midpoint and endpoints from each subinterval defined by the partition above, rather than just the endpoints. However, Simpson's rule requires the continuity of the fourth derivative $f^{(4)}(x)$ on $[a,b]$.

Specifically, on a given subinterval, say $[a, a + \Delta x]$ for simplicity, the Simpson's rule approximation is defined as

$$\int_a^{a+\Delta x} f(x)\, dx \approx \frac{1}{6}\left[f(a) + 4f\left(\frac{2a + \Delta x}{2}\right) + f(a + \Delta x)\right]\Delta x,$$

where $\Delta x = \frac{b-a}{n}$. Adding over all subintervals produces Simpson's rule:

$$\int_a^b f(x)\, dx \approx \frac{1}{6}\sum_{i=1}^{n}\left[f(x_{i-1}) + 4f\left(\frac{x_{i-1} + x_i}{2}\right) + f(x_i)\right]\Delta x.$$

In terms of the resulting coefficients of the function values, a simple calculation produces the following:

$$\int_a^b f(x)\, dx \approx \frac{1}{6}\left[f(x_0) + 2\sum_{i=1}^{n-1} f(x_i) + 4\sum_{i=1}^{n} f\left(\frac{x_{i-1} + x_i}{2}\right) + f(x_n)\right]\Delta x. \tag{10.44}$$

The development of the error in this approximation follows that above for the trapezoidal rule but utilizes the Taylor approximation up to the $f^{(4)}(x)$ term. We present the final result of these calculations without proof (see exercise 12):

Proposition 10.64 *If $f(x)$ is a four times differentiable function with continuous $f^{(4)}(x)$ on the bounded interval $[a, b]$, with partition given by $\{x_i\}_{i=0}^n = \{a + i\Delta x\}_{i=0}^n$ and $\Delta x = \frac{b-a}{n}$, then the error in Simpson's rule defined in (10.44) is given by*

$$I - I^S = -\frac{1}{180} f^{(4)}(w) \frac{(b-a)^5}{(2n)^4} \tag{10.45}$$

for some $w \in [a, b]$. If $|f^{(4)}(x)| \le M_4$ on $[a, b]$, the absolute error bound is given by

$$|I - I^S| \le \frac{M_4(b-a)^5}{180(2n)^4}. \tag{10.46}$$

Remark 10.65 *Note that the error estimate of Simpson's rule is $O(\Delta x^4)$. Specifically, since $\Delta x = \frac{b-a}{n}$, we have from (10.46) that*

$$|I - I^T| \le \frac{M_4(b-a)}{2880} \Delta x^4. \tag{10.47}$$

10.11   Continuous Probability Theory

10.11.1   Probability Space and Random Variables

Recall that chapter 6 on series provided the needed tools to develop most of a discrete probability theory. Exceptions included (7.67), which required some chapter 9 tools, and the statement that the moment-generating function or characteristic function uniquely characterize a probability density, addressed somewhat in section 8.1. Similarly the tools of Riemann integration in this chapter 10 are sufficient to develop most of the "continuous" counterpart to this theory.

As in chapter 7, we begin with a sample space, S, which we do not require to be finite or discrete as was the case for discrete probability theory. We might imagine S to be the real numbers $\mathbb{R}$ or Euclidean space $\mathbb{R}^n$, for example. The critical observation in this generalization is that we can no longer rely on the restriction that S has a countable collection of sample points.

In this section we introduce many of the relevant aspects of this continuous theory, and provide a more general "mixed" discrete and continuous model in exercises 40 through 42. The more formal and mathematically more complete development, which provides a framework for an even more general probability theory which encompasses both the discrete and continuous theories, and more, requires the tools of real analysis.

Given $\mathcal{S}$, we define the **complete collection of events** as in definition 7.2, but begin to emphasize the alternate terminology noted in remark 7.3.

Definition 10.66   *Given a sample space, $\mathcal{S}$, a collection of events, $\mathcal{E} = \{A \mid A \subset \mathcal{S}\}$, is called **complete**, or is a **sigma algebra**, if it satisfies the following properties:*

1. $\emptyset, \mathcal{S} \in \mathcal{E}$.

2. *If $A \in \mathcal{E}$, then $\tilde{A} \in \mathcal{E}$.*

3. *If $A_j \in \mathcal{E}$ for $j = 1, 2, 3, \ldots$, then $\bigcup_j A_j \in \mathcal{E}$.*

In other words, we require that a sigma algebra of events contain the null event $\emptyset$, the certain event $\mathcal{S}$, the complement of all events, and that it be closed under countable unions. However, while item 3 is stated only for countable unions, it is also true for countable intersections because of property 2 and De Morgan's laws. Hence it is also the case that $\bigcap_j A_j \in \mathcal{E}$. Similarly, if $A, B \in \mathcal{E}$, then $A \sim B \in \mathcal{E}$, where $A \sim B \equiv \{x \in \mathcal{S} \mid x \in A \text{ and } x \notin B\}$, since $A \sim B = A \cap \tilde{B}$.

Remark 10.67

1. *In the discrete sample spaces of chapter 7, $\mathcal{E}$ usually contained each of the sample points and all subsets of $\mathcal{S}$, and was consequently always complete. In a general sample space that is uncountably infinite, the collection of events will virtually always be a proper subset of the collection of all subsets. Consequently the structure of the collection of events implied by the sigma algebra definition is all that we will have to work with, and hence all that can be assumed about $\mathcal{E}$ in the development of this theory.*

2. *It was noted in chapter 7 that the use of the term "complete" was not standard, but was introduced there for simplicity. The three conditions in the definition above are general requirements for $\mathcal{E}$ to be a **sigma algebra**, and this is a more natural language here given the greater generality of this collection.*

3. *Although perhaps not formally, but at least intuitively, it should be clear that the generality of the definition of a sigma algebra of events implies that on any given sample space, any number of sigma algebras can be defined as long as they satisfy the conditions above. Once that is contemplated, it becomes clear that one might find two sigma algebras, $\mathcal{E}$ and $\mathcal{E}'$ where $\mathcal{E} \subset \mathcal{E}'$ in the sense that every event in $\mathcal{E}$ is an event in $\mathcal{E}'$. In that sense, $\mathcal{E}'$ is a **finer sigma algebra** because it contains more events, and $\mathcal{E}$ a **coarser sigma algebra** because it contains fewer events. One also imagines that there* *may be two sigma algebras where neither $\mathcal{E} \subset \mathcal{E}'$ nor $\mathcal{E}' \subset \mathcal{E}$ is true. Outstanding questions to be pursued in more advanced treatments using the tools of real analysis are:*

- *If $\mathcal{D}$ is any collection of subsets of $\mathcal{S}$, is there a sigma algebra $\mathcal{E}$ that contain the sets in $\mathcal{D}$ so that $\mathcal{D} \subset \mathcal{E}$?*

- *If yes, is there a smallest such sigma algebra?*
  For example, if $\mathcal{E}$ and $\mathcal{E}'$ are two sigma algebras on $\mathcal{S}$, is there a sigma algebra $\mathcal{E}''$ so that $\mathcal{E} \cup \mathcal{E}' \subset \mathcal{E}''$? For another example, if $X : \mathcal{S} \to \mathbb{R}$ is a given function, is there a sigma algebra that contains all sets of the form $X^{-1}(a, b)$ for all open intervals $(a, b) \subset \mathbb{R}$?

The notion of a probability measure on $\mathcal{E}$ is identical to that in chapter 7. Because of the generality of the event space and the fact that $\mathcal{E}$ does not contain sample points as events, we use the general notation μ, which is standard in the theory, rather than the notation for the discrete theory of Pr.

Definition 10.68   *Given a sample space $\mathcal{S}$, and a sigma algebra of events $\mathcal{E} = \{A \mid A \subset \mathcal{S}\}$, a **probability measure** is a function $\mu : \mathcal{E} \to [0, 1]$, which satisfies the following properties:*

1. $\mu(\mathcal{S}) = 1$.

2. *If $A \in \mathcal{E}$, then $\mu(A) \geq 0$ and $\mu(A) = 1 - \mu(A)$.*

3. *If $A_j \in \mathcal{E}$ for $j = 1, 2, 3, \ldots$ are **mutually exclusive events**, that is, with $A_j \cap A_k = \emptyset$ for all $j \neq k$, then*

$$\mu\left(\bigcup_j A_j\right) = \sum \mu(A_j).$$

*In this case the triplet: $(\mathcal{S}, \mathcal{E}, \mu)$ is called a **probability space**.*

Definition 10.69   *An event $A \in \mathcal{E}$ is a **null event** under μ if $\mu(A) = 0$. If A is a null event and every $A' \subset A$ satisfies:*

1. $A' \in \mathcal{E}$,

2. $\mu(A') = 0$,

*then the triplet, $(\mathcal{S}, \mathcal{E}, \mu)$ is called a **complete probability space**.*

Remark 10.70   *Questions on probability spaces to be pursued in more advanced treatments using the tools of real analysis are:*

1. *If $(\mathcal{S}, \mathcal{E}, \mu)$ is a probability space that is not complete, can $\mathcal{E}$ be expanded to a sigma algebra $\mathcal{E}'$ that is complete and hence includes all subsets of null sets?*

2. *If* $(\mathcal{S}, \mathcal{E}, \mu)$ *is a probability space which is not complete, and* $\mathcal{E}$ *is expanded to include all subsets of null sets, can the definition of* μ *be expanded to* $\mathcal{E}'$ *without changing its values on* $\mathcal{E}$?

3. *Given* $(\mathcal{S}, \mathcal{E}, \mu)$, *define* n-**trial sample space**, *denoted* $\mathcal{S}^n$, *by*

$$\mathcal{S}^n = \{(s_1, s_2, \ldots, s_n) \mid s_j \in \mathcal{S}\}.$$

How can an associated sigma algebra of events $\mathcal{E}^n$ *be defined and reflective of the sigma algebra* $\mathcal{E}$? *Also, how can a probability measure* $\mu_n(A)$ *be defined for* $A \in \mathcal{E}^n$ *in a way that allows the identification of up to n-events in* $\mathcal{E}$ *with n-independent events in* $\mathcal{E}^n$, *which have the same probability measures?*

So far not too much of what has been defined is materially different from the discrete setting of chapter 7. What really distinguishes the discrete and continuous models is the nature of a random variable defined on $\mathcal{S}$.

Definition 10.71   *Given a sample space* $\mathcal{S}$, *and a sigma algebra of events* $\mathcal{E} = \{A \mid A \subset \mathcal{S}\}$, *a* **continuously distributed random variable** *is a function*

$$X : \mathcal{S} \to \mathbb{R},$$

so that:

1. *For any bounded or unbounded interval* $\{a, b\} \subset \mathbb{R}$, *where* $\{a, b\}$ *denotes that this interval may be open, closed, or half-open,*

$$X^{-1}\{a, b\} \in \mathcal{E}.$$

2. *There is a* **continuous function**, *denoted* f *or* f_X, *with* $f(x) \geq 0$, *the* **probability function (p.f.)**, *or* **probability density function (p.d.f.) of** X *so that given any interval* $\{a, b\}$,

$$\int_a^b f(x)\, dx = \mu[X^{-1}\{a, b\}]. \tag{10.48}$$

The **distribution function (d.f.)**, *or* **cumulative distribution function (c.d.f.)** *associated with* X, *denoted* F *or* F_X, *is then defined on* $\mathbb{R}$ *by*

$$F(x) = \mu[X^{-1}(-\infty, x]] \tag{10.49a}$$

$$= \int_{-\infty}^x f(y)\, dy. \tag{10.49b}$$

Note that for any point $a \in \mathbb{R}$, $X^{-1}[a] \in \mathcal{E}$, since $X^{-1}\left[a, a + \frac{1}{n}\right] \in \mathcal{E}$ for all n, and by the property of a sigma algebra,

$$\bigcap_n X^{-1}\left[a, a + \frac{1}{n}\right] = X^{-1}[a] \in \mathcal{E}.$$

Of course, if $a \notin \mathrm{Rng}[X]$, then $X^{-1}[a] = \emptyset$. Also, by (10.48), it must be the case that

$$\mu[X^{-1}[a]] = 0 \qquad \text{for all } a \in \mathbb{R}.$$

Consequently the pre-image of any point under X has probability measure 0 and is a null event in $\mathcal{E}$. In addition the pre-image of any countable collection of points is again a null event, since given $\{a_j\}_{j=1}^{\infty}$, the collection of events $\{X^{-1}[a_j]\}_{j=1}^{\infty}$ are mutually exclusive since X is a function, and so by definition 10.68,

$$\mu\left[\bigcup_j X^{-1}[a_j]\right] = \sum_j \mu[X^{-1}[a_j]] = 0.$$

Remark 10.72 *These conclusions highlight a stark contrast between continuous and discrete probability theory. In chapter 7, given any random variable X, there is a finite or countable collection $\{a_j\}_{j=1}^{\infty} \subset \mathbb{R}$ so that $\{X^{-1}[a_j]\}_{j=1}^{\infty} \subset \mathcal{E}$ are mutually independent events, and $\Pr[\bigcup_j X^{-1}[a_j]] = 1$. In continuous probability theory, for any collection $\{a_j\}_{j=1}^{\infty} \subset \mathbb{R}$, while it is still true that $\{X^{-1}[a_j]\}_{j=1}^{\infty} \subset \mathcal{E}$ and are mutually independent events, we now have $\mu[\bigcup_j X^{-1}[a_j]] = 0$.*

Note that since the c.d.f. is the integral of a continuous function, we have from the Fundamental Theorem of Calculus version II that $F(x)$ is a differentiable function and

$$F'(x) = f(x). \tag{10.50}$$

Note also from the definition of $F(x)$ that:

1. $F(\infty) = 1$, since $F(x) = \mu[X^{-1}(-\infty, x)]$, and as $x \to \infty$, $\mu[X^{-1}(-\infty, x)] \to \mu(\mathcal{S})$ $= 1$.

2. $F(x)$ is nondecreasing, that is, $x < x' \Rightarrow F(x) \leq F(x')$, since $f(x) \geq 0$ and by (10.49),

$$F(x') - F(x) = \int_x^{x'} f(y)\, dy \geq 0.$$

3. $F(-\infty) = 0$, since for any x we have $\mathcal{S} = X^{-1}(-\infty, x] \cup X^{-1}(x, \infty)$. Consequently

$$F(x) = \mu[X^{-1}(-\infty, x]] = 1 - \mu[X^{-1}(x, \infty)],$$

and by item 1 above, $F(x) \to 0$ as $x \to -\infty$.

Remark 10.73 *Note that no comment has been made about the random variable X being a continuous function defined on $\mathcal{S}$. It would seem natural that a continuous probability theory ought to be the probability theory of continuous random variables. But to do so would require that $\mathcal{S}$ has more structure than is guaranteed by the sigma algebra of events. Specifically, in order to be able to define that X is a continuous function requires either that:*

1. *$\mathcal{S}$ is a metric space so X can be defined to be continuous in the usual $\epsilon - \delta$ sense, or equivalently, by the condition that $X^{-1}[G]$ is open for any open $G \subset \mathbb{R}$, or*

2. *$\mathcal{S}$ is a topological space, so X can be defined to be continuous by the condition that $X^{-1}[G]$ is open in $\mathcal{S}$ for any open $G \subset \mathbb{R}$.*

*Since it also must be the case that $X^{-1}[G] \subset \mathcal{E}$ for all intervals and hence all open sets, the sigma algebra must then be defined to contain all the open sets in $\mathcal{S}$. In other words, in order to be able to define X to be a continuous random variable requires that $\mathcal{S}$ have a topology of sets, and that the sigma algebra $\mathcal{E}$ contain all these open sets. In more advanced treatments based on the tools of real analysis, the minimal sigma algebra with this property will be called a **Borel sigma algebra**, and the associated events called **Borel sets**, after **Émile Borel** (1871–1956).*

While this extra structure is needed to define the notion of a continuous random variable, it would not be enough to ensure, without a lot of math tools we have not yet developed, that there is a continuous function $f(x)$ so that (10.48) is satisfied. So this development is circumvented and continuous probability theory is, in effect, defined as the probability theory of random variables with continuous probability density functions.

10.11.2 Expectations of Continuous Distributions

The general structure of the formulas below will be seen to be analogous to those in section 7.5.1. These formulas again represent what are known as **expected value calculations**, and sometimes referred to as **taking expectations**. The general structure of this calculation is defined first and then specific examples are presented.

Definition 10.74 *Given a continuously distributed random variable $X : \mathcal{S} \to \mathbb{R}$ with continuous probability density function $f(x)$, and a continuous function $g(x)$ defined on* the range of X, $\mathrm{Rng}[X] \subset \mathbb{R}$, **the expected value of** $g(X)$, denoted $E[g(X)]$, is defined as

$$E[g(X)] = \int_{-\infty}^{\infty} g(x) f(x)\, dx, \tag{10.51}$$

as long as the associated integral is absolutely convergent. In other words, since $f(x) \geq 0$, it is required that

$$\int_{-\infty}^{\infty} |g(x)| f(x)\, dx < \infty. \tag{10.52}$$

If (10.52) is not satisfied, we say that $E[g(X)]$ does not exist.

Remark 10.75

1. If there is a small disappointment in the above definition vis-a-vis the discrete case in definition 7.35, it is that there is no natural counterpart to formula (7.35):

$$E[g(X)] = \sum_{s_j \in S} g(X(s_j)) \Pr(s_j),$$

where $\{x_j\} \subset \mathbb{R}$ denotes the range of X. In other words, in chapter 7 expected values could be equivalently defined as a calculation on S using the probability measure $\Pr$, or as a calculation on $\mathbb{R}$ using the probability density function $f(x)$. At the moment, without the more general tools of real analysis, there is no way to define $E[g(X)]$ as a calculation on S using the probability measure μ. If there was, we might expect that this definition would look something like:

$$E[g(X)] = \int_S g(X(s))\, d\mu(s),$$

although some amount of work needs to be done to define exactly what such an integral means.

2. The condition in (10.52) is automatically satisfied if $g(x)$ is a bounded function on the range of X, $|g(x)| \leq K$, since then

$$\int_{-\infty}^{\infty} |g(x)| f(x)\, dx \leq K \int_{-\infty}^{\infty} f(x)\, dx = K.$$

So this restriction is "only" important for unbounded functions. That said, in practice we are primarily interested in the expected value of unbounded functions, so this condition cannot in general be assumed to be valid.

*10.11.3   Discretization of a Continuous Distribution

The goal of this section is to better link the notions of expected value in the discrete and continuous contexts. At the moment it might appear that the summations of definition 7.35 of chapter 7 were simply converted to integrals. To see why this is the correct answer, and not just a notational trick, we begin with a somewhat long definition. The idea is simple and natural, but it takes a lot of words to convey.

Definition 10.76   *Given a continuously distributed random variable, $X : S \to \mathbb{R}$, a **discretization** of X **of mesh size** δ, denoted X_δ, is a discrete random variable defined on the **discretization of the sample space** S, denoted S_δ, constructed as follows:*

1. *A **partition** of $\mathbb{R}$ is defined with mesh size, δ. In other words, there is given $\{x_i\}_{i=0}^{\infty}, \{y_i\}_{i=0}^{\infty} \subset \mathbb{R}$, with*

$$\cdots < y_2 < y_1 < y_0 = x_0 < x_1 < x_2 < \cdots,$$

with $x_{i+1} - x_i \leq \delta$ and $y_i - y_{i+1} \leq \delta$ for all i, and the partition is defined as $\{[x_i, x_{i+1})\}_{i=0}^{\infty} \cup \{[y_{i+1}, y_i)\}_{i=0}^{\infty}$.

2. *From each partition interval is chosen a point, or **interval tag**,*

$$\tilde{x}_i \in [x_i, x_{i+1}), \quad \tilde{y}_i \in [y_{i+1}, y_i), \quad i \geq 0.$$

3. *Mutually exclusive events are defined in $\mathcal{E}$ by*

$$A_i^+ = X^{-1}[x_i, x_{i+1}), \quad A_i^- = X^{-1}[y_{i+1}, y_i), \quad i \geq 0.$$

*Then S_δ is defined as the **discrete sample space** in which these events are sample points*

$$S_\delta = \{A_i^+\}_{i=0}^{\infty} \cup \{A_i^-\}_{i=0}^{\infty}, \tag{10.53}$$

*with the **complete collection of events**, denoted $\mathcal{E}_\delta$, defined as these sample points plus all unions and complements of unions of these sample points.*

   *The **probability measure**, $\Pr_\delta$, is then defined on sample points by*

$$\Pr_\delta[A_i^+] = \mu[X^{-1}[x_i, x_{i+1})], \tag{10.54a}$$

$$\Pr_\delta[A_i^-] = \mu[X^{-1}[y_{i+1}, y_i)], \tag{10.54b}$$

and extended additively to all events, where in this definition, $X^{-1}[x_i, x_{i+1})$ and $X^{-1}[y_{i+1}, y_i)$ are considered as events in S.

*Finally, the **discrete random variable** $X_\delta : S_\delta \to \mathbb{R}$ is defined by*

$$X_\delta(A_i^+) = \tilde{x}_i, \quad X_\delta(A_i^-) = \tilde{y}_i, \tag{10.55}$$

*with associated **probability density function**, $f_\delta(x)$, defined by*

$$f_\delta(\tilde{x}_i) \equiv \mathrm{Pr}_\delta[X_\delta^{-1}[\tilde{x}_i]] = F(x_{i+1}) - F(x_i), \tag{10.56a}$$

$$f_\delta(\tilde{y}_i) \equiv \mathrm{Pr}_\delta[X_\delta^{-1}[\tilde{y}_i]] = F(y_i) - F(y_{i+1}). \tag{10.56b}$$

Example 10.77 *Other continuous distributions are introduced below, but the **unit normal distribution** was introduced in section 8.6 and can be discretized as follows. In the background is $(S, \mathcal{E}, \mu)$, representing a sample space, sigma algebra, and probability measure, and also a continuously distributed random variable $X : S \to \mathbb{R}$. So for any interval $\{a, b\}$, which we take to be closed for definitiveness, we have that $X^{-1}[a, b] \in \mathcal{E}$ and*

$$\mu[X^{-1}[a,b]] = \frac{1}{\sqrt{2\pi}} \int_a^b e^{-x^2/2} \, dx.$$

Although not necessary, it is natural to define a discretization that is symmetric in terms of the collection of interval tags since $\phi(x) = \frac{1}{\sqrt{2\pi}} e^{-x^2/2}$ is symmetric about $x = 0$. To this end, and with mesh size $\delta = \frac{1}{n}$, it is notationally convenient to eliminate y_0 and x_0, and define

$$x_i = \frac{2i-1}{2n}, \qquad y_i = -x_i, \, i = 1, 2, 3, \ldots,$$

and associated events in S by

$$A_0 = X^{-1}\left[-\frac{1}{2n}, \frac{1}{2n}\right),$$

$$A_i^+ = X^{-1}\left[\frac{2i-1}{2n}, \frac{2i+1}{2n}\right), \quad A_i^- = X^{-1}\left[-\frac{2i+1}{2n}, -\frac{2i-1}{2n}\right), \qquad i \ge 1.$$

The discrete sample space S_δ is then defined as the collection of sample points as in (10.53) with probability measure Pr_δ as in (10.54). In this case, note that with $\Phi(x)$ denoting the normal cumulative distribution function as defined in (10.49):

$$\mathrm{Pr}_\delta[A_0] = \Phi\left(\frac{1}{2n}\right) - \Phi\left(-\frac{1}{2n}\right),$$

$$\Pr_\delta[A_i^+] = \Phi\left(\frac{2i+1}{2n}\right) - \Phi\left(\frac{2i-1}{2n}\right),$$

$$\Pr_\delta[A_i^-] = \Phi\left(-\frac{2i-1}{2n}\right) - \Phi\left(-\frac{2i+1}{2n}\right).$$

Finally, with interval tags $\{\tilde{x}_i, \tilde{y}_i\}$ defined as interval midpoints,

$$\tilde{x}_0 = 0, \quad \tilde{x}_i = \frac{i}{n}, \quad \tilde{y}_i = -\frac{i}{n}, \qquad i \geq 1,$$

the discretized normal random variable X_δ is defined as

$$X_\delta(A_0) = 0, \quad X_\delta(A_i^+) = \frac{i}{n}, \quad X_\delta(A_i^-) = -\frac{i}{n}, \quad i \geq 1,$$

with probability density function given in (10.56).

We can compare the cumulative distribution functions of the normal and its discretization with $\delta = 0.5$ in **figure 10.4**. *Note that using midpoint tags produced a balanced discretization, in that within each interval of the partition, for example, $[-0.25, 0.25)$ the discretized normal c.d.f. is below $\Phi(x)$ on $[-0.25, 0)$ and above on $[0, 0.25)$.*

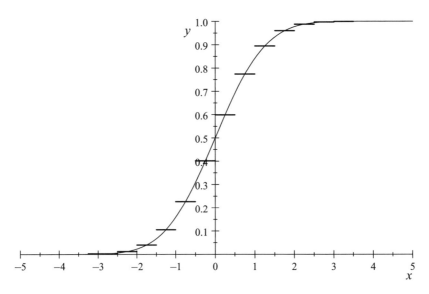

Figure 10.4
$\Phi(x)$ and $\Phi_\delta(x)$ compared for $\delta = 0.5$, midpoint tags

Analogously, left- and right-endpoint tagging produces discretized c.d.f.s that are almost always above, or below, the continuous c.d.f. $\Phi(x)$.

The connection between expected values in a discrete and continuous context can now be formulated by the next result. For notational ease, assume $y_i = -x_i$.

Proposition 10.78   *Given a continuously distributed random variable,* $X : S \to \mathbb{R}$, *and discretizations of* X *of mesh size* δ, X_δ, *defined on* S_δ, *then for* $g(x)$ *a continuous function for which (10.52) holds:*

$$E[g(X_\delta)] \to E[g(X)] \qquad as \ \delta \to 0. \tag{10.57}$$

Proof   By (7.36) applied with (10.56),

$$E[g(X_\delta)] = \sum_{i=0}^{\infty} g(x_i^+)[F(x_{i+1}) - F(x_i)] + \sum_{i=0}^{\infty} g(x_i^-)[F(-x_i) - F(-x_{i+1})].$$

We detail the convergence of the first summation, and leave the analogous derivation for the second summation as an exercise. Now, since $F(x)$ is differentiable, the mean value theorem in (9.22) yields that

$$F(x_{i+1}) - F(x_i) = F'(x_i')\Delta x_i,$$

where $\Delta x_i = x_{i+1} - x_i$ and $x_i' \in (x_i, x_{i+1})$. Hence, because $F'(x_i') = f(x_i')$ by (10.50),

$$\sum_{i=0}^{\infty} g(x_i^+)[F(x_{i+1}) - F(x_i)] = \sum_{i=0}^{\infty} g(x_i^+) f(x_i')\Delta x_i.$$

As $g(x)f(x)$ is a continuous function, it achieves its maximum and minimum in every compact set, and hence in the closure of every interval in the partition. Consequently for every i there exists $x_i^{max}, x_i^{min} \in [x_i, x_{i+1}]$ so that

$$f(x_i^{min})g(x_i^{min}) \le g(x_i^+)f(x_i') \le f(x_i^{max})g(x_i^{max}).$$

As $g(x)f(x)$ is assumed absolutely integrable, it is certainly integrable, and so the Riemann sums defined by either x_i^{max} or x_i^{min} converge to this integral as $\Delta x_i \to 0$. Consequently, as $\delta \to 0$, we have by definition that $\Delta x_i \to 0$ and can conclude that

$$\sum_{i=0}^{\infty} g(x_i^+) f(x_i')\Delta x_i \to \int_0^{\infty} g(x)f(x)\,dx.$$

The same argument can be applied to the second summation in the definition of $E[g(X_\delta)]$, which together produces

$$E[g(X_\delta)] \rightarrow \int_{-\infty}^{\infty} g(x)f(x)\,dx.$$

Finally, this last integral equals $E[g(X)]$ since by assumption, $g(x)$ satisfies (10.52).

■

Remark 10.79   *The proposition above was stated with the relatively strong assumption that $g(x)$ is a continuous function. A review of the proof provides the insight that all that was needed was that: 1) $g(x)$ be continuous except on a set of measure 0 so that $\int_{-\infty}^{\infty} g(x)f(x)\,dx$ is defined, and, 2) $g(x)$ is bounded on every bounded interval, so that we could produce an upper and lower bound for $g(x_i^+)f(x_i')$ on each subinterval.*

10.11.4   Common Expectation Formulas

We now list a collection of expectation formulas which includes the moments of X. As noted above, in each case the expectation is defined only when (10.52) is satisfied. The notation is consistent with that in section 7.5.1.

nth Moment

$$\mu_n' \equiv \int_{-\infty}^{\infty} x^n f(x)\,dx, \qquad n = 1, 2, 3, \ldots \tag{10.58}$$

Mean

$$\mu \equiv \mu_1' = \int_{-\infty}^{\infty} xf(x)\,dx \tag{10.59}$$

nth Central Moment

$$\mu_n \equiv \int_{-\infty}^{\infty} (x - \mu)^n f(x)\,dx, \qquad n = 1, 2, 3, \ldots \tag{10.60}$$

Variance

$$\sigma^2 \equiv \mu_2 = \int_{-\infty}^{\infty} (x - \mu)^2 f(x)\,dx \tag{10.61}$$

Standard Deviation

$$\sigma = \sqrt{\int_{-\infty}^{\infty} (x - \mu)^2 f(x)\, dx} \tag{10.62}$$

Moment-Generating Function
$M_X(t)$ is defined only when the integral is convergent for t in an interval about 0,

$$M_X(t) \equiv \int_{-\infty}^{\infty} e^{xt} f(x)\, dx \tag{10.63}$$

Characteristic Function

$$C_X(t) \equiv \int_{-\infty}^{\infty} e^{ixt} f(x)\, dx \tag{10.64}$$

$C_X(t)$ is defined for all t since by (10.10),

$$|C_X(t)| \leq \int_{-\infty}^{\infty} |e^{ixt}| f(x)\, dx$$

$$= \int_{-\infty}^{\infty} f(x)\, dx = 1,$$

because by Euler's formula in (2.5), $|e^{ixt}| = |\cos xt + i \sin xt| = 1$.

Example 10.80   *All of the many formulas in section 7.5.1 involving expectations can now be shown to be valid in this continuous probability model, except for those that involve probability density functions of two or more variables. As we do not yet have either a differential or an integral calculus for these functions, the continuous counterparts to the formulas relating to the joint, conditional, and marginal probability densities, the law of total probability, sample statistics, or sums of independent and identically distributed random variables must be deferred as an application of multivariate calculus. However, once these tools are developed, these discrete results will again prove to be applicable in this continuous and in even more general settings.*

Examples of formulas that can be derived now follow (see exercises 13 and 32).

1. *As in (7.45):*

$$\sigma^2 = \mathrm{E}[X^2] - \mathrm{E}[X]^2. \tag{10.65}$$

2. *As in exercise* 12 *in chapter* 7:

$$\mu_n = \sum_{j=0}^{n}(-1)^{n-j}\binom{n}{j}\mu_j'\mu^{n-j},$$ (10.66a)

$$\mu_n' = \sum_{j=0}^{n}\binom{n}{j}\mu_j\mu^{n-j}.$$ (10.66b)

3. *As in (7.64) and (7.65):*

$$M_X(t) = \sum_{n=0}^{\infty}\frac{\mu_n' t^n}{n!},$$ (10.67)

$$\mu_n' = M_X^{(n)}(0),$$ (10.68)

if all moments exist.

4. *As in (7.71) and (7.72):*

$$C_X(t) = \sum_{n=0}^{\infty}\frac{\mu_n'(it)^n}{n!},$$ (10.69)

$$\mu_n' = \frac{1}{i^n}C_X^{(n)}(0),$$ (10.70)

if all moments exist.

10.11.5   Continuous Probability Density Functions

As was the case in section 7.6, there are infinitely many continuous probability density functions in theory. Specifically, if $h(x)$ is any continuous function with absolutely convergent integral,

$$0 < \int_{-\infty}^{\infty} |h(x)|\, dx = C < \infty,$$

then a p.d.f. can be defined by

$$f(x) = \frac{|h(x)|}{\int_{-\infty}^{\infty}|h(x)|}.$$

While true in theory, it is another question altogether whether this p.d.f. will be found to be useful or reflective of the probability density of a random variable of interest.

In this section we identify several of the common continuous probability distributions, and some of their properties.

Continuous Uniform Distribution

Perhaps the simplest continuous probability density that can be imagined is one which assumes the same value on every point. The domain of this distribution is arbitrary, and is conventionally denoted as the interval $[a, b]$. The p.d.f. of the **continuous uniform distribution**, sometimes called the **continuous rectangular distribution**, is defined on $[a, b]$ by the density function

$$f_U(x) = \begin{cases} \frac{1}{b-a}, & x \in [a, b], \\ 0, & x \notin [a, b]. \end{cases} \tag{10.71}$$

It is an easy calculation to derive the mean and variance of this distribution:

$$\mu_U = \frac{1}{2}(b + a), \tag{10.72a}$$

$$\sigma_U^2 = \frac{1}{12}(b - a)^2. \tag{10.72b}$$

Similarly the moment-generating function can be calculated from the integral of e^{xt}, producing

$$M_U(t) = \frac{e^{bt} - e^{at}}{t(b - a)}, \qquad t \in \mathbb{R}. \tag{10.73}$$

Although $M_U(t)$ has an apparent singularity at $t = 0$, the numerator can be expanded in a Taylor series, and one finds that

$$M_U(t) = 1 + \sum_{n=2}^{\infty} \left(\frac{b^n - a^n}{b - a} \right) \frac{t^{n-1}}{n!},$$

which converges for all t.

Letting $a = 0$ and $b = 1$ in these formulas produces the limiting results as $n \to \infty$ of the discrete rectangular distribution developed in section 7.6.1. Moreover the discrete rectangular distribution can be seen to be a discretization of this continuous distribution, with $\delta = \frac{1}{n}$ and right-endpoint tags.

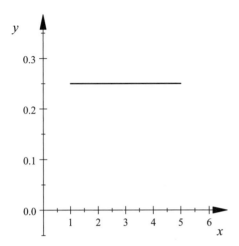

Figure 10.5
$f_U(x) = \frac{1}{4}, 1 \le x \le 5$

An example of this density function is seen in **figure 10.5**.

Beta Distribution

The **beta distribution** contains two shape parameters $v > 0$ and $w > 0$, and it is defined on the interval $[0, 1]$ by the density function

$$f_\beta(x) = \frac{x^{v-1}(1-x)^{w-1}}{B(v, w)}. \tag{10.74}$$

Here the **beta function** $B(v, w)$ is one of the "special functions" in mathematics defined by a definite integral that, in general, requires numerical evaluation:

$$B(v, w) = \int_0^1 y^{v-1}(1-y)^{w-1} \, dy. \tag{10.75}$$

By definition, therefore $\int_0^1 f_\beta(x) \, dx = 1$.

If v or w or both parameters are less than 1, the beta density is unbounded at $x = 0$ or $x = 1$, or both, and this integral converges as an improper integral discussed in section 10.6 because the exponent of both x and $1 - x$ exceeds -1. If both parameters are greater than 1, this density function is 0 at the interval endpoints, and by the methods of section 9.5 has a unique maximum at $x = \frac{v-1}{v+w-2}$. Examples

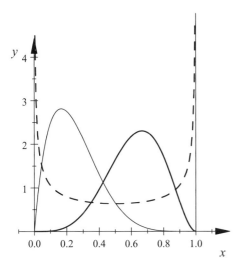

Figure 10.6
$$f_\beta(x) = \frac{x^{v-1}(1-x)^{w-1}}{B(v,w)}$$

of this density function are displayed in **figure 10.6**. In this figure the parameters are
defined by

$$(v,w) = \begin{cases} (0.5, 0.5), & \text{dashed line,} \\ (2,6), & \text{light line,} \\ (5,3), & \text{bold line.} \end{cases}$$

By definition, one has for any positive integer n,

$$E[x^n] = \frac{B(v+n, w)}{B(v, w)}.$$

Now the beta function $B(v, w)$ satisfies an important identity that is useful in evaluating moments of this distribution:

$$B(v+1, w) = \frac{v}{v+w} B(v, w), \tag{10.76}$$

which is exercise 14.

Applying the iterative formula in (10.76), we have

$$\mu_\beta = \frac{v}{v+w}, \tag{10.77a}$$

$$\mu'_{n\beta} = \prod_{i=0}^{n-1} \left(\frac{v+i}{v+w+i} \right), \tag{10.77b}$$

$$\sigma_\beta^2 = \frac{vw}{(v+w)^2(v+w+1)}. \tag{10.77c}$$

Using this same iterative formula, we derive by mathematical induction that if n, m are positive integers,

$$B(n,m) = \frac{(n-1)!(m-1)!}{(n+m-1)!}, \tag{10.78}$$

which is exercise 33.

Exponential Distribution

The **exponential distribution** is defined on $[0, \infty)$, and with a single scale parameter $\lambda > 0$, by the density function:

$$f_E(x) = \lambda e^{-\lambda x}. \tag{10.79}$$

It is apparent that $\int_0^\infty f_E(x)\, dx = 1$ as an improper integral for any $\lambda > 0$, that $f_E(0) = \lambda$ and that $f_E(x)$ is strictly decreasing over $[0, \infty)$. This distribution is a special case of the gamma distribution discussed next.

Gamma Distribution

The **gamma distribution** is defined on $[0, \infty)$, reflects a scale parameter $b > 0$ and a shape parameter $c > 0$, and is given by the density function

$$f_\Gamma(x) = \frac{1}{b} \left(\frac{x}{b} \right)^{c-1} \frac{e^{-x/b}}{\Gamma(c)}. \tag{10.80}$$

As in the case of the beta distribution, the **gamma function** $\Gamma(c)$ is another "special function" defined by the integral

$$\Gamma(c) = \int_0^\infty y^{c-1} e^{-y}\, dy, \qquad c > 0. \tag{10.81}$$

When $c = 1$ and $b = \frac{1}{\lambda}$, the gamma density function is the **exponential density** function noted above.

The gamma function exists as an improper integral, and for its evaluation both the unboundedness of the interval and, in the case of $c < 1$, the unboundedness of the

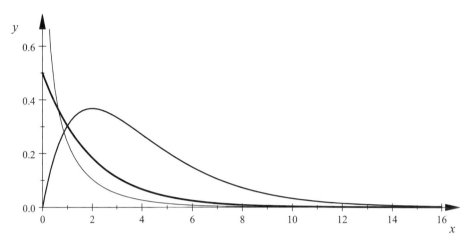

Figure 10.7
$f_\Gamma(x) = \frac{1}{b}\left(\frac{x}{b}\right)^{c-1}\frac{e^{-x/b}}{\Gamma(c)}$

integrand near $x = 0$ must be considered. The fact that $\int_0^\infty f_\Gamma(x)\,dx = 1$ then follows from the substitution $y = \frac{x}{b}$ and (10.81).

When $c \le 1$, the gamma density is a strictly decreasing function, since $f_\Gamma'(x) < 0$, whereas for $c > 1$, the gamma density has a unique maximum at $x = b(c-1)$. Also, as noted above, when $c < 1$ the gamma density is unbounded at $x = 0$. Gamma densities are displayed in **figure 10.7** for various parameters. In particular, the density displayed with a bold line is an exponential, with $\lambda = 0.5$:

$$(b,c) = \begin{cases} (2,0.5), & \text{light line,} \\ (2,2), & \text{medium line,} \\ (2,1), & \text{bold line.} \end{cases}$$

The gamma function $\Gamma(c)$ satisfies an iterative formula that is useful for generating moments of this distribution, and which follows from an integration by parts:

$$\Gamma(c) = (c-1)\Gamma(c-1). \tag{10.82}$$

From the substitution $y = \frac{x}{b}$,

$$E[x^n] = \frac{b^n \Gamma(c+n)}{\Gamma(c)},$$

and this iterative formula produces

$$\mu_\Gamma = bc, \tag{10.83a}$$

$$\mu'_{n\Gamma} = b^n \prod_{j=0}^{n-1} (c + j), \tag{10.83b}$$

$$\sigma_\Gamma^2 = b^2 c. \tag{10.83c}$$

The moment-generating function can also be calculated (see exercise 15):

$$M_\Gamma(t) = (1 - bt)^{-c}, \qquad |t| < \frac{1}{b}. \tag{10.84}$$

As noted above, when $c = 1$ and $b = \frac{1}{\lambda}$, the gamma density function becomes the exponential density function, and so the moment and m.g.f. formulas above are easily converted to that case.

The gamma function $\Gamma(c)$ satisfies $\Gamma(1) = 1$ by direct integration, so mathematical induction can be used with (10.82) to prove that for any positive integer n,

$$\Gamma(n) = (n - 1)!, \tag{10.85}$$

and so $\Gamma(c)$ can be seen to be a continuous generalization of the discrete **factorial function** for $c > 0$. This factorial identity is also the motivation behind defining $0! = 1$, which makes perhaps little sense directly. However, considering $\Gamma(c)$ as a generalization of this discrete function, the statement $0! = 1$ really means that by (10.85), $0!$ is defined in terms of the gamma function, and so

$$0! \equiv \Gamma(1) = 1.$$

Cauchy Distribution

The **Cauchy distribution**, named for **Augustin Louis Cauchy** (1789–1857), is of interest as an example of a p.d.f. that has no finite moments. This density function is defined on $\mathbb{R}$ as a function of a location parameter, $x_0 \in \mathbb{R}$, and a scale parameter $\lambda > 0$, by

$$f_C(x) = \frac{1}{\pi\lambda} \frac{1}{1 + \left(\frac{x - x_0}{\lambda}\right)^2}. \tag{10.86}$$

This function is symmetric about $x = x_0$, at which point $f_C(x_0) = \frac{1}{\pi\lambda}$, the density's maximum value. The parameter λ is a scaling parameter that determines how quickly (λ small) or how slowly (λ large) $f_C(x)$ decreases from this maximum as $|x - x_0| \to \infty$.

When $x_0 = 0$ and $\lambda = 1$, this function is the probability density of a ratio of independent unit normal random variables, but we do not derive this.

That $\int_{-\infty}^{\infty} f(x)\,dx = 1$ as an improper integral follows from two substitutions. First off, substituting $y = \frac{x - x_0}{\lambda}$ produces

$$\int_{-\infty}^{\infty} f_C(x)\,dx = \frac{1}{\pi} \int_{-\infty}^{\infty} \frac{1}{1 + y^2}\,dy.$$

The second substitution is $y = \tan z$, which produces $1 + y^2 = \sec^2 z$. Since $\tan z = \frac{\sin z}{\cos z}$, this function can then be differentiated using the tools of chapter 9, and in particular (9.16), to produce that $(\tan z)' = \sec^2 z$. Finally, this substitution changes the limits of integration from $y \in (-\infty, \infty)$ to $z \in \left(-\frac{\pi}{2}, \frac{\pi}{2}\right)$, so

$$\int_{-\infty}^{\infty} f_C(x)\,dx = \frac{1}{\pi} \int_{-\pi/2}^{\pi/2} dz = 1.$$

This function has no finite moments, even though it would appear by a cancellation argument that $\mu = x_0$. But recall that in order for an expectation to be defined, the associated integral must be absolutely convergent. Simplifying the calculation to $x_0 = 0$ and $\lambda = 1$, which is equivalent to making a substitution of $y = \frac{x - x_0}{\lambda}$, consider

$$\int_{-\infty}^{\infty} |y| f_C(y)\,dy = 2 \int_{0}^{\infty} y f_C(y)\,dy$$

$$= \frac{2}{\pi} \int_{0}^{\infty} \frac{y}{1 + y^2}\,dy.$$

This integral can be explicitly evaluated by the substitution, $z = 1 + y^2$, producing

$$\int_{-\infty}^{\infty} |y| f_C(y)\,dy = \lim_{N \to \infty} \frac{1}{\pi} \int_{1}^{N} \frac{dz}{z}$$

$$= \frac{1}{\pi} \lim_{N \to \infty} [\ln z]\big|_1^N$$

$$= \frac{1}{\pi} \lim_{N \to \infty} \ln N = \infty.$$

So the Cauchy distribution has no finite mean nor higher moments, and hence no moment-generating function. It does have a characteristic function, although (10.70)

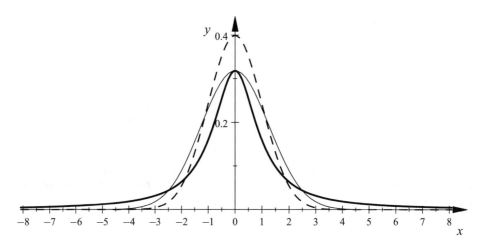

Figure 10.8
$f_C(x) = \frac{1}{\pi} \frac{1}{1+x^2}$, $\phi(x) = \frac{1}{\sigma\sqrt{2\pi}} \exp\left(-\frac{x^2}{2\sigma^2}\right)$

can not be valid here. As it turns out, $C_C(t)$ is not differentiable at $t = 0$ since it is a function of $|t|$.

This density is graphed in bold in **figure 10.8** with $x_0 = 0$ and $\lambda = 1$. For comparison, also graphed are the standard unit normal density (dashed line) and another normal but with $\sigma = \sqrt{\frac{\pi}{2}} \approx 1.2533$ (light line) to have the same maximum value as the Cauchy. The "fat tails" of the Cauchy density are evident.

Normal Distribution

The **normal distribution** is defined on $(-\infty, \infty)$, and depends on a location parameter, $\mu \in \mathbb{R}$, and a scale parameter, $\sigma^2 > 0$, and is defined by the density function, where we use $\exp A \equiv e^A$ to simplify notation:

$$f_N(x) = \frac{1}{\sigma\sqrt{2\pi}} \exp\left(-\frac{(x - \mu)^2}{2\sigma^2}\right). \tag{10.87}$$

When $\mu = 0$ and $\sigma^2 = 1$, this is known as the **unit normal distribution**, and often denoted $\phi(x)$,

$$\phi(x) = \frac{1}{\sqrt{2\pi}} \exp\left(-\frac{x^2}{2}\right), \tag{10.88}$$

introduced in (8.26) with the De Moivre theorem.

The normal density is displayed in **figure 10.8** with $\mu = 0$, for $\sigma = 1$ (dashed line) and $\sigma = 1.2533$ (light line).

The substitution $y = \frac{x-\mu}{\sigma}$ into the integral of $f_N(x)$ shows that this integral equals the integral of $\phi(y)$. Unfortunately, there is no approach to demonstrating that this latter integral has value 1 with the tools currently at our disposal, so a formal proof will be deferred as an application, surprisingly, of multivariate calculus. We simply state the result:

$$\int_{-\infty}^{\infty} \phi(y)\,dy = 1.$$

However, since $\exp\left(-\frac{x^2}{2}\right) < x^{-N}$ as $x \to \infty$ for any N, it is easy to validate that $\int_{-\infty}^{\infty} y^n \phi(y)\,dy < \infty$ for any $n \geq 0$ using the results from section 10.6 on improper integrals.

In general, it is easiest to calculate the central moments μ_{nN} and to use (10.66b) if the corresponding moments μ'_{nN} are needed. To this end, note that using the substitution $y = \frac{x-\mu}{\sigma}$ produces

$$\int_{-\infty}^{\infty} (x - \mu)^n f_N(x)\,dx = \sigma^n \int_{-\infty}^{\infty} y^n \phi(y)\,dy.$$

For n odd, it is apparent that $\int_{-\infty}^{\infty} y^n \phi(y)\,dy = 0$, since with the substitution of $z = -y$ in the second integral,

$$\int_{-\infty}^{\infty} y^n \phi(y)\,dy = \int_{0}^{\infty} y^n \phi(y)\,dy + \int_{-\infty}^{0} y^n \phi(y)\,dy$$

$$= \int_{0}^{\infty} y^n \phi(y)\,dy - \int_{\infty}^{0} (-z)^n \phi(-z)\,dz$$

$$= \int_{0}^{\infty} y^n \phi(y)\,dy - \int_{0}^{\infty} z^n \phi(z)\,dz$$

$$= 0.$$

Here $(-z)^n = -z^n$ since n is odd, $\phi(-z) = \phi(z)$ from (10.88), and the interchange of limits follows from (10.17).

The mean of the normal is easily calculated from this result with the same substitution:

$$\int_{-\infty}^{\infty} x f_N(x)\, dx = \int_{-\infty}^{\infty} (\sigma y + \mu)\phi(y)\, dy$$

$$= \mu,$$

so $\mu_N = \mu$.

For $n = 2m$ even, exercise 34 develops the iterative formula:

$$\int_{-\infty}^{\infty} y^{2m}\phi(y)\, dy = (2m - 1)\int_{-\infty}^{\infty} y^{2m-2}\phi(y)\, dy,$$

and this plus mathematical induction will prove that

$$\int_{-\infty}^{\infty} y^{2m}\phi(y)\, dy = \frac{(2m)!}{2^m m!}.$$

Combining with the above, we derive

$$\mu_{nN} = \begin{cases} 0, & n = 2m + 1, \\ \frac{\sigma^{2m}(2m)!}{2^m m!}, & n = 2m, \end{cases} \tag{10.89a}$$

$$\mu_N = \mu, \tag{10.89b}$$

$$\mu_{2N} \equiv \sigma_N^2 = \sigma^2. \tag{10.89c}$$

So predictably the parameters μ and σ^2 equal the mean and the variance of this distribution.

The final derivation is for the moment-generating function:

$$M_N(t) = \int_{-\infty}^{\infty} e^{tx} f_N(x)\, dx$$

$$= \frac{1}{\sigma\sqrt{2\pi}} \int_{-\infty}^{\infty} \exp\left(-\frac{(x - \mu)^2 - 2\sigma^2 tx}{2\sigma^2} \right) dx.$$

Now **completing the square** produces

$$(x - \mu)^2 - 2\sigma^2 tx = [x - (\mu + \sigma^2 t)]^2 - 2\sigma^2 t\left(\mu + \frac{1}{2}\sigma^2 t \right),$$

and so

$$M_N(t) = \frac{1}{\sigma\sqrt{2\pi}} \exp\left(\mu t + \frac{1}{2}\sigma^2 t^2\right) \int_{-\infty}^{\infty} \exp\left(-\frac{[x - (\mu + \sigma^2 t)]^2}{2\sigma^2}\right) dx.$$

The substitution $y = \frac{x - (\mu + \sigma^2 t)}{\sigma}$ in this integral produces $\int_{-\infty}^{\infty} \phi(y)\, dy$, which equals $\sqrt{2\pi}$, and so

$$M_N(t) = \exp\left(\mu t + \frac{1}{2}\sigma^2 t^2\right). \tag{10.90}$$

Correspondingly for the m.g.f. of the unit normal,

$$M_\Phi(t) = \exp\left(\frac{1}{2}t^2\right). \tag{10.91}$$

An analogous derivation produces the following results for the characteristic function:

$$C_N(t) = \exp\left(i\mu t - \frac{1}{2}\sigma^2 t^2\right), \tag{10.92}$$

$$C_\Phi(t) = \exp\left(-\frac{1}{2}t^2\right). \tag{10.93}$$

Lognormal Distribution

The **lognormal distribution** is defined on $[0, \infty)$, depends on a location parameter $\mu \in \mathbb{R}$ and a shape parameter $\sigma^2 > 0$, and unsurprisingly is intimately related to the normal distribution introduced in section 8.6 and discussed above. However, to some the name "lognormal" appears to be opposite of the relationship that exists. Stated one way, a random variable X is lognormal with parameters (μ, σ^2) if $X = e^Z$ where Z is normal with the same parameters. So X can be understood as an exponentiated normal. Stated another way, a random variable X is lognormal with parameters (μ, σ^2) if $\ln X$ is normal with the same parameters. The name comes from the second statement, in that the log of a lognormal is normal.

The probability density function of the lognormal is defined as follows, again using $\exp A \equiv e^A$ to simplify notation:

$$f_L(x) = \frac{1}{\sigma x \sqrt{2\pi}} \exp\left(-\frac{(\ln x - \mu)^2}{2\sigma^2}\right). \tag{10.94}$$

The substitution $y = \frac{\ln x - \mu}{\sigma}$ produces

$$\int_0^\infty f_L(x)\,dx = \frac{1}{\sqrt{2\pi}} \int_{-\infty}^\infty \exp\left(-\frac{y^2}{2}\right) dy$$

$$= \int_{-\infty}^\infty \phi(y)\,dy.$$

In other words, the integral of the lognormal density over $[0, \infty)$ equals 1.

This density function is well defined at $x = 0$, and $f_L(0) = 0$. To see this, let $x = e^{-y}$ and consider $y \to \infty$. With this transformation,

$$f_L(e^{-y}) = \frac{e^y}{\sigma\sqrt{2\pi}} \exp\left(-\frac{(y+\mu)^2}{2\sigma^2}\right)$$

$$= \frac{1}{\sigma\sqrt{2\pi}} \exp\left(y - \frac{(y+\mu)^2}{2\sigma^2}\right).$$

As $y \to \infty$, it is apparent that $\left[y - \frac{(y+\mu)^2}{2\sigma^2}\right] \to -\infty$, and so $f_L(e^{-y}) \to 0$.

Also the density function $f_L(x)$ has a unique critical point, a maximum, and this is found at $x = \exp(\mu - \sigma^2)$. In **figure 10.9** is displayed the lognormal (bold line) with $\mu = 0$ and $\sigma = 1$. A gamma density is also displayed (thin line), and was engineered to have the same critical point as the lognormal, namely to have a maximum at

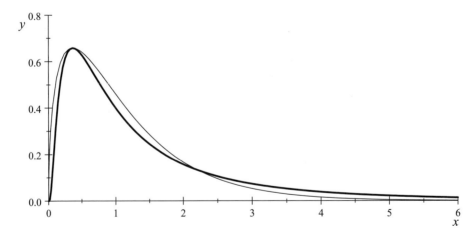

Figure 10.9
$$f_L(x) = \frac{1}{x\sqrt{2\pi}} \exp\left(-\frac{(\ln x)^2}{2}\right), \quad f_\Gamma(x) = \frac{1}{b}\left(\frac{x}{b}\right)^{c-1} \frac{e^{-x/b}}{\Gamma(c)}$$

$x = e^{-1}$ equal to $\frac{1}{e^{-1}\sqrt{2\pi}} \exp\left(-\frac{(\ln e^{-1})^2}{2}\right) \approx 0.65774$. From the analysis above of the gamma, the location of the maximum requires $b(c-1) = e^{-1}$, and the parameters were numerically estimated as

$$c \approx 1.5, \quad b \approx 2e^{-1} = 0.73576.$$

Finally, $f_L(x)$ has moments of all orders. Specifically, using the substitution $y = \frac{\ln x - \mu}{\sigma}$, we write

$$\mu'_{nL} = \int_0^\infty x^n f_L(x) \, dx$$

$$= \int_{-\infty}^\infty \exp(n\sigma y + n\mu)\phi(y) \, dy$$

$$= e^{n\mu} M_\Phi(n\sigma).$$

In other words, the moments of the lognormal can be calculated from the moment-generating function of the unit normal. Specifically, using (10.91), we obtain

$$\mu'_{nL} = e^{n\mu + (n\sigma)^2/2}, \tag{10.95a}$$

$$\mu_L = e^{\mu + \sigma^2/2}, \tag{10.95b}$$

$$\sigma_L^2 = e^{2\mu + \sigma^2}(e^{\sigma^2} - 1). \tag{10.95c}$$

Surprisingly, although the lognormal distribution has moments of all orders, it does not have a convergent moment-generating function. To see this, assume that the m.g.f. exists and by (10.67),

$$M_L(t) = \sum_{n=0}^\infty \frac{\mu'_n t^n}{n!}$$

$$= \sum_{n=0}^\infty \frac{e^{n\mu + (n\sigma)^2/2} t^n}{n!}, \qquad |t| < R.$$

Then as a power series, its interval of convergence is related to the limits superior and inferior of the coefficient ratios as noted in proposition 6.24 on the ratio test.

Letting $c_n = \frac{e^{n\mu + (n\sigma)^2/2}t^n}{n!}$, we have $\limsup_{n\to\infty} \left|\frac{c_{n+1}}{c_n}\right| = \liminf_{n\to\infty} \left|\frac{c_{n+1}}{c_n}\right|$, and so

$$L = \lim_{n\to\infty} \left|\frac{c_{n+1}}{c_n}\right|$$

$$= \lim_{n\to\infty} \left|\frac{e^{\mu + (2n+1)\sigma^2/2}}{n+1}t\right|$$

$$= \infty,$$

for all $t \neq 0$. So by the ratio test this series is divergent for $t \neq 0$, and $M_L(t)$ only exists at $t = 0$. The moments simply grow to fast to allow convergence for any $|t| > 0$.

10.11.6 Generating Random Samples

In chapter 7 was introduced a general approach to generating independent and identically distributed random samples given any discrete probability density function. The proof of this result depended on the structure of the n-**trial sample space**, denoted $\mathcal{S}^n$, which was associated with the original sample space $\mathcal{S}$ on which this random variable was defined. This sample space was endowed with a complete collection of events, denoted $\mathcal{E}^n$, and associated probability measure P_n, each intimately related to the respective notions on $\mathcal{S}$. An independent and identically distributed (i.i.d.) sample of the random variable X could then be defined as stated in proposition 7.60, which we repeat for completeness, with additional clarifying references.

Proposition 10.81 *Let X be a discrete random variable on a sample space $\mathcal{S}$, with range $\{x_k\} \subset \mathbb{R}$, and distribution function $F(x)$. Then, if $\{r_j\}_{j=1}^n \subset [0, 1]$ is a uniformly distributed random sample in the sense of (7.117), then $\{F^{-1}(r_j)\}_{j=1}^n$ is a random sample of X in the sense of (7.7), where $F^{-1}(r_j)$ is defined in (7.118). In other words, if $\{x_{k_j}\}_{j=1}^n \subset \mathrm{Rng}[X]$, then*

$$f(x_{k_1}, x_{k_2}, \ldots, x_{k_n}) = \prod_{j=1}^n f(x_{k_j}).$$

Unfortunately, we do not yet have the necessary tools to generalize this result to the continuous distribution case. However, the discretization result above provides a useful approach which is nearly identical with the theoretical result in practice.

To develop this application, suppose that we are given a continuously distributed random variable X for which we wish to generate i.i.d. random samples of size n. To simplify notation, we assume that $\mathrm{Rng}[X]$ is unbounded in only one direction, say

Rng$[X] \subset [a, \infty)$. Then for any $\delta > 0$ a discretization of X with mesh size δ can be constructed, denoted X_δ, with range $\{\tilde{x}_i\} \subset \mathbb{R}$, with $\tilde{x}_i \in [x_i, x_{i+1})$, and with probability density $f_\delta(x)$ defined as in (10.56) by $f_\delta(\tilde{x}_i) \equiv F(x_{i+1}) - F(x_i)$. Recall that the significance of δ is that $x_{i+1} - x_i \leq \delta$ for all i.

The result above for discrete random variables then assures us that for a uniformly distributed random sample $\{r_j\}_{j=1}^n \subset [0, 1]$, that $\{F_\delta^{-1}(r_j)\}_{j=1}^n$ is independent and identically distributed, so that for any $\{\tilde{x}_{k_j}\}_{j=1}^n \subset$ Rng$[X_\delta]$,

$$f_\delta(\tilde{x}_{k_1}, \tilde{x}_{k_2}, \ldots, \tilde{x}_{k_n}) = \prod_{j=1}^n f_\delta(\tilde{x}_{k_j}).$$

On the other hand, since $f_\delta(\tilde{x}_i) \equiv F(x_{i+1}) - F(x_i)$, we conclude that

$$f_\delta(\tilde{x}_{k_1}, \tilde{x}_{k_2}, \ldots, \tilde{x}_{k_n}) = \prod_{j=1}^n [F(x_{k_j+1}) - F(x_{k_j})]$$

$$= \prod_{j=1}^n \Pr[X \in [x_{k_j}, x_{k_j+1})].$$

In other words, for any discretization of the random variable X, the procedure above provides a methodology for generating i.i.d. random samples of size n that have the correct probability structures. The one compromise in this procedure is that for any interval $[x_i, x_{i+1})$ defined by the discretization, the only value of X that can be sampled is the tagged point $\tilde{x}_i \in [x_i, x_{i+1})$.

In practice, this is of little consequence, since the discretization can be made as fine as desired. For example, in theory one can define a discretization for which δ is smaller that the precision we wish to use in the measurement of the sample points $\tilde{x}_i$. For example, if one wants a random sample with one decimal accuracy, one could choose $\delta = 0.05$, say, or smaller.

10.12 Applications to Finance

10.12.1 Continuous Discounting

A common application of integrals in finance is for interest manipulations with continuous compounding. Given an annual rate, r, the equivalent rate based on compounding m times per year, denoted $r^{(m)}$, is defined in (2.14) by

$$1 + r = \left(1 + \frac{r^{(m)}}{m}\right)^m .$$
(10.96)

The **continuous rate** of compounding is defined as

$$r^{(\infty)} \equiv \lim_{m \to \infty} r^{(m)}.$$

This limit is easily calculated as follows, where we substitute $m = \frac{1}{\Delta x}$ and evaluate the result as $\Delta x \to 0$:

$$r^{(m)} = m[(1 + r)^{1/m} - 1]$$

$$= \frac{(1 + r)^{\Delta x} - 1}{\Delta x}.$$

For $\Delta x \to 0$, we recognize this expression from (9.8) as the derivative of the function $f(x) = (1 + r)^x$ at $x = 0$, which from (9.12) is

$$r^{(\infty)} = \ln(1 + r), \quad \text{or}$$
(10.97a)

$$1 + r = e^{r^{(\infty)}}.$$
(10.97b)

Put another way, the **present value function** with continuous compounding for \$1 at time t is given by $e^{-\delta t}$, while the **accumulated value function** at time t of \$1 at time 0 is $e^{\delta t}$, using the simplifying notation, $\delta \equiv r^{(\infty)}$. This follows from (10.96) by raising each side to $\pm t$, then taking the limit as $m \to \infty$ as above.

An alternative approach to this notion of continuous compounding is to denote by $A(t)$ the value at time t of \$1 invested at time 0, assuming continuous compounding. Then, using an annual rate, $A(t + \Delta t) = A(t)(1 + r)^{\Delta t}$, we conclude that

$$\frac{A(t + \Delta t) - A(t)}{\Delta t} = \left(\frac{(1 + r)^{\Delta t} - 1}{\Delta t}\right) A(t),$$

and from the calculation above conclude that $A(t)$ is a differentiable function and that $A'(t) = \delta A(t)$.

Then from $\frac{A'(t)}{A(t)} = \delta$, and $\frac{A'(t)}{A(t)} = \frac{d}{dt}[\ln A(t)]$, we derive

$$\frac{d}{dt}[\ln A(t)] = \delta,$$

$$\int_0^T \frac{d}{dt}[\ln A(t)]\,dt = \delta T,$$

$$A(T) = A(0)e^{\delta T},$$

where the last step comes from the Fundamental Theorem of Calculus version I:

$$\int_0^T \frac{d}{dt}[\ln A(t)]\,dt = \ln A(T) - \ln A(0) = \ln \frac{A(T)}{A(0)}.$$

Naturally one does not need continuous compounding for discrete cash flows, but this provides a framework for considering the value of a continuously paid cash flow stream. A continuous function $C(t)$ represents a **continuously payable cash flow** stream if over any interval of time $[a,b]$ the total cash paid is given by

$$C(a,b) = \int_a^b C(t)\,dt.$$

The function $C(t)$ represents the "annualized" rate of payment at time t, in that the amount of cash payable over $[t, t + \Delta t]$ is approximately $C(t)\Delta t$. This follows from the first MVT for integrals in (10.12), which can be restated so that for $t' \in [t, t + \Delta t]$,

$$\int_t^{t+\Delta t} C(s)\,ds = C(t')\Delta t.$$

Also this integral is approximated by $C(t)\Delta t$, a single term of a Riemann sum for Δt small.

The present value at time a, or accumulated value at time b, given continuous compounding at rate δ, then proceeds by starting with a discrete approximation, and recognizing the Riemann integral in the limit. For example, the present value calculation requires cash flow over $[t, t + \Delta t]$, which equals $C(t')\Delta t$, to be discounted to time a, and this is approximated by a factor of $e^{-\delta(t'-a)}$. So with $\Delta t = \frac{b-a}{n}$, we have a partition defined by $\{a + j\Delta t\}_{j=0}^n$ and subinterval tags denoted $t'_j \in (a + (j-1)\Delta t, a + j\Delta t)$:

$$PV_{[a,b]}[C(t)] = \lim_{\Delta t \to 0} \sum_{j=0}^{n-1} C(t'_j)e^{-\delta(t'_j - a)}\Delta t.$$

In other words,

$$PV_{[a,b]}[C(t)] = \int_a^b C(t)e^{-\delta(t-a)}\, dt. \tag{10.98}$$

When $C(t) = C$, a constant cash flow stream, we get

$$PV_{[a,b]}[C] = C\left[\frac{1 - e^{-\delta(b-a)}}{\delta}\right]. \tag{10.99}$$

Similarly the accumulated value of this cash flow stream requires cash flow over $[t, t + \Delta t]$ to be accumulated to time b, and this is approximated by a factor of $e^{\delta(b-s')}$:

$$AV_{[a,b]}[C(t)] = \lim_{\Delta t \to 0} \sum_{j=0}^{n-1} C(s_j')e^{\delta(b-s_j')}\Delta t,$$

which is to say

$$AV_{[a,b]}[C(t)] = \int_a^b C(t)e^{\delta(b-t)}\, dt. \tag{10.100}$$

When $C(t) = C$, a constant cash flow stream, we get

$$AV_{[a,b]}[C] = C\left[\frac{e^{\delta(b-a)} - 1}{\delta}\right]. \tag{10.101}$$

Note that in general,

$$AV_{[a,b]}[C(t)] = e^{\delta(b-a)}PV_{[a,b]}[C(t)], \tag{10.102}$$

a formula that simply adjusts the valuation from $t = a$ to $t - b$.

10.12.2   Continuous Term Structures

In chapter 3 was introduced discrete interest rate term structure models, whereby based on market observations, one calculates the term structure in one or all of the available bases of bond yields, spot rates, or forward rates. In this section this model is generalized to a continuous framework.

Bond Yields
Although mathematically possible, the continuous counterpart to the bond yield structure is rarely used in practice, since to be meaningful, a continuous bond yield at each time t, denoted i_t say, would represent the bond yield on a t-period

bond that paid coupons continuously at rate r_t say. Generalizing (3.36) using (10.99), we obtain the price of this bond, P_t, for a par amount of F_t, is given by

$$P_t = F_t r_t \left[\frac{1 - e^{-i_t t}}{i_t} \right] + F_t e^{-i_t t}. \tag{10.103}$$

Note that the annuity symbol

$$\bar{a}_{t; i_t} \equiv \frac{1 - e^{-i_t t}}{i_t} \tag{10.104}$$

appears to be the continuous counterpart to the discrete formula in chapter 2 in (2.11) for continuous interest rates. But it is important to understand that the continuity of the cash flows is explicitly reflected in (10.104), and that this formula is not equivalent to the formula in (2.11) under the assumption that the rate alone is continuous. Indeed, by (10.97), if $r = i_t$ denotes a continuous rate, and $n = t$ is assumed an integer, the formula from chapter 2 becomes

$$a_{n; i_t} \equiv \frac{1 - e^{-i_t t}}{e^{i_t} - 1}.$$

Both annuity factors reflect the present value of payment streams of 1 per year using a continuous rate of interest. But $\bar{a}_{t; i_t}$ treats this payment as made continuously, while $a_{n; i_t}$ treats this payment as a lump sum at the end of each year. So intuitively $\bar{a}_{t; i_t} > a_{n; i_t}$, since cash is received earlier. More formally, the continuous cash flow underlying $\bar{a}_{t; i_t}$ can be accumulated to the end of each year with (10.101), producing

$$AV[1] = \int_0^1 e^{i_t(1-t)} \, dt = \frac{e^{i_t} - 1}{i_t}.$$

Logically $\bar{a}_{t; i_t}$ should then equal the value of an annual payment annuity, which pays $AV[1]$ at the end of each year, and not surprisingly, we have

$$\bar{a}_{t; i_t} = \frac{e^{i_t} - 1}{i_t} a_{n; i_t}.$$

Forward Rates

It is often convenient to assume that continuous spot rates and forward rates are continuously denominated in time, and denoted s_t and f_t respectively. This is motivated by an interest in developing models of future rates that evolve "stochastically" in continuous time, which is to say randomly, and an interest in what these models can tell us about today's pricing of bonds. These stochastic pricing topics are quite advanced for the tools developed up to this point. But we can develop the relationship between the continuous forward term structure model at a given point in time and the prices of bonds.

Imagine a model specification for **continuous forward interest rates** f_t for $t > 0$. Intuitively this means that the present value at time t, of 1 payable at time $t + \Delta t$, is approximately $e^{-\Delta t f_t}$. Extending this idea, the present value at time 0 of 1 payable at time T is approximately

$$Z_T \approx \exp\left[-\sum_{j=0}^{n-1} f_{j\Delta t}\Delta t\right],$$

where Z_T denotes the price of this T-period zero coupon bond, and with $\Delta t = \frac{T}{n}$.

It is apparent that if the f_t model is continuous, which is more than is needed but often the case in models in practice, this approximate price converges as $\Delta t \to 0$. Specifically, the price of a T-period **zero-coupon bond**, given a continuous forward rate specification, satisfies

$$Z_T = \exp\left[-\int_0^T f_t\, dt\right]. \tag{10.105}$$

Since a fixed cash flow coupon bond is just a portfolio of zero-coupon bonds, the formula in (10.105) can also be used for these bonds, generalizing (3.39).

So given a model specification for forward rates, one can price fixed cash flow bonds using this formula. Of course, in practice, such models do not produce a single specification of this structure, since if that is all one wants, that can generally be observed in today's financial markets. The goal of such models is to produce randomly generated collections of future forward rate "paths," a sample space of such paths, on which one can then interpret Z_T as a random variable. Once done, an entire theory exists for translating these statistical distributions of rate paths and prices into logical prices for today's fixed and variable cash flow securities that rely on these future rates. This is an advanced subject that requires the tools of stochastic processes.

Fixed Income Investment Fund

A model of interest rates can also be interpreted in the context of providing investment returns in a fund, such as a money market fund, in which the forward rate f_t

is earned from time t to time $t + \Delta t$. Again, assuming that such a rate path is continuous, if A_t denotes the fund balance at time t, then

$$A_{t+\Delta t} \approx A_t e^{f_t \Delta t}.$$

Consequently, since $\frac{A_T}{A_0} = \prod_{j=0}^{n-1} \frac{A_{(j+1)\Delta t}}{A_{j\Delta t}}$, where $\Delta t = \frac{T}{n}$, we conclude that

$$\frac{A_T}{A_0} \approx \exp\left[\sum_{j=0}^{n-1} f_{j\Delta t} \Delta t\right].$$

As above, if f_t is a continuous function, this summation converges as $\Delta t \to 0$, producing the **investment fund model**

$$A_T = A_0 \exp\left[\int_0^T f_t\, dt\right]. \tag{10.106}$$

This model makes sense in both a "deterministic" setting, where a forward rate path is specified, or in a statistical context, where various rate paths are generated and the resulting fund balance, A_T for fixed T, is treated as a random variable on the sample space of rate paths.

Because of the Fundamental Theorem of Calculus version II, A_T is a differentiable function of T when f_t is continuous, and we have that

$$\frac{dA_T}{dT} = f_T A_T.$$

In this interpretation the instantaneous change in the fund balance at time T is proportional to the fund balance, with proportionality factor of f_T. This formula is sometimes expressed in the differential notation:

$$dA_t = f_t A_t\, dt. \tag{10.107}$$

This notation is best understood in the context of integration theory, as was seen in section 10.7.1 on integration by the substitution method. In other words, a differential of a function is a mathematical object one integrates to determine how a function changes. Now, if we simply integrate both sides of this equation, we get

$$\int_0^T dA_t = \int_0^T f_t A_t\, dt,$$

which doesn't appear very promising. Logically, the left-hand side is the integral of 1, and so

$$\int_0^T dA_t = A_t\big|_{t=0}^T = A_T - A_0.$$

But the right-hand side is not readily evaluated.

But if we first divide the equation in (10.107) by A_t, which is justified since $A_t > 0$, and then integrate, we get

$$\int_0^T \frac{dA_t}{A_t} = \int_0^T f_t\, dt.$$

The left-hand integral is now

$$\int_0^T \frac{dA_t}{A_t} = \ln A_t\big|_{t=0}^T = \ln \frac{A_T}{A_0},$$

and when equated to the right-hand integral, (10.106) is reproduced.

Spot Rates

If s_T denotes the **continuous spot rate** for term T, it must be the case that in addition to (10.105), we have by definition,

$$Z_T = \exp[-T s_T], \tag{10.108}$$

and hence from (10.105),

$$s_T = \frac{1}{T} \int_0^T f_t\, dt. \tag{10.109}$$

Recall the first mean value theorem for integrals in proposition 10.27. The continuous spot rate at time T is seen to equal the average value of the continuous forward rates over the interval $[0, T]$.

This continuous spot-forward relationship can be reversed with the help of the Fundamental Theorem of Calculus version II above. First, note that if f_t is continuous, then s_T is a differentiable function of T for $T > 0$, since it is the product of $\frac{1}{T}$, which is differentiable for $T \neq 0$, and $\int_0^T f_t\, dt$, which is differentiable by this theorem. Also

$$\frac{ds_T}{dT} = \frac{-1}{T^2} \int_0^T f_t\, dt + \frac{1}{T} f_T,$$

which can be rewritten as

$$\frac{ds_T}{dT} = \frac{1}{T}(f_T - s_T) \tag{10.110}$$

and also

$$f_T = \frac{d(Ts_T)}{dT}. \tag{10.111}$$

This analysis allows some easy conclusions based on (10.110) and chapter 9 tools:

1. Spot rates increase as a function of t if and only if $\frac{ds_t}{dt} > 0$ for all t, which occurs if and only if $f_t > s_t$ for all t.

2. Spot rates decrease as a function of t if and only if $\frac{ds_t}{dt} < 0$ for all t, which occurs if and only if $f_t < s_t$ for all t.

3. If spot rates increase then decrease, or conversely, there is a time t_0 so that $\frac{ds_t}{dt}\big|_{t_0} = 0$, and hence $f_{t_0} = s_{t_0}$.

Note further that from (10.111) we can conclude that $f_T > 0$ for all T if and only if the function $g(T) = Ts_T$ is a strictly increasing function of T. It is not necessary to have s_T an increasing function in order for $f_T > 0$. Indeed, $\frac{d(Ts_T)}{dT} > 0$ simply implies that $\frac{ds_T}{dT} > -\frac{s_T}{T}$.

10.12.3   Continuous Stock Dividends and Reinvestment

The analysis above for a fixed income fund carries over readily to the context of an equity fund. Specifically, if R_t denotes the equity fund return at time t, then with the same derivation, we have that with E_T denoting the fund balance at time T:

$$E_T = E_0 \exp\left[\int_0^T R_t\, dt\right]. \tag{10.112}$$

As before, E_T is a differentiable function of T when R_t is a continuous function, so this composite function can be differentiated and expressed in differential notation as

$$dE_t = R_t E_t\, dt. \tag{10.113}$$

Now, it is often assumed that such an equity will pay continuous cash dividends, and that these dividends are continuously reinvested in more equity. By continuous dividends is meant that if D_t denotes the rate of dividend payout at time t, the total change in value to the investor at time t is approximately

Total return $\approx (R_t E_t + D_t E_t)\Delta t.$

The investor receives $R_t E_t \Delta t$ as appreciation/depreciation in the fund, and $D_t E_t \Delta t$ in cash dividends.

In this form it is difficult to model total investor wealth at some time in the future. Although this cash could be invested in a risk free asset like a T-bill, the position in the T-bill is not risk free, since the principal flows into that fund, $D_t E_t \Delta t$, reflect the riskiness of the equity fund. Because it is common to want to partition total investments between risk assets and risk-free assets, for example when one is replicating a option, there is a motivation to reinvest these dividends in stock, rather than to accumulate this risky asset in T-bills.

With that goal, we now seek to determine the total value of the fund when dividends are so reinvested. To this end, let E_t again denote the value of the equity fund when dividends are disbursed in cash to the investor, and let F_t denote the value of this fund when all dividends are continuously reinvested in more stock. Logically the change in the total fund, $F_{t+\Delta t} - F_t$, reflects two components:

1. An increment or decrement based on the performance of the equities, as implied by R_t, which can be captured by the return in the E fund, scaled to reflect the assets in the F fund:

$$\frac{[E_{t+\Delta t} - E_t]}{E_t} F_t.$$

2. An increment, since $D_t \geq 0$, based on the payment of continuous cash dividends on the total fund balance of F_t, equal to

$$F_t D_t \Delta t,$$

that are then reinvested in more equities in the F fund.

Combining, we derive that

$$F_{t+\Delta t} - F_t = \frac{[E_{t+\Delta t} - E_t]}{E_t} F_t + F_t D_t \Delta t,$$

or

$$\frac{F_{t+\Delta t} - F_t}{\Delta t F_t} = \frac{E_{t+\Delta t} - E_t}{\Delta t E_t} + D_t.$$

As $\Delta t \to 0$, the limit on the right side of the equation exists because E_t is differentiable as noted above. Consequently F_t is also differentiable and

$$\frac{F_t'}{F_t} = \frac{E_t'}{E_t} + D_t.$$

Now $\frac{E_t'}{E_t} = R_t$ is continuous by assumption, as is D_t, and hence so too is $\frac{F_t'}{F_t}$. Integrating this expression from $t = 0$ to $t = T$, and recalling that $\frac{d \ln f(x)}{dx} = \frac{f'(x)}{f(x)}$, we obtain

$$\ln\left[\frac{F_T}{F_0}\right] = \ln\left[\frac{E_T}{E_0}\right] + \int_0^T D_t \, dt.$$

Finally, assuming that $F_0 = E_0$, so both funds begin with the same level of assets, we obtain that

$$F_T = E_T \exp\left[\int_0^T D_t \, dt\right]. \tag{10.114}$$

When $D_t = D$ is constant, this simplifies to

$$F_T = E_T e^{DT}. \tag{10.115}$$

Combining (10.114) with (10.112), we obtain

$$F_T = E_0 \exp\left[\int_0^T (R_t + D_t) \, dt\right]. \tag{10.116}$$

10.12.4 Duration and Convexity Approximations

In section 9.8.5 Taylor series approximations were applied to model the price sensitivity of a fixed income security or portfolio. Using the tools of this chapter, we develop an alternative price sensitivity model.

Recall the definition of the duration of the price function in (9.57):

$$D(r) = -\frac{P'(r)}{P(r)}.$$

Assuming continuity of $D(r)$ and $P(r) > 0$, we can integrate this expression from i_0 to i, obtaining

$$\int_{i_0}^{i} D(r)\, dr = -\int_{i_0}^{i} \frac{P'(r)}{P(r)}\, dr$$

$$= -\ln\left[\frac{P(i)}{P(i_0)}\right].$$

A little algebra then provides the identity

$$P(i) = P(i_0)e^{-\int_{i_0}^{i} D(r)\, dr}, \tag{10.117}$$

which can be transformed to an approximation formula with a one-step Riemann sum:

$$P(i) \approx P(i_0)e^{-D(i_0)(i-i_0)}. \tag{10.118}$$

This approximation can then be improved by analyzing the function in the exponential in (10.117):

$$f(i) = \int_{i_0}^{i} D(r)\, dr,$$

and applying the Fundamental Theorem of Calculus version II in (10.20), and then (9.65), to obtain

$$f'(i) = D(i), \quad f''(i) = D^2(i) - C(i).$$

Expanding the second-order Taylor series of $f(i)$ about i_0, and noting that $f(i_0) = 0$, we obtain an improvement to (10.118):

$$P(i) \approx P(i_0)e^{-D(i_0)(i-i_0)-(1/2)[D^2(i_0)-C(i_0)](i-i_0)^2}. \tag{10.119}$$

It is interesting to compare the approximations above to those developed in chapter 9. To this end, if we apply the formula for an exponential power series in (7.63) to the approximation in (10.118), we obtain

$$P(i) \approx P(i_0)\left[1 - D(i_0)(i - i_0) + \frac{1}{2}D^2(i_0)(i - i_0)^2\right] + O(\Delta i^3).$$

Expanding (10.119) in the same way obtains

$$P(i) \approx P(i_0)\left[1 - D(i_0)(i - i_0) + \frac{1}{2}C(i_0)(i - i_0)^2\right] + O(\Delta i^3).$$

So to an error of $O(\Delta i^3)$, (10.119) provides the same result as the second-order Taylor approximation in (9.61). The same can be said for (10.118) and (9.60) to an error of $O(\Delta i^2)$. However, for price functions with positive convexity, (10.118) will generally provide better approximations using only $D(i_0)$ than will (9.60) because of the $\frac{1}{2}D^2(i_0)(i-i_0)^2$ adjustment above.

Finally, as an application of Riemann sums, we demonstrate that by partitioning the interval $[i_0, i]$, and applying the simple approximation in (9.60) to each sub-interval, that in the limit the identity in (10.117) is produced. To this end, define $i_j = i_0 + \frac{j}{n}\Delta i$ for $j = 0, 1, \ldots, n$, where $\Delta i = i - i_0$. Apparently,

$$\frac{P(i)}{P(i_0)} = \prod_{j=1}^{n} \frac{P(i_j)}{P(i_{j-1})},$$

and each factor in this product can be approximated by (9.60):

$$\frac{P(i_j)}{P(i_{j-1})} = 1 - D(i_{j-1})\frac{\Delta i}{n} + O\left(\frac{1}{n^2}\right).$$

Consequently

$$\prod_{j=1}^{n}\left[\frac{P(i_j)}{P(i_{j-1})}\right] = \prod_{j=1}^{n}\left[1 - D(i_{j-1})\frac{\Delta i}{n} + O\left(\frac{1}{n^2}\right)\right],$$

and by assuming that $D(i)$ is continuous and hence bounded on this interval, we conclude that all factors in this product are positive for n large enough, justifying the taking of natural logarithms. This produces

$$\ln\left[\prod_{j=1}^{n}\left[1 - D(i_{j-1})\frac{\Delta i}{n} + O\left(\frac{1}{n^2}\right)\right]\right] = \sum_{j=1}^{n}\ln\left[1 - D(i_{j-1})\frac{\Delta i}{n} + O\left(\frac{1}{n^2}\right)\right]$$

$$= -\sum_{j=1}^{n} D(i_{j-1})\frac{\Delta i}{n} + O\left(\frac{1}{n}\right),$$

using (8.20). Note that in this calculation, although the error in each logarithmic power series is $O\left(\frac{1}{n^2}\right)$, this error increases to $O\left(\frac{1}{n}\right)$ because there are n terms in the summation.

Letting $n \to \infty$, the last expression converges as a Riemann sum to the integral of the continuous function, $D(i)$. In other words,

$$\ln\left[\prod_{j=1}^{n}\left[1 - D(i_{j-1})\frac{\Delta i}{n} + O\left(\frac{1}{n^2}\right)\right]\right] \to -\int_{i_0}^{i} D(r)\,dr.$$

Now, since $g(x) = e^x$ is a continuous function, and hence sequentially continuous, we can exponentiate this sequence and limit to obtain that as $n \to \infty$:

$$\frac{P(i)}{P(i_0)} = \prod_{j=1}^{n}\left[1 - D(i_{j-1})\frac{\Delta i}{n} + O\left(\frac{1}{n^2}\right)\right] \to e^{-\int_{i_0}^{i} D(r)\,dr}.$$

10.12.5 Approximating the Integral of the Normal Density

The unit normal density function was introduced in section 8.6 and studied in more detail in section 10.11.5. As given in (10.88), it is defined as

$$\phi(x) = \frac{1}{\sqrt{2\pi}}e^{-x^2/2}.$$

It was shown to be a critically important function in chapter 8 due to the De Moivre–Laplace theorem, and the more general central limit theorem, and this is even more so because the statement and proof of the latter result can be generalized much further. The tools that will be needed to generalize this result relate to properties of multivariate functions that will be used in the study of independent, identically distributed random variables, as well as a more general integration theory and probability theory.

In this section we apply some of the results studied above to the question of approximating integrals of $\phi(x)$. Of course, if X is a random variable with density function $\phi(x)$, then

$$\Pr[a \le X \le b] = \int_a^b \phi(x)\,dx.$$

On the other hand, if Y is a random variable with density function equal to the more general $f_N(y)$, then by a substitution $x = \frac{y-\mu}{\sigma}$,

$$\Pr[c \le Y \le d] = \int_c^d f_N(y)\,dy = \int_a^b \phi(x)\,dx,$$

where $a = \dfrac{c - \mu}{\sigma}$, $b = \dfrac{d - \mu}{\sigma}$.

Consequently, all probability statements about Y can be translated to probability statements about X, and so it is the integral of $\phi(x)$ that is addressed in this section.

As was noted in section 8.6, the most common probability values to develop are of the form

$$\Phi(b) = \int_{-\infty}^{b} \phi(x)\,dx, \qquad b > 0,$$

since all other statements can be derived from these as seen in (8.32). However, if $b > 0$, it is apparent since $\Phi(0) = 0.5$, that

$$\Phi(b) = 0.5 + \int_{0}^{b} \phi(x)\,dx, \qquad b > 0,$$

and only integrals of the form $\int_{0}^{b} \phi(x)\,dx$, for $b > 0$ need be addressed.

Power Series Method

As was seen in section 10.9.5 on the integrability of power series, a power series can be integrated term by term over any subinterval of its interval of convergence. Since $\phi(x)$ is an analytic function which converges for all x, this approach can be utilized over any interval. To this end, recall that by the Taylor series expansion of the exponential function,

$$\phi(x) = \frac{1}{\sqrt{2\pi}} \sum_{j=0}^{\infty} \frac{\left(-\frac{1}{2}x^2\right)^j}{j!}$$

$$= \frac{1}{\sqrt{2\pi}} \sum_{j=0}^{\infty} \frac{(-1)^j x^{2j}}{2^j j!}.$$

Integrating term by term, as noted in (10.39), we get for $b > 0$,

$$\int_{0}^{b} \phi(x)\,dx = \frac{1}{\sqrt{2\pi}} \sum_{j=0}^{\infty} \frac{(-1)^j b^{2j+1}}{(2j+1)2^j j!}. \tag{10.120}$$

This expression converges absolutely by the above noted section, and as an alternating series, we have an error estimate associated with any partial summation from section 6.1.5. Specifically, by the alternating series convergence test for n large to ensure that series terms decrease,

$$\left| \int_0^b \phi(x)\, dx - \frac{1}{\sqrt{2\pi}} \sum_{j=0}^{n-1} \frac{(-1)^j b^{2j+1}}{(2j+1)2^j j!} \right| \leq \frac{1}{\sqrt{2\pi}} \frac{b^{2n+1}}{(2n+1)2^n n!}.$$

This error term decreases to 0 quickly as n increases, as can be seen by an application of Stirling's formula from (8.24):

$$e^{1/(12n+1)} < \frac{n!}{\sqrt{2\pi n}\, n^{n+(1/2)} e^{-n}} < e^{1/12n},$$

which produces an error estimate after a bit of algebra:

$$\left| \int_0^b \phi(x)\, dx - \frac{1}{\sqrt{2\pi}} \sum_{j=0}^{n-1} \frac{(-1)^j b^{2j+1}}{(2j+1)2^j j!} \right| \leq \frac{b e^{-1/(12n+1)}}{2\pi\sqrt{n}(2n+1)} \left(\frac{b^2 e}{2n} \right)^n. \tag{10.121}$$

Upper and Lower Riemann Sums
Because $\phi(x)$ is a strictly decreasing function on $[0, b]$ for $b > 0$, the upper Riemann sums are defined by the left subinterval endpoints, while the lower sums are defined by the right endpoints. Consequently, defining the partition of $[0, b]$, with $\Delta x = \frac{b}{n}$, we obtain

$$\frac{1}{\sqrt{2\pi}} \sum_{j=1}^{n} \Delta x \exp\left(-\frac{1}{2}[j\Delta x]^2 \right) \leq \int_0^b \phi(x)\, dx$$

$$\leq \frac{1}{\sqrt{2\pi}} \sum_{j=0}^{n-1} \Delta x \exp\left(-\frac{1}{2}[j\Delta x]^2 \right).$$

This can be simplified by defining the sum $S = \sum_{j=1}^{n-1} \Delta x \exp\left(-\frac{1}{2}[j\Delta x]^2\right)$, to produce

$$\frac{1}{\sqrt{2\pi}} \left(S + \Delta x \exp\left(-\frac{1}{2}b^2 \right) \right) \leq \int_0^b \phi(x)\, dx \leq \frac{1}{\sqrt{2\pi}} (S + \Delta x). \tag{10.122}$$

With the given definition of S, the range between the upper and lower bounds is $\frac{\Delta x}{\sqrt{2\pi}} \left(1 - \exp\left(-\frac{1}{2}b^2\right) \right)$, and so a midpoint estimate gives half this error. Defining $I^{U/L}$ as this midpoint value produces

$$I^{U/L} = \frac{1}{\sqrt{2\pi}} \left(S + \frac{\Delta x}{2} \left(1 + \exp\left(-\frac{1}{2}b^2 \right) \right) \right),$$

and we get

$$\left| \int_0^b \phi(x)\,dx - I^{U/L} \right| \leq \frac{\Delta x}{2\sqrt{2\pi}} \left(1 - \exp\left(-\frac{1}{2} b^2 \right) \right). \tag{10.123}$$

So given $b > 0$, the error on this approach is $O(\Delta x) = O\!\left(\frac{1}{n}\right)$.

Trapezoidal Rule

The trapezoidal rule is defined as the average of the Riemann sums defined with the left endpoints and the right endpoints, as is clear from (10.40). Consequently in the case of a monotonic function like $\phi(x)$, the trapezoidal approximation can also be defined as the average of the upper and lower Riemann sums. So here the trapezoidal approximation I^T equals $I^{U/L}$ above.

However, the error estimate for the trapezoidal rule reflects higher order derivative values of $\phi(x)$. Specifically, we have from (10.42),

$$\left| \int_0^b \phi(x)\,dx - I^T \right| \leq \frac{M_2 b}{12} (\Delta x)^2, \tag{10.124}$$

where M_2 is an upper bound for $|\phi^{(2)}(x)|$ on $[0, b]$. Taking these derivatives, we obtain

$$\phi'(x) = -\frac{x}{\sqrt{2\pi}} e^{-x^2/2}, \quad \phi^{(2)}(x) = \frac{(x^2 - 1)}{\sqrt{2\pi}} e^{-x^2/2}.$$

To estimate M_2, we can locate the critical points of $f(x) \equiv \phi^{(2)}(x)$ by the methods of section 9.5.1. Evaluating $f'(x)$, we have that

$$f'(x) = \frac{[x(3 - x^2)]}{\sqrt{2\pi}} e^{-x^2/2},$$

so it is apparent that the critical points of $f'(x)$ occur at $x = 0, \pm\sqrt{3}$. Also, as $|x| \to \infty$, $\phi^{(2)}(x) \to 0$, and so since $\phi^{(2)}(0) < 0$ and $\phi^{(2)}(\pm\sqrt{3}) > 0$,

$$M_2 = \max[|\phi^{(2)}(0)|, \phi^{(2)}(\pm\sqrt{3})] \approx 0.3989.$$

Consequently we obtain the trapezoidal error estimate:

$$\left| \int_0^b \phi(x)\,dx - I^T \right| \leq 0.03325 b (\Delta x)^2, \tag{10.125}$$

which is considerably better than the estimate in (10.123), despite using the same approximation. This is due to the use of information on this function's second

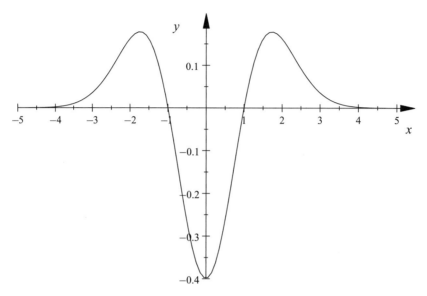

Figure 10.10
$\varphi^{(2)}(x) = \frac{1}{\sqrt{2\pi}}(x^2 - 1)e^{(-x^2/2)}$

derivative which is ignored above. This information reduces the error to $O(\Delta x^2)$ where $\Delta x = \frac{b}{n}$. The graph of $\phi^{(2)}(x)$ is seen in **figure 10.10**.

Note that this graph also indicates that the normal density has second derivative that is negative on the interval $[-1, 1]$, which is the interval $[-\sigma, \sigma]$ in the general case, implying that this function is concave on this interval and changes to convex outside this interval. The points $x = \pm 1$, or more generally $x = \pm \sigma$, are therefore **inflection points**, or **points of inflection**, of the normal density function, as noted in section 9.6. It is also the case that in this example, these inflection points are exactly the points where $\phi^{(2)}(x) = 0$.

Simpson's Rule

Simpson's rule, as can be observed in (10.44), requires three Riemann sums, the same two used for the trapezoidal rule defined in terms of the subinterval left and right endpoints, as well as a third Riemann sum defined by the subinterval midpoints. The weight put on the endpoint Riemann sums is $\frac{1}{6}$ each, while the weight on the midpoint Riemann sum is $\frac{4}{6}$. Denoting by I^S the Simpson approximation using the same partition as above with $\Delta x = \frac{b}{n}$, we obtain by (10.46) that

$$|I - I^S| \le \frac{M_4 b}{2880} (\Delta x)^4, \tag{10.126}$$

where M_4 is an upper bound for $|\phi^{(4)}(x)|$ on $[0, b]$.

Continuing to take these derivatives, we have

$$\phi^{(3)}(x) = \frac{[x(3 - x^2)]}{\sqrt{2\pi}} e^{-x^2/2}, \quad \phi^{(4)}(x) = \frac{[x^4 - 6x^2 + 3]}{\sqrt{2\pi}} e^{-x^2/2}.$$

Again seeking the critical points of $f(x) = \phi^{(4)}(x)$, we obtain

$$f'(x) = \frac{-x[x^4 - 10x^2 + 15]}{\sqrt{2\pi}} e^{-x^2/2},$$

and the critical points are obtained by the quadratic formula applied to $y^2 - 10y + 15$ with $y = x^2$, producing

$$x = 0, \pm\sqrt{3 + \sqrt{6}}, \pm\sqrt{3 - \sqrt{6}}.$$

The upper bound for $|\phi^{(4)}(x)|$ is again seen to occur at $x = 0$ by substitution, producing

$$M_4 = \phi^{(4)}(0) \approx 1.1968.$$

Consequently the Simpson error estimate becomes

$$|I - I^S| \le 0.00042b(\Delta x)^4. \tag{10.127}$$

This error is significantly better than the trapezoidal estimate above due to the error being $O(\Delta x^4)$ compared to $O(\Delta x^2)$, although it could be argued that the comparison is not quite fair since Simpson's rule uses midpoints of subintervals, and hence a finer partition.

Had the trapezoidal estimate been implemented using this same number of interval evaluation points, the implied partition would be $\Delta x' = \frac{b}{2n} = \frac{\Delta x}{2}$, and the revised trapezoidal estimate, denoted $I^{T'}$, stated in terms of the original Δx, would have an error:

$$\left| \int_0^b \phi(x)\, dx - I^{T'} \right| \le 0.00832b(\Delta x)^2.$$

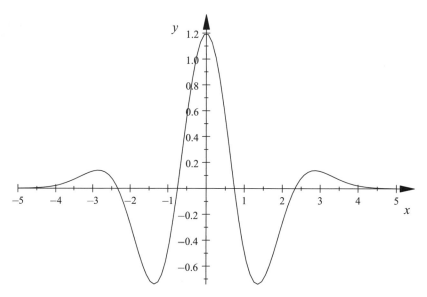

Figure 10.11
$\varphi^{(4)}(x) = \frac{1}{\sqrt{2\pi}}(x^4 - 6x^2 + 3)e^{(-x^2/2)}$

So, despite using the same interval points in both estimates, Simpson's rule is seen to be far superior because of the way it weighted the various points, and not because it included more points. The graph of $\phi^{(4)}(x)$ can be seen in **figure 10.11**.

*10.12.6   Generalized Black–Scholes–Merton Formula

This section develops a generalization of the classical Black–Scholes–Merton option pricing formulas. It is a generalization in that is applies the earlier methodology to a more general **European-style derivative** than a European put or call option. By European-style derivative is meant a financial contract with a general payoff function at time T that depends on the value of an underlying investment asset at time T, and that does not allow early exercise. As always, we use the language that S is a common stock, but as noted in section 7.8.6, we really only need to assume that S is an **investment asset** to justify the replicating portfolio pricing argument.

   It is important to note that this section does not generalize these famous formulas in the sense that these are "new" and would have been unknown to the authors. Indeed the mathematical tools used in the original papers certainly handle the payoff functions considered here, and the authors certainly knew this. But we have not yet developed the tools used by these authors, so we will solve these problems, the specific and the general, with the tools of this chapter.

The goals of this section are as follows:

1. Derive a general integration formula for the price of a European-style derivative, based on a replicating portfolio approach, that reduces to the classical B-S-M formulas when the payoff function is a European put or call.

2. Demonstrate that this evaluation only utilizes the risk-neutral probability distribution of stock prices.

3. Develop this formula using the tools of integration theory studied in this chapter, because the tools of chapter 9, while adequate for a put or call option, do not readily apply to this general situation.

To begin with, recall the chapter 7 price of a European derivative with expiry at time T, in (7.147) as generalized in section 8.8.3. The lattice-based price of a European option or other European-style derivative security on an investment asset, exercisable in n periods and derived based on a replicating portfolio argument, is

$$\Lambda_0(S_0) = e^{-nr} \sum_{j=0}^{n} \binom{n}{j} q^j (1-q)^{n-j} \Lambda(S_n^j),$$

$$S_n^j = S_0 e^{ju + (n-j)d}.$$

In the generalized chapter 8 setting where T is fixed, and time-step periods are defined by $\Delta t = \frac{T}{n}$, this formula is applicable with risk-neutral probability as in (8.52):

$$q(\Delta t) = \frac{e^{r(\Delta t)} - e^{d(\Delta t)}}{e^{u(\Delta t)} - e^{d(\Delta t)}},$$

where by (8.53) and (8.54), the binomial asset period returns and risk-free rate are given by

$$u(\Delta t) = \mu \Delta t + \sqrt{\frac{p'}{p}} \sigma \sqrt{\Delta t},$$

$$d(\Delta t) = \mu \Delta t - \sqrt{\frac{p}{p'}} \sigma \sqrt{\Delta t},$$

$$r(\Delta t) = r\Delta t,$$

where $0 < p < 1$ is the real world probability of $u(\Delta t)$ and $p' \equiv 1 - p$.

We state the main result of this section in proposition 10.84 below. The only requirement on the payoff function at time T, $\Lambda(S_T)$, is that it is bounded and **piecewise continuous with limits**. The notion of piecewise continuous was encountered in section 10.2.2 and will be generalized here with a definition.

Definition 10.82   *A function $f(x)$ is **piecewise continuous with limits** on $\mathbb{R}$ if there exist points*

$$\cdots < a_{-2} < a_{-1} < a_0 < a_1 < a_2 < \cdots.$$

so that:

1. *On each open interval, (a_j, a_{j+1}), $f(x)$ is bounded and continuous*

2. *For each a_j, $\lim_{x \to a_j^+} f(x)$ and $\lim_{x \to a_j^-} f(x)$ exist, and $f(a_j)$ is defined as one of these limits*

3. *The collection $\{a_j\}$, if infinite, has no accumulation points, so that $\min[a_{j+1} - a_j] = m > 0$*

*A function $f(x)$ is **piecewise continuous with limits** on $[a, b]$ if there exist points*

$$a \leq a_0 < a_1 < a_2 < \cdots < a_n \leq b$$

with the same properties.

For the purposes of the existence of the Riemann integral below, as was seen in section 10.2.2 on the Riemann integral without continuity, only constraint 1 in this definition is needed, which is the typical definition of piecewise continuous. In order for such a function to make sense as a payoff function for a European-style derivative, we add constraint 2.

Besides being logical in the financial markets, constraint 2 will also allow us to express $f(x)$ as a continuous function on each closed interval, $[a_j, a_{j+1}]$, redefining $f(x)$ at the endpoints in terms of its one-sided limits. This partitioning of $f(x)$ will not change the value of its integral, of course, and will be seen to provide for a needed technicality in the proof of the proposition below.

Example 10.83   *A **European binary call option**, with an expiry of T and strike price of K, is defined with the payoff function*

$$\Lambda(S_T) = \begin{cases} A, & S_T > S_0, \\ 0, & S_T \leq S_0, \end{cases} \tag{10.128}$$

*for some fixed amount $A > 0$. A **European binary put option** is defined with*

$$\Lambda(S_T) = \begin{cases} 0, & S_T \geq S_0, \\ A, & S_T < S_0. \end{cases} \tag{10.129}$$

European binary options are the simplest examples of derivative securities with payoff functions that are piecewise continuous with limits.

The main result is stated next for bounded payoff functions, and will be proved over the next few sections. In the real world every payoff function is of necessity a bounded function, say $|\Lambda(S_T)| \leq M$, where M represents global gross domestic product, say. Strictly speaking, we do not need to assume boundedness for the statement of this result. However, if not bounded, there is the question of the existence of the integral in the statement of the proposition, and this generalization will create some unnecessary technical difficulties since this boundedness assumption is not a restriction in any real world application.

Proposition 10.84   *For any bounded payoff function, $\Lambda(S_T)$, that is piecewise continuous with limits, we have that as $\Delta t \to 0$,*

$$\Lambda_0(S_0) \to e^{-rT} \int_{-\infty}^{\infty} \Lambda(S_0 e^x) f(x)\, dx, \tag{10.130}$$

where $f(x)$ is the probability density function for $N\left(\left(r - \frac{1}{2}\sigma^2\right)T, \sigma^2 T\right)$, and $\Lambda_0(S_0)$ is the binomial summation defined in (7.147) generalized for $\Delta t = T/n$.

Remark 10.85   *Note that by the section 9.8.10 analysis, $N\left(\left(r - \frac{1}{2}\sigma^2\right)T, \sigma^2 T\right)$ is the limiting distribution of the log-ratio of equity prices under the risk-neutral probability as developed in (9.93). In other words, (10.130) states that the price of a European-style derivative, based on a replicating portfolio on a binomial lattice, converges to the expected present value of the payoff function values. This expectation is calculated under the assumption that future stock prices are lognormally distributed, with mean returns consistent with the assumption that investors are risk neutral. Here, as in (7.144), risk neutrality means that investors will pay S_0 for the security, equal to the expected present value of future stock prices:*

$$S_0 = e^{-rT} \int_{-\infty}^{\infty} (S_0 e^x) f(x)\, dx, \tag{10.131}$$

where $f(x)$ is the probability density function for $N\left(\left(r - \frac{1}{2}\sigma^2\right)T, \sigma^2 T\right)$. This identity follows because the right-hand expression equals $S_0 e^{-rT} M_Z(1)$, where $Z \sim N\left(\left(r - \frac{1}{2}\sigma^2\right)T, \sigma^2 T\right)$, which reduces to S_0.

We now develop the tools needed to prove this general result.

The Piecewise "Continuitization" of the Binomial Distribution
In section 10.11.3 the discretization of a continuously distributed random variable
was introduced. Here we introduce the first step in the opposite concept, and that is
for the continuitization of a discrete random variable, where we note that pronunci-
ation of this term is facilitated with meter: "ba-ba-boom-ba-ba-boom-ba". We con-
centrate on the binomial distribution of equity returns, as this is the application in
hand, but it will be clear from the construction that this approach is more generally
applicable.

We first define a **piecewise continuitization** of the probability density function of
the binomial $\text{Bin}(n, q)$ used in the derivative pricing formula above. Specifically,
given $f_B(j) = \binom{n}{j} q^j (1-q)^{n-j}$ for $j = 0, 1, \ldots, n$, and **interval tags** equal to the stock
returns on a binomial lattice after n time steps,

$$x_j = nd + (u - d)j, \qquad j = 0, 1, \ldots, n+1,$$

the piecewise continuitization of $f_B(j)$ is defined on the interval $[x_0, x_{n+1})$ by

$$\tilde{f}_n(x) = \frac{1}{u - d} f_B(j), \qquad x_j \le x < x_{j+1}, \tag{10.132}$$

and is defined to be 0 outside the interval $[x_0, x_{n+1})$.

With the formulas in (8.53) for $u(\Delta t)$ and $d(\Delta t)$, and recalling that $n \equiv \frac{T}{\Delta t}$, we have

$$x_0 = T\left[\mu - \sqrt{\frac{p}{p'}} \frac{\sigma}{\sqrt{\Delta t}}\right],$$

$$x_{n+1} = T\left[\mu + \sqrt{\frac{p'}{p}} \frac{\sigma}{\sqrt{\Delta t}}\right] + \frac{\sigma\sqrt{\Delta t}}{\sqrt{pp'}}.$$

So the interval $[x_0, x_{n+1})$ grows without bound as $\Delta t \to 0$.

In **figure 10.12** the piecewise continuitization of the binomial distribution with
$n = 6$, $q = 0.55$, $u = 0.05$ and $d = -0.04$ is represented by the seven horizontal lines
in bold. On an n period binomial lattice, only $\{x_j\}_{j=0}^n$ are produced as equity returns,
of course. Here x_{n+1} is defined consistently to simplify the model. To avoid this ex-
traneous return, each of the horizontal bars in this figure could have been "centered"
on $\{x_j\}_{j=0}^n$, and defined as

$$\hat{f}_n(x) = \frac{1}{u - d} f_B(j), \quad x_j - \frac{1}{2}(u - d) \le x < x_j + \frac{1}{2}(u - d),$$

but this will make the subsequent work a bit messier with little apparent payoff.

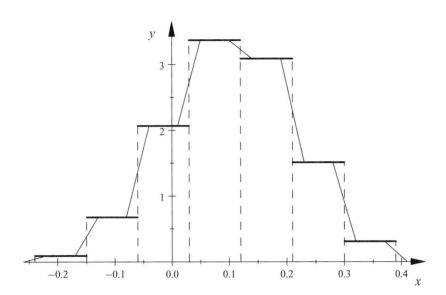

Figure 10.12
Piecewise continuitization and continuitization of the binomial $f(x)$

Note that $\tilde{f}_n(x)$ is piecewise continuous and has integral 1, since $x_{j+1} - x_j = u - d$. In this case the integral is just the area of rectangles:

$$\int_{x_0}^{x_{n+1}} \tilde{f}_n(x)\, dx = \sum_{j=0}^{n} \int_{x_j}^{x_{j+1}} \tilde{f}_n(x)\, dx$$

$$= (u - d) \sum_{j=0}^{n} \frac{1}{u - d} f_B(j) = 1.$$

In exercise 19 is assigned the development of the following expectation formulas, with $\tilde{X}_n$ denoting the piecewise continuously distributed random variable with probability density $\tilde{f}_n(x)$, and X_n^B the discrete random variable with probability density $f_B(j)$ and domain $\{x_j\}_{j=0}^{n}$:

$$E[\tilde{X}_n] = E[X_n^B] + \frac{1}{2}[u - d], \tag{10.133a}$$

$$E[(\tilde{X}_n)^2] = E[(X_n^B)^2] + [u - d]E[X_n^B] + \frac{[u - d]^2}{3}, \tag{10.133b}$$

$$\text{Var}[\tilde{X}_n] = \text{Var}[X_n^B] + \frac{[u-d]^2}{12},$$ (10.133c)

$$M_{\tilde{X}_n}(t) = \frac{e^{t(u-d)} - 1}{t(u-d)} M_{X_n^B}(t).$$ (10.133d)

Recall that $u - d = \frac{\sigma\sqrt{\Delta t}}{\sqrt{pp'}}$, and by a Taylor series expansion, $\frac{e^{t(u-d)}-1}{t(u-d)} = 1 + O(u-d)$. Hence, as $\Delta t \to 0$, the moments above and the moment-generating function for $\tilde{X}_n$ approach the respective values for X_n^B, which we recall from section 9.8.10 approach the respective values for $Z \sim N\big((r - \frac{1}{2}\sigma^2)T, \sigma^2 T\big)$.

The "Continuitization" of the Binomial Distribution

Next we define the continuitization of $f_B(j)$ in a way that makes integrating this function easy. In **figure 10.12** this continuous function is made up of the the light diagonal curves and the connecting portions of the bold horizontal lines. As can be seen, this function, denoted $f_n(x)$, is defined so that $\int f_n(x)\,dx = \int \tilde{f}_n(x)\,dx$, since the difference in functions is simply a summation of offsetting triangles.

To formally define $f_n(x)$, we choose $\{\epsilon_j\}_{j=0}^{n+1}$ with the restriction that $0 < \epsilon_j < \frac{u-d}{2}$, although the goal will later be to better specify the rate of convergence of $\epsilon_j \to 0$. This **continuitization** is now defined for $x \in [x_0 - \epsilon_0, x_{n+1} + \epsilon_n]$ by

$$f_n(x) = \begin{cases} \tilde{f}_n(x), & x \in [x_j + \epsilon_j, x_{j+1} - \epsilon_j], \\ (1-t)\tilde{f}_n(x_j - \epsilon_j) + t\tilde{f}_n(x_j + \epsilon_j), & x = (1-t)(x_j - \epsilon_j) + t(x_j + \epsilon_j). \end{cases}$$

(10.134)

We define $f_n(x) = 0$ outside the interval $[x_0 - \epsilon_0, x_{n+1} + \epsilon_n]$. For this definition, $j = 0, 1, \ldots, n$, in the first line, which defines the horizontal portions of $f_n(x)$, and $j = 0, 1, \ldots, n+1$ and $0 < t < 1$ for the second line, which defines the diagonal portions of $f_n(x)$. Also recall that $\tilde{f}_n(x) = 0$ outside $[x_0, x_{n+1}]$, and so in particular, this is used in the second line for $\tilde{f}_n(x_0 - \epsilon_0)$ and $\tilde{f}_n(x_{n+1} + \epsilon_{n+1})$.

In **figure 10.12** is displayed the continuitization of the binomial with $n = 6$, $q = 0.55$, $u = 0.05$ and $d = -0.04$ using $\epsilon_j = 0.02$ for all j. Note that $f_n(x)$ is continuous, since at each point $x_j - \epsilon_j$, we have that $f_n(x_j - \epsilon_j) = \tilde{f}_n(x_j - \epsilon_j) = \tilde{f}_n(x_{j-1})$ by the first line, while by the second, $t = 0$ at this point, and hence the same result is produced. The same analysis shows continuity at all $x_j + \epsilon_j$.

In order to show that $f_n(x)$ is a probability density function, we show that $\int [f_n(x) - \tilde{f}_n(x)]\,dx = 0$. To do this, we only need to demonstrate that this equation holds over each of the diagonal portions of $f_n(x)$, since this integral will of course equal 0 on the horizontal portions.

Consider the interval $[x_j - \epsilon_j, x_j + \epsilon_j]$ for $j = 0, 1, \ldots, n+1$. We have that with $x = (1-t)(x_j - \epsilon_j) + t(x_j + \epsilon_j)$,

$$f_n(x) - \tilde{f}_n(x) = \begin{cases} t[\tilde{f}_n(x_j + \epsilon_j) - \tilde{f}_n(x_j - \epsilon_j)], & x_j - \epsilon_j \leq x < x_j, \\ -(1-t)[\tilde{f}_n(x_j + \epsilon_j) - \tilde{f}_n(x_j - \epsilon_j)], & x_j \leq x \leq x_j + \epsilon_j. \end{cases}$$

Now for $\int_{x_j - \epsilon_j}^{x_j + \epsilon_j} [f_n(x) - \tilde{f}_n(x)]\, dx$, we first need to express $f_n(x) - \tilde{f}_n(x)$ explicitly as a function of x, rather than implicitly in terms of t. To do this, we have from $x = (1-t)(x_j - \epsilon_j) + t(x_j + \epsilon_j)$ that

$$t = \frac{x - (x_j - \epsilon_j)}{2\epsilon_j}, \quad 1 - t = \frac{(x_j + \epsilon_j) - x}{2\epsilon_j},$$

and so with an algebraic step,

$$f_n(x) - \tilde{f}_n(x) = \begin{cases} \left[\frac{x - (x_j - \epsilon_j)}{2\epsilon_j}\right][\tilde{f}_n(x_j + \epsilon_j) - \tilde{f}_n(x_j - \epsilon_j)], & x \in [x_j - \epsilon_j, x_j), \\ -\left[\frac{(x_j + \epsilon_j) - x}{2\epsilon_j}\right][\tilde{f}_n(x_j + \epsilon_j) - \tilde{f}_n(x_j - \epsilon_j)], & x \in [x_j, x_j + \epsilon_j]. \end{cases}$$

Factoring out the common terms of $[\tilde{f}_n(x_j + \epsilon_j) - \tilde{f}_n(x_j - \epsilon_j)]$ and $2\epsilon_j$, and splitting the integral due to the discontinuity at $x = x_j$, we derive for $j = 0, 1, \ldots, n+1$,

$$\frac{2\epsilon_j}{[\tilde{f}_n(x_j + \epsilon_j) - \tilde{f}_n(x_j - \epsilon_j)]} \int_{x_j - \epsilon_j}^{x_j + \epsilon_j} [f_n(x) - \tilde{f}_n(x)]\, dx$$

$$= \int_{x_j - \epsilon_j}^{x_j} [x - (x_j - \epsilon_j)]\, dx - \int_{x_j}^{x_j + \epsilon_j} [(x_j + \epsilon_j) - x]\, dx$$

$$= 0.$$

This approach also provides an efficient way to evaluate the moments and moment-generating function for X_n, the continuously distributed random variable with density function $f_n(x)$, in terms of the respective values for $\tilde{X}_n$ identified in (10.133). In other words, for any function $g(x)$,

$$\int g(x) f_n(x)\, dx = \int g(x) \tilde{f}_n(x)\, dx + \sum_{j=0}^{n+1} \int_{x_j - \epsilon_j}^{x_j + \epsilon_j} g(x)[f_n(x) - \tilde{f}_n(x)]\, dx. \qquad (10.135)$$

In exercise 38 is assigned the application of (10.135) to $g(x) = x$ and x^2 to produce the following formulas:

$$\mathrm{E}[X_n] = \mathrm{E}[\tilde{X}_n] + \frac{1}{6}\sum_{j=0}^{n} f_B(j)\left(\frac{\epsilon_{j+1}^2 - \epsilon_j^2}{u-d}\right), \tag{10.136a}$$

$$\mathrm{E}[X_n^2] = \mathrm{E}[(\tilde{X}_n)^2] + \frac{1}{3}\sum_{j=0}^{n} f_B(j)\left[\left(\frac{\epsilon_{j+1}^2 - \epsilon_j^2}{u-d}\right)x_j + \epsilon_{j+1}^2\right], \tag{10.136b}$$

$$\mathrm{Var}[X_n] = \mathrm{Var}[\tilde{X}_n] + \frac{1}{3}\sum_{j=0}^{n} f_B(j)\left[\left(\frac{\epsilon_{j+1}^2 - \epsilon_j^2}{u-d}\right)(x_j - \mathrm{E}[\tilde{X}_n]) + \epsilon_{j+1}^2\right]$$

$$+ \frac{1}{36}\left[\sum_{j=0}^{n} f_B(j)\left(\frac{\epsilon_{j+1}^2 - \epsilon_j^2}{u-d}\right)\right]^2. \tag{10.136c}$$

Note that if $\epsilon_j^2 = \epsilon^2$ for all j, these messy formulas simplify greatly to

$$\mathrm{E}[X_n] = \mathrm{E}[\tilde{X}_n],$$

$$\mathrm{Var}[X_n] = \mathrm{Var}[\tilde{X}_n] + \frac{\epsilon^2}{3}.$$

If $\{\epsilon_j^2\}$ are not constant, some care is needed to ensure that these summations converge as $\Delta t \to 0$, since $n = O[(\Delta t)^{-1}]$. For example, the first moment formula suggests that in order for this summation to converge as $\Delta t \to 0$, since $\sum_{j=0}^{n} f_B(j) = 1$, it is simply necessary that $\left\{\frac{\epsilon_{j+1}^2 - \epsilon_j^2}{u-d}\right\}$ must converge to 0 uniformly in j. As $u - d = O[(\Delta t)^{1/2}]$, if $\epsilon_{j+1}^2 - \epsilon_j^2 = O[(\Delta t)^{(1/2)+\delta}]$ for some $\delta > 0$, the resulting summation will be $O[(\Delta t)^\delta]$ and converge to 0 with Δt. This condition on $\epsilon_{j+1}^2 - \epsilon_j^2$ is generally stronger than the original defining condition that $0 < \epsilon_j < \frac{u-d}{2} = O[(\Delta t)^{1/2}]$.

For the second moment and variance, because $\max\{|x_j|\} = O[(\Delta t)^{-1/2}]$, which follows from the definition of x_j, we need $\epsilon_{j+1}^2 - \epsilon_j^2 = O[(\Delta t)^{1+\delta}]$ as well as $\epsilon_{j+1}^2 = O[(\Delta t)^\delta]$ to ensure that the terms involving $\{\epsilon_{j+1}^2 - \epsilon_j^2\}$ and those involving $\{\epsilon_j^2\}$ converge to 0 as $\Delta t \to 0$.

In the next section, $\{\epsilon_j^2\}$ will be chosen to do more than stabilize the limit of these two moments of X_n as $n \to \infty$. The goal will be to ensure that the moment-generating function of X_n converges to that of $\tilde{X}_n$ as $n \to \infty$.

The Limiting Distribution of the "Continuitization"
The goal of this section is to show that as $\Delta t \to 0$, the moment-generating function for the continuitization of this binomial converges to the m.g.f. of the $N\left((r - \frac{1}{2}\sigma^2)T, \sigma^2 T\right)$. This will be demonstrated by showing that the moment-

generating function of this continuitization converges with the m.g.f. of the original binomial distribution, which, as was demonstrated in section 9.8.10, converges to the m.g.f. of the $N\left((r - \frac{1}{2}\sigma^2)T, \sigma^2 T\right)$ as $\Delta t \to 0$.

To this end, and to avoid a messy integral with $f_n(x)$, we again apply (10.135):

$$\int e^{tx} f_n(x)\, dx = \int e^{tx} \tilde{f}_n(x)\, dx + \sum_{j=0}^{n+1} \int_{x_j - \epsilon_j}^{x_j + \epsilon_j} e^{tx}[f_n(x) - \tilde{f}_n(x)]\, dx.$$

We now show in two steps that the first integral produces the desired result, and that $\{\epsilon_j^2\}$ can be chosen so that for all t the second term converges to 0 as $n \to \infty$, or equivalently, as $\Delta t \to 0$.

1. As noted in (10.133),

$$\int e^{tx} \tilde{f}_n(x)\, dx = \frac{e^{t(u-d)} - 1}{t(u - d)} M_B(t),$$

where $M_B(t)$ is the moment generating function of the binomial random variable denoted X_n^B above, which takes values $\{x_j\}$.

Recall that in section 9.8.10 it was demonstrated that $M_B(t) \to M_Z(t)$ as $\Delta t \to 0$, with $Z \sim N\left((r - \frac{1}{2}\sigma^2)T, \sigma^2 T\right)$. Also, by expanding $e^{t(u-d)}$ as a Taylor series, and using that $u - d = \frac{\sigma \sqrt{\Delta t}}{\sqrt{pp'}}$, we have

$$\frac{e^{t(u-d)} - 1}{t(u - d)} = 1 + O((u - d))$$

$$= 1 + O[(\Delta t)^{1/2}],$$

and so

$$\int e^{tx} \tilde{f}_n(x)\, dx = (1 + O[(\Delta t)^{1/2}]) M_B(t)$$

$$\to M_Z(t) \text{ as } \Delta t \to 0.$$

2. For the second integral, note that by the analysis in the previous section, only the subintervals $[x_j - \epsilon_j, x_j + \epsilon_j]$ need to be evaluated, since $f_n(x) = \tilde{f}_n(x)$ elsewhere. As noted above,

$$f_n(x) - \tilde{f}_n(x) = \begin{cases} \left[\frac{x - (x_j - \epsilon_j)}{2\epsilon_j}\right][\tilde{f}_n(x_j + \epsilon_j) - \tilde{f}_n(x_j - \epsilon_j)], & x \in [x_j - \epsilon_j, x_j), \\ -\left[\frac{(x_j + \epsilon_j) - x}{2\epsilon_j}\right][\tilde{f}_n(x_j + \epsilon_j) - \tilde{f}_n(x_j - \epsilon_j)], & x \in [x_j, x_j + \epsilon_j]. \end{cases}$$

Now note that the coefficient functions of x are bounded in absolute value by $\frac{1}{2}$, and since $\tilde{f}_n(x) = \frac{1}{u-d} f_B(j)$ for $x \in [x_j, x_{j+1})$ and $f_B(j) \le 1$ for all j, we conclude that by the triangle inequality,

$$|f_n(x) - \tilde{f}_n(x)| \le \frac{1}{u-d}.$$

So by (10.10),

$$\left| \int e^{tx} [f_n(x) - \tilde{f}_n(x)] \, dx \right| \le \frac{1}{u-d} \sum_{j=0}^{n} \int_{x_j - \epsilon_j}^{x_j + \epsilon_j} e^{tx} \, dx$$

$$= \frac{1}{u-d} \sum_{j=0}^{n} \frac{e^{t(x_j+\epsilon_j)} - e^{t(x_j-\epsilon_j)}}{t}.$$

Now, using a Taylor series expansion, we derive as $\epsilon_j \to 0$,

$$\frac{e^{t(x_j+\epsilon_j)} - e^{t(x_j-\epsilon_j)}}{2\epsilon_j t} = e^{tx_j}(1 + O[(\epsilon_j t)^2]).$$

From this we conclude that

$$\left| \int e^{tx} [f_n(x) - \tilde{f}_n(x)] \, dx \right| \le \frac{2}{u-d} \sum_{j=0}^{n} \epsilon_j e^{tx_j}(1 + O[(\epsilon_j t)^2])$$

$$= \frac{2}{u-d} \sum_{j=0}^{n} \epsilon_j e^{tnd} e^{j(u-d)}(1 + O[(\epsilon_j t)^2]).$$

We are free to choose $\{\epsilon_j\}$ at will, subject to the constraints above to preserve moments, and so we set

$$\epsilon_j = \sqrt{\Delta t}\, e^{-j(u-d)} = \sqrt{\Delta t} \exp\left[\frac{-j\sigma\sqrt{\Delta t}}{\sqrt{pp'}} \right]. \tag{10.137}$$

Then, since $0 \le j \le n = \frac{T}{\Delta t}$, we have that $\epsilon_j \to 0$ as $\Delta t \to 0$:

$$\sqrt{\Delta t} \exp\left[\frac{-\sigma}{\sqrt{\Delta t}\sqrt{pp'}} \right] \le \epsilon_j \le \sqrt{\Delta t},$$

and it can be checked that these ϵ_j values also satisfy the necessary moment conditions above.

Substituting $nd = \frac{T}{\Delta t}\left[\mu \Delta t - \frac{p\sigma\sqrt{\Delta t}}{\sqrt{pp'}}\right]$ and $u - d = O(\sqrt{\Delta t})$, we derive with constants $C, c > 0$, since $O[(t\epsilon_j)^2] = t^2 O(\Delta t)$ and $n = \frac{T}{\Delta t}$,

$$\left|\int e^{tx}[f_n(x) - \tilde{f}_n(x)]\, dx\right| \leq C(1 + t^2 O(\Delta t)) \sum_{j=0}^{n} e^{-ct/\sqrt{\Delta t}}$$

$$= C(1 + t^2 O(\Delta t))\left(\frac{T}{\Delta t} + 1\right)e^{-ct/\sqrt{\Delta t}}.$$

That is because there are $n + 1$ constant terms in this summation.

To see that as $\Delta t \to 0$ this integral converges to 0 for all t, substitute $s = \frac{1}{\sqrt{\Delta t}}$, and consider the limit of this upper bound as $s \to \infty$:

$$C\left(1 + t^2 O\left(\frac{1}{s^2}\right)\right)(Ts^2 + 1)e^{-cts} \to 0.$$

The Generalized Black–Scholes–Merton Formula

We are now in a position to address the result quoted above in (10.130). To simplify notation, we ignore the e^{-rT} term, which is simply a multiplicative factor in both the discrete and limiting continuous pricing formulas. The major steps in this demonstration are:

1. With $f_n(x)$ defined as in (10.134), and $f(x)$ the normal distribution in (10.130), we first show that as $n \to \infty$, or equivalently, $\Delta t \to 0$,

$$\int_{-\infty}^{\infty} \Lambda(S_0 e^x) f_n(x)\, dx \to \int_{-\infty}^{\infty} \Lambda(S_0 e^x) f(x)\, dx.$$

As shown above, $M_{X_n}(t) \to M_X(t)$ pointwise for all t as $\Delta t \to 0$. Restricting to any compact interval $[-N, N]$, this pointwise convergence of analytic functions is therefore uniform.

Also, by (10.136), the collection of variances, $\{\sigma_n^2\}$ is bounded, and by the Chebyshev inequality, for any $\epsilon > 0$ there is an N so that

$$\Pr[|X| > N] < \epsilon,$$

$$\Pr[|X_n| > N] < \epsilon \qquad \text{for all } n.$$

As noted previously but not proved, the convergence of moment-generating functions also implies the pointwise convergence of $f_n(x) \to f(x)$, and as continuous functions, this convergence is uniform on any compact interval, $[-N, N]$. On this interval, splitting $\Lambda(S_0 e^x)$ into its finite number of piecewise continuous functions on the subintervals $[a_j, a_{j+1}] \subset [-N, N]$, we have $\Lambda(S_0 e^x) f_n(x) \to \Lambda(S_0 e^x) f(x)$ uniformly on each subinterval, and consequently as well as on $[-N, N]$. Hence, by proposition 10.55,

$$\int_{a_j}^{a_{j+1}} \Lambda(S_0 e^x) f_n(x)\, dx \to \int_{a_j}^{a_{j+1}} \Lambda(S_0 e^x) f(x)\, dx,$$

for all $[a_j, a_{j+1}] \subset [-N, N]$, and the same is then true for the integrals over $[-N, N]$.

Putting this all together, we can split the integral over $(-\infty, \infty)$ into integrals over $[-N, N]$, $(-\infty, N]$, and $[N, \infty)$. We then have by the triangle inequality and (10.10), the Chebyshev bounds above, and the assumption that $\Lambda(S_0 e^x)$ is bounded and hence $|\Lambda(S_0 e^x)| < M$ for some M,

$$\left| \int_{-\infty}^{\infty} \Lambda(S_0 e^x) f_n(x)\, dx - \int_{-\infty}^{\infty} \Lambda(S_0 e^x) f(x)\, dx \right|$$

$$\leq \left| \int_{-N}^{N} \Lambda(S_0 e^x) f_n(x)\, dx - \int_{-N}^{N} \Lambda(S_0 e^x) f(x)\, dx \right| + 2M\epsilon.$$

Since the difference of integrals over $[-N, N]$ converges to 0 as $n \to \infty$, we have shown that the difference of integrals over $(-\infty, \infty)$ can be made as small as desired, proving the result.

2. Next we convert the integrals with $f_n(x)$ into a summation with binomial probabilities, where we begin with the observation

$$\int_{-\infty}^{\infty} \Lambda(S_0 e^x) f_n(x)\, dx = \sum_j \int_{a_j}^{a_{j+1}} \Lambda(S_0 e^x) f_n(x)\, dx.$$

On each interval $[a_j, a_{j+1}]$ the integrand $\Lambda(S_0 e^x) f_n(x)$ is continuous by defining $\Lambda(S_0 e^x)$ at the endpoints in terms of its limiting values. Also $f_n(x)$ is identically 0 outside the interval $[x_0, x_{n+1}] = [nd, (n+1)u - d]$. With $\Delta x = \frac{x_{n+1} - x_0}{n+1} = u - d$, and interval partition defined with $x_j = nd + (u - d)j$, for $j = 0, 1, \ldots, n+1$, each integral in the summation above can be expressed as follows, where $a_j \leq x_k < x_{k+1} < \cdots < x_l \leq a_{j+1}$:

$$\int_{a_j}^{a_{j+1}} \Lambda(S_0 e^x) f_n(x)\, dx = \int_{a_j}^{x_k} \Lambda(S_0 e^x) f_n(x)\, dx + \sum_j \int_{x_j}^{x_{j+1}} \Lambda(S_0 e^x) f_n(x)\, dx$$

$$+ \int_{x_l}^{a_{j+1}} \Lambda(S_0 e^x) f_n(x)\, dx.$$

Now by the first mean value theorem for integrals in (10.12), there is $\hat{x}_j \in (x_j, x_{j+1})$ with

$$\int_{x_j}^{x_{j+1}} \Lambda(S_0 e^x) f_n(x)\, dx = \Lambda(S_0 e^{\hat{x}_j}) f_n(\hat{x}_j)(x_{j+1} - x_j),$$

and similarly for the first and last integrals. For the integrals in the summation, since $\hat{x}_j \in (x_j, x_{j+1})$, and this interval's value of ϵ_j can be chosen smaller than defined in (10.137), we can assume that $\hat{x}_j \in (x_j + \epsilon_j, x_{j+1} - \epsilon_{j+1})$ and so $f_n(\hat{x}_j) = \tilde{f}_n(\hat{x}_j) = \frac{1}{u-d} f_B(j)$. Then, since $x_{j+1} - x_j = u - d$,

$$\int_{x_j}^{x_{j+1}} \Lambda(S_0 e^x) f_n(x)\, dx = \Lambda(S_0 e^{\hat{x}_j}) \tilde{f}_n(\hat{x}_j)(x_{j+1} - x_j)$$

$$= \Lambda(S_0 e^{\hat{x}_j}) \binom{n}{j} q^j (1-q)^{n-j}.$$

Now, for the integrals that involve a given a_j, say $x_k < a_j < x_{k+1}$, we combine the integral over $[x_k, a_j]$ and the integral over $[a_j, x_{k+1}]$, and a similar argument produces the following, where $\hat{x}_{k1} \in (x_k + \epsilon_k, a_j)$, $\hat{x}_{k2} \in (a_j, x_{k+1} - \epsilon_{k+1})$, $\lambda_{k1} = \frac{a_j - x_k}{x_{k+1} - x_k}$, and $\lambda_{k2} = 1 - \lambda_{k1} = \frac{x_{k+1} - a_j}{x_{k+1} - x_k}$:

$$\int_{x_k}^{x_{k+1}} \Lambda(S_0 e^x) f_n(x)\, dx$$

$$= \binom{n}{k} q^k (1-q)^{n-k} [\lambda_{k1} \Lambda(S_0 e^{\hat{x}_{k1}}) + \lambda_{k2} \Lambda(S_0 e^{\hat{x}_{k2}})]$$

$$= \binom{n}{k} q^k (1-q)^{n-k} \Lambda(S_0 e^{\hat{x}_{k1}}) + \binom{n}{k} q^k (1-q)^{n-k} \lambda_{k2} [\Lambda(S_0 e^{\hat{x}_{k2}}) - \Lambda(S_0 e^{\hat{x}_{k1}})].$$

Combining all integrals, we have that

$$\int_{-\infty}^{\infty} \Lambda(S_0 e^x) f_n(x) \, dx$$

$$= \sum_{j=0}^{n} \binom{n}{j} q^j (1-q)^{n-j} \Lambda(S_0 e^{\hat{x}_j})$$

$$+ \sum_{a_k \in (x_j, x_{j+1})} \binom{n}{j} q^j (1-q)^{n-j} \lambda_{j2} [\Lambda(S_0 e^{\hat{x}_{j2}}) - \Lambda(S_0 e^{\hat{x}_{j1}})], \qquad (10.138)$$

where the second summation includes only those values of j for which $a_k \in (x_j, x_{j+1})$ for some k.

3. The final step is to show that the summations in (10.138) converge to the binomial summation represented by $\Lambda_0(S_0)$ in (10.130). To this end, we show that the first summation converges to $\Lambda_0(S_0)$, and the second converges to 0 as $n \to \infty$. First off,

$$\Lambda_0(S_0) - \sum_{j=0}^{n} \binom{n}{j} q^j (1-q)^{n-j} \Lambda(S_0 e^{\hat{x}_j})$$

$$= \sum_{j=0}^{n} \binom{n}{j} q^j (1-q)^{n-j} [\Lambda(S_0 e^{x_j}) - \Lambda(S_0 e^{\hat{x}_j})],$$

where by construction, $\hat{x}_j \in (x_j + \epsilon_j, x_{j+1} - \epsilon_{j+1})$. Also $\Lambda(S_0 e^x)$ can be assumed to be continuous at each x_j, perhaps not for a fixed n for which it may happen that $x_j = a_k$ for some j and k, but as $n \to \infty$, which is our concern. Consequently $\hat{x}_j \to x_j$ as $n \to \infty$ for each j. Now, because the binomial density in this summation has bounded variance for all n, we again apply the Chebyshev inequality to derive that for any $\epsilon > 0$ there is an interval $[-N, N]$ so that $\Pr[X_n^B \in [-N, N]] \geq 1 - \epsilon$ for all n. On this interval, since $\Lambda(S_0 e^x)$ is piecewise continuous with limits, and there are only a finite number of intervals, $[a_k, a_{k+1}] \subset [-N, N]$, we conclude that as $n \to \infty$,

$$\max_{x_j \in [-N, N]} |\Lambda(S_0 e^{x_j}) - \Lambda(S_0 e^{\hat{x}_j})| \to 0.$$

Hence summing over all j for which $x_j \in [-N, N]$ produces

$$\sum_{x_j \in [-N, N]} \binom{n}{j} q^j (1-q)^{n-j} |\Lambda(S_0 e^{x_j}) - \Lambda(S_0 e^{\hat{x}_j})| \to 0.$$

Now for all j for which $x_j \notin [-N, N]$, we apply the triangle inequality

$$\sum_{x_j \notin [-N,N]} \binom{n}{j} q^j (1-q)^{n-j} |\Lambda(S_0 e^{x_j}) - \Lambda(S_0 e^{\hat{x}_j})| \le 2M\epsilon,$$

since $\Lambda(S_0 e^x)$ is bounded by M and $\Pr[X_n^B \notin [-N,N]] < \epsilon$. Consequently the first summation in (10.138) converges to $\Lambda_0(S_0)$ as claimed.

For the second summation in (10.138), by the triangle inequality,

$$\sum_{a_k \in (x_j, x_{j+1})} \binom{n}{j} q^j (1-q)^{n-j} \lambda_{j2} |\Lambda(S_0 e^{\hat{x}_{j2}}) - \Lambda(S_0 e^{\hat{x}_{j1}})|$$

$$\le 2M \sum_{a_k \in (x_j, x_{j+1})} \binom{n}{j} q^j (1-q)^{n-j},$$

since $\Lambda(S_0 e^x)$ is bounded by M and $0 \le \lambda_{j2} \le 1$. We can split this summation into the finite collection of $\{a_k\} \subset [-N,N]$, and the rest, and obtain

$$\sum_{a_k \in (x_j, x_{j+1})} \binom{n}{j} q^j (1-q)^{n-j} < \sum_{a_k \in [-N,N]} \binom{n}{j} q^j (1-q)^{n-j} + \epsilon.$$

Now, since this summation includes only those values of j for which $a_k \in (x_j, x_{j+1})$ for some k, this finite summation converges to 0 as $n \to \infty$, completing the derivation.

Exercises

Practice Exercises

1. Demonstrate by explicit evaluation of the Riemann sums, the following integrals for $c \in \mathbb{R}$, where for simplicity assume that $0 \le a < b$:

(a) $\int_a^b c \, dx = (b-a)c$

(b) $\int_a^b cx \, dx = \frac{c}{2}(b^2 - a^2)$ (*Hint:* $\sum_{j=1}^n j = \frac{n(n+1)}{2}$.)

(c) $\int_a^b cx^2 \, dx = \frac{c}{3}(b^3 - a^3)$ (*Hint:* $\sum_{j=1}^n j^2 = \frac{n(n+1)(2n+1)}{6}$.)

2. For the function,

$$f(x) = \begin{cases} x^2, & 0 \le x < 1 \\ x^2 + 5, & 1 \le x \le 2 \end{cases}$$

(a) Verify explicitly that

$$\int_0^2 f(x)\,dx = \int_0^1 x^2\,dx + \int_1^2 (x^2 + 5)\,dx = \frac{23}{3},$$

by demonstrating that the contribution of the terms in the Riemann sums containing the point $x = 1$ converge to 0.

(b) Confirm that this conclusion is independent of the definition of $f(1)$.

3. Consider a collection of intervals containing a point x': $\{I_j\} = \{(x' - a_j, x' + b_j)\}$, where $\{a_j\}$ and $\{b_j\}$ are positive sequences which converge to 0. Prove that for a given function, $f(x)$, with M_j and m_j defined as in (10.2), that $M_j - m_j \to 0$ if and only if $f(x)$ is continuous at x'.

4. For each of the functions in exercise 1, determine the value of d as promised by the mean value theorem for which

$$\int_a^b f(x)\,dx = f(d)(b - a).$$

5. Using the Fundamental Theorem of Calculus version I in (10.15):

(a) Confirm the formulas in exercise 1.

(b) Generalize exercise 1 to show that for $a, b \in \mathbb{R}$:

$$\int_a^b cx^n\,dx = \frac{c}{n + 1}(b^{n+1} - a^{n+1}), \qquad n \in \mathbb{R},\ n \neq -1.$$

(c) Confirm that for part (b), if $n = -1$,

$$\int_a^b cx^{-1}\,dx = c\ln\left(\frac{b}{a}\right), \qquad b > a > 0.$$

(d) Generalize part (c) if $a < b < 0$. (*Hint:* Compare $\int_a^b cx^{-1}\,dx$ to $-\int_a^b cx^{-1}\,dx$ to $-\int_{-b}^{-a} cx^{-1}\,dx$.)

6. Use the integral test in the following analyses:

(a) Show that $\sum_{n=1}^{\infty} e^{-n}$ converges and estimate its value. (*Note:* This can also be summed exactly as a geometric series, of course, but that is not what is to be done here.)

(b) Show that $\sum_{n=1}^{\infty} n^m$, for $m \geq -1$, diverges, and estimate the rate of growth of the partial sums.

(c) For $0 < q < 1$, determine whether $\sum_{n=1}^{\infty} nq^n$ converges or diverges, and correspondingly estimate its summation value, or the growth rate of its partial sums. (*Hint*: Integrate $f(x) = xq^x = xe^{x\ln q}$ using integration by parts.)

7. Evaluate the following definite integrals using the method of substitution, and then identify an antiderivative of the integrand:

(a) $\int_0^{\infty} xe^{-x^2}\,dx$ (*Hint*: First consider $\int_0^N xe^{-x^2}\,dx$ as a definite integral.)

(b) $\int_0^{\infty} (4z^3 + 6z)(z^4 + 3z^2 + 5)^{-2}\,dz$

(c) $\int_0^{10} \frac{e^{2x}\,dx}{4e^{2x}-1}$

8. Evaluate the following definite integrals using integration by parts, and then identify an antiderivative of the integrand. (*General hint*: Once a potential antiderivative is found, this formula can be verified by differentiation.)

(a) $\int_0^{10} x^m e^x\,dx$ for positive integer m (*Hint*: Implement two or three integration by parts steps and observe the pattern.)

(b) $\int_3^{20} x^m e^{x^2}\,dx$ for positive odd integer $m = 2n + 1$ (*Hint*: Implement two or three integration by parts steps and observe the pattern, using xe^{x^2}.)

9. Show using a Taylor series expansion that if $f(y) = \frac{1}{1+y}$, for $|y| < 1$, that $\int_0^x f(y)\,dy - \ln(1+x)$. Justify integrating term by term as well as the convergence of the final series to the desired answer.

10. Using the definite integrals over bounded intervals in exercise 7(c), 8(a), and 8(b) (use $m = 5$ for exercise 8):

(a) Implement both the trapezoidal rule and Simpson's rule for several values of n and compare the associated errors. (*Hint*: Try $n = 5, 10, 25$, and 100, say.)

(b) For each approximation, evaluate the error as n increases significantly, to see if the respective orders of convergence, $O\left(\frac{1}{n^2}\right)$ and $O\left(\frac{1}{n^4}\right)$, are apparent. (*Hint*: If $\epsilon_n^T = |I - I^T|$ for $\Delta x = \frac{b-a}{n}$, the error $\epsilon_n^T = O\left(\frac{1}{n^2}\right)$ means that $n^2\epsilon_n^T \leq C^T$ for some constant C^T as $n \to \infty$, and similarly for $\epsilon_n^S = |I - I^S|$, that $n^4\epsilon_n^S \leq C^S$ as $n \to \infty$. Attempt to verify that the C^T and C^S values obtained are no bigger than the values predicted in theory using the maxima of the derivatives of the given functions.)

11. Evaluate $\Pr[-1 \leq X \leq 2]$ for the Cauchy distribution with $x_0 = 1$, and a scale parameter $\lambda = 2$, by:

(a) Trapezoidal rule with $n = 30$

(b) Simpson's rule with $n = 30$

(c) Evaluate the error in each approximation

12. Derive the error estimate for Simpson's rule over the subinterval $[a, a + \Delta x]$. (*Hint*: Use the Taylor approximation:

$$f(x) = f(a) + f'(a)(x - a) + \frac{1}{2}f^{(2)}(a)(x - a)^2$$

$$+ \frac{1}{3!}f^{(3)}(a)(x - a)^3 + \frac{1}{4!}f^{(4)}(y)(x - a)^4,$$

for some $y \in [a, a + \Delta x]$. Calculate $\int_a^{a+\Delta x} f(x)\, dx$, using the second MVT for integrals in (10.35), and also evaluate the expression for I^S over this interval, and subtract, recalling the intermediate value theorem in (9.1).)

13. Prove the following identities:

(a) As in (10.65): $\sigma^2 = E[X^2] - E[X]^2$.

(b) As in (10.66):

i. $\mu_n = \sum_{j=0}^{n}(-1)^{n-j}\binom{n}{j}\mu_j'\mu^{n-j}$ (*Hint*: Use the binomial theorem.)

ii. $\mu_n' = \sum_{j=0}^{n}\binom{n}{j}\mu_j\mu^{n-j}$ (*Hint*: $X = [X - \mu] + \mu$.)

14. Prove the iterative formula for the beta function in (10.76):

$$B(v + 1, w) = \frac{v}{v + w}B(v, w).$$

(*Hint*: Integrate by parts to first show: $B(v + 1, w) = \frac{v}{w}B(v, w + 1)$. Then by expressing $(1 - x)^w = (1 - x)(1 - x)^{w-1}$, and simplifying, that $B(v, w + 1) = B(v, w) - B(v + 1, w)$.)

15. Derive the moment-generating function formula for the gamma distribution:

$$M_\Gamma(t) = (1 - bt)^{-c}, \qquad |t| < \frac{1}{b}.$$

(*Hint*: $\int e^{tx}f_\Gamma(x)\, dx = \frac{1}{\Gamma(c)}\int \frac{1}{b}\left(\frac{x}{b}\right)^{c-1}e^{-((1-tb)/b)x}\, dx$; substitute $y = \frac{x}{b}$, then $z = (1 - tb)y$, or do this in one step.)

16. Evaluate the present value of a 50 year annuity, payable continuously at the rate of $1000 per year, at the continuous rate of 6%.

17. Repeat exercise 16, in the case where the annuity is continuously payable, and continuously increasing, so that the annualized rate of payment at time t is $C(t) = 1000(1.08)^t$. (*Hint*: Consider converting the 8% annual rate to another basis.)

18. Repeat exercise 18 of chapter 9 using the price function approximations in (10.118) and (10.119).

19. Derive (10.133). (*Hint*: Split each integral, such as

$$\int_{x_0}^{x_{n+1}} x\tilde{f}_n(x)\,dx = \sum_{j=0}^{n} \int_{x_j}^{x_{j+1}} x\tilde{f}_n(x)\,dx.)$$

20. Assume that the price of a t-period zero-coupon bond is given by $Z_t = \frac{1}{1+t}$ for all $t \geq 0$.

(a) Evaluate the implied continuous forward rates, f_t, and spot rates, s_t for all $t \geq 0$.

(b) Confirm (10.111).

21. With $r = 0.03$ on a continuous basis, $S_0 = 100$, and $\ln\left[\frac{S_{t+1}}{S_t}\right] \sim N(0.12, (0.18)^2)$ over annual periods:

(a) Determine the value of a 0.5-year binary call option on a stock with payoff function

$$\Lambda(S_{0.5}) = \begin{cases} 10, & S_{0.5} > 105, \\ 0, & S_{0.5} \leq 105. \end{cases}$$

(b) Evaluate the corresponding price for a binary put option, with payoff function

$$\Lambda(S_{0.5}) = \begin{cases} 0, & S_{0.5} \geq 105, \\ 10, & S_{0.5} < 105. \end{cases}$$

(c) Derive put-call parity for these binary options:

$$\Lambda^P(S_0) + \Lambda^C(S_0) = 10e^{-0.015}.$$

Assignment Exercises

22. Repeat exercise 1 in the cases where:

(a) $a < 0 < b$

(b) $a < b < 0$

(*Hint*: Consider $\int_a^b = \int_a^0 + \int_0^b$ for part (a), and identify the relationship between $\int_a^0$ and $\int_0^{-a}$ of the given functions. For part (b), consider the relationship between $\int_a^b$ and $\int_{-b}^{-a}$ of the given functions. In both cases keep track of the sign of $f(x)$.)

23. Show that if $f(x)$ is continuous on bounded $[a, b]$, so too is $|f(x)|$. In other words, show that $f(x) \to f(x_0)$ implies that $|f(x)| \to |f(x_0)|$. (*Hint*: To show this, prove that

$$||a| - |b|| \leq |a - b|.)$$

(10.139)

(a) Give a different example from what is in the text of where $|f(x)|$ continuous does not imply that $f(x)$ is continuous.

(b) Give a second example where the continuity of $f(x)^2$ does not imply the continuity of $f(x)$.

24. For the functions in exercise 5(b) and 5(c), explicitly determine the value of d as promised by the mean value theorem for which

$$\int_a^b f(x)\,dx = f(d)(b - a).$$

25. Use the integral test in the following analyses:

(a) Show that $\sum_{n=1}^{\infty} n^2 e^{-n}$ converges and estimate its value. (*Hint*: integrate by parts.)

(b) Show that $\sum_{n=1}^{\infty} \frac{n}{n^2+10}$ diverges, and estimate the rate of growth of the partial sums.

(c) For $0 < q < 1$, determine whether $\sum_{n=1}^{\infty} n^2 q^n$ converges or diverges, and correspondingly estimate its summation value, or the growth rate of its partial sums. (*Hint*: Integrate $f(x) = x^2 q^x = x^2 e^{x \ln q}$ using integration by parts.)

26. Evaluate the following definite integrals using the method of substitution, and then identify an antiderivative of the integrand:

(a) $\int_{-\infty}^{\infty} y e^{-y^2}\,dy$ (*Hint*: First consider $\int_{-M}^{N} y e^{-y^2}\,dy$ as a definite integral.)

(b) $\int_2^{20} \frac{\ln\sqrt{w}}{w}\,dw$ (*Hint*: Focus on $\ln\sqrt{w}$.)

(c) $\int_0^{10} (8x^3 + 10x - 3)(2x^4 + 5x^2 - 3x)^{-1/2}\,dx$ (*Hint*: First consider $\int_a^{10} f(x)\,dx$ for $a > 0$.)

27. Evaluate the following definite integrals using integration by parts, and then identify an antiderivative of the integrand (*General hint*: Once a potential antiderivative is found, this formula can be verified by induction on n.):

(a) $\int_0^{20} x^n e^{-rx}\,dx$ for positive integer n, positive real r

(b) $\int_0^{10} x^n e^{-x^2}\,dx$ for positive integer odd $n = 2m + 1$

28. Show using a Taylor series expansion that if $f(y) = e^y$, then $\int_0^x f(y)\,dy = e^x - 1$. Justify integrating term by term as well as the convergence of the final series to the desired answer.

29. Assume that the value of a t-period continuous forward rate is given by $f_t = \frac{0.03}{1+0.1t}$ for all $t \geq 0$.

(a) Evaluate the implied continuous spot rates, s_t, and zero-coupon bond prices, Z_t, for all $t \geq 0$.

(b) Confirm (10.111).

30. Using the definite integrals over bounded intervals in exercises 26(b) and 26(c), and 27(a) and 27(b) (use $n = 10$ and $r = 0.10$ in exercise 27):

(a) Implement both the trapezoidal rule and Simpson's rule for several values of n and compare the associated errors. (*Hint*: Try $n = 5, 10, 25$, and 100, say.)

(b) For each, evaluate the error as n increases significantly, to see if the respective orders of convergence, $O\left(\frac{1}{n^2}\right)$ and $O\left(\frac{1}{n^4}\right)$, are apparent. (*Hint*: If $\epsilon_n^T = |I - I^T|$ for $\Delta x = \frac{b-a}{n}$, the error $\epsilon_n^T = O\left(\frac{1}{n^2}\right)$ means that $n^2\epsilon_n^T \leq C^T$ for some constant C^T as $n \to \infty$, and similarly for $\epsilon_n^S = |I - I^S|$, that $n^4\epsilon_n^S \leq C^S$ as $n \to \infty$. Attempt to verify that the C^T and C^S values obtained are no bigger than the values predicted in theory using the maxima of the derivatives of the given functions.)

31. Evaluate $\Pr[1 \leq X \leq 5]$ for the gamma distribution with $b = 1$ and shape parameter $c = 3$, by:

(a) Trapezoidal rule with $n = 100$

(b) Simpson's rule with $n = 100$

(c) Evaluate the error in each approximation

32. Prove the following identities:

(a) As in (10.67) that $M_X(t) = \sum_{n=0}^{\infty} \frac{\mu_n' t^n}{n!}$ (*Hint*: Compare to the discrete derivation in chapter 9, using section 10.7.2 on convergence of a sequence of integrals.)

(b) As in (10.68) that $\mu_n' = M_X^{(n)}(0)$ (*Note*: Justify term by term differentiation and substitution of $t = 0$.)

33. Show directly that for the beta function: $B(1, 1) = 1$, and then using the same hint as in exercise 14, show that

$$B(v, w) = \frac{(v - 1)(w - 1)}{(v + w - 1)(v + w - 2)} B(v - 1, w - 1),$$

and that with this and mathematical induction, derive (10.78).

34. Derive the iterative formulas for moments of the unit normal:

(a) For $m = 1, 2, 3, \ldots$:

$$\int_{-\infty}^{\infty} y^{2m} \phi(y) \, dy = (2m - 1) \int_{-\infty}^{\infty} y^{2m-2} \phi(y) \, dy.$$

(*Hint*: Try integration by parts, splitting the integrand into y^{2m-1} and $y\phi(y)$, and note the latter can be integrated by substitution.)

(b) For $m = 1, 2, 3, \ldots$:

$$\int_{-\infty}^{\infty} y^{2m-1} \phi(y) \, dy = 0.$$

(*Hint*: Consider $f(y)$ and $f(-y)$, then Riemann sums.)

35. Evaluate the present value of a perpetuity, payable continuously at the rate of $10,000 per year, at the continuous rate of 10%.

36. Repeat exercise 35 in the case where the annuity is continuously payable, and continuously increasing, so that the annualized rate of payment at time t is $C(t) = 10,000(1 + 2t)$.

37. Repeat exercise 41 of chapter 9 using the price function approximations in (10.118) and (10.119).

38. Derive (10.136). (*Hint*: Use (10.135) and recall that by (10.132), for $j = 0, 1, \ldots,$ $n + 1$,

$$\tilde{f}_n(x_j + \epsilon_j) - \tilde{f}_n(x_j - \epsilon_j) = \frac{1}{u - d} [f_B(j) - f_B(j - 1)].)$$

39. Derive the Black–Scholes–Merton formulas for the price of a European put or call using (10.130). (*Hint*: Use a substitution in the integral.)

40. The notion of Riemann integral can be generalized to become a **Riemann–Stieltjes integral**, in recognition of the work of **Thomas Joannes Stieltjes** (1856–1894).

Definition 10.86 *Given a function, $g(x)$, a function $f(x)$ is **Riemann–Stieltjes integrable with respect to** $g(x)$ **on an interval** $[a, b]$ if as $\mu \to 0$, with μ defined as in (10.3), we have that*

$$\left[\sum_{i=1}^{n} M_i \Delta g_i - \sum_{i=1}^{n} m_i \Delta g_i \right] \to 0, \tag{10.140}$$

*where M_i and m_i are defined in (10.2). Here $\Delta g_i = g(x_i^-) - g(x_{i-1}^+)$, where $g(x_i^-) = \lim_{x \to x_i^-} g(x)$ and $g(x_{i-1}^+) = \lim_{x \to x_{i-1}^+} g(x)$, are defined as one-sided limits from the left $(-)$ and right $(+)$. In this case we define the **Riemann–Stieltjes integral of** $f(x)$ **with respect to** $g(x)$ **over** $[a, b]$, by*

$$\int_a^b f(x)\, dg = \lim_{\mu \to 0} \sum_{i=1}^n f(\tilde{x}_i)\Delta g_i,$$

which exists and is independent of the choice of $\tilde{x}_i \in [x_{i-1}, x_i]$ by (10.140).

(a) Show that if $g(x)$ and $f(x)$ are continuous on $[a, b]$, and $g(x)$ is differentiable on (a, b) with $g'(x)$ a continuous function with limits as $x \to a$ and $x \to b$, then

$$\int_a^b f(x)\, dg = \int_a^b f(x) g'(x)\, dx, \tag{10.142}$$

where the integral on the right is a Riemann integral. (*Hint*: Consider the mean value theorem from chapter 9.)

(b) Generalize part (a) to the case where there is a partition of $[a, b]$:

$$a = y_0 < y_1 < \cdots < y_{m+1} = b$$

so that $g(x)$ satisfies the conditions of part (a) on each subinterval, $[y_j, y_{j+1}]$ but has "jumps" at $\{y_j\}_{j=1}^m$:

$$\lim_{x \to y_j+} g(x) \neq \lim_{x \to y_j-} g(x), \qquad j = 1, 2, \ldots, m.$$

Show that in this case

$$\int_a^b f(x)\, dg = \sum_{j=0}^m \int_{y_j}^{y_{j+1}} f(x) g'(x)\, dx + \sum_{j=1}^m f(y_j)[g(y_j^+) - g(y_j^-)]. \tag{10.143}$$

41. Evaluate $\int_0^{10} x^2\, dg$ with:

(a) $g(x) = e^{-0.04x}$

(b) $g(x) = \begin{cases} e^{-0.04x}, & 0 \le x < 2 \\ e^{-0.04x} - 4, & 2 \le x < 6 \\ e^{-0.04x} + 4, & 6 \le x \le 10 \end{cases}$

42. This exercise investigates the application of Riemann–Stieltjes integration to probability theory.

(a) Show that if $f(x)$ is a continuous probability density function with distribution function $F(x)$, then for any function $g(x)$ for which $E[g(x)]$ exists,

$$E[g(x)] = \int g(x)\, dF. \tag{10.144}$$

(*Hint*: Use (10.142).)

(b) Show that if $f(x)$ is a discrete probability density function with domain $\{x_j\}$ assumed to have no accumulation points, and with distribution function $F(x)$, that for any function $g(x)$ for which $E[g(x)]$ exists, (10.144) remains valid. (*Hint:* Use (10.143).)

Remark 10.87 *A probability density function can be **mixed**, meaning both with continuous and discrete components. The distribution function $F(x)$ then is nondecreasing, $0 \leq F(x) \leq 1$, has the structure required in exercise 40.b, and $E[g(x)]$ is again defined as in (10.144) using (10.143).*

(c) Evaluate the mean and variance of the random variable with mixed distribution function defined by

$$
F(x) = \begin{cases}
0, & x < 0, \\
0.25, & x = 0, \\
\frac{1}{4}\left(1 + \frac{x}{100}\right), & 0 < x < 50, \\
0.5, & x = 50, \\
\frac{1}{3}\left(1 + \frac{x}{100}\right), & 50 < x < 100, \\
0.75, & x = 100, \\
1 - \frac{1}{4}e^{100-x}, & x > 100.
\end{cases}
$$

References

I have listed in this section a number of textbook references for the mathematics and finance presented in this book. All these textbooks provide theoretical and applied materials in their respective areas beyond their development here, and they are worth pursuing by the reader interested in gaining greater depth or breadth of knowledge. This list is by no means complete and is intended only as a guide to further study.

The reader will no doubt observe that the mathematics references are somewhat older than the finance references and, upon web searching, will find that some of the older texts in each category have been updated to newer editions, sometimes with additional authors. Since I own and use the editions listed below, I decided to present these rather than reference the newer editions that I have not reviewed. As many of these older texts are considered "classics," they are also likely to be found in university and other libraries. That said, there are undoubtedly many very good new texts by both new and established authors with similar titles that are also worth investigating.

My rules of thumb for a textbook, whether recommended by a colleague or newly discovered, are as follows:

1. If it provides a clear and complete exposition that makes it easy to understand both simple and deep connections, it is a very good textbook.

2. If it provides compelling derivations and applications that motivate the reader to want to read on and learn more, it is an excellent textbook.

3. If it is difficult to understand and does not motivate the reader, it is either poorly written or ahead of the reader's current state of knowledge, and in either case the reader should seek another reference text.

Topic Mapping

Numbers refer to the numbered references that follow.

Finance

Investment markets: 2, 3, 5, 6, 8, 11, 12, 14
Fixed income pricing: 1, 2, 3, 5, 6, 7, 8, 9, 10, 11, 12, 13, 14
Equity pricing: 1, 2, 3, 5, 7, 12, 14
Portfolio theory: 1, 2, 3, 5, 7, 12, 14
Insurance finance: 4, 10, 12
Utility theory: 4, 5, 7, 12
Option pricing: 1, 2, 3, 5, 6, 7, 8, 11, 12, 13, 14
Risk analysis: 1, 4, 6, 8, 9, 11, 12

Mathematics

Logic: 25, 31
Number systems: 15, 26, 28, 30, 31
Functions: 15, 20, 28, 30, 31, 32
Euclidean and metric spaces: 16, 18, 19, 27, 31
Set theory: 16, 21, 28, 31
Topology: 16, 19, 29, 30, 31
Sequences and series: 15, 20, 26, 30
Probability theory: 17, 22, 24, 29
Calculus: 15, 20, 23, 30, 31, 32
Approximation theory: 15, 23

Bibliography

Finance

1. Benninga, Simon. *Financial Modeling*, 3rd ed. Cambridge: MIT Press, 2008.

2. Bodie, Zvi, Alex Kane, and Alan J. Marcus. *Investments*, 7th ed. New York: McGraw-Hill/Irwin, 2008.

3. Bodie, Zvi, and Robert C. Merton. *Finance*. Upper Saddle River, NJ: Prentice Hall, 2000.

4. Bowers, Newton L., Jr., Hans U. Gerber, James C. Hickman, Donald A. Jones, and Cecil J. Nesbitt. *Actuarial Mathematics*. Itasca, IL: Society of Actuaries, 1986.

5. Copeland, Thomas E., J. Fred Weston, and Kupdeep Shastri. *Financial Theory and Corporate Policy*, 4th ed. Boston: Pearson Addison-Wesley, 2005.

6. Fabozzi, Frank J. *Bond Markets, Analysis, and Strategies*, 6th ed. Upper Saddle River, NJ: Pearson Prentice Hall, 2007.

7. Huang, Chi-fu, and Robert H. Litzenberger. *Foundations for Financial Economics*. Upper Saddle River, NJ: Prentice Hall, 1988.

8. Hull, John C. *Options, Futures, and Other Derivatives*, 7th ed. Upper Saddle River, NJ: Pearson Prentice Hall, 2009.

9. Hull, John C. *Risk Management and Financial Institutions*. Upper Saddle River, NJ: Pearson Prentice Hall, 2006.

10. Kellison, Stephen G. *The Theory of Interest*. Homewood, IL: Irwin, 1970.

11. McDonald, Robert L. *Derivatives Markets*, 2nd ed. Boston: Pearson Addison-Wesley, 2006.

12. Panjer, Harry H., ed. *Financial Economics*. Schaumburg, IL: Actuarial Foundation, 1998.

13. Shreve, Steven E. *Stochastic Calculus for Finance I: The Binomial Asset Pricing Model.* New York: Springer, 2000.

14. Sharpe, William F. *Investments*, 3rd ed. Englewood Cliffs, NJ: Prentice-Hall, 1985.

Mathematics

15. Courant, Richard, and Fritz John. *Introduction to Calculus and Analysis*, Vol. 1. New York: Interscience Publishers, 1965.

16. Dugundji, James. *Topology*. Boston: Allyn and Bacon, 1970.

17. Feller, William. *An Introduction to Probability Theory and Its Applications*, Vol. 1. New York: Wiley, 1968.

18. Gel'fand, I. [Izrail'] M. *Lectures on Linear Algebra.* New York: Dover, 1989.

19. Gemignani, Michael C. *Elementary Topology*. Reading, MA: Addison-Wesley Publishing, 1967.

20. Goldberg, Richard R. *Methods of Real Analysis.* Waltham, MA: Xerox College Publishing, 1964.

21. Halmos, Paul R. *Naive Set Theory.* New York: Van Nostrand Reinhold, 1960.

22. Hoel, Paul G. *Introduction to Mathematical Statistics*, 4th ed. New York: Wiley, 1971.

23. Kellison, Stephen G. *Fundamentals of Numerical Analysis.* Homewood, IL: Irwin, 1975.

24. Lindgren, Bernard W. *Statistical Theory*, 3rd ed. New York: Macmillan, 1976.

25. Margaris, Angelo. *First Order Mathematical Logic.* Waltham, MA: Xerox College Publishing, 1967.

26. Maor, Eli. *To Infinity and Beyond*. Princeton: Princeton University Press, 1991.

27. Paige, Lowell J., and J. Dean Swift. *Elements of Linear Algebra.* Waltham, MA: Blaisdell, 1961.

28. Pinter, Charles C. *Set Theory.* Reading, MA: Addison-Wesley, 1971.

29. Ross, Sheldon. *A First Course in Probability*. New York: Macmillan, 1976.

30. Rudin, Walter. *Principals of Mathematical Analysis*, 3rd ed. New York: McGraw-Hill, 1976.

31. Sentilles, Dennis. *A Bridge to Advanced Mathematics*. Baltimore: Williams and Wilkins, 1975.

32. Thomas, George B., Jr. *Calculus and Analytic Geometry*, 4th ed, Part 1. Reading, MA: Addison-Wesley, 1968.

Index